Heterocyclic Compounds in Medicinal Chemistry

Heterocyclic Compounds in Medicinal Chemistry

Valentina Noemi Madia
Davide Ialongo

Basel • Beijing • Wuhan • Barcelona • Belgrade • Novi Sad • Cluj • Manchester

Editors

Valentina Noemi Madia
Dipartimento di Chimica e
Tecnologie del Farmaco
Sapienza University of Rome
Rome
Italy

Davide Ialongo
Dipartimento di Chimica e
Tecnologie del Farmaco
Sapienza University of Rome
Rome
Italy

Editorial Office
MDPI AG
Grosspeteranlage 5
4052 Basel, Switzerland

This is a reprint of articles from the Special Issue published online in the open access journal *Pharmaceuticals* (ISSN 1424-8247) (available at: www.mdpi.com/journal/pharmaceuticals/special_issues/GC66SFS95K).

For citation purposes, cite each article independently as indicated on the article page online and as indicated below:

Lastname, A.A.; Lastname, B.B. Article Title. *Journal Name* **Year**, *Volume Number*, Page Range.

ISBN 978-3-7258-2292-8 (Hbk)
ISBN 978-3-7258-2291-1 (PDF)
doi.org/10.3390/books978-3-7258-2291-1

© 2024 by the authors. Articles in this book are Open Access and distributed under the Creative Commons Attribution (CC BY) license. The book as a whole is distributed by MDPI under the terms and conditions of the Creative Commons Attribution-NonCommercial-NoDerivs (CC BY-NC-ND) license.

Contents

Mladenka Jurin, Ana Čikoš, Višnja Stepanić, Marcin Górecki, Gennaro Pescitelli and Darko Kontrec et al.
Synthesis, Absolute Configuration, Biological Profile and Antiproliferative Activity of New 3,5-Disubstituted Hydantoins
Reprinted from: *Pharmaceuticals* 2024, 17, 1259, doi:10.3390/ph17101259 1

Md Ali Asif Noor, Md Mazedul Haq, Md Arifur Rahman Chowdhury, Hilal Tayara, HyunJoo Shim and Kil To Chong
In Silico Exploration of Novel EGFR Kinase Mutant-Selective Inhibitors Using a Hybrid Computational Approach
Reprinted from: *Pharmaceuticals* 2024, 17, 1107, doi:10.3390/ph17091107 32

Tebyan O. Mirgany, Hanadi H. Asiri, A. F. M. Motiur Rahman and Mohammed M. Alanazi
Discovery of 1H-benzo[d]imidazole-(halogenated) Benzylidenebenzohydrazide Hybrids as Potential Multi-Kinase Inhibitors
Reprinted from: *Pharmaceuticals* 2024, 17, 839, doi:10.3390/ph17070839 52

Daria Klimoszek, Małgorzata Jeleń, Małgorzata Dołowy and Beata Morak-Młodawska
Study of the Lipophilicity and ADMET Parameters of New Anticancer Diquinothiazines with Pharmacophore Substituents
Reprinted from: *Pharmaceuticals* 2024, 17, 725, doi:10.3390/ph17060725 70

Noha Fathallah, Wafaa M. Elkady, Sara A. Zahran, Khaled M. Darwish, Sameh S. Elhady and Yasmin A. Elkhawas
Unveiling the Multifaceted Capabilities of Endophytic *Aspergillus flavus* Isolated from *Annona squamosa* Fruit Peels against *Staphylococcus* Isolates and HCoV 229E—In Vitro and In Silico Investigations
Reprinted from: *Pharmaceuticals* 2024, 17, 656, doi:10.3390/ph17050656 91

Yechan Lee, Sunhee Lee, Younho Lee, Doona Song, So-Hyeon Park and Jieun Kim et al.
Anticancer Evaluation of Novel Benzofuran–Indole Hybrids as Epidermal Growth Factor Receptor Inhibitors against Non-Small-Cell Lung Cancer Cells
Reprinted from: *Pharmaceuticals* 2024, 17, 231, doi:10.3390/ph17020231 132

Baskar Nammalwar and Richard A. Bunce
Recent Advances in Pyrimidine-Based Drugs
Reprinted from: *Pharmaceuticals* 2024, 17, 104, doi:10.3390/ph17010104 159

Eid E. Salama, Mohamed F. Youssef, Ahmed Aboelmagd, Ahmed T. A. Boraei, Mohamed S. Nafie and Matti Haukka et al.
Discovery of Potent Indolyl-Hydrazones as Kinase Inhibitors for Breast Cancer: Synthesis, X-ray Single-Crystal Analysis, and In Vitro and In Vivo Anti-Cancer Activity Evaluation
Reprinted from: *Pharmaceuticals* 2023, 16, 1724, doi:10.3390/ph16121724 209

Kiran Shahzadi, Syed Majid Bukhari, Asma Zaidi, Tanveer A. Wani, Muhammad Saeed Jan and Seema Zargar et al.
Novel Coumarin Derivatives as Potential Urease Inhibitors for Kidney Stone Prevention and Antiulcer Therapy: From Synthesis to In Vivo Evaluation
Reprinted from: *Pharmaceuticals* 2023, 16, 1552, doi:10.3390/ph16111552 225

Lizeth Arce-Ramos, Juan-Carlos Castillo and Diana Becerra
Synthesis and Biological Studies of Benzo[b]furan Derivatives: A Review from 2011 to 2022
Reprinted from: *Pharmaceuticals* 2023, 16, 1265, doi:10.3390/ph16091265 245

Diana Tzankova, Hristina Kuteva, Emilio Mateev, Denitsa Stefanova, Alime Dzhemadan and Yordan Yordanov et al.
Synthesis, DFT Study, and In Vitro Evaluation of Antioxidant Properties and Cytotoxic and Cytoprotective Effects of New Hydrazones on SH-SY5Y Neuroblastoma Cell Lines
Reprinted from: *Pharmaceuticals* **2023**, *16*, 1198, doi:10.3390/ph16091198 339

Aisha A. Alsfouk, Hanan M. Alshibl, Najla A. Altwaijry, Ashwag Alanazi, Omkulthom AlKamaly and Ahlam Sultan et al.
New Imadazopyrazines with CDK9 Inhibitory Activity as Anticancer and Antiviral: Synthesis, In Silico, and In Vitro Evaluation Approaches
Reprinted from: *Pharmaceuticals* **2023**, *16*, 1018, doi:10.3390/ph16071018 359

Article

Synthesis, Absolute Configuration, Biological Profile and Antiproliferative Activity of New 3,5-Disubstituted Hydantoins

Mladenka Jurin [1], Ana Čikoš [2], Višnja Stepanić [3], Marcin Górecki [4], Gennaro Pescitelli [5,*], Darko Kontrec [1], Andreja Jakas [1], Tonko Dražić [6] and Marin Roje [1,*]

[1] Laboratory for Chiral Technologies, Division of Organic Chemistry and Biochemistry, Ruđer Bošković Institute, Bijenička Cesta 54, 10000 Zagreb, Croatia; mladenka.jurin@irb.hr (M.J.); darko.kontrec@irb.hr (D.K.); andreja.jakas@irb.hr (A.J.)
[2] NMR Centre, Ruđer Bošković Institute, Bijenička Cesta 54, 10000 Zagreb, Croatia; ana.cikos@irb.hr
[3] Laboratory for Machine Learning and Knowledge Representation, Division of Electronics, Ruđer Bošković Institute, Bijenička Cesta 54, 10000 Zagreb, Croatia; visnja.stepanic@irb.hr
[4] Institute of Organic Chemistry, Polish Academy of Sciences, 01-224 Warsaw, Poland; marcin.gorecki@icho.edu.pl
[5] Department of Chemistry and Industrial Chemistry, University of Pisa, Via Giuseppe Moruzzi 13, 56124 Pisa, Italy
[6] Laboratory for Biocolloids and Surface Chemistry, Division of Physical Chemistry, Ruđer Bošković Institute, Bijenička Cesta 54, 10000 Zagreb, Croatia; tdrazic@gmail.com
* Correspondence: gennaro.pescitelli@unipi.it (G.P.); marin.roje@irb.hr (M.R.)

Citation: Jurin, M.; Čikoš, A.; Stepanić, V.; Górecki, M.; Pescitelli, G.; Kontrec, D.; Jakas, A.; Dražić, T.; Roje, M. Synthesis, Absolute Configuration, Biological Profile and Antiproliferative Activity of New 3,5-Disubstituted Hydantoins. *Pharmaceuticals* 2024, 17, 1259. https://doi.org/10.3390/ph17101259

Academic Editors: Valentina Noemi Madia and Davide Ialongo

Received: 16 August 2024
Revised: 16 September 2024
Accepted: 20 September 2024
Published: 24 September 2024

Copyright: © 2024 by the authors. Licensee MDPI, Basel, Switzerland. This article is an open access article distributed under the terms and conditions of the Creative Commons Attribution (CC BY) license (https:// creativecommons.org/licenses/by/ 4.0/).

Abstract: Hydantoins, a class of five-membered heterocyclic compounds, exhibit diverse biological activities. The aim of this study was to synthesize and characterize a series of novel 3,5-disubstituted hydantoins and to investigate their antiproliferative activity against human cancer cell lines. The new hydantoin derivatives **5a–i** were prepared as racemic mixtures of *syn-* and *anti-*isomers via a base-assisted intramolecular amidolysis of C-3 functionalized β-lactams. The enantiomers of *syn-***5a** and *anti-*hydantoins **5b** were separated by preparative high-performance liquid chromatography (HPLC) using *n*-hexane/2-propanol (90/10, *v/v*) as the mobile phase. The absolute configuration of the four allyl hydantoin enantiomers **5a** was assigned based on a comparison of the experimental electronic circular dichroism (ECD) and vibrational circular dichroism (VCD) spectra with those calculated using density functional theory (DFT). The antiproliferative activity evaluated *in vitro* against three different human cancer cell lines: HepG2 (liver hepatocellular carcinoma), A2780 (ovarian carcinoma), and MCF7 (breast adenocarcinoma), and on the non-tumor cell line HFF1 (normal human foreskin fibroblasts) using the MTT cell proliferation assay. In silico drug-like properties and ADMET profiles were estimated using the ADMET Predictor ver. 9.5 and the online server admetSAR. Eighteen new 3,5-disubstituted hydantoins were synthesized and characterized. The compound *anti-***5c** showed potent cytotoxic activity against the human tumor cell line MCF7 (IC$_{50}$ = 4.5 µmol/L) and the non-tumor cell line HFF1 (IC$_{50}$ = 12.0 µmol/L). In silico analyzes revealed that the compounds exhibited moderate water solubility and membrane permeability and are likely substrates for CYP3A4 and P-glycoprotein and have a high probability of antiarthritic activity.

Keywords: 3,5-disubstituted hydantoins; preparative HPLC separation; NMR analysis; absolute configuration; CD/VCD; antiproliferative activity; in silico biological profiling

1. Introduction

Imidazolidine-2,4-diones, commonly referred to as hydantoins, are five-membered heterocyclic compounds characterized by the presence of two nitrogen atoms at positions one and three, and two carbonyl functions at positions two and four within the hydantoin ring [1–4]. The hydantoin moiety is an important structural element found in various bioactive medicinal compounds. The biological activity of these compounds depends on the nature of the substituents at positions N-1, N-3, and C-5 of the hydantoin ring. The hydantoin core structure is found in marketed drugs for the treatment of epileptic seizures (phenytoin, ethotoine, and norantoine) [5,6], and metastatic prostate cancer (nilutamide) [7,8], muscle relaxants drugs, and drugs for the prevention of malignant hyperthermia (dantrolene) [9] (Figure 1). BMS-587101 is a potent and orally active antagonist of leukocyte function associated antigen-1 (LFA-1) [10], while BMS-564929 is a highly potent, orally active, nonsteroidal tissue selective androgen receptor (AR) modulator [11]. In addition, hydantoin derivatives exhibit diverse biological and pharmacological activities in medicine, such as antimicrobial [12,13], antiviral [14–16], anticonvulsant [17–21], antitumor [7,22], antiarrhythmic [23,24], antihypertensive [25], antithrombotic, anti-inflammatory [26], antidiabetic [27,28], and agrochemical applications, such as herbicidal and fungicidal [29]. Hydantoins also have an inhibitory effect on some enzymes (human aldose reductase and leucocyte elastase) [30,31]. Furthermore, the hydantoin moiety is frequently found in natural products, predominantly isolated from a diverse range of marine organisms and bacteria [32]. For example, hemimycallins A and B (Figure 1) were extracted from the marine sponge *Hemimycale arabica* [31], mukanadine B (Figure 1) was extracted from the marine sponge *Agelas nakamurai* [33], midpacamide (Figure 1) was extracted from the Fijian sponge *Agelas mauritiana* [34], and parazoanthines A–J (Figure 1) were extracted from the Mediterranean anemone *Parazoanthus axinellae* [35,36]. From the viewpoint of organic synthesis, hydantoins have been widely used as precursors for the synthesis of some optically pure natural and unnatural amino acids [37]. In addition, the optically pure hydantoins have been widely used as chiral auxiliaries [38] and metal ligands in asymmetric catalysis [39].

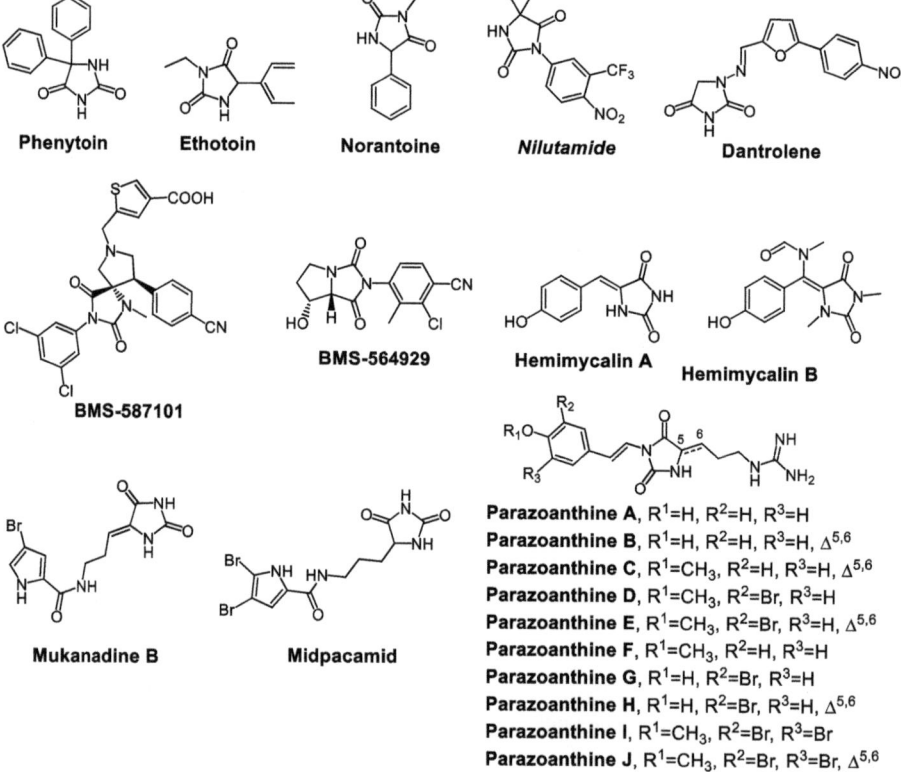

Figure 1. Chemical structures of some biologically active compounds and natural products containing a hydantoin moiety.

Numerous methods for the synthesis of the hydantoin derivatives have been described in the literature [10,40]. Hydantoins can be prepared either in solution by classical methods, on the solid phase, under microwave irradiation, by mechanochemistry, or by the continuous flow method [10,40–52]. Zhang et al. prepared enantiomerically pure hydantoins by the cyclization of optically pure α-amino amides with triphosgene [41]. The same group reported that the use of 1,1′-carbonyldiimidazole (CDI) led to a racemization of the stereogenic center. They suggest that the imidazole carbamate intermediate is responsible for the observed racemization with CDI in this type of reaction. Chen et al. recently prepared a library of enantiomerically pure 3,5-disubstituted hydantoins by a direct cyclization of the corresponding ureas with sodium hydride [42]. The obtained hydantoins were isolated in good to high yields and with excellent enantioselectivities. Tanwar et al. obtained 3,5-disubstituted hydantoins, including the anticonvulsant ethotoin, in a one-pot synthesis of α-amino methyl ester hydrochlorides with carbamates [43]. This reaction led to the formation of the corresponding ureido derivatives, which subsequently cyclized under basic conditions to give the corresponding hydantoins in good yields. Liu et al. used a simple activation strategy mediated by trifluoromethanesulfonic anhydride to prepare highly substituted chiral hydantoins from simple Boc-protected dipeptides in a single step under mild conditions [44]. Mehra and Kumar performed a base-promoted intramolecular amidolysis of C-3 functionalized azetidin-2-ones, prepared from racemic cis-β-lactam carbamates and amines, with sodium methoxide to afford the 3,5-disubstituted hydantoins [45]. The basic condition led not only to the formation of the desired hydantoins but also to the formation of their dimers. Rajić et al. synthesized 3,5-disubstituted hydantoins by a base-induced cyclization of the corresponding N-(1-benzotriazolecarbonyl)-L- and D-amino acid amides and investigated their antiviral and antitumor activity [14]. Of all of the hydantoins synthesized, cyclohexyl-5-phenyl hydantoin showed antitumor activity against HeLa and MCF7 cell lines but also cytotoxic effects on human non-tumor fibroblasts WI 38, whereas 3-benzhydryl-5-phenyl hydantoin showed moderate antitumor activity against HeLa, MCF7, MiaPaCa-2, H 460, and SW 620 but no cytotoxic effects on normal cells. In 2016, Konnert et al. reported the mechanochemical reaction of α-amino methyl esters with CDI or with substituted alkyl isocyanates to give a series of 3,5 disubstituted hydantoins, including ethotoin, with or without poly(ethylene glycol)s as grinding assisting agents [46]. Konnert et al. also developed a mechanochemical solvent-free method to prepare various 3,5-disubstituted hydantoin derivatives from amino esters or dipeptides via a CDI-mediated one-pot/two-step cyclization reaction involving unsymmetrical urea or carboxy-imidazolyl dipeptide ester intermediates [47]. Recently, Mascitti et al. optimized the mechanochemical synthesis of highly functionalized 3,5-disubstituted hydantoins from α-amino esters by liquid-assisted grinding (LAG) in the presence of various poly(ethylene) glycols [48]. The first reported solid-phase synthesis of 3,5-disubstituted hydantoins was described by De Witt and coworkers [49]. A series of polystyrene Wang resin, ester-linked amino acids were exposed to isocyanates to give resin-linked α-ureido acids, which were then treated with 6 M hydrochloric acid. The strongly acidic conditions promoted N-cyclisation and the simultaneous cleavage of hydantoin from the resin. Colacino et al. reported a microwave-assisted solid-phase synthesis of 3,5 hydantoin derivatives in which resin-bound amino acids with a free N-terminal moiety reacted with phenyl isocyanate [37]. The ureido derivatives formed were then simultaneously cyclized and released from the resin in the presence of triethylamine to afford the corresponding hydantoins. Colacino et al. discussed [50] innovative mechanochemical methods for the synthesis of hydantoin-based active pharmaceutical ingredients (APIs) and emphasized their environmentally friendly nature and advantages over conventional synthetic approaches. The authors emphasize the improved yields, the simplified reaction conditions and the lower environmental impact and advocate mechanochemistry as a sustainable alternative in drug development. Furthermore, the hydantoins can also be synthesized using flow chemistry techniques. Monteiro et al. [51] presented a new continuous flow method for the synthesis of hydantoins using the Bucherer–Bergs reaction. The authors successfully optimized the reaction conditions and achieved high yields for a variety of substrates. They also investigated the selective N(3)-monoalkylation of the hydantoins, which opens up possibilities for further modifications and potentially new therapeutic applications. Vukelić et al. [52] synthesized hydantoins from commercially available amines using a semi-continuous flow process involving photooxidative cyanation followed by carbon dioxide introduction at elevated temperatures. Using this method, various hydantoins were produced in good yields without the need for extensive purification steps, demonstrating an efficient and sustainable approach to hydantoin synthesis.

In the present work, a series of novel 3,5 disubstituted hydantoins **5a–i** were prepared as racemic mixtures of *syn*- and *anti*-isomers via a base-assisted intramolecular amidolysis of C-3 functionalized β-lactams [45]. Each synthesized hydantoin **5a–i** has two stereogenic carbon centers, one at the C-5 position of the hydantoin ring and the other in the side chain. The absolute configuration of the four isolated enantiomers of allyl hydantoin **5a** was determined by a comparison of experimental and DFT calculated ECD and VCD spectra. All of the synthesized compounds **5a–i** were evaluated in vitro for their antiproliferative activity against liver hepatocellular carcinoma (HepG2), ovarian carcinoma (A2780), breast adenocarcinoma (MCF7), and untransformed human fibroblasts (HFF-1). In silico drug-like properties and ADMET (Absorption, Distribution, Metabolism, Excretion, and Toxicity) profiles of the synthesized hydantoin compounds **5a–i** were estimated using the commercial software ADMET Predictor ver. 9.5 [53] and the free online server admetSAR [54]. The biological activities of each hydantoin **5a–i** were predicted by the commercial software PASS 2020 [55] and the open access web server SwissTargetPrediction [56].

2. Results and Discussion

2.1. Synthesis and Separation of Syn- and Anti-(±)-3,5-Disubstituted Hydantoins **5a–i**

The synthesis of the new 3,5-disubstituted hydantoins **5a–i** was carried out as shown in Scheme 1. This study demonstrates the in situ cyclization of racemic *trans*-β-lactam ureas **4a–i** under basic conditions by intramolecular amidolysis leading to the formation of a five-membered hydantoin ring. In the first step, imine **1** was prepared by a condensation reaction between 4-methoxybenzaldehyde and 4-fluoroaniline in an anhydrous dichloromethane medium, using powdered molecular sieves 4 Å (Scheme 1) [57]. The isolated imine **1** was obtained in a yield of 73% after recrystallization from a mixture of ethyl acetate and *n*-hexane.

Following the formation of imine **1**, a reaction was carried out with N-phthalimidoglycine in the presence of triethylamine and 2-chloro-1-methylpyridinium iodide, a reagent known as Mukaiyama reagent. This step led to the preparation of a racemic mixture of *cis*- and *trans*-3-phthalimido-β-lactam **2**, as shown in Scheme 1 [58,59]. The choice of starting materials, 4-methoxybenzaldehyde and 4-fluoroaniline, lays in the already demonstrated bioactivity of these pharmacophores [59]. The [2 + 2]-cycloaddition reaction of imine **1** with in situ-generated ketene from acid and triethylamine, also known as the Staudinger reaction, is characterized by the sequential formation of the covalent N1–C2 and C3–C4 bonds within the β-lactam ring [60,61]. To gain a deeper insight into the nature of the reaction, the crude reaction mixture was analyzed by ^{1}H NMR spectroscopy. The analysis revealed the presence of two isomeric β-lactams, with a remarkable *cis*/*trans* ratio of approximately 1:5. This finding was subsequently validated by HPLC analysis, which provided a definitive representation of the isomer distribution. In the ^{1}H NMR spectra of the *cis* stereoisomer, the two protons of the β-lactam ring, H-3 and H-4, were observed as two doublets at 5.42 ppm and 5.65 ppm, respectively, with a coupling constant $^3J_{H3,H4}$ of 5.5 Hz. In contrast, the *trans*-isomer exhibited a coupling constant of J = 2.6 Hz, with the corresponding protons appearing as doublets at 5.28 and 5.31 ppm. These spectral data enabled the accurate determination of the structures of the expected diastereomeric β-lactams, namely (±)-*cis*-**2** and (±)-*trans*-**2**, as shown in Scheme 1. Flash column chromatography was used to separate the *cis*- and *trans*-isomers, yielding pure *trans*-**2** (75%) and pure *cis*-**2** (11%). Given that the ketene generated in situ from the reaction between N-phthalimidoglycine and triethylamine corresponds to a Sheehan ketene, it exhibits a preferential formation of the *trans*-β-lactam [62].

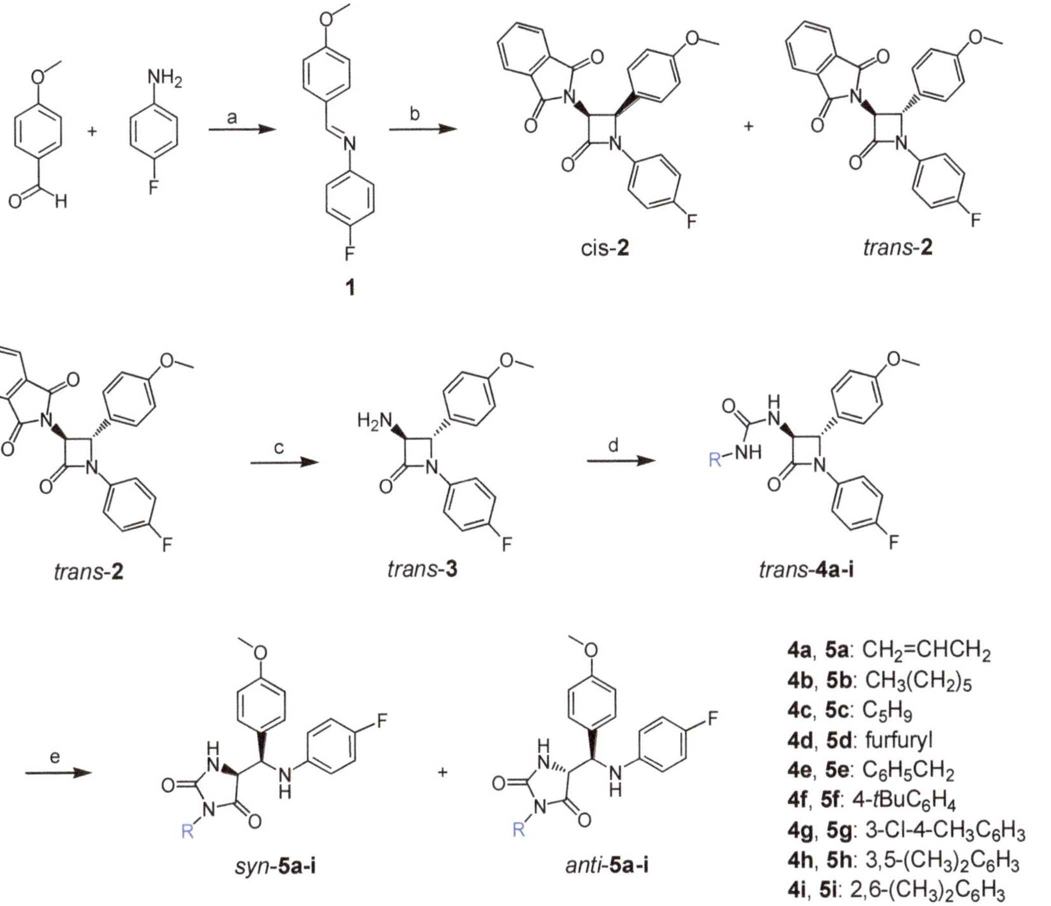

Scheme 1. Reagents and conditions: (**a**) CH$_2$Cl$_2$, room temperature, 20 h; (**b**) phthalimidoacetic acid, 2-chloro-1-methylpyridinium iodide, triethylamine, CH$_2$Cl$_2$, 0 °C, 2 h, room temperature, 20 h; (**c**) ethylenediamine, EtOH, 65 °C, 1 h; (**d**) R-NCO, acetonitrile, 0 °C, 2 h, room temperature, 18 h; and (**e**) 25% NaOMe in MeOH, MeOH, 65 °C, 1 h.

The protecting group, the phthalimide group, was removed with ethylenediamine in anhydrous ethanol, yielding the (±)-3-amino-β-lactam **3** in a 67% yield (Scheme 1) [58].

In the next reaction step, compound **3** was reacted with various aliphatic and aromatic isocyanates in dry acetonitrile at room temperature, resulting in the formation of (±)-*trans*-β-lactam ureas **4a–i** (Scheme 1) [63]. The resulting products **4a–i** were obtained in yields ranging from 75 to 98%. ^{1}H and ^{13}C NMR, HRMS, and FTIR were used to characterize these compounds **4a–i**. In ^{1}H NMR, the appearance of proton signals at about δ = 4.40 ppm and about 5.00 ppm, respectively, as doublets, correspond to the protons of the C-3 and C-4 of the β-lactam ring, respectively. The signal of the NH proton bound to the ring system is observed as a doublet in the chemical shift range of 5.78 to 6.55 ppm. The signal of the second NH proton is observed as a singlet, doublet, or triplet, depending on which alkyl or aryl substituent is bound to this NH proton. The ^{13}C NMR spectra also showed the peaks for the β-lactam and urea carbonyl carbons at about 166 and 156 ppm, respectively, confirming the synthesis of the β-lactam ureas. The appearance of peaks at 3311–3396 cm^{-1} (N-H stretching), 1732–1756 cm^{-1} (C=O stretching), and 1426 cm^{-1} (C-N stretching), in the FTIR spectra reconfirmed the β-lactam and urea moieties.

Subsequently, the treatment of *trans*-β-lactam urea **4a–i** with sodium methoxide in dry methanol at 65 °C for 1 h [45] resulted in the formation of hydantoins **5a–i** in good to excellent yields but with poor-to-modest diastereoselectivities (Scheme 1 and Table 1). The diastereomeric ratio was determined by RP-HPLC analysis and ^{1}H NMR spectroscopy of reaction mixtures. For example, the diastereomeric ratio for the allyl derivative **5a** (Entry 1) is 52.5:47.5, indicating a slight preference for the *syn*-isomer. However, the diastereomeric ratio for the 2,6-dimethylphenyl derivative **5i** (Entry 9) is 58.8:41.2, indicating a much stronger preference for the *syn*-isomer. The diastereomeric ratio for 3-chloro-4-methylphenyl **5g** indicates a slight preference for the *anti*-isomer, but it is not a highly diastereoselective reaction (48.3:51.7).

Table 1. Structures, yields, and diastereomeric ratio of 3,5-disubstituted hydantoins **5a–i**.

Entry	β-Lactam Urea	Hydantoin	R	η (%)	HPLC Syn:Anti
1	4a	5a	allyl	85.8	52.5:47.5
2	4b	5b	hexyl	91.8	50.1:49.9
3	4c	5c	cylopentyl	79.5	51.8:48.2
4	4d	5d	furfuryl	81.5	57.1:42.9
5	4e	5e	benzyl	86.4	52.2:47.8
6	4f	5f	4-*tert*-butylphenyl	88.8	66.0:34.0
7	4g	5g	3-chloro-4-methylphenyl	75.3	48.3:51.7
8	4h	5h	3,5-dimethylphenyl	58.8	64.3:35.7
9	4i	5i	2,6-dimethylphenyl	75.8	58.8:41.2

According to the reaction mechanism proposed by Mehra and Kumar (Scheme 2), the formation of hydantoin **5** from β-lactam urea **4** can involve methoxide-assisted tandem intermolecular amidolysis–intramolecular cyclization (pathway A, Scheme 2) or a base-assisted formation of a ureido anion followed by intramolecular amidolysis (pathway B, Scheme 2) [45,64]. If the reaction was carried out under mild reaction conditions, the second reaction pathway is more likely. In the presence of a base, such as sodium methoxide, optically active hydantoin tautomerizes easily, indicating the existence of an enol tautomer **8** [65].

The racemic *trans*-β-lactam ureas **4a–i** were used to prepare 3,5-disubstituted hydantoins, and as a result, a racemic mixture of *syn*- and *anti*-hydantoins **5a–i** was obtained. The diastereomerically pure hydantoins **5a–i** were obtained using preparative HPLC with a linear gradient of water (mobile phase A) and acetonitrile (mobile phase B).

The isolated *syn*- and *anti*-diastereomers **5a–i** were characterized using various analytical techniques, including melting point determination, thin-layer chromatography (TLC), infrared (IR) spectroscopy, nuclear magnetic resonance (NMR) spectroscopy, high-performance liquid chromatography (HPLC), and high-resolution mass spectrometry (HRMS), as detailed in the Materials and Methods section.

The IR spectroscopy study confirms the structure of the compounds mentioned above. The IR spectra of the hydantoin derivatives **5a–i** show absorption peaks in the range of 3200 to 3300 cm^{-1}, corresponding to the N-H stretching vibrations of the imide and amide functionalities. Absorption bands observed between 1770 and 1705 cm^{-1} indicate the asymmetric and symmetric stretching of the carbonyl groups in the hydantoin ring. In addition, the IR spectra of compounds **5a–i** show absorption bands around 1610 cm^{-1} (amide I) and around 1510 cm^{-1} (amide II), which are characteristic of amide groups.

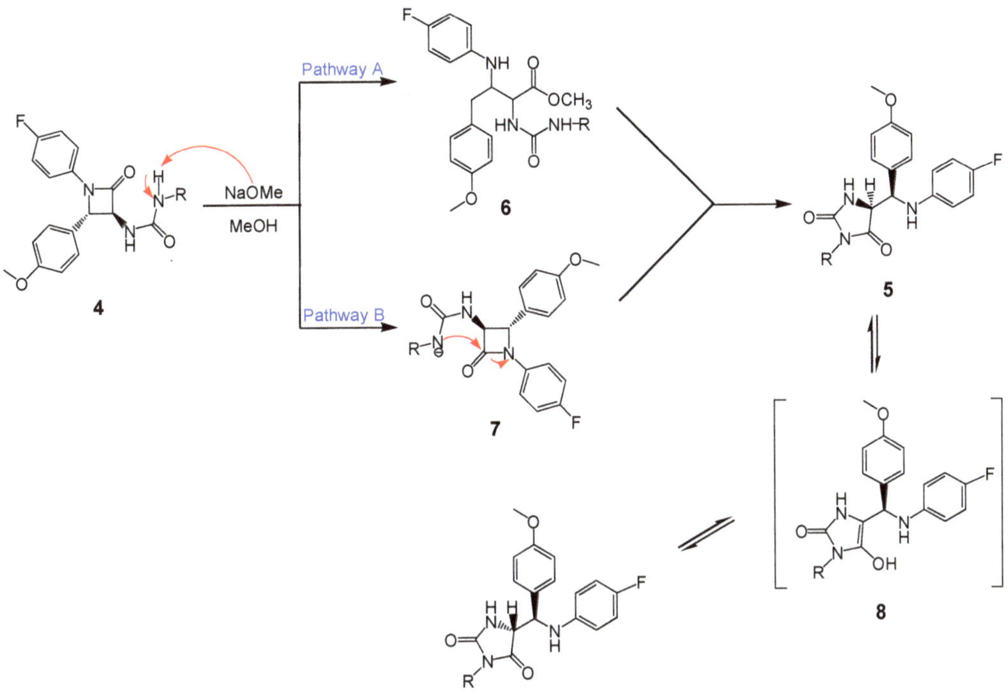

Scheme 2. Plausible mechanism for the formation of hydantoin **5a–i** from β-lactam urea **4a–i**.

2.2. NMR Analysis of Syn- and Anti-5a

A comparison of chemical shifts, NOE interactions, and coupling constants was performed to obtain more information about the conformation of the two diastereoisomers. Structures, stereochemistry, and numbering for the first eluted (Peak 1) and the second eluted (Peak 2) diastereomer are shown in Figure 2. A comparison of the chemical proton shifts (Table S1) revealed the largest difference in the allylic protons CH_2-7, H-8 and CH_2-9, which does not reflect differences in the stereochemistry of the two isomers, but is most likely a consequence of the different positioning of the allylic moiety, possibly due to a different conformation of the hydantoin ring in the two diastereoisomers. A comparison of the carbon chemical shifts (Table S2) revealed the largest difference in the chemical shifts of C-6 and C-17, which is a direct consequence of the different stereochemistry at position C-6. A comparison of the NOE interactions (Table S3) revealed only one significant difference. Two weak interactions exist in the first eluted diastereoisomer (Peak 1), H-5/H12,16 and H-5/H18,22, indicating that in this molecule both aromatic rings are positioned equally close to the hydantoin ring proton H-5. In the second eluted diastereoisomer (Peak 2), however, there is only one medium interaction H-5/H18,22 which means that in this molecule the conformation is different and the methoxybenzene is close to H-5, but the fluorobenzene is not. A comparison of the coupling constants (Table S4) showed the largest difference in $^3J_{\text{H-6,NH-10}}$, which was 6.4 Hz for Peak 1 (*anti*-**5a**) and 10.6 Hz for Peak 2 (*syn*-**5a**). This suggests that the angle between H-9 and NH-10 has changed with different stereochemistry. Although it was not possible to unambiguously determine the stereochemistry of **5a** using NMR data alone, the combination with molecular modelling proved to be successful.

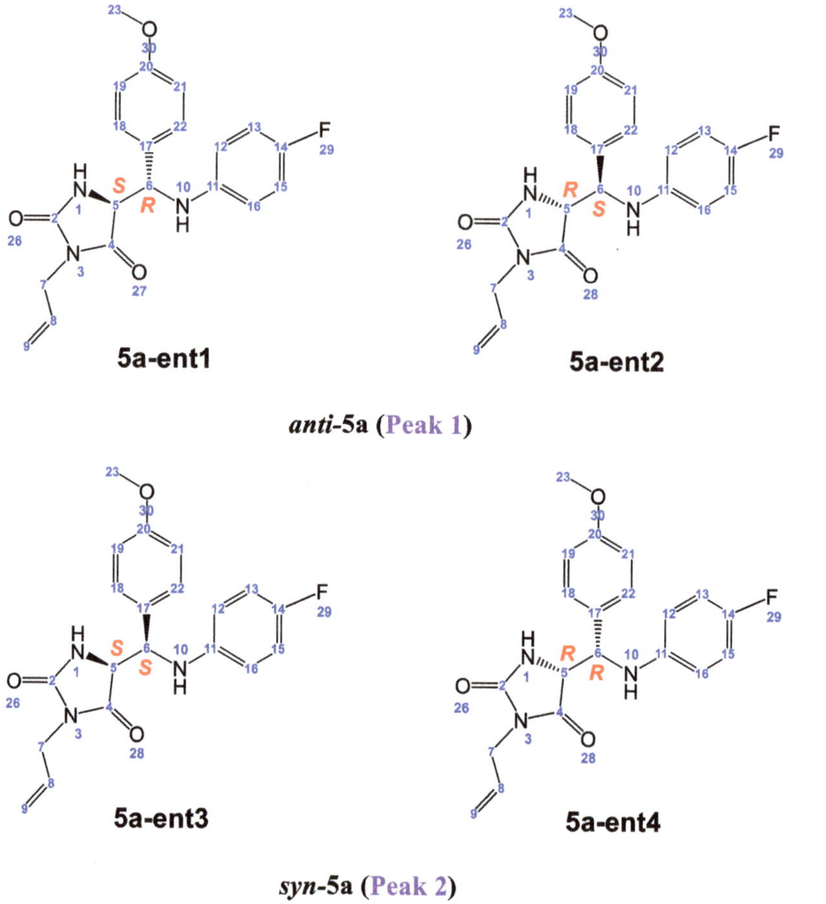

Figure 2. Structures, stereochemistry, and numbering for Peak 1 and Peak 2.

2.3. ECD and VCD Analysis and Absolute Configuration Determination of Syn-5a and Anti-5a Allyl Hydantoin

The racemic *syn-* and *anti-*allyl hydantoins, *syn-***5a** and *anti-***5a**, were selected for this study and purified to enantiomeric purity (e.e. > 99%) by preparative HPLC on a chiral semi-preparative CHIRAL ART Amylose-SA column using *n*-hexane/2-PrOH (90/10, *v/v*) as the mobile phase. Electronic circular dichroism (ECD) and vibrational circular dichroism (VCD) spectra of the enantiomers **5a-ent1**, **5a-ent2** (*anti-***5a**, Peak 1), **5a-ent3**, and **5a-ent4** (*syn-***5a**, Peak 2) were recorded in acetonitrile and calculated by density functional theory (DFT). Figure 3 reports the UV and ECD spectra of the four isomers of **5a**. As expected, the UV spectra are all very similar, while the ECD spectra appear paired in mirror-image couples for the two enantiomers of each diastereomer of **5a**. It was anticipated that ECD is dominated by the reciprocal arrangement between the two aromatic chromophores, and therefore, could be especially sensitive to the configuration at C-6; however, major differences also emerge between the diastereomeric pairs, related to the configuration at C-5.

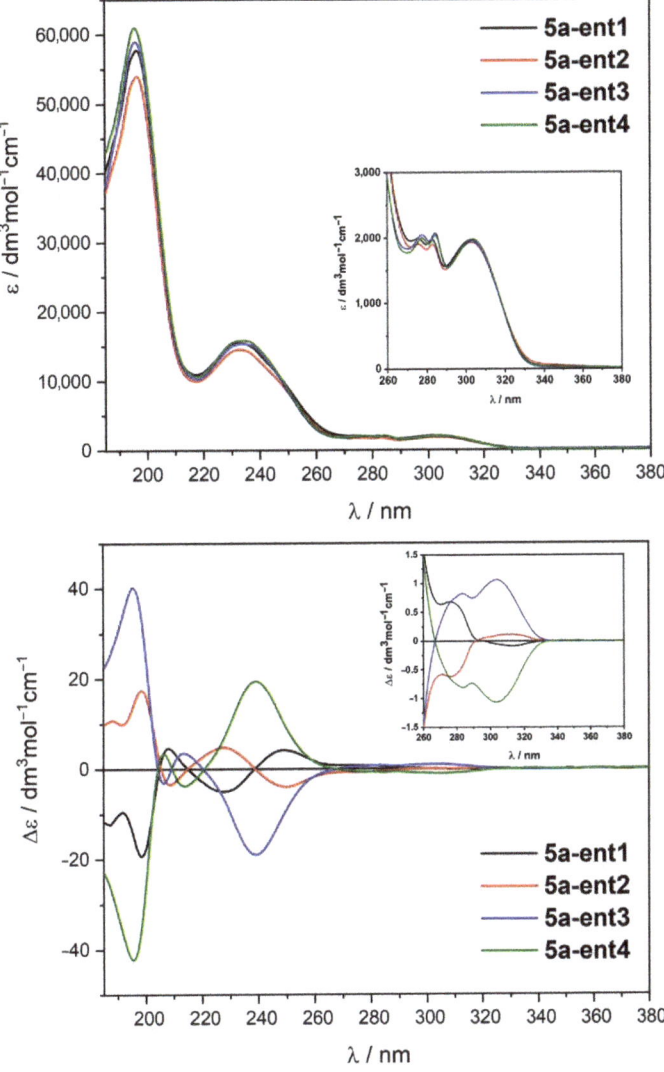

Figure 3. Experimental UV and ECD spectra of hydantoins **5a-ent1** (black line), **5a-ent2** (red line), **5a-ent3** (blue line), and **5a-ent4** (green line) recorded in CH_3CN.

The computational procedure employed for ECD calculations followed a well-established workflow [66,67]. The input structures had (5R,6S) and (5S,6S) configurations for the *syn*- and *anti*-isomers of **5a**, respectively. The conformational space was sampled by means of a systematic search with molecular mechanics force field (MMFF). All conformers thus obtained were first optimized by DFT at B3LYP-D3/6-31G(d) level in vacuo, then reoptimized at B3LYP-D3BJ/6-311+G(d,p) level in vacuo, and final energies were estimated at the same level including the PCM solvent model for acetonitrile. In this way, five different conformers were obtained for (5R,6S)-**5a** and eight for (5S,6S)-**5a** with Boltzmann populations > 3% at 300 K (see Supporting Information, Tables S5 and S6). Although comparing experimental NOE data with optimized geometries is not trivial for compounds with multiple fast-exchanging conformers, we notice that the (5S,6S)-**5a** isomer yields low-energy structures with H-2 directed toward the fluorobenzene ring, while this is not true for (5*R*,6*S*)-**5a**. According to NOE results (see above), this suggests an *anti* stereochemistry for Peak 1 and *syn* stereochemistry for Peak 2, corresponding, respectively, to configuration (5*S**,6*S**) and (5*R**,6*S**). ECD calculations were then run with time-dependent DFT (TD-DFT) at CAM-B3LYP/def2-TZVP/PCM and B3LYP/def2-TZVP/PCM levels. Final spectra were generated as Boltzmann averages at 300 K using internal energies and plotted as a sum of Gaussians with best-fit bandwidth. The comparison between the calculated and experimental ECD spectra (Figure 4) suggests that the isomer **5a-ent1** has a (5*S*,6*S*) configuration and the isomer **5a-ent4** has a (5*R*,6*S*) configuration. The assignment was substantiated by evaluating similarity factors between experimental and calculated spectra (Supporting Information, Table S7) [68]. Furthermore, to confirm the assignment for **5a-ent1/5a-ent2**, the uncertainty related to the conformational averaging was reduced by considering a truncated model where the *N*-allyl group was replaced by *N*-methyl. This truncation approach [69] led to the same assignment as above (Supporting Information, Tables S8 and S9 and Figure S147).

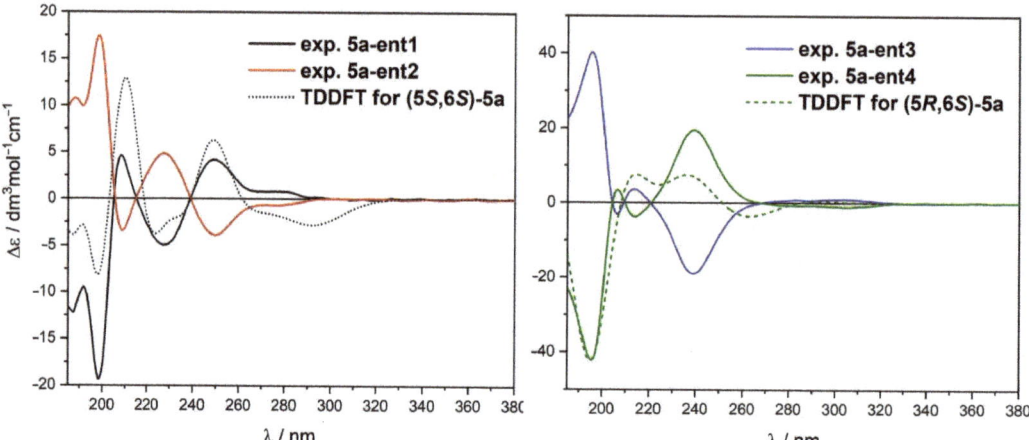

Figure 4. Comparison between experimental ECD spectra of hydantoins **5a-ent1/5a-ent2** (black/red lines) and **5a-ent3/5a-ent4** (blue/green lines) recorded in CH_3CN with the spectra calculated for (5*S*,6*S*)-**5a** (black dotted line) and (5*R*,6*S*)-**5a** (green dashed line) at TD B3LYP/def2-TZVP/PCM//B3LYP-D3BJ/6-311+G(d,p) level as Boltzmann averages over eight and five conformers, respectively. Plotting parameters for (5*S*,6*S*)-**5a**: UV shift +15 nm, σ = 0.24 eV. Plotting parameters for (5*R*,6*S*)-**5a**: UV-shift +4 nm, σ = 0.38 eV.

Finally, following the advice to use multiple chiroptical techniques to avoid potentially incorrect assignments [70,71], we extended our study to vibrational CD (VCD). The VCD spectra measurements for **5a-ent3** and **5a-ent4** were especially good in terms of signal-to-noise ratio thanks to the availability of a larger amount of enantiopure samples and looked like mirror images throughout the fingerprint region (Figure 5). DFT calculations run at B3LYP-D3BJ/6-311+G(d,p)/PCM level on (5*R*,6*S*)-**5a** reproduced satisfactorily the experimental VCD spectrum for **5a-ent4**, thus confirming the assignment obtained by ECD (Figure 5; similarity factors are provided in the Supporting Information, Table S7). In conclusion, the chiroptical analysis afforded the following assignments: (5*S*,6*S*)-**5a-ent1**, (5*R*,6*R*)-**5a-ent2**, (5*S*,6*R*)-**5a-ent3**, and (5*R*,6*S*)-**5a ent4**.

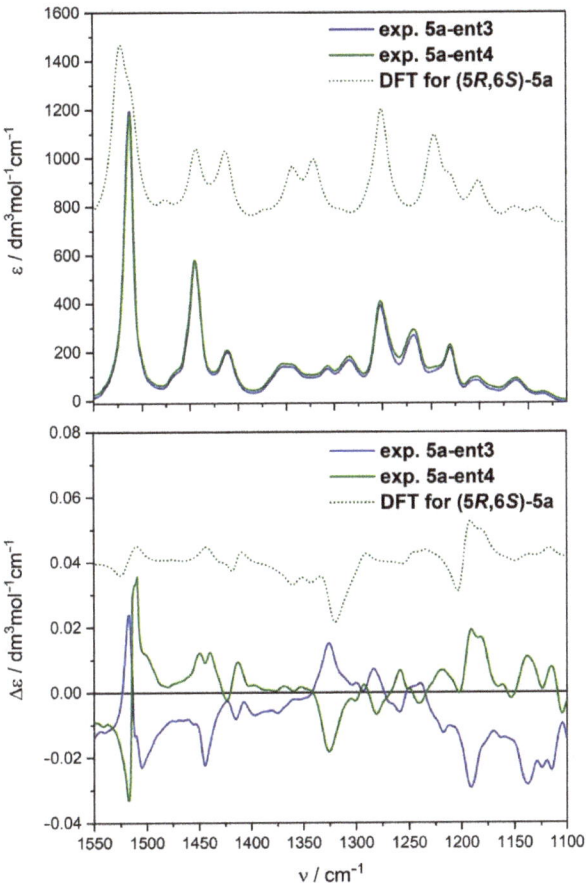

Figure 5. Comparison between experimental IR (top) and VCD spectra (bottom) of hydantoins **5a-ent3** (blue line) and **5a-ent4** (green line) recorded in CD$_3$CN with the spectrum calculated for (5R,6S)-**5a** at B3LYP-D3BJ/6-311+G(d,p)/PCM level as Boltzmann averages over 8 conformers. Plotting parameters: scaling factor 0.985, σ = 8 cm^{-1}.

2.4. Antiproliferative Activity of Syn- and Anti-**5a–i**

In the present work, the eighteen synthesized hydantoin derivatives, *syn*-**5a–i** and *anti*-**5a–i**, were evaluated in vitro for their antiproliferative activity by MTT assay. Three human cancer cell lines, including liver hepatocellular carcinoma cells (HepG2), ovarian carcinoma cells (A2780), and breast adenocarcinoma cells (MCF7), and a human non-tumor foreskin fibroblasts cell line HFF-1, were selected to determine in vitro antiproliferative activity. The results are summarized in Table 2. The following compounds, *anti*-**5b**, *syn*-**5f**, *anti*-**5f**, and *anti*-**5g**, showed moderate activity against liver cancer cell lines HepG2 with IC$_{50}$ values in the range 15–35 μmol/L, while the other *syn*- and *anti*-hydantoin compounds showed insignificant inhibitory potential on HepG2 cell lines. Most of the tested compounds showed moderate antiproliferative activity, while the two hydantoins *anti*-**5a** and *syn*-**5d** showed no cytotoxicity against A2780 cells. Of the eighteen tested hydantoins, eleven exhibited moderate antiproliferative potential against MCF7 cells with IC$_{50}$ values in the range 20–55 μmol/L. Compounds *syn*-**5a**, *anti*-**5a**, *syn*-**5d**, *anti*-**5d**, and *syn*-**5e** showed no cytotoxicity against MCF7 cells. Interestingly, the compound *anti*-**5c** with a cyclopentyl group at the N-3 position of the hydantoin ring showed significant antiproliferative effects on human breast carcinoma cell line MCF7 with an IC_{50} value of 4.5 µmol/L. However, its diastereomer *syn*-**5c** only showed moderate antiproliferative effect against the same cell line (IC_{50} = 41 µmol/L). The hydantoin compounds *anti*-**5b**, *anti*-**5c**, *syn*-**5f**, *syn*-**5g**, *anti*-**5g**, *syn*-**5h**, and *anti*-**5h** showed moderate cytotoxic effects, whereas the other hydantoins had less or no toxic effect on the HFF-1 healthy cells.

Table 2. Antiproliferative activities of hydantoins *syn*-**5a–i** and *anti*-**5a–i** in vitro (IC_{50}, µmol/L).

Compound	HepG2 *	A2780 *	MCF7 *	HFF-1 *
syn-**5a**	>100	71 ± 34	>100	>100
anti-**5a**	>100	>100	>100	>100
syn-**5b**	>100	93 ± 123	55 ± 1.8	>100
anti-**5b**	35 ± 12	33 ± 12	34 ± 9	52 ± 15
syn-**5c**	>100	35 ± 7.9	41 ± 5.4	>100
anti-**5c**	>100	42 ± 57	4.5 ± 0.45	12 ± 13
syn-**5d**	>100	>100	>100	>100
anti-**5d**	>100	79 ± 5.4	>100	>100
syn-**5e**	>100	59 ± 17	>100	>100
anti-**5e**	>100	58 ± 7.7	55 ± 2.4	>100
syn-**5f**	15 ± 1.2	15 ± 1.7	20 ± 0.15	20 ± 0.48
anti-**5f**	33 ± 1.4	22 ± 5.9	39 ± 9.5	>100
syn-**5g**	>100	28 ± 13	21 ± 0.35	33 ± 18
anti-**5g**	30 ± 3.0	43 ± 13	27 ± 1.4	35 ± 55
syn-**5h**	>100	43 ± 4.0	27 ± 2.2	21 ± 49
anti-**5h**	>100	59 ± 2.4	48 ± 14	18 ± 27
syn-**5i**	>100	77 ± 3.0	87 ± 32	>100
anti-**5i**	>100	48 ± 10	52 ± 21	94 ± 11

* The human cancer cell lines used in this study were HepG2 (derived from liver), A2780 (derived from ovarian tissue), and MCF-7 (derived from breast tissue). The normal human cell line used was HFF1 (human fibroblasts). Cell proliferation was assessed after 72 h using the MTT assay, as delineated in the Materials and Methods section.

2.5. In Silico Physicochemical and Biological Profiling of Syn/Anti-5a–i

The ADMET profile for the synthesized derivatives **5a–i** was estimated using the program ADMET Predictor [53] and the web server admetSAR [54]. All synthesized 3,5-disubstituted hydantoins **5a–i**, with the exception of **5f**, are drug-like molecules compliant with Lipinski's and Veber's rules (Table S12). Orally administered molecules are likely to have logP ≤ 5, MW ≤ 500, HBD ≤ 5, and HBA ≤ 10 according to Lipinski's rule of five [72,73], and/or TPSA ≤ 140 Å^2 (or 12 or fewer HBD and HBA), and a number of rotatable bonds equal to or less than 10 according to Veber's rule [74]. The derivative **5f** with a 4-*tert*-butylphenyl substituent violates Lipinski's rule in terms of lipophilicity (logP = 5.01). While all 3,5-disubstituted hydantoins **5a–i** are nonionizable compounds and most of them have logP ≥ 3 (Table S10), they are predicted to be moderately water soluble (Table S11). All molecules are also predicted to have moderate membrane permeability according to the ADMET predictor's Peff and MDCK values, which indicates their good oral bioavailability. The membrane permeability of a compound may be reduced by its interaction with efflux pumps such as P-glycoprotein (Pgp). The compounds **5b–i** are estimated to be Pgp substrates and many of them are not predicted to be Pgp inhibitors (Table S11). Furthermore, hydantoins **5a–i** are estimated to not penetrate the BBB (BBB_filter) and have low retention in brain (logBB, Table S11), and hence it is very likely that they do not enter the brain [75]. Regarding distribution in the body, they are predicted to be transported by plasma proteins (Table S11, low hum_fup% values). Considering metabolism, the web server admetSAR is commonly used to estimate the probability that the compounds are substrates and inhibitors of cytochrome P450 (CYP450) oxidoreductases. According to admetSAR, all hydantoins **5a–i** should be substrates only for CYP3A4 and most of them are also predicted to be CYP3A4 inhibitors (Table S11). In terms of toxicity, hydantoins **5f–i** are classified as cardiotoxic (hERG_Filter). All synthesized hydantoins **5a–i** are predicted to be non-mutagenic and non-carcinogenic to rats or mice (Table S11).

The software PASS 2020 [55] and the online server SwissTargetPrediction [56] were used to predict biological activities of the new 3,5-disubstituted hydantoins **5a–i**. The PASS 2020 simultaneously predicts 1945 recommended biological activities, including pharmacological effects, mechanisms of action, interaction with drug-metabolizing enzymes and transporters, toxic and adverse effects, influence on gene expression, etc., using built-in machine-learning models. The results consist of Pa and Pi values, which are computed probabilities that a compound is active and inactive, respectively, for each type of activity from the biological activity spectrum. The Pa and Pi values range from 0.0 to 1.0. Considering only the activities predicted with Pa > Pi and Pa > 0.5, all synthesized hydantoins **5a–i** should have antiarthritic activity (Pa $\geq$ 0.727). In addition, only the compound syn/anti-**5d** is predicted to inhibit thioredoxin glutathione reductase (Pa = 0.665), beta lactamase (Pa = 0.554), DNA-directed DNA polymerase (Pa = 0.553), and DNA polymerase beta (Pa = 0.542). In contrast to PASS, SwissTargetPrediction estimated protein targets based on the structural similarity of hydantoins **5a–i** with 327,719 already known actives on 2092 human proteins. No high-confidence predictions were found indicating substantial novelty in the structures of the synthesized 3,5-disubstituted hydantoins.

3. Materials and Methods

3.1. Chemistry

3.1.1. General

All required chemicals were purchased from commercial suppliers, including Sigma-Aldrich (Munich, Germany), Merck (Darmstadt, Germany), Fluka (Buchs, Switzerland), and Kemika (Zagreb, Croatia). Prior to use, the solvents dichloromethane, ethanol, and acetonitrile were subjected to the required drying procedures according to established protocols. The HPLC-grade solvents, acetonitrile, 2-propanol, and *n*-hexane, were purchased from Honeywell (Seelze, Germany). The reactions were monitored by thin-layer chromatography (TLC), which was performed on 0.25 mm silica gel plates (50F-254, DC-Alufolien-Kieselgel F254, Sigma Aldrich, Merck KGaA, Darmstadt, Germany). The TLC plates were observed under ultraviolet light (254 nm) or stained with ninhydrin. Flash column chromatography was performed with silica gel Si50 (particle size 0.04–0.053 mm, 230–400 mesh, Sigma Aldrich, Munich, Germany).

The determination of melting points was conducted using an Olympus BX51 polarizing microscope (Olympus Corporation, Tokyo, Japan), which was equipped with a Linkam TH500 hot stage and a PR500 temperature controller (Technical Manufacturing Corporation, Peabody, Massachusetts, USA). Images were captured using the Olympus C5050 ZOOM digital camera.

FTIR-ATR spectra were obtained using a PerkinElmer UATR Two spectrometer (PerkinElmer Inc., Waltham, MA, USA), operating within a range of 450 cm^{-1} to 4000 cm^{-1}.

All solvents used for NMR sample preparation were purchased from EurIsotop (Saint-Aubin, France). ^{1}H and ^{13}C NMR spectra were recorded on a Brucker AV 300 or AV 500 (1H 300 or 500 MHz and ^{13}C 75 or 151 MHz) spectrometers (Bruker Technologies, Ettlingen and Leipzig, Germany) in chloroform (CDCl$_3$) or acetonitrile (CD$_3$CN) at ambient temperature. All solvents utilized for the preparation of NMR samples were obtained from EurIsotop (Saint-Aubin, France). The ^{1}H and ^{13}C NMR spectra were recorded on Brucker AV 300 or AV 500 spectrometers (^{1}H 300 or 500 MHz and ^{13}C 75 or 151 MHz) (Bruker Technologies, Ettlingen and Leipzig, Germany), utilizing chloroform (CDCl$_3$) or acetonitrile (CD$_3$CN) at ambient temperature. The chemical shifts (δ) were expressed in parts per milion (ppm) relative to tetramethylsilane (TMS), which was used as an internal standard. All coupling constants (*J*) are given in hertz (Hz). The splitting patterns were reported as follows: s for singlet; d for doublet; t for triplet; p for pentet; h for heptet; m for multiplet; bs for broad singlet; dd for a doublet of doublets; and tdd for a triplet of a doublet of doublets.

An Agilent 1290 Infinity II/5550 Q-TOF instrument (Agilent Technologies, Waldbronn, Germany) with electrospray ionization (ESI) quadrupole-time-of-flight (Q-TOF) mass spectrometry with high-resolution was used as the analytical technique to determine the molecular weight of the synthesized compounds.

RP-HPLC studies were performed with the aim of determining the *cis/trans* ratio of 3-phthalimido-β-lactam **2** and determining the purity of compound 3-amino-β-lactam **3**. The Symmetry C18 column (150 × 4.6 mm, 5 µm, Waters, Milford, MA, USA) was used for these analyses. The analyses were carried out on an Agilent 1200 Series System (Agilent Technologies, Waldbronn, Germany), which was equipped with a vacuum degasser, a quaternary pump, a thermostatically controlled column compartment, an autosampler, and a variable wavelength detector. The methodology used was as follows: water served as eluent A, while acetonitrile constituted eluent B. The gradient for the first 0–15 min went from 30% to 70% eluent B. From 15–18 min, an isocratic phase with 70% eluent B was applied. From 18.01 to 21 min, an additional isocratic phase with 30% B was introduced. The flow rate was maintained at 1 mL/min, the detection wavelength was set to 254 nm, the column temperature was set to 30 °C, and the injection volume was set to 20 µL.

Trans-(±)-β-lactam ureas **4a–i** were analyzed by RP-HPLC using a Synergi Polar-RP 80A column (150 mm × 4.6 mm, 4 µm, Phenomenex, Torrance, CA, USA) with a gradient elution setup consisting of (A) water + 0.1% trifluoroacetic acid and (B) acetonitrile. The elution program was as follows: 0–10 min, 50–100% B; 10–13 min, 100% B; and 13.01–17 min, 50% B. The flow rate was maintained at 1.0 mL/min. The column temperature was maintained at 30 °C, and the sample injection volume was 20 µL. UV detection was performed at a wavelength of 254 nm.

The *syn/anti* ratio of hydantoins was determined by HPLC on an Agilent 1200 series system using a Zorbax Extend-C18 column (250 × 4.6 mm, 5 µm, Agilent Technologies, Milford, MA, USA). The flow rate was set to 0.8 mL/min, the column temperature was maintained at 30 °C. A sample volume of 20 µL was injected and UV detection was performed at a wavelength of 254 nm. A gradient elution of two solvents was used—Solvent A (water) and Solvent B (acetonitrile). The mobile phase gradient elutions used in the three methods were as follows: For Method A (for **5a, 5d, 5e**), the gradient program was as follows: 0–60 min, 35–48% B; 60–63 min, 48% B; and 63.01–67 min, 35% B. For Method B (for **5c, 5h, 5i**), the gradient program was as follows: 0–60 min, 35–60% B; 60–63 min, 60% B; and 63.01–67 min, 35% B. For Method C (for **5b, 5f, 5g**), the gradient program was as follows: 0–60 min, 45 58% B; 60–63 min, 58% B; and 63.01–67 min, 45% B. The preparative HPLC was carried out on an Agilent 1260 Infinity II HPLC system, which was equipped with an Agilent MWD detector. A Zorbax Extend C-18 PrepHT preparative column (250 mm × 21.2 mm i. d., 5-µm particle size) was used, and the gradient elution program with water (phase A) and acetonitrile (phase B) was delivered at 17 mL/min. The chromatograms were monitored at 254 nm. In addition, the injection volume was 500 µL. The mobile phase gradient elutions used in the three methods were as follows: For Method D (for **5a, 5d, 5e**), the gradient program was as follows: 0–35.98 min, 35–48% B; 35.98–37.78 min, 48% B; and 37.79–40.18 min, 35% B. For Method E (for **5c, 5h, 5i**), the gradient program was as follows: 0–35.98 min, 35–60% B; 35.98–37.78 min, 60% B; and 37.78–40.18 min, 35% B. For Method F (for **5b, 5f, 5g**), the gradient program was as follows: 0–35.98 min, 45–58% B; 35.98–37.78 min, 58% B; and 37.78–40.18, min 45% B.

The chiral separation of *syn*- and *anti*-allyl hydantoin enantiomers **5a** was carried out on the CHIRAL ART Amylose-SA semi-preparative column (250 mm × 8.0 mm i. d., 10-µm particle size). An isocratic elution was performed with *n*-hexane/PrOH (90/10, *v/v*) at a flow rate of 5 mL/min. The chromatograms were monitored at 254 nm and the injection volume was 500 µL. The 10 µm bulk immobilized amylose-based chiral stationary phase CHIRAL ART Amylose-SA S-10 µm used in the HPLC column was purchased from YMC (Kyoto, Japan) and packed in house using empty stainless steel HPLC columns (dimensions: 250 mm × 4.6 mm i. d. and 250 mm × 8.0 mm i. d.) from Knauer GmbH (Berlin, Germany).

The enantioseparation of the hydantoins *syn*-**5a–i** and *anti*-**5a–i** was carried out on an Agilent 1200 Series HPLC System (Agilent Technologies, Waldbronn, Germany). The system included a vacuum degasser, a quaternary pump, a thermostatically controlled column compartment, an autosampler, and a variable wavelength detector. The mobile phase was a mixture of *n*-hexane and 2-PrOH (90/10, v/v). All analyses were performed with a flow rate of 1.0 mL/min and at a column temperature of 30 °C. An injection volume of 20 μL was employed.

ECD spectra were recorded using a J-815 spectrometer (Jasco, Tokyo, Japan) at room temperature in spectroscopic grade acetonitrile. Solutions with concentrations in the range of 0.25 to 0.27 mM were measured in quartz cells with a path length of 1 and 0.1 cm. All spectra were carried out in the range from 380 to 180 nm using a scanning speed of 100 nm/min, a step size of 0.2 nm, a bandwidth of 1 nm, a response time of 0.5 s, and an accumulation of 5 scans. The spectra were background-corrected using acetonitrile as the solvent, and they were recorded under the same measurement conditions.

VCD spectra were obtained using an FVS-6000 VCD spectrometer (Jasco, Tokyo, Japan) with a resolution of 4 cm^{-1} within the 2000–850 cm^{-1} range in spectroscopic grade CD$_3$CN for 2 h. Solutions with concentrations in the range of 0.2–0.15 M in a BaF$_2$ cell with a path length of 200 μm were placed on a rotating holder. Baseline correction was made by subtracting the spectrum of CD$_3$CN recorded under identical conditions.

3.1.2. Synthesis of (4-Fluorophenyl)[(4-methoxyphenyl)methylene]amine (**1**)

To a solution containing 4-methoxybenzaldehyde (6 mL, 49.313 mmol) in anhydrous dichloromethane (30 mL) maintained at room temperature, 4-fluoroaniline (4.67 mL, 49.313 mmol) was then added. This was followed by the addition of activated 4 Å molecular sieve powder (2 g). The resulting reaction mixture was stirred at room temperature for 20 h. The mixture was then filtered through a short layer of Celite, which had previously been rinsed with dichloromethane, to remove the molecular sieve powder. The filtrate was concentrated under reduced pressure, yielding a crude imine in the form of an oil. After recrystallization from a solvent mixture of ethyl acetate and *n*-hexane, pure imine **1** (8.1 g, 73%) was isolated as white crystals. m.p. 68–70 °C. FTIR (ATR, cm^{-1}): 2925, 2847, 1626, 1604, 1512, 1458, 1308, 1253, 1208, 1179, 1159, 1021, 841. ^{1}H NMR (300 MHz, CDCl$_3$), δ/ppm: 8.35 (s, 1H), 7.83 (d, J = 8.7 Hz, 2H), 7.21–7.12 (m, 2H), 7.06 (t, J = 8.7 Hz, 2H), 6.97 (d, J = 8.8 Hz, 2H), 3.86 (s, 3H). ^{13}C NMR (150 MHz, CDCl$_3$), δ/ppm: 162.7, 161.0 (d, J = 244.1 Hz), 159.6, 148.5 (d, J = 2.7 Hz), 130.6, 129.3, 122.3 (d, J = 8.1 Hz), 115.9 (d, J = 22.4 Hz), 114.3, 55.6. C$_{14}$H$_{12}$FNO (229.25): calcd. C 73.35, H 5.28, N 6.11; found C 73.06, H 5.40, N 6.12.

3.1.3. Cis/trans-(±)-1-(4-fluorophenyl)-2-(4-methoxyphenyl)-4-oxoazetidin-3-yl-isoindoline-1,3-dione (**2**)

A suspension containing phthalimidoacetic acid (4.03 g, 19.529 mmol) in anhydrous dichloromethane (150 mL) was cooled to 0 °C under an inert argon atmosphere. 2-Chloro-1-methylpyridinium iodide (6.69 g, 39.258 mmol) and triethylamine (10.9 mL, 78.516 mmol) were added to this solution. The resulting reaction mixture was stirred for 2 h to 0 °C. After that, a solution of imine **1** (3.0 g, 13.086 mmol) in anhydrous dichloromethane (25 mL) was added dropwise and stirring was then continued for a further 2 h, during which the temperature rose from 0 °C to room temperature. The mixture was then refluxed for 20 h and progress was monitored by TLC and HPLC. After the completion of the reaction, the reaction mixture was washed with a solution of saturated sodium bicarbonate (150 mL), a solution of saturated sodium chloride (150 mL), and deionized water (150 mL). The organic layer was then dried over anhydrous sodium sulfate, filtered, and concentrated under reduced pressure, yielding the crude product. The *cis/trans* ratio of the crude product was determined to be 1:5. The crude product was purified by column chromatography over silica gel in a chloroform/dichloromethane/ethyl acetate mixture (7/1/1, $v/v/v$). The pure forms of the *cis*- and *trans*-3-phthalimido-β-lactams were obtained as a white solid. R_f (*cis*-**2**) = 0.59, R_f (*trans*-**2**) = 0.71 (chloroform/dichloromethane/ethyl acetate = 7/1/1, $v/v/v$).

Cis-**2**

White solid (600 mg, 11%), m.p. 194.1−198.8 °C. FTIR (ATR, cm^{-1}): 2895, 1781, 1756, 1721, 1611, 1510, 1467, 1442, 1385, 1254, 1175, 1115, 1053, 956, 832, 814, 750, 716. ^{1}H NMR (600 MHz, CDCl$_3$), δ/ppm: 7.77−7.61 (m, 4H), 7.48−7.38 (m, 2H), 7.17 (d, J = 8.7 Hz, 2H), 7.02 (t, J = 8.7 Hz, 2H), 6.72 (d, J = 8.8 Hz, 2H), 5.65 (d, J = 5.5 Hz, 1H), 5.42 (d, J = 5.5 Hz, 1H), 3.66 (s, 3H). ^{13}C NMR (151 MHz, CDCl$_3$), δ/ppm: 166.9, 160.9, 159.8, 159.5 (d, J = 244.2 Hz), 134.4, 133.8 (d, J = 2.7 Hz), 131.3, 128.7, 123.7, 119.0 (d, J = 7.9 Hz), 116.1 (d, J = 22.8 Hz), 114.2, 61.0, 59.3, 55.3. HR-MS (ESI-QTOF) m/z: calcd. for C$_{24}$H$_{17}$FN$_2$O$_4$, [M + H]$^+$, 417.1251; found 417.1255.

Trans-**2**

White solid (4.1 g, 75%), m.p. 197.3−198.8 °C. FTIR (ATR, cm^{-1}): 3007, 2932, 2836, 1778, 1754, 1713, 1612, 1507, 1467, 1369, 1311, 1253, 1229, 1180, 1148, 1120, 1105, 1034, 970, 941, 833, 717, 530. ^{1}H NMR (600 MHz, CDCl$_3$), δ/ppm: 7.91−7.84 (m, 2H), 7.80−7.73 (m, 2H), 7.36−7.27 (m, 4H), 7.01−6.88 (m, 4H), 5.31 (d, J = 2.6 Hz, 1H), 5.28 (d, J = 2.6 Hz, 1H), 3.81 (s, 3H). ^{13}C NMR (151 MHz, CDCl$_3$), δ/ppm: 166.9, 162.1, 160.5, 159.5 (d, J = 244.1 Hz), 134.7, 133.6 (d, J = 2.8 Hz), 131.8, 127.7, 127.4, 124.0, 119.3 (d, J = 8.0 Hz), 116.1 (d, J = 22.8 Hz), 115.0, 63.1, 61.3, 55.5. HR-MS (ESI-QTOF) m/z: calcd. for C$_{24}$H$_{17}$FN$_2$O$_4$, [M + H]$^+$, 417.1251; found 417.1248.

3.1.4. Trans-3-amino-1-(4-fluorophenyl)-4-(4-methoxyphenyl)-2-azetidinone (**3**)

To a suspension of *trans*-3-phthalimido-β lactam **2b** (3.78 g, 9.078 mmol) in dry ethanol (100 mL) was added ethylenediamine (1.21 mL, 18.156 mmol), and the reaction mixture was stirred at 65 °C under an argon atmosphere. The progress of the reaction was monitored by TLC. After one hour, the ethanol was evaporated in vacuo and the resulting crude product was dissolved in ethyl acetate (200 mL). The solution was then washed with a solution of saturated sodium chloride (100 mL) and deionized water (100 mL). The organic layer was dried over anhydrous sodium sulfate, filtered and concentrated under reduced pressure to give the crude product which was then purified by SiO$_2$ flash column chromatography with ethyl acetate as an eluent to give compound **3** (1.73 g, 67%) as a white solid. R_f = 0.39 (ethyl acetate), m.p. 98.7−101.2 °C. FTIR (ATR, cm^{-1}): 3384, 3341, 3177, 2933, 2840, 1732, 1611, 1510, 1483, 1380, 1246, 1229, 1154, 1115, 1029, 965, 832, 793, 717. ^{1}H NMR (600 MHz, CDCl$_3$), δ/ppm: 7.30−7.20 (m, 4H), 6.98−6.84 (m, 4H), 4.60 (d, J = 2.2 Hz, 1H), 4.04 (d, J = 2.2 Hz, 1H), 3.80 (s, 3H), 1.82 (bs, 2H). ^{13}C NMR (151 MHz, CDCl$_3$), δ/ppm: 168.0, 160.0, 159.2 (d, J = 243.6 Hz), 133.8 (d, J = 2.8 Hz), 128.6, 127.3, 119.0 (d, J = 7.9 Hz), 116.0 (d, J = 22.6 Hz), 114.7, 70.2, 66.6, 55.5. HR-MS (ESI-QTOF) m/z: calcd. for C$_{16}$H$_{15}$FN$_2$O$_2$, [M + H]$^+$, 287.1191; found 287.1183.

3.1.5. General Procedure for the Preparation of (±)-*trans*-β-lactam Ureas **4a**–**i**

To a solution of *trans*-3-amino-β-lactam **3** (1.0 equiv.) in dry acetonitrile (5 mL) was added the appropriate isocyanate (1.5 equiv.) and the resulting reaction mixture was stirred for 20 h at room temperature. Acetonitrile was then concentrated under reduced pressure and the obtained crude product was purified by SiO$_2$ flash column chromatography with *n*-hexane/ethyl acetate (4/1, v/v). The pure *trans*-β-lactam urea was finally obtained by trituration with *n*-hexane.

Allyl-3-[(±)-*trans*-1-(4-fluorophenyl)-2-(4-methoxyphenyl)-4-oxoazetidin-3-yl]urea (**4a**)

Compound **4a** was prepared from **3** (50 mg, 0.175 mmol) and allyl isocyanate (23.2 μL, 0.263 mmol) according to the general synthetic procedure, producing a white solid (57.7 mg, 89%). R_f = 0.22 (chloroform/ethyl acetate = 4/1, v/v), m.p. 86.1−88.5 °C. FTIR (ATR, cm^{-1}): 3346, 2933, 2838, 1745, 1636, 1612, 1560, 1507, 1426, 1288, 1247, 1224, 1175, 1137, 1102, 923, 830. ^{1}H NMR (600 MHz, CDCl$_3$), δ/ppm: 7.25−7.17 (m, 4H), 6.92−6.84 (m, 4H), 6.04 (d, J = 6.8 Hz, 1H), 5.84−5.74 (m, 1H), 5.37 (t, J = 5.8 Hz, 1H), 5.16 (dd, J = 17.2, 1.6 Hz, 1H), 5.07 (dd, J = 10.3, 1.5 Hz, 1H), 5.01 (d, J = 2.3 Hz, 1H), 4.42 (dd, J = 6.8, 2.3 Hz,

1H), 3.79 (s, 3H), 3.78–3.67 (m, 2H). ^{13}C NMR (151 MHz, CDCl$_3$), δ/ppm: 166.3, 160.1, 159.4 (d, J = 244.5 Hz), 157.5, 135.1, 133.6 (d, J = 2.7 Hz), 128.2, 127.6, 119.3 (d, J = 7.9 Hz), 116.0 (d, J = 22.8 Hz), 115.9, 114.8, 67.2, 63.9, 55.5, 42.9. HR-MS (ESI-QTOF) m/z: calcd. for C$_{20}$H$_{20}$FN$_3$O$_3$, [M + H]$^+$, 370.1567; found 370.1579.

1-Hexyl-3-[(±)-*trans*-1-(4-fluorophenyl)-2-(4-methoxyphenyl)-4-oxoazetidin-3-yl]urea (**4b**)

Compound **4b** was prepared from 3 (50 mg, 0.175 mmol) and hexyl isocyanate (38.2 μL, 0.263 mmol) according to the general synthetic procedure, producing a white solid (64.3 mg, 89%). R$_f$ = 0.32 (chloroform/ethyl acetate = 4/1, v/v), m.p. 58.3–62.4 °C. FTIR (ATR, cm^{-1}): 3352, 2956, 2929, 2858, 1748, 1639, 1563, 1507, 1386, 1247, 1226, 1175, 1136, 1031, 830. ^{1}H NMR (600 MHz, CDCl$_3$), δ/ppm: 7.25–7.18 (m, 4H), 6.91–6.85 (m, 4H), 5.95 (d, J = 6.7 Hz, 1H), 5.22 (t, J = 5.6 Hz, 1H), 5.01 (d, J = 2.2 Hz, 1H), 4.40 (dd, J = 6.7, 2.3 Hz, 1H), 3.79 (s, 3H), 3.16–3.00 (m, 2H), 1.43 (p, J = 7.2 Hz, 2H), 1.32–1.19 (m, 6H), 0.90–0.83 (m, 3H). ^{13}C NMR (151 MHz, CDCl$_3$), δ/ppm: 166.5, 160.0, 159.4 (d, J = 244.0 Hz), 157.5, 133.6 (d, J = 2.7 Hz), 128.3, 127.6, 119.2 (d, J = 7.8 Hz), 116.0 (d, J = 22.7 Hz), 114.7, 67.3, 64.0, 55.4, 40.6, 31.7, 30.2, 26.7, 22.7, 14.2. HR-MS (ESI-QTOF) m/z: calcd. for C$_{23}$H$_{28}$FN$_3$O$_3$, [M + H]$^+$, 414.2193; found 414.2158.

1-Cyclopentyl-3-[(±)-*trans*-1-(4-fluorophenyl)-2-(4-methoxyphenyl)-4-oxoazetidin 3-yl]urea (**4c**)

Compound **4c** was prepared from 3 (50 mg, 0.175 mmol) and cyclopentyl isocyanate (29.6 μL, 0.263 mmol) according to the general synthetic procedure, producing a white solid (59.5 mg, 86%). R$_f$ = 0.30 (chloroform/ethyl acetate = 4/1, v/v), m.p. 138.7–143.7 °C. FTIR (ATR, cm^{-1}): 3342, 2959, 1747, 1633, 1556, 1509, 1385, 1289, 1250, 1226, 1175, 1131, 1101, 1031, 830. ^{1}H NMR (600 MHz, CDCl$_3$), δ/ppm: 7.28–7.19 (m, 4H), 6.93–6.85 (m, 4H), 5.74 (d, J = 6.5 Hz, 1H), 5.09 (d, J = 7.3 Hz, 1H), 4.99 (d, J = 2.2 Hz, 1H), 4.43 (dd, J = 6.6, 2.2 Hz, 1H), 3.97 (h, J = 6.8 Hz, 1H), 3.79 (s, 3H), 1.90 (tdd, J = 13.2, 10.1, 6.2 Hz, 2H), 1.67–1.58 (m, 2H), 1.48–1.58 (m, 2H), 1.40–1.28 (m, 2H). ^{13}C NMR (151 MHz, CDCl$_3$), δ/ppm: 166.2, 160.1, 159.4 (d, J = 244.0 Hz), 157.1, 133.7 (d, J = 2.4 Hz), 128.3, 127.6, 119.2, 116.0 (d, J = 22.6 Hz), 114.7, 67.3, 64.1, 55.5, 52.3, 33.6, 33.5, 23.7. HR MS (ESI-QTOF) m/z: calcd. for C$_{22}$H$_{24}$FN$_3$O$_3$, [M + H]$^+$, 398.1880; found 398.1871.

1-[(±)-*Trans*-1-(4-fluorophenyl)-2-(4-methoxyphenyl)-4-oxoazetidin-3-yl)-3-(furan-2 ylmethyl]urea (**4d**)

Compound **4d** was prepared from 3 (50 mg, 0.175 mmol) and furfuryl isocyanate (28.2 μL, 0.263 mmol) according to the general synthetic procedure, producing a white solid (64.4 mg, 90%). R$_f$ = 0.22 (chloroform/ethyl acetate = 4/1, v/v), m.p. 188.8–189.9 °C. FTIR (ATR, cm^{-1}): 3333, 1732, 1650, 1611, 1556, 1508, 1427, 1388, 1426, 1247, 1218, 1176, 1146, 1104, 1076, 1028, 1010, 833, 743. ^{1}H NMR (600 MHz, CDCl$_3$), δ/ppm: 7.28–7.27 (m, 1H), 7.23–7.17 (m, 4H), 6.91–6.84 (m, 4H), 6.26 (dd, J = 3.2, 1.8 Hz, 1H), 6.17 (d, J = 2.9 Hz, 1H), 5.99 (d, J = 6.7 Hz, 1H), 5.60 (t, J = 5.7 Hz, 1H), 5.00 (d, J = 2.2 Hz, 1H), 4.40 (dd, J = 6.7, 2.3 Hz, 1H), 4.32 (dd, J = 15.6, 5.8 Hz, 1H), 4.25 (dd, J = 15.6, 5.5 Hz, 1H), 3.79 (s, 3H). ^{13}C NMR (151 MHz, CDCl$_3$), δ/ppm: 166.2, 160.1, 159.4 (d, J = 244.3 Hz), 157.2, 152.2, 142.2, 133.5 (d, J = 2.5 Hz), 128.1, 127.6, 119.3 (d, J = 7.9 Hz), 116.0 (d, J = 22.8 Hz), 114.8, 110.6, 107.2, 67.2, 63.9, 55.5, 37.4. HR-MS (ESI-QTOF) m/z: calcd. for C$_{22}$H$_{20}$FN$_3$O$_4$, [M + H]$^+$, 410.1516; found 410.1509.

1-Benzyl-3-[(±)-*trans*-1-(4-fluorophenyl)-2-(4-methoxyphenyl)-4-oxoazetidin-3-yl]urea (**4e**)

Compound **4e** was prepared from 3 (50 mg, 0.175 mmol) and benzyl isocyanate (32.5 μL, 0.263 mmol) according to the general synthetic procedure, producing a white solid (65.9 mg, 90%). R$_f$ = 0.28 (chloroform/ethyl acetate = 4/1, v/v), m.p. 164.1–167.3 °C. FTIR (ATR, cm^{-1}): 3358, 3264, 3035, 2965, 2929, 1735, 1646, 1555, 1506, 1425, 1384, 1243, 1176, 1155, 1105, 1022, 841, 748, 706. ^{1}H NMR (600 MHz, CDCl$_3$), δ/ppm: 7.30–7.24 (m, 2H), 7.25–7.14 (m, 7H), 6.89–6.83 (m, 4H), 5.97 (d, J = 6.6 Hz, 1H), 5.57 (t, J = 5.8 Hz, 1H), 4.97 (d, J = 2.3 Hz, 1H), 4.38 (dd, J = 6.7, 2.3 Hz, 1H), 4.30 (dd, J = 14.7, 5.8 Hz, 1H), 4.23

(dd, *J* = 14.9, 5.6 Hz, 1H), 3.78 (s, 3H). ^{13}C NMR (151 MHz, CDCl$_3$), δ/ppm: 166.2, 160.1, 159.4 (d, *J* = 244.0 Hz), 157.5, 139.0, 133.5 (d, *J* = 2.6 Hz), 128.7, 128.1, 127.6, 127.4, 119.2 (d, *J* = 7.9 Hz), 116.0 (d, *J* = 22.8 Hz), 114.7, 67.2, 63.9, 55.5, 44.4. HR-MS (ESI-QTOF) m/z: calcd. for C$_{24}$H$_{22}$FN$_3$O$_3$, [M + H]$^+$, 420.1723; found 420.1700.

1-(4-(Tert-Butyl)phenyl)-3-[(±)-*trans*-1-(4-fluorophenyl)-2-(4-methoxyphenyl)-4-oxoazetidin-3-yl]urea (**4f**)

Compound **4f** was prepared from **3** (50 mg, 0.175 mmol) and 4-tert-butyl isocyanate (46.7 µL, 0.263 mmol) according to the general synthetic procedure, producing a white solid (73.7 mg, 91%). R$_f$ = 0.47 (chloroform/ethyl acetate = 4:1, *v/v*), m.p. 195.4–196.0 °C. FTIR (ATR, cm^{-1}): 3332, 2961, 1748, 1657, 1603, 1542, 1507, 1387, 1292, 1249, 1228, 1176, 1138, 1102, 1032, 831. ^{1}H NMR (600 MHz, CDCl$_3$), δ/ppm: 7.35 (s, 1H), 7.23–7.18 (m, 6H), 7.18–7.13 (m, 2H), 6.90–6.84 (m, 4H), 6.18 (d, *J* = 6.8 Hz, 1H), 5.06 (d, *J* = 2.3 Hz, 1H), 4.46 (dd, *J* = 6.8, 2.3 Hz, 1H), 3.78 (s, 3H), 1.25 (s, 9H). ^{13}C NMR (151 MHz, CDCl$_3$), δ/ppm: 166.2, 160.1, 159.5 (d, *J* = 244.5 Hz), 155.5, 147.1, 135.4, 133.5 (d, *J* = 2.6 Hz), 128.0, 127.7, 126.1, 121.0, 119.4 (d, *J* = 8.0 Hz), 116.0 (d, *J* = 22.8 Hz), 114.8, 67.1, 63.8, 55.5, 34.4, 31.5. HR-MS (ESI-QTOF) m/z: calcd. for C$_{27}$H$_{28}$FN$_3$O$_3$, [M + H]$^+$, 462.2193; found 462.2183.

1-(3-Chloro-4-methylphenyl)-3-[(±)-*trans*-1-(4-fluorophenyl)-2-(4-methoxyphenyl)-4-oxoazetidin-3-yl]urea (**4g**)

Compound **4g** was prepared from **3** (50 mg, 0.175 mmol) and 3-chloro-4-methyl isocyanate (36.0 µL, 0.263 mmol) according to the general synthetic procedure, producing a white solid (69.1 mg, 87%). R$_f$ = 0.41 (chloroform/ethyl acetate = 4/1, *v/v*), m.p. 174.2–178.8 °C. FTIR (ATR, cm^{-1}): 3327, 2931, 1741, 1658, 1588, 1542, 1508, 1384, 1290, 1225, 1175, 1136, 1102, 1030, 830. ^{1}H NMR (600 MHz, CDCl$_3$), δ/ppm: 7.62 (s, 1H), 7.27 (d, *J* = 2.0 Hz, 1H), 7.23–7.15 (m, 4H), 6.85–6.82 (m, 6H), 6.48 (d, *J* = 6.9 Hz, 1H), 5.12 (d, *J* = 2.3 Hz, 1H), 4.39 (dd, *J* = 6.9, 2.3 Hz, 1H), 3.79 (s, 3H), 2.20 (s, 3H). ^{13}C NMR (151 MHz, CDCl$_3$), δ/ppm: 166.8, 160.4, 160.2, 159.6 (d, *J* = 244.7 Hz), 155.1, 137.2, 134.4, 133.3 (d, *J* = 2.6 Hz), 130.9, 130.7, 127.7, 127.6, 120.2, 119.5 (d, *J* = 7.9 Hz), 118.2, 116.1 (d, *J* = 22.7 Hz), 114.8, 66.9, 63.7, 55.5, 19.4. HR-MS (ESI-QTOF) m/z: calcd. for C$_{24}$H$_{21}$ClFN$_3$O$_3$, [M + H]$^+$, 454.1333; found 454.1257.

1-(3,5-Dimethylphenyl)-3-[(±)-*trans*-1-(4 fluorophenyl)-2-(4-methoxyphenyl)-4-oxoazetidin 3-yl]urea (**4h**)

Compound **4h** was prepared from **3** (50 mg, 0.175 mmol) and 3,5-dimethylphenyl isocyanate (37.0 µL, 0.263 mmol) according to the general synthetic procedure, producing a white solid (68.3 mg, 90%). R$_f$ = 0.44 (chloroform/ethyl acetate = 4/1, *v/v*), m.p. 163.2–164.2 °C. FTIR (ATR, cm^{-1}): 3320, 2918, 2851, 1753, 1645, 1612, 1566, 1508, 1387, 1279, 1247, 1224, 1175, 1153, 1102, 1031, 831. ^{1}H NMR (600 MHz, CDCl$_3$), δ/ppm: 7.40 (s, 1H), 7.32–7.06 (m, 4H), 6.94–6.75 (m, 6H), 6.61 (s, 1H), 6.32 (d, *J* = 6.9 Hz, 1H), 5.08 (d, *J* = 2.3 Hz, 1H), 4.40 (dd, *J* = 6.8, 2.3 Hz, 1H), 3.78 (s, 3H), 2.15 (s, 6H). ^{13}C NMR (151 MHz, CDCl$_3$), δ/ppm: 166.6, 160.1, 159.5 (d, *J* = 244.2 Hz), 155.4, 138.8, 138.1, 133.4 (d, *J* = 2.7 Hz), 128.0, 127.6, 125.5, 119.4 (d, *J* = 8.0 Hz), 118.3, 116.0 (d, *J* = 22.8 Hz), 114.7, 67.0, 63.7, 55.4, 21.4. HR-MS (ESI-QTOF) m/z: calcd. for C$_{25}$H$_{24}$FN$_3$O$_3$, [M +H]$^+$, 434.1880; found 434.1871.

1-(2,6-Dimethylphenyl)-3-[(±)-*trans*-1-(4-fluorophenyl)-2-(4-methoxyphenyl)-4-oxoazetidin 3-yl]urea (**4i**)

Compound **4i** was prepared from **3** (50 mg, 0.175 mmol) and 2,6-dimethylphenyl isocyanate (36.6 µL, 0.263 mmol) according to the general synthetic procedure, producing a white solid (67.2 mg, 89%). R$_f$ = 0.30 (chloroform/ethyl acetate = 4/1, *v/v*), m.p. 127.2–128.9 °C. FTIR (ATR, cm^{-1}): 3311, 2961, 1753, 1638, 1612, 1548, 1508, 1384, 1248, 1226, 1175, 1134, 1101, 1030, 831. ^{1}H NMR (600 MHz, CDCl$_3$), δ/ppm: 7.25–7.17 (m, 4H), 7.16–7.03 (m, 3H), 6.92–6.82 (m, 3H), 6.39 (s, 1H), 4.96 (d, *J* = 2.3 Hz, 1H), 4.44 (dd, *J* = 6.8, 2.3 Hz, 1H), 3.77 (s, 3H), 2.29 (s, 6H). ^{13}C NMR (151 MHz, CDCl$_3$), δ/ppm: 166.0, 160.1, 159.3 (d, *J* = 243.7 Hz), 156.2, 133.7 (d, *J* = 2.6 Hz), 128.9, 128.2, 127.5, 119.1 (d, *J* = 7.8 Hz),

115.9 (d, J = 22.8 Hz), 114.8, 67.2, 63.9, 55.5, 18.5. HR-MS (ESI-QTOF) m/z: calcd. for $C_{25}H_{24}FN_3O_3$, [M + H]$^+$, 434.1880; found 434.1859.

3.1.6. General Procedure for the Preparation of *Syn/anti*-(±)-3,5-disubstituted Hydantoins **5a–i**

To a stirred solution of corresponding β-lactam urea **4a–i** (1.0 mmol) in dry methanol at 65 °C was added a solution of 25% sodium methoxide in methanol (1.1 mmol). The reaction mixture was allowed to reflux for 1 h (the progress of the reaction was monitored by TLC) and the resulting reaction mixture was then treated with brine and extracted with chloroform (3 × 15 mL). The organic layer was dried over anhydrous sodium sulfate, filtered and concentrated in vacuo to give a residue. The residue was then purified by SiO$_2$ flash column chromatography with chloroform/ethyl acetate (2:1 or 4:1) as the eluent solvent. The *syn/anti*-ratio in the product was determined by RP-HPLC and ^{1}H NMR spectroscopy. The mixture of diastereomeric hydantoins was then fractionated by preparative RP-HPLC using a Zorbax Extend C18 PrepHT column (250 × 9.4 mm I.D., 5-μm particle size, 300 Å pore size, Agilent Technologies). A linear gradient AB (method D–F) was used at a flow rate of 17 mL/min, with mobile phase A consisting of water and mobile phase B of acetonitrile. The chromatograms were monitored at 254 nm. Collected fractions of *syn*- and *anti*-hydantoins were extracted with chloroform (3 × 20 mL). The purity of the *syn*- and *anti*-isomers fraction was verified by analytical RP-HPLC as described in the Materials and Methods section. The organic phase was dried over anhydrous sodium sulfate and filtered, and the resulting filtrate was concentrated in vacuo to obtain the pure *syn*- and *anti*-isomers.

Syn/anti-3-allyl-5-{[(4-fluorophenyl)amino](4-methoxyphenyl)methyl}-imidazolidine-2,4-dione (**5a**)

Compound *syn/anti*-**5a** was prepared from *trans*-β-lactam urea **4a** (40 mg, 0.108 mmol) and a 25 wt.% solution of sodium methoxide in methanol (24.8 μL, 0.119 mmol) according to the general synthetic procedure, producing a white solid (34.3 mg, 85.8%). The diastereomeric ratio of *syn/anti*-isomers (52.5:47.5) was determined by HPLC (retention time: *anti*-isomer 29.5 min, *syn*-isomer 31.3 min, method A) of the crude reaction mixture. The diastereomeric mixture of the *syn/anti*-isomer **5a** was further separated by preparative HPLC (retention time: *anti*-isomer 18.0 min, *syn*-isomer 19.6 min, method D) to obtain isomers *syn*-**5a** (10 mg) and *anti*-**5a** (11 mg). R_f (*syn/anti*-**5a**) = 0.42 (chloroform/ethyl acetate = 2/1, v/v).

Syn-**5a**: m.p. 168.4–170.7 °C. FTIR (ATR, cm^{-1}): 3351, 2933, 2837, 1774, 1702, 1610, 1508, 1445, 1409, 1344, 1249, 1216, 1177, 1108, 989, 930, 823. ^{1}H NMR (600 MHz, CD$_3$CN), δ/ppm: 7.28 (d, J = 8.8 Hz, 2H), 6.89 (d, J = 8.8 Hz, 2H), 6.81 (t, J = 8.8 Hz, 2H), 6.61 (dd, J = 8.8, 4.8 Hz, 2H), 6.22 (br s, 1H), 5.71 (ddt, J = 17.1, 10.4, 5.0 Hz, 1H), 5.26 (d, J = 10.6 Hz, 1H), 5.04 (dq, J = 10.5, 1.4 Hz, 1H), 5.02 (dq, J = 17.3, 1.6 Hz, 1H), 4.91 (dd, J = 10.6, 3.3 Hz, 1H), 4.38 (dd, J = 3.3, 2.2 Hz, 1H), 3.98 (ddt, J = 16.5, 4.8, 1.8 Hz, 1H), 3.93 (ddt, J = 16.5, 4.8 Hz, 1.8 Hz, 1H), 3.75 (s, 3H). ^{13}C NMR (151 MHz, CD$_3$CN), δ/ppm: 173.2, 160.3, 158.2, 156.7 (d, J = 233,6 Hz), 144.4 (d, J = 1,7 Hz), 132.9, 131.5, 129.4, 116.6, 116.3 (d, J = 22.6 Hz), 116.0 (d, J = 7.5 Hz), 115.0, 63.3, 57.4, 56.0, 41.0. HR-MS (ESI-QTOF) m/z: calcd. for $C_{20}H_{20}FN_3O_3$, [M + H]$^+$, 370.1567; found 370.1555.

Anti-**5a**: m.p. 155.0–156.9 °C. FTIR (ATR, cm^{-1}): 3351, 2933, 2837, 1774, 1702, 1610, 1508, 1445, 1409, 1344, 1249, 1216, 1177, 1108, 989, 930, 823. ^{1}H NMR (600 MHz, CD$_3$CN), δ/ppm: 7.26 (d, J = 8.8 Hz, 2H), 6.84 (t, J = 9.0 Hz, 2H), 6.84 (d, J = 8.4 Hz, 2H), 6.62 (dd, J = 8.8, 4.4 Hz, 2H), 6.38 (s, 1H), 5.43 (ddt, J = 17.1, 10.5, 4.9 Hz, 1H), 5.00 (d, J = 7.3 Hz, 1H), 4.85 (dq, J = 10.3, 1.1 Hz, 1H), 4.80 (dd, J = 6.4, 3.9 Hz, 1H), 4.52 (dd, J = 3.7, 1.8 Hz, 1H), 4.52 (dq, J = 17.2, 1.5 Hz, 1H), 3.82 (ddt, J = 16.1, 5.1, 1.5 Hz, 1H), 3.74 (s, 3H), 3.73 (ddt, J = 16.2, 5.1, 1.8 Hz, 1H). ^{13}C NMR (151 MHz, CD$_3$CN), δ/ppm: 172.5, 160.7, 157.6, 156.9 (d, J = 233,6 Hz), 144.4 (d, J = 1,7 Hz), 132.6, 130.4, 129.6, 116.5, 116.4 (d, J = 22.6 Hz), 116.0 (d, J = 7.5 Hz), 114.6, 62.1, 59.5, 56.0, 40.8. HR-MS (ESI QTOF) m/z: calcd. for $C_{20}H_{20}FN_3O_3$, [M + H]$^+$, 370.1567; found 370.1555.

Syn/anti-{[(4-fluorophenyl)amino](4-methoxyphenyl)methyl}-3-hexylimidazolidine-2,4-dione (**5b**)

Compound *syn/anti*-**5b** was prepared from *trans*-β-lactam urea **4b** (40 mg, 0.097 mmol) and a 25 wt.% solution of sodium methoxide in methanol (24.5 µL, 0.107 mmol) according to the general synthetic procedure, producing a white solid (36.7 mg, 91.8%). The diastereomeric ratio of *syn/anti* isomers (50.1:49.9) was determined by HPLC (retention time: *anti*-isomer 40.5 min, *syn*-isomer 41.9 min, method C) of the crude reaction mixture. The diastereomeric mixture of the *syn/anti*-isomer **5b** was further separated by preparative HPLC (retention time: *anti*-isomer 29.8 min, *syn*-isomer 30.6 min, method F) to obtain isomers *syn*-**5b** (15 mg) and *anti*-**5b** (13 mg). R_f (*syn/anti*-**5b**) = 0.39 (chloroform/ethyl acetate = 4/1, *v/v*).

Syn-**5b**: m.p. 137.4–139.3 °C. FTIR (ATR, cm^{-1}): 3387, 3359, 2950, 2929, 2856, 1762, 1709, 1612, 1581, 1506, 1455, 1422, 1359, 1256, 1178, 1113, 1030, 985, 820. ^{1}H NMR (600 MHz, CD$_3$CN), δ/ppm: 7.25 (d, *J* = 8.7 Hz, 2H), 6.84–6.77 (m, 4H), 6.60–6.57 (m, 2H), 6.15 (s, 1H), 5.25 (d, *J* = 10.5 Hz, 1H), 4.88 (dd, *J* = 10.7, 3.2 Hz, 1H), 4.32 (dd, *J* = 3.2, 2.1 Hz, 1H), 3.74 (s, 3H), 3.40–3.33 (m, 1H), 3.33–3.27 (m, 1H), 1.45–1.37 (m, 2H), 1.25–1.02 (m, 6H), 0.86–0.83 (m, 3H). ^{13}C NMR (151 MHz, CD$_3$CN), δ/ppm: 172.5, 160.1, 158.4, 156.4 (d, *J* = 235.0 Hz), 142.7 (d, *J* = 2.1 Hz), 131.2, 129.2, 116.0 (d, *J* = 22.5 Hz), 115.7 (d, *J* = 7.4 Hz), 114.6, 62.7, 59.2, 55.7, 38.8, 32.0, 28.5, 26.8, 23.0, 14.1. HR-MS (ESI-QTOF) *m/z*: calcd. for C$_{23}$H$_{28}$FN$_3$O$_3$, [M + H]$^+$, 414.2193; found 414.2184.

Anti-**5b**: m.p. 164.2–167.5 °C. FTIR (ATR, cm^{-1}): 3387, 3359, 2950, 2929, 2856, 1762, 1709, 1612, 1581, 1506, 1455, 1422, 1359, 1256, 1178, 1113, 1030, 985, 820. ^{1}H NMR (600 MHz, CD$_3$CN), δ/ppm: 7.27 (d, *J* = 8.7 Hz, 2H), 6.89–6.44 (m, 4H), 6.67–6.60 (m, 2H), 6.34 (s, 1H), 4.98 (d, *J* = 7.5 Hz, 1H), 4.79 (dd, *J* = 7.9, 3.6 Hz, 1H), 4.47 (dd, *J* = 3.6, 1.7 Hz, 1H), 3.72 (s, 3H), 3.23–3.16 (m, 1H), 3.14–3.07 (m, 1H), 1.25–1.02 (m, 6H), 0.95–0.86 (m, 2H), 0.86–0.83 (m, 3H). ^{13}C NMR (151 MHz, CD$_3$CN), δ/ppm: 173.3, 160.3, 157.9, 156.8 (d, *J* = 235.8 Hz), 142.1 (d, *J* = 2.1 Hz), 130.1, 129.3, 116.3 (d, *J* = 22.5 Hz), 115.9 (d, *J* = 7.4 Hz), 114.2, 61.6, 57.3, 55.6, 38.6, 31.9, 28.2, 26.7, 22.9, 14.1. HR-MS (ESI-QTOF) *m/z*: calcd. for C$_{23}$H$_{28}$FN$_3$O$_3$, [M + H]$^+$, 414.2193; found 414.2184.

Syn/anti-3-cyclopentyl-5-[[(4-fluorophenyl)amino](4-methoxyphenyl)methyl}-imidazolidine-2,4-dione (**5c**)

Compound *syn/anti*-**5c** was prepared from *trans*-β-lactam urea **4c** (40 mg, 0.101 mmol) and a 25 wt.% solution of sodium methoxide in methanol (25.4 µL, 0.111 mmol) according to the general synthetic procedure, producing a white solid (31.8 mg, 79.5%). The diastereomeric ratio of *syn/anti* isomers (51.8:48.2) was determined by HPLC (retention time: *anti*-isomer 41.2 min, *syn*-isomer 42.3 min, method B) of the crude reaction mixture. The diastereomeric mixture of the *syn/anti*-isomer **5c** was further separated by preparative HPLC (retention time: *anti*-isomer 23.2 min, *syn*-isomer 24.1 min, method F) to obtain isomers *syn*-**5c** (14 mg) and *anti*-**5c** (13 mg). R_f (*syn/anti*-**5c**) = 0.41 (chloroform/ethyl acetate = 4/1, *v/v*).

Syn-**5c**: m.p. 211.0–211.8 °C. FTIR (ATR, cm^{-1}): 3347, 3231, 2959, 1762, 1699, 1613, 1510, 1432, 1352, 1308, 1282, 1258, 1221, 1131, 1157, 1111, 965, 832. ^{1}H NMR (600 MHz, CD$_3$CN), δ/ppm: 7.26 (d, *J* = 8.8 Hz, 2H), 6.87 (d, *J* = 8.8 Hz, 2H), 6.81 (t, *J* = 8.9 Hz, 2H), 6.61–6.57 (m, 2H), 6.15 (s, 1H), 5.26 (d, *J* = 10.5 Hz, 1H), 4.84 (dd, *J* = 10.5, 3.3 Hz, 1H), 4.26 (dd, *J* = 3.4, 2.2 Hz, 1H), 4.10 4.02 (m, 1H), 3.74 (s, 3H), 1.90–1.75 (m, 4H), 1.66–1.57 (m, 2H), 1.57–1.49 (m, 2H). ^{13}C NMR (151 MHz, CD$_3$CN), δ/ppm: 173.5, 160.3, 158.4, 156.6 (d, *J* = 233.2 Hz), 144.4 (d, *J* = 1.6 Hz), 131.3, 129.5, 116.2 (d, *J* = 22.5 Hz), 115.9 (d, *J* = 7.4 Hz), 114.8, 62.1, 59.7, 57.8, 52.2, 29.6, 29.5, 25.8. HR-MS (ESI-QTOF) *m/z*: calcd. for C$_{22}$H$_{24}$FN$_3$O$_3$, [M + H]$^+$, 398.1880; found 398.1870.

Anti-**5c**: m.p. 141.1–142.5 °C. FTIR (ATR, cm^{-1}): 3347, 3231, 2959, 1762, 1699, 1613, 1510, 1432, 1352, 1308, 1282, 1258, 1221, 1131, 1157, 1111, 965, 832. ^{1}H NMR (600 MHz, CD$_3$CN), δ/ppm: 7.24 (d, *J* = 8.7 Hz, 2H), 6.86–6.81 (m, 4H), 6.64 6.60 (m, 2H), 6.32 (s, 1H), 4.99 (d, *J* = 7.0 Hz, 1H), 4.77 (dd, *J* = 7.3, 3.6 Hz, 1H), 4.39 (dd, *J* = 3.7, 1.6 Hz, 1H), 4.29–4.15

(m, 1H), 3.73 (s, 3H), 1.75–1.66 (m, 4H), 1.57–1.42 (m, 2H), 1.50–1.42 (m, 2H). ^{13}C NMR (151 MHz, CD$_3$CN), δ/ppm: 172.8, 160.6, 158.0, 156.9 (d, J = 233.6 Hz), 144.4 (d, J = 1.7 Hz), 130.3, 129.5, 116.3 (d, J = 22.5 Hz), 116.0 (d, J = 7.6 Hz), 114.5, 61.3, 57.8, 56.0, 52.0, 29.5, 29.3, 25.7. HR-MS (ESI-QTOF) m/z: calcd. for C$_{22}$H$_{24}$FN$_3$O$_3$, [M + H]$^+$, 398.1880; found 398.1870.

Syn/anti-5-{[(4-fluorophenyl)amino](4-methoxyphenyl)methyl}-3-(furan-2-ylmethyl)imidazolidine-2,4-dione (**5d**)

Compound *syn/anti*-**5d** was prepared from *trans*-β-lactam urea **4d** (40 mg, 0.098 mmol) and a 25 wt.% solution of sodium methoxide in methanol (24.7 μL, 0.108 mmol) according to the general synthetic procedure, producing a white solid (32.6 mg, 81.5%). The diastereomeric ratio of *syn/anti*-isomers (57.1:42.9) was determined by HPLC (retention time: *anti*-isomer 37.7 min, *syn*-isomer 40.3 min, method A) of the crude reaction mixture. The diastereomeric mixture of the *syn/anti*-isomer **5d** was further separated by preparative HPLC (retention time: *anti*-isomer 23.2 min, *syn*-isomer 25.2 min, method D) to obtain isomers *syn*-**5d** (13 mg) and *anti*-**5d** (11 mg). R_f (*syn/anti*-**5d**) = 0.43 (chloroform/ethyl acetate = 4/1, v/v).

Syn-**5d**: m.p. 161.0–163.9 °C. FTIR (ATR, cm^{-1}): 3356, 1776, 1705, 1610, 1508, 1441, 1348, 1248, 1215, 1178, 1129, 1031, 1010, 937, 822. ^{1}H NMR (600 MHz, CD$_3$CN), δ/ppm: 7.37 (dd, J = 1.8 Hz, 0.9 Hz, 1H), 7.26 (d, J = 8.6 Hz, 2H), 6.87–6.75 (m, 6H), 6.31 (dd, J = 3.2, 1.9 Hz, 1H), 6.25 (s, 1H), 6.16 (dd, J = 3.2, 0.8 Hz, 1H), 5.21 (d, J = 10.6 Hz, 1H), 4.89 (dd, J = 10.6, 3.2 Hz, 1H), 4.55 (d, J = 15.9 Hz, 1H), 4.49 (d, J = 15.5 Hz, 1H), 4.39 (dd, J = 3.3, 2.1 Hz, 1H), 3.74 (s, 3H). ^{13}C NMR (151 MHz, CD$_3$CN), δ/ppm: 172.8, 160.2, 157.7, 156.6 (d, J = 233.5 Hz), 150.6, 144.3 (d, J = 1.4 Hz), 143.3, 131.3, 129.3, 116.2 (d, J = 22.4 Hz), 115.9 (d, J = 7.4 Hz), 114.9, 111.4, 108.8, 63.1, 57.4, 55.9, 35.7. HR-MS (ESI-QTOF) m/z: calcd. for C$_{22}$H$_{20}$FN$_3$O$_4$, [M + H]$^+$, 410.1516; found 410.1514.

Anti-**5d**: m.p. 165.8–168,9 °C. FTIR (ATR, cm^{-1}): 3356, 1776, 1705, 1610, 1508, 1441, 1348, 1248, 1215, 1178, 1129, 1031, 1010, 937, 822. ^{1}H NMR (600 MHz, CD$_3$CN), δ/ppm: 7.28 (dd, J = 1.8, 0.9 Hz, 1H), 7.20 (d, J = 8.7 Hz, 2H), 6.61–6.55 (m, 6H), 6.40 (s, 1H), 6.23 (dd, J = 3.3, 1.8 Hz, 1H), 5.80 (dd, J = 3.3, 0.9 Hz, 1H), 4.97 (d, J = 7.3 Hz, 1H), 4.80 (dd, J = 7.4, 3.7 Hz, 1H), 4.52 (dd, J = 3.8, 1.7 Hz, 1H), 4.38 (d, J = 15.0 Hz, 1H), 4.30 (d, J = 15.8 Hz, 1H), 3.73 (s, 3H). ^{13}C NMR (151 MHz, CD$_3$CN), δ/ppm: 172.2, 160.4, 157.2, 156.8 (d, J = 233.5 Hz), 150.3, 144.3 (d, J = 1.5 Hz), 143.2, 131.3, 130.1, 116.2 (d, J = 22.4 Hz), 115.9 (d, J = 7.4 Hz), 114.5, 111.2, 108.6, 62.0, 59.3, 55.8, 35.4. HR-MS (ESI-QTOF) m/z: calcd. for C$_{22}$H$_{20}$FN$_3$O$_4$, [M + H]$^+$, 410.1516; found 410.1514.

Syn/anti-3-benzyl-5-{[(4-fluorophenyl)amino](4-methoxyphenyl)methyl}imidazolidine-2,4-dione (**5e**)

Compound *syn/anti*-**5e** was prepared from *trans*-β-lactam urea **4e** (47 mg, 0.112 mmol) and a 25 wt.% solution of sodium methoxide in methanol (28.1 μL, 0.123 mmol) according to the general synthetic procedure, producing a white solid (40.6 mg, 86.4%). The diastereomeric ratio of *syn/anti*-isomers (52.2:47.8) was determined by HPLC (retention time: *anti*-isomer 51.5 min, *syn*-isomer 54.4 min, method A) of the crude reaction mixture. The diastereomeric mixture of the *syn/anti*-isomer **5e** was further separated by preparative HPLC (retention time: *anti*-isomer 32.3 min, *syn*-isomer 33.8 min, method E) to obtain isomers *syn*-**5e** (16 mg) and *anti*-**5e** (14 mg). R_f (*syn/anti*-**5e**) = 0.46 (chloroform/ethyl acetate = 2/1, v/v).

Syn-**5e**: m.p. 187.3–191.0 °C. FTIR (ATR, cm^{-1}): 3359, 1771, 1706, 1611, 1509, 1446, 1251, 1217, 1179, 10301, 822. ^{1}H NMR (600 MHz, CD$_3$CN), δ/ppm: 7.42–7.37 (m, 1H), 7.28 (d, J = 8.7 Hz, 2H), 7.21–7.10 (m, 3H), 6.96–6.73 (m, 8H), 6.67–6.53 (m, 2H), 6.36–6.30 (m, 1H), 6.21–6,15 (m, 1H), 6.30 (s, 1H), 4.91 (d, J = 7.5 Hz, 1H), 4.84 (dd, J = 10.8, 3.2 Hz, 1H), 4.45 (dd, J = 3.2, 2.1 Hz, 1H), 4.63 (d, J = 15.5 Hz, 1H), 4.52 (d, J = 15.5 Hz, 1H), 3.77 (s, 3H). ^{13}C NMR (151 MHz, CD$_3$CN), δ/ppm: 174.0, 160.9, 158.9, 157.4 (d, J = 233.3 Hz), 144.9 (d, J = 1.6 Hz), 138.2, 131.9, 130.1, 130.0, 128.8, 128.9, 116.9 (d, J = 22.7 Hz), 116.5 (d, J = 7.4 Hz),

115.50, 63.9, 58.0, 56.5, 43.0. HR MS (ESI-QTOF) m/z: calcd. for $C_{24}H_{22}FN_3O_3$, [M + H]$^+$, 420.1723; found 420.1715.

Anti-5e: m.p. 154.3–156.1 °C. FTIR (ATR, cm^{-1}): 3359, 1771, 1706, 1611, 1509, 1446, 1251, 1217, 1179, 10301, 822. ^{1}H NMR (600 MHz, CD$_3$CN), δ/ppm: 7.24 (d, J = 8.8 Hz, 2H), 7.21–7.10 (m, 3H), 6.90–6.74 (m, 4H), 6.74–6.67 (m, 2H), 6.66–6.58 (m, 2H), 6.54 (s, 1H), 5.21 (d, J = 7.5 Hz, 1H), 4.76 (dd, J = 7.5, 3.2 Hz, 1H), 4.60 (dd, J = 3.5, 1.8 Hz, 1H), 4.45 (d, J = 15.5 Hz, 1H), 4.29 (d, J = 14.8 Hz, 1H), 3.75 (s, 3H). ^{13}C NMR (151 MHz, CD$_3$CN), δ/ppm: 173.2, 161.2, 158.1, 157.6 (d, J = 233.8 Hz), 144.9 (d, J = 1.6 Hz), 137.8, 131.9, 131.0, 129.9, 128.8, 128.4, 116.9 (d, J = 22.4 Hz), 116.6 (d, J = 7.5 Hz), 115.2, 62.8, 59.8, 56.4, 42.7. HR-MS (ESI-QTOF) m/z: calcd. for $C_{24}H_{22}FN_3O_3$, [M + H]$^+$, 420.1723; found 420.1715.

Syn/anti-3-[4-(tert-butyl)phenyl]-5-{[(4-fluorophenyl)amino](4-methoxyphenyl)methyl} imidazoli-dine-2,4-dione (**5f**)

Compound *syn/anti*-**5f** was prepared from *trans*-β-lactam urea **4f** (40 mg, 0.087 mmol) and a 25 wt.% solution of sodium methoxide in methanol (22.0 μL, 0.096 mmol) according to the general synthetic procedure, producing a white solid (35.5 mg, 88.8%). The diastereomeric ratio of *syn/anti*-isomers (66.0:34.0) was determined by HPLC (retention time: *anti*-isomer 50.6 min, *syn*-isomer 52.6 min, method C) of the crude reaction mixture. The diastereomeric mixture of the *syn/anti*-isomer **5f** was further separated by preparative HPLC (retention time: *anti*-isomer 31.9 min, *syn*-isomer 33.6 min, method F) to obtain isomers *syn*-**5f** (13 mg) and *anti*-**5f** (8 mg). R$_f$ (*syn/anti*-**5f**) = 0.61 (chloroform/ethyl acetate = 4/1, v/v).

Syn-5f: m.p. 233.9–234.6 °C. FTIR (ATR, cm^{-1}): 3356, 3237, 2963, 1770, 1716, 1610, 1509, 1462, 1364, 1253, 1223, 1178, 1109, 1031, 828. ^{1}H NMR (600 MHz, CD$_3$CN), δ/ppm: 7.49 (d, J = 8.6 Hz, 2H), 7.42 (d, J = 8.6 Hz, 2H), 7.11 (d, J = 8.6 Hz, 2H), 6.94–6.88 (m, 2H), 6.68–6.62 (m, 2H), 6.48 (s, 1H), 5.39 (d, J = 10.5 Hz, 1H), 4.98 (dd, J = 10.5, 3.1 Hz, 1H), 4.50 (dd, J = 3.4, 2.0 Hz, 1H), 3.76 (s, 3H), 1.32 (s, 9H). ^{13}C NMR (151 MHz, CD$_3$CN), δ/ppm: 172.7, 160.4, 157.5, 156.7 (d, J = 233.5 Hz), 152.5, 144.3 (d, J = 1.5 Hz), 131.2, 129.5, 129.4, 127.3, 126.9, 116.3 (d, J = 22.5 Hz), 116.0 (d, J = 7.4 Hz), 114.6, 61.9, 59.8, 56.0, 31.5. HR-MS (ESI-QTOF) m/z: calcd. for $C_{27}H_{28}FN_3O_3$, [M + H]$^+$, 462.2193; found 462.2187.

Anti-5f: m.p. 87.5–90.5 °C. FTIR (ATR, cm^{-1}): 3356, 3237, 2963, 1770, 1716, 1610, 1509, 1462, 1364, 1253, 1223, 1178, 1109, 1031, 828. ^{1}H NMR (600 MHz, CD$_3$CN), δ/ppm: 7.36–7.29 (m, 4H), 6.92–6.88 (m, 4H), 6.84–6.76 (m, 4H), 6.69 (s, 1H), 5.08 (d, J = 7.0 Hz, 1H), 4.89 (dd, J = 7.1, 3.4 Hz, 1H), 4.64 (dd, J = 3.5, 1.4 Hz, 1H), 3.76 (s, 3H), 1.30 (s, 9H). ^{13}C NMR (151 MHz, CD$_3$CN), δ/ppm: 172.1, 160.7, 157.1, 156.9 (d, J = 233.8 Hz), 152.52, 144.4 (d, J = 1.5 Hz), 130.5, 130.3, 130.2, 127.4, 126.9, 116.3 (d, J = 22.5 Hz), 116.0 (d, J = 7.4 Hz), 114.9, 63.0, 57.9, 55.9, 31.4. HR MS (ESI-QTOF) m/z: calcd. for $C_{27}H_{28}FN_3O_3$, [M + H]$^+$, 462.2193; found 462.2187.

Syn/anti-3-(3-chloro-4-methylphenyl)-5{[(4-fluorophenyl)amino](4-methoxyphenyl)methyl} imidazolidine-2,4-dione (**5g**)

Compound *syn/anti*-**5g** was prepared from *trans*-β-lactam urea **4g** (42 mg, 0.093 mmol) and a 25 wt.% solution of sodium methoxide in methanol (23.3 μL, 0.102 mmol) according to the general synthetic procedure, producing a white solid (30.1 mg, 75.3%). The diastereomeric ratio of *syn/anti*-isomers (48.3:51.7) was determined by HPLC (retention time: *anti*-isomer 36.8 min, *syn*-isomer 38.3 min, method C) of the crude reaction mixture. The diastereomeric mixture of the *syn/anti*-isomer **5g** was further separated by preparative HPLC (retention time: *anti*-isomer 23.2 min, *syn*-isomer 24.4 min, method F) to obtain isomers *syn*-**5g** (11 mg) and *anti*-**5g** (12 mg). R$_f$ (*syn/anti*-**5g**) = 0.37 (chloroform/ethyl acetate = 2/1, v/v).

Syn-5g: m.p. 207.3–208.9 °C. FTIR (ATR, cm^{-1}): 3370, 1761, 1708, 1610, 1509, 1419, 1252, 1252, 1223, 1177, 1108, 1029, 819. ^{1}H NMR (600 MHz, CD$_3$CN), δ/ppm: 7.31–7.26 (m, 3H), 7.19 (d, J = 2.0 Hz, 1H), 7.09 (dd, J = 8.1, 2.0 Hz, 1H), 6.84 6.76 (m, 4H), 6.68–6.63 (m, 2H), 6.50 (s, 1H), 5.34 (d, J = 10.7 Hz, 1H), 4.98 (dd, J = 10.6, 3.1 Hz, 1H), 4.51 (dd, J = 3.3, 2.1 Hz, 1H), 3.76 (s, 3H), 2.34 (s, 3H). ^{13}C NMR (151 MHz, CD$_3$CN), δ/ppm: 171.7, 160.8,

157.0, 156.7 (d, J = 233.3 Hz), 144.2 (d, J = 1.8 Hz), 137.4, 134.6, 132.2, 131.5, 131.1, 130.3, 127.7, 126.0, 116.3 (d, J = 22.4 Hz), 116.1 (d, J = 7.6 Hz), 114.9, 63.0, 57.9, 55.9, 19.8. HR MS (ESI-QTOF) m/z: calcd. for $C_{24}H_{21}ClFN_3O_3$, [M + H]$^+$, 454.1333; found 454.1326.

Anti-5g: m.p. 128.1–131.6 °C. FTIR (ATR, cm^{-1}): 3370, 1761, 1708, 1610, 1509, 1419, 1252, 1252, 1223, 1177, 1108, 1029, 819. ^{1}H NMR (600 MHz, CD$_3$CN), δ/ppm: 7.38–7.29 (m, 3H), 6.95–6.91 (m, 2H), 6.91–6.84 (m, 4H), 6.69 (s, 1H), 6.69–6.64 (m, 2H), 5.05 (d, J = 7.3 Hz, 1H), 4.88 (dd, J = 7.3, 3.6 Hz, 1H), 4.64 (dd, J = 3.6, 1.5 Hz, 1H), 3.76 (s, 3H), 2.37 (s, 3H). ^{13}C NMR (151 MHz, CD$_3$CN), δ/ppm: 172.3, 160.4, 156.5, 156. 9 (d, J = 233.8 Hz), 144.3 (d, J = 1.7 Hz), 137.3, 134.6, 132.24, 131.8, 129.4, 129.30, 127.7, 126.1, 116.3 (d, J = 22.3 Hz), 116.1 (d, J = 7.6 Hz), 114.6, 61.9, 59.8, 56.0, 19.9. HR-MS (ESI-QTOF) m/z: calcd. for $C_{24}H_{21}ClFN_3O_3$, [M + H]$^+$, 454.1333; found 454.1326.

Syn/anti-3-(3,5-dimethylphenyl)-5-{[(4-fluorophenyl)amino](4-methoxyphenyl)methyl} imidazolidine-2,4-dione (5h)

Compound *syn/anti*-5h was prepared from *trans*-β-lactam urea 4h (40 mg, 0.092 mmol) and a 25 wt.% solution of sodium methoxide in methanol (23.1 μL, 0.101 mmol) according to the general synthetic procedure, producing a white solid (23.5 mg, 58.8%). The diastereomeric ratio of *syn/anti*-isomers (64.3:35.7) was determined by HPLC (retention time: *anti*-isomer 45.0 min, *syn*-isomer 46.5 min, method B) of the crude reaction mixture. The diastereomeric mixture of the *syn/anti*-isomer 5h was further separated by preparative HPLC (retention time: *anti*-isomer 31.0 min, *syn*-isomer 32.3 min, method E) to obtain isomers *syn*-5h (9 mg) and *anti*-5h (8 mg). R_f (*syn/anti*-5h) = 0.39 (chloroform/ethyl acetate = 2/1, v/v).

Syn-5h: m.p. 202.3–206.4 °C. FTIR (ATR, cm^{-1}): 3353, 3248, 1771, 1706, 1611, 1509, 1438, 1411, 1358, 1279, 1254, 1221, 1173, 1108, 1029, 824. ^{1}H NMR (600 MHz, CD$_3$CN), δ/ppm: 7.33 (dd, J = 8.7, 5.4 Hz, 4H), 7.04 (s, 1H), 6.75 (s, 2H), 6.68 6.63 (m, 4H), 6.41 (s, 1H), 5.37 (d, J = 10.6 Hz, 1H), 4.97 (dd, J = 10.6, 3.2 Hz, 1H), 4.49 (dd, J = 3.2, 2.1 Hz, 1H), 3.77 (s, 3H), 2.30 (s, 6H). ^{13}C NMR (151 MHz, CD$_3$CN), δ/ppm: 172.7, 160.4, 157.5, 156.7 (d, J = 233.5 Hz), 144.4 (d, J = 1.7 Hz), 139.9, 133.0, 131.2, 130.8, 129.5, 125.5, 116.3 (d, J = 22.3 Hz), 115.5 (d, J = 7.5 Hz), 114.9, 62.9, 57.9, 56.0, 21.2. HR-MS (ESI-QTOF) m/z: calcd. for $C_{25}H_{24}FN_3O_3$, [M + H]$^+$, 434.1880; found 434.1876.

Anti-5h: m.p. 101.4–104.0 °C. FTIR (ATR, cm^{-1}): 3353, 3248, 1771, 1706, 1611, 1509, 1438, 1411, 1358, 1279, 1254, 1221, 1173, 1108, 1029, 824. ^{1}H NMR (600 MHz, CD$_3$CN), δ/ppm: 6.92 (dd, J = 8.7, 5.2 Hz, 4H), 6.99 (s, 1H), 6.85 (dt, J = 19.8, 8.9 Hz, 4H), 6.61 (s, 1H), 6.40 (s, 2H), 5.05 (d, J = 7.3 Hz, 1H), 4.88 (dd, J = 7.4, 3.6 Hz, 1H), 4.63 (dd, J = 3.6, 1.5 Hz, 1H), 3.77 (s, 3H), 2.25 (s, 6H). ^{13}C NMR (151 MHz, CD$_3$CN), δ/ppm: 172.0, 160.8, 157.1, 156.9 (d, J = 233.6 Hz), 144.3 (d, J = 1.8 Hz), 139.8, 132.7, 130.8, 129.4, 130.4, 125.5, 116.3 (d, J = 22.4 Hz), 115.0 (d, J = 7.5 Hz), 114.6, 61.9, 59.8, 55.9, 21.1. HR-MS (ESI-QTOF) m/z: calcd. for $C_{25}H_{24}FN_3O_3$, [M + H]$^+$, 434.1880; found 434.1876.

Syn/anti-3-(2,6-Dimethylphenyl)-5-{[(4-fluorophenyl)amino](4-methoxyphenyl)methyl} imidazolidine-2,4-dione (5i)

Compound *syn/anti*-5i was prepared from *trans*-β-lactam urea 4i (45 mg, 0.104 mmol) and a 25 wt.% solution of sodium methoxide in methanol (26.1 μL, 0.114 mmol) according to the general synthetic procedures, producing a white solid (34.1 mg, 75.8%). The diastereomeric ratio of *syn/anti*-isomers (58.8:41.2) was determined by HPLC (retention time: *anti*-isomer 41.7 min, *syn*-isomer 43.1 min, method B) of the crude reaction mixture. The diastereomeric mixture of the *syn/anti*-isomer 5i was further separated by preparative HPLC (retention time: *anti*-isomer 28.8 min, *syn*-isomer 30.1 min, method F) to obtain isomers *syn*-5i (14 mg) and *anti*-5i (14 mg). R_f (*syn/anti*-5i) = 0.56 (chloroform/ethyl acetate = 4/1, v/v).

Syn-5i: m.p. 168.6–170.8 °C. FTIR (ATR, cm^{-1}): 3346, 2929, 1781, 1712, 1611, 1508, 1479, 1442, 1404, 1250, 1177, 1216, 1109, 1031, 823. ^{1}H NMR (600 MHz, CD$_3$CN), δ/ppm: 7.37 (d, J = 7.4 Hz, 2H), 7.24 (t, J = 7.6 Hz, 1H), 7.12 (t, J = 7.5 Hz, 2H), 6.93 (d, J = 7.4 Hz, 2H), 6.88 (d, J = 8.5 Hz, 2H), 6.69–6.62 (m, 2H), 6.75 (s, 1H), 5.42 (d, J = 11.2 Hz, 1H), 5.04 (dd,

J = 10.3 Hz, 2.9 Hz, 1H), 4.66 (dd, J = 2.9, 2.1 Hz, 1H), 3.77 (s, 3H), 2.13 (s, 3H), 1.98 (s, 3H). ^{13}C NMR (151 MHz, CD$_3$CN), δ/ppm: 171.5, 160.9, 156.9 (d, J = 233.9 Hz), 156.7, 144.2 (d, J = 1.8 Hz), 138.0, 137.9, 131.5, 130.3, 129.5, 129.4, 129.2, 129.1, 116.3 (d, J = 22.5 Hz), 116.0 (d, J = 7.5 Hz), 115.0, 64.1, 57.3, 55.9, 18.0. HR-MS (ESI-QTOF) m/z: calcd. for C$_{25}$H$_{24}$FN$_3$O$_3$, [M + H]$^+$, 434.1880; found 434.1872.

Anti-**5i**: m.p. 115.7–117.3 °C. FTIR (ATR, cm^{-1}): 3346, 2929, 1781, 1712, 1611, 1508, 1479, 1442, 1404, 1250, 1177, 1216, 1109, 1031, 823. ^{1}H NMR (600 MHz, CD$_3$CN), δ/ppm: 7.35 (d, J = 8.8 Hz, 2H), 7.18 (dd, J = 7.6, 3.2 Hz, 2H), 7.00 (ddt, J = 7.6, 1.5, 0.7 Hz, 1H), 6.88–6.80 (m, 4H), 6.69–6.62 (m, 2H), 6.50 (s, 1H), 5.02 (d, J = 7.4 Hz, 1H), 4.95 (dd, J = 7.9, 3.5 Hz, 1H), 4.86 (dd, J = 3.5, 1.8 Hz, 1H), 3.75 (s, 3H), 2.17 (s, 3H), 2.08 (s, 3H). ^{13}C NMR (151 MHz, CD$_3$CN), δ/ppm: 171.5, 160.9, 156.9 (d, J = 233.9 Hz), 156.7, 144.2 (d, J = 1.8 Hz), 138.0, 137.9, 131.2, 130.9, 130.8, 130.20, 129.1, 129.0, 116.3 (d, J = 22.5 Hz), 115.1 (d, J = 7.6 Hz), 114.9, 62.3, 59.1, 55.9, 16.7. HR-MS (ESI QTOF) m/z: calcd. for C$_{25}$H$_{24}$FN$_3$O$_3$, [M + H]$^+$, 434.1872; found 434.1872.

3.1.7. Computational Section

Conformational searches and preliminary DFT calculations were run with Spartan'20 (Wavefunction, Irvine, CA, USA) with default parameters, default grids, and convergence criteria. Geometry optimizations and frequency calculations, and TD-DFT calculations, were run with Gaussian'16 (Revision C.01) [76] with default grids and convergence criteria. The conformational search was run with the Monte Carlo algorithm implemented in Spartan'20 using Merck molecular force field. All structures thus obtained were pre-optimized with DFT at the B3LYP-D3/6-31G(d) level in vacuo. The structures were then reoptimized at B3LYP-D3BJ/6-311+G(d,p) in vacuo, and final energies were estimated at the same level including the PCM solvent model for acetonitrile. ECD calculations were run with TD-DFT at the CAM-B3LYP/def2-TZVP/PCM and B3LYP/def2-TZVP/PCM level, including the IEF-PCM solvent model for acetonitrile. The number of excited states (roots) was 40 for both functionals. VCD (frequency) calculations were run with DFT at the B3LYP-D3BJ/6-311+G(d,p)/PCM level, using geometries optimized at the same level of theory. All structures had only real (positive) frequencies. UV/ECD and IR/VCD spectra were plotted using the program SpecDis (version 1.71, Berlin, Germany, http://specdis-software.jimdo.com, accessed on 23 September 2024).

3.1.8. Antiproliferative Activity of *Syn/anti*-(±)-3,5 Disubstituted Hydantoins 5a–i

The antiproliferative activity of compounds **5a–i** was evaluated across three distinct human cell lines. The following cell lines were utilized in this study: HepG2 (liver hepatocellular carcinoma, HB 8055, from ATCC, Manassas, VA, USA), A2780 (ovarian carcinoma, 93112519, from ECACC, Porton Down, Wiltshire, UK), MCF7 (breast adenocarcinoma, HTB-22, from ATCC, Manassas, VA, USA), and one human non-tumor cell line, HFF1 (untransformed human fibroblasts, SCRC-1041, from ATCC, Manassas, VA, USA), employing the MTT assay [77,78]. The above cells were cultured as monolayers and maintained in various media, specifically Dulbecco's modified Eagle's medium (DMEM) or Roswell Park Memorial Institute (RPMI-1549) medium, which were enriched with differing concentrations of fetal bovine serum (FBS). Specifically, the A2780 cancer cells were cultured in RPMI-1540 medium (ATCC, Manassas, VA, USA) supplemented with 10% FBS. The MCF7 carcinoma cells were cultured in DMEM supplemented with 10% FBS and insulin (0.01 mg/mL). The HepG2 cells were cultured in DMEM supplemented with 10% FBS, whereas the HFF1 cells were cultured in DMEM containing 15% FBS. All cell lines were incubated in a humidified atmosphere with 5% CO$_2$ at 37 °C. The growth inhibition activity was assessed according to the slightly modified procedure of the National Cancer Institute, Developmental Therapeutics Program [79]. On day 0, the cells were inoculated onto standard 96-well microtiter plates. The cell concentrations were calibrated based on the population doubling time (PDT). The cell concentrations were 9.6 × 10^3 per mL for HepG2, 1.3 × 10^4 per mL for A2780, 1.6 × 10^3 per mL for MCF7, and 1.6 × 10^3 per mL for

HFF1. All compounds were prepared as stock solutions in DMSO at a concentration of 0.04 mol/L. On the subsequent day, the cells were treated with a range of concentrations of compounds **5a–i** (from 10^{-8} to 10^{-4} mol/L in duplicate), and were then incubated for an additional 72 h at 37 °C in a 5% CO_2 environment. The working dilutions in DMSO were freshly generated on the day of testing. Following a 72 h incubation period, the growth rate of the cells was evaluated via the MTT assay, which quantifies dehydrogenase activity in viable cells. The MTT cell proliferation assay is a colorimetric assay system that quantifies the conversion of the tetrazolium compound (MTT) into an insoluble formazan product by the mitochondria of viable cells. For this analysis, the medium containing the substance was removed, and 40 µL of MTT reagent was added to each well at a concentration of 0.5 mg/mL. Subsequently, the precipitates were dissolved in 160 µL of DMSO following a four-hour incubation period. The absorbance (A) was determined using a microplate reader set at a wavelength of 595 nm (Hidex Chameleon V, Hidex, Turku, Finland). The percentage of growth (PG) for the cell lines was calculated using one of two subsequent expressions: Each experiment was conducted in quadruplicate across two independent trials.

$$\text{If (mean } A_{test} - \text{mean } A_{tzero}) \geq 0, \text{ then} \tag{1}$$

$$PG = 100 \times (\text{mean } A_{test} - \text{mean } A_{tzero}) / (\text{mean } A_{ctrl} - \text{mean } A_{tzero}), \tag{2}$$

$$\text{and if (mean } A_{test} - \text{mean } A_{tzero}) < 0, \text{ then} \tag{3}$$

$$PG = 100 \times (\text{mean } A_{test} - \text{mean } A_{tzero}) / A_{tzero} \tag{4}$$

where the mean A_{tzero} is the average of the optical density measurements before the exposure of the cells to the test compound, the mean A_{test} is the average of the optical density measurements after the desired period, and the mean A_{ctrl} is the average of the optical density measurements after the desired period with no exposure of the cells to the test compound. Each experimental trial was conducted in quadruplicate across two separate experiments. The results are expressed as IC_{50}, which indicates the concentration required for 50% inhibition. The IC_{50} values for each compound are calculated from concentration–response curves using linear regression analysis by fitting the test concentrations that give PG values above and below the reference value (i.e., 50%). However, in instances where all tested concentrations for a specific cell line yield PGs that exceed the corresponding reference effect level (e.g., a PG value of 50), the highest tested concentration is designated as the default value, prefixed by a ">" symbol.

3.1.9. In Silico Drug-Likeness and Biological Profiling of 3,5-Disubstituted Hydantoins **5a–i**

The commercial program ADMET Predictor ver. 9.5 (Simulation Plus Inc., Lancaster, CA, USA) [53] was used to predict various simple structural features, physicochemical properties, and pharmacokinetic- and toxicity-related parameters of novel hydantoin derivatives. These were the topological polar surface area (TPSA) and numbers of hydrogen bond donating (HBD) and accepting (HBA) atoms, water solubility (Sw), lipophilicity coefficient (logP), human effective jejunal permeability (Peff), permeability to Madin-Darby canine kidney cells (MDCK), percent unbound to human blood plasma proteins (hum_fup%), percent unbound in human liver microsomes (fumic%), high/low classification of the probability of passing through the blood–brain barrier (BBB), brain/blood partition coefficient (logBB), likeness of P-glycoprotein efflux (Pgp sub) and inhibition (Pgp inh), rodent chronic toxicity Rat_ and Mouse_TD50 values estimating the oral dose (in mg kg^{-1} per day) of a compound required to induce a tumor in 50% of a rat/mouse population after exposure over a standard lifetime, yes/no likelihood of the hERG potassium channel inhibition (hERG_filter), and potential toxicity liabilities summarizing all ADMET risk models (Tox_Code). The web server admetSAR (https://lmmd.ecust.edu.cn/admetsar2/, accessed on 17 February 2020) was used to predict whether a compound may be an inhibitor (CYPs 3A4, 2D6, 2C19, 2C9, and 1A2) and a substrate (CYPs 3A4, 2D6, and 3A4) of the CYP enzymes [54].

Potential biological activities were screened in silico by using the commercial software PASS 2020 [55] and freely available web server SwissTargetPrediction [56].

4. Conclusions

In summary, a series of novel drug-like (±)-3,5-disubstituted hydantoins syn/anti-**5a–i** were synthesized by a base-assisted intramolecular amidolysis of trans-β-lactams ureas, with yields from good to excellent. The analysis of the NMR spectra revealed significant differences in carbon chemical shifts of two diastereoisomers. In addition, the NOE interactions, coupled with molecular modelling calculations, helped determine the relative configuration of the two diastereoisomers. The absolute configuration of the four enantiomers of allyl hydantoin syn/anti-**5a** was assigned using TD-DFT calculations of ECD and VCD spectra. The biological potential of novel hydantoins was evaluated in vitro by testing on antiproliferative activities and in silico using the commercial programs PASS and ADMET Predictor and the free online servers SwissTargetPrediction and admetSAR. Most of the tested compounds **5a–i** showed moderate antiproliferative activity on the tested tumor cell lines A2780 and MCF7. Four hydantoins, anti-**5b**, syn-**5f**, anti-**5f**, and anti-**5g**, showed moderate activity against the liver cancer cell line HepG2. The hydantoin compounds anti-**5b**, anti-**5c**, syn-**5f**, syn-**5g**, anti-**5g**, syn-**5h**, and anti-**5h** showed a moderate cytotoxic effect, while the other hydantoins had less or no toxic effect on the HFF-1 healthy cells. Among the synthesized hydantoins, compound anti-**5c**, with a cyclopentyl group at the N-3 position of the hydantoin ring, exhibited a strong cytotoxic effect against human tumor cell line MCF7 (IC_{50} = 4.5 µmol/L) and non-tumor cell line HFF-1 (IC_{50} = 12.0 µmol/L). In silico ADMET profiling describes hydantoins **5a–i** as aqueous soluble and membrane permeable molecules that tend to be metabolized by CYP3A4 and eliminated by the Pgp pump. According to PASS, synthesized hydantoins **5a–i** have a high probability of antiarthritic activity ($Pa \geq 0.727$). The hydantoin syn/anti-**5d**, with a furfuryl group at the N-3 position of the hydantoin ring, is also predicted to be an inhibitor of thioredoxin glutathione reductase, beta lactamase, DNA-directed DNA polymerase, and DNA polymerase beta. The SwissTargetPrediction tool did not predict significant biological targets for compounds **5a–i**, consistent with the novelty of the structure of the synthesized compounds.

Supplementary Materials: The following supporting information can be downloaded at: https://www.mdpi.com/article/10.3390/ph17101259/s1, Figure S1: ^{1}H NMR (300 MHz; $CDCl_3$) spectra of compound **1**; Figure S2: ^{13}C NMR (75 MHz; $CDCl_3$) spectra of compound **1**; Figure S3: ^{1}H NMR (300 MHz, $CDCl_3$) spectra of β-lactam protons H-3 and H-4 of the cis/trans-3-phthalimido-β-lactam mixture **2a/2b**; Figure S4: ^{1}H NMR (600 MHz; $CDCl_3$) spectra of compound **2a**; Figure S5: ^{13}C NMR (151 MHz; $CDCl_3$) spectra of compound **2a**; Figure S6: ^{1}H NMR (600 MHz; $CDCl_3$) spectra of compound **2b**; Figure S7: ^{13}C NMR (151 MHz; $CDCl_3$) spectra of compound **2b**; Figure S8: ^{1}H NMR (600 MHz; $CDCl_3$) spectra of compound **3**; Figure S9: ^{13}C NMR (151 MHz; $CDCl_3$) spectra of compound **3**; Figure S10: ^{1}H NMR (600 MHz; $CDCl_3$) spectra of compound **4a**; Figure S11: ^{13}C NMR (151 MHz; $CDCl_3$) spectra of compound **4a**; Figure S12: ^{1}H NMR (600 MHz; $CDCl_3$) spectra of compound **4b**; Figure S13: ^{13}C NMR (151 MHz; $CDCl_3$) spectra of compound **4b**; Figure S14: ^{1}H NMR (600 MHz; $CDCl_3$) spectra of compound **4c**; Figure S15: ^{13}C NMR (151 MHz; $CDCl_3$) spectra of compound **4c**; Figure S16: ^{1}H NMR (600 MHz; $CDCl_3$) spectra of compound **4d**; Figure S17: ^{13}C NMR (151 MHz; $CDCl_3$) spectra of compound **4d**; Figure S18: ^{1}H NMR (600 MHz; $CDCl_3$) spectra of compound **4e**; Figure S19: ^{13}C NMR (151 MHz; $CDCl_3$) spectra of compound **4e**; Figure S20: ^{1}H NMR (600 MHz; $CDCl_3$) spectra of compound **4f**; Figure S21: ^{13}C NMR (151 MHz; $CDCl_3$) spectra of compound **4f**; Figure S22: ^{1}H NMR (600 MHz; $CDCl_3$) spectra of compound **4g**; Figure S23: ^{13}C NMR (151 MHz; $CDCl_3$) spectra of compound **4g**; Figure S24: ^{1}H NMR (600 MHz; $CDCl_3$) spectra of compound **4h**; Figure S25: ^{13}C NMR (151 MHz; $CDCl_3$) spectra of compound **4h**; Figure S26: ^{1}H NMR (600 MHz; $CDCl_3$) spectra of compound **4i**; Figure S27: ^{13}C NMR (151 MHz; $CDCl_3$) spectra of compound **4i**; Figure S28: ^{1}H NMR (600 MHz; CD_3CN) spectra of compound **5a**; Figure S29: ^{13}C NMR (151 MHz; CD_3CN) spectra of compound **5a**; Figure S30: ^{1}H NMR (600 MHz; CD_3CN) spectra of compound **5b**; Figure S31: ^{13}C NMR (151 MHz; CD_3CN) spectra of compound **5b**; Figure S32: ^{1}H NMR (600 MHz; CD_3CN) spectra of compound **5c**; Figure S33: ^{13}C NMR (151 MHz;

CD$_3$CN) spectra of compound **5c**; Figure S34: ^{1}H NMR (600 MHz; CD$_3$CN) spectra of compound **5d**; Figure S35: ^{13}C NMR (151 MHz; CD$_3$CN) spectra of compound **5d**; Figure S36: ^{1}H NMR (600 MHz; CD$_3$CN) spectra of compound **5e**; Figure S37: ^{13}C NMR (151 MHz; CD$_3$CN) spectra of compound **5e**; Figure S38: ^{1}H NMR (600 MHz; CD$_3$CN) spectra of compound **5f**; Figure S39: ^{13}C NMR (151 MHz; CD$_3$CN) spectra of compound **5f**; Figure S40: ^{1}H NMR (600 MHz; CD$_3$CN) spectra of compound **5g**; Figure S41: ^{13}C NMR (151 MHz; CD$_3$CN) spectra of compound **5g**; Figure S42: ^{1}H NMR (600 MHz; CD$_3$CN) spectra of compound **5h**; Figure S43: ^{13}C NMR (151 MHz; CD$_3$CN) spectra of compound **5h**; Figure S44: ^{1}H NMR (600 MHz; CD$_3$CN) spectra of compound **5i**; Figure S45: ^{13}C NMR (151 MHz; CD$_3$CN) spectra of compound **5i**; **Figure S46**: ^{1}H spectrum, structure, numbering, and full assignment of Peak 1 in acetonitrile-d$_3$ at 25 °C; **Figure S47**: ^{13}C spectrum, structure, numbering, and full assignment of Peak 1 in acetonitrile-d$_3$ at 25 °C; Figure S48: Fully assigned ^{1}H-^{1}H COSY spectra of Peak 1 in acetonitrile-d$_3$ at 25 °C; Figure S49: Fully assigned ^{1}H-^{13}C HSQC spectrum of Peak 1 in acetonitrile-d$_3$ at 25 °C; Figure S50: Fully assigned ^{1}H-^{13}C HMBC spectrum of Peak 1 in acetonitrile-d$_3$ at 25 °C; Figure S51: Fully assigned ^{1}H-^{1}H NOESY spectrum of Peak 1 in acetonitrile-d$_3$ at 25 °C; Figure S52: ^{1}H spectrum, structure, numbering, and full assignment of Peak 2 in acetonitrile-d$_3$ at 25 °C; Figure S53: ^{13}C spectrum, structure, numbering, and full assignment of Peak 2 in acetonitrile-d$_3$ at 25 °C; Figure S54: Fully assigned ^{1}H-^{1}H COSY spectrum of Peak 2 in acetonitrile-d$_3$ at 25 °C; Figure S55: Fully assigned ^{1}H-^{13}C HSQC spectrum of Peak 2 in acetonitrile-d$_3$ at 25 °C; Figure S56: Fully assigned ^{1}H-^{13}C HMBC spectrum of Peak 2 in acetonitrile-d$_3$ at 25 °C; Figure S57: Fully assigned ^{1}H-^{1}H NOESY spectrum of Peak 2 in acetonitrile-d$_3$ at 25 °C; Figure S58: RP-HPLC chromatogram of compound **2a**. Figure S59: RP-HPLC chromatogram of compound **2b**; Figure S60: RP-HPLC chromatogram of compound **3**; Figure S61: RP-HPLC chromatogram of compound **4a**; Figure S62: RP-HPLC chromatogram of compound **4b**; Figure S63: RP-HPLC chromatogram of compound **4c**; Figure S64: RP-HPLC chromatogram of compound **4d**; Figure S65: RP-HPLC chromatogram of compound **4e**; Figure S66: RP-HPLC chromatogram of compound **4f**; Figure S67: RP-HPLC chromatogram of compound **4g**; Figure S68: RP-HPLC chromatogram of compound **4h**; Figure S69: RP-HPLC chromatogram of compound **4i**; Figure S70: RP-HPLC chromatogram of compound *syn/anti*-**5a**; Figure S71: RP-HPLC chromatogram of compound *anti*-**5a**; Figure S72: RP-HPLC chromatogram of compound *syn*-**5a**; Figure S73: RP-HPLC chromatogram of compound *syn/anti*-**5b**; Figure S74: RP-HPLC chromatogram of compound *anti*-**5b**; Figure S75: RP-HPLC chromatogram of compound *syn*-**5b**; Figure S76: RP-HPLC chromatogram of compound *syn/anti*-**5c**, Figure S77: RP-HPLC chromatogram of compound *anti*-**5c**; Figure S78: RP-HPLC chromatogram of compound *syn*-**5c**; Figure S79: RP-HPLC chromatogram of compound *syn/anti*-**5d**; Figure S80: RP-HPLC chromatogram of compound *anti*-**5d**; Figure S81: RP-HPLC chromatogram of compound *syn*-**5d**; Figure S82: RP-HPLC chromatogram of compound *syn/anti*-**5e**; Figure S83: RP-HPLC chromatogram of compound *anti*-**5e**; Figure S84: RP-HPLC chromatogram of compound *syn*-**5e**; Figure S85: RP-HPLC chromatogram of compound *syn/anti*-**5f**; Figure S86: RP-HPLC chromatogram of compound *anti*-**5f**; Figure S87: RP-HPLC chromatogram of compound *syn*-**5f**; Figure S88: RP-HPLC chromatogram of compound *syn/anti*-**5g**; Figure S89: RP-HPLC chromatogram of compound *anti*-**5g**; Figure S90: RP-HPLC chromatogram of compound *syn*-**5g**; Figure S91: RP-HPLC chromatogram of compound *syn/anti*-**5h**; Figure S92: RP-HPLC chromatogram of compound *anti*-**5h**; Figure S93: RP-HPLC chromatogram of compound *syn*-**5h**; Figure S94: RP-HPLC chromatogram of compound *syn/anti*-**5i**; Figure S95: RP-HPLC chromatogram of compound *anti*-**5i**; Figure S96: RP-HPLC chromatogram of compound *syn*-**5i**; Figure S97: Chromatogram of *syn/anti*-**5a** on preparative Zorbax Extend C-18 PrepHT preparative column; Figure S98: Chromatogram of *anti*-**5a** (Peak 1) on analytical CHIRAL ART Amylose SA column; Figure S99: Chromatogram of *syn*-**5a** (Peak 2) on analytical CHIRAL ART Amylose SA column; Figure S100: Chromatogram of enantiomer **5a-ent1** (*anti*-**5a**, Peak 1); Figure S101: Chromatogram of **5a-ent2** (*anti*-**5a**, Peak 1); Figure S102: Chromatogram of **5a-ent3** (*syn*-**5a**, Peak 2). Figure S103: Chromatogram of **5a-ent4** (*syn*-**5a**, Peak 2); Figure S104: Chromatogram of *anti*-**5a** (Peak 1) on the semi-preparative CHIRAL ART Amylose SA column; Figure S105: Chromatogram of *syn*-**5a** (Peak 2) on the semi-preparative CHIRAL ART Amylose SA column; Figure S106: HR-MS (ESI-QTOF) spectra of compound **2a**; Figure S107: HR-MS (ESI-QTOF) spectra of compound **2b**; Figure S108: HR-MS (ESI-QTOF) spectra of compound **4a**; Figure S109: HR-MS (ESI-QTOF) spectra of compound **4b**; Figure S110: HR-MS (ESI-QTOF) spectra of compound **4c**; Figure S111: HR-MS (ESI-QTOF) spectra of compound **4d**; Figure S112: HR-MS (ESI-QTOF) spectra of compound **4e**; Figure S113: HR-MS (ESI-QTOF) spectra of compound **4f**; Figure S114: HR-MS (ESI-QTOF) spectra of compound **4g**; Figure S115: HR-MS (ESI-QTOF) spectra of compound **4h**; Figure S116:

HR-MS (ESI-QTOF) spectra of compound **4i**; Figure S117: HR-MS (ESI-QTOF) spectra of compound *syn/anti*-**5a**; Figure S118: HR-MS (ESI-QTOF) spectra of compound *syn/anti*-**5b**; Figure S119: HR-MS (ESI-QTOF) spectra of compound *syn/anti*-**5c**; Figure S120: HR-MS (ESI-QTOF) spectra of compound *syn/anti*-**5d**; Figure S121: HR-MS (ESI-QTOF) spectra of compound *syn/anti*-**5e**; Figure S122: HR-MS (ESI-QTOF) spectra of compound *syn/anti*-**5f**; Figure S123: HR-MS (ESI-QTOF) spectra of compound *syn/anti*-**5g**; Figure S124: HR-MS (ESI-QTOF) spectra of compound *syn/anti*-**5h**; Figure S125: HR-MS (ESI-QTOF) spectra of compound *syn/anti*-**5i**; Figure S126: IR spectra of compound **2a**; Figure S127: IR spectra of compound **2b**; Figure S128: IR spectra of compound **3**; Figure S129: IR spectra of compound **4a**; Figure S130: IR spectra of compound **4b**; Figure S131: IR spectra of compound **4c**; Figure S132: IR spectra of compound **4d**; Figure S133: IR spectra of compound **4e**; Figure S134: IR spectra of compound **4f**; Figure S135: IR spectra of compound **4g**; Figure S136: IR spectra of compound **4h**; Figure S137: IR spectra of compound **4i**; Figure S138: IR spectra of compound *syn/anti*-**5a**; Figure S139: IR spectra of compound *syn/anti*-**5b**; Figure S140: IR spectra of compound *syn/anti*-**5c**; Figure S141: IR spectra of compound *syn/anti*-**5d**; Figure S142: IR spectra of compound *syn/anti*-**5e**; Figure S143: IR spectra of compound *syn/anti*-**5f**; Figure S144: IR spectra of compound *syn/anti*-**5g**; Figure S145: IR spectra of compound *syn/anti*-**5h**; Figure S146: IR spectra of compound *syn/anti*-**5i**; Figure S147: ECD spectra calculated for the truncated analogs of (5*S*,6*S*)-**5a** and (5*R*,6*S*)-**5a** (with the allyl group replaced by a methyl) at the TD-B3LYP/def2-TZVP/PCM//B3LYP-D3BJ/6-311 + G(d,p) level. Plotting parameters: UV shift + 15 nm, σ = 0.3 eV; Table S1: Proton chemical shifts comparison for Peak 1 and Peak 2 in acetonitrile-d_3 at 25 °C, with major differences marked in red; Table S2: Carbon chemical shifts comparison for Peak 1 and Peak 2 in acetonitrile-d_3 at 25 °C, with major differences marked in red; Table S3: Comparison of NOE interactions for Peak 1 in acetonitrile-d_3 at 25 °C, with major differences marked in red, where s = strong, m = medium, and w = weak; Table S4: Comparison of coupling constants for Peak 1 and Peak 2 in acetonitrile-d_3 at 25 °C, with major differences marked in red; Table S5: Structures, relative energy (kcal/mol), and Boltzmann populations of low-energy minima calculated for (5*S*,6*S*)-**5a** at the B3LYP-D3BJ/6-311+G(d,p)/PCM level; Table S6: Structures, relative energy (kcal/mol), and Boltzmann populations of low-energy minima calculated for (5*R*,6*S*)-**5a** at the B3LYP-D3BJ/6-311+G(d,p)/PCM level; Table S7: ECD similarity factors calculated for the range 190–290 nm, and VCD similarity factors calculated for the range 1500–1200 cm^{-1}; Table S8: Structures, relative energy (kcal/mol), and Boltzmann populations of low-energy minima calculated for the truncated model of (5*S*,6*S*)-**5a** at the B3LYP-D3BJ/6-311+G(d,p)/PCM level; Table S9: Structures, relative energy (kcal/mol), and Boltzmann populations of low-energy minima calculated for the truncated model of (5*R*,6*S*)-**5a** at the B3LYP-D3BJ/6-311+G(d,p)/PCM level; Table S10: Prediction of Lipinski's rule of five properties for the hydantoins **5a–i**; Table S11: ADMET properties of the hydantoins **5a–i** calculated by ADMET Predictor and admetSAR.

Author Contributions: Conceptualization, M.J. and T.D.; methodology, M.J., A.Č., M.G. and G.P.; investigation, M.J., A.Č., V.S., M.G. and D.K.; writing—original draft preparation, M.J., A.Č. and G.P.; writing—reviewing and editing, V.S., D.K., A.J. and M.R.; visualization, M.J., M.G. and G.P.; supervision, M.R. All authors have read and agreed to the published version of the manuscript.

Funding: This study was supported by the Croatian Government and the European Union through the European Regional Development Fund—the Competitiveness and Cohesion Operational Programme (KK.01.1.1.01) through the project Bioprospecting of the Adriatic Sea (KK.01.1.1.01.0002) granted to The Scientific Centre of Excellence for Marine Bioprospecting—BioProCro.

Institutional Review Board Statement: Not applicable.

Informed Consent Statement: Not applicable.

Data Availability Statement: Data is contained within the article and Supplementary Materials.

Acknowledgments: Calculations have been carried out using resources provided by Wroclaw Centre for Networking and Supercomputing in Poland and by computing@unipi, a Computing Service provided by the University of Pisa. We also thank the Center for NMR for recording spectra.

Conflicts of Interest: The authors declare no conflicts of interest.

References

1. Dapporto, P.; Paoli, P.; Rossi, P.; Altamura, M.; Perrotta, E.; Nannicini, R. Structural characterisation of a tetrasubstituted hydantoin by experimental and theoretical approaches: X-ray and ab initio studies. *J. Mol. Struct. THEOCHEM* **2000**, *532*, 195–204. [CrossRef]
2. Šmit, B.M.; Pavlović, R.Z. Three-step synthetic pathway to fused bicyclic hydantoins involving a selenocyclization step. *Tetrahedron* **2015**, *71*, 1101–1108. [CrossRef]
3. Akpan, E.D.; Dagdag, O.; Ebenso, E.E. Progress on the coordination chemistry and application of hydantoins and its derivatives as anticorrosive materials for steel: A review. *Coord. Chem. Rev.* **2023**, *489*, 215207. [CrossRef]
4. Gawas, P.P.; Ramakrishna, B.; Veeraiah, N.; Nutalapati, V. Multifunctional hydantoins: Recent advances in optoelectronics and medicinal drugs from Academia to the chemical industry. *J. Mater. Chem. C* **2021**, *9*, 16341–16377. [CrossRef]
5. Velázquez-Macías, R.F.; Aguilar-Patiño, S.; Cortez-Betancourt, R.; Rojas-Esquivel, I.; Fonseca-Reyes, G.; Contreras-González, N. Evaluation of efficacy of buserelin plus nilutamide in Mexican Male patients with advanced prostate cancer. *Rev. Mex. Urol.* **2016**, *76*, 346–351. [CrossRef]
6. Ito, Y.; Sadar, M.D. Enzalutamide and blocking androgen receptor in advanced prostate cancer: Lessons learnt from the history of drug development of antiandrogens. *Res. Rep. Urol.* **2018**, *10*, 23–32. [CrossRef]
7. Anderson, J. The role of antiandrogen monotherapy in the treatment of prostate cancer. *BJU Int.* **2003**, *91*, 455–461. [CrossRef]
8. Kassouf, W.; Tanguay, S.; Aprikian, A.G. Nilutamide as Second Line Hormone Therapy for Prostate Cancer After Androgen Ablation Fails. *J. Urol.* **2003**, *169*, 1742–1744. [CrossRef]
9. Krause, T.; Gerbershagen, M.U.; Fiege, M.; Weißhorn, R.; Wappler, F. Dantrolene—A review of its pharmacology, therapeutic use and new developments. *Anaesthesia* **2004**, *59*, 364–373. [CrossRef]
10. Konnert, L.; Lamaty, F.; Martinez, J.; Colacino, E. Recent Advances in the synthesis of hydantoins: The state of the art of a valuable scaffold. *Chem. Rev.* **2017**, *117*, 13757–13809. [CrossRef]
11. Ostrowski, J.; Kuhns, J.E.; Lupisella, J.A.; Manfredi, M.C.; Beehler, B.C.; Krystek, S.R.; Bi, Y.; Sun, C.; Seethala, R.; Golla, R.; et al. Pharmacological and X-Ray structural characterization of a novel selective androgen receptor modulator: Potent hyperanabolic stimulation of skeletal muscle with hypostimulation of prostate in rats. *Endocrinol.* **2007**, *148*, 4–12. [CrossRef] [PubMed]
12. Ali, O.M.; El-Sayed, S.A.; Eid, S.A.; Abdelwahed, N.A.M. Antimicrobial activity of new synthesized [(oxadiazolyl)methyl]phenytoin derivatives. *Acta Pol. Pharm.-Crug Res.* **2012**, *69*, 657–667.
13. Oliveira, S.M.D.; Silva, J.B.P.D.; Hernandes, M.Z.; Lima, M.D.C.A.D.; Galdino, S.L.; Pitta, I.D.R. Estrutura, reatividade e propriedades biológicas de hidantoínas. *Quím. Nova* **2008**, *31*, 614–622. [CrossRef]
14. Rajic, Z.; Zorc, B.; Raic-Malic, S.; Ester, K.; Kralj, M.; Pavelic, K.; Balzarini, J.; De Clercq, E.; Mintas, M. Hydantoin derivatives of L- and D-amino acids: Synthesis and evaluation of their antiviral and antitumoral activity. *Molecules* **2006**, *11*, 837–848. [CrossRef]
15. Kim, D.; Wang, L.; Caldwell, C.G.; Chen, P.; Finke, P.E.; Oates, B.; MacCoss, M.; Mills, S.G.; Malkowitz, L.; Gould, S.L.; et al. Discovery of human CCR5 antagonists containing hydantoins for the treatment of HIV-1 infection. *Bioorg. Med. Chem. Lett.* **2001**, *11*, 3099–3102. [CrossRef]
16. Verlinden, Y.; Cuconati, A.; Wimmer, E.; Rombaut, E.B. The antiviral compound 5-(3,4 dichlorophenyl) methylhydantoin inhibits the post synthetic cleavages and the assembly of poliovirus in a cell-free system. *Antiviral Res.* **2000**, *48*, 61–69. [CrossRef]
17. Botros, S.; Khalil, N.A.; Naguib, B.H.; El Dash, Y. Synthesis and anticonvulsant activity of new phenytoin derivatives. *Eur. J. Med. Chem.* **2013**, *60*, 57–63. [CrossRef]
18. Deodhar, M.; Sable, P.; Bhosale, A.; Juvale, K.; Dumbare, R.; Sakpal, P. Synthesis and evaluation of phenytoin derivatives as anticonvulsant agents. *Turk. J. Chem.* **2009**, *33*, 367–373. [CrossRef]
19. Thenmozhiyal, J.C.; Wong, P.T.-H.; Chui, W.-K. Anticonvulsant activity of phenylmethylenehydantoins: A structure–activity relationship study. *J. Med. Chem.* **2004**, *47*, 1527–1535. [CrossRef]
20. LeTiran, A.; Stables, J.P.; Kohn, H. Functionalized amino acid anticonvulsants: Synthesis and pharmacological evaluation of conformationally restricted analogues. *Bioorg. Med. Chem.* **2001**, *9*, 2693–2708. [CrossRef]
21. Anger, T.; Madge, D.J.; Mulla, M.; Riddall, D. Medicinal Chemistry of Neuronal Voltage Gated Sodium Channel Blockers. *J. Med. Chem.* **2001**, *44*, 115–137. [CrossRef] [PubMed]
22. Nakabayashi, M.; Regan, M.M.; Lifsey, D.; Kantoff, P.W.; Taplin, M.; Sartor, O.; Oh, W.K. Efficacy of nilutamide as secondary hormonal therapy in androgen-independent prostate cancer. *BJU Int.* **2005**, *96*, 783–786. [CrossRef] [PubMed]
23. Kiec-Kononowicz, K. Synthesis, structure and antiarrhythmic properties evaluation of new basic derivatives of 5,5-diphenylhydantoin. *Eur. J. Med. Chem.* **2003**, *38*, 555–566. [CrossRef]
24. Ciechanowicz-Rutkowska, M.; Stadnicka, K.; Kiec-Kononowicz, K.; Byrtus, H.; Filipek, B.; Zygmunt, M.; Maciag, D. Structure-Activity Relationship of Some New Anti-Arrhythmic Phenytoin Derivatives. *Arch. Pharm. Pharm. Med. Chem. (Weinheim)* **2000**, *333*, 357–364. [CrossRef]
25. Edmunds, J.J.; Klutchko, S.; Hamby, J.M.; Bunker, A.M.; Connolly, C.J.C.; Winters, R.T.; Quin, J.; Sircar, I.; Hodges, J.C. Derivatives of 5-[[1-4(4 carboxybenzyl)imidazolyl]methylidene]hydantoins as orally active angiotensin II receptor antagonists. *J. Med. Chem.* **1995**, *38*, 3759–3771. [CrossRef]
26. Lu, H.; Kong, D.; Wu, B.; Wang, S.; Wang, Y. Synthesis and Evaluation of Anti-Inflammatory and Antitussive Activity of Hydantion Derivatives. *Lett. Drug Des. Discov.* **2012**, *9*, 638–642. [CrossRef]

27. Somsák, L.; Kovács, L.; Tóth, M.; Ősz, E.; Szilágyi, L.; Györgydeák, Z.; Dinya, Z.; Docsa, T.; Tóth, B.; Gergely, P. Synthesis of and a comparative study on the inhibition of muscle and liver glycogen phosphorylases by epimeric pairs of D-gluco- and D-xylopyranosylidene-spiro(thio)hydantoins and N-(D-glucopyranosyl) amides. *J. Med. Chem.* **2001**, *44*, 2843–2848. [CrossRef] [PubMed]
28. Oka, M.; Matsumoto, Y.; Sugiyama, S.; Tsuruta, N.; Matsushima, M. A potent aldose reductase inhibitor, (2S,4S)-6-fluoro-2',5'-dioxospiro[chroman-4,4'-imidazolidine]-2 carboxamide (Fidarestat): Its absolute configuration and interactions with the aldose reductase by X-ray crystallography. *J. Med. Chem.* **2000**, *43*, 2479–2483. [CrossRef]
29. Mizuno, T.; Kino, T.; Ito, T.; Miyata, T. Synthesis of aromatic urea herbicides by the selenium-assisted carbonylation using carbon monoxide with sulfur. *Synth. Commun.* **2000**, *30*, 1675–1688. [CrossRef]
30. Fiallo, M.M.L.; Kozlowski, H.; Garnier-Suillerot, A. Mitomycin antitumor compounds. *Eur. J. Pharm. Sci.* **2001**, *12*, 487–494. [CrossRef]
31. Youssef, D.; Shaala, L.; Alshali, K. Bioactive hydantoin alkaloids from the Red Sea marine sponge *Hemimycale arabica*. *Mar. Drugs* **2015**, *13*, 6609–6619. [CrossRef] [PubMed]
32. Kalník, M.; Gabko, P.; Bella, M.; Koóš, M. The Bucherer–Bergs multicomponent synthesis of hydantoins—Excellence in simplicity. *Molecules* **2021**, *26*, 4024. [CrossRef]
33. Uemoto, H.; Tsuda, M.; Kobayashi, J. Mukanadins A–C, new bromopyrrole alkaloids from marine sponge *Agelas nakamurai*. *J. Nat. Prod.* **1999**, *62*, 1581–1583. [CrossRef] [PubMed]
34. Jiménez, C.; Crews, P. Mauritamide A and accompanying oroidin alkaloids from the sponge *Agelas mauritiana*. *Tetrahedron Lett.* **1994**, *35*, 1375–1378. [CrossRef]
35. Cachet, N.; Genta-Jouve, G.; Regalado, E.L.; Mokrini, R.; Amade, P.; Culioli, G.; Thomas, O.P. Parazoanthines A–E, hydantoin alkaloids from the Mediterranean Sea anemone *Parazoanthus axinellae*. *J. Nat. Prod.* **2009**, *72*, 1612–1615. [CrossRef]
36. Audoin, C.; Cocandeau, V.; Thomas, O.; Bruschini, A.; Holderith, S.; Genta-Jouve, G. Metabolome consistency: Additional parazoanthines from the Mediterranean zoanthid *Parazoanthus Axinellae*. *Metabolites* **2014**, *4*, 421–432. [CrossRef] [PubMed]
37. Colacino, E.; Lamaty, F.; Martinez, J.; Parrot, I. Microwave-assisted solid-phase synthesis of hydantoin derivatives. *Tetrahedron Lett.* **2007**, *48*, 5317–5320. [CrossRef]
38. Lu, G.-J.; Nie, J.-Q.; Chen, Z.-X.; Yang, G.-C.; Lu, C.-F. Synthesis and evaluation of a new non-cross-linked polystyrene supported hydantoin chiral auxiliary for asymmetric aldol reactions. *Tetrahedron Asymmetry* **2013**, *24*, 1331–1335. [CrossRef]
39. Metallinos, C.; John, J.; Zaifman, J.; Emberson, K. Diastereoselective synthesis of N-substituted planar chiral ferrocenes using a proline hydantoin-derived auxiliary. *Adv. Synth. Catal.* **2012**, *354*, 602–606. [CrossRef]
40. Meusel, M.; Gütschow, M. Recent developments in hydantoin chemistry. A review. *Org. Prep. Proced. Int.* **2004**, *36*, 391–443. [CrossRef]
41. Zhang, D.; Xing, X.; Cuny, G.D. Synthesis of hydantoins from enantiomerically pure α-amino amides without epimerization. *J. Org. Chem.* **2006**, *71*, 1750–1753. [CrossRef] [PubMed]
42. Chen, Y.; Su, L.; Yang, X.; Pan, W.; Fang, H. Enantioselective synthesis of 3,5-disubstituted thiohydantoins and hydantoins. *Tetrahedron* **2015**, *71*, 9234–9239. [CrossRef]
43. Tanwar, D.; Ratan, A.; Gill, M. Facile one pot synthesis of substituted hydantoins from carbamates. *Synlett* **2017**, *28*, 2285–2290. [CrossRef]
44. Liu, H.; Yang, Z.; Pan, Z. Synthesis of highly substituted imidazolidine-2,4-dione (hydantoin) through Tf$_2$O-mediated dual activation of Boc-protected dipeptidyl compounds. *Org. Lett.* **2014**, *16*, 5902–5905. [CrossRef]
45. Mehra, V.; Kumar, V. Facile diastereoselective synthesis of functionally enriched hydantoins via base-promoted intramolecular amidolysis of C-3 functionalized azetidin-2-ones. *Tetrahedron Lett.* **2013**, *54*, 6041–6044. [CrossRef]
46. Konnert, L.; Dimassi, M.; Gonnet, L.; Lamaty, F.; Martinez, J.; Colacino, E. Poly(ethylene) glycols and mechanochemistry for the preparation of bioactive 3,5-disubstituted hydantoins. *RSC Adv.* **2016**, *6*, 36978–36986. [CrossRef]
47. Konnert, L.; Gonnet, L.; Halasz, I.; Suppo, J.-S.; De Figueiredo, R.M.; Campagne, J.-M.; Lamaty, F.; Martinez, J.; Colacino, E. Mechanochemical preparation of 3,5-disubstituted hydantoins from dipeptides and unsymmetrical ureas of amino acid derivatives. *J. Org. Chem.* **2016**, *81*, 9802–9809. [CrossRef]
48. Mascitti, A.; Lupacchini, M.; Guerra, R.; Taydakov, I.; Tonucci, L.; d'Alessandro, N.; Lamaty, F.; Martinez, J.; Colacino, E. Poly(ethylene glycol)s as grinding additives in the mechanochemical preparation of highly functionalized 3,5 disubstituted hydantoins. *Beilstein J. Org. Chem.* **2017**, *13*, 19–25. [CrossRef]
49. DeWitt, S.H.; Kiely, J.S.; Stankovic, C.J.; Schroeder, M.C.; Cody, D.M.; Pavia, M.R. "Diversomers": An approach to nonpeptide, nonoligomeric chemical diversity. *Proc. Natl. Acad. Sci.* **1993**, *90*, 6909–6913. [CrossRef]
50. Colacino, E.; Porcheddu, A.; Charnay, C.; Delogu, F. From Enabling Technologies to Medicinal Mechanochemistry: An Eco-Friendly Access to Hydantoin-Based Active Pharmaceutical Ingredients. *React. Chem. Eng.* **2019**, *4*, 69–80. [CrossRef]
51. Monteiro, J.L.; Pieber, B.; Corrêa, A.G.; Kappe, C.O. Continuous Synthesis of Hydantoins: Intensifying the Bucherer-Bergs Reaction. *Synlett* **2016**, *27*, 83–87. [CrossRef]
52. Vukelić, S.; Koksch, B.; Seeberger, P.H.; Gilmore, K. A Sustainable, Semi-Continuous Flow Synthesis of Hydantoins. *Chem. Eur. J.* **2016**, *22*, 13451–13454. [CrossRef] [PubMed]
53. Lawless, M.S.; Waldman, M.; Fraczkiewicz, R.; Clark, R.D. Using cheminformatics in drug discovery. In *New Approaches to Drug Discovery*; Nielsch, U., Fuhrmann, U., Jaroch, S., Eds.; Springer International Publishing: Cham, Switzerland, 2015; pp. 139–168. [CrossRef]

54. Yang, H.; Lou, C.; Sun, L.; Li, J.; Cai, Y.; Wang, Z.; Li, W.; Liu, G.; Tang, Y. admetSAR 2.0: Web-service for prediction and optimization of chemical ADMET properties. *Bioinformatics* **2019**, *35*, 1067–1069. [CrossRef] [PubMed]
55. Filimonov, D.; Poroikov, V. Probabilistic approaches in activity prediction. In *Chemoinformatics Approaches to Virtual Screening*; Varnek, A., Tropsha, A., Eds.; The Royal Society of Chemistry: London, UK, 2008; pp. 182–216. [CrossRef]
56. Daina, A.; Michielin, O.; Zoete, V. SwissTargetPrediction: Updated data and new features for efficient prediction of protein targets of 30 small molecules. *Nucleic Acids Res.* **2019**, *47*, W357–W364. [CrossRef]
57. Dražić, T.; Roje, M.; Jurin, M.; Pescitelli, G. Synthesis, separation and absolute configuration determination by ECD spectroscopy and TDDFT calculations of 3-amino-β-lactams and derived guanidines. *Eur. J. Org. Chem.* **2016**, *2016*, 4189–4199. [CrossRef]
58. Bandyopadhyay, D.; Cruz, J.; Banik, B.K. Novel synthesis of 3-pyrrole substituted β-lactams via microwave-induced bismuth nitrate-catalyzed reaction. *Tetrahedron* **2012**, *68*, 10686–10695. [CrossRef]
59. Jurin, M.; Stepanić, V.; Bojanić, K.; Vadlja, D.; Kontrec, D.; Dražić, T.; Roje, M. Novel (±) trans-β-lactam ureas: Synthesis, *in silico* and *in vitro* biological profiling. *Acta Pharm.* **2024**, *74*, 37–59. [CrossRef] [PubMed]
60. Hosseyni, S.; Jarrahpour, A. Recent advances in β-lactam synthesis. *Org. Biomol. Chem.* **2018**, *16*, 6840–6852. [CrossRef]
61. Cossío, F.P.; De Cózar, A.; Sierra, M.A.; Casarrubios, L.; Muntaner, J.G.; Banik, B.K.; Bandyopadhyay, D. Role of imine isomerization in the stereocontrol of the Staudinger reaction between ketenes and imines. *RSC Adv.* **2022**, *12*, 104–117. [CrossRef]
62. Deketelaere, S.; Van Nguyen, T.; Stevens, C.V.; D'Hooghe, M. Synthetic approaches toward monocyclic 3-amino-β-lactams. *ChemistryOpen* **2017**, *6*, 301–319. [CrossRef]
63. Habuš, I.; Radolović, K.; Kralj, B. New thiazolidinone and triazinethione conjugates derived from amino-β-lactams. *Heterocycles* **2009**, *78*, 1729. [CrossRef]
64. Mehra, V.; Singh, P.; Manhas, N.; Kumar, V. β-Lactam-synthon-interceded facile synthesis of functionally decorated thiohydantoins. *Synlett* **2014**, *25*, 1124–1126. [CrossRef]
65. Trisovic, N.; Uscumlic, G.; Petrovic, S. Hydantoins: Synthesis, properties and anticonvulsant activity. *Hem. Ind.* **2009**, *63*, 17–31. [CrossRef]
66. Srebro-Hooper, M.; Autschbach, J. Calculating natural optical activity of molecules from first principles. *Annu. Rev. Phys. Chem.* **2017**, *68*, 399–420. [CrossRef]
67. Pescitelli, G.; Bruhn, T. Good computational practice in the assignment of absolute configurations by TDDFT calculations of ECD spectra. *Chirality* **2016**, *28*, 466–474. [CrossRef]
68. Bruhn, T.; Schaumlöffel, A.; Hemberger, Y.; Bringmann, G. SpecDis: Quantifying the comparison of calculated and experimental electronic circular dichroism spectra. *Chirality* **2013**, *25*, 243–249. [CrossRef]
69. Iwahana, S.; Iida, H.; Yashima, E.; Pescitelli, G.; Di Bari, L.; Petrovic, A.G.; Berova, N. Absolute stereochemistry of a 4 a-hydroxyriboflavin analogue of the key intermediate of the FAD-Monooxygenase cycle. *Chem. Eur. J.* **2014**, *20*, 4386–4395. [CrossRef]
70. Mándi, A.; Kurtán, T. Applications of OR/ECD/VCD to the structure elucidation of natural products. *Nat. Prod. Rep.* **2019**, *36*, 889–918. [CrossRef]
71. Superchi, S.; Scafato, P.; Gorecki, M.; Pescitelli, G. Absolute configuration determination by quantum mechanical calculation of chiroptical spectra: Basics and applications to fungal metabolites. *Curr. Med. Chem.* **2018**, *25*, 287–320. [CrossRef] [PubMed]
72. Lipinski, C.A. Drug-like properties and the causes of poor solubility and poor permeability. *J. Pharmacol. Toxicol. Methods* **2000**, *44*, 235–249. [CrossRef]
73. Lipinski, C.A.; Lombardo, F.; Dominy, B.W.; Feeney, P.J. Experimental and computational approaches to estimate solubility and permeability in drug discovery and development settings. *Adv. Drug Deliv. Rev.* **1997**, *23*, 3–25. [CrossRef]
74. Veber, D.F.; Johnson, S.R.; Cheng, H.-Y.; Smith, B.R.; Ward, K.W.; Kopple, K.D. Molecular properties that influence the oral bioavailability of drug candidates. *J. Med. Chem.* **2002**, *45*, 2615–2623. [CrossRef] [PubMed]
75. Haddad-Tóvolli, R.; Dragano, N.R.V.; Ramalho, A.F.S.; Velloso, L.A. Development and function of the blood-brain barrier in the context of metabolic control. *Front. Neurosci.* **2017**, *11*, 224. [CrossRef]
76. Frisch, M.J.; Trucks, G.W.; Schlegel, H.B.; Scuseria, G.E.; Robb, M.A.; Cheeseman, J.R.; Scalmani, G.; Barone, V.; Petersson, G.A.; Nakatsuji, H.; et al. *Gaussian 16*; Revision C.01; Gaussian, Inc.: Wallingford, CT, USA, 2016.
77. Hranjec, M.; Kralj, M.; Piantanida, I.; Sedić, M.; Šuman, L.; Pavelić, K.; Karminski-Zamola, G. Novel cyano- and amidino-substituted derivatives of styryl-2-benzimidazoles and benzimidazo[1,2 a]quinolines. synthesis, photochemical synthesis, DNA binding and antitumor evaluation, Part 3. *J. Med. Chem.* **2007**, *50*, 5696–5711. [CrossRef] [PubMed]
78. Hranjec, M.; Piantanida, I.; Kralj, M.; Šuman, L.; Pavelić, K.; Karminski-Zamola, G. Novel amidino-substituted thienyl- and furylvinyl benzimidazole derivatives and their photochemical conversion into corresponding diaza cyclopenta[c]fluorenes. Synthesis, interactions with DNA and RNA and antitumor evaluation. *J. Med. Chem.* **2008**, *51*, 4899–4910. [CrossRef]
79. Boyd, M.R.; Paull, K.D. Some practical considerations and applications of the national cancer institute *in vitro* anticancer drug discovery screen. *Drug Dev. Res.* **1995**, *34*, 91–109. [CrossRef]

Disclaimer/Publisher's Note: The statements, opinions and data contained in all publications are solely those of the individual author(s) and contributor(s) and not of MDPI and/or the editor(s). MDPI and/or the editor(s) disclaim responsibility for any injury to people or property resulting from any ideas, methods, instructions or products referred to in the content.

Article

In Silico Exploration of Novel EGFR Kinase Mutant-Selective Inhibitors Using a Hybrid Computational Approach

Md Ali Asif Noor [1], Md Mazedul Haq [2], Md Arifur Rahman Chowdhury [2], Hilal Tayara [3], HyunJoo Shim [4,*] and Kil To Chong [1,*]

[1] Department of Electronics and Information Engineering, Jeonbuk National University, Jeonju 54896, Republic of Korea; 202155519@jbnu.ac.kr
[2] Research Center of Bioactive Materials, Department of Bioactive Material Sciences, Division of Life Sciences (Molecular Biology Major), Jeonbuk National University, Jeonju 54896, Republic of Korea; haqmazed@jbnu.ac.kr (M.M.H.); 201855347@jbnu.ac.kr (M.A.R.C.)
[3] School of International Engineering and Science, Jeonbuk National University, Jeonju 54896, Republic of Korea; hilaltayara@jbnu.ac.kr
[4] School of Pharmacy, Jeonbuk National University, Jeonju 54896, Republic of Korea
* Correspondence: shimhj@jbnu.ac.kr (H.S.); kitchong@jbnu.ac.kr (K.T.C.)

Citation: Noor, M.A.A.; Haq, M.M.; Chowdhury, M.A.R.; Tayara, H.; Shim, H.; Chong, K.T. In Silico Exploration of Novel EGFR Kinase Mutant-Selective Inhibitors Using a Hybrid Computational Approach. *Pharmaceuticals* 2024, 17, 1107. https://doi.org/10.3390/ph17091107

Academic Editors: Roberta Rocca, Valentina Noemi Madia and Davide Ialongo

Received: 28 June 2024
Revised: 12 August 2024
Accepted: 19 August 2024
Published: 23 August 2024

Copyright: © 2024 by the authors. Licensee MDPI, Basel, Switzerland. This article is an open access article distributed under the terms and conditions of the Creative Commons Attribution (CC BY) license (https://creativecommons.org/licenses/by/4.0/).

Abstract: Targeting epidermal growth factor receptor (EGFR) mutants is a promising strategy for treating non-small cell lung cancer (NSCLC). This study focused on the computational identification and characterization of potential EGFR mutant-selective inhibitors using pharmacophore design and validation by deep learning, virtual screening, ADMET (Absorption, distribution, metabolism, excretion and toxicity), and molecular docking-dynamics simulations. A pharmacophore model was generated using Pharmit based on the potent inhibitor JBJ-125, which targets the mutant EGFR (PDB 5D41) and is used for the virtual screening of the Zinc database. In total, 16 hits were retrieved from 13,127,550 molecules and 122,276,899 conformers. The pharmacophore model was validated via DeepCoy, generating 100 inactive decoy structures for each active molecule and ADMET tests were conducted using SWISS ADME and PROTOX 3.0. Filtered compounds underwent molecular docking studies using Glide, revealing promising interactions with the EGFR allosteric site along with better docking scores. Molecular dynamics (MD) simulations confirmed the stability of the docked conformations. These results bring out five novel compounds that can be evaluated as single agents or in combination with existing therapies, holding promise for treating the EGFR-mutant NSCLC.

Keywords: NSCLC; JBJ-125; deep learning; pharmacophore; virtual screening; molecular docking; molecular dynamics; ADMET

1. Introduction

Lung cancer is a major global health concern owing to its high mortality rates. Non-small cell lung cancer (NSCLC) is the most prevalent, accounting for over 80% of lung cancer cases [1]. However, the effectiveness of early-stage treatment with chemotherapeutic agents targeting wild-type epidermal growth factor receptor (EGFR) in NSCLC remains uncertain, as previous research has suggested limited benefits for patient survival [2]. EGFR is a transmembrane protein belonging to the ERBB (erythroblastic leukemia viral oncogene homologue) family of receptor tyrosine kinases (RTKs) [3]. The EGFR family comprises four members: ERBB1, ERBB2, ERBB3, and ERBB4 [4]. EGFR contains extracellular ligand-attachment domains and is divided into four subdomains (I, II, III, and IV). Subdomains I and III, also known as L1 and L2, are responsible for binding growth factors, whereas subdomains II and IV, or CR1 and CR2, facilitate protein dimerization [5]. First- and second-generation EGFR tyrosine kinase inhibitors (TKIs) were initially designed to target the ATP binding site but faced challenges due to resistance [6]. Although third-generation

TKIs have been developed to address this issue, they face challenges, such as C797S mutation emergence [6]. Consequently, alternative and effective treatment strategies should be urgently explored. Allosteric site targeting is a promising approach in this regard [7]. It involves binding to regions other than the active site, thereby influencing protein conformation and downstream signaling pathways [8]. By targeting allosteric sites, EGFR activity and downstream signaling may be inhibited, thereby hindering cancer cell proliferation. In addition, adverse effects associated with the existing treatments for NSCLC may be overcome [9]. Previous studies have suggested that Leu747, Met766, Leu777, Leu788, Ile789, Met790, Phe856, and Asp855 constitute an EGFR allosteric site [8]. Another study by Singh et al. identified Lys745, Leu788, Thr854, Asp855, and Phe856 as the amino acids that interact with potential allosteric inhibitors [10]. So, targeting these residues can be a way to develop potential allosteric inhibitors. Recent studies have investigated compounds such as JBJ-125 (Figure 1A) and JBJ-063 (Figure 1B), which show promise for overcoming resistance mutations like L858R/T790M/C797S. Beyettet. al. reported the synergistic effect of JBJ-125 and osimertinib against TKI resistance [11].

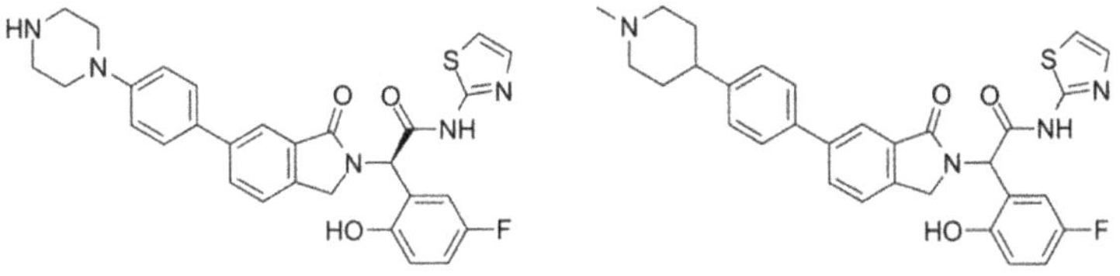

Figure 1. (**A**) JBJ-125. (**B**) JBJ-063.

Now, to investigate potential new compounds, in silico techniques like molecular docking are often applied, which helps in observing the interactions such as hydrogen bonds, hydrophobic bonds, pi-pi stacking, etc. [12]. This study aimed to identify distinguished derivatives, referencing JBJ-125. By targeting the allosteric sites, we aimed to contribute to the development of effective therapies for NSCLC, particularly for overcoming drug resistance.

2. Result

2.1. Pharmacophore Model Generation

The pharmacophore model was constructed based on the structural features of the reference ligand JBJ-125, a known potent EGFR mutant selective inhibitor (PDB 5D41), using the Pharmit tool. Ten pharmacophoric features (Table 1), including three aromatic rings, one hydrogen bond donor, three hydrogen bond acceptors, and three hydrophobic rings (Figure 2) were identified. By employing the default parameters within the Pharmit server, the Zinc database was screened using this pharmacophore model.

Table 1. Pharmacophore model features along with X, Y, Z coordination and radius.

Feature	X	Y	Z	Radious
Aromatic Ring 1	8.6	−0.7	−0.1	1
Aromatic Ring 2	15.4	−3.9	−0.2	1
Aromatic Ring 3	17.6	2.2	−0.1	1
Hydrogen Bond Donor	15.9	0.1	0.5	1
Hydrogen Bond Acceptor 1	11.8	1.3	0.2	1
Hydrogen Bond Acceptor 2	14.1	1	−0.6	1
Hydrogen Bond Acceptor 3	12.7	−4.3	0.2	1
Hydrophobic Bond 1	8.6	−0.7	−0.1	1
Hydrophobic Bond 2	15.4	−3.9	−0.2	1
Hydrophobic Bond 3	17.6	2.2	−0.1	1

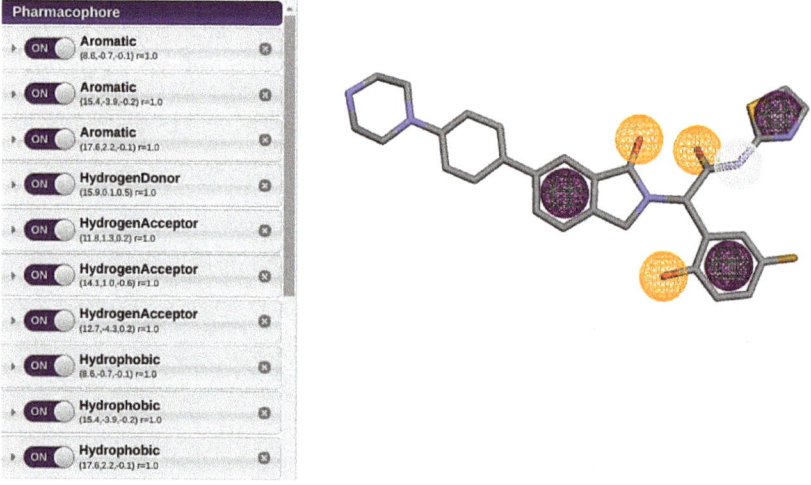

Figure 2. Pharmacophore model with its features.

2.2. Virtual Screening

The Zinc database was virtually screened using the developed pharmacophore model and the default protocol embedded within Pharmit. This screening process yielded a collection of 16 hits that were retrieved from the database of 13,127,550 molecules and 122,276,899 conformers. In addition to these 16 hits identified as possible new compounds, JBJ-125 was included as a reference compound for further analysis (Table 2). Subsequently, all compounds were rigorously evaluated against Lipinski's rule of five, ADMET analysis, and other pertinent in silico investigations.

Table 2. List of pharmacophore-derived compounds and reference compound with code numbers.

Scheme	Compound ID	Code Number	SMILE
1	ZINC000012638703	BNS1	COc1cc(OC)cc(C(=O)Nc2ccccc2-c2nnn(CC(=O)N3CCCc4ccccc43)n2)c1
2	ZINC000016694801	BNS2	COc1ccccc1N(CC(=O)Nc1ccccc1Oc1ccccc1)S(=O)(=O)c1cccs1
3	ZINC000012777271	BNS3	COC(=O)c1occc1CSc1nc(NC(=O)c2ccccc2)c2c(C)c(C)oc2n1
4	ZINC000033067859	BNS4	COc1ccc(C(=O)Nc2ccccc2OCc2cc(=O)n3cccc(C)c3n2)cc1OC
5	ZINC000020617126	BNS5	COc1ccc(C)cc1NC(=O)c1cn(C)nc1C(=O)Nc1cc(C)ccc1OC
6	ZINC000020617150	BNS6	COc1ccccc1NC(=O)c1cn(C)nc1C(=O)Nc1ccccc1OC
7	ZINC000059488018	BNS7	Oc1ccc(Br)cc1/C=N/N=C1/c2ccccc2-c2nc3ccccc3nc21
8	ZINC000059488022	BNS8	O=[N+]([O-])c1ccc(O)c(/C=N/N=C2/c3ccccc3-c3nc4ccccc4nc32)c1
9	ZINC000059488016	BNS9	Oc1cc(Cl)ccc1/C=N/N=C1/c2ccccc2-c2nc3ccccc3nc21
10	ZINC000013577005	BNS10	COc1ccc(NC(=O)C[C@H]2C(=O)N(c3ccccc3)C(=S)N2CCc2ccccc2OC)cc1
11	ZINC000021535964	BNS11	Cc1cn2c(=O)cc(CSc3ccccc3NC(=O)COc3ccc(Cl)cc3)nc2s1
12	ZINC000059488021	BNS12	COc1ccc(O)c(/C=N/N=C2/c3ccccc3-c3nc4ccccc4nc32)c1
13	ZINC000229934991	BNS13	O=C(Nc1ccc(Cl)cc1)[C@@H]1[C@H](c2ccccc([N+](=O)[O-])c2)C2(C(=O)c3ccccc3C2=O)[C@H]2c3ccccc3C=NN12
14	ZINC000041077159	BNS14	COc1ccc(C(=O)Nc2ccccc2OCc2cc(=O)n3c(ncn3C(C)C)n2)c1
15	ZINC000000831474	BNS15	COc1ccccc1NC(=O)c1nc1[nH]c1C(=O)Nc1ccccc1OC
16	ZINC000033067751	BNS16	COc1ccc(C(=O)Nc2ccccc2OCc2cc(=O)n3cccc3n2)cc1OC
17	Pubchem CID 124173751	JBJ-125	ClCN(CCN1)C2=CC=C(C=C2)C3=CC4=C(CN(C4=O)C(C5=C(C=CC(=C5)F)O)C(=O)NC6=NC=CS6)C=C3

2.3. Pharmacophore Validation

Pharmacophore validation was performed by deep decoy. The overall process assesses the hypothesis to discriminate among the active compounds and inactive decoys. The final dataset was prepared with 16 known active molecules, screened from 13,127,550 molecules and 122,276,899 conformers. From the assessment, we found an ROC (Receiver Operating Characteristic) value of 0.778, indicating better model quality and effectiveness. An Area Under the Receiver Operating Characteristic Curve (AUC-ROC) value of 0.5 suggests no discriminative power (equivalent to random guessing), while 1.0 indicates perfect classification [13]. Three types of metrics were used to evaluate the model. Performance metrics -which evaluates the ability of machine learning models to correctly classify compounds as actives or decoys. Higher values indicate better performance in distinguishing between actives and decoys [14]. Property matching metrics- which evaluates how well the properties of decoys match those of the actives, to ensure that decoys are chosen based on similar properties to actives. Thus, these metrics ensure that the decoys are appropriate controls by having similar properties to the actives. This is important for the validity of the screening process [15]. Finally, the structural similarity metrics, which evaluates the structural similarity between actives and decoys. These metrics ensure that decoys structurally resemble actives, which is important for the validity of structure-based screening methods [15].

2.4. ADMET Properties

To evaluate the ADMET properties, all pharmacophore-derived compounds were assessed alongside the reference compound JBJ-125, which served as the standard. Various parameters including bioavailability radar and fundamental physicochemical properties such as molecular weight, lipophilicity, water solubility, metabolic characteristics, and drug likeliness were examined. The consistent bioavailability pattern of the pharmacophore-derived compounds BNS1–BNS6, BNS10–BNS11, and BNS14–BNS16 is comparable to that of the reference compounds. Figure 3 illustrates the distribution of these compounds with respect to their bioavailability.

Analysis of the basic physicochemical properties (Table S1) revealed that the molecular weights of all the compounds except BNS13 were below 500 g/mol. As indicated by the consensus Log P values (Table S2), the lipophilicity of the screened compounds was 2.22 to 4.32. In comparison, the Log P value of JBJ-125 was 3.42.

Assessment of water solubility patterns (Table S3) revealed that, based on the Log S(ESOL) class categorization, all compounds were soluble, with most being moderately soluble. However, according to the Ali solubility classification, most pharmacophore-derived compounds exhibited poor water solubility, a trend consistent with the SILICOS-IT class category, in which all compounds, including the reference, demonstrated poor water solubility. All compounds, along with the reference compound except BNS2, BNS3, BNS11, and BNS13 demonstrated high absorption in the gastrointestinal tract (GI, Table S4). Notably, none of these compounds permeated across the blood–brain barrier (BBB). Most compounds, including BNS1, BNS4, BNS6, BNS10, and BNS13–BNS16, along with the reference compounds, are potential permeability glycoprotein (P-gp) substrates. BNS8 inhibited only one and BNS14 inhibited two out of five cytochrome P450 (CYP) isoforms. Conversely, all other compounds inhibited a minimum of three out of the five isoforms, whereas reference compounds inhibited four isoforms. Moreover, all compounds, except for BNS13 and reference compound JBJ-125, adhered to Lipinski's rule of five (Table 3). Additionally, all compounds exhibited a bioavailability score of 0.55. Among the compounds, BNS1-BNS6, BNS11, and BNS14–BNS16 showed no PAIN and BRENK alerts, whereas others displayed one or both alerts. The reference compound JBJ-125 presented one PAIN alert.

In the PROTOX study, hepatotoxicity, respiratory toxicity, carcinogenicity, immunotoxicity, mutagenicity, etc., were analyzed (Table 4). PROTOX works on the similarity method, which is based on the fact that structurally similar molecules are likely to exhibit similar toxic profiles [16]. In toxicity classification, most compounds (BNS1, BNS5 to BNS7, BNS10, BNS11, and BNS13–BNS16) were categorized into level IV toxicity classes. As for

BNS3, BNS8, BNS9, and BNS12, they were classified as level V toxicity classes, BNS4 as level III, BNS2 as level VI, and JBJ-125 had toxicity level IV.

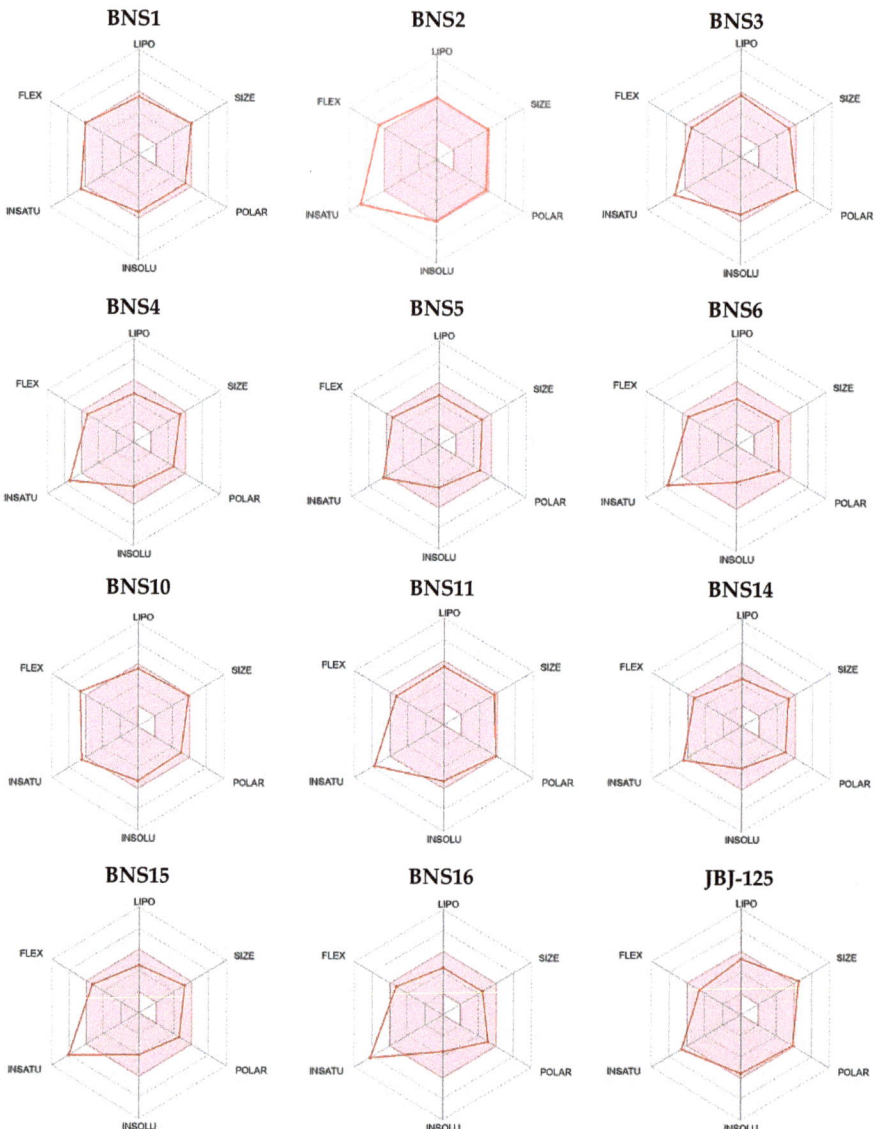

Figure 3. The radar of bioavailability prediction is displayed in the pink area. It shows the range of physicochemical properties optimum for oral bioavailability. Lipophilicity (−0.7 < XLOGP3 < +0.5), size (150 < MW < 500 g/mol), polarity (20A2 < TPSA < 130A2), insolubility (0 < Log S (ESOL) < 6), instauration (0.25 < Fraction Csp3 < 1), and flexibility (number of rotatable bonds < 9).

Table 3. Predicted drug likeliness properties of compounds.

Compounds	Lipinski #Violations	Ghose #Violations	Veber #Violations	Egan #ViolAtions	Muegge #Violations	Bioavailability Score	PAINS #Alert	Brenk #Alert	Lead Likeness #Violations	Synthetic Accessibility
BNS1	Yes, 0 violations	No#2	yes	yes	yes	0.55	0	0	No#3	3.75
BNS2	Yes, 0 violations	No#3	yes	No	No	0.55	0	0	No#3	3.93
BNS3	Yes, 0 Violation	Yes	Yes	No	Yes	0.55	0	0	No#3	3.73
BNS4	Yes, 0 Violations	Yes	Yes	Yes	Yes	0.55	0	0	No#2	3.32
BNS5	Yes, 0 Violations	Yes	Yes	Yes	Yes	0.55	0	0	No#2	3.04
BNS6	Yes, 0 Violations	Yes	Yes	Yes	Yes	0.55	0	0	No#2	2.81
BNS7	Yes, 0 Violations	Yes	Yes	Yes	Yes	0.55	1	1	No#2	3.3
BNS8	Yes, 0 Violations	Yes	Yes	Yes	Yes	0.55	1	3	No#2	3.37
BNS9	Yes, 0 violations	Yes	Yes	Yes	Yes	0.55	1	1	No#2	3.25
BNS10	Yes, 0 Violations	No#2	Yes	Yes	Yes	0.55	0	1	No#3	3.88
BNS11	Yes, 0 Violations	Yes	Yes	Yes	Yes	0.55	0	0	No#3	3.5
BNS12	Yes, 0 Violations	Yes	Yes	Yes	Yes	0.55	1	1	No#2	3.37
BNS13	Yes, 1 Violation	No#2	Yes	Yes	No#1	0.55	1	3	No#2	5.32
BNS14	Yes,0 Violations	Yes	Yes	Yes	Yes	0.55	0	0	No#2	3.34
BNS15	Yes, 0 Violations	Yes	Yes	Yes	Yes	0.55	0	0	No#2	2.66
BNS16	Yes, 0 Violations	Yes	Yes	Yes	Yes	0.55	0	0	No#2	3.17
JBJ-125	Yes, 1 Violations	No#2	No#2	Yes	Yes	0.55	1	0	No#2	4.26

Table 4. Toxicity prediction by PROTOX.

Compounds	Hepato-Toxicity	Neuro Toxicity	Respiratory Toxicity	Carcino Genicity	Immuno Toxicity	Muta Genicity	Cyto Toxicity	Toxicity Class
BNS1	Inactive	Active	Active	inactive	inactive	Moderately active	Moderately active	IV
BNS2	Inactive	Inactive	Active	Moderately inactive	inactive	inactive	inactive	VI
BNS3	Moderately active	Inactive	Moderately active	Moderately active	Moderately active	Moderately inactive	Inactive	V
BNS4	Moderately inactive	Moderately active	Active	Moderately inactive	Moderately active	Moderately active	Moderately inactive	III
BNS5	Moderately active	Moderately active	Moderately inactive	Moderately inactive	Inactive	Moderately inactive	Inactive	IV
BNS6	Moderately active	Moderately active	Moderately inactive	Moderately inactive	Inactive	Moderately inactive	Inactive	IV
BNS7	Moderately active	Moderately active	Moderately inactive	Moderately inactive	Moderately active	Moderately inactive	Moderately inactive	IV
BNS8	Moderately active	Moderately inactive	Moderately inactive	Active	Moderately inactive	Active	inactive	V
BNS9	Moderately active	Moderately active	Moderately inactive	Moderately inactive	inactive	Moderately inactive	inactive	V
BNS10	Moderately inactive	Active	Active	Moderately inactive	inactive	inactive	inactive	IV
BNS11	Moderately active	Moderately active	Moderately inactive	Moderately inactive	Moderately inactive	Moderately inactive	Moderately inactive	IV
BNS12	Moderately active	Moderately active	Moderately inactive	Moderately active	Active	Moderately active	inactive	V
BNS13	Moderately active	Moderately inactive	Moderately inactive	Moderately active	inactive	Active	inactive	IV
BNS14	Moderately inactive	Moderately active	Moderately active	Moderately inactive	Active	Moderately active	inactive	IV
BNS15	Moderately inactive	Moderately inactive	Moderately inactive	Moderately active	inactive	Moderately inactive	inactive	IV
BNS16	Moderately inactive	Moderately active	Active	Moderately inactive	Moderately active	Moderately active	Moderately inactive	IV
JBJ-125	Moderately inactive	Active	Active	Moderately inactive	Active	Moderately inactive	Moderately inactive	IV

Therefore, from the absorption, distribution, metabolism, and excretion (ADME) and PAIN (Pan Assay Interference) and BRENK alert analysis, we filtered BNS1-BNS6, BNS11, and BNS14–BNS16 as they were likely to possess a structurally promising moiety by not eliciting false-positive responses (PAIN alert). Usually, PAIN alert holding molecules contain substructures, which are likely to produce false positive biological results regardless of the target protein [17], thereby, reducing the likelihood of putative toxicity or metabolic instability (BRENK alert). The bioavailability ranges of these compounds were similar compared to that of the reference compound. Upon comparing the 10 compounds obtained after filtering through the PAIN and BRENK alert analysis with the 11 compounds identified from the bioavailability radar, we found overlapping compounds, except BNS10. Further scrutiny revealed that BNS10 possessed a BRENK alert, leading to its exclusion. Consequently, we selected 10 compounds that matched both the bioavailability radar and drug likeliness criteria and from the Protox toxicity class classification, which were in a considerable range; thus, they were considered for further evaluation, specifically molecular docking.

2.5. Molecular Docking Validation

Molecular docking validation was conducted using Glide by docking the extracted native ligand (57N) to the EGFR protein (PDB ID: 5D41) and superimposing the docked ligand. Superimposition revealed that the docked ligand conformation was nearly identical to that of the native co-crystallized ligand (Figure 4), with a 0.998 root mean square deviation (RMSD) value.

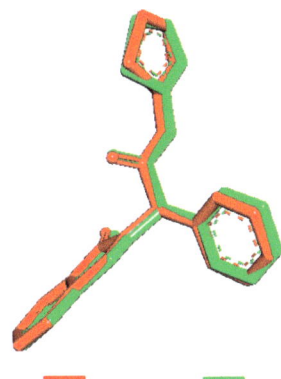

Reference Pose   Glide Pose

Figure 4. Superimposing of the Glide pose over the reference pose.

2.6. Molecular Docking

The molecular docking results of the 10 compounds identified from the ADMET tests are presented in Table 5.

From the docking score, pharmacophore-derived compounds ranged from −9.692 to −11.625 (Table 5) and for the reference compound JBJ-125 it was −11.119. So, compared to the reference, BNS1 (−11.625) and BNS16 (−11.237) showed better docking scores. Also, considering the total amino acid interactions, JBJ-125 had 26 interactions, whereas BNS1 and BNS2 both had a maximum of 29 amino acid interactions. Apart from this, BNS3, BNS4, and BNS11 showed 27 amino acid interactions, which are more than reference JBJ-125 (Table 5). So, based on the docking score and number of amino acid interactions, BNS1–BNS4, BNS11, and BNS16 were selected for molecular dynamics simulation.

Table 5. Molecular docking (Glide score, IFD score, and total amino acid interactions) result.

Compound	Glide Score (Kcal/mol)	IFD Score	Total Amino Acid Interaction
BNS1	−11.625	−667.73	29
BNS2	−10.313	−663.75	29
BNS3	−9.874	−666.74	27
BNS4	−10.408	−671.39	27
BNS5	−10.217	−664.96	25
BNS6	−9.853	−664.44	23
BNS11	−10.193	−665.93	27
BNS14	−10.442	−671.61	25
BNS15	−9.692	−663.93	21
BNS16	−11.237	−673.11	24
57N	−10.388	−662.43	20
JBJ-125	−11.119	−659.97	26

From the interaction category (Figure 5), it was observed that all compounds including the reference have shown hydrophobic interaction with "LEU747, ILE759, MET766, LEU777, LEU788, MET790, PHE856, and LEU858". Regarding hydrogen bond interactions we found that all pharmacophore-derived compounds (except BNS4) showed a hydrogen bond interaction with LYS745. Polar interaction was observed with THR854 among all the pharmacophore-derived and reference compounds. In negative charge interactions, JBJ-125 showed interaction with ASP800, ASP855 and GLU762 but all the other pharmacophore-derived compounds formed interaction with ASP855, GLU762, and GLU866. Therefore, from the molecular docking experiment, we found that our compounds interacting with amino acids mostly matched with the earlier discussed amino acids from the previous findings [8,10]. Additionally, it was observed that JBJ-125 formed a salt bridge interaction with GLU762. Pi cation interactions were observed among BNS1 and BNS4 with LYS745 and pi-pi stacking was observed with PHE856 among BNS2, BNS3, and BNS16.

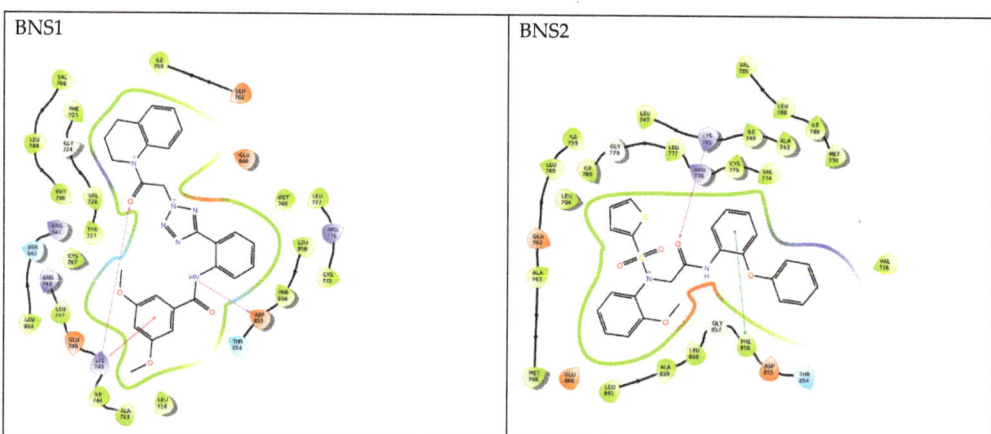

Figure 5. *Cont.*

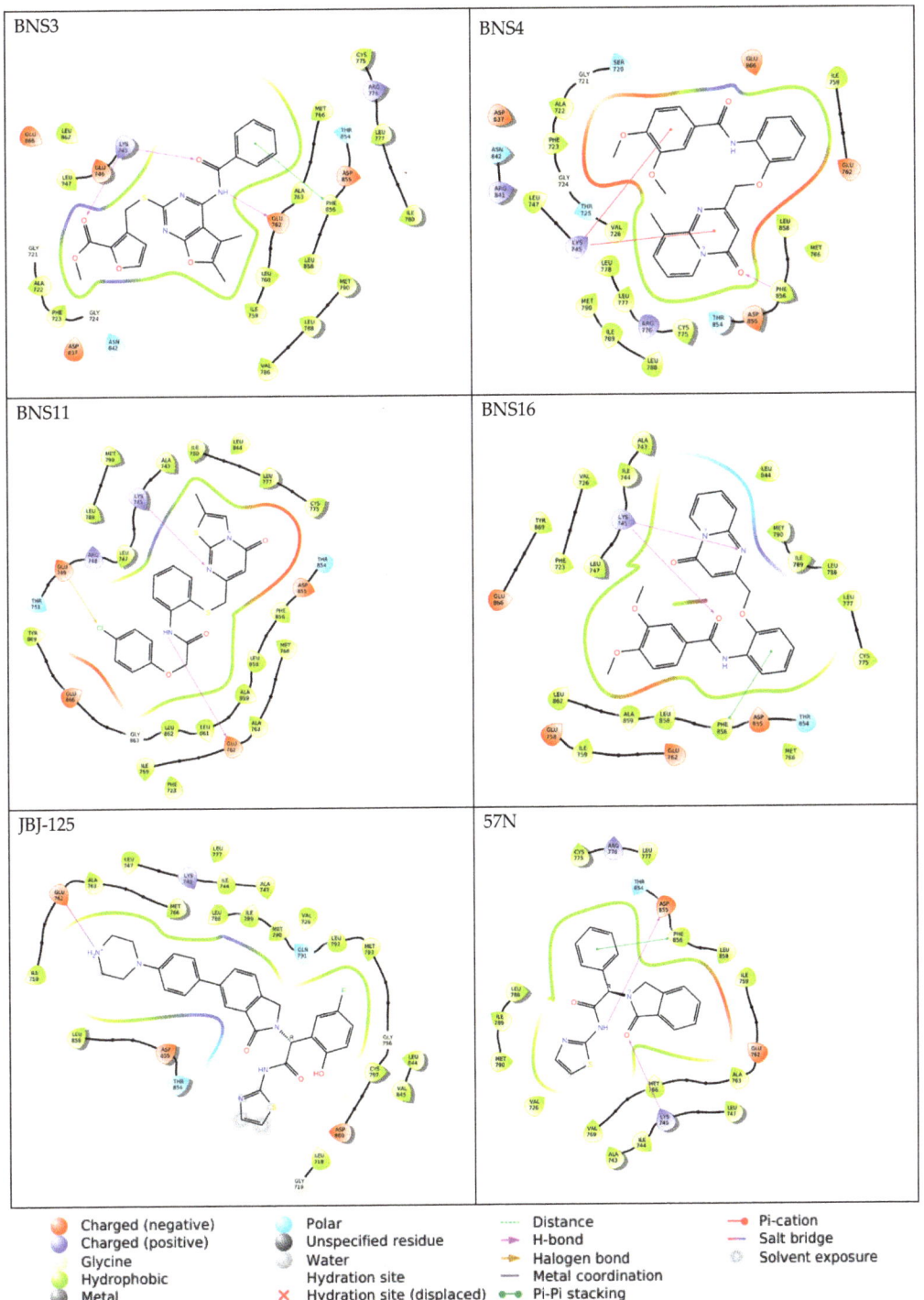

Figure 5. Molecular docking interactions. Specific types of interactions shown with specific marks.

Upon further analysis, it was observed that the pharmacophore-derived compounds BNS1, BNS3, BNS4, and BNS16 had the presence of a benzamide group in common and interesting interactions were formed with the benzamide group. For example, the benzene ring of the benzamide group was found to form a pi-cation interaction with LYS745 in BNS1 and BNS4 and pi-pi stacking interactions with BNS3 (Table S5). Considering the carbonyl group present in benzamide, BNS3 and BNS16 had a hydrogen bond interaction with LYS745 (Table S5). Again, from the amide group, hydrogen bond interactions were observed with ASP855 in BNS1 and with GLU762 in BNS3. On the other hand, BNS2 and BNS11 had an acetamide group in common. Here, with the acetamide group, LYS745 had a hydrogen bond interaction with the carbonyl group of BNS2 and BNS11.

2.7. Induced Fit Docking

By using Induced Fit Docking (IFD), it is possible to create multiple poses for the ligand–protein complex. Here, multiple conformational alterations matching the receptor–ligand position are conducted, followed by ranking the poses based on the IFD score for the identification of the ideal structure of the docked complex. In this study, using IFD, we compared the IFD scores of pharmacophore-derived compounds with the reference compound JBJ-125 to investigate the ideal ligand posture. All scores are listed in Table 5. From Table 5, we can find that the IFD score of JBJ-125 was -659.97 whereas, from the compounds obtained from pharmacophore, the IFD score range varied from -663.75 to -673.11, indicating the better performance of IFD of the pharmacophore-derived compounds. The highest IFD score was observed in BNS16 (-673.11) and the lowest in BNS2 (-663.75). With induced-fit docking, it is possible to generate multiple ligand-receptor structures along with certain conformational changes made by the receptor to receive a ligand. Therefore, this comprehensive technique helps in identifying promising ligand-receptor combinations for additional studies.

2.8. Molecular Dynamics Simulation

Molecular dynamics (MD) simulations were performed for 100 nanoseconds (ns) to evaluate the stability of ligand binding with proteins along with complex flexibility [18]. The tested compounds' binding stability and protein–ligand complex flexibility were observed utilizing the root mean square deviation (RMSD) and root mean square fluctuation (RMSF), respectively. In addition, the radius of gyration (rGyr) and molecular surface area (MolSA) were also analyzed to observe the nature of ligand extendedness and molecular surface calculation respectively.

From the molecular dynamics simulation, for JBJ-125, the average RMSD of the protein backbone atom was 2.204 Å (Figure 6A), with a maximum of 2.92 Å at 75.40 ns. The backbone RMSD was almost stable and slight fluctuation was observed within the 68 to 76 ns range. The average RMSD of ligand fit to protein was 2.421 Å with a maximum value of 3.44 Å at 18.80; other than this, the RMSD was almost stable. Overall, the average protein–ligand complex RMSD was below 3.00 Å, indicating a good stability pattern. The protein RMSF value average was 1.032 Å, the average rGyr value was 6.063, and MolSA was 480.72 Å^2 (Figure 7).

For BNS1, the average RMSD of the protein backbone was 3.144 Å (Figure 6A) and the average RMSD of ligand fit to protein was 2.97 Å. For the protein backbone, it took around 20 ns to reach a stable point and after that, it was almost stable through the 100 ns run; an almost similar pattern was also observed in ligand RMSD. The overall ligand protein RMSD was close to 3.00 Å. The protein RMSF value average was 1.210 Å (Figure 6B). The rGyr value average was 5.27 Å, which was better than JBJ-125, and the average MolSA value was 466.93 Å^2 (Figure 7).

For BNS2, the protein backbone RMSD value average was 2.821 Å (Figure 6A). The RMSD distribution pattern of the BNS2 protein backbone was almost similar to BNS1. The ligand RMSD average was 2.10 Å with minor fluctuations. This indicates the overall considerable RMSD of the protein–ligand complex. The average protein RMSF (Figure 6B) value was observed at 1.171 Å. The rGyr was observed at 4.688 Å, which was better than the reference compound. The average MolSA value was 426.87 Å² (Figure 7).

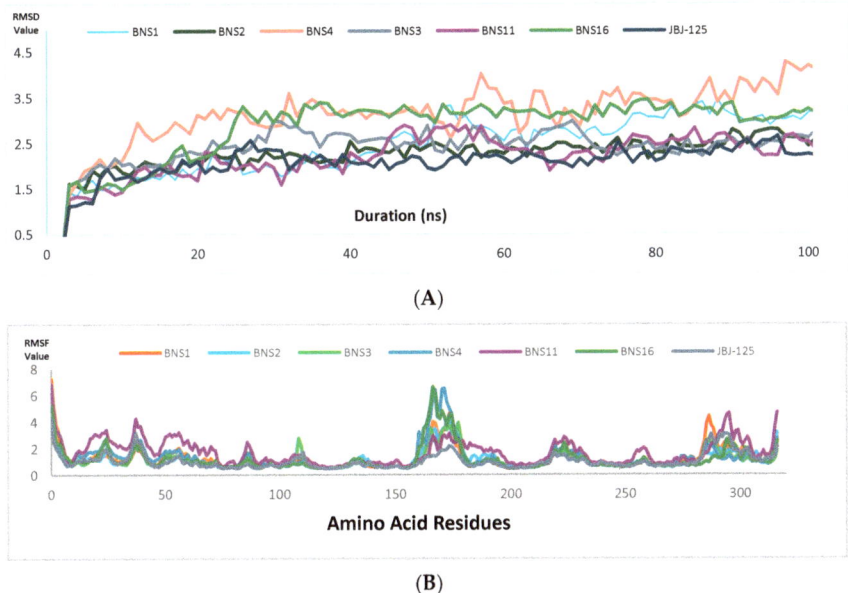

Figure 6. (**A**) Protein backbone RMSD graph. (**B**) Protein RMSF graph.

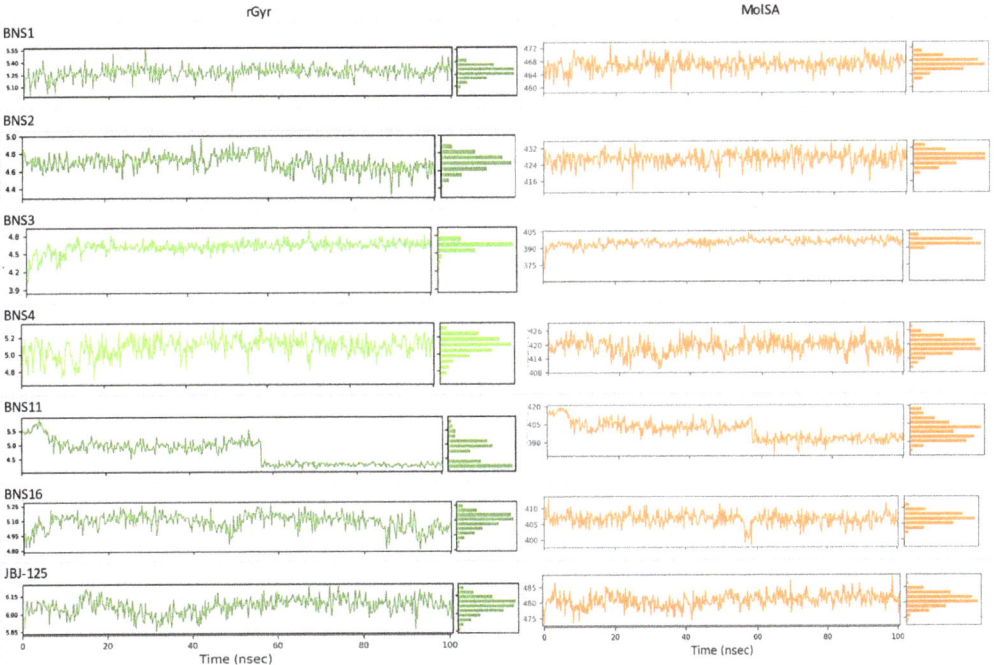

Figure 7. rGyr and MolSA of the pharmacophore-derived compounds and reference compound.

In BNS3, the observed protein backbone RMSD average was 2.805 Å (Figure 6A), and ligand RMSD had an average of 4.155 Å. For protein backbone RMSD, initial fluctuation was seen within 20 ns; after that, similar distribution was observed. In ligand RMSD, after 40 ns, stable distribution was seen. Here, the ligand RMSD value observed was slightly higher than the JBJ-125. The protein RMSF value average was 1.09 Å (Figure 6B). Also, the average rGyr was 4.605 Å, which was better than JBJ-125, and the MolSA average was 394.94 Å^2 (Figure 7).

In BNS4, the average protein backbone RMSD value was 4.014 Å (Figure 6A) and the ligand RMSD was 4.713 Å. The average protein RMSF value was 1.309 Å (Figure 6B) but the highest fluctuation crossed 5 Å. The rGyr value observed was 5.08 Å, whereas MolSA was 419.194 Å^2 (Figure 7).

In BNS11, the protein backbone RMSD average was observed at 2.95 Å (Figure 6A) but in the protein backbone RMSD, among the overall distribution, fluctuation was frequently raised above 3.0 Å. The ligand RMSD average was 4.706 Å but after 10 ns to the rest, the overall distribution was above 4.0 Å, indicating less ligand protein binding compared to JBJ-125. The average protein backbone RMSF was 1.68 Å (Figure 6B). The rGyr was 4.71 Å and the MolSA average was 399.08 Å^2 (Figure 7).

For BNS16, the protein backbone RMSD average was 2.67 Å (Figure 6A); after 30 ns to the rest, the average distribution was below 3.0 Å. The ligand RMSD average was 2.280 Å; after initial fluctuation, the RMSD graph declined to 2.50 Å till the first 50 ns. During the last 50 ns, distribution was observed below 2.5 Å. The protein RMSF average was 1.180 Å^2 (Figure 6B). The average rGyr was 5.08 Å and the average MolSA was 406.49 Å^2 (Figure 7).

For all the compounds obtained from the pharmacophore, the rGyr value score was lower than the reference compound, indicating that the pharmacophore-derived compounds will undergo less conformational change within the active site than the reference one [19]. The MolSA value of the compounds indicates the polarity of the compounds, which is competitive toward the reference [19]. The post-MD simulation interaction is presented in the Supplementary Figure S1 and the data are presented in Tables S6–S10.

Compared to the reference compound, BNS2 and BNS16 had a similar protein–ligand RMSD value average, within the range below 3.0 Å, indicating a stable complex [20]. For BNS1, the ligand RMSD average was 2.97 Å and the protein RMSD value average was 3.144 Å, which is not significantly higher than the acceptable range. In BNS3 and BNS11, the average ligand RMSD value was slightly higher than 3.0 Å (i.e., 4.155 Å and 4.706 Å) but not significantly different; thus, BNS1, BNS3, and BNS11 can also be considered for further evaluation. But, in BNS4, both the protein and ligand RMSD value average was over 4.00 Å; thus, excluding this, we have BNS2, BNS16, BNS1, BNS3, and BNS11 (Figure S2) as potential candidates for further evaluation.

3. Discussion

After pharmacophore modeling and virtual screening, we validated the pharmacophore model using deep learning techniques and found considerable results. The physicochemical attributes of pharmacophore-derived compounds were comprehensibly analyzed utilizing SWISS ADME. Lipophilicity, a fundamental determinant of drug absorption, was meticulously evaluated from the Log $P_{o/w}$ (ranging from 2.22–4.32 across the compound set). These positive values signify favorable lipophilic characteristics, indicating potential gastrointestinal absorption [21,22]. Furthermore, the assessment of solubility, a pivotal parameter governing drug bioavailability, revealed a collective trend toward poor water solubility among the pharmacophore-derived compounds and reference standards. Despite this, the observed lipophilicity suggests the prospect of substantial oral absorption, facilitating systemic distribution and eventual therapeutic action [23,24]. This intricate interplay between lipophilicity and solubility underscores the nuanced pharmacokinetic profile of the identified compounds, warranting further exploration to elucidate their therapeutic potential with precision and depth.

The assessment of Absorption, Distribution, Metabolism, and Excretion (ADME) properties constitutes a pivotal aspect in delineating the pharmacological behavior of potential drug candidates [25]. Particularly, for orally administered drugs, efficient absorption within the gastrointestinal tract (GIT) is of paramount importance for optimizing pharmacokinetic parameters. Conversely, the blood–brain barrier (BBB) serves as a pivotal physiological barrier that selectively regulates the entry of substances into the central nervous system (CNS) [26]. A comprehensive analysis of ADME properties revealed compelling insights into the pharmacokinetic profile of the identified compounds. Notably, except for BNS2, BNS3, BNS11, and BNS13, the evaluated compounds exhibited high gastrointestinal absorption rates, indicating favorable oral bioavailability. Furthermore, the absence of blood–brain barrier permeation among the investigated compounds suggests a reduced likelihood of adverse effects within the CNS, thus augmenting their safety profile for potential therapeutic applications. These findings underscore the potential utility of the identified compounds as orally administered agents with favorable pharmacokinetic attributes and minimal CNS-related side effects.

P-glycoprotein (P-gp) serves as a pivotal efflux transporter, facilitating substrate translocation from intracellular to extracellular compartments, thereby mitigating the potential toxic effects of compounds [27,28]. In our in silico investigation, we tested the P-gp substrate affinity of the identified compounds to elucidate their potential pharmacokinetic interactions. Notably, JBJ-125 exhibited P-gp substrate positivity, indicating their propensity to interact with this efflux transporter. Similarly, BNS1, BNS4, BNS6, BNS10, and BNS13-BNS16 also demonstrated P-gp substrate positivity. These observations shed light on the potential pharmacokinetic behavior of the identified compounds, particularly their interaction with P-gp and subsequent implications for drug disposition and efficacy. These insights will be instrumental in guiding further pharmacological evaluation and therapeutic applications of the identified compounds.

Understanding the intricate interactions between compounds and the cytochrome P450 (CYP) system is crucial to elucidating the pharmacokinetic profiles of potential drugs. These interactions play a pivotal role in mediating the biotransformation and elimination of drugs from the systemic circulation [17]. In this study, we examined the inhibitory potential of the identified compounds against various CYP isoforms to elucidate their pharmacokinetic implications. Our findings revealed that BNS8 and BNS14 inhibited only one and two CYP isoforms, respectively. Interestingly, the remaining compounds inhibited a minimum of three CYP isoforms. Comparative analysis of the reference compound JBJ-125 demonstrated a striking similarity in the inhibitory patterns, underscoring the consistency in pharmacokinetic behavior across the compounds. These observations underscore the importance of assessing CYP-mediated drug interactions in predicting the pharmacokinetic profile and potential drug–drug interactions of novel compounds. Such insights are invaluable for guiding further pharmacological investigations and optimizing therapeutic strategies.

The identification of structurally promising moieties and assessment of potential toxicity are critical steps in the preclinical evaluation of novel compounds. In the study, we used PAIN and BRENK to identify the structural motifs associated with false-positive responses in silico and putative toxicity, chemical reactivity, and metabolic instability [29,30]. Our analysis revealed that BNS1- BNS6, BNS11, and BNS14–BNS16 exhibited no PAIN or BRENK alerts. Conversely, JBJ-125 exhibited one PAIN alert. Furthermore, the PROTOX study indicated that all other pharmacophore-derived compounds, except BNS4, demonstrated toxicity levels below III, suggesting their potential safety for further evaluation [31]. Based on the bioavailability radar and drug-likeness properties, we selected 10 compounds (BNS1-BNS6, BNS11, and BNS14–BNS16) for subsequent molecular docking studies.

From the docking studies, we found that compounds BNS1 to BNS4 and BNS11 showed more amino acid interactions and BNS1 and BNS16 had docking scores higher than reference JBJ-125. From the induced fit docking; we found all the pharmacophore-derived compounds had higher IFD scores. Compared with JBJ-125, we can observe that, like JBJ-

125, BNS1, BN2, BNS3, BNS4, BNS11, and BNS16 had common positive charge interactions with LYS745. Here, we do not see any additional interaction between JBJ-125 and LYS745 but in BNS1 there is one hydrogen bond interaction and one pi cation interaction with LYS745, one hydrogen bond interaction with BNS2 and BNS3, two pi cation interactions with BNS4, BNS11, and BNS16 present, which shows the stronger bond formation of these compounds compared to JBJ-125. JBJ-125 had negative charge interactions with GLU762, so was seen among the other compounds also. Notably, JBJ-125 had a salt bridge interaction with GLU762. Additionally, BNS3 and BNS11 had formed one hydrogen bond interaction with GLU762, indicating their competitiveness with JBJ-125. Also, in the introduction part, THR854, Asp855, and Phe856 were discussed as key important amino acid residues for the allosteric site [8,10]. Here, JBJ-125, THR854, Asp855, and Phe856 formed polar, charge-negative, and hydrophobic interactions, respectively. A similar pattern was observed among all the other compounds also as well. Additionally, we observed that in BNS1, a hydrogen bond interaction was formed with ASP855 and a pi-pi stacking interaction was formed with PHE856 in BNS2, BNS11, and BNS16, stating stronger interaction of these compounds than JBJ-125. Previous studies have shown that afatinib and erlotinib showed docking scores of −7.69 and −7.37, respectively, against EGFR [32,33], whereas, our pharmacophore-derived compounds showed better docking scores than them indicating their better binding affinity and selectivity.

Compound BNS1 bears a 1,2,3,4-tetrahydroquinoline scaffold. In previous studies, quinazoline derivatives containing the 1,2,3,4-tetrahydroquinoline moiety demonstrated significant inhibitory activity against EGFR kinase, comparable to the positive control, afatinib [34]. This suggests that BNS1 could potentially exhibit strong EGFR inhibitory effects, making it a promising candidate for further experiments.

Also, Compound BNS3 features a thiazolo[3,2-a]pyrimidine scaffold. Another study reported that a novel series of naphtho[2′,3′:4,5]thiazolo[3,2-a]pyrimidine hybrids were synthesized and evaluated for their topo IIα/EGFR inhibitory activities [35]. Compounds 6i, 6a, and 6c from this series showed superior cytotoxic activity compared to doxorubicin and erlotinib against tested cancer cell lines. Molecular docking studies revealed that compound 6a forms the same hydrogen bond interaction with LYS 745 as observed with BNS3 in our study. This structural similarity and interaction suggest that BNS3 may also exhibit potent EGFR inhibition and could offer enhanced efficacy in treating cancers with EGFR involvement. Both BNS1 and BNS3 show potential for strong EGFR inhibitory activity due to their structural resemblance to compounds that have demonstrated efficacy in preclinical studies. This enhances their prospects as effective EGFR inhibitors.

Along with these tests, the considerable MD simulation pattern increases the acceptance of our compounds. Moreover, all these compounds showed interactions with the key important amino acid residues regarded as potential allosteric sites as mentioned earlier. Thus, our final compounds can be considered for further experiments as a better therapeutic choice compared to JBJ-125.

4. Material and Methods

Virtual experimentation was started with pharmacophore design and virtual screening using Pharmit [36], followed by the ADMET test using SWISS ADME and PROTOX 3.0 [17,31]. For docking, the glide function was used to perform a systematic search for the conformational, orientation, and positional space of the ligand in the binding pocket [37]. A molecular dynamics study was performed using Desmond in the Schrodinger molecular modeling suite [18].

4.1. Pharmacophore Designing/Modeling

A pharmacophore is an exposure of the drug-likeness of a molecule to its steric and electronic features, which are required to ensure optimal intermolecular interactions with a specific biological target, that is, a protein or enzyme, and inhibit or block its activity [38]. The pharmacophore technique can be used to facilitate drug development while searching large libraries or databases. In this study, the structure-based pharmacophore for the allosteric site of PDB ID:5D41 was generated using JBJ-125, the active inhibitor. The pharmacophore was generated using the free online server Pharmit, an open tool available at (http://pharmit.csb.pitt.edu, accessed on 13 June 2024). The value specification we used here are aromatic ring 1 (X: 8.6, Y: −0.7, Z: −0.1), aromatic ring 2 (X: 15.4, Y: −3.9, Z: −0.2), aromatic ring 3 (X: 17.6, Y: 2.2, Z: −0.1), hydrogen bond donor (X: 15.9, Y: 0.1, Z: 0.5), hydrogen bond acceptor 1 (X: 11.8, Y: 1.3, Z: 0.2) hydrogen bond acceptor 2 (X: 14.1, Y: 1.0, Z: −0.6), hydrogen bond acceptor 3 (X: 12.7, Y: −4.3, Z: 0.2), hydrophobic bond 1 (X: 8.6, Y: −0.7, Z: −0.1), hydrophobic bond 2 (X: 15.4, Y: −3.9, Z: −0.2), and hydrophobic bond 3 (X: 17.6, Y: 2.2, Z: −0.1).

4.2. Pharmacophore-Based Virtual Screening

In computational drug development and discovery processes, pharmacophore-based virtual screening is one of the most important steps for searching large libraries to identify LEADS against specific targets. Several tools and servers are available for pharmacophore-based virtual screening. Here, we used Pharmit, a free online server with an algorithm that can screen compound libraries based on the pharmacophore model or molecular shape and rank the results by energy minimization [36]. Using Pharmit, large databases of compounds can be screened based on their pharmacophoric features or molecular shapes. In this study, we screened the zinc database (https://zinc20.docking.org/, accessed on 13 June 2024) [39] based on JBJ-125 using Pharmit, and the top hits generated from the model are given in Table 6.

Table 6. Top hits generated from the Pharmit model.

Compound ID	RMSD	Mass	RBnds
ZINC000012777271	0.617	437	8
ZINC000013577005	0.642	490	10
ZINC000229934991	0.687	577	5
ZINC000033067751	0.699	431	8
ZINC000012638703	0.701	499	9
ZINC000041077159	0.714	433	8
ZINC000020617150	0.750	380	8
ZINC000020617126	0.751	408	8
ZINC000033067859	0.754	445	8
ZINC000000831474	0.762	366	8
ZINC000059488016	0.797	385	2
ZINC000059488018	0.798	429	2
ZINC000059488021	0.798	380	3
ZINC000059488022	0.798	395	3
ZINC000016694801	0.807	495	10
ZINC000021535964	0.821	472	8

4.3. Pharmacophore Validation

To determine the model accuracy of our pharmacophore model in predicting active chemicals, pharmacophore validation was performed. Here, we used the Deep decoy dataset (https://github.com/oxpig/DeepCoy, accessed on 18 June 2024), which generates property—matching decoy molecules, using a deep learning strategy called deep coy [15]. Here, we took the active molecules SMILE and generated 100 inactive decoy structures for each active molecule.

4.4. ADME Profile

The absorption, distribution, metabolism, and excretion (ADME) profile of the selected compounds was determined using SwissADME. The freely accessible SwissADME web tool (http://www.swissadme.ch/, accessed on 19 June 2024) is the most relevant computational method for providing a global appraisal of the pharmacokinetic profiles of small molecules.

These methods were selected by web tool designers for robustness and ease of interpretation to enable efficient translation into medicinal chemistry [17]. Additionally, hepatotoxicity, neurotoxicity, carcinogenicity, immune-toxicity, mutagenicity, cytotoxicity, and toxicity were predicted using PROTOX 3.0 (ProTox-3.0-Prediction of Toxicity of chemicals) available in (https://tox.charite.de/protox3/, accessed on 19 June 2024) [31].

4.5. Ligand Preparation

For ligand preparation, we designed a structure-based pharmacophore, targeting the allosteric site of PDB ID-5D41 focusing on JBJ-125, its active inhibitor. The pharmacophore was generated using the free online server Pharmit. Using the pharmacophore, we screened 16 compounds having structural similarity with JBJ-125 from the zinc database. The selected compounds were processed for energy minimization via the LigPrep module of Schrodinger using the OPLS3e force field [40]. The ZINC and PubChem ID of the pharmacophore-derived and reference compounds, respectively, are presented in Table 1.

4.6. Protein Preparation

We selected a mutant-selective EGFR protein structure targeting T790M and C797S mutations (PDB ID-5D41). The PDB structure was downloaded from RCSB PDB [41]. After the protein structure was retired from RCSB PDB, it underwent protein preparation processes available in Schrodinger [42]. During the protein preparation, water molecules were removed, missing side chains were added using Prime, and all co-crystallized ligands except 57N were deleted because they represent an allosteric inhibitor. 57N was used later for generating the receptor grid. The protein energy minimization was performed using the OPLS3e force field. The Van der Waals radius scaling factor was kept at 1.0 with a partial cutoff value of 0.25. For receptor grid generation, the centroid of the workspace ligand (57N) was selected and the grid box was generated accordingly (X: −23.71, Y: 31.37, Z: 12.3).

4.7. Docking Simulation Validation

Docking simulation was validated by re-docking the native ligand to the receptor binding site, to validate docking analysis, reproducibility, and reliability.

4.8. Molecular Docking

The GLIDE operational ligand docking tool in Maestro was used to generate molecular docking [43]. In GLIDE, compounds having atom numbers more than 500 and rotatable bonds more than 100 were set to reject [44]. As the number of designed analogs and generated tautomer was less, they were screened using the standard precision (SP) method, which uses descriptors and explicit water technology. The SP method eliminates false positives and employs a protocol with a refined growth strategy [37] and for ligand sampling, the flexible option was chosen along with nitrogen inversion and ring conformation in consideration. The application of sample bias was performed to all torsions presented with attached functional groups. Also, the Epik tool was enabled to enhance the docking score [45]. Minimization of post-docking was also performed, where the number of ten poses per ligand was evaluated to report the most effective conformation.

4.9. Induced Fit Docking

For induced fit docking using Schrodinger, the induced fit docking module was utilized [19,28]. Here, we used the previously used receptor grid box. For conformational sampling, sample ring conformation was kept with an energy window of 2.5 kcal/mol; additionally, receptor van der Waals scaling was kept at 0.50 along with ligand van der Waals scaling at 0.50. Residue refinement was kept within 5.0 Å of the ligand poses. Glide redocking of structures was kept within 30.0 kcal/mol of the best structure with standard precision mode. Table 5 presents the outcome of the induced fit docking score.

4.10. Molecular Dynamics

Target–ligand complex flexibility was studied via molecular dynamics (MD) to mimic biological systems. MD simulations were performed using the Desmond tool of the Schrödinger Drug Design Suite. Based on the docking score, the ligands were subjected to MD simulations for 100 ns to study their stability. The three steps performed for the MD simulation were building the system, minimization, and MD simulation. The docked ligand–protein complex was selected and the system was modeled by a predefined solvent system—TIP3P under orthorhombic boundary conditions. System neutralization was conducted by adding counter ions and salt was added as a concentration of 0.15 M Na^+ and Cl^- ions for reaching physiological circumstances and the system building was performed using OPLS3e force field. In a 100 ns run of molecular dynamics, trajectory data were taken every 50 picoseconds, energy data were captured at 1.2 picoseconds intervals, and the approximate number of frames was 500. NPT ensemble class was selected and 300 K temperature followed by 1.01325 pressures (bar) was carried out for MD simulation. Later, utilization of the simulation interaction diagram function was used for generating figures and plots to present the results. Any negative charges on the model were neutralized with sodium ions and the model was subjected to energy minimization until 25 kcal/mol/Å gradient thresholds were achieved at 300 K and 1 bar pressure via the NPT ensemble class. When conducting the MD simulation, the trajectory was recorded at 50 ps with approximately 500 frames. The complex stability was evaluated by protein and ligand RMSD (Root-Mean-Square Deviation) fluctuations, protein–ligand interactions, and contacts with various amino acids using the Simulation Event Analysis tool of Desmond [18].

5. Conclusions

In our study, the aim was to identify potential allosteric inhibitors to overcome the mutations that happen in EGFR NSCLC. To do it, we considered compound JBJ−125 as a reference and developed a pharmacophore-based on the features of JBJ-125 and performed a deep learning-based method to validate the pharmacophore model, followed by virtual screening. After that, we evaluated their toxicity via Swiss ADME and Protox. The screened compounds from ADMET tests undergo molecular docking, induced fit docking, and molecular dynamics studies. We found that BNS1, BNS2, BNS3, BNS11, and BNS16 have better interactions and docking scores than JBJ-125, and interactions with previously reported amino acid residues as allosteric sites were also observed among them. Recent studies indicate the capacity of JBJ-125 as a promising one to overcome resistance as a single agent or in combination with Osimertinib; hence, we believe we have potential outcomes and in vitro studies need to be performed to fully discover their therapeutic potential.

Supplementary Materials: The following supporting information can be downloaded at https://www.mdpi.com/article/10.3390/ph17091107/s1. The supporting information of physicochemical properties (Table S1), lipophilicity (Table S2), water solubility (Table S3), metabolism properties (Table S4), key interactions (Table S5), protein backbone RMSD (Table S6), ligand fit to protein RMSD (Table S7), protein RMSF (Table S8), rGyr value (Table S9), and MolSA (Table S10) and Figures post MD simulation interactions (Figure S1) and final compounds (Figure S2) are provided.

Author Contributions: Conceptualization, M.A.A.N., methodology, M.A.A.N. and M.M.H., software, M.A.A.N., M.M.H. and H.T., formal analysis M.A.A.N. and M.M.H.; investigation and writing-original draft preparation, M.A.A.N., writing—review and editing, M.A.A.N., M.M.H., M.A.R.C. and H.T.; supervision, H.T. and K.T.C.; funding acquisition, K.T.C. and H.S. All authors have read and agreed to the published version of the manuscript.

Funding: This work was supported in part by the National Research Foundation of Korea (NRF) grant funded by the Korea government (MSIT) (No. 2020R1A2C2005612) and in part by the Technology Innovation Program (20012892) funded by the Ministry of Trade, Industry and Energy (MOTIE, Korea) and the Institute of New Drug Development.

Institutional Review Board Statement: Not applicable.

Informed Consent Statement: Not applicable.

Data Availability Statement: Data are contained within the article and Supplementary Materials.

Conflicts of Interest: The authors declare no conflicts of interest.

References

1. Fawwaz, M.; Mishiro, K.; Nishii, R.; Sawazaki, I.; Shiba, K.; Kinuya, S.; Ogawa, K. Synthesis and Fundamental Evaluation of RadioiodinatedRociletinib (CO-1686) as a Probe to Lung Cancer with L858R/T790M Mutations of Epidermal Growth Factor Receptor (EGFR). *Molecules* **2020**, *25*, 2914. [CrossRef]
2. Huang, Y.; Cuan, X.; Zhu, W.; Yang, X.; Zhao, Y.; Sheng, J.; Zi, C.; Wang, X. An EGCG Derivative in Combination with Nimotuzumab for the Treatment of Wild-Type EGFR NSCLC. *Int. J. Mol. Sci.* **2023**, *24*, 14012. [CrossRef]
3. Roskoski, R., Jr. The ErbB/HER family of protein-tyrosine kinases and cancer. *Pharmacol. Res.* **2014**, *79*, 34–74. [CrossRef] [PubMed]
4. Schechter, A.; Stern, D.; Vaidyanathan, L.; Decker, S.J.; Drebin, J.A.; Greene, M.I.; Weinberg, R.A. Neu Oncogene: An erb-B-related gene encoding a 185,000-Mr tumour antigen. *Nature* **1984**, *312*, 513–516. [CrossRef]
5. Amelia, T.; Kartasasmita, R.E.; Ohwada, T.; Tjahjono, D.H. Structural Insight and Development of EGFR Tyrosine Kinase Inhibitors. *Molecules* **2022**, *27*, 819. [CrossRef] [PubMed]
6. Saini, R.; Kumari, S.; Bhatnagar, A.; Singh, A.; Mishra, A. Discovery of the allosteric inhibitor from actinomyces metabolites to target EGFRCSTMLR mutant protein: Molecular modeling and free energy approach. *Sci. Rep.* **2023**, *13*, 8885. [CrossRef]
7. Hubbard, P.A.; Moody, C.; Murali, R. Allosteric modulation of Ras and the PI3K/AKT/mTOR pathway: Emerging therapeutic opportunities. *Front. Physiol.* **2014**, *5*, 478. [CrossRef] [PubMed]
8. Maity, S.; Pai, K.S.R.; Nayak, Y. Advances in targeting EGFR allosteric site as anti-NSCLC therapy to overcome the drug resistance. *Pharmacol. Rep.* **2020**, *72*, 799–813. [CrossRef] [PubMed] [PubMed Central]
9. Roskoski, R., Jr. Classification of small molecule protein kinase inhibitors based upon the structures of their drug-enzyme complexes. *Pharmacol. Res.* **2016**, *103*, 26–48. [CrossRef] [PubMed]
10. Purba, E.; Saita, E.; Maruyama, I. Activation of the EGF receptor by ligand binding and oncogenic mutations: The "rotation model". *Cell* **2017**, *6*, 13. [CrossRef]
11. Beyett, T.S.; To, C.; Heppner, D.E.; Rana, J.K.; Schmoker, A.M.; Jang, J.; De Clercq, D.J.H.; Gomez, G.; Scott, D.A.; Gray, N.S.; et al. Molecular basis for cooperative binding and synergy of ATP-site and allosteric EGFR inhibitors. *Nat. Commun.* **2022**, *13*, 2530. [CrossRef]
12. Simonyan, H.; Palumbo, R.; Petrosyan, S.; Mkrtchyan, A.; Galstyan, A.; Saghyan, A.; Scognamiglio, P.L.; Vicidomini, C.; Fik-Jaskólka, M.; Roviello, G.N. BSA Binding and Aggregate Formation of a Synthetic Amino Acid with Potential for Promoting Fibroblast Proliferation: An In Silico, CD Spectroscopic, DLS, and Cellular Study. *Biomolecules* **2024**, *14*, 579. [CrossRef] [PubMed]
13. Hanley, J.A. Receiver Operating Characteristic (ROC) Curves. In *Wiley StatsRef:Statistics Reference Online*; Balakrishnan, N., Colton, T., Everitt, B., Piegorsch, W., Teugels, F.R.J.L., Eds.; John Wiley & Sons: Hoboken, NJ, USA, 2014. [CrossRef]
14. Rácz, A.; Bajusz, D.; Héberger, K. Multi-level comparison of machine learningclassifiers and their performance metrics. *Molecules* **2019**, *24*, 2811. [CrossRef]
15. Imrie, F.; Bradley, A.R.; Deane, C.M. Generating property-matched decoy molecules using deep learning. *Bioinformatics* **2021**, *37*, 2134–2141. [CrossRef]
16. Drwal, M.N.; Banerjee, P.; Dunkel, M.; Wettig, M.R.; Preissner, R. ProTox: A web server for the in silico prediction of rodent oral toxicity. *Nucleic Acids Res.* **2014**, *42*, W53–W58. [CrossRef]
17. Daina, A.; Michielin, O.; Zoete, V. SwissADME: A free web tool to evaluate pharmacokinetics, drug-likeness and medicinal chemistry friendliness of small molecules. *Sci. Rep.* **2017**, *7*, 42717. [CrossRef] [PubMed]
18. Bowers, K.J.; Chow, E.; Xu, H.; Dror, R.O.; Eastwood, M.P.; Gregersen, B.A.; Klepeis, J.L.; Kolossvary, I.; Moraes, M.A.; Sacerdoti, F.D.; et al. Scalable algorithms for molecular dynamics simulations on commodity clusters. In Proceedings of the 2006 ACM/IEEE Conference on Supercomputing (SC '06), Tampa, FL, USA, 11–17 November 2006.
19. Vemula, D.; Maddi, D.R.; Bhandari, V. Homology modeling, virtual screening, molecular docking, and dynamics studies for discovering Staphylococcus epidermidis FtsZ inhibitors. *Front. Mol. Biosci.* **2023**, *10*, 1087676. [CrossRef]
20. Nyambo, K.; Tapfuma, K.I.; Adu-Amankwaah, F.; Julius, L.; Baatjies, L.; Niang, I.S.; Smith, L.; Govender, K.K.; Ngxande, M.; Watson, D.J.; et al. Molecular docking, molecular dynamics simulations and binding free energy studies of interactions between Mycobacterium tuberculosis Pks13, PknG and bioactive constituents of extremophilic bacteria. *Sci. Rep.* **2024**, *14*, 6794. [CrossRef] [PubMed]
21. Rai, M.; Singh, A.V.; Paudel, N.; Kanase, A.; Falletta, E.; Kerkar, P.; Heyda, J.; Barghash, R.F.; Pratap Singh, S.; Soos, M. Herbal concoction Unveiled: A computational analysis of phytochemicals' pharmacokinetic and toxicological profiles using novel approach methodologies (NAMs). *Curr. Res. Toxicol.* **2023**, *5*, 100118. [CrossRef] [PubMed] [PubMed Central]
22. Arnott, J.A.; Planey, S.L. The influence of lipophilicity in drug discovery and design. *Expert Opin. Drug Discov.* **2012**, *7*, 863–875. [CrossRef]
23. Lipinski, C.A.; Lombardo, F.; Dominy, B.W.; Feeney, P.J. Experimental and computational approaches to estimate solubility and permeability in drug discovery and development settings. *Adv. Drug Deliv. Rev.* **2012**, *64*, 4–17. [CrossRef]

24. Savjani, K.T.; Gajjar, A.K.; Savjani, J.K. Drug solubility: Importance and enhancement techniques. *Int. Sch. Res. Not.* **2012**, *2012*, 195727. [CrossRef] [PubMed]
25. Abbott, N.J. Prediction of blood–brain barrier permeation in drug discovery from in vivo, in vitro and in silico models. *Drug Discov. Today Technol.* **2004**, *1*, 407–416. [CrossRef]
26. Kim, R.B. Drugs as P-glycoprotein substrates, inhibitors, and inducers. *Drug Metab. Rev.* **2002**, *34*, 47–54. [CrossRef] [PubMed]
27. Farid, R.; Day, T.; Friesner, R.A.; Pearlstein, R.A. New insights about HERG blockade obtained from protein modeling, potential energy mapping, and docking studies. *Bioorg. Med. Chem.* **2006**, *14*, 3160–3173. [CrossRef] [PubMed]
28. Sherman, W.; Day, T.; Jacobson, M.P.; Friesner, R.A.; Farid, R. Novel procedure for modeling ligand/receptor induced fit effects. *J. Med. Chem.* **2006**, *49*, 534–553. [CrossRef]
29. Baell, J.B.; Holloway, G.A. New substructure filters for removal of pan assay interference compounds (PAINS) from screening libraries and for their exclusion in bioassays. *J. Med. Chem.* **2010**, *53*, 2719–2740. [CrossRef]
30. Brenk, R.; Schipani, A.; James, D.; Krasowski, A.; Gilbert, I.H.; Frearson, J.; Wyatt, P.G. Lessons Learnt from Assembling Screening Libraries for Drug Discovery for Neglected Diseases. *ChemMedChem Chem. Enabling Drug Discov.* **2007**, *3*, 435–444. [CrossRef]
31. Swain, S.S.; Singh, S.R.; Sahoo, A.; Hussain, T.; Pati, S. Anti-HIV-drug and phyto-flavonoid combination against SARS-CoV-2: A molecular docking-simulation base assessment. *J. Biomol. Struct. Dyn.* **2021**, *40*, 6463–6476. [CrossRef]
32. Patil, S.; Randive, V.; Mahadik, I.; Asgaonkar, K. Design, In Silico Molecular Docking, and ADMET Prediction of Amide Derivatives of Chalcone Nucleus as EGFR Inhibitors for the Treatment of Cancer. *Curr. Drug Discov. Technol.* **2024**, *21*, 9–19. [CrossRef]
33. Yang, H.; Zhang, Z.; Liu, Q.; Yu, J.; Liu, C.; Lu, W. Identification of Dual-Target Inhibitors for Epidermal Growth Factor Receptor and AKT: Virtual Screening Based on Structure and Molecular Dynamics Study. *Molecules* **2023**, *28*, 7607. [CrossRef]
34. OuYang, Y.; Zou, W.; Peng, L.; Yang, Z.; Tang, Q.; Chen, M.; Jia, S.; Zhang, H.; Lan, Z.; Zheng, P.; et al. Design, Synthesis, Antiproliferative Activity and Docking Studies of Quinazoline Derivatives Bearing 2,3-dihydro-indole or 1,2,3,4-tetrahydroquinoline As Potential EGFR Inhibitors. *Eur. J. Med. Chem.* **2018**, *154*, 29–43. [CrossRef] [PubMed]
35. Mourad, M.A.; Abo Elmaaty, A.; Zaki, I.; Mourad, A.A.; Hofni, A.; Khodir, A.E.; Elkamhawyf, A.; Roh, E.J.; Al-Karmalawy, A.A. Novel topoisomerase II/EGFR dual inhibitors: Design, synthesis and docking studies of naphtho[2′,3′:4,5]thiazolo[3,2-a]pyrimidine hybrids as potential anticancer agents with apoptosis inducing activity. *J. Enzym. Inhib. Med. Chem.* **2023**, *38*, 2205043. [CrossRef] [PubMed]
36. Sunseri, J.; Koes, D.R. Pharmit: Interactive exploration of chemical space. *Nucleic Acids Res.* **2016**, *44*, W442–W448. [CrossRef]
37. Kumar, A.; Agarwal, P.; Rathi, E.; Kini, S.G. Computer-aided identification of human carbonic anhydrase isoenzyme VII inhibitors 211as potential antiepileptic agents. *J. Biomol. Struct. Dyn.* **2022**, *40*, 4850–4865. [CrossRef] [PubMed]
38. Yang, S. Pharmacophore modeling and applications in drug discovery: Challenges and recent advances. *Drug Discov. Today* **2010**, *15*, 444–450. [CrossRef] [PubMed]
39. Irwin, J.J.; Tang, K.G.; Young, J.; Dandarchuluun, C.; Wong, B.R.; Khurelbaatar, M.; Moroz, Y.S.; Mayfield, J.; Sayle, R.A. ZINC20—A Free Ultralarge-Scale Chemical Database for Ligand Discovery. *J. Chem. Inf. Model.* **2020**, *60*, 6065–6073. [CrossRef]
40. Dixon, S.L.; Smondyrev, A.M.; Rao, S.N. PHASE: A novel approach to pharmacophore modeling and 3D database searching. *Chem. Biol. Drug Des.* **2006**, *67*, 370–372. [CrossRef]
41. Jia, Y.; Yun, C.H.; Park, E.; Ercan, D.; Manuia, M.; Juarez, J.; Xu, C.; Rhee, K.; Chen, T.; Zhang, H.; et al. Overcoming EGFR(T790M) and EGFR(C797S) resistance with mutant-selective allosteric inhibitors. *Nature* **2016**, *534*, 129–132. [CrossRef] [PubMed] [PubMed Central]
42. Schödinger Release, S. *2: Protein Preparation Wizard*; Epik, Schrödinger, LLC: New York, NY, USA, 2021.
43. Friesner, R.A.; Banks, J.L.; Murphy, R.B.; Halgren, T.A.; Klicic, J.J.; Mainz, D.T.; Repasky, M.P.; Knoll, E.H.; Shelley, M.; Perry, J.K.; et al. Glide: A new approach for rapid, accurate docking and scoring. 1. Method and assessment of docking accuracy. *J. Med. Chem.* **2004**, *47*, 1739–1749. [CrossRef]
44. Tripathi, S.K.; Muttineni, R.; Singh, S.K. Extra precision docking, free energy calculation and molecular dynamics simulation studies of CDK2 inhibitors. *J. Theor. Biol.* **2013**, *334*, 87–100. [CrossRef] [PubMed]
45. Shelley, J.C.; Cholleti, A.; Frye, L.L.; Greenwood, J.R.; Timlin, M.R.; Uchimaya, M. Epik: A software program for pKa prediction and protonation state generation for drug-like molecules. *J. Comput.-Aided Mol. Des.* **2007**, *21*, 681–691. [CrossRef] [PubMed]

Disclaimer/Publisher's Note: The statements, opinions and data contained in all publications are solely those of the individual author(s) and contributor(s) and not of MDPI and/or the editor(s). MDPI and/or the editor(s) disclaim responsibility for any injury to people or property resulting from any ideas, methods, instructions or products referred to in the content.

Article

Discovery of 1*H*-benzo[*d*]imidazole-(halogenated) Benzylidenebenzohydrazide Hybrids as Potential Multi-Kinase Inhibitors

Tebyan O. Mirgany, Hanadi H. Asiri, A. F. M. Motiur Rahman *[iD] and Mohammed M. Alanazi *[iD]

Department of Pharmaceutical Chemistry, College of Pharmacy, King Saud University, Riyadh 11451, Saudi Arabia; tmirgany@ksu.edu.sa (T.O.M.); hasiri@ksu.edu.sa (H.H.A.)
* Correspondence: afmrahman@ksu.edu.sa (A.F.M.M.R.); mmalanazi@ksu.edu.sa (M.M.A.)

Citation: Mirgany, T.O.; Asiri, H.H.; Rahman, A.F.M.M.; Alanazi, M.M. Discovery of 1*H*-benzo[*d*]imidazole-(halogenated)Benzylidenebenzohydrazide Hybrids as Potential Multi-Kinase Inhibitors. *Pharmaceuticals* **2024**, *17*, 839. https://doi.org/10.3390/ph17070839

Academic Editors: Valentina Noemi Madia and Davide Ialongo

Received: 2 June 2024
Revised: 20 June 2024
Accepted: 21 June 2024
Published: 26 June 2024

Copyright: © 2024 by the authors. Licensee MDPI, Basel, Switzerland. This article is an open access article distributed under the terms and conditions of the Creative Commons Attribution (CC BY) license (https://creativecommons.org/licenses/by/4.0/).

Abstract: In an effort to develop improved and effective targeted tyrosine kinase inhibitors (TKIs), a series of twelve novel compounds with the structural motif "(*E*)-4-(((1*H*-benzo[*d*]imidazol-2-yl)methyl)amino)-*N*′-(halogenated)benzylidenebenzohydrazide" were successfully synthesized in three steps, yielding high product yields (53–97%). Among this new class of compounds, **6c** and **6h-j** exhibited excellent cytotoxic effects against four different cancer cell lines, with half-maximal inhibitory concentration (IC$_{50}$) values ranging from 7.82 to 21.48 µM. Notably, compounds **6h** and **6i** emerged as the most potent inhibitors, demonstrating significant activity against key kinases such as EGFR, HER2, and CDK2. Furthermore, compound **6h** displayed potent inhibitory activity against AURKC, while **6i** showed potent inhibitory effects against the mTOR enzyme, with excellent IC$_{50}$ values comparable with well-established TKIs. The mechanistic study of lead compound **6i** revealed its ability to induce cell cycle arrest and apoptosis in HepG2 liver cancer cells. This was accompanied by upregulation of pro-apoptotic caspase-3 and Bax and downregulation of anti-apoptotic Bcl-2. Additionally, molecular docking studies indicated that the binding interactions of compounds **6h** and **6i** with the target enzymes give multiple interactions. These results underscore the ability of compound **6i** as a compelling lead candidate warranting further optimization and development as a potent multi-targeted kinase inhibitor, which could have significant implications for the treatment of various cancers. The detailed structural optimization, mechanism of action, and in vivo evaluation of this class of compounds warrant further investigation to assess their therapeutic potential.

Keywords: 1*H*-benzo[*d*]imidazole; tyrosine kinase inhibitor; multiple kinase inhibitor; cancer; apoptosis

1. Introduction

Cancer therapy continues to be a formidable task, demanding the discovery and development of effective treatments. Among the emerging classes of anticancer agents, halogenated TKIs have garnered significant attention owing to their unique chemical structures and enhanced pharmacological properties. Incorporating halogen atoms, such as fluorine, chlorine, bromine, or iodine, into the molecular backbone of TKIs offers a promising avenue to improve their therapeutic efficacy, selectivity, and potency [1]. The addition of halogen substituents in TKIs has been shown to exert a profound influence on their binding affinity to target kinases. Through their strategic placement, halogen atoms modulate the interactions between the TKIs and their kinase targets, potentially leading to enhanced therapeutic effects. These modifications can optimize the TKIs' ability to bind to the kinase active site, resulting in increased potency and selectivity [1–3]. The halogenated TKI gefitinib, developed to precisely target the epidermal growth factor receptor (EGFR), has demonstrated its clinical utility in the care of non-small cell lung cancer (NSCLC) patients, leading to its adoption as a first-line standard of care therapy for advanced NSCLC cases harboring activating EGFR mutations. Furthermore, gefitinib offers advantages such as oral bioavailability and a comparatively lower toxicity profile, in contrast to conventional chemotherapy agents. Notably, it has shown promising outcomes in patients with EGFR mutations, which are frequently observed in Asian populations and allied with higher response rates to TKIs (Figure 1) [4–8]. Dasatinib, another halogenated TKI, exerts its therapeutic effects by targeting multiple tyrosine kinases, including BCR-ABL, SRC family kinases, and c-KIT. The incorporation of halogen substituents in dasatinib has been demonstrated to enhance its potency and selectivity, rendering it a valuable therapeutic option for various types of cancer. Notably, dasatinib has demonstrated efficacy in treating chronic myeloid leukemia (CML) and acute lymphoblastic leukemia (ALL), warranting its approval as a first-line or second-line therapy for these conditions. Additionally, dasatinib has demonstrated the capacity to overcome resistance to other TKIs, such as imatinib, which are commonly employed in the treatment of CML. However, it is crucial to recognize the potential adverse effects associated with dasatinib, which may include fluid retention, bleeding, and pulmonary arterial hypertension (PAH). Furthermore, the higher cost linked to the use of dasatinib is an important consideration [6]. Afatinib, a halogenated TKI with targets including EGFR, HER2, and HER4, has emerged as an effective therapy for NSCLC and head and neck cancer. Halogen substituents incorporated into afatinib have demonstrated the potential to enhance its potency and selectivity, thus improving its effectiveness in specific patient populations. Afatinib's broad spectrum of activity against multiple members of the HER family of receptor tyrosine kinases positions it as a valuable treatment option for diverse cancer types. It has shown efficacy, particularly in patients with EGFR mutations, leading to its approval as a first-line therapy in this population. Additionally, afatinib has shown the capability of overwhelming resistance to other EGFR TKIs, including gefitinib and erlotinib. However, it is essential to consider potential adverse effects, such as diarrhea, rash, and mucositis, as well as the higher cost associated with afatinib [7,8]. Several halogenated TKIs have shown promising results in clinical trials and preclinical studies. Entrectinib, a halogenated TKI, has demonstrated efficacy in targeting ROS1 and NTRK fusions in solid tumors, including lung cancer and neuroblastoma [9]. Avitinib, another halogenated TKI, has shown potential in inhibiting the EGFR T790M mutation, which is allied with battling first-generation EGFR TKIs [10,11]. Saracatinib, a halogenated Src/Abl kinase inhibitor, has exhibited promising activity in solid tumors, particularly in pancreatic cancer [12,13]. Ponatinib, a halogenated TKI with potent activity against BCR-ABL and other tyrosine kinases, has been effective in treating chronic myeloid leukemia (CML) and Philadelphia chromosome-positive acute lymphoblastic leukemia (Ph+ ALL) [6,14] Foretinib, a halogenated TKI targeting MET, VEGFR, and other kinases, has shown efficacy in various solid tumors, including renal cell carcinoma and hepatocellular carcinoma [15]. Vandetanib, flumatinib, vemurafenib, sorafenib, cabozantinib, nilotinib, lapatinib, and selumetinib, all halogenated TKIs, have demonstrated efficacy in different cancer types by targeting specific kinases involved in tumor growth and progression [16–24]. These include VEGFR, BRAF, c-Met, RET, and HER2, among others. However, it is important to consider the potential adverse effects associated with these halogenated TKIs. These halogenated tyrosine kinase inhibitors may cause debilitating fatigue, cognitive deficits, and even life-threatening lung disease. Patients may also experience gastrointestinal issues, blood cell depletion, heart rhythm abnormalities, high blood pressure, painful skin, nail changes, and heightened photosensitivity. Vigilant monitoring and active management of these diverse adverse effects are crucial when prescribing these halogenated TKI therapies. Close monitoring and management of these side effects are crucial for ensuring patient safety and treatment effectiveness. Moreover, it is worth noting that some of these halogenated TKIs may have limitations, including high costs and limited accessibility in certain healthcare systems. The affordability and availability of these drugs are important factors to consider when evaluating their clinical utility and impact on patient care. However, halogenated TKIs have emerged as a promising class of anticancer agents, with multiple compounds demonstrating efficacy in various cancer types. While these drugs offer significant therapeutic potential, it is essential to carefully assess their adverse effects and limitations to optimize their clinical utility and ensure equitable access to effective treatments in cancer care. Further research and clinical investigations are warranted to elucidate the full potential of halogenated TKIs and their role in improving patient outcomes in oncology.

On the other hand, benzimidazole compounds have gained attention in drug development owing to their versatility and diverse applications. They exhibit antibacterial properties [25], potential as anti-tubercular agents [26], antifungal activities [27], and antiprotozoal effects [28]. Benzimidazole derivatives also show promise as antiviral agents [29] and as protein kinase inhibitors [30,31]. Studies have demonstrated the therapeutic action of benzimidazole-based compounds, such as nazartinib, in inhibiting EGFR and HER2 proteins [32–35]. Our previous research focused on hydrazone derivatives, which exhibited potent multi-kinase inhibition activity, including EGFR and HER2 [36–39]. To further explore their potential, we designed benzimidazole and (halogenated)benzylidenebenzohydrazide hybrids, namely, (E)-4-(((1H-benzo[d]imidazol-2-yl)methyl)amino)-N'-(halogenated)benzylidenebenzohydrazides, aiming for multi-kinase inhibitory activities. The synthesized compounds (6a-l) were evaluated for their in-vitro cytotoxicity against cancerous/normal cell lines, and their ability to inhibit multiple tyrosine kinases, investigated for their apoptosis-inducing ability, cell cycle-suppressing effects, drug-likeness properties, and in silico molecular docking, shed light on their selective mechanism of action and substantiated their potential as promising anticancer agents.

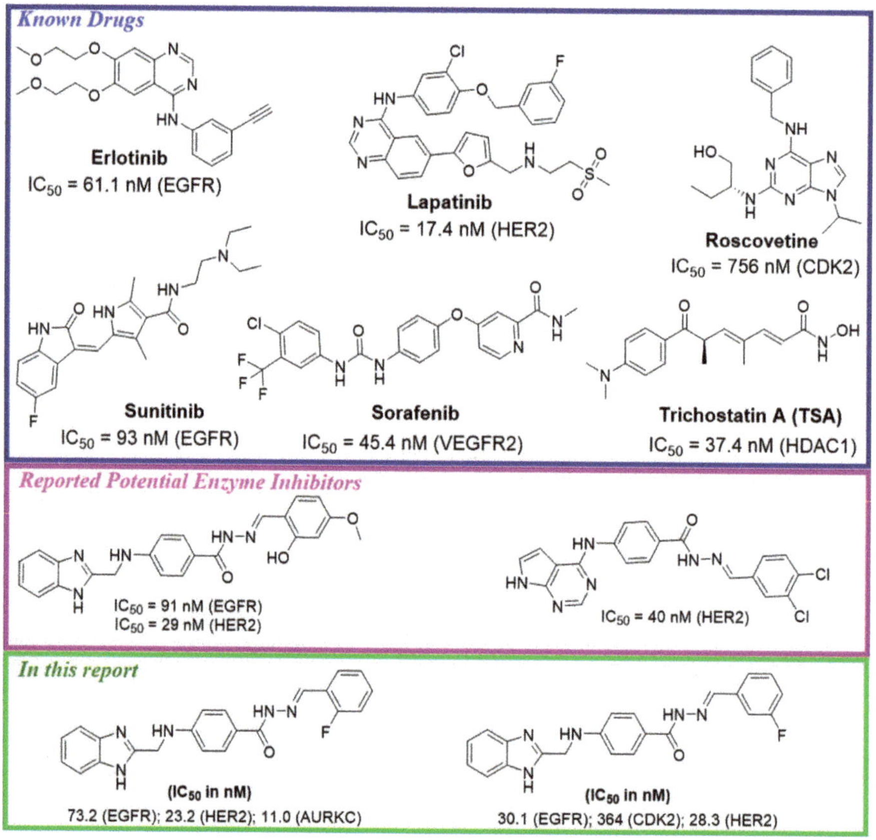

Figure 1. An example of some known TKIs [39], recently reported TKIs [40], and synthesized potential TKIs.

2. Results and Discussion

2.1. Chemistry

The synthesis of (E)-4-(((1H-benzo[d]imidazol-2-yl)methyl)amino)-N'-(halogenated) benzylidenebenzohydrazide derivatives (6a-l) was carried out using a straightforward method, as described in Scheme 1, following the multi-step procedure adopted for our previously reported method [39]. The use of a previously reported synthetic methodology, along with the straightforward nature of the individual steps, allowed for the efficient preparation of the desired series of halogenated benzylidenebenzohydrazide derivatives (6a-l) bearing the benzo[d]imidazole moiety. The structural elucidation of the synthesized compounds was carried out using a comprehensive analytical approach, employing a suite of spectroscopic techniques. These included infrared (IR) spectroscopy, mass spectrometry (MS), and proton and carbon nuclear magnetic resonance (^{1}H-NMR and ^{13}C-NMR) analyses (please see the Supplementary Materials for the spectra). Additionally, we also reported the physical properties of the compounds 6a-l, specifically their color and melting point ranges. The ^{1}H-NMR spectroscopic analysis of the synthesized compounds 6a-l revealed characteristic signals for the secondary aromatic amine protons located at the 4-position of the benzohydrazide moiety. These amine proton signals were observed in the 7.01 to 7.06 ppm region of the ^{1}H-NMR spectra as triplets, with coupling constant values ranging from 5.7 to 6.1 Hz. The observed chemical shift and coupling pattern of these amine proton signals provided evidence that the aromatic amine functionality was directly bonded to an aliphatic (CH2) carbon, specifically the methylene group derived from the condensation of compounds 1 and 2 during the synthetic sequence. These ^{1}H-NMR data thus confirmed the successful formation of the desired compound 6a-l structures, with the secondary aromatic amine protons serving as a characteristic spectroscopic signature for the benzohydrazide moiety within the target molecules. The ^{1}H-NMR spectra of the synthesized compounds 6a-l also exhibited characteristic signals for the benzylidene protons. These benzylidene proton signals were observed in the 8.35 to 8.38 ppm region of the ^{1}H-NMR spectra, except for the compounds with substituents at the 2-position of the benzene ring. Specifically, compound 6b, which contained a bromo substituent at the 2-position, displayed a downfield shift of 0.38 ppm compared with the non-substituted compound 6a. Similarly, the 2-chloro substituted compound 6e exhibited a 0.43 ppm downfield shift, while the 2-fluoro substituted compound 6h showed a 0.27 ppm downfield shift relative to 6a. Furthermore, the compounds with 2,4-dihalogen substitution, 6k (2,4-dichloro) and 6l (2,5-difluoro), also demonstrated downfield shifts of 0.38 ppm and 0.22 ppm, respectively, compared with the unsubstituted analog 6a. These observed downfield shifts in the benzylidene proton signals for the 2-substituted and 2,4-disubstituted compounds can be attributed to the electronic effects of the halogen substituents, which influenced the chemical environment and deshielding of the benzylidene protons.

The structures of the synthesized compounds 6a-l were further proved by the observation of distinctive IR spectral signatures, such as the sharp peaks observed in the range of 1604–1613 cm^{-1}, which were characteristic of the benzylidene hydrazone (-N=CH-) functional group present in all the compounds. N-H stretching vibration bonds of amide -NH (-CO-NH-) appeared between 3044 and 3464 cm^{-1} and around 1502–1568 cm^{-1}. Carbonyl (C=O) vibrational bonds appeared at around 1642–1656 cm^{-1}. Furthermore, each compound displayed distinct IR spectral features. Notably, the bromo-substituted analogs (6b-d) exhibited characteristic stretching bands in the range of 741–748 cm^{-1}. In contrast, the chloro- and fluoro-substituted compounds (6e-l) showed distinctive stretching bands between 504 and 758 cm^{-1}. Finally, the structures of compounds 6e-l were further confirmed through their mass spectral analysis data.

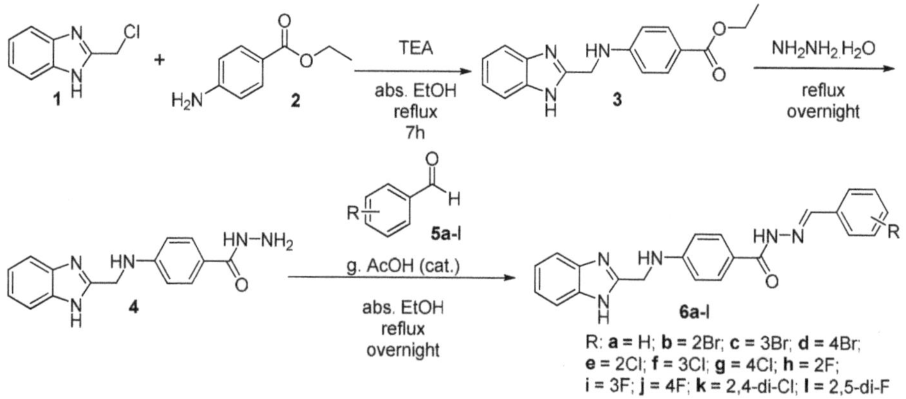

Scheme 1. Synthesis of (E)-4-(((1H-benzo[d]imidazol-2-yl)methyl)amino)-N'-(substituted)benzy-lidenebenzohydrazide (**6a-l**).

2.2. Biological Evaluation

2.2.1. In Vitro Cytotoxicity

The cytotoxic activities of compounds **6a-l** were assessed using a standard MTT method, which was performed across three different cancer cell line models, namely human colon cancer (HCT-116), hepatocellular carcinoma (HepG2), mammary gland cancer (MCF-7), and normal (WI-38) cell lines. The results presented in Table 1 show that the cytotoxicity of the tested compounds was quantified by the concentration required to induce 50% inhibition of cancer cell viability. The cytotoxicity results provide a comprehensive evaluation of the synthesized compounds **6a-6l** in comparison with the standard drugs sorafenib, doxorubicin, and sunitinib. Among the synthesized compounds, **6c** (3-Br substituted) and **6i** (3-F substituted) exhibited the most potent cytotoxic activity, with IC_{50} values ranging from 7.82 to 10.21 µM across the tested cancer cell lines. These values are comparable to the standard drugs, which have IC_{50} values between 4.17 and 24.06 µM. Compounds **6h** (2-F), **6j** (4-F), and **6d** (4-Br) also demonstrated relatively strong cytotoxicity, with IC_{50} values generally below 30 µM in the cancer cell lines. The remaining compounds, including **6a, 6b, 6e, 6f, 6g, 6k**, and **6l**, exhibited moderate to weaker cytotoxic activity, with IC_{50} values mostly above 20 µM. When considering the selectivity toward cancer cells over the normal WI-38 fibroblast cell line, compounds **6c, 6i**, and **6j** stood out, showing significantly lower IC_{50} values in the cancer cell lines compared with the WI-38 cells, indicating a higher degree of selectivity. Other compounds, such as **6h** and **6d**, also exhibited relatively better selectivity toward the cancer cell lines. The structure–activity relationship analysis suggested that the presence of bromo (Br) or fluoro (F) substituents at the 3-position of the phenyl ring is favorable for potent and selective cytotoxic activity, as observed with compounds **6c** and **6i**. Substitution at the 4-position, as in the case of compounds **6d, 6g**, and **6j**, also seems beneficial, with fluoro (**6j**) showing the best results. In contrast, the disubstituted compounds **6k** and **6l** generally exhibited lower cytotoxic potency and selectivity compared with their monosubstituted counterparts. Overall, the cytotoxicity data provide valuable insights into the structure–activity relationships of the synthesized compounds and can guide further optimization efforts to develop potential anticancer agents with improved potency and selectivity.

Table 1. In vitro cytotoxicity of compounds **6a-l** against selected cancer cell lines.

Compound		In Vitro Cytotoxicity IC$_{50}$ (µM)						
No.	R	HCT-116	SI	HepG2	SI	MCF-7	SI	WI-38
6a	H	24.62 ± 1.9	2.95	31.76 ± 2.2	2.29	26.31 ± 1.9	2.76	72.60 ± 4.0
6b	2-Br	42.90 ± 2.5	1.96	26.16 ± 1.9	3.22	31.82 ± 2.1	2.64	84.15 ± 4.3
6c	3-Br	10.21 ± 0.8	4.72	8.90 ± 0.6	5.41	7.82 ± 0.6	6.16	48.17 ± 2.7
6d	4-Br	30.26 ± 2.1	1.79	17.78 ± 1.3	3.05	25.18 ± 1.8	2.16	54.30 ± 3.1
6e	2-Cl	46.67 ± 2.7	1.33	28.80 ± 1.7	2.15	36.16 ± 2.3	1.71	61.86 ± 3.5
6f	3-Cl	63.72 ± 3.5	1.57	34.79 ± 2.1	2.87	45.61 ± 2.5	2.19	>100
6g	4-Cl	35.29 ± 2.3	2.02	24.90 ± 1.7	2.86	22.51 ± 1.6	3.16	71.19 ± 3.5
6h	2-F	21.48 ± 1.6	4.33	12.94 ± 1.0	7.19	16.31 ± 1.2	5.70	92.98 ± 4.6
6i	3-F	13.44 ± 1.2	4.20	9.39 ± 0.8	6.01	11.64 ± 0.9	4.85	56.46 ± 3.0
6j	4-F	18.72 ± 1.4	2.08	14.02 ± 1.2	2.78	8.31 ± 0.7	4.70	39.03 ± 2.3
6k	2,4-di-Cl	67.83 ± 3.8	1.47	44.78 ± 2.4	2.23	48.64 ± 2.7	2.06	>100
6l	2,5-di-F	39.84 ± 2.2	1.95	20.02 ± 1.5	3.88	29.81 ± 2.0	2.60	77.61 ± 4.1
Sorafenib		5.47 ± 0.3	1.95	9.18 ± 0.6	1.16	7.26 ± 0.3	1.47	10.65 ± 0.8
Doxorubicin		5.23 ± 0.3	1.28	4.50 ± 0.2	1.49	4.17 ± 0.2	1.61	6.72 ± 0.5
Sunitinib		17.91 ± 1.3	3.11	8.38 ± 0.5	6.64	24.06 ± 2.0	2.31	55.63 ± 3.3

IC$_{50}$ values are the mean ± SD of triplicate measurements; SI = Selectivity Index = IC$_{50}$ of normal cell line (WI-38)/IC$_{50}$ of cancer cell line [41].

The Selectivity Index (SI) values shown in Table 1 provide important information about the selectivity of the synthesized compounds (**6a-l**) against the various cancer cell lines tested. The SI values represent the ratio of the IC$_{50}$ values for the normal WI-38 cell line compared with the respective cancer cell lines, with higher SI values indicating greater selectivity for the cancer cells over normal cells. Compounds **6c**, **6h**, and **6i** exhibited the highest SI values against the SI cancer cell line, with SI values of 4.20–4.72 (HCT-116), 5.41–7.19 (HepG2), and 4.85–6.16 (MCF-7), respectively. The SI values for the standard drugs were 1.16–1.95 (sorafenib), 1.28–1.61 (doxorubicin), and 2.31–3.11 (sunitinib), except 6.64 against the HepG2 cell line for sunitinib 6.64, respectively. This suggests these compounds (**6c**, **6h**, **6i**) demonstrated the greatest selectivity for the SI cancer cells compared with the normal WI-38 cells.

2.2.2. In Vitro Protein Kinase Inhibition Assays

Based on the cytotoxicity analysis of the synthesized compounds (**6a-l**), four compounds (**6c**, **6h-j**) were chosen for further enzymatic activity assessment against a range of kinase enzymes, including EGFR, Her2, VEGFR2, CDK2, AURKC, HDAC1, and mTOR (Table 2). To benchmark the activity of these compounds, several well-known kinase inhibitors were used as reference standards for the tested kinases, such as erlotinib for EGFR, lapatinib for Her2, sorafenib for VEGFR2, roscovetine for CDK2, TSA for AURKC and HDAC1, and rapamycin for mTOR. The synthesized compounds **6c** and **6h-j** demonstrated varying degrees of potent inhibition across the different protein kinases tested. For instance, compound **6h** exhibited very high potency against EGFR (IC$_{50}$ = 73.2 nM), which is almost 1-fold higher than the standard erlotinib (IC$_{50}$ = 61.1 nM), and against Her2 (IC$_{50}$ = 23.2 nM), which is also almost 1-fold higher than lapatinib (IC$_{50}$ = 17.4 nM). Additionally, **6h** showed high potency against CDK2 (IC$_{50}$ = 284 nM), which is 2.5-fold higher than roscovetine (IC$_{50}$ = 756 nM), and very high potency against AURKC (IC$_{50}$ = 11 nM), which is 3-fold higher than the standard TSA (IC$_{50}$ = 30.4 nM). Compound **6i** showed excellent potency against EGFR (IC$_{50}$ = 30.1 nM), which is almost 2-fold higher than the standard erlotinib, and against Her2 (IC$_{50}$ = 28.3 nM), which is almost 1-fold higher than lapatinib. Additionally, **6i** exhibited high potency against CDK2 (IC_{50} = 364 nM), which is 2-fold higher than roscovetine, and against mTOR (IC_{50} = 152 nM), which is 1-fold higher than the standard rapamycin (IC_{50} = 208 nM). In contrast, compound **6c** gave disappointing results compared with the expectations based on the cytotoxicity against various cancer cell lines. Additionally, compound **6j** showed very little inhibitory activity against the tested kinases. Compared with the standard drugs, the synthesized compounds **6h** and **6i** exhibited superior potency across multiple kinase targets. The low IC_{50} values for the synthesized compounds indicate their high binding affinity and inhibitory activity toward these clinically relevant protein kinases, suggesting their potential to be developed as potent and selective anticancer agents targeting multiple dysregulated signaling pathways in cancer. These findings underscore the promising potential of compounds **6h** and **6i** as candidates for further development as kinase inhibitors.

Table 2. In vitro protein kinase inhibition of compounds **6c** and **6h-j** against EGFR, Her2, VEGFR2, CDK2, AURKC, HDAC1, and mTOR.

Compound	In Vitro Protein Kinase Inhibition IC_{50} (nM)						
	EGFR	Her2	VEGFR2	CDK2	AURKC	HDAC1	mTOR
6c	125.2 ± 0.041	55.6 ± 0.023	604.5 ± 0.022	938 ± 0.039	94.4 ± 0.036	2263 ± 0.077	1461 ± 0.05
6h	73.2 ± 0.004	23.2 ± 0.001	194.5 ± 0.007	284 ± 0.012	11 ± 0.004	151.1 ± 0.005	413 ± 0.014
6i	30.1 ± 0.03	28.3 ± 0.001	172.2 ± 0.006	364 ± 0.011	74.5 ± 0.003	96.6 ± 0.003	152 ± 0.005
6j	166.4 ± 0.008	204.7 ± 0.009	307.2 ± 0.011	1448 ± 0.062	589.4 ± 0.022	473.3 ± 0.016	1305 ± 0.044
Erlotinib	61.1 ± 0.002	-	-	-	-	-	-
Lapatenib	-	17.4 ± 0.001	-	-	-	-	-
Sorafenib	-	-	45.4 ± 0.002	-	-	-	-
Roscovetine	-	-	-	756 ± 0.032	-	-	-
TSA	-	-	-	-	30.4 ± 0.001	37.4 ± 0.001	-
Rapamycin	-	-	-	-	-	-	208 ± 0.007

IC_{50} values are the mean ± SD of triplicate measurements.

2.2.3. Cell Cycle Analysis

The cytotoxicity assay, selectivity index (SI) values, and enzyme inhibitory activities of the synthesized compounds **6a-l** were evaluated, and based on these results, compound **6i** was selected for further investigation to assess its impact on cell cycle progression in HepG2 cells. To examine the effect of compound **6i** on the cell cycle, HepG2 cells were treated with the compound at its IC_{50} concentration for 24 h. Following the treatment, the cells were stained with propidium iodide and analyzed using flow cytometry to determine the cell cycle phase distribution. The cell cycle analysis results, presented in Table 3 and Figure 2, revealed that treatment with compound **6i** elicited a distinct cell cycle arrest effect. Specifically, the data showed that the DNA content of cells treated with **6i** increased in the G1 phase while decreasing in the S and G2/M phases. This indicates that compound **6i** induced a significant cell cycle arrest at the G1 phase. The quantitative data further substantiated this observation. The percentage of cells in the G0-G1 phase increased from 52.39% in the untreated control to 72.13% upon treatment with compound **6i**. Conversely, the proportion of cells in the S phase decreased from 34.77% to 25.19%, and the percentage of cells in the G2/M phase declined from 12.84% to 2.68%. These findings suggest that treatment with compound **6i** substantially altered the cell cycle distribution of the HepG2 cancer cells. It led to a marked accumulation of cells in the G1 phase, accompanied by a concomitant reduction in the proportion of cells in the S and G2/M phases. The cell cycle arrest effect observed with compound **6i** may be a key mechanistic aspect contributing to its anticancer potential. The observed cell cycle arrest at the G1 phase is a common mechanism of action for many anti-cancer agents, as it can prevent cell division and proliferation, ultimately leading to cell cycle arrest and potentially inducing apoptosis or other forms of cell death. These results, along with the previously reported cytotoxicity and kinase inhibitory activities of compound **6i**, further highlight its potential as a promising anticancer candidate that warrants further investigation and optimization for targeted cancer therapy.

Table 3. Results of cell cycle analysis of HepG2 cells treated with compound **6i**.

Compound/Cell Line	DNA Content (%)			Cell Cycle Distribution Index (CDI)
	%G0-G1	%S	%G2/M	
Cont. HepG2	52.39	34.77	12.84	0.91
Compound 6i/HepG2	72.13	25.19	2.68	0.39

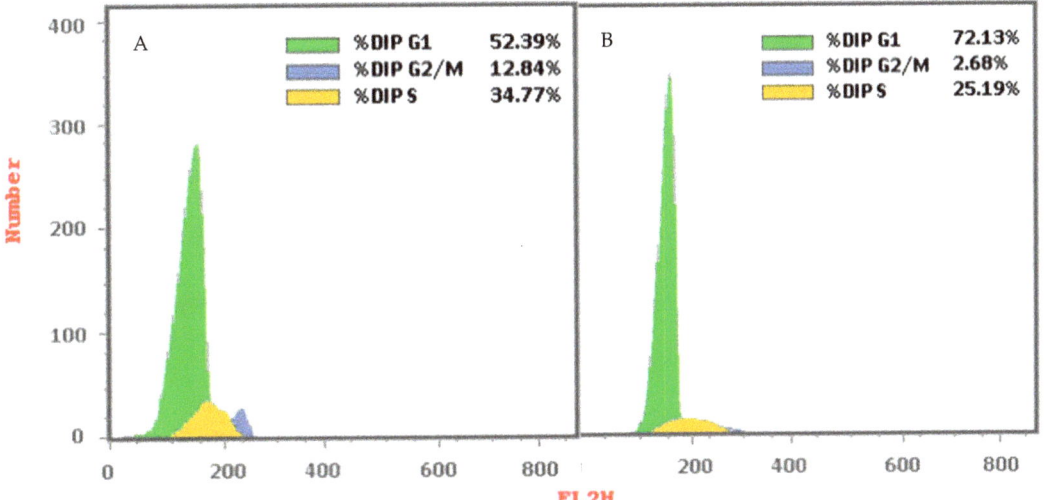

Figure 2. Cell cycle phases of HepG2 cells following treatment with compound **6i**: (**A**) Control HepG2 cell lines; (**B**) HepG2 cells treated with compound **6i**.

The cell cycle distribution index (CDI) is a metric that quantifies the rate of cell proliferation, calculated as CDI = (G2/M + S)/(G0 − G1), where the values represent the percentages of cells in each phase. In this study, the CDI values for the control and compound **6i**-treated HepG2 cells were 0.91 and 0.39, respectively. A decrease in CDI indicates a reduced rate of cell proliferation, suggesting cell cycle arrest. Compound **6i** was further evaluated for apoptosis analysis in HepG2 cell lines.

2.2.4. Apoptosis Analysis

Annexin-V/Propidium Iodide (PI) Staining Assay

To investigate the mode of cell death induced by the synthesized compound **6i**, flow cytometry analysis was performed. HepG2 cancer cells were treated with compound **6i** at their respective IC$_{50}$ concentrations and incubated for 24 h. Following the treatment, the cells were double-stained with Annexin V and propidium iodide (PI) to assess the levels of apoptosis and necrosis. The data presented in Table 4 reveal a significant decline in the percentage of viable cells in the compound **6i**-treated group (63.93%) compared with the untreated control group (97.58%). In the early stages of the cell death process, the percentage of cells undergoing apoptosis was 0.61% in the control group, whereas it increased to 22.07% in the compound **6i**-treated group. Additionally, in the late stages of apoptosis, the percentage of apoptotic cells rose from 0.21% in the control group to 9.98% in the compound **6i**-treated group. These findings suggest that the antiproliferative effects of compound **6i** against HepG2 cancer cells are at least partially mediated through the induction of apoptosis, particularly during the early stages of the cell death process.

The flow cytometry data indicate that treatment with compound **6i** leads to a substantial increase in the proportion of HepG2 cells undergoing both early and late stages of apoptosis. The observed increase in both early and late apoptotic cell populations upon treatment with compound **6i** indicates that this compound may trigger the apoptotic signaling cascades, leading to programmed cell death. Additionally, the data show a slight increase in the percentage of necrotic cells from 1.6% in the control group to 4.02% in the compound **6i**-treated group, suggesting that the compound may also induce some degree of necrosis in HepG2 cells. Taken together, these results highlight the pro-apoptotic and potentially cytotoxic effects of compound **6i** on HepG2 cells, which could have important implications for its further development as a potential anticancer agent. The apoptosis levels in the control HepG2 cell lines and the cells treated with compound **6i** are shown in the graphical representation provided in Figure 3.

Table 4. Apoptotic cell distributions of HepG2 cells treated with compound **6i**.

Sample	Alive Cell (%)	Apoptosis		Necrosis
		Early	Late	
Cont. HepG2	97.58	0.61	0.21	1.6
Compound 6i/HepG2	63.93	22.07	9.98	4.02

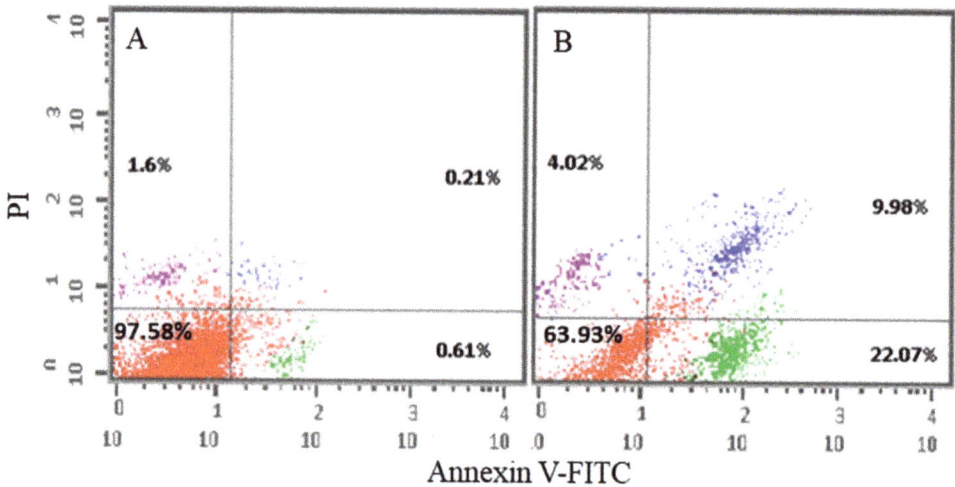

Figure 3. Apoptotic cell distributions: (**A**) Control HepG2 cells; (**B**) HepG2 cells treated with compound **6i**. The colors red, green, blue and purple are representing alive cells, early apoptosis, late apoptosis and necrosis, respectively.

Determination of Apoptotic Protein Levels

To further elucidate the mechanism by which compound **6i** induces cell death, the expression levels of key apoptotic proteins were assessed in the HepG2 cell line. As shown in Table 5, treatment with compound **6i** resulted in significant changes in the levels of critical proteins involved in the apoptotic signaling cascade in HepG2 cells. Specifically, the levels of the pro-apoptotic proteins caspase-3 and Bax were markedly higher in the compound **6i**-treated group compared with the control. Caspase-3 levels were approximately 3.9-fold higher, while Bax levels were 7.22-fold higher in the treated cells. In contrast, the level of the anti-apoptotic protein Bcl-2 was 7.5-fold lower in the compound **6i**-treated group compared with the control. These findings strongly suggest that the induction of apoptosis by compound **6i** in HepG2 cells is mediated, at least in part, by the upregulation of the pro-apoptotic proteins caspase-3 and Bax, coupled with the downregulation of the anti-apoptotic protein Bcl-2. The magnitude of the changes observed in the levels of these key apoptotic regulators upon treatment with compound **6i** was comparable to, or even exceeded, the effects seen with the known apoptosis-inducing agent staurosporine. These results further substantiate the pro-apoptotic mechanism of action of compound **6i** and highlight its potential as a promising candidate for the development of novel therapies targeting diseases involving dysregulated cell death, such as cancer. Additional studies are warranted to elucidate the detailed molecular mechanisms underlying the apoptosis-inducing effects of compound **6i** and to assess its therapeutic potential in relevant in vivo models.

Table 5. Expression changes in apoptotic protein following treatment with compound **6i**.

Kinase Protein	Protein Expression (Pg/mL) (Folds)		
	Caspase-3	Bax	Bcl-2
Control HepG2	99.904 ± 3.88 (1)	71.075 ± 2.762 (1)	15.668 ± 0.53 (1)
6i/HepG2	388.497 ± 15.09 (3.9)	513.731 ± 19.96 (7.22)	2.073 ± 0.07 (0.132)
Saurosporine/HepG2	541.162 ± 21.02 (5.4)	386.743 ± 15.03 (5.44)	3.336 ± 0.11 (0.212)

Protein expression values are the mean ± SD of triplicate measurements.

2.3. In Silico Studies

Molecular Docking

In order to predict the potential binding interactions between compounds **6h** and **6i** and the investigated protein kinase enzymes, compound **6h** was docked into the active site of Her2 (PDB: 7PCD) and AURKC (PDB: 6GR8), and **6i** was docked into the active side of EGFR (PDB: 4HJO) and Her2 (PDB: 7PCD), respectively.

Based on the docking data of compound **6h** with the HER2 kinase enzyme (PDB: 7PCD), the hydrogen bond network with Lys753, Asp863, and Thr862 through water molecules stabilized the compound inside the active site of the target enzyme (Figure 4A). Additionally, binding interactions between the benzylidene benzene moiety of compound **6h** and the amino acid residues Leu726 and Leu852 were observed, involving H-π bonds (Figure 4A). However, when compound **6h** was co-crystallized with the AURKC kinase enzyme (PDB: 6GR8), three hydrogen bonds were formed between the key amino acid residues and the hydrazide moiety, facilitating the fitting of the compound within the active site (Figure 4B). Specifically, one hydrogen bond was between the side chain of Lys35 and the carbonyl group, the second hydrogen bond was between Lys72 and a nitrogen atom, and the third hydrogen bond was part of a network involving Glu91 and Ala183, mediated by a water molecule (Figure 4B).

The docking study results of compound **6i** with the EGFR (PDB: 4HJO) and the active site of HER2 kinase enzyme (PDB: 7PCD) are depicted in Figure 5. For the EGFR kinase enzyme, the benzimidazole moiety of compound **6i** formed a hydrogen bonding network, mediated by water molecules, with side chains of Thr766 and Thr830 (Figure 5A). Additionally, the nitrogen atom of the imidazole ring participated in a hydrogen bond with the key amino acid residue Asp831, while another important hydrogen bond was formed between the backbone of Met769 and the carbonyl oxygen of the compound, facilitating its fitting within the active site. Furthermore, the presence of a fluorine atom in the meta position of the benzene ring enhanced its interaction with Lys704 through the formation of a cation-π bond compared with other derivatives (Figure 5A). For the Her2 active site, compound **6i** made a hydrogen bond network with Lys753, Asp863, and Met501, stabilized by water molecules inside the active site (Figure 5B). Additionally, binding interactions between the benzylidene benzene moiety of compound **6i** and the amino acid residues Leu726 and Leu852 were also observed, similar to the interactions seen for compound **6h**.

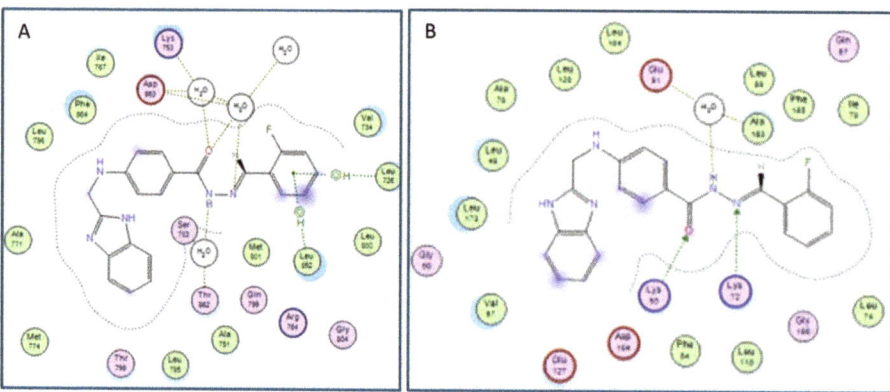

Figure 4. Docking study of compound **6h** showing different types of binding interaction with the key amino acid residues: (**A**) **6h** with Her2; and (**B**) **6h** with AURKC kinase enzymes.

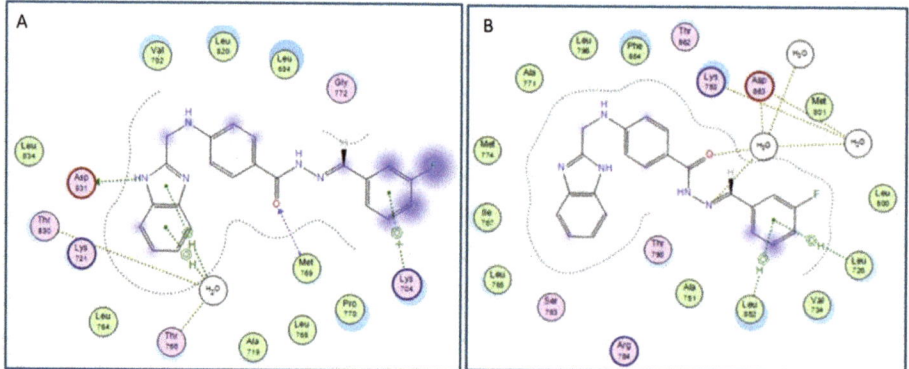

Figure 5. Docking study of compound **6i** showing different types of binding interaction with the key amino acid residues: (**A**) compound **6i** with EGFR; and (**B**) compound **6i** with Her2 kinase enzymes.

3. Materials and Methods

3.1. General

All reagents and solvents used in the experiments were of commercial grade and employed without further purification. Barnstead electrothermal digital melting point apparatus (model IA9100, BIBBY scientific limited, Staffordshire, UK) was used to determine the melting points. A Jasco FT/IR-6600 spectrometer (Tokyo, Japan) was used for recording IR data. Bruker 700 MHz NMR spectrometry (Zurich, Switzerland) was used to obtain the NMR data. Mass spectra were taken using an Agilent 6320 ion trap mass spectrometer equipped with an ESI ion source (Agilent Technologies, Palo Alto, CA, USA).

3.2. Chemistry

3.2.1. Synthesis of Ethyl 4-(((1*H*-benzo[d]imidazol-2-yl)methyl)amino)benzoate (**3**)

Ethyl 4-(((1*H*-benzo[*d*]imidazol-2-yl)methyl)amino)benzoate (**3**) was prepared using a previously reported method [39]. White powder (65%), mp. 255 °C (lit. [39] mp. = 255 °C).

3.2.2. Synthesis of 4-(((1*H*-benzo[*d*]imidazol-2-yl)methyl)amino)benzohydrazide (**4**)

Additionally, 4-(((1*H*-Benzo[*d*]imidazol-2-yl)methyl)amino)benzohydrazide (**4**) was prepared using our previously reported method [39]. White powder (80%), p. 240 °C (lit. [39] mp. = 240 °C).

3.3. General Procedure for the Preparation of (E)-4-(((1H-benzo[d]imidazol-2-yl)methyl)amino)-N′-(substitutedbenzylidene)benzohydrazide (6a-l)

An equimolar mixture of hydrazide **4** and halogen-substituted benzaldehyde (**5**) was reacted to obtain the desired 1*H*-benzo[*d*]imidazole-(halogenated)benzylidenebenzohydrazide (**6a-l**) following the reported procedure [39,40,42].

3.3.1. (*E*)-4-(((1*H*-benzo[d]imidazol-2-yl)methyl)amino)-*N*′-benzylidenebenzohydrazide (**6a**)

White solid (53.43%), mp. 278 °C (lit. [39] mp. = 278 °C; lit [43] mp. = 246 °C), CAS registry number 76321-88-5.

3.3.2. (*E*)-4-(((1*H*-benzo[d]imidazol-2-yl)methyl)amino)-*N*′-(2-bromobenzylidene)benzohydrazide (**6b**)

White solid (105 mg, 0.234 mmol, 66%), mp. 284 °C. FT-IR (KBr): ν (cm^{-1}) = 3437, 3174, 3135, 3044, 2976, 2829, 1926, 1893, 1649, 1604, 1558, 1522, 1431, 1350, 1294, 1268, 1187,1057, 1014, 832, 761, 739, 689, 647, and 608 cm^{-1}. ^{1}H-NMR (700 MHz, DMSO-d_6), δ ^{1}H NMR (700 MHz, DMSO), δ 12.35 (s, 1H), 11.76 (s, 1H), 8.75 (s, 1H), 7.99 (s, 1H), 7.75 (d, *J* = 8.3 Hz, 2H), 7.69 (d, *J* = 8.2 Hz, 1H), 7.58 (s, 1H), 7.46 (t, *J* = 7.7 Hz, 2H), 7.35 (t, *J* = 7.6 Hz, 1H), 7.15 (s, 2H), 7.04 (t, *J* = 6.1 Hz, 1H), 6.74 (d, *J* = 8.3 Hz, 2H), and 4.59 (d, *J* = 5.8 Hz, 2H) ppm. ^{13}C-NMR (176 MHz, DMSO-d_6), δ 163.54, 153.45, 152.11, 144.83, 143.78, 134.90, 133.98, 133.68, 131.93, 129.87, 128.62, 127.67, 123.87, 122.38, 121.66, 120.71, 118.98, 111.98, 111.80, and 41.85 ppm. Mass (ESI): *m/z* 448 [^{79}(Br)M+H]$^+$, 450 [^{81}(Br)M+H]$^+$.

3.3.3. (*E*)-4-(((1*H*-benzo[d]imidazol-2-yl)methyl)amino)-*N*′-(3-bromobenzylidene)benzohydrazide (**6c**)

White solid (97 mg, 0.216 mmol, 61%), mp. 278 °C. FT-IR (KBr): ν (cm^{-1}) = 3388, 3096, 2963, 2907, 2845, 2790, 1963, 1656, 1606, 1522, 1438, 1255, 1073, 1021, 895, 822, 741,692, 640, and 478 cm^{-1}. ^{1}H-NMR (700 MHz, DMSO-d_6), δ ^{1}H NMR (700 MHz, DMSO), δ 12.35 (s, 1H), 11.62 (s, 1H), 8.35 (s, 1H), 7.89 (s, 1H), 7.73 (d, *J* = 8.3 Hz, 2H), 7.68 (d, *J* = 8.0 Hz, 1H), 7.60 (d, *J* = 8.2 Hz, 1H), 7.57 (s, 1H), 7.45 (s, 1H), 7.41 (t, *J* = 8.0 Hz, 1H), 7.15 (s, 2H), 7.04 (t, *J* = 6.1 Hz, 1H), 6.73 (d, *J* = 8.5 Hz, 2H), and 4.58 (d, *J* = 5.7 Hz, 2H) ppm. ^{13}C-NMR (176 MHz, DMSO-d_6), δ 163.46, 153.35, 151.97, 144.67, 143.67, 137.64, 134.79, 132.65, 131.45, 129.73, 129.29, 126.49, 122.62, 122.27, 121.56, 120.68, 118.87, 111.87, 111.70, and 41.75 ppm. Mass (ESI): *m/z* 448 [^{79}(Br)M+H]$^+$, 450 [^{81}(Br)M+H]$^+$.

3.3.4. (*E*)-4-(((1*H*-benzo[d]imidazol-2-yl)methyl)amino)-*N*′-(4-bromobenzylidene)benzohydrazide (**6d**)

White solid (124 mg, 0.276 mmol, 63%), mp. 287 °C (lit. [43] mp. = 230 °C), CAS registry number 76321-87-4. FT-IR (KBr): ν (cm^{-1}) =3389, 3060, 2957, 2906, 2796, 1655, 1610, 1568, 1526, 1487, 1455, 1438, 1357, 1335, 1296, 1276, 1260, 1212, 1190, 1150, 1132, 1068, 1026, 1009, 956, 934, 825, 748, 695, 638, 515, and 503 cm^{-1}. ^{1}H-NMR (700 MHz, DMSO-d_6), δ 12.35 (s, 1H), 11.56 (s, 1H), 8.36 (s, 1H), 7.72 (d, *J* = 8.6 Hz, 2H), 7.64 (s, 4H), 7.58 (s, 1H), 7.45 (s, 1H), 7.15 (s, 2H), 7.03 (s, 1H), 6.73 (d, *J* = 8.7 Hz, 2H), and 4.58 (d, *J* = 5.8 Hz, 2H) ppm. ^{13}C-NMR (176 MHz, DMSO-d_6), δ 163.40, 153.37, 151.93, 145.29, 143.68, 134.80, 134.43, 132.27, 129.70, 129.17, 123.31, 122.29, 121.56, 120.79, 118.88, 111.88, 111.71, and 41.76 ppm. Mass (ESI): *m/z* 448 [^{79}(Br)M+H]$^+$, 450 [^{81}(Br)M+H]$^+$.

3.3.5. (*E*)-4-(((1*H*-benzo[d]imidazol-2-yl)methyl)amino)-*N*′-(2-chlorobenzylidene)benzohydrazide (**6e**)

White solid (140 mg, 0.346 mmol, 97%), mp. 289 °C. FT-IR (KBr): ν (cm^{-1}) = 1656, 1606, 1522, 1431, 1360, 1330, 1298, 1258, 1187, 1148, 1132, 1050, 1024, 936, 826, 744, 713, 692, 640, and 498 cm^{-1}. ^{1}H-NMR (700 MHz, DMSO-d_6), δ ^{1}H NMR (700 MHz, DMSO), δ 12.35 (s, 1H), 11.73 (s, 1H), 8.80 (s, 1H), 8.00 (s, 1H), 7.74 (d, *J* = 8.8 Hz, 2H), 7.58 (d, *J* = 7.6 Hz, 1H), 7.52 (d, *J* = 4.4 Hz, 1H), 7.43 (t, *J* = 8.4 Hz, 3H), 7.15 (s, 2H), 7.04 (t, *J* = 6.1 Hz, 1H), 6.74 (d, *J* = 8.7 Hz, 2H), and 4.58 (d, *J* = 5.7 Hz, 2H) ppm. ^{13}C-NMR (176 MHz, DMSO-d_6), δ 163.39, 158.77, 153.34, 152.01, 143.68, 142.37, 134.79, 133.37, 132.38, 131.57, 130.35, 129.75,

128.04, 127.18, 122.29, 121.55, 120.60, 118.87, 111.88, 111.70, and 41.74 ppm. Mass (ESI): m/z 404 [^{35}(Cl)M+H]$^+$, 406 [^{37}(Cl)M+H]$^+$.

3.3.6. (E)-4-(((1H-benzo[d]imidazol-2-yl)methyl)amino)-N'-(3-chlorobenzylidene) benzohydrazide (6f)

White solid (136 mg, 0.336 mmol, 94%), mp. 271 °C. FT-IR (KBr): ν (cm^{-1}) = 3398, 3057, 3025, 2871, 2836, 1609, 1525, 1473, 1434, 1334, 1281, 1184, 1132, 1024, 920, 822, 739, 696, 621, and 556 cm^{-1}. ^{1}H-NMR (700 MHz, DMSO-d_6), δ ^{1}H NMR (700 MHz, DMSO), δ 12.35 (s, 1H), 11.62 (s, 1H), 8.37 (s, 1H), 7.74 (s, 1H), 7.73 (d, J = 8.8 Hz, 2H), 7.65 (s, 1H), 7.58 (d, J = 7.6 Hz, 1H), 7.47 (d, J = 5.2 Hz, 2H), 7.44 (d, J = 7.5 Hz, 1H), 7.18–7.12 (m, 2H), 7.04 (t, J = 5.8 Hz, 1H), 6.74 (d, J = 8.8 Hz, 2H), and 4.58 (d, J = 5.8 Hz, 2H) ppm. ^{13}C-NMR (176 MHz, DMSO-d_6), δ 163.49, 153.35, 151.97, 144.78, 137.41, 134.80, 134.09, 131.17, 129.77, 126.46, 126.06, 122.29, 121.55, 120.70, 118.87, 111.87, 111.70, and 41.76 ppm. Mass (ESI): m/z 404 [^{35}(Cl)M+H]$^+$, 406 [^{37}(Cl)M+H]$^+$.

3.3.7. (E)-4-(((1H-benzo[d]imidazol-2-yl)methyl)amino)-N'-(4-chlorobenzylidene) benzohydrazide (6g)

White solid (100 mg, 0.247 mmol, 69%), mp. 290 °C (lit. [43] mp. = 238 °C), CAS registry number 76321-91-0. FT-IR (KBr): ν (cm^{-1}) = 3385, 3057, 2952, 2911, 2787, 2514, 1656, 1606, 1522, 1434, 1334, 1294, 1255, 1210, 1184, 1086, 1026, 934, 822, 741, 637, 504, and 436 cm^{-1}. ^{1}H-NMR (700 MHz, DMSO-d_6), δ 12.35 (s, 1H), 11.55 (s, 1H), 8.38 (s, 1H), 7.72 (d, J = 8.5 Hz, 4H), 7.57 (s, 1H), 7.51 (d, J = 8.3 Hz, 2H), 7.44 (s, 1H), 7.14 (s, 2H), 7.02 (t, J = 5.8 Hz, 1H), 6.73 (d, J = 8.6 Hz, 2H), and 4.58 (d, J = 5.8 Hz, 2H) ppm. ^{13}C-NMR (176 MHz, DMSO-d_6), δ 163.38, 153.37, 151.92, 145.20, 143.69, 134.80, 134.54, 134.09, 129.69, 129.36, 128.92, 122.28, 121.55, 120.79, 118.87, 111.87, 111.70, and 41.76 ppm. Mass (ESI): m/z 404 [^{35}(Cl)M+H]$^+$, 406 [^{37}(Cl)M+H]$^+$.

3.3.8. (E)-4-(((1H-benzo[d]imidazol-2-yl)methyl)amino)-N'-(2-fluorobenzylidene) benzohydrazide (6h)

White solid (100 mg, 0.258 mmol, 72%), mp. 289 °C. FT-IR (KBr): ν (cm^{-1}) = 3388, 3096, 2963, 2907, 2845, 2790, 1963, 1655, 1606, 1521, 1437, 1255, 1073, 1021, 894, 822, 741, 692, 640, and 478 cm^{-1}. ^{1}H-NMR (700 MHz, DMSO-d_6), δ 12.35 (s, 1H), 11.62 (s, 1H), 8.64 (s, 1H), 7.92 (s, 1H), 7.73 (d, J = 8.7 Hz, 2H), 7.58 (d, J = 8.0 Hz, 1H), 7.47 (d, J = 6.1 Hz, 1H), 7.45 (d, J = 7.6 Hz, 1H), 7.29 (t, J = 8.5 Hz, 2H), 7.15 (s, 2H), 7.04 (s, 1H), 6.74 (d, J = 8.7 Hz, 2H), and 4.58 (d, J = 5.8 Hz, 2H) ppm. ^{13}C-NMR (176 MHz, DMSO-d_6), δ 163.37, 160.61 ($^1J_{C-F}$ = 250 Hz), 153.36, 151.97, 143.67, 139.15, 134.79, 132.03 ($^3J_{C-F}$ = 8.4 Hz), 129.71, 126.63 ($^4J_{C-F}$ = 2.8 Hz), 125.37 ($^3J_{C-F}$ = 3Hz), 122.66 ($^2J_{C-F}$ = 9.8 Hz), 122.30, 121.56, 120.66, 118.88, 116.43 ($^2J_{C-F}$ = 20.8 Hz), 111.90, 111.70, and 41.74 ppm. Mass (ESI): m/z 388 [(^{18}F)M+H]$^+$; 389 [(^{19}F)M+H]$^+$.

3.3.9. (E)-4-(((1H-benzo[d]imidazol-2-yl)methyl)amino)-N'-(3-fluorobenzylidene) benzohydrazide (6i)

Beige solid (151 mg, 0.309 mmol, 87%), mp. 278 °C. FT-IR (KBr): ν (cm^{-1}) =3464, 3405, 3206, 3066, 2924, 2843, 1613, 1525, 1443, 1353, 1294, 1265, 1187, 1021, 936, 898, 872, 826, 758, 735, 683, and 521 cm^{-1}. ^{1}H-NMR (700 MHz, DMSO-d_6), δ 12.35 (s, 1H), 11.60 (s, 1H), 8.39 (s, 1H), 7.73 (d, J = 8.4 Hz, 2H), 7.58 (d, J = 4.4 Hz, 1H), 7.52 (s, 1H), 7.49 (d, J = 7.6 Hz, 1H), 7.44 (d, J = 8.0 Hz, 1H), 7.25 (t, J = 8.6 Hz, 1H), 7.15 (s, 2H), 7.03 (s, 1H), 6.73 (d, J = 8.3 Hz, 2H), and 4.58 (d, J = 5.6 Hz, 2H) ppm. ^{13}C-NMR (176 MHz, DMSO-d_6), δ 163.49, 162.69 ($^1J_{C-F}$ = 245 Hz), 153.36, 151.95, 145.13, 143.67, 137.76 ($^3J_{C-F}$ = 7.8Hz), 134.79, 131.34 ($^3J_{C-F}$ = 8.5 Hz), 129.72, 123.67, 122.29, 121.55, 120.72, 118.87, 116.88 ($^2J_{C-F}$ = 21 Hz), 113.20 ($^2J_{C-F}$ = 22 Hz), 111.87, 111.70, and 41.75 ppm. Mass (ESI): m/z 388 [(^{18}F)M+H]$^+$; 389 [(^{19}F)M+H]$^+$.

3.3.10. (E)-4-(((1H-benzo[d]imidazol-2-yl)methyl)amino)-N'-(4-fluorobenzylidene)benzohydrazide (6j)

White solid (70 mg, 0.180 mmol, 53%), mp. 291 °C. FT-IR (KBr): ν (cm^{-1}) =3392, 2956, 2898, 2790, 2693, 1649,1604, 1502, 1438, 1298, 1187, 1145, 832, 751, 696, and 514 cm^{-1}. ^{1}H-NMR (700 MHz, DMSO-d_6), δ 12.34 (s, 1H), 11.49 (s, 1H), 8.38 (s, 1H), 7.75 (s, 2H), 7.71 (d, J = 8.6 Hz, 2H), 7.55 (s, 1H), 7.46 (s, 1H), 7.28 (t, J = 8.7 Hz, 2H), 7.15 (s, 2H), 7.01 (t, J = 6.0 Hz, 1H), 6.73 (d, J = 8.6 Hz, 2H), and 4.57 (d, J = 5.8 Hz, 2H) ppm. ^{13}C-NMR (176 MHz, DMSO-d_6), δ ^{13}C NMR (176 MHz, DMSO), δ 163.37, 163.35 ($^1J_{C-F}$ = 247 Hz), 153.38, 151.87, 145.44, 131.75, 129.62, 129.42 ($^1J_{C-F}$ = 8.5 Hz), 122.24, 121.58, 120.89, 118.87, 116.31 ($^2J_{C-F}$ = 22 Hz), 111.87, 111.68, and 41.76 ppm. Mass (ESI): m/z 388 [(^{18}F)M+H]$^+$; 389 [(^{19}F)M+H]$^+$.

3.3.11. (E)-4-(((1H-benzo[d]imidazol-2-yl)methyl)amino)-N'-(2,4-dichlorobenzylidene)benzohydrazide (6k)

White solid (134 mg, 0.305 mmol, 86%), mp. 294 °C. FT-IR (KBr): ν (cm^{-1}) = 3385, 3096, 2956, 2790, 1656, 1604, 1522l, 1441, 1330, 1298, 1258, 1184, 1044, 970, 822, 739, 692, 637, and 514 cm^{-1}. ^{1}H-NMR (700 MHz, DMSO-d_6), δ ^{1}H NMR (700 MHz, DMSO), δ 12.35 (s, 1H), 11.78 (s, 1H), 8.75 (s, 1H), 7.99 (s, 1H), 7.74 (d, J = 8.7 Hz, 2H), 7.71 (s, 1H), 7.57 (s, 1H), 7.51 (d, J = 8.8 Hz, 1H), 7.45 (s, 1H), 7.15 (s, 2H), 7.06 (t, J = 5.8 Hz, 1H), 6.74 (d, J = 8.7 Hz, 2H), and 4.59 (d, J = 5.7 Hz, 2H) ppm. ^{13}C-NMR (176 MHz, DMSO-d_6), δ 163.47, 153.42, 152.17, 143.77, 141.38, 135.20, 134.88, 134.11, 131.62, 129.89, 128.55, 128.46, 122.36, 121.68, 120.58, 118.99, 111.99, 111.80, and 41.83 ppm. Mass (ESI): m/z 438 [^{35}(Cl)M+H]$^+$, 440 [^{37}(Cl)M+H]$^+$.

3.3.12. (E)-4-(((1H-benzo[d]imidazol-2-yl)methyl)amino)-N'-(2,5-difluorobenzylidene)benzohydrazide (6l)

Cream solid (110 mg, 0.271 mmol, 84%), mp. 160 °C. FT-IR (KBr): ν (cm^{-1}) =3444, 3206, 3139, 2963, 2826, 1642, 1606, 1519, 1486, 1428, 1353, 1298, 1184, 1090, 1060, 1014, 940, 898, 836, 813, 739, and 481 cm^{-1}. ^{1}H-NMR (700 MHz, DMSO-d_6) δ ^{1}H NMR (700 MHz, DMSO) δ 12.36 (s, 1H), 11.72 (s, 1H), 8.59 (s, 1H), 7.73 (d, J = 8.5 Hz, 2H), 7.63–7.59 (m, 1H), 7.51 (dt, J = 9.8, 4.8 Hz, 1H), 7.39–7.36 (m, 1H), 7.32 (dt, J = 8.8, 4.3 Hz, 1H), 7.15 (dt, J = 7.5, 3.7 Hz, 2H), 7.06 (t, J = 5.7 Hz, 1H), 6.74 (d, J = 8.4 Hz, 2H), and 4.58 (d, J = 5.7 Hz, 2H) ppm. ^{13}C-NMR (176 MHz, DMSO-d_6), δ 164.49, 158.00 (dd, $^1J_{C-F}$ = 277, 243 Hz), 157 (ddd, $^1J_{C-F}$ = 245, 50, 2.4Hz), 153.32, 152.07, 138.07, 129.77, 124.35 (dd, $^1J_{C-F}$ = 12.1, 7.8 Hz), 121.94, 121.75 (dd, $^1J_{C-F}$ = 24, 9 Hz), 120.45, 119.34 (dd, $^1J_{C-F}$ = 26, 8.4 Hz), 118.53 (dd, $^1J_{C-F}$ = 24.6, 9 Hz), 118.34 (dd, $^1J_{C-F}$ = 24, 8.4 Hz), 118.20, 118.05, 112.04, 112.03, 111.90, 111.88, and 41.72 ppm. Mass (ESI): m/z 406 [(^{18}F)M+H]$^+$; 407 [(^{19}F)M+H]$^+$.

3.4. Biological Evaluation

3.4.1. In Vitro Cytotoxicity Assay

The cytotoxic potential of compounds **6a-l** was evaluated against a panel of cancer cell lines, including HCT-116, HepG2, and MCF-7, as well as a normal cell line, WI-38, using the reported methodology [44]. The cell lines utilized in this study were obtained from the American Type Culture Collection (ATCC) through the Holding Company for Biological Products and Vaccines (VACSERA) in Cairo, Egypt. These included: 1. A human colon cancer cell line (HCT-116); 2. Hepatocellular carcinoma (HepG2)—a liver cancer cell line; 3. Mammary gland breast cancer (MCF-7); and 4. A human lung fibroblast (WI-38)—a normal cell line.

3.4.2. In Vitro Enzyme Inhibitory Assays

The enzyme inhibitory activities of compounds **6c** and **6h-j** against EGFR, HER2, VEGFR2, CDK2, AURKC, HDAC1, and mTOR enzymes were assessed using recently reported methods [44].

3.4.3. Cell Cycle Analysis

Cell cycle distribution was evaluated using ab139418 Propidium Iodide (PI) flow cytometry kit/BD following a previously reported method [44,45] for the most potent compound **6i**.

3.4.4. Annexin-V/Propidium Iodide (PI) Double Staining Assay

The apoptosis effect of compound **6i** was studied using a previously reported method [44,45].

3.4.5. Determination of Apoptotic Protein Levels

Apoptotic protein levels for compound **6i** were determined using an ELISA technique for caspase-3, BAX, and Bcl-2, applying a previously reported method [44,46,47].

3.5. In Silico Molecular Docking Studies

Molecular docking analysis of compounds **6h** and **6i** were conducted using the active sites of EGFR (PDB: 4HJO), Her2 (PDB: 7PCD), and AURKC (PDB: 6GR8) kinase enzymes. For docking studies, MOE 2020 software programs were employed.

4. Conclusions

The findings of this study highlight the significant potential of the novel class of halogenated compounds with the structural motif "(E)-4-(((1H-Benzo[d]imidazol-2-yl)methyl)amino)-N'-(halogenated)benzohydrazide (**6a-l**)" as promising candidates for the development of highly potent and effective targeted kinase inhibitors (TKIs). The in vitro evaluation of this compound series demonstrated excellent cytotoxic effects against a panel of diverse cancer cell lines, with select analogs, such as **6c** and **6h-j**, exhibiting remarkably low IC$_{50}$ values in the range of 7.82 to 21.48 µM. Of particular note, compounds **6h** and **6i** emerged as the most potent inhibitors, displaying significant inhibitory activity against critical oncogenic kinases, including EGFR, HER2, CDK2, AURKC, and mTOR. The in-depth mechanistic investigation of lead compound **6i** uncovered its remarkable capability to induce cell cycle arrest and programmed cell death (apoptosis) within HepG2 liver cancer cells. This was accompanied by a notable upregulation of pro-apoptotic proteins caspase-3 and Bax, coupled with a downregulation of the anti-apoptotic protein Bcl-2. Furthermore, computational molecular docking analyses suggested favorable binding interactions between compounds **6h** and **6i** and the relevant target enzymes. These findings collectively underscore the immense potential of this class of halogenated-1H-benzo[d]imidazol-2-yl)methyl)amino)-N'-benzohydrazide compounds as a promising avenue for the development of next-generation multi-targeted kinase inhibitors with enhanced potency and therapeutic efficacy. The ability of these compounds to modulate multiple oncogenic signaling pathways simultaneously presents a compelling opportunity to address the inherent complexity and adaptability of cancer cells, potentially overcoming the limitations associated with single-target kinase inhibitors. Moving forward, further optimization of the lead compounds, including structural modifications, in-depth mechanistic studies, and rigorous in vivo evaluations, will be crucial to fully elucidate the therapeutic potential of this compound class. Successful translation of these findings into clinical settings could pave the way for the development of innovative treatment strategies, ultimately benefiting patients with a wide range of cancers and other diseases characterized by deregulated kinase signaling.

Supplementary Materials: The following supporting information can be downloaded at https://www.mdpi.com/article/10.3390/ph17070839/s1: ^{1}H-NMR, ^{13}CNMR, and Mass spectra.

Author Contributions: Conceptualization, M.M.A. and A.F.M.M.R.; methodology, H.H.A. and T.O.M.; validation, A.F.M.M.R.; formal analysis, H.H.A. and T.O.M.; investigation, M.M.A. and A.F.M.M.R.; resources, M.M.A.; data curation, T.O.M. and A.F.M.M.R.; writing—original draft preparation, M.M.A. and A.F.M.M.R.; writing—review and editing, M.M.A. and A.F.M.M.R.; supervision, M.M.A. and A.F.M.M.R.; funding acquisition, M.M.A. All authors have read and agreed to the published version of the manuscript.

Funding: This work was funded by the Researchers Supporting Project number (RSPD2024R628), King Saud University, Riyadh, Saudi Arabia.

Institutional Review Board Statement: Not applicable.

Informed Consent Statement: Not applicable.

Data Availability Statement: Data are given in the Supplementary Materials and are available upon request.

Acknowledgments: The authors extend their appreciation to the Researchers Supporting Project number (RSPD2024R628), King Saud University, Riyadh, Saudi Arabia, for funding this work.

Conflicts of Interest: The authors declare no conflicts of interest.

References

1. Hernandes, M.Z.; Cavalcanti, S.M.; Moreira, D.R.; de Azevedo Junior, W.F.; Leite, A.C. Halogen atoms in the modern medicinal chemistry: Hints for the drug design. *Curr. Drug Targets* **2010**, *11*, 303–314. [CrossRef] [PubMed]
2. Tsang, J.E.; Urner, L.M.; Kim, G.; Chow, K.; Baufeld, L.; Faull, K.; Cloughesy, T.F.; Clark, P.M.; Jung, M.E.; Nathanson, D.A. Development of a Potent Brain-Penetrant EGFR Tyrosine Kinase Inhibitor against Malignant Brain Tumors. *ACS Med. Chem. Lett.* **2020**, *11*, 1799–1809. [CrossRef] [PubMed]
3. Poznanski, J.; Winiewska-Szajewska, M.; Czapinska, H.; Poznańska, A.; Shugar, D. Halogen bonds involved in binding of halogenated ligands by protein kinases. *Acta Biochim. Pol.* **2016**, *63*, 203–214. [CrossRef] [PubMed]
4. Nanda, R. Targeting the human epidermal growth factor receptor 2 (HER2) in the treatment of breast cancer: Recent advances and future directions. *Rev. Recent. Clin. Trials* **2007**, *2*, 111–116. [CrossRef] [PubMed]
5. Zhou, C.; Wu, Y.L.; Chen, G.; Feng, J.; Liu, X.Q.; Wang, C.; Zhang, S.; Wang, J.; Zhou, S.; Ren, S.; et al. Erlotinib versus chemotherapy as first-line treatment for patients with advanced EGFR mutation-positive non-small-cell lung cancer (OPTIMAL, CTONG-0802): A multicentre, open-label, randomised, phase 3 study. *Lancet Oncol.* **2011**, *12*, 735–742. [CrossRef] [PubMed]
6. Cortes, J.E.; Kim, D.W.; Pinilla-Ibarz, J.; le Coutre, P.; Paquette, R.; Chuah, C.; Nicolini, F.E.; Apperley, J.F.; Khoury, H.J.; Talpaz, M.; et al. A phase 2 trial of ponatinib in Philadelphia chromosome-positive leukemias. *N. Engl. J. Med.* **2013**, *369*, 1783–1796. [CrossRef] [PubMed]
7. Sequist, L.V.; Yang, J.C.; Yamamoto, N.; O'Byrne, K.; Hirsh, V.; Mok, T.; Geater, S.L.; Orlov, S.; Tsai, C.M.; Boyer, M.; et al. Phase III Study of Afatinib or Cisplatin Plus Pemetrexed in Patients With Metastatic Lung Adenocarcinoma With EGFR Mutations. *J. Clin. Oncol.* **2023**, *41*, 2869–2876. [CrossRef] [PubMed]
8. Cárdenas-Fernández, D.; Soberanis Pina, P.; Turcott, J.G.; Chávez-Tapia, N.; Conde-Flores, E.; Cardona, A.F.; Arrieta, O. Management of diarrhea induced by EGFR-TKIs in advanced lung adenocarcinoma. *Ther. Adv. Med. Oncol.* **2023**, *15*, 17588359231192396. [CrossRef] [PubMed]
9. Drilon, A.; Laetsch, T.W.; Kummar, S.; DuBois, S.G.; Lassen, U.N.; Demetri, G.D.; Nathanson, M.; Doebele, R.C.; Farago, A.F.; Pappo, A.S.; et al. Efficacy of Larotrectinib in TRK Fusion-Positive Cancers in Adults and Children. *N. Engl. J. Med.* **2018**, *378*, 731–739. [CrossRef]
10. Yang, J.C.; Shih, J.Y.; Su, W.C.; Hsia, T.C.; Tsai, C.M.; Ou, S.H.; Yu, C.J.; Chang, G.C.; Ho, C.L.; Sequist, L.V.; et al. Afatinib for patients with lung adenocarcinoma and epidermal growth factor receptor mutations (LUX-Lung 2): A phase 2 trial. *Lancet Oncol.* **2012**, *13*, 539–548. [CrossRef]
11. Wu, Q.; Jiang, H.; Wang, S.; Dai, D.; Chen, F.; Meng, D.; Geng, P.; Tong, H.; Zhou, Y.; Pan, D.; et al. Effects of avitinib on the pharmacokinetics of osimertinib in vitro and in vivo in rats. *Thorac. Cancer* **2020**, *11*, 2775–2781. [CrossRef] [PubMed]
12. Foran, J.; Ravandi, F.; Wierda, W.; Garcia-Manero, G.; Verstovsek, S.; Kadia, T.; Burger, J.; Yule, M.; Langford, G.; Lyons, J.; et al. A phase I and pharmacodynamic study of AT9283, a small-molecule inhibitor of aurora kinases in patients with relapsed/refractory leukemia or myelofibrosis. *Clin. Lymphoma Myeloma Leuk.* **2014**, *14*, 223–230. [CrossRef] [PubMed]
13. Thompson, S.L.; Gianessi, C.A.; O'Malley, S.S.; Cavallo, D.A.; Shi, J.M.; Tetrault, J.M.; DeMartini, K.S.; Gueorguieva, R.; Pittman, B.; Krystal, J.H.; et al. Saracatinib Fails to Reduce Alcohol-Seeking and Consumption in Mice and Human Participants. *Front. Psychiatry* **2021**, *12*, 709559. [CrossRef] [PubMed]
14. Massaro, F.; Molica, M.; Breccia, M. Ponatinib: A Review of Efficacy and Safety. *Curr. Cancer Drug Targets* **2018**, *18*, 847–856. [CrossRef] [PubMed]
15. Fujino, T.; Suda, K.; Koga, T.; Hamada, A.; Ohara, S.; Chiba, M.; Shimoji, M.; Takemoto, T.; Soh, J.; Mitsudomi, T. Foretinib can overcome common on-target resistance mutations after capmatinib/tepotinib treatment in NSCLCs with MET exon 14 skipping mutation. *J. Hematol. Oncol.* **2022**, *15*, 79. [CrossRef] [PubMed]
16. Wells, S.A., Jr.; Robinson, B.G.; Gagel, R.F.; Dralle, H.; Fagin, J.A.; Santoro, M.; Baudin, E.; Elisei, R.; Jarzab, B.; Vasselli, J.R.; et al. Vandetanib in patients with locally advanced or metastatic medullary thyroid cancer: A randomized, double-blind phase III trial. *J. Clin. Oncol.* **2012**, *30*, 134–141. [CrossRef] [PubMed]

17. Zhang, L.; Meng, L.; Liu, B.; Zhang, Y.; Zhu, H.; Cui, J.; Sun, A.; Hu, Y.; Jin, J.; Jiang, H.; et al. Flumatinib versus Imatinib for Newly Diagnosed Chronic Phase Chronic Myeloid Leukemia: A Phase III, Randomized, Open-label, Multi-center FESTnd Study. *Clin. Cancer Res.* **2021**, *27*, 70–77. [CrossRef] [PubMed]
18. Chapman, P.B.; Hauschild, A.; Robert, C.; Haanen, J.B.; Ascierto, P.; Larkin, J.; Dummer, R.; Garbe, C.; Testori, A.; Maio, M.; et al. Improved survival with vemurafenib in melanoma with BRAF V600E mutation. *N. Engl. J. Med.* **2011**, *364*, 2507–2516. [CrossRef]
19. Cheng, A.L.; Kang, Y.K.; Chen, Z.; Tsao, C.J.; Qin, S.; Kim, J.S.; Luo, R.; Feng, J.; Ye, S.; Yang, T.S.; et al. Efficacy and safety of sorafenib in patients in the Asia-Pacific region with advanced hepatocellular carcinoma: A phase III randomised, double-blind, placebo-controlled trial. *Lancet Oncol.* **2009**, *10*, 25–34. [CrossRef]
20. Yakes, F.M.; Chen, J.; Tan, J.; Yamaguchi, K.; Shi, Y.; Yu, P.; Qian, F.; Chu, F.; Bentzien, F.; Cancilla, B.; et al. Cabozantinib (XL184), a novel MET and VEGFR2 inhibitor, simultaneously suppresses metastasis, angiogenesis, and tumor growth. *Mol. Cancer Ther.* **2011**, *10*, 2298–2308. [CrossRef]
21. Kantarjian, H.; Giles, F.; Wunderle, L.; Bhalla, K.; O'Brien, S.; Wassmann, B.; Tanaka, C.; Manley, P.; Rae, P.; Mietlowski, W.; et al. Nilotinib in imatinib-resistant CML and Philadelphia chromosome-positive ALL. *N. Engl. J. Med.* **2006**, *354*, 2542–2551. [CrossRef] [PubMed]
22. Geyer, C.E.; Forster, J.; Lindquist, D.; Chan, S.; Romieu, C.G.; Pienkowski, T.; Jagiello-Gruszfeld, A.; Crown, J.; Chan, A.; Kaufman, B.; et al. Lapatinib plus capecitabine for HER2-positive advanced breast cancer. *N. Engl. J. Med.* **2006**, *355*, 2733–2743. [CrossRef] [PubMed]
23. Xu, B.; Yan, M.; Ma, F.; Hu, X.; Feng, J.; Ouyang, Q.; Tong, Z.; Li, H.; Zhang, Q.; Sun, T.; et al. Pyrotinib plus capecitabine versus lapatinib plus capecitabine for the treatment of HER2-positive metastatic breast cancer (PHOEBE): A multicentre, open-label, randomised, controlled, phase 3 trial. *Lancet Oncol.* **2021**, *22*, 351–360. [CrossRef] [PubMed]
24. Carvajal, R.D.; Sosman, J.A.; Quevedo, J.F.; Milhem, M.M.; Joshua, A.M.; Kudchadkar, R.R.; Linette, G.P.; Gajewski, T.F.; Lutzky, J.; Lawson, D.H.; et al. Effect of selumetinib vs chemotherapy on progression-free survival in uveal melanoma: A randomized clinical trial. *Jama* **2014**, *311*, 2397–2405. [CrossRef] [PubMed]
25. Tunçbilek, M.; Kiper, T.; Altanlar, N. Synthesis and in vitro antimicrobial activity of some novel substituted benzimidazole derivatives having potent activity against MRSA. *Eur. J. Med. Chem.* **2009**, *44*, 1024–1033. [CrossRef] [PubMed]
26. Shingalapur, R.V.; Hosamani, K.M.; Keri, R.S. Synthesis and evaluation of in vitro anti-microbial and anti-tubercular activity of 2-styryl benzimidazoles. *Eur. J. Med. Chem.* **2009**, *44*, 4244–4248. [CrossRef] [PubMed]
27. Zhang, H.Z.; Damu, G.L.; Cai, G.X.; Zhou, C.H. Design, synthesis and antimicrobial evaluation of novel benzimidazole type of Fluconazole analogues and their synergistic effects with Chloromycin, Norfloxacin and Fluconazole. *Eur. J. Med. Chem.* **2013**, *64*, 329–344. [CrossRef] [PubMed]
28. Hernández-Luis, F.; Hernández-Campos, A.; Castillo, R.; Navarrete-Vázquez, G.; Soria-Arteche, O.; Hernández-Hernández, M.; Yépez-Mulia, L. Synthesis and biological activity of 2-(trifluoromethyl)-1H-benzimidazole derivatives against some protozoa and Trichinella spiralis. *Eur. J. Med. Chem.* **2010**, *45*, 3135–3141. [CrossRef]
29. Luo, Y.; Yao, J.P.; Yang, L.; Feng, C.L.; Tang, W.; Wang, G.F.; Zuo, J.P.; Lu, W. Design and synthesis of novel benzimidazole derivatives as inhibitors of hepatitis B virus. *Bioorg Med. Chem.* **2010**, *18*, 5048–5055. [CrossRef]
30. Garuti, L.; Roberti, M.; Bottegoni, G. Benzimidazole derivatives as kinase inhibitors. *Curr. Med. Chem.* **2014**, *21*, 2284–2298. [CrossRef]
31. Singla, P.; Luxami, V.; Paul, K. Benzimidazole-biologically attractive scaffold for protein kinase inhibitors. *RSC Adv.* **2014**, *4*, 12422–12440. [CrossRef]
32. Woolley, D.W. Some biological effects produced by benzimidazole and their reversal by purines. *J. Biol. Chem.* **1944**, *152*, 225–232. [CrossRef]
33. Demirel, S.; Ayhan Kilcigil, G.; Kara, Z.; Güven, B.; Onay Beşikci, A. Synthesis and Pharmacologic Evaluation of Some Benzimidazole Acetohydrazide Derivatives as EGFR Inhibitors. *Turk. J. Pharm. Sci.* **2017**, *14*, 285–289. [CrossRef] [PubMed]
34. Akhtar, M.J.; Siddiqui, A.A.; Khan, A.A.; Ali, Z.; Dewangan, R.P.; Pasha, S.; Yar, M.S. Design, synthesis, docking and QSAR study of substituted benzimidazole linked oxadiazole as cytotoxic agents, EGFR and erbB2 receptor inhibitors. *Eur. J. Med. Chem.* **2017**, *126*, 853–869. [CrossRef] [PubMed]
35. Jia, Y.; Juarez, J.; Li, J.; Manuia, M.; Niederst, M.J.; Tompkins, C.; Timple, N.; Vaillancourt, M.T.; Pferdekamper, A.C.; Lockerman, E.L.; et al. EGF816 Exerts Anticancer Effects in Non-Small Cell Lung Cancer by Irreversibly and Selectively Targeting Primary and Acquired Activating Mutations in the EGF Receptor. *Cancer Res.* **2016**, *76*, 1591–1602. [CrossRef]
36. Brown, J.R. 3 Adriamycin and Related Anthracycline Antibiotics. In *Progress in Medicinal Chemistry*; Ellis, G.P., West, G.B., Eds.; Elsevier: Amsterdam, The Netherlands, 1978; Volume 15, pp. 125–164.
37. Daunorubicin. Available online: https://go.drugbank.com/drugs/DB00694 (accessed on 11 December 2023).
38. Zorubicin. Available online: https://en.wikipedia.org/wiki/Zorubicin (accessed on 11 December 2023).
39. Mirgany, T.O.; Rahman, A.F.M.M.; Alanazi, M.M. Design, synthesis, and mechanistic evaluation of novel benzimidazole-hydrazone compounds as dual inhibitors of EGFR and HER2: Promising candidates for anticancer therapy. *J. Mol. Struct.* **2024**, *1309*, 138177. [CrossRef]
40. Indrayanto, G.; Putra, G.S.; Suhud, F. Chapter Six—Validation of in-vitro bioassay methods: Application in herbal drug research. In *Profiles of Drug Substances, Excipients and Related Methodology*; Al-Majed, A.A., Ed.; Academic Press: Cambridge, MA, USA, 2021; Volume 46, pp. 273–307.

41. Alotaibi, A.A.; Asiri, H.H.; Rahman, A.F.M.M.; Alanazi, M.M. Novel pyrrolo[2,3-d]pyrimidine derivatives as multi-kinase inhibitors with VEGFR-2 selectivity. *J. Saudi Chem. Soc.* **2023**, *27*, 101712. [CrossRef]
42. Alotaibi, A.A.; Alanazi, M.M.; Rahman, A.F.M.M. Discovery of New Pyrrolo[2,3-d]pyrimidine Derivatives as Potential Multi-Targeted Kinase Inhibitors and Apoptosis Inducers. *Pharmaceuticals* **2023**, *16*, 1324. [CrossRef]
43. Bahadur, S.; Pandey, K.K. Synthesis of p-alkyl-(2-benzimidazolyl)-methyl-aminobenzoates and corresponding hydrazides as possible antimalarial agents. *J. Indian. Chem. Soc.* **1980**, *57*, 447–448.
44. Alanazi, A.S.; Mirgany, T.O.; Alsfouk, A.A.; Alsaif, N.A.; Alanazi, M.M. Antiproliferative Activity, Multikinase Inhibition, Apoptosis- Inducing Effects and Molecular Docking of Novel Isatin–Purine Hybrids. *Medicina* **2023**, *59*, 610. [CrossRef]
45. Eldehna, W.M.; Hassan, G.S.; Al-Rashood, S.T.; Al-Warhi, T.; Altyar, A.E.; Alkahtani, H.M.; Almehizia, A.A.; Abdel-Aziz, H.A. Synthesis and in vitro anticancer activity of certain novel 1-(2-methyl-6-arylpyridin-3-yl)-3-phenylureas as apoptosis-inducing agents. *J. Enzym. Inhib. Med. Chem.* **2019**, *34*, 322–332. [CrossRef] [PubMed]
46. Sabt, A.; Abdelhafez, O.M.; El-Haggar, R.S.; Madkour, H.M.F.; Eldehna, W.M.; El-Khrisy, E.; Abdel-Rahman, M.A.; Rashed, L.A. Novel coumarin-6-sulfonamides as apoptotic anti-proliferative agents: Synthesis, in vitro biological evaluation, and QSAR studies. *J. Enzym. Inhib. Med. Chem.* **2018**, *33*, 1095–1107. [CrossRef] [PubMed]
47. Alanazi, M.M.; Aldawas, S.; Alsaif, N.A. Design, Synthesis, and Biological Evaluation of 2-Mercaptobenzoxazole Derivatives as Potential Multi-Kinase Inhibitors. *Pharmaceuticals* **2023**, *16*, 97. [CrossRef] [PubMed]

Disclaimer/Publisher's Note: The statements, opinions and data contained in all publications are solely those of the individual author(s) and contributor(s) and not of MDPI and/or the editor(s). MDPI and/or the editor(s) disclaim responsibility for any injury to people or property resulting from any ideas, methods, instructions or products referred to in the content.

Article

Study of the Lipophilicity and ADMET Parameters of New Anticancer Diquinothiazines with Pharmacophore Substituents

Daria Klimoszek [1], Małgorzata Jeleń [2,*], Małgorzata Dołowy [1] and Beata Morak-Młodawska [2]

[1] Department of Analytical Chemistry, Faculty of Pharmaceutical Sciences in Sosnowiec, Medical University of Silesia in Katowice, Jagiellońska Street 4, 41-200 Sosnowiec, Poland; d201204@365.sum.edu.pl (D.K.); mdolowy@sum.edu.pl (M.D.)

[2] Department of Organic Chemistry, Faculty of Pharmaceutical Sciences in Sosnowiec, Medical University of Silesia in Katowice, Jagiellońska Street 4, 41-200 Sosnowiec, Poland; bmlodawska@sum.edu.pl

* Correspondence: manowak@sum.edu.pl

Abstract: Lipophilicity is one of the principal parameters that describe the pharmacokinetic behavior of a drug, including its absorption, distribution, metabolism, elimination, and toxicity. In this study, the lipophilicity and other physicochemical, pharmacokinetic, and toxicity properties that affect the bioavailability of newly synthesized dialkylaminoalkyldiquinothiazine hybrids as potential drug candidates are presented. The lipophilicity, as R_{M0}, was determined experimentally by the RP-TLC method using RP18 plates and acetone–TRIS buffer (pH 7.4) as the mobile phase. The chromatographic parameters of lipophilicity were compared to computationally calculated partition coefficients obtained by various types of programs such as iLOGP, XLOGP3, WLOGP, MLOGP, SILCOS-IT, LogP, logP, and milogP. In addition, the selected ADMET parameters were determined in silico using the SwissADME and pkCSM platforms and correlated with the experimental lipophilicity descriptors. The results of the lipophilicity study confirm that the applied algorithms can be useful for the rapid prediction of logP values during the first stage of study of the examined drug candidates. Of all the algorithms used, the biggest similarity to the chromatographic value (R_{M0}) for certain compounds was seen with iLogP. It was found that both the SwissADME and pkCSM web tools are good sources of a wide range of ADMET parameters that describe the pharmacokinetic profiles of the studied compounds and can be fast and low-cost tools in the evaluation of examined drug candidates during the early stages of the development process.

Keywords: lipophilicity; diquinothiazines; ADME; chromatography; phenothiazines; anticancer agents

Citation: Klimoszek, D.; Jeleń, M.; Dołowy, M.; Morak-Młodawska, B. Study of the Lipophilicity and ADMET Parameters of New Anticancer Diquinothiazines with Pharmacophore Substituents. *Pharmaceuticals* **2024**, *17*, 725. https://doi.org/10.3390/ph17060725

Academic Editors: Valentina Noemi Madia and Davide Ialongo

Received: 16 May 2024
Revised: 28 May 2024
Accepted: 30 May 2024
Published: 3 June 2024

Copyright: © 2024 by the authors. Licensee MDPI, Basel, Switzerland. This article is an open access article distributed under the terms and conditions of the Creative Commons Attribution (CC BY) license (https://creativecommons.org/licenses/by/4.0/).

1. Introduction

Heterocyclic compounds are some of the best-known and most important structural components of drugs. Of these, nitrogen-containing heterocycles are particularly important. As the FDA (Food and Drug Administration) data show, 59% of all unique small-molecule drugs contain a nitrogen atom, and it should also be noted that 4 of the 10 most commonly used nitrogen heterocycles also contain a sulfur atom [1]. Phenothiazine is considered to be the third most commonly used six-membered nonaromatic nitrogen heterocycle and is present in 16 unique small-molecule drugs with various effects ranging from antihistaminic, sedative, and antipsychotic effects to anti-neurodegenerative (i.e., Parkinson's and Alzheimer's diseases) effects [1,2].

Classical phenothiazines, mainly used as neuroleptics, are substituted at position 10 with dialkylaminoalkyl groups and additionally at position 2 with small groups. These substances have significant neuroleptic, antiemetic, antihistaminic, antipruritic, analgesic, and anthelmintic effects. Continuing research in new directions on the activity of neuroleptic phenothiazines and on the modification of their structures provides information on their anticancer, antiviral (including anti-SARS-CoV-2), antibacterial, and anti-inflammatory activities, and the reversal of multidrug resistance. These substances have antioxidant and antihyperlipidemic effects [3–10].

The lipophilicity of medicinal substances has a significant impact on their ADMET parameters (absorption, distribution, metabolism, excretion, toxicity), which refer to absorption, distribution, metabolism, elimination, and toxicity.

The assessment and consideration of the lipophilicity of medicinal substances are important during drug design, as it can have a significant impact on their pharmacokinetic properties and toxicity [11]. Medicinal substances with moderate lipophilicity tend to be better absorbed through cell membranes, which can affect their rate and absorption efficiency from the gastrointestinal tract or through the skin. Lipophilic substances can more easily penetrate cell membranes and migrate to lipid-rich tissues, which can affect their distribution in the body. Substances with increased lipophilicity may be more susceptible to metabolism in the liver through oxidation, reduction, and conjugation reactions. The impact of lipophilicity on metabolism can have consequences for pharmacological activity and toxicity. The lipophilicity of drugs can influence their excretion, as substances with increased lipophilicity can be stored in fatty tissues and exhibit a prolonged presence in the body. The lipophilicity of drugs may be related to their toxicity, as this can affect their accumulation in tissues and interactions with receptors and proteins in the body [12–17]. Therefore, to better understand the behavior of biologically active compounds, including new drug candidates, their lipophilic properties should be assessed. Theoretical and experimental methods are commonly used to describe the lipophilicity of compounds. Calculation methods are used to estimate the lipophilicity parameter quantified as P (partition coefficient) or its decimal logarithm (logP). The extensive development of chemoinformatics has an influence on the number of programs available for the online prediction (in silico) of this important parameter and other ADMET properties, and such platforms include ADMETlab, pkCSM platform, SwissADME, and MetaTox [18]. Calculation approaches are useful to rapidly predict logP values, especially during the early stages of drug development; thus, next, these values should be complemented with experimental data.

Among the experimental techniques, the classic shake-flask method and liquid chromatography play an important role in determining lipophilicity. The lipophilicity chromatographic parameters (R_{M0}) that are obtained by RP-TLC (reversed-phase thin-layer chromatography) and $logk_0$ and assessed by reversed-phase high-performance liquid chromatography (RP-HPLC) are commonly used to assess the lipophilic nature of compounds [19]. The standard shake-flask procedure recommended by the Organization for Economic Co-operation and Development involves the direct measurement of the partition coefficient [20]. It allows for the accurate measurement of the logP values in the range of −2 to 4 but requires relatively large amounts of pure compounds compared to other methods. The main disadvantage of this method is that it is time-consuming and requires the control of many parameters affecting the equilibrium state of the tested system, usually lasting from 1 h to 24 h [21]. Therefore, currently, most lipophilicity tests are conducted by means of chromatographic techniques. Chromatographic approaches in reversed-phase systems (RP-TLC and RP-HPLC) are the most widely used indirect methods to experimentally determine lipophilicity. Both of these chromatographic methods need a smaller amount of sample and a relatively shorter time for analysis compared to the classical shake-flask method. The obtained results are repeatable, and the accuracy of the partition coefficient values can be within ±1 unit in relation to the shake-flask value [22].

A comprehensive review of chromatographic procedures dedicated to the determination of the lipophilicity parameters of different drug substances as an essential tool in medicinal chemistry was performed by Soares and co-workers [23]. Taking into account the importance of lipophilicity parameters as key factors in drug chemistry, namely in the design of new drugs, the aim of this study was to assess the lipophilicity of a newly synthesized group of diquinothiazines by means of both the RP-TLC and calculation methods.

All diquinothiazines that were the subject of this study were tested early for their antiproliferative activity using cultured glioblastoma SNB-19, colorectal carcinoma Caco-2, breast cancer MDA-MB-231, and lung cancer A549 cell lines and NHDF normal fibroblasts [24]. They can therefore be considered as bioactive compounds. Most of the compounds were very active against at least one cancer cell line, with an IC_{50} value < 3 µM being more active than cisplatin. The most tested diquinothiazines showed higher activity against the A549 lung cancer cell line. Compounds **1–3**, **5–8**, and **11–14** were very active against all cancer cells. As was stated, the most active were the dimethylaminopropyldiquinothiazine **2** against the A549 cell line and the pyrrolidinylethyldiquinothiazine **8** against the SNB-19 cell line, with an IC_{50} value of 0.3 µM. The mechanism of the antiproliferative effect was examined using the RT-QPCR method. It caused a significant reduction in CDKN1A expression in the MDA-MB-231, A549, and SNB-19 tumor lines. Compound **8** markedly reduced the expression of *BCL-2* in A549 and SNB-19 and the expression of *BAX* in cancer cell lines [24].

In another study, the diquinothiazine **2** demonstrated significant in vitro anticancer activity against the human lung carcinoma A549 and non-small lung carcinoma H1299 lines and protective potential for the healthy cell lines BEAS-2B and NHDF. Using the 72 h MTT, strong cytotoxic activity was observed in the viability test (Promega). The weak lethal effect observed in NHDF or BEAS-2B cells at IC_{50} doses against A549 or H1299 cells confirms promising cancer selectivity. The cell cycle revealed that substance **2** activated the necrosis phase [25].

Continuing from our previous studies, the purpose of this work was to determine the lipophilicity parameters of fifteen newly developed anticancer, angularly condensed diquinothiazines, **1–15**, with pharmacophore dialkylaminoalkyl substituents using combined computational and chromatographic approaches as $logP_{calcd}$, R_{M0}, and $logP_{TLC}$ (Figure 1). The full spectral characteristics of these compounds and their synthesis have been well described previously [24]. The current work aimed to discuss the influence of the nature of the substituents and the method of condensation of rings in a five-ring molecule system on the value of lipophilicity indices, determined by both calculation and RP-TLC methods, as well as on other drug-likeness and ADME properties predicted by in silico studies that are key for describing the pharmacokinetic behavior of these drugs.

The usefulness of the RP-TLC technique as well as logP predictions using different computational software for the design of promising drug candidates belonging to the studied diquinothiazines with pharmacophore dialkylaminoalkyl substituents was evaluated.

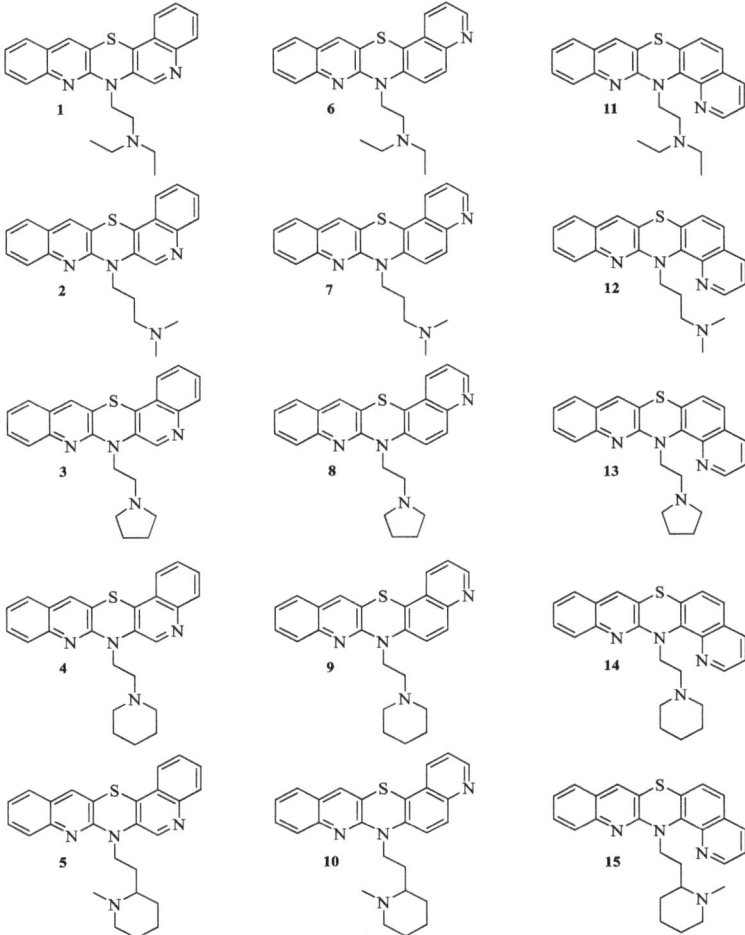

Figure 1. Structures of angularly condensed *N*-dialkyloaminoalkylodiquinothiazines **1–15**.

2. Results

2.1. Lipophilicity Studies

Both the computational and chromatographic values of the lipophilicity parameters logP$_{calcd}$, R$_{M0}$, and logP$_{TLC}$ were determined for fifteen 7- and 14-substituted angularly condensed diquinothiazines with pharmacophore dialkylaminoalkyl substituents on the thiazine nitrogen atom. The tested diquinothiazines were divided into three groups, differing in the way the quinoline rings are connected to the 1,4-thiazine ring: 7-substituted diquino[3,2-b;3′,4′-e]thiazines, **1–5**; 7-substituted diquino[3,2-b;6′,5′-e]thiazines, **6–10**; and 14-substituted diquino[3,2-b;8′,7′-e]thiazines, **11–15** (Figure 1).

Firstly, the computational method was chosen to determine the lipophilicity parameters of the studied compounds. For this purpose, popular computational programs were used based on various mathematical algorithms available on the following platforms: SwissADME [26], pkCSM [27], Molinspiration [28], and the ChemDraw Ultra program [29].

The calculated logP values (logP$_{calcd}$) for the angularly fused diquinothiazines **1–15** are shown in Table 1 and differed depending on the substituents on the thiazine nitrogen atom, the shape of the five-ring diquinothiazine system, and on the calculation program.

Table 1. The computed lipophilicity parameters (logP$_{calcd}$) for diquinothiazines **1–15** using the following Internet databases: SwissADME [26], pkCSM [a] [27], Molinspiration [b] [28], and ChemDraw [c] [29].

No. of Compound	iLOGP	XLOGP3	WLOGP	MLOGP	SILICOS-IT	LogP [a]	milogP [b]	logP [c]	logP$_{average}$
1	4.21	5.39	5.35	4.24	4.66	5.73	5.61	5.17	5.05 (±0.60)
2	3.98	5.01	4.96	4.03	4.27	5.34	5.13	4.60	4.67 (±0.52)
3	4.09	5.14	4.72	4.24	4.51	5.48	5.26	4.81	4.91 (±0.61)
4	4.17	5.50	5.11	4.45	4.74	5.87	5.77	5.22	5.10 (±0.61)
5	4.28	5.93	5.50	4.65	4.84	6.26	6.03	5.31	5.35 (±0.71)
6	4.08	5.39	5.35	4.24	4.66	5.73	5.44	5.17	5.01 (±0.61)
7	3.83	5.01	4.96	4.03	4.27	5.34	4.96	4.60	4.63 (±0.53)
8	4.06	5.14	4.72	4.24	4.51	5.48	5.09	4.81	4.76 (±0.48)
9	4.01	5.50	5.11	4.45	4.74	5.87	5.60	5.22	5.06 (±0.63)
10	4.28	5.93	5.50	4.65	4.84	6.26	5.86	5.31	5.33 (±0.69)
11	3.99	5.39	5.35	4.24	4.66	5.73	5.14	5.17	4.96 (±0.60)
12	3.67	5.01	4.96	4.03	4.27	5.34	4.66	4.60	4.57 (±0.55)
13	4.12	5.14	4.72	4.24	4.51	5.48	4.79	4.81	4.73 (±0.45)
14	3.77	5.50	5.11	4.45	4.74	5.87	5.30	5.22	5.00 (±0.66)
15	3.97	5.93	5.50	4.65	4.84	6.26	5.55	5.31	5.25 (±0.74)

The next step of our study focused on determining the more reliable lipophilicity parameters of the tested compounds. In order to obtain the relative lipophilicity, expressed as R_{M0}, the chromatographic behavior of the fifteen tested diquinothiazines, **1–15**, was investigated under proper RP-TLC conditions. RP-18 plates were used as the stationary phase, while organic modifiers containing acetone were used as the mobile phase. The linear relationship between R_{M0} and acetone concentration was determined on the basis of Equation (2) (Table 2). In addition to this, thanks to the relationship observed between R_{M0} and the slope of these linear plots (b), the lipophilicity parameter C_0 was also determined (Table 2).

Table 2. Data for linear correlation ($R_M = R_{M0} + bC$) for compounds **1–15**.

No. of Compound	R_{M0}	b	r	C_0
1	3.45	−4.77	0.9932	0.7233
2	3.01	−4.09	0.9961	0.7359
3	3.05	−4.23	0.9916	0.7210
4	3.48	−4.89	0.9953	0.7117
5	3.64	−5.14	0.9980	0.7082
6	3.68	−4.85	0.9951	0.7588
7	3.31	−4.34	0.9938	0.7627
8	3.48	−4.60	0.9941	0.7565
9	3.69	−4.88	0.9962	0.7562
10	3.83	−5.04	0.9962	0.7599
11	3.60	−4.75	0.9927	0.7579
12	3.17	−4.25	0.9935	0.7458
13	3.23	−4.21	0.9907	0.7672
14	3.61	−4.69	0.9952	0.7697
15	3.82	−5.01	0.9942	0.7625

Then, the relative lipophilicity parameter R_{M0} was converted to an absolute value lipophilicity parameter, logP$_{TLC}$, using a calibration curve determined under the same measurement conditions for a set of standards, **I–V**, with the literature values of logP$_{lit}$ in the range of 1.21–6.38 (Table 3). The obtained values of the R_{M0} coefficient of the tested compounds were in the range of 3.01–3.83. The correlation between the logP$_{lit}$ values and the experimental R_{M0} values for standards **I–V** gave the following calibration equation:

$$\log P_{TLC} = 1.2838 \, R_{M0} + 0.2138$$

$(r = 0.9967; s = 0.1920; F = 459.32; p < 0.001)$

Table 3. The R_{M0} and $logP_{lit}$ values and b (slope) and r (correlation coefficient) values of the equation $R_M = R_{M0} + bC$ for standards **I–V**.

Lipophilicity Parameters	Standards				
	I	II	III	IV	V
$logP_{lit}$	1.21 [30]	1.87 [31]	3.18 [31]	4.45 [31]	6.38 [32]
R_{M0}	0.78	1.16	2.51	3.33	4.69
-b	0.0162	0.0247	0.0328	0.0412	0.0564
r	0.9923	0.9937	0.9971	0.9982	0.9977
$logP_{TLC}$	1.21	1.70	3.43	4.49	6.24

The standard curve equation was used to obtain the $logP_{TLC}$ parameter for compounds **1–15**, and the results are presented in Table 4.

Table 4. The experimental lipophilicity parameters ($logP_{TLC}$ values) for compounds **1–15**.

No. of Compound	$logP_{TLC}$	No. of Compound	$logP_{TLC}$	No. of Compound	$logP_{TLC}$
1	4.64	6	4.94	11	4.84
2	4.08	7	4.46	12	4.28
3	4.13	8	4.68	13	4.36
4	4.68	9	4.95	14	4.85
5	4.89	10	5.13	15	5.12

2.2. Molecular Descriptors

For all tested compounds, **1–15**, selected molecular descriptors such as molar mass (M), molar volume (V_M), molar refraction (Ref_M), and surface area were calculated to check how they correlate with the experimentally determined lipophilicity parameter R_{M0} (Table 5).

Table 5. The molecular descriptors for compounds **1–15**.

No. of Compound	Molar Mass (M) [g/mol]	Molar Volume (V_M) [cm^3]	Molar Refractivity (Ref_M) [cm^3/mol]	Surface Area [Å]
1	400.55	322.1	121.077	175.12
2	386.52	305.6	116.446	168.75
3	398.54	304.6	119.121	174.11
4	412.56	322.4	123.722	180.48
5	426.59	342.3	128.258	186.84
6	400.55	322.1	121.077	175.12
7	386.52	305.6	116.446	168.75
8	398.54	304.6	119.121	174.11
9	412.56	322.4	123.722	180.48
10	426.59	342.3	128.258	186.84
11	400.55	322.1	121.077	175.12
12	386.52	305.6	116.446	168.75
13	398.54	304.6	119.121	174.11
14	412.56	322.4	123.722	180.48
15	426.59	342.3	128.258	186.84

2.3. In Silico ADME Prediction

Physicochemical parameters used to predict drug-like properties, i.e., Lipinski's rule of five and the Ghose, Veber, Egan, and Muegge rules, were also calculated (Table 6).

Table 6. The drug-likeness and ADME properties predicted by in silico studies using SwissADME.

Predicted Parameter	Compound No.				
	1, 6, 11	2, 7, 12	3, 8, 13	4, 9, 14	5, 10, 15
Physicochemical Properties					
Num. heavy atoms	29	28	29	30	31
Num. arom. heavy atoms	20	20	20	20	20
Hydrogen bond acceptors	3	3	3	3	3
Hydrogen bond donors	0	0	0	0	0
Number of rotatable bonds	5	4	3	3	3
Topological polar surface area [Å^2]	57.56	57.56	57.56	57.56	57.56
Drug-Likeness Prediction					
Rule of Lipinski	+	+	+	+	+
Rule of Ghose	+	+	+	−(MR>130)	−(MR>130)
Rule of Veber	+	+	+	+	+
Rule of Egan	+	+	+	+	+
Rule of Muegge	−(XLOGP3>5)	−(XLOGP3>5)	−(XLOGP3>5)	−(XLOGP3>5)	−(XLOGP3>5)
Bioavailability					
Bioactivity score	0.55	0.55	0.55	0.55	0.55

ADME parameters were obtained using the pkCSM and PreADMET servers (Tables 7–9). As shown in Tables 7–9, the compounds studied showed significant differences in the molecular descriptors as well as in their ADME parameters.

Table 7. The absorption descriptors for compounds **1–15**.

No. of Compound	Water Solubility [log mol/L]	Caco-2 Permeability [log Papp in 10^{-6} cm/s]	Intestinal Absorption [% Absorbed]	Skin Permeability [log Kp]
1	−5.871	1.003	92.241	−2.697
2	−5.783	1.028	92.931	−2.701
3	−4.551	1.145	92.725	−2.733
4	−4.660	1.143	92.337	−2.733
5	−5.820	1.021	92.309	−2.711
6	−5.793	1.032	95.220	−2.694
7	−5.737	1.057	96.401	−2.700
8	−4.517	1.226	94.349	−2.723
9	−4.652	1.224	93.960	−2.723
10	−5.829	1.049	95.779	−2.712
11	−5.373	1.043	95.329	−2.691
12	−5.263	1.068	96.510	−2.697
13	−3.965	1.211	93.469	−2.744
14	−4.061	1.209	93.080	−2.743
15	−5.329	1.060	95.888	−2.709

Table 8. The distribution descriptors for compounds **1–15**.

No. of Compound	VDss [log L/kg]	Unbound Fraction [Fu]	BBB Permeability [log BB]	CNS Permeability [log PS]
1	0.986	0.259	0.478	−1.464
2	0.864	0.253	0.483	−1.402
3	0.868	0.193	0.424	−1.394
4	0.923	0.189	0.437	−1.381
5	1.041	0.255	0.535	−1.422
6	1.178	0.268	0.537	−1.493
7	1.062	0.262	0.484	−1.478
8	1.269	0.206	0.559	−1.318
9	1.328	0.200	0.572	−1.304
10	1.244	0.257	0.536	−1.498
11	1.332	0.267	0.387	−1.475
12	1.200	0.261	0.357	−1.483
13	1.170	0.200	0.425	−1.370
14	1.222	0.196	0.438	−1.356
15	1.368	0.262	0.410	−1.503

Table 9. The excretion and toxicity for compounds **1–15**.

No. of Compound	Total Clearance [log ml/min/kg]	Max. Tolerated Dose [log mg/kg/day]	Oral Rat Acute Toxicity [mol/kg]	Oral Rat Chronic Toxicity [log mg/kg bw/day]	*Tetrahymena pyriformis* Toxicity [log µg/L]	Minnow Toxicity [log mM]
1	0.672	0.672	2.277	0.449	0.299	1.983
2	0.526	0.647	2.272	0.490	0.300	1.944
3	0.824	0.537	2.734	0.827	0.291	−0.279
4	0.779	0.550	2.756	0.848	0.291	−0.396
5	0.599	0.568	2.363	0.542	0.291	1.736
6	0.714	0.293	2.361	0.773	0.327	0.423
7	0.584	0.258	2.312	0.807	0.331	0.597
8	0.832	0.211	3.157	1.074	0.300	−0.872
9	0.787	0.228	3.184	1.096	0.299	−0.989
10	0.657	0.188	2.413	0.606	0.304	0.389
11	0.707	0.594	2.513	0.590	0.299	0.542
12	0.562	0.557	2.485	0.610	0.300	0.460
13	0.826	0.543	2.949	1.020	0.292	−0.848
14	0.781	0.557	2.972	1.041	0.291	−0.965
15	0.634	0.483	2.588	0.663	0.291	0.252

2.3.1. Absorption

The potential absorption of 15 tested angularly condensed diquinothiazines, **1–15**, was evaluated using the parameters of water solubility, Caco-2 cell permeability (human colon adenoma cells), absorption in the human intestine, and skin permeability, obtained using the pkCSM platform. The obtained results are summarized in Table 7. Water solubility was measured using the logS parameter (S—solubility, expressed in mol/L). When it comes to the absorption of administered drugs, the most frequently used method to determine this parameter is to test the permeability of potential drugs through a monolayer of Caco-2 cells. This is due to the similarity in structure and function of Caco-2 cells to the human intestinal epithelium. Since the main site of absorption of an orally administered drug is usually the intestine, it is important to determine the amount of the compound that is absorbed here. This method predicts the percentage of the compound that was absorbed. The skin permeability parameter is also significant for the effectiveness of some products. This parameter is expressed as logKp, and a compound is considered to have relatively low skin permeability if logKp > −2.5 [33,34].

2.3.2. Distribution

The potential distribution of the tested diquinothiazines **1–15** was assessed using the parameters of VDss, fraction unbound, BBB permeability, and CNS permeability, obtained using the pkCSM platform. The results obtained are summarized in Table 8. The volume of distribution (VDss) provides an indication of the distribution of the drug in the body and is a pharmacokinetic parameter representing the volume into which the dose of drug would have to be distributed to give rise to the same concentration observed in the blood plasma. A low VDss indicates high water solubility or high plasma protein binding because more of the drug remains in the plasma; a high VDss suggests significant concentration in tissues, for example, due to tissue binding or high lipid solubility [35,36]. According to the model used in the pkCMS software, VDss is considered low if log VDss < −0.15 and high if log VDss > 0.45.

Fu (unbound fraction) is also an important pharmacokinetic parameter because it affects various factors of drug effectiveness and side effects (including glomerular filtration in the kidneys, total clearance, and hepatic metabolism). Therefore, it is important to accurately predict Fu during drug development. Unbound drugs in plasma may exhibit pharmacological activity by interacting with targets such as proteins, enzymes, receptors, and channels; hence, the plasma unbound fraction (Fu) of a drug is an important factor in determining drug efficacy [37].

When considering a substance as a drug candidate, it is also important to determine the extent to which it will cross the blood–brain barrier. This parameter is measured in vivo as logBB, the logarithmic ratio of brain to plasma drug concentrations. According to the computational model used in pkCSM, if the value of the logBB parameter is greater than 0.3, the substance crosses the blood–brain barrier, while if logBB is lower than −1, the substance is distributed to the brain to a small extent [38].

2.3.3. Excretion and Toxicity

The potential excretion and toxicity of the tested diquinothiazines **1–15** were evaluated using the parameters of total clearance, maximum tolerated dose, acute oral toxicity of rats, *Tetrahymena pyriformis* toxicity, and minnow toxicity, obtained using the pkCSM platform. The results obtained are summarized in Table 9. The expected total clearance, expressed as log (ml/min/kg), is the volume of plasma completely cleared of the drug per unit of time by the organ eliminating the drug from the body. Knowledge of this parameter is necessary to determine the maintenance dose of the drug [39]. The maximum recommended tolerated dose (MRTD) allows us to initially determine the toxic dose of a chemical substance to humans. The model used in pkCSM was developed using 1222 experimental data points from human clinical trials. The obtained results of the MRTD parameter values are given as log (mg/kg/day). Calculating this parameter is helpful in determining the maximum recommended starting dose for drugs under investigation. If the calculated value is less than or equal to 0.477 log (mg/kg/day), it is considered low, and if it is higher, it is considered high. A model based on *Tetrahymena pyriformis*, a protozoal bacterium whose toxicity is often considered a toxic endpoint, was used to determine in silico toxicity in the pkCSM program. This method was built using the concentrations of 1571 compounds required to inhibit 50% of growth. This parameter is designated as $pIGC_{50}$ (negative logarithm of the concentration required to inhibit 50% growth in log μg/L), and a compound can be assessed as toxic if the value of this parameter is calculated to be greater than −0.5 log μg/L. The minnow toxicity parameter is based on measurements of the LC_{50} parameter, i.e., the concentration of the substance necessary to cause the death of 50% of flathead minnows, which were used as an animal model in this study. The model was based on tests of the LC_{50} parameter for over 550 substances. According to this model, an LC_{50} value below 0.5 mM ($logLC_{50} < -0.3$) is considered to indicate high acute toxicity.

3. Discussion

It is known that lipophilicity is one of the most frequently studied physicochemical parameters of new substances that are drug candidates [19,40]. It is most often defined to support quantitative structure–activity relationships (QSARs), including absorption, distribution in tissues, and drug transport across the biological barrier [41–43]. This parameter is characterized by the distribution of a dissolved substance in two-phase liquid–liquid or solid–liquid systems. In silico methods have also been proposed to assess lipophilicity. Several programs have been developed to calculate the value of the logP parameter [26–29]. It is still important to use experimental methods because most programs determining the logP parameter calculate lipophilicity using atomic methods that do not take into account some structural parameters. There may be discrepancies between the calculated data and those determined experimentally, which can be explained by the fact that newly synthesized drug candidates may contain substructures or heterocycle systems that are not covered in the software development training set [44,45].

The present research on lipophilicity began with in silico analyses using various computer programs available on the SwissADME [26], pkCSM [27], and Molinspiration [28] platforms and the ChemDraw program [29]. These platforms use various mathematical modules described on the website of the above-mentioned servers. The obtained results of the calculated lipophilicity are within a wide range of values. This is most likely due to differences in the calculation models. The comparison of the chromatographic parameters (logP) of the tested angularly condensed N-dialkylaminoalkyldiquinothiazines **1–15** is shown in Figure 2. As observed (Table 1), the lowest value of the logP$_{calc}$ parameter was obtained for the 14-(3′-dimethylaminopropyl)diquino[3,2-b;8′,7′-e]thiazine **12** according to calculations with the iLOGP program (SwissADME) and the highest one was obtained for all three isomeric diquinothiazines **5**, **10**, and **15** with a piperidinylethyl substituent based on calculations with the logP program (pkCSM). This program also calculated the same values of the logP parameter for the remaining isomers (i.e., for diquinothiazines **1**, **6**, and **11**, logP = 5.17; for **2**, **7**, and **12**, logP = 4.60; for **3**, **8**, and **13**, logP = 4.81; and for **4**, **9**, and **14**, logP = 5.22).

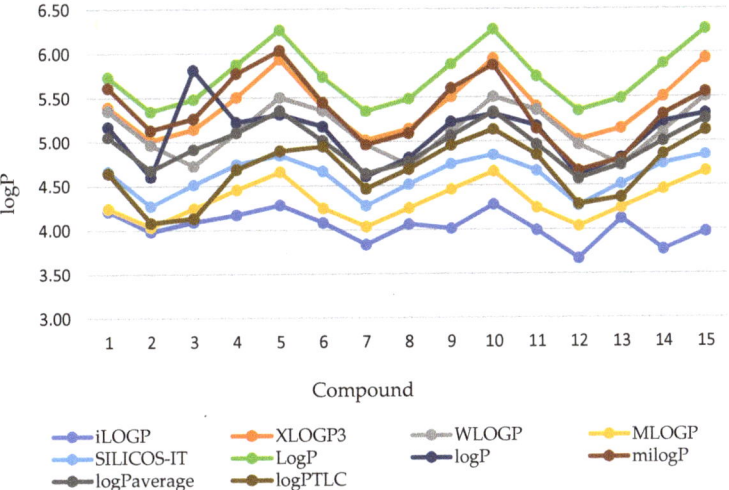

Figure 2. Comparison of the theoretical lipophilicity parameters (the logP values) of the examined angularly condensed N-dialkyloaminoalkyldiquinothiazines **1–15**.

In order to compare all obtained theoretical values of the partition coefficient of the examined drugs expressed in the form of logP, and to then estimate the lipophilic character of the studied N-dialkyloaminoalkyldiquinothiazines, a chemometric approach, i.e., cluster analysis (CA) with Euclidean distance, was conducted. The results are presented in Figures 3 and 4.

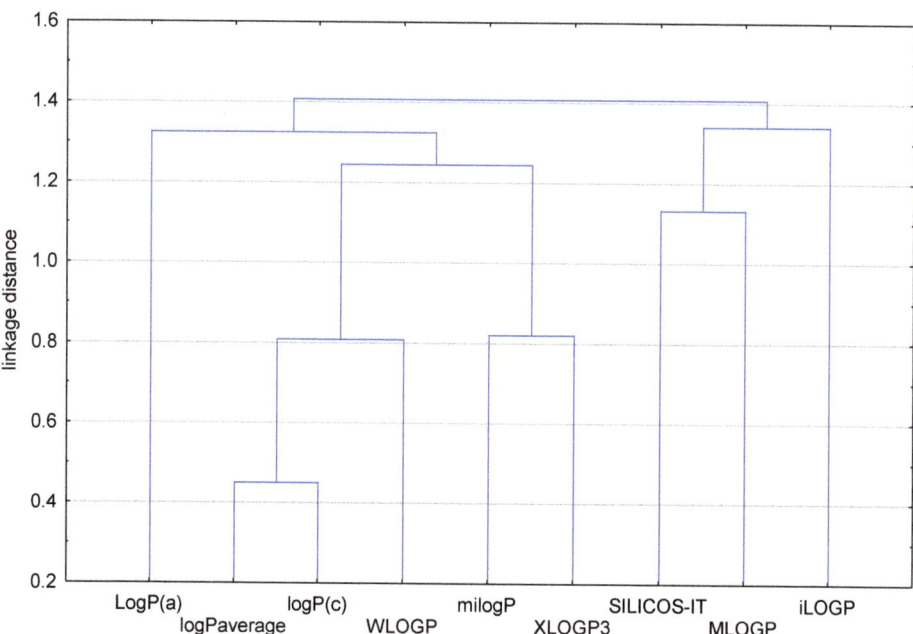

Figure 3. Dendrogram of partition coefficients of the examined angularly condensed *N*-dialkyloaminoalkyldiquinothiazines **1–15**.

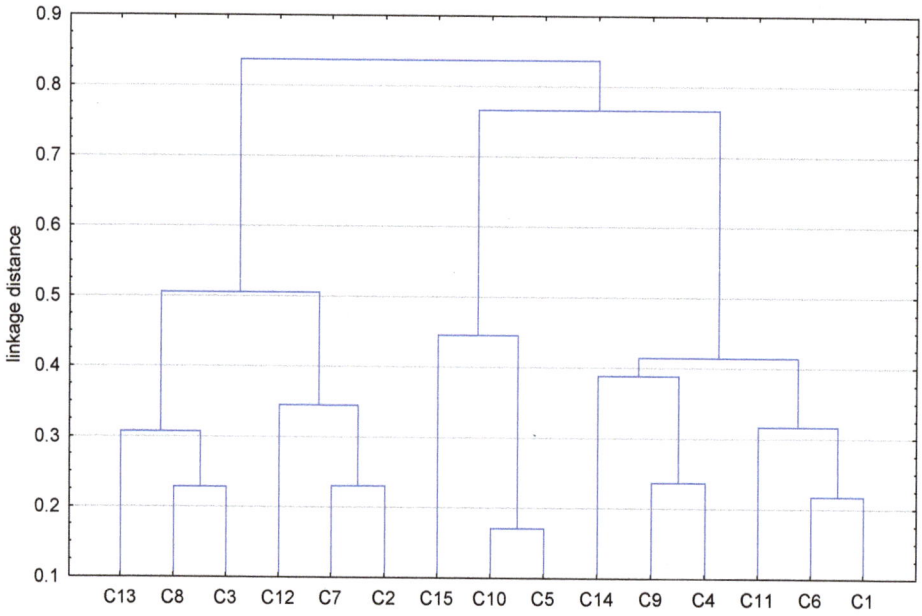

Figure 4. Dendrogram of the examined angularly condensed *N*-dialkyloaminoalkyldiquinothiazines **1–15** based on their partition coefficients.

Figure 3 shows the dendrogram of all the theoretical lipophilicity parameters of the examined angularly condensed N-dialkyloaminoalkyldiquinothiazines **1–15**. The first cluster contains the partition coefficients $LogP^a$, $logP_{average}$, $logP^c$, WLOGP, milogP, and XLOGP3. In the second cluster, the rest of the computed partition coefficients are placed: SILICOS-IT, MLOGP, and iLOGP. The reason for these differences is the prediction power of each software. This grouping of partition coefficient values shows the biggest similarity (the smallest distance on the dendrogram) of $logP^c$ with the $logP_{average}$ value calculated on the basis of all theoretical logP values. Therefore, the parameter $logP^c$, may be a good alternative for estimation of the lipophilicity of the examined group of N–dialkyloaminoalkyldiquinothiazines.

Figure 4 shows the similarity analysis of the examined compounds based on their partition coefficients. According to the computed lipophilicity parameters (the logP values), the studied compounds can be divided into two groups based on variation in their structural particularities, i.e., depending on the kind of substituent that may influence the lipophilic character of these compounds. The first group consists of the compounds C13, C8, and C3 in the first subgroup, with C12, C7, and C2 in the second subgroup. Similarly, the second group is divided into two smaller visible subgroups, with C15, C10, and C5 in the first subgroup and C14, C9, C4, C11, C6, and C1 in the second subgroup.

The cluster analysis shown in Figure 4 confirms the greatest similarity in lipophilic character of derivatives C10 and C5.

In the next steps of this study, the calculated logP values using different computational methods, summarized in Table 1, were correlated with the experimental lipophilicity parameter R_{M0}, presented in Table 2. The correlation coefficients and the equations obtained for the correlation of the R_{M0} parameter with the logP values obtained by means of an appropriate program are summarized in Table 10, which shows the strongest relationships between logP and R_{M0}.

In order to obtain the linear equations listed in Table 10, RP-TLC was chosen as the method for experimental determination of the lipophilicity parameter of the tested diquinothiazines **1–15**. Using this technique, the relative lipophilicity parameter R_{M0} was obtained. It was then converted to the relative parameter $logP_{TLC}$ (as described in Sections 2 and 4). For all investigated derivatives, in a wide range of organic modifier concentrations in the mobile phase, high values of correlation coefficients ($r = 0.99$) made it possible to determine the lipophilicity parameter R_{M0} by extrapolation (Table 2). The analysis of the chromatographic parameter R_{M0} of the examined angularly condensed N-dialkyloaminoalkyldiquinothiazines **1–15** presented in Table 2 shows that they are relatively smaller compared to the theoretical logP values obtained by means of different computer software and their mean value ($logP_{average}$). Generally, a difference in the obtained R_{M0} within one unit to logP was noted in the case of all chromatographic descriptors. This fact confirms that calculation approaches are useful only for the rapid prediction of logP values during the first stage of study of new drug candidates, after which they should be complemented with experimental data such as chromatographic descriptors. Of all the algorithms, iLogP showed the biggest similarity to the chromatographic value, especially for compounds **2, 7, 11, 12, 14**, and **15**. This shows the potential utility of iLogP for the rapid estimation of the lipophilic character of these compounds.

In the next stage, satisfactory relationships between the R_{M0} values and b (slope) values of the linear equations allowed the calculation of the next lipophilicity parameter, namely C_0. The results of these data confirm that all compounds studied belong to a congeneric group. However, as shown in Figure 5, the dissimilarity of the C_0 values with the partition coefficients as well as the chromatographic parameter (R_{M0}) indicates that both parameters cannot be fully replaced by C_0.

Table 10. Linear correlations between the partition coefficients and the chromatographic parameters of lipophilicity of the tested compounds **1–15** ($p < 0.05$).

Compounds	Lipophilicity Parameter	Equation	r
1–5	iLOGP	iLOGP = 0.3877 R_{M0} + 2.8566	0.9403
1–15	XLOGP3	XLOGP3 = 1.0578 R_{M0} + 1.7233	0.8426
1–5		XLOGP3 = 1.1972 R_{M0} + 1.4122	0.9384
6–10		XLOGP3 = 1.6538 R_{M0} − 0.5565	0.9427
11–15		XLOGP3 = 1.2383 R_{M0} + 1.0773	0.9577
1–15	WLOGP	WLOGP = 0.8350 R_{M0} + 2.2304	0.7683
1–5		WLOGP = 0.9794 R_{M0} + 1.8705	0.8868
11–15		WLOGP = 0.9985 R_{M0} + 1.6473	0.8919
1–15	MLOGP	MLOGP = 0.6500 R_{M0} + 2.0664	0.7840
6–10		MLOGP = 1.0545 R_{M0} + 0.5279	0.9101
11–15		MLOGP = 0.7511 R_{M0} + 1.7035	0.8795
1–15	SILCOS-IT	SILCOS-IT = 0.6630 R_{M0} + 2.3034	0.8486
1–5		SILCOS-IT = 0.7442 R_{M0} + 2.1287	0.9374
6–10		SILCOS-IT = 1.0799 R_{M0} + 0.7184	0.9892
11–15		SILCOS-IT = 0.7588 R_{M0} + 1.9588	0.9432
1–15	LogP	LogP = 1.0697 R_{M0} + 2.0243	0.8484
1–5		LogP = 1.2120 R_{M0} + 1.7050	0.9460
6–10		LogP = 1.6739 R_{M0} − 0.2866	0.9500
11–15		LogP = 1.2499 R_{M0} + 1.3788	0.9624
6–10		logP = 1.4693 R_{M0} − 0.2644	0.9877
11–15		logP = 1.0652 R_{M0} + 1.3086	0.9714
1–15	milogP	milogP = 0.9755 R_{M0} + 1.9610	0.6499
1–5		milogP = 1.2903 R_{M0} + 1.2684	0.9813
6–10		milogP = 1.7585 R_{M0} − 0.9371	0.9725
11–15		milogP = 1.3037 R_{M0} + 0.5431	0.9867
1–15	logP$_{average}$	logP$_{average}$ = 0.7369 R_{M0} + 2.4017	0.7854
1–5		logP$_{average}$ = 0.8307 R_{M0} + 2.2530	0.9292
6–10		logP$_{average}$ = 1.3129 R_{M0} + 0.2343	0.9791
11–15		logP$_{average}$ = 0.9300 R_{M0} + 1.6601	0.9819

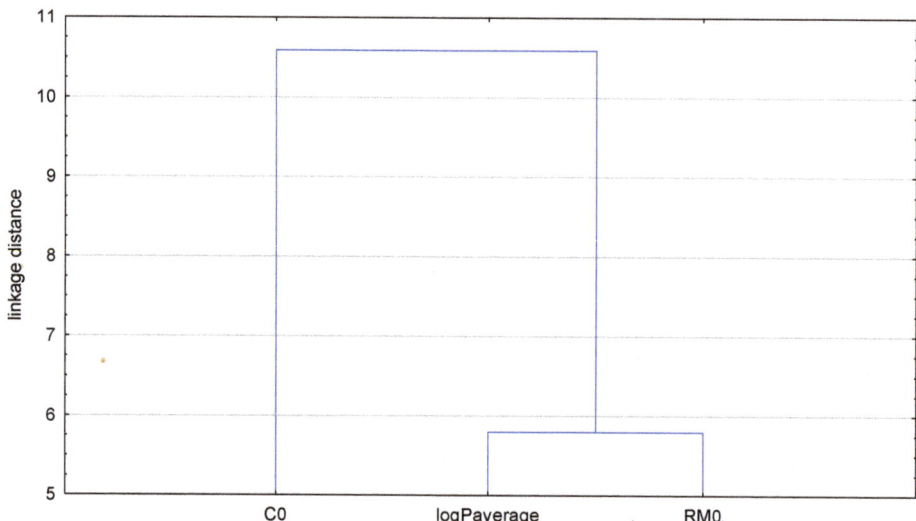

Figure 5. Dendrogram of theoretical (logP$_{averge}$) and chromatographic parameters (R_{M0}, C_0) of examined angularly condensed *N*-dialkyloaminoalkyldiquinothiazines **1–15**.

The linear equations obtained between the theoretical partition coefficients and chromatographic parameters of lipophilicity of the tested compounds **1–15** (Table 10) confirmed the usefulness of iLOGP, XLOGP3, WLOGP, MLOGP, LOGP, miLOGP, logP$_{average}$, and SILICOS-IT for predicting the experimental value of this parameter in the form of R_{M0}.

Analyzing the experimentally obtained values of the logP$_{TLC}$ parameter, it can be observed that in each of the three groups of isomeric angularly condensed diquinothiazines, the lowest lipophilicity was found in derivatives with a dimethylaminopropyl substituent (**2**, **7**, and **12**), while the highest was found in derivatives with an N-methylpiperidine substituent (**5**, **10**, and **15**). Comparing the three tested groups of isomeric angularly condensed diquinothiazines, it can be concluded that the compounds with the 7-substituted diquino[3,2-b;3′,4′-e]thiazine structure, **1–5**, are characterized by the lowest lipophilicity, while those with the 7-substituted diquino[3,2-b;6′,5′-e]thiazine structure, **6–10**, show the highest values of the logP$_{TLC}$ parameter.

Selected molecular descriptors were also calculated for all tested diquinothiazines. The values are presented in Table 5. As shown by the results in Table 5, the same values of these parameters were obtained for isomeric angularly condensed diquinothiazines with the same substituents. The results obtained correlated with the experimentally obtained R_{M0} parameter (Table 11). Good correlations were obtained for all descriptors ($r \geq 0.8$). Correlations in subgroups of individual isomers resulted in improved correlations.

Table 11. Linear correlations between experimentally determined R_{M0} and the molecular descriptors' values ($p < 0.05$).

Compounds	Molecular Descriptor	Equation	r
1–15	Molar mass	M = 42.779 R_{M0} + 256.508	0.8000
1–5		M = 47.859 R_{M0} + 245.773	0.8807
6–10		M = 68.869 R_{M0} + 157.161	0.9216
11–15		M = 49.570 R_{M0} + 232.152	0.8999
1–15	Molar volume	V_M = 45.782 R_{M0} + 160.711	0.8444
1–5		V_M = 52.381 R_{M0} + 145.180	0.9518
6–10		V_M = 69.514 R_{M0} + 69.288	0.9182
11–15		V_M = 53.970 R_{M0} + 131.262	0.9674
1–15	Molar refractivity	Ref$_M$ = 13.202 R_{M0} + 75.913	0.8311
1–5		Ref$_M$ = 14.871 R_{M0} + 72.263	0.9213
6–10		Ref$_M$ = 20.942 R_{M0} + 46.375	0.9434
11–15		Ref$_M$ = 15.363 R_{M0} + 68.170	0.9389
1–15	Surface area	Surface area = 19.960 R_{M0} + 109.882	0.8017
1–5		Surface area = 21.667 R_{M0} + 105.000	0.8831
6–10		Surface area = 31.142 R_{M0} + 65.010	0.9229
11–15		Surface area = 22.437 R_{M0} + 98.843	0.9021

In Silico ADME Prediction

Potential drug candidates usually have similar physicochemical properties. Therefore, based on analysis of the simple molecular properties of existing drugs and/or drug candidates, simple filters defining acceptable limits for these properties are being developed. In 2007, Lipinski proposed the 'Rule of Five' [46], the most famous drug similarity filter which defines four rules that determine whether a molecule is well absorbed or not after oral administration. These parameters include molecular weight (MW $\leq$ 500), octanol/water partition coefficient (ClogP $\leq$ 5), number of hydrogen bond donors (HBD $\leq$ 5), and number of hydrogen bond acceptors (HBA $\leq$ 10). If the relationship violates two or more bases, it may not be active when administered orally. Data in the literature indicate that 85.4% of FDA-approved drugs meet the rule of five [47]. The results obtained on the SwissADME platform show that all tested diquinothiazines meet Lipiński's rule.

Since then, various principles have been developed similar to the Rule of Five. For example, after parsing a file of 6304 molecules in a database, Ghose et al. found that over 80% of compounds from the Comprehensive Medicinal Chemistry (CMC) database meet the following qualification ranges: AlogP between −0.4 and 5.6, MW between 160 and 480, molar refractive index from 40 to 130, and integer atoms from 20 to 70 [48]. From the current group of 15 tested diquinothiazines, according to the SwissADME platform, derivatives with piperidine (**4**, **9**, and **14**) and *N*-methylpiperidine (**5**, **10**, and **15**) substituents do not meet this rule because their molar refractive index is greater than 130. However, the value of this parameter obtained with the program Chem3D is less than 130 (Table 5).

All tested substances, **1–15**, also meet Veber and Egan's rules. According to Veber's rule, compounds that satisfy only the following two criteria will most likely have good oral bioavailability in rats: have 10 or fewer rotatable bonds and a polar surface area equal to or less than 140 Å^2 (or 12 or fewer hydrogen bond donors and acceptors). This may be because a reduced polar area correlates better with an increased permeation rate than lipophilicity, and an increased number of rotatable bonds has a negative effect on the permeation rate [49,50]. In turn, Egan's rule is based on two descriptors: PSA and AlogP98.

It has been shown that molecular weight, although often used in passive absorption models, is unnecessary because it is already a component of both PSA and AlogP98. Following extensive model validation on hundreds of known orally administered drugs, "drug-like" molecules that have been tested for Caco-2 cell permeability have shown a good rate of successful predictions (74–92%) [51]. However, none of the tested substances meet Muegge's rule due to the calculated XLOGP value being greater than 5.

The bioavailability score calculated using the SwissADME platform for all tested substances was 0.55. Substances with a bioavailability score ≥ 0.55 are considered very well absorbed by the body. A bioavailability score of 0.55 and a TPSA score between 60 and 140 Å^2 indicate optimal absorption [52].

When developing new drugs, it is also important to determine and evaluate ADME properties (absorption, distribution, metabolism, and excretion). Poor ADME properties are a significant cause of failure and increased drug design costs [53]. It was considered acceptable to determine these parameters using in silico methods at the early stages of substance evaluation for drug suitability. This allows, without incurring high research costs, to eliminate compounds with undesirable ADME properties at an early stage of research [54]. For the discussed substances, selected parameters responsible for absorption (water solubility, Caco-2 permeability, intestinal absorption, and skin permeability), distribution (VDss, unbound fraction, BBB permeability, and CNS permeability), and excretion and toxicity (total clearance, max. tolerated dose, oral rat acute toxicity, oral rat chronic toxicity, *Tetrahymena pyriformis* toxicity, and minnow toxicity) were also determined in silico. The pkCSM platform was used to determine these parameters. All 15 tested compounds are characterized by poor solubility in water due to their chemical structure (five six-membered fused rings and no hydrophilic groups). The obtained parameter values range from −5.871 for diquinothiazine **1** to −3.965 for diquinothiazine **13**. Caco-2 permeability obtained using the pkCSM platform is given as the logarithm of the apparent permeability coefficient (log Papp). Compounds with a predicted log Papp at 10^{-6} cm/s greater than 0.9 are considered to have high Caco-2 permeability. According to the analysis results, all the synthesized compounds show high Caco-2 cell permeability (1.003–1.226). The calculated absorption values in the human intestine show that all compounds have a similar, very high probability of intestinal absorption (92.24–96.51%). The lowest values were obtained for derivatives with the 7-substituted diquino[3,2-b;3′,4′-e]thiazine structure, i.e., **1–5**. For the other two groups of isomers, the values were similar, which shows the influence of the shape of the molecule on this parameter. The calculated skin permeability values are in the range of −2.744 to −2.694 and indicate poor permeability (Table 7). Almost all tested angularly condensed diquinothiazines can be substrates for P-glycoprotein (except for substance **1**), and all of them are predicted to inhibit P-glycoproteins I and II (Table S1).

VDss is a pharmacokinetic parameter that determines half-life and represents the degree of drug distribution in tissues. The predicted log VDss values ranged from 0.86 to 1.37, which means that the compounds have a high constant distribution in plasma and tissues. The unbound fraction in plasma (Fu) is also an important pharmacokinetic parameter which determines the amount of drug that is "free" in the plasma, and therefore the fraction capable of diffusing from the plasma into tissues. From the results obtained for the tested diquinothiazines **1–15**, the unbound fraction had low values ranging from 0.189 to 0.268. The BBB permeability parameter was also determined using the pkCSM platform. The blood–brain barrier controls the transfer of essential substances needed for a proper functioning brain and participates in the removal of cellular metabolites and toxins. The expected permeability through the blood–brain barrier for the tested substances ranged from 0.357 to 0.572, which means that they will pass through the BBB barrier. All tested diquinothiazines may also be able to penetrate the central nervous system because the logPS parameter values obtained for them were in the range of -1.304 to -1.503, and substances with logPS > -2 are considered to penetrate the CNS, while those with logPS < -3 do not penetrate the CNS (Table 8).

The possible interaction of the tested diquinothiazines with cytochrome P450 was also calculated. The obtained results show that all tested substances may be CYP1A2 inhibitors, and almost all of them may be CYP2D6 (except substances **3** and **4**) and CYP3A4 (except substance **2**) inhibitors; however, none of the tested compounds may be CYP2C9 inhibitors (Table S1). Although the abundance of CYP3A4 is poor, it contributes to nearly 50% of drug metabolism, so drug–drug interactions need to be considered in the subsequent development of compounds.

The excretion and toxicity parameters determined in silico are presented in Table 9. Total clearance of the tested diquinothiazines ranged from 0.526 to 0.832. Total clearance measures the efficiency of drug elimination from the entire body and is useful in determining the rate of drug dosing. The maximum tolerated dose value for the tested diquinothiazines was calculated in the range from 0.188 to 0.672. A dose equal to or less than 0.477 is considered low, and such values were obtained for diquinothiazines **6–10**, while dose values higher than 0.477, which are considered high, were obtained for the remaining compounds tested. The values of the parameters of oral rat acute toxicity and chronic toxicity were calculated in the range from 2.272 to 3.184 and from 0.449 to 1.096, respectively. The determined values of the *Tetrahymena pyriformis* toxicity parameter were above -0.5 (0.291–0.331), which means that the compounds may be toxic, while the minnow toxicity parameter values ranged from -0.989 to 1.944. Values of this parameter of less than -0.3 may indicate the toxicity of the substance and were calculated for diquinothiazines **4, 8, 9, 13,** and **14**. It was also checked whether the tested substances could be substrates of Organic Cation Transporter 2 (Renal OCT2). Renal OCT2 is a renal uptake transporter that plays an important role in renal clearance and the clearance of drugs and endogenous compounds. Diquinothiazines **8** and **9** may have this effect (Table S2). However, none of the substances discussed should show AMES toxicity or irritate the skin, but they may be hepatotoxic. Calculations using the pkCSM program also showed that the tested substances may be hERG II inhibitors but should not show any inhibition against hERG I (Table S1).

Absorption from the gastrointestinal tract and access to the brain are two pharmacokinetic parameters important to estimate at different stages of the drug discovery process. For this purpose, an estimated method of penetration into the brain or intestines has been developed—boiled egg [55]. It is an accurate predictive model that works by calculating the lipophilicity and polarity of small molecules. At the same time, penetration into both the brain and intestines is predicted based on the same two physicochemical descriptors. This computational method was also used to analyze the tested diquinothiazines **1–15** (Figure 6). Molecules located in the white region are expected to be passively absorbed in the gastrointestinal tract, while molecules in the "yolk" region are expected to passively cross the BBB. It can be observed that none of the analyzed compounds can be passively absorbed in the intestines, but all of them can pass through the BBB. This shows that the dosing compliance may be poor due to the inability of all the compounds to be absorbed through the intestines. Additionally, all tested compounds can become substrates for p-glycoprotein (blue points).

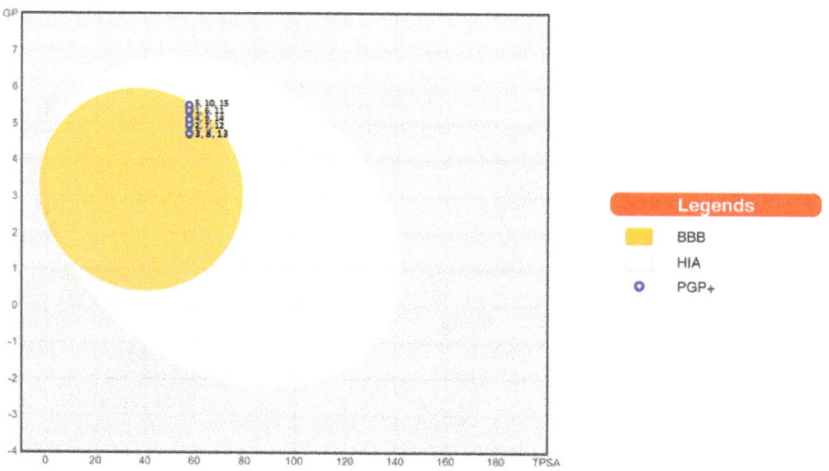

Figure 6. The boiled egg representation of the intestinal absorption and permeation through the blood–brain barrier for diquinothiazines **1–15**.

4. Materials and Methods

4.1. Solvents and Reference Standards

The studied compounds, i.e., 7-substituted diquino[3,2-b;3′,4′-e]thiazines **1–5**, 7-substituted diquino[3,2-b;6′,5′-e]thiazines **6–10**, and 14-substituted diquino[3,2-b;8′,7′-e]thiazines **11–15**, were previously designed and synthesized [24]. The chemical structures of the examined compounds are shown in Figure 1. The summary formulas, molecular weights, melting points, and physical appearances of the tested diquinothiazines **1–15** are listed in Table S3.

Analytical grade acetone (POCh, Gliwice, Poland) and TRIS (tris(hydroxymethyl)aminomethane, Fluka, Buchs, Switzerland) were used in RP-TLC studies as the mobile phase components. To prepare the calibration curve, the reference standards of five chemical compounds with the described lipophilicity parameter ($logP_{lit}$) were used: acetanilide (**I**, 1.21 [30], POCh, Gliwice, Poland), benzoic acid (**II**, 1.87 [31], POCh, Gliwice, Poland), benzophenone (**III**, 3.18 [31], Fluka, Buchs, Switzerland), anthracene (**IV**, 4.45 [31], POCh, Gliwice, Poland), and p,p′-DDT (1,1,1-trichloro-2,2-bis(4-chlorophenyl, Merck, Darmstadt, Germany), **V**, 6.38 [32]).

4.2. RP-TLC Analysis

All chromatographic experiments were performed on a silica gel 60 RP-18 F_{254S} (10 cm × 10 cm) RP-TLC plate (Merck, Darmstadt, Germany). The mobile phase was prepared by mixing the respective amount of aqueous TRIS (tris(hydroxymethyl)aminomethane) buffer, pH = 7.4 (ionic strength 0.2 M), to meet physiological conditions and acetone in a range from 50 to 80% (v/v) in increments of 5%. Ethanol (96%) was used to prepare the solutions of angularly condensed diquinothiazines **1–15** and the standards **I–V** at a concentration of 2.0 mg/mL. Next, solutions (2 mL) of the analyzed compounds were applied to the plates 5 mm apart and 10 mm from the lower edge and sides of the plates. Before plate development, the chromatographic chambers were saturated with the mobile phase for 0.5 h. After the development of the plates and drying in a stream of air, the chromatograms were observed in ultraviolet light at 254 nm. Each chromatographic experiment was run in triplicate, and then mean values of the retardation factor, i.e., R_F values, were used to determine the R_{MW} values of the compounds.

The R_M values calculated from the experimental R_F values by using Equation (1) were linearly dependent on the concentration of acetone:

$$R_M = \log\left(\frac{1}{R_F} - 1\right) \quad (1)$$

The R_{M0} values were obtained by extrapolating to zero acetone concentration by using Equation (2):

$$R_M = R_{M0} + b \cdot C \quad (2)$$

where C is the volume fraction of the organic modifier in the mobile phase and b is the change in the R_M value due to the 1% increase in the organic modifier in the mobile phase (associated with the specific hydrophobic surface area) [56].

Next, the chromatographic data in the form of R_{M0} and coefficient b were used to calculate the lipophilicity parameter C_0 [56] from the following Equation (3):

$$C_0 = -\frac{R_{M0}}{b} \quad (3)$$

The C_0 parameter is interpreted as the molecular hydrophobicity per unit of the specific hydrophobic surface, i.e., corresponding to the specific hydrophobic surface area of the substance in contact with the stationary phase.

4.3. In Silico Calculated Descriptors

Theoretical values of the partition coefficients (logP) for each compound were determined using various Internet servers such as SwissADME [26], pkCSM [27], Molinspiration [28], and the ChemDraw program [29], including iLOGP, XLOGP3, WLOGP, MLOGP, SILCOS-IT, LogP, logP, and milogP. The logP$_{average}$ is an arithmetic mean of the values predicted by the algorithms. The software used are based on different methods of logP calculation [52,57,58]. The following lipophilicity descriptors were obtained using the web tool SwissADME: iLOGP, a physics-based method that relies on free energies of solvation in n-octanol and water calculated by the generalized-born and solvent accessible surface area (GB/SA) model developed by Daina and coworkers; WLOGP, an atomistic method stationed on fragments and topological descriptors; XLOGP3, an atomistic method including corrective factors and a knowledge-based library; MLOGP, based on topological indices and the linear relationship of structure–logP; and SILICOS-IT, a hybrid fragment/topological approach based on 27 fragments and 7 topological descriptors [26,52,57,58]. The next method used for the calculation of miLogP is a fragment-based approach developed by Molinspiration, and it calculates log P from the sum of group or fragment contributions and correction factors [28]. To compute all other descriptors, such as LogP[a] and logP[c], the pkCSM and ChemDraw programs were used, respectively [27,29].

The molecular descriptors and the ADME parameters were also obtained using the SwissADME and pkCSM platform [26,27].

The cluster analysis procedure was conducted using Statistica software v. 13.3. Calculations were performed on Euclidean distances and a single linkage distance [59].

5. Conclusions

In the search for the most promising drug-like substances, the key parameter that describes the pharmacokinetic behavior of a drug is lipophilicity. Therefore, in the present work, fifteen newly synthesized anticancer, angularly condensed diquinothiazines with pharmacophore dialkylaminoalkyl substituents were studied with the goal of predicting selected pharmacokinetic parameters influencing their ADMET properties, including the lipophilicity descriptors logP and R_{M0}. Lipophilicity indices were determined experimentally by the RP-TLC method, with RP$_{18}$ as the stationary phase and an acetone–TRIS buffer (pH 7.4) mixture as the mobile phase, as well as theoretically using computer approaches. In addition to this, the selected ADMET parameters were determined in silico using the

SwissADME and pkCSM platforms and then correlated with the experimental lipophilicity parameters of the examined compounds. The results of the lipophilicity study confirm that the applied software can be useful for the rapid prediction of logP values during the first stage of study of the examined drug candidates, and next they should be complemented with experimental data such as chromatographic descriptors. Of all the algorithms used, iLogP showed the biggest similarity to the chromatographic value (R_{M0}), especially for compounds **2**, **7**, **12**, **14**, and **15**. In addition to this, it was stated that both the SwissADME and pkCSM web tools are good sources of a wide range of ADMET parameters that describe the pharmacokinetic profiles of the studied compounds and can be fast and low-cost tools in the evaluation of drug candidates during the early stages of the development process.

The satisfactory linear correlation equations of the chromatographic parameter of lipophilicity (R_{M0}) with the theoretical logP values and other physicochemical properties of the tested compounds, such as the molar mass, molar volume, molar refractive index, and surface area, and with the results of other pharmacokinetics parameters that are key in the ADMET profile demonstrate the utility of the computational approaches in estimating the physicochemical properties of new drug candidates in the development process.

Supplementary Materials: The following supporting information can be downloaded at: https://www.mdpi.com/article/10.3390/ph17060725/s1, Table S1: The absorption and metabolism descriptors for compounds **1–15**; Table S2: The excretion and toxicity descriptors for compounds **1–15**; Figure S1: The bioavailability radars for compounds **1–15**; Table S3: The summary formulas, molecular weights, melting points, and appearance of the tested diquinothiazines **1–15**.

Author Contributions: Conceptualization, M.J.; methodology, M.J. and D.K.; software, M.J., D.K. and B.M.-M.; validation, M.J. and M.D.; formal analysis, M.J. and D.K.; investigation, M.J. and D.K.; writing—original draft preparation, M.J. and M.D.; writing—review and editing, M.J., M.D., D.K. and B.M.-M.; visualization, B.M.-M.; supervision, M.J.; project administration, M.J. All authors have read and agreed to the published version of the manuscript.

Funding: This research was funded by the Medical University of Silesia in Katowice, grant number BNW-1-013/N/3/F.

Institutional Review Board Statement: Not applicable.

Informed Consent Statement: Not applicable.

Data Availability Statement: Data from the research described in the manuscript are available from the authors.

Conflicts of Interest: The authors declare no conflicts of interest.

References

1. Vitaku, E.; Smith, D.T.; Njardarson, J.T. Analysis of the structural diversity, substitution patterns, and frequency of nitrogen heterocycles among US FDA approved pharmaceuticals. *J. Med. Chem.* **2014**, *57*, 10257–10274. [CrossRef]
2. Ohlow, M.J.; Moosmann, B. Foundation Review: Phenothiazine: The seven lives of pharmacology's first lead structure. *Drug Discov. Today* **2011**, *16*, 119–131. [CrossRef]
3. Edinoff, A.N.; Armistead, G.; Rosa, C.A.; Anderson, A.; Patil, R.; Cornett, E.M.; Murnane, K.S.; Kaye, A.M.; Kaye, A.D. Phenothiazines and their evolving roles in clinical practice: A narrative review. *Health Psychol. Res.* **2022**, *10*, 38930. [CrossRef]
4. Posso, M.C.; Domingues, F.C.; Ferreira, S.; Silvestre, S. Development of phenothiazine hybrids with potential medicinal interest: A Review. *Molecules* **2022**, *27*, 276. [CrossRef]
5. Pluta, K.; Morak-Młodawska, B.; Jeleń, M. Recent progress in biological activities of synthesized phenothiazines. *Eur. J. Med. Chem.* **2011**, *46*, 3179–3189. [CrossRef]
6. Spengler, G.; Csonka, Á.; Molnár, J.; Amaral, L. The anticancer activity of the old neuroleptic phenothiazine-type drug thioridazine. *Anticancer Res.* **2016**, *36*, 5701–5706. [CrossRef]
7. Zubas, A.; Ghinet, A.; Farce, A.; Dubois, J.; Bîcu, E. Phenothiazineand carbazole-cyanochalcones as dual inhibitors of tubulin polymerization and human farnesyltransferase. *Pharmaceuticals* **2023**, *16*, 888. [CrossRef]
8. Mosnaim, A.D.; Ranade, V.V.; Wolf, M.E.; Puente, J.; Valenzuela, A.M. Phenothiazine molecule provides the basic chemical structure for various classes of pharmacotherapeutic agents. *Am. J. Ther.* **2006**, *13*, 261–273. [CrossRef]

9. Chuang, S.T.; Papp, H.; Kuczmog, A.; Eells, R.; Condor Capcha, J.M.; Shehadeh, L.A.; Jakab, F.; Buchwald, P. Methylene blue is a nonspecific protein–protein interaction inhibitor with potential for repurposing as an antiviral for COVID-19. *Pharmaceuticals* **2022**, *15*, 621. [CrossRef]
10. Boyd-Kimball, D.; Gonczy, K.; Lewis, B.; Mason, T.; Siliko, N.; Wolfe, J. Classics in chemical neuroscience: Chlorpromazine. *ACS Chem. Neurosci.* **2019**, *10*, 79–88. [CrossRef]
11. Ginex, T.; Vazquez, J.; Gilbert, E.; Herrero, E.; Luque, F.J. Lipophilicity in drug design: An overview of lipophilicity descriptors in 3D-QSAR studies. *Future Med. Chem.* **2019**, *11*, 1177–1193. [CrossRef]
12. Johnson, T.W.; Dress, K.R.; Edwards, M. Using the Golden Triangle to optimize clearance and oral absorption. *Bioorg. Med. Chem. Lett.* **2009**, *19*, 5560–5564. [CrossRef]
13. Johnson, T.W.; Gallego, R.A.; Edwards, M.P. Lipophilic efficiency as an important metric in drug design. *J. Med. Chem.* **2018**, *61*, 6401–6420. [CrossRef]
14. Ditzinger, F.; Price, D.J.; Ilie, A.R.; Köhl, N.J.; Jankovic, S.; Tsakiridou, G.; Aleandri, S.; Kalantzi, L.; Holm, R.; Nair, A.; et al. Lipophilicity and hydrophobicity considerations in bioenabling oral formulations approaches—A PEARRL review. *J. Pharm. Pharmacol.* **2019**, *71*, 464–482. [CrossRef]
15. Leeson, P.D.; Davis, A.M. Time-related differences in the physical property profiles of oral drugs. *J. Med. Chem.* **2004**, *47*, 6338–6348. [CrossRef]
16. Markovic, M.; Ben-Shabat, S.; Keinan, S.; Aponick, A.; Zimmermann, E.M.; Dahan, A. Lipidic prodrug approach for improved oral drug delivery and therapy. *Med. Res. Rev.* **2019**, *39*, 579–607. [CrossRef]
17. Di, L.; Kerns, E.H. Profiling drug-like properties in discovery research. *Curr. Opin. Chem. Biol.* **2003**, *7*, 402–408. [CrossRef]
18. Dulsat, J.; López-Nieto, B.; Estrada-Tejedor, R.; Borrell, J.I. Evaluation of free online ADMET tools for academic or small biotech environments. *Molecules* **2023**, *28*, 776. [CrossRef]
19. Rutkowska, E.; Pająk, K.; Jóźwiak, K. Lipophilicity—Methods of determination and its role in medicinal chemistry. *Acta Pol. Pharm. Drug Res.* **2013**, *70*, 3–18.
20. OECD. *Test No. 117: Partition Coefficient (n-Octanol/Water), HPLC Method*; OECD Publishing: Paris, France, 2022.
21. Roman, I.P.; Mastromichali, A.; Tyrovola, K.; Canals, A.; Psillakis, E. Rapid determination of octanol-water partition coefficient using vortex-assisted liquid-liquid microextraction. *J. Chromatogr. A* **2014**, *1330*, 1–5. [CrossRef]
22. Eadsforth, C.V.; Moser, P. Assessment of reverse-phase chromatographic methods for determining partition coefficients. *Chemosphere* **1983**, *12*, 1459–1475. [CrossRef]
23. Soares, J.X.; Santos, Á.; Fernandes, C.; Pinto, M.M.M. Liquid chromatography on the different methods for the determination of lipophilicity: An essential analytical tool in medicinal chemistry. *Chemosensors* **2022**, *10*, 340. [CrossRef]
24. Jeleń, M.; Pluta, K.; Latocha, M.; Morak-Młodawska, B.; Suwińska, K.; Kuśmierz, D. Evaluation of angularly condensed diquinothiazines as potential anticancer agents. *Bioorg. Chem.* **2019**, *87*, 810–820. [CrossRef]
25. Skonieczna, M.; Kasprzycka, A.; Jeleń, M.; Morak-Młodawska, B. Tri- and pentacyclic azaphenothiazine as pro-apoptotic agents in lung carcinoma with a protective potential to healthy cell lines. *Molecules* **2022**, *27*, 5255. [CrossRef]
26. Available online: http://swissadme.ch (accessed on 10 January 2024).
27. Available online: https://biosig.unimelb.edu.au/pkcsm/prediction (accessed on 5 January 2024).
28. Available online: https://www.molinspiration.com/services/logp.html (accessed on 5 January 2024).
29. *ChemDraw: ChemDraw Ultra*, Version 22.2.0; CambridgeSoft, PerkinElmer Informatics: Austin, TX, USA, 2022.
30. Bodor, N.; Gabanyi, Z.; Wong, C.K. A new method for the estimation of partition coefficient. *J. Am. Chem. Soc.* **1989**, *111*, 3783–3786. [CrossRef]
31. Mannhold, R.; Cruciani, G.; Dross, K.; Rekker, R. Multivariate analysis of experimental and computational descriptors of molecular lipophilicity. *J. Comput. Mol. Des.* **1998**, *12*, 573–581. [CrossRef]
32. Brooke, D.; Dobbs, J.; Williams, N. Octanol:water partition coefficients (P): Measurement, estimation, and interpretation, particularly for chemicals with $P > 10^5$. *Ecotoxicol. Environ. Saf.* **1986**, *11*, 251–260. [CrossRef]
33. Azman, M.; Sabri, A.H.; Anjani, Q.K.; Mustaffa, M.F.; Hamid, K.A. Intestinal absorption study: Challenges and absorption enhancement strategies in improving oral drug delivery. *Pharmaceuticals* **2022**, *15*, 975. [CrossRef]
34. Tsakovska, I.; Pajeva, I.; Al Sharif, M.; Alov, P.; Fioravanzo, E.; Kovarich, S.; Worth, A.P.; Richarz, A.N.; Yang, C.; Mostrag-Szlichtyng, A.; et al. Quantitative structure-skin permeability relationships. *Toxicology* **2017**, *387*, 27–42. [CrossRef]
35. Gombar, V.K.; Hall, S.D. Quantitative structure−activity relationship models of clinical pharmacokinetics: Clearance and volume of distribution. *J. Chem. Inf. Model.* **2013**, *53*, 948–957. [CrossRef]
36. Poulin, P.; Theil, F.P. Prediction of pharmacokinetics prior to in vivo studies. 1. Mechanism-based prediction of volume of distribution. *J. Pharm. Sci.* **2002**, *91*, 129–156. [CrossRef]
37. Mulpuru, V.; Mishra, N. In silico prediction of fraction unbound in human plasma from chemical fingerprint using automated machine learning. *ACS Omega* **2021**, *10*, 6791–6797. [CrossRef]
38. Sweeney, M.D.; Zhao, Z.; Montagne, A.; Nelson, A.R.; Zlokovic, B.V. Blood-Brain Barrier: From physiology to disease and back. *Physiol. Rev.* **2019**, *99*, 21–78. [CrossRef]
39. Horde, G.W.; Gupta, V. Drug Clearance. 20 June 2023. In *StatPearls [Internet]*; StatPearls Publishing: Treasure Island, FL, USA, 2024. [PubMed]

40. Das, A.; Kumar, S.; Persoons, L.; Daelemans, D.; Schols, D.; Alici, H.; Tahtaci, H.; Karki, S.S. Synthesis, in silico ADME, molecular docking and in vitro cytotoxicity evaluation of stilbene linked 1,2,3-triazoles. *Heliyon* **2021**, *7*, e05893. [CrossRef]
41. Goetz, G.H.; Shalaeva, M. Leveraging chromatography based physicochemical properties for efficient drug design. *ADMET DMPK* **2018**, *6*, 85. [CrossRef]
42. Kaliszan, R. QSRR: Quantitative Structure-(Chromatographic) Retention Relationships. *Chem. Rev.* **2007**, *107*, 3212–3246. [CrossRef]
43. Kempińska, D.; Chmiel, T.; Kot-Wasik, A.; Mróz, A.; Mazerska, Z.; Namie'snik, J. State of the art and prospects of methods for determination of lipophilicity of chemical compounds. *Trends Anal. Chem.* **2019**, *113*, 54–73. [CrossRef]
44. Di, L.; Kerns, E.H. *Drug-Like Properties*; Academic Press: New York, NY, USA, 2008.
45. Petrauskas, A.A.; Kolovanov, E.A. ACD/Log P method description. *Perspect. Drug Discov. Des.* **2000**, *19*, 99–116. [CrossRef]
46. Lipinski, C.A.; Lombardo, F.; Dominy, B.W.; Feeney, P.J. Experimental and computational approaches to estimate solubility and permeability in drug discovery and development settings. *Adv. Drug Deliv. Rev.* **1997**, *23*, 3–25. [CrossRef]
47. Law, V.; Knox, C.; Djoumbou, Y.; Jewison, T.; Guo, A.C.; Liu, Y.; Maciejewski, A.; Arndt, D.; Wilson, M.; Neveu, V.; et al. DrugBank 4.0: Shedding new light on drug metabolism. *Nucleic Acids Res.* **2014**, *42*, D1091–D1097. [CrossRef]
48. Ghose, A.K.; Viswanadhan, V.N.; Wendoloski, J.J. A knowledge-based approach in designing combinatorial or medicinal chemistry libraries for drug discovery. A qualitative and quantitative characterization of known drug databases. *J. Comb. Chem.* **1999**, *1*, 55–68. [CrossRef]
49. Veber, D.F.; Johnson, S.R.; Cheng, H.-Y.; Smith, B.R.; Ward, K.W.; Kopple, K.D. Molecular properties that influence the oral bioavailability of drug candidates. *J. Med. Chem.* **2002**, *45*, 2615–2623. [CrossRef]
50. Egan, W.J.; Merz, K.M.; Baldwin, J.J. Prediction of drug absorption using multivariate statistics. *J. Med. Chem.* **2000**, *43*, 3867–3877. [CrossRef]
51. Martin, Y.C. A Bioavailability Score. *J. Med. Chem.* **2005**, *48*, 3164–3170. [CrossRef]
52. Daina, A.; Michielin, O.; Zoete, V. SwissADME: A free web tool to evaluate pharmacokinetics, drug-likeness and medicinal chemistry friendliness of small molecules. *Sci. Rep.* **2017**, *7*, 42717. [CrossRef]
53. Lobo, S. Is there enough focus on lipophilicity in drug discovery? *Expert Opin. Drug Discov.* **2020**, *15*, 261–263. [CrossRef]
54. Pantaleao, S.Q.; Fernandes, P.O.; Goncalves, J.E.; Maltarollo, V.G.; Honorio, K.M. Recent advances in the prediction of pharmacokinetics properties in drug design studies: A review. *Chem. Med. Chem.* **2022**, *17*, e202100542. [CrossRef]
55. Daina, A.; Zoete, V. A BOILED-Egg to predict gastrointestinal absorption and brain penetration of small molecules. *Chem. Med. Chem.* **2016**, *11*, 1117–1121. [CrossRef]
56. Šegan, S.; Penjišević, J.; Šukalović, V.; Andrić, D.; Milojković-Opsenica, D.; Kostić-Rajačić, S. Investigation of lipophilicity and pharmacokinetic properties of 2-(methoxy)phenylpiperazine dopamine D2 ligands. *J. Chromatogr. B Anal. Technol. Biomed. Life Sci.* **2019**, *1124*, 146–153. [CrossRef]
57. Moriguchi, I.; Shuichi, H.; Liu, Q.; Nakagome, I.; Matsushita, Y. Simple method of calculating octanol/water partition coefficient. *Chem. Pharm. Bull.* **1992**, *40*, 127–130. [CrossRef]
58. Moriguchi, I.; Shuichi, H.; Nakagome, I.; Hirano, H. Comparison of reliability of log P values for drugs calculated by several methods. *Chem. Pharm. Bull.* **1994**, *42*, 976–978. [CrossRef]
59. Stanisz, A. *Przystępny Kurs Statystyki z Zastosowaniem STATISTICA PL na Przykładach Medycyny; Analizy Wielowymiarowe*; Stat Soft Polska: Kraków, Poland, 2007; Volume 3.

Disclaimer/Publisher's Note: The statements, opinions and data contained in all publications are solely those of the individual author(s) and contributor(s) and not of MDPI and/or the editor(s). MDPI and/or the editor(s) disclaim responsibility for any injury to people or property resulting from any ideas, methods, instructions or products referred to in the content.

Article

Unveiling the Multifaceted Capabilities of Endophytic *Aspergillus flavus* Isolated from *Annona squamosa* Fruit Peels against *Staphylococcus* Isolates and HCoV 229E—In Vitro and In Silico Investigations

Noha Fathallah [1,*,†], Wafaa M. Elkady [1,†], Sara A. Zahran [2], Khaled M. Darwish [3], Sameh S. Elhady [4,5,6] and Yasmin A. Elkhawas [1,*]

1. Department of Pharmacognosy and Medicinal Plants, Faculty of Pharmacy, Future University in Egypt, Cairo 11835, Egypt; welkady@fue.edu.eg
2. Department of Microbiology and Immunology, Faculty of Pharmacy, Future University in Egypt, Cairo 11835, Egypt; sara.zahran@fue.edu.eg
3. Department of Medicinal Chemistry, Faculty of Pharmacy, Suez Canal University, Ismailia 41522, Egypt; khaled_darwish@pharm.suez.edu.eg
4. King Abdulaziz University Herbarium, Faculty of Science, King Abdulaziz University, Jeddah 21589, Saudi Arabia; ssahmed@kau.edu.sa
5. Department of Biological Sciences, Faculty of Science, King Abdulaziz University, Jeddah 21589, Saudi Arabia
6. Center for Artificial Intelligence in Precision Medicines, King Abdulaziz University, Jeddah 21589, Saudi Arabia
* Correspondence: noha.mostafa@fue.edu.eg (N.F.); yasmien.alaa@fue.edu.eg (Y.A.E.)
† These authors contributed equally to this work.

Citation: Fathallah, N.; Elkady, W.M.; Zahran, S.A.; Darwish, K.M.; Elhady, S.S.; Elkhawas, Y.A. Unveiling the Multifaceted Capabilities of Endophytic *Aspergillus flavus* Isolated from *Annona squamosa* Fruit Peels against *Staphylococcus* Isolates and HCoV 229E—In Vitro and In Silico Investigations. *Pharmaceuticals* 2024, 17, 656. https://doi.org/10.3390/ph17050656

Academic Editors: Valentina Noemi Madia and Davide Ialongo

Received: 9 April 2024
Revised: 14 May 2024
Accepted: 16 May 2024
Published: 19 May 2024

Copyright: © 2024 by the authors. Licensee MDPI, Basel, Switzerland. This article is an open access article distributed under the terms and conditions of the Creative Commons Attribution (CC BY) license (https://creativecommons.org/licenses/by/4.0/).

Abstract: Recently, there has been a surge towards searching for primitive treatment strategies to discover novel therapeutic approaches against multi-drug-resistant pathogens. Endophytes are considered unexplored yet perpetual sources of several secondary metabolites with therapeutic significance. This study aims to isolate and identify the endophytic fungi from *Annona squamosa* L. fruit peels using morphological, microscopical, and transcribed spacer (ITS-rDNA) sequence analysis; extract the fungus's secondary metabolites by ethyl acetate; investigate the chemical profile using UPLC/MS; and evaluate the potential antibacterial, antibiofilm, and antiviral activities. An endophytic fungus was isolated and identified as *Aspergillus flavus* L. from the fruit peels. The UPLC/MS revealed seven compounds with various chemical classes. The antimicrobial activity of the fungal ethyl acetate extract (FEA) was investigated against different Gram-positive and Gram-negative standard strains, in addition to resistant clinical isolates using the agar diffusion method. The CPE-inhibition assay was used to identify the potential antiviral activity of the crude fungal extract against low pathogenic human coronavirus (HCoV 229E). Selective Gram-positive antibacterial and antibiofilm activities were evident, demonstrating pronounced efficacy against both methicillin-resistant *Staphylococcus aureus* (MRSA) and methicillin-sensitive *Staphylococcus aureus* (MSSA). However, the extract exhibited very weak activity against Gram-negative bacterial strains. The ethyl acetate extract of *Aspergillus flavus* L exhibited an interesting antiviral activity with a half maximal inhibitory concentration (IC_{50}) value of 27.2 µg/mL against HCoV 229E. Furthermore, in silico virtual molecular docking-coupled dynamics simulation highlighted the promising affinity of the identified metabolite, orienting towards three MRSA biotargets and HCoV 229E main protease as compared to reported reference inhibitors/substrates. Finally, ADME analysis was conducted to evaluate the potential oral bioavailability of the identified metabolites.

Keywords: *Annona squamosa* L.; endophytic fungi; MRSA; antiviral; ADME prediction; public health; drug discovery; molecular docking-coupled dynamics simulation

1. Introduction

Drug-resistant bacteria and fungi are thought to pose a global health risk. Microorganisms that produce biofilms present one of the challenges that scientists face today due to their unique capacity to modify their immediate environs through intriguing phenotypic plasticity that involves changes in their physiology and their resistance to antimicrobial agents [1]. Since the late 1970s, (MRSA) infections have been linked to multiple hospital outbreaks and are a major cause of morbidity and mortality among hospitalized patients worldwide. In comparison to other African nations as well as countries in the southern and eastern Mediterranean, Egypt had the highest MRSA rates among clinical isolates of *S. aureus* [2]. Another major problem facing the healthcare system in Egypt is acute respiratory infections (ARIs) which are a chief cause of morbidity and mortality among children under five, which also causes absenteeism due to respiratory symptoms among primary and preparatory school students. At the end of 2022, numerous governmental surveillances detected a surge of respiratory viruses including coronavirus [3].

Coronavirus species are known to cause human infection, one of which, HCoV 229 E, typically causes cold symptoms in immunocompetent individuals [4]; it causes mild to severe enteric, respiratory, and systemic disease in animals, poultry, and rodents, and causes common cold or pneumonia in humans. Thus, it was deemed necessary to search for potential new and promising antimicrobial and antiviral non-conventional drugs. Since the dawn of human civilization, plants have been a significant source of medicinal compounds [5,6]. Current demand for new and potent medications and other plant-based items is rising.

Drug-resistant bacteria are thought to pose a global health problem. Biofilm-forming bacteria are among the issues facing scientists today, with their special ability to alter their immediate environs by unusual phenotypical plasticity that encompasses changes in their physiology and resistance to antimicrobial treatments; Singab et al. reported that endophytes have been identified as a hidden treasure for secondary metabolites. According to previously reported data, various compounds isolated from *Aspergillus flavus* showed antimicrobial, anti-biofilm activity [7]. Khattak et al. reported that the *Aspergillus flavus* isolated compound demonstrated significant antibacterial activity against *S. aureus* [8].

It has been established after more than a century of research that the majority of plants in ecosystems —if not all of them—are symbiotic with fungal endophytes, including grass, trees, algae, and herbaceous plants [9,10]. The expression of host plant diseases can be significantly altered by non-pathogenic fungi found within plants, also known as endophytes ("endo" = within, "phyte" = plant), according to recent studies [11,12]. These fungi are valuable sources of bioactive secondary metabolites that can produce broad-spectrum antimicrobial substances [6,13].

Annona squamosa Linn tree, commonly known as the sugar apple, is endogenous to Egypt [14]. It yields edible fruits and is used to make both industrial and therapeutic items. *A. squamosa* Linn is currently employed as an anti-inflammatory [15], cytotoxic [16], antitumor, hepatoprotective [17], antidiabetic [18], and anti-lice agent [19]. It is associated with the presence of alkaloids, carbohydrates, tannins, fixed oils, and phenolics [20–22]. It was previously evaluated for its antimicrobial activity [5,23,24] and was established as a plant with a potential wide spectrum of antimicrobial activity. Due to the genetic transmission or co-evolution between the endophyte and host, some of the fruits' therapeutic benefits may be attributed to the endophytes [9].

On one hand, the current study aimed to isolate the endophytic fungi associated with *A. squamosa* fruits, and to identify the metabolites that may be useful by employing the UPLC/MS analytical technique, which is a rapid and affordable method of identification. On the other hand, in vitro experiments to evaluate the ethyl acetate's antibacterial and antiviral potentials followed by a rapid prediction using preliminary computational in silico and ex silico studies are undertaken to assess the drug-like properties of those lead compounds.

2. Results and Discussion

2.1. Isolation and Identification of Endophytic Fungi

In the present study, the particular fungus under investigation was the only one being successfully sub-cultured and purified through repeated culturing of the crushed *A. squamosa* L. fruits. Notably, the mother culture revealed various endophytes, yet they failed to grow upon sub-culturing. Using the morphological and microscopical features listed in Table 1 and Figure 1, the isolated purified endophyte would belong to the *Aspergillus* species. The identification was confirmed using amplification and sequencing of the internal transcribed spacer ribosomal RNA (ITS rRNA) gene. Sequence analysis showed a 99% identity with *Aspergillus flavus* (*A. flavus*) as seen in Figure 2. Upon submission, a GenBank accession number OM095472 was assigned to the ITS rRNA gene sequence.

Table 1. Morphological and microscopical description of *A. flavus*.

Morphological Characters		Microscopic Characters	
Surface	Yellowish-black	Hyphae	Thread-like septate branched
Margins	Entire	Conidia	Olive green (4 to 7 μm), roughened
Reverse side	Greenish-yellow	Phialides	uniseriate and biseriate phialides
Growth	Moderate		
Elevations	Umbonate		

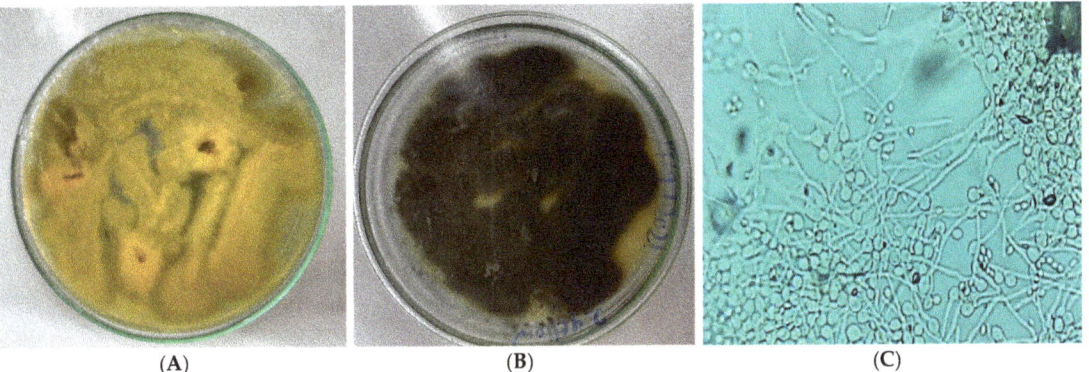

(A)   (B)   (C)

Figure 1. Photos of *Aspergillus flavus* fungus; (**A**,**B**) colony morphology on potato dextrose agar after 2 days of incubation; (**A**) front view, (**B**) back view. (**C**) Under microscope (1000×).

Patil et al., Liu et al., and Ola et al. [25–27] previously covered the significance of the endophytic *A. flavus* isolated from several plant species. They demonstrated how it might be valuable as an antibacterial and anticancer agent. This provided us with a clue to design a study that would investigate the contribution of endophytic fungus to the previously reported activity of *A. squamosa*, as well as identify the chemicals responsible for antibacterial, antibiofilm, and antiviral activity.

Aflatoxins are common toxic active metabolites usually produced by *A. flavus*. They are known to appear in the media as yellow pigments, which could be easily visualized on the reverse side of a coconut-agar medium colony [28]; their products turn pink/plum red when exposed to ammonia vapor and usually give blue fluorescence on CAM when exposed to UV light (365 nm). Interestingly, our isolate did not produce any yellow pigments, any pink color, or any blue fluorescence upon applying the three tests; thus, it was concluded that it is a non-aflatoxigenic isolate [29].

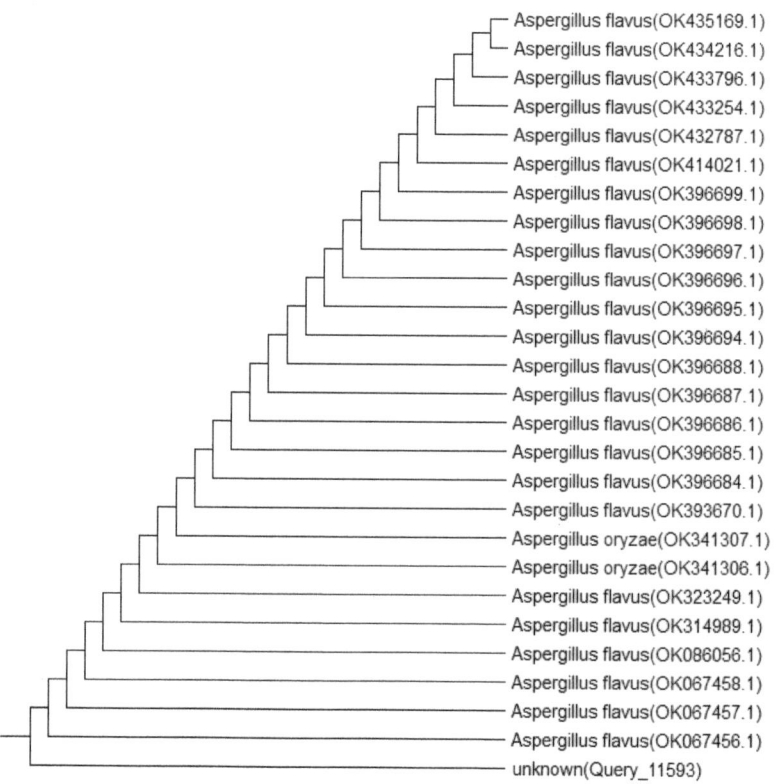

Figure 2. Phylogenetic tree of the isolated endophyte *Aspergillus flavus*.

2.2. Metabolic Profiling of the Ethyl Acetate Extract

To investigate the active metabolites present in the identified fungus, metabolic profiling using UPLC/MS was conducted. This approach was used as it is a sensitive and accurate method of analysis, allows for separation in a shorter development and analysis time than conventional LC/UV, and it provides a comprehensive profile of the compounds present in the extract [30]. The chromatogram represented in Figure S1 revealed seven compounds as illustrated in Table 2 and Figure 3. Most of the identified metabolites have been previously reported from various endophyte species. It was noted that the extract revealed several classes of active metabolites such as sesquiterpenoids, phenolics, fatty acids, flavonoids, and pyrones. Heptelidic acid, ferulic acid, and oleic acid were the dominant active metabolites identified with areas of 26.8%, 25.3%, and 23.2%, respectively.

On the one hand, heptelidic acid was the most dominant compound. This compound was reported before by Tanaka et al., Itoh et al., and Kim et al. as an antimalarial, antibiotic, and anticancer agent [31–33]. On the other hand, ferulic acid is known for its broad antimicrobial activities [34,35]. Nevertheless, the rest of the identified compounds are known for their remarkable array of biochemical and pharmacological actions [36,37], suggesting that they may significantly affect the function of various mammalian cellular systems. These results encouraged additional research on the FEA's antiviral, antibiofilm, and antimicrobial properties.

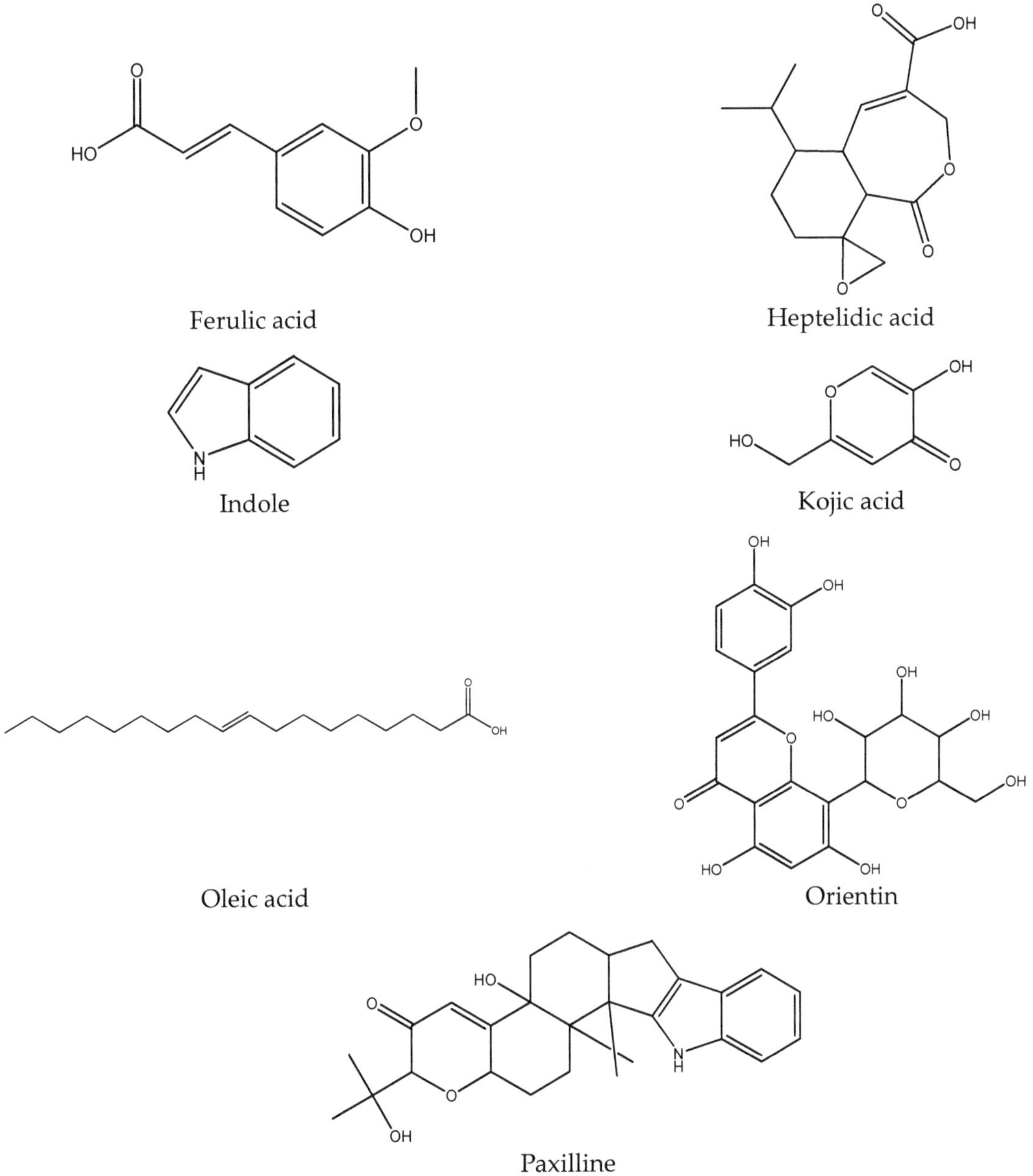

Ferulic acid

Heptelidic acid

Indole

Kojic acid

Oleic acid

Orientin

Paxilline

Figure 3. Two-dimensional structures of the identified compounds by UPLC/MS analysis (Chemdraw ultra-version 14).

Table 2. Peak assignments of the ethyl acetate extract of *A. flavus*, L via UPLC-ESI-MS/MS in negative ionization mode. (N.B. the compounds numbered according to their abundance.)

No.	Compound	Chemical Class	Molecular Formula	[M-H]$^-$	Abundance	M. Weight	Ref.
1	Heptelidic acid	Sesquiterpene	$C_{15}H_{20}O_5$	279	26.8%	280	[38,39]
2	Ferulic acid	Phenolic	$C_{10}H_{10}O_4$	193	25.3%	194	[40,41]
3	Oleic acid	Fatty acid	$C_{18}H_{34}O_2$	281	23.2%	282	[42]
4	Paxilline	Diterpene indole polycyclic alkaloid	$C_{27}H_{33}NO_4$	432	8.3%	435	[43,44]
5	Indole	Alkaloid	C_8H_7N	116	7.4%	117	[45]
6	Orientin	Flavonoid	$C_{21}H_{20}O_{11}$	446	6.4%	447	[46]
7	Kojic acid	Pyrone	$C_6H_6O_4$	141	2%	142	[47,48]

2.3. Antimicrobial Potential

Endophytes' interactions with the plants vary from antagonism to mutualism. Usually, the host plant provides the endophytes with food and protection while the latter increases the host's resistance to herbivores, infections, as well as different abiotic stressors [49], thus it is now considered as a promising approach for discovering new potent antimicrobial agents.

The antimicrobial activity of FEA (20% w/v) was evaluated using the disc diffusion technique against a diverse panel of microbes. Notably, FEA demonstrated maximum inhibition zones against Gram-positive bacteria, specifically *S. aureus* ATCC 25923 (which is MSSA) and MRSA ATCC-700788. The inhibition zones were measured at (15 ± 0.4 mm) and (11 ± 0.7 mm), respectively, approaching the efficacy of the standard drug vancomycin, which displayed zones of inhibition at (18 ± 0.2 mm) and (13 ± 0.3 mm) against the same strains (Table 3). Interestingly, no observable inhibition zones were detected when testing FEA against Gram-negative isolates such as *Escherichia coli* ATCC 25922 and *Pseudomonas aeruginosa* ATCC 9027, as well as the *Candida albicans* ATCC 10231 strains. The results unveil a distinct selective antibacterial activity of FEA, particularly targeting Gram-positive bacteria.

Table 3. Antimicrobial activity—Zones of inhibition for FEA of *Aspergillus flavus* against tested isolates.

Bacterial Strains	Inhibition Zone Diameter (mm)				
	Negative Control		Positive Control		
	FEA	DMSO	Vancomycin	Gentamicin	Nystatin
S. aureus ATCC 25923 (MSSA)	15 ± 0.4	0	18 ± 0.2	–	–
MRSA ATCC-700788	11 ± 0.7	0	13 ± 0.3	–	–
E. coli ATCC 25922	0	0	–	19 ± 0.7	–
P. aeruginosa ATCC9027	0	0	–	25 ± 1.1	–
C. albicans ATCC 10231	0	0	–	–	15 ± 0.5

All measurements were conducted in triplicate, and the results are presented as means ± standard deviation (SD).

2.4. Minimum Inhibitory Concentration

Figure 4 presents a summary of the Minimum Inhibitory Concentration (MIC) values for FEA against sensitive Gram-positive strains, specifically *S. aureus* ATCC 25923 (MSSA) and MRSA ATCC-700788. The results reveal that FEA exhibited potent antimicrobial activity, with the lowest MIC recorded at 50 mg/mL for MRSA ATCC-700788. In contrast, a higher MIC value of 100 mg/mL was observed for *S. aureus* ATCC 25923 (MSSA), indicating a higher susceptibility of MRSA to the extract. Despite the higher MIC for MSSA, FEA remains effective against both *S. aureus* strains. These results highlight the potential of FEA as a natural antimicrobial agent, particularly against problematic bacterial strains such as MRSA. This could be a valuable approach to conquer *S. aureus* as this organism when compared to other microorganisms can serve as an example of the adaptive evolution of bacteria during the antibiotic era. That is because it has shown a remarkable capacity to rapidly adapt to new antibiotics by developing resistance mechanisms. Not only does the resistance mechanism involve the antibiotic's enzymatic deactivation but also it forms biofilm which is considered a major virulence factor [50,51].

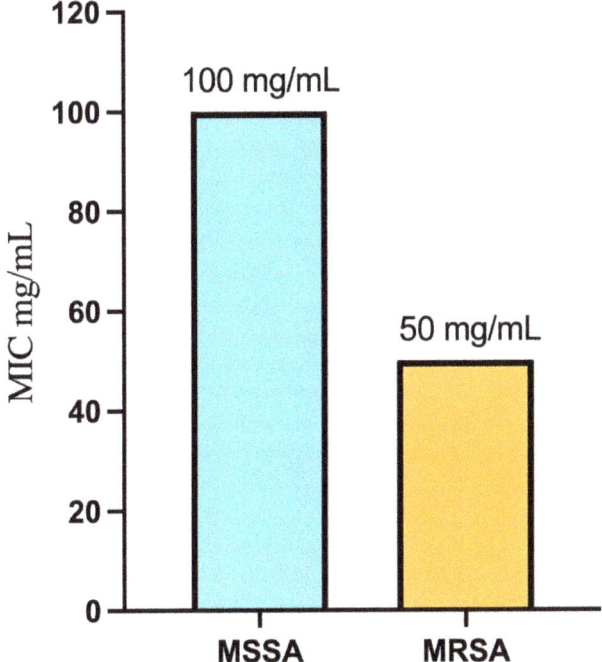

Figure 4. Minimum inhibitory concentration of FEA against sensitive Gram-positive strains.

2.5. Antibiofilm Activity/Anti-Adhesion

2.5.1. Prevention of Cell Attachment

The effect of Sub-Minimum Inhibitory Concentration (Sub-MIC) of FEA on biofilm formation by *S. aureus* ATCC 25923 (MSSA) and MRSA ATCC-700788 is illustrated in Figure 5A. As per established criteria [52], percentage inhibition values ranging from 0 to 100% are indicative of biofilm inhibition, while values below 0% suggest the enhancement of biofilm formation. Activities surpassing the 50% inhibition threshold are considered good, while those falling between 0 and 49% are deemed poor [53]. The fungal extract displayed notable activity in preventing biofilm attachment, and the observed effects were found to be dosage-dependent. Notably, FEA exhibited effective prevention of biofilm attachment for *S. aureus* ATCC 25923 (MSSA) at concentrations of 75 and 50 mg/mL (75 and 50% MIC), surpassing the significant 50% inhibition threshold. However, for MRSA ATCC-700788, the observed suppression remained below the 50% inhibition threshold, even at the highest tested concentration of 37.5 mg/mL (75% MIC).

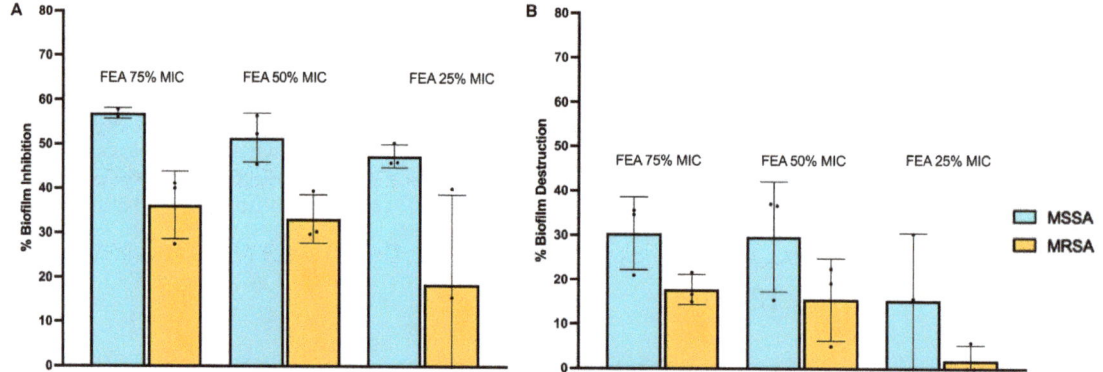

Figure 5. Antibiofilm activity of FEA at its sub-MIC concentrations (75, 50, and 25%) against both MSSA and MRSA isolates. (**A**) Inhibition of biofilm formation and (**B**) biofilm mass destruction. Values ranging from 0% to 50% indicate low activity, and values exceeding 50% indicate high activity. All measurements were conducted in triplicate, and the results are presented as means ± standard deviation (SD).

2.5.2. Evaluating Biofilm Mass Destruction

Figure 5B illustrates the effects of the fungal extracts on destroying or reducing further development in 24 h preformed biofilms. Once again, a dose-dependent antibiofilm activity was evident. However, it is noteworthy that the ability to destroy an already-formed biofilm is not as powerful as the prevention of attachment. In this context, all observed activities exhibited poor biofilm inhibition, falling below 50%. Across all concentrations of FEA, the inhibitory effects were consistently more pronounced against *S. aureus* ATCC 25923 (MSSA) compared to MRSA ATCC-700788.

These findings suggest that while the fungal extract may effectively prevent initial biofilm attachment, particularly against MSSA, its ability to eradicate established biofilms is less potent.

2.6. Antiviral Activity of Crude Extract

FEA demonstrated noteworthy antiviral activity, as evidenced by a CC_{50} value of 46.38 µg/mL, and (IC_{50}) value of 27.2 µg/mL against low pathogenic coronavirus (HCoV 229E), indicating that the extract effectively inhibits viral replication at a relatively low concentration. However, the calculated selectivity index (SI = CC_{50}/IC_{50}) of approximately 2 implies a narrow therapeutic window for the extract, raising concerns about its safety profile [54]. It was reported by Hasöksüz et al. [4] that the virulence and pathophysiology mechanisms of CoVs may be attributed to nonstructural proteins which block the host's innate immune response and structural proteins that play a crucial role in promoting viral assembly and release. FEA established a distinct potency against the HCoV 229E virus which may indicate that its compounds interfere with the function of the nonstructural protein or affect the envelop formation by hindering the structural proteins. Overall, while the antiviral potential of the FEA is promising, further studies are needed to optimize its safety profile and evaluate its efficacy in vivo before considering it as a potential antiviral agent for clinical use.

2.7. Online Software Swiss ADME Prediction (Boiled Egg Method and Lipinski's Rule of Five)

As discussed by [55], it is commonly known that ADME data, whether computationally predicted or empirically observed, provide important information about how a drug will eventually be absorbed, distributed, metabolized, or excreted by the body. While there are other ways to administer drugs, oral dosage is strongly recommended for patient comfort and compliance. An important criterion for decision making at different stages of the discovery process is the early calculation of oral bioavailability, which is defined as the fraction of the dose that enters the bloodstream following oral administration

Identified compounds' physicochemical properties were assessed using Lipinski's rule of five and ADME, which aid in the approval process for prospective compounds for use in biological systems [54,56]. As can be seen in Table 4, most of the compounds met Lipinski's requirements to become an oral medication. However, Orientin exhibited two violations in the number of hydrogen bond donors (>5) and acceptors (>10). Nevertheless, as can be seen in Figure 6, the radar plot bioavailability technique predicted that two compounds, namely heptelidic acid and paxilline, can become completely orally bioavailable as all their parameters were found in the pink bioavailable area. Yet five compounds exhibited deviation in one parameter. Ferulic acid, indole, and kojic acid showed INSATU parameter deviancy while oleic acid and orientin were offshoots of the vertex in flexibility and polarity parameters, respectively. The EGG-BOILED model facilitates the intuitive assessment of the white part of passive gastrointestinal absorption (HIA) and the yellow part of brain penetration (BBB). The physicochemical zone containing chemicals expected to have significant intestinal absorption is known as the "grey region". Regarding the compounds, as observed in Figure 7, two of them were found in the yolk area, namely ferulic acid and indole, while four were in the white zone, namely kojic acid, paxilline, oleic acid, and heptelidic acid. Orientin TPSA 201.28 Å^2 was out of the threshold area [57]. Additionally, most of the compounds were predicted by software as non-substrates (PGP−) of the permeability glycoprotein (PGP) being shown in red circles. Contrarily, only paxilline was shown as a blue circle corresponding to a substrate (PGP +) of glycoprotein permeability.

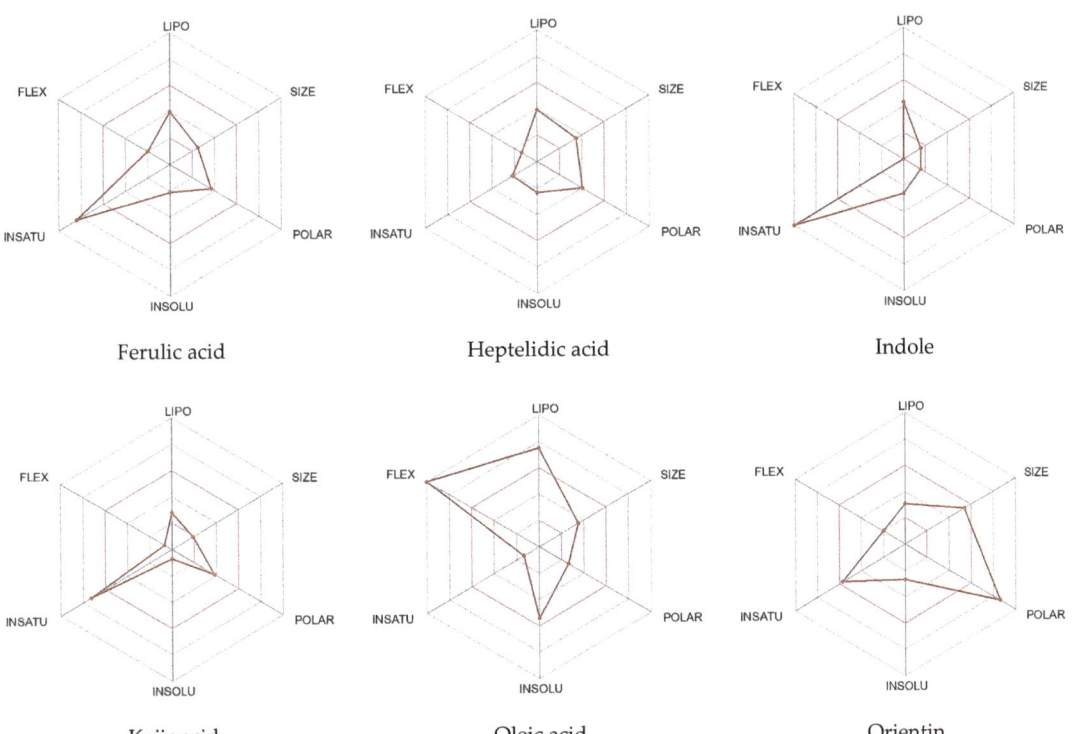

Figure 6. *Cont.*

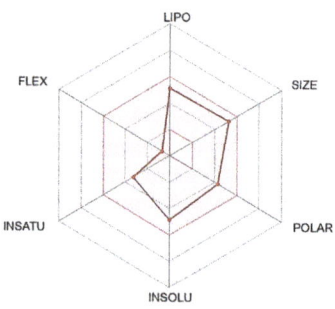

Paxilline

Figure 6. Compounds' polarity, lipophilicity, solubility, flexibility, and saturation are represented on the radar map as POLAR, LIPO, INSOLU, and IN-SATU, respectively. The magenta area represents the optimal range for each molecular property. Solubility: log S < 6; sizes: MW between 150 and 500 g/mol; saturation: fraction of carbons in the sp3 hybridization > 0.25; polarity: TPSA between 20 and 130 Å; flexibility: <9 rotatable bonds. Lipophilicity: XLOGP3 between −0.7 and +5.0.

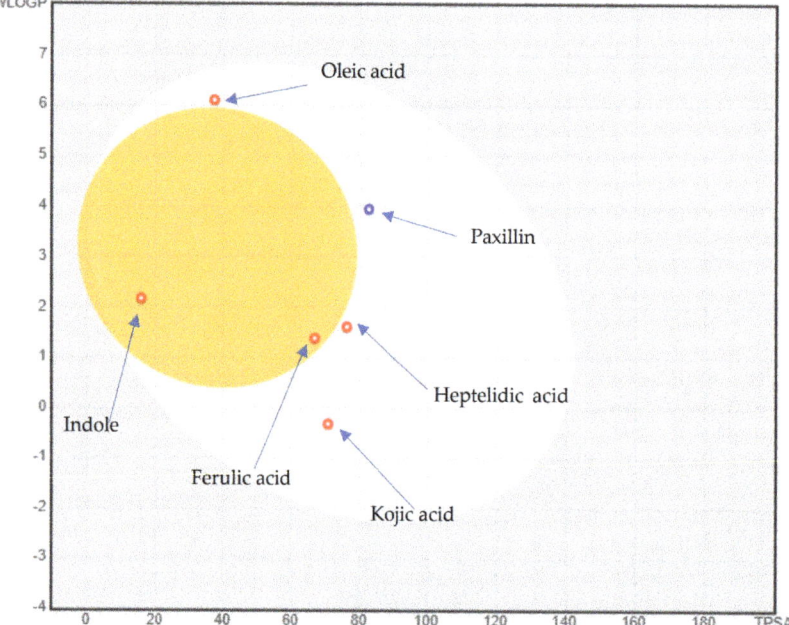
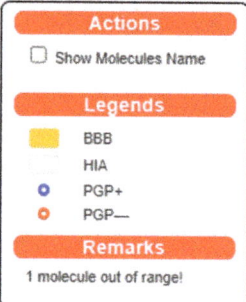

Figure 7. Compounds evaluation using Boiled Egg Method for BBB and GIT absorption. N.B: Boiled egg 2D graphical representation; the yolk area represents the molecules that can passively permeate through the blood–brain barrier (BBB); the molecules located in the white region are predicted to be passively absorbed by the gastrointestinal (GI) tract. Ferulic acid TPSA: 66.76 Å^2, WLOGP 1.39; heptelidic acid TPSA: 76.13 Å^2, WLOGP 1.62; indole TPSA: 15.79 Å^2, WLOGP 2.17; kojic acid TPSA: 70.67 Å^2, WLOGP −0.31; oleic acid TPSA: 37.30 Å^2, WLOGP 6.11; paxilline 82.55 Å^2, WLOGP 3.96.

Table 4. Lipinski's rule of five for ADME analysis of the identified compounds.

No.	Compound	M. wt.	Lipophilicity Log Po/w(MLOGP)	Hydrogen Bond Donors	Hydrogen Bond Acceptors	No. of Rule Violations	Drug Likeness
		Less than 500 g/mol	Less than 5	Less than 5	Less than 10	Less than 2 Violations	Lipinski's Rule Follows Rule
1	Heptelidic acid	280	1.60	1	5	0	Yes
2	Ferulic acid	194	1.00	2	4	0	Yes
3	Oleic acid	282	4.57	1	2	0	Yes
4	Paxilline	435	2.58	3	4	0	Yes
5	Indole	117	1.57	1	0	0	Yes
6	Orientin	448	−2.51	8	11	2	No
7	Kojic acid	142	−1.69	2	4	0	Yes

2.8. In Silico Investigation: Molecular Docking Simulation

In silico studies are performed as they are considered an effective approach for determining drug protein-bound structures and binding affinities down to their molecular levels. Driven by the obtained antibacterial and antiviral activities, we were interested in investigating such activities down to the molecular levels. The identified phytochemicals were evaluated for their binding affinities and interactions towards several potential biotargets highlighting their antibacterial and antiviral activities. In terms of activity against *S. aureus* and its methicillin-resistant strain (MRSA), the molecular aspects of the identified phytochemicals' binding affinity with several targets involved in peptidoglycan biosynthesis were investigated. Most of the marketed drugs commonly applied for managing *S. aureus* and MRSA are those designed for hampering its peptidoglycan biosynthesis, the crucial component of the bacterial cell wall [58]. Typically, peptidoglycans confer the bacterial cell wall's flexibility and robustness and thus interfering with their biosynthesis would mediate bactericidal actions [59]. The presented study explored the potential of identified phytochemicals to block relevant steps across peptidoglycan biosynthesis. The bacterial MurE ligase is typically involved within the cytosolic biosynthesis of peptidoglycan's starting units: UDP-*N*-acetylglucosamine for producing the UDP-*N*-acetylmuramyl-multi-peptide product [60]. The final stage involved the formation of linear peptidoglycans via the DD-transpeptidase catalytic activity of penicillin-binding proteins (e.g., PBP2a) following the transfer of the disaccharide pentapeptides to the cell membrane's outer surface [61]. The development of multi-target drugs has been considered advantageous for circumventing the most common antibiotic resistance mechanism which is the target mutations [62,63].

Out of an evolutionary concept, targeting multiple independent paths for inhibitions is unlikely to allow bacteria to develop resistance over time that would circumvent the pipeline of antimicrobial drug discovery [64]. In these terms, additional targeting of the *S. aureus* teichoic acid-associated β-glycosyltransferase enzyme (TarS) has been considered beneficial to hamper methicillin resistance [65]. This transferase enzyme has been reported to involve several mechanistic aspects enrolled with *S. aureus*'s ability to cope with microenvironmental stresses, biofilm formation, evasion of immune responses, lysozyme resistances, and triggering inflammatory responses [66–71]. The enzyme is chiefly responsible for beta-acylation of bacterial cell wall teichoic acid via *N*-acetylglucosamine being implicated within the IgG-mediated opsonophagocytosis and complement activations at clinical-isolated *S. aureus* strains [72]. Furthermore, resensitization of MRSA strains towards β-lactam antibiotics was reported to be achieved following TarS deletion [73].

For investigating the herein reported anti-human coronavirus 229E (HCoV-229E) activity, targeting the virus main protease (Mpro) has been considered ideal for developing broad-spectrum targeted therapeutics [74]. Owing to the high structural conservation in general across different coronavirus lineages and their integral role within the virus life cycle, these proteases represent promising targets for hampering the virus activities [75,76]. Three coronavirus lineages are identified and classified according to the degree of strain pathogenicity towards humans (lower, HCoV-229E; moderate, HCoV-OC43; and higher, SARS-CoV-2). Thus, targeting singular lineage's Mpro would harbor the potentiality to target, at least to a great extent, the other coronavirus lineages [77–79]. Moreover, these viral proteases lack homologous assemblies within the human biological systems, which highlights the potential safety profile for Mpro inhibitors to any other coronavirus biotargets [80].

Throughout our docking investigation and across the above four designated targets, two identified phytochemicals were depicted with the highest promising affinities (Table 5). Showing high negative docking binding energies (ΔG), both orientin and heptelidic acid were considered promising as multi-target drugs against relevant antibacterial and antiviral target molecules. Docking energies for both phytochemical compounds were higher at *S. aureus* targets MurE and *HCoV-229E* Mpro as compared to their respective ones towards *S. aureus* PBP2a and TarS. On the contrary, both ferulic acid and kojic acid were modest binding energies against all investigated targets conferring their lower relevant affinities for these targets. To our delight, both top-docked phytochemicals depict higher docking scores and predicted affinity towards TarS as compared to a reference target inhibitor. On the other hand, both top-docked compounds had just lower binding energy than the reference inhibitor at the other *S. aureus* targets, MurE and PBP2a, while being only ~0.25-fold lower than the Mpro reference compound.

Table 5. CDOCKER interaction energies for the identified phytochemicals at the binding sites of both bacterial and viral biotargets.

Compounds	Designated Targets			
	S. aureus MurE (PDB; 4c12)	*S. aureus* PBP2a (PDB; 3zg0)	*S. aureus* TarS (PDB; 5tzj)	HCoV-229E Mpro (PDB; 7yrz)
Orientin	−51.75	−49.69	−37.48	−50.15
Heptelidic acid	−49.58	−42.93	−34.54	−39.71
Paxilline	−43.45	−40.51	−31.21	−26.44
Ferulic acid	−39.22	−38.56	−20.76	−25.21
Kojic acid	−22.18	−14.11	−13.23	−19.49
Oleic acid	−21.24	−36.54	−20.15	−21.23
Reference	−54.87	−51.58	−33.67	−69.75

Reference compounds were adopted throughout the docking investigation to ensure the clinical significance of the docking findings. Applying the same docking protocol and algorithm for the reported target inhibitor and/or relevant co-crystalline ligand serves as positive control references permitting comparative docking findings with reported experimental data [81,82]. Herein, a thiazolidinylidene-based compound (T26) was adopted as a positive control as a MurE inhibitor. The reference compound was reported with high dual inhibition activities towards MurE and MurD from *S. aureus* (IC_{50} = 17.0 µM and 6.4 µM, respectively) [83]. Reported studies highlighted close similarity between the MurE secondary structure originating from MRSA and *E. Coli* microorganisms (RMSD 1.48 Å along > 450 Cα-atoms and Z-score 21.2) [60,84]. Furthermore, T26 highlighted great antibacterial activity against MRSA and its wildtype strain with a minimum inhibition concentration of 9.0 µg/mL [83]. Concerning PBP2a, the co-crystallized cephalosporin antibiotic, ceftaroline [85], was a relevant positive control. The novel 5th generation β-lactam drug exhibits broad-spectrum activities, particularly towards the gram-negative bacteria and highly resistant microorganisms, including MRSA, vancomycin-resistant, -intermediate and heteroresistant vancomycin-intermediate *S. aureus* strains [86]. The co-crystallized TarS substrate (UDP-GluNAc) was the suitable comparator, where the ability of investigated phytochemicals to achieve higher docking energies would confer their ability to competitively displace the substrate and hamper the enzyme machinery [65]. Finally, the co-crystallized HCoV-229E Mpro ligand, nirmatrelvir, was adopted as a relevant reference compound where this peptide-like small molecule served as a potent inhibitor of the SARS-CoV2 Mpro enzyme with IC_{50} = 0.79 nM and Ki = 3 nM [87]. The superior docking score of nirmatrelvir can be highly rationalized to its reported great inhibition activity down to the low nanomolar concentrations.

To highlight the differential binding affinities for the top-docked phytochemicals, a comprehensive evaluation of the ligand's orientation/conformation and residue-wise interactions were undertaken at each target. Interestingly, molecular docking of orientin and heptelidic acid at *S. aureus* MurE revealed preferential anchoring of the hypertensive compound at the binding domain of co-crystallized product UDP-N-acetylmuramyl-tripeptide (UNAM-tripeptide). Typically, the product binds predominantly across the central domain of the ligase protein near the ATP-binding site (Figure 8A). Several MRSA MurE key residues have been reported as important, including Asp406, Ser456, and Glu460, for product/substrate binding and recognition [60], as well as an affinity for promising inhibitors [88–91]. Both the negatively charged sidechains of Asp406 and Glu460 as well as the polar mainchain of Ser456 served as the electrostatic trap mediating the stability of UNAM-tripeptide at the binding site. Validation of the docking protocol was highlighted through redocking the co-crystallized ligand under the same adopted parameter, highlighting great superimposed alignment for the co-crystallized and redocked conformation (RMSD = 1.8 Å) (Figure 8A). Furnishing RMSD below 2.0 Å for the co-crystallized ligand to its reference conformation/orientation signifies that both the adopted docking parameters and algorithms were efficient for predicting relevant binding poses, highlighting respective biological significance and, in turn, the docking energies [92].

Residue-wise interactions for orientin and heptelidic acid depicted a wide polar network with surrounding residues. Orientin predicted polar interactions with magnesium ions, besides hydrogen bonding at (hydrogen bond-Donor–Acceptor at angles/distances) with Tyr462 sidechain –OH (2.8Å/135.2°), Thr111 mainchain C=O (2.1Å/129.1°), Lys114 sidechain –N$^+$H$_3$ (2.5Å/128.2° and 2.1Å/144.2°), His205 sidechain NHτ (2.3Å/124.2°), Asn407 sidechain –NH$_2$ (2.4Å/126.7° and 2.3Å/121.1°), and Glu460 sidechain OH (2.4Å/128.0°). Furthermore, several hydrophobic contacts with surrounding non-polar residues are also shown in Figure 8B. Hydrophobic π-mediated contacts with Tyr351 further stabilized orientin at the catalytic site. Owing to its smaller size, heptelidic acid predicted a preferential orientation towards domain II of the active site furnishing several polar contacts with Mg^{2+}, Lys114 sidechain –N$^+$H$_3$ (2.1Å/155.7° and 2.7Å/133.8°), Thr152 (sidechain OH; 2.3Å/174.1° and mainchain –NH; 3.0Å/144.2°), His205 sidechain NHτ (2.5Å/152.6°), and Arg383 sidechain–NHτ– (3.1Å/124.8°). The latter could rationalize the inferior docking score of heptelidic acid as compared to orientin. The sesquiterpene lactone phytochemical predicted favored van der Waal contacts via its hydrophobic cage-like structure with the surrounding pocket residues including Ala150, His181, His353, and Met379 (Figure 8C). Finally, docking of the reference positive control, T26, at MurE highlighted dominant electrostatic potentiality guiding its anchoring at the substrate site with extended orientation/conformation across the domain I/II/III of the active site explaining its relatively higher docking energy as compared to docked phytochemicals. Interactions with Mg^{2+}, Thr46 sidechain OH (2.6 Å/159.1°), Asp406 sidechain C=O (2.6 Å/123.3°), Asn407 sidechain –NH$_2$ (2.6 Å/118.4° and 2.8 Å/115.0°), and Glu460 sidechain –OH (2.0 Å/140.5°) residues were highlighted (Figure 8D).

Exploring the final stage of peptidoglycan synthesis, targeting PBP2a has been considered beneficial for hampering MRSA survival. Generally, the catalytic active site of PBP2a resides at the transpeptidase domain residing within an open groove on the protein surface readily accessible to ligands (Figure 9A) [85]. Notably, three conserved motifs have been reported to cluster around the active sites while harboring the catalytic serine and all the residues required to activate the catalytic hydroxyl group for a nucleophilic attack. The first motif comprises the S-X-X-K (Ser403-Thr404-Gln405-Lys406) tetrad where the catalytic serine resides and its sidekick, lysine amino acid, can exhibit their vital role for organizing the nearby residues as well as minimizing the pKa of the catalytic serine-OH [93]. The second and third conserved motifs are composed of the S-X-N (Ser462-Asp463-Asn464) and K-X-G (Lys570-Ser571-Gly572) triads. The characteristic tetrad and triad motifs adopt strikingly similar conformations in a way that makes all active sites within the serine-based PBPs appear just the same [94]. Interestingly, the β-lactam-inhibiting enzymes (β-lactamases), which are responsible for bacteria resistance through β-lactam catalytic hydrolysis, exhibit the same three conserved motifs making them evolutionary and mechanistically related to all PBPs [94,95]. Such observations explained how penicillins, cephalosporins, and carbapenems exhibit affinity for several PBPs and β-lactamases, where the latter can confer bacterial resistance. Therefore, introducing non-β-lactam-based antimicrobial agents, like propranolol, to circumvent the overgrowing resistance against β-lactam antibiotics is considered highly rationalized [95].

Redocking the co-crystallized cephalosporin antibiotic, ceftaroline [85], provided a validation tool for the adopted docking protocol and algorithm. At the depicted aligned RMSD of 0.5 Å, the redocked ceftaroline managed to replicate its co-crystallized conformation/orientation and residue-wise patterns (Figure 9A). Polar interaction with Ser462 sidechain –OH (2.2 Å/157.4°), Thr600 mainchain C=O (2.1 Å/169.6°), and Glu602 sidechain –NH (1.8 Å/140.3°) were conserved towards the ligand's polar functionalities of the opened β-lactam ring, amidic sidechain, and thiadiazole ring substitution (Figure 9B). Stacking between the ligand's thiazole ring and Tyr446 sidechain through close range π-π hydrophobic contact (4.1 Å) provides extra stability near the conserved S-X-N motif. Docking orientin at PBP2a was dominant through polar interaction with the Tyr519 mainchain C=O (2.1 Å/133.2°), Gln521 sidechain C=O (2.9 Å/127.4°), Ser462 sidechain –OH (2.4 Å/121.0°), and Asn464 sidechain (–NH$_2$; 2.8 Å/162.4° and C=O; 2.6 Å/124.7°). Orientin stability was further mediated through non-polar contacts with surrounding residues (Ala601 and Met641) as well as hydrophobic π–π interaction between the compound's resorcinol ring and the Tyr446 sidechain (Figure 9C). For heptelidic acid, limited polar interactions were depicted at the PBP2a binding site since few polar networks were seen with the Lys406 sidechain –N$^+$H$_3$ (2.8 Å/162.4°), Ser462 sidechain –OH (2.8 Å/162.4°), and Asn464 sidechain –NH$_2$ (2.8 Å/162.4°) (Figure 9D). The latter docking observation could be related to less inherited structural flexibility of heptelidic acid, the thing that limits its conformational maneuvers conferring a lower docking score to orientin. The lack of the compound's aromaticity could provide a reason for the fewer hydrophobic interactions depicted by heptelidic acid towards the pocket lining residues.

Investigating the compounds' residue-wise interactions at the TarS catalytic site would provide valuable insights regarding the ability of top-docked phytochemicals to interfere with bacterial virulence and biofilm production [66–71]. The enzyme catalytic site is settled at the carboxy-terminal domain exhibiting the canonical GTA folding (double α/β/α sandwiched Rossman motifs) (Figure 10A) [65]. The binding site is enclosed within two key loops: (a) the catalytic site loop (CS-loop; Glu171–Asp178) encompassing the putative base catalytic residue Asp178; (b) the substrate access loop (SA-loop; Lys205–Tyr215). Several pocket residues including Tyr10, Arg75, Asp91, Glu177, Asp178, His210, and Ser212 have been reported as important for recognizing and binding the enzyme's substrate (Uridine diphosphate N-acetylglucosamine; UDP-GluNAc) as well as small molecule TarS inhibitors [65,96–98]. Preliminary redocking of the co-crystallized substrate revealed the validity of the adopted docking protocol where UDP-GluNAc achieved low RMSD (0.9 Å) to its co-crystalline orientation/conformation (Figure 10A). The redocked substrate recaptured the co-crystallized residue-wise interaction patterns including salt bridges with vicinal residues including Arg75 sidechain (sidechain =N$^+$H$_2$; 1.9 Å/150.1° and sidechain –N$^+$H$_2$; 2.3 Å/134.7°), Glu177 sidechain (C=O; 1.8 Å/148.0° and –O$^-$; 3.0 Å/128.8°), Arg206 sidechain –N$^+$H$_2$ (2.2 Å/123.6° and 2.5 Å/124.7°), and Ser212 (mainchain –NH; 1.9 Å/164.0° and sidechain –OH; 2.7 Å/127.3°). Hydrophobic contacts with several prolines (Pro8, Pro71, Pro74, and Pro153) as well as π–π stacking for the pyrimidindione ring with Tyr10 were also relevant at close proximities (Figure 10B). Docking interactions for the identified top-docked phytochemicals were mostly differentiated based on polar contacts with surrounding residues. Owing to the higher number of hydrogen bond donors and acceptors for orientin as compared to heptelidic acid, the earlier depicted a wider polar network towards pocket-lining residues. Hydrogen bonds with Arg126 sidechain =N$^+$H$_2$ (2.3 Å/129.4°), His210 sidechain NHτ (2.9 Å/125.5°), Ser213 mainchain –NH (2.3 Å/142.9°), and Ala214 mainchain –NH (2.6 Å/123.9°) were predicted for orientin at optimum angles and distances (Figure 10C). Displaced face-to-face π–π stacking was depicted between orientin's resorcinol ring and Tyr10 at a close distance. Fewer polar contacts were depicted for heptelidic acid (Ser92 sidechain –OH; 2.0 Å/152.7° and Met211 mainchain –NH; 2.3 Å/124.7°) with limited hydrophobic contacts (Figure 10D).

Moving towards the compound's differential Mpro-related affinities, a comprehensive examination of the ligand's residue-wise interaction was conducted. Typically, the target's substrate pocket illustrated that the co-crystallized binary complex comprises four sub-sites (S$_1'$-to-S$_4$) for anchoring the four peptido-partitions (P$_1'$-to-P$_4$) of its substrate (Figure 11A) [74]. Within the current literature, several Mpro pocket amino acids are identified as important for binding different ligands [77,80,99–101]. Binding to the S1' sub-site, especially towards the Mpro catalytic dyads His41 and Cys144, has been identified as important for strong ligand–protein interactions and enzyme hydrolytic activity blockage [102]. Significant non-polar contacts with the sidechains of either Glu165 or Asn189 at the S3 sub-site, as well as S2 sub-site residues (Ala49 and Leu190), can serve as hydrophobic grips for anchoring different small molecules in the enzyme's pocket [74]. Regarding polar binding interactions, both carbonyl and nitrogen of the Glu165 mainchain were highlighted as crucial for providing relevant ligand–Mpro binding at the S1 sub-site. Several other amino acids including Asn24, Thr25, Ser168, His171, Phe184, and Ala195 were reported in literature as being relevant for preferential ligand binding [99–101]. Initially, the furnished docking poses and energies were considered valid since preliminary redocking of co-crystallized nirmatrelvir showed a root-mean-squared deviation: RMSD = 1.72 Å (Figure 11A). Redocked nirmatrelvir was able to replicate the co-crystallized ligand–Mpro binding interactions showing double polar hydrogen bonds with the S1 pocket Glu165 mainchain (–NH; 1.9 Å/171.8° and C=O; 2.0 Å/168.9°) via the ligand's amide moiety (Figure 11B). The ligand's pyrrolidinyl moiety mediated polar interaction with the Glu165 sidechain oxyanion (2.5 Å/128.2°) and the Phe139 mainchain C=O (2.5 Å/141.6°). The ring further mediated the hydrogen bond with the S1 pocket His162 sidechain NHτ (1.8 Å/168.4°). Additional hydrogen bonds between the compound's central amide linker and the S2 pocket Gln163 mainchain C=O (2.2 Å/163.8°) were also depicted. Close-range hydrophobic interactions (π-CH) towards His41 and the ligand's bicyclic ring (5.0 Å) served to further the ligand's stability at the S1' sub-site. Further van der Waal contacts with non-polar pocket lining residues Ile51, Ala143, Ile164, Leu166, and Pro188 were observed. Altogether, these favored ligand–target interactions would be translated into superior docking binding scores corresponding to the reported experimental in vitro Mpro inhibition assay (IC$_{50}$ at low-range nanomolar concentration). To our delight, the identified phytochemicals revealed relevant ligand accommodation at the Mpro binding site. Orientin depicted extended orientation across the four sub-sites, S1'–S3 (Figure 11C). Lodging at the S1' sub-site was solely relevant for the orientin as compared to the other identified phytochemicals through polar interaction with Gly142 mainchain –NH (2.2 Å/163.8°). Further T-shaped π–π hydrophobic contact was shown between the compound's resorcinol ring and S1's pocket His41 sidechain. Further polar interactions were predicted for orientin including residues of pockets S1 Glu165 mainchain (–NH; 2.2 Å/163.8° and C=O; 2.1 Å/146.5°), His162 sidechain NHτ (2.2 Å/163.8°), as well as vicinal residue Thr25 sidechain –OH (2.2 Å/163.8°). Non-polar van der Waal contacts with Ile51, Phe139, Ile140, Ala143, and Pro188 were also depicted. Moving to heptelidic acid, a lower extent of polar interaction was depicted. Interactions with Pocket S1 His162 sidechain NHτ (2.2 Å/156.4°) and S1' Cys144 sidechain –SH (2.9 Å/140.5°) were only depicted for the sesquiterpene lactone derivative (Figure 11D). Such differential ligand's residue-wise interactions confer higher docking energy for orientin regarding heptelidic acid. Despite limited polar interactions, heptelidic acid is predicted to mediate several non-polar contacts with several residues (His41, Ile164, Leu166, and Pro188) owing to its cage-like architecture and isopropyl arm chain. This could partially compensate for the limited electrostatic interactions predicted by this smaller-sized phytochemical compound.

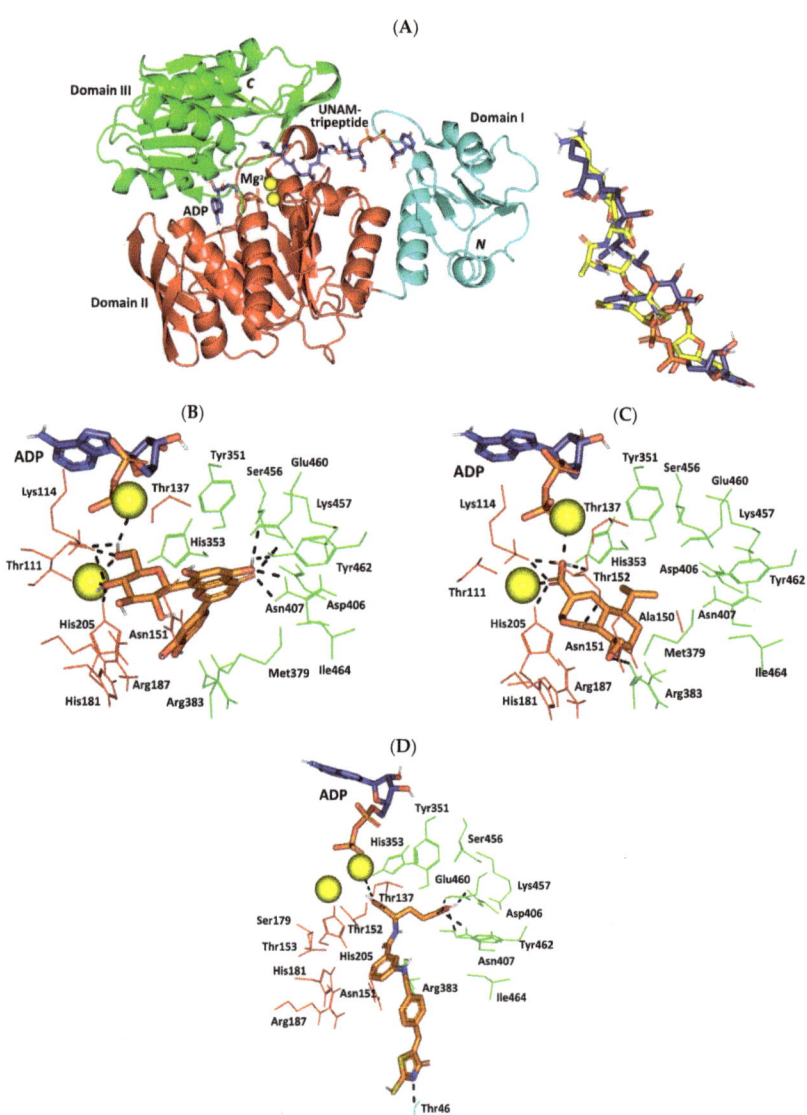

Figure 8. The architecture of *S. aureus* MurE and depicted molecular docking poses. (**A**) **Left panel**: Cartoon 3D-representation of *S. aureus* MurE (PDB; 4c12) ligase showing structural domains; I, II, and III (cyan, red, and green, respectively) bound to two magnesium ions (yellow) and the co-crystallized adenosine diphosphate (ADP) and product UDP-*N*-acetylmuramyl-tripeptide (UNAM-tripeptide) as blue sticks. Bold *C* and *N* letters denote the carboxy and amino terminals. **Right Panel**: Aligned redocked MurE product (UNAM-tripeptide; yellow) over its co-crystalline state (blue). Predicted binding mode of (**B**) orientin, (**C**) heptelidic acid, and (**D**) antibacterial T26 as reference ligand. Only surrounding residues within 5 Å radius as lines are shown and polar interactions are illustrated as black dash lines.

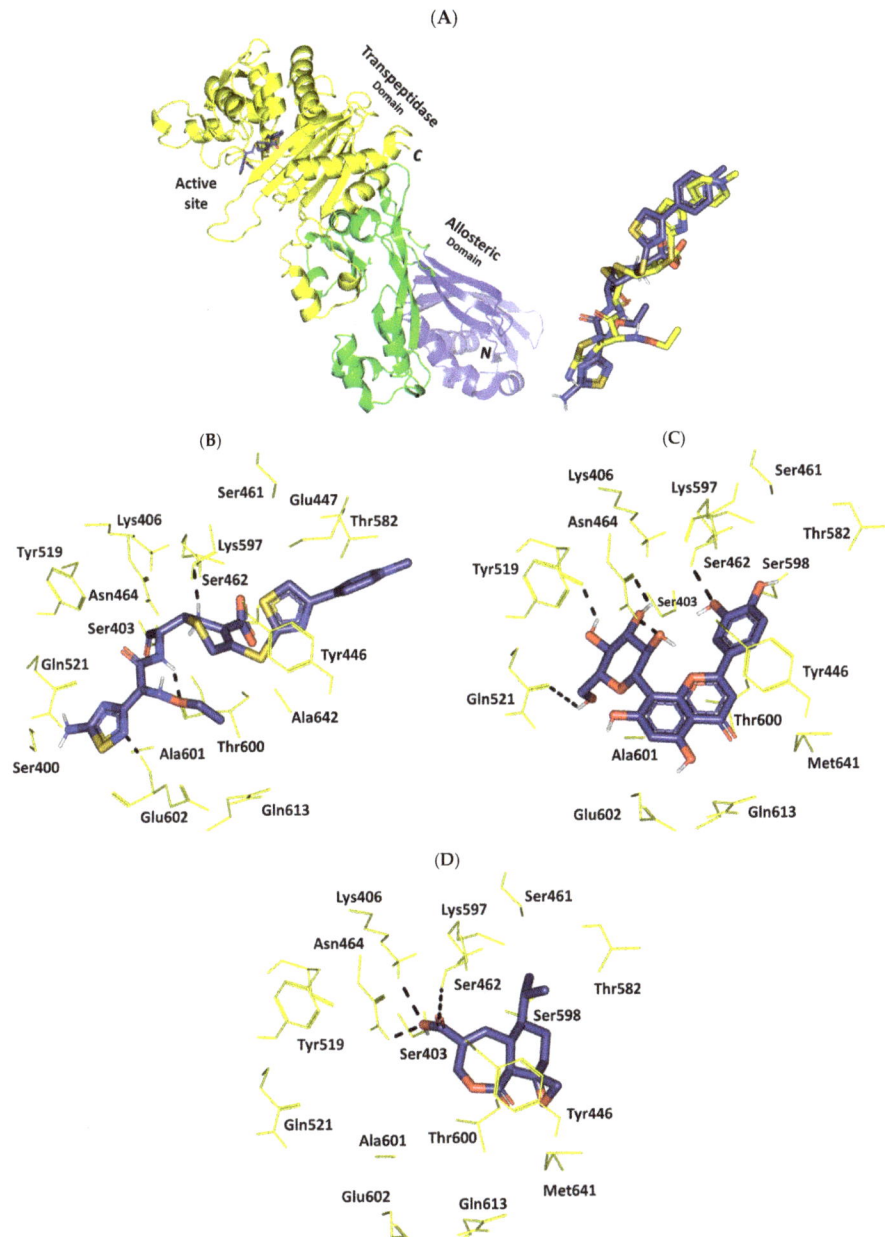

Figure 9. The architecture of *S. aureus* PBP2a and depicted molecular docking poses. (**A**) **Left Panel**: Cartoon 3D-representation of *S. aureus* PBP2a (PDB; 3zg0) transpeptidase enzyme in complex with ceftaroline co-crystalline ligand (blue sticks) and showing structural domains; transpeptidase domain (yellow) and allosteric domain (blue green). Bold letters *C* and *N* denote carboxy and amino terminals. **Right Panel**: Aligned redocked ceftaroline (yellow) over its co-crystalline state (blue); predicted binding mode of (**B**) ceftaroline as positive reference control, (**C**) orientin, and (**D**) heptelidic acid. Only surrounding residues within a 5 Å radius as lines are shown and colored as per constituting domains. Polar interactions are illustrated as black dash lines.

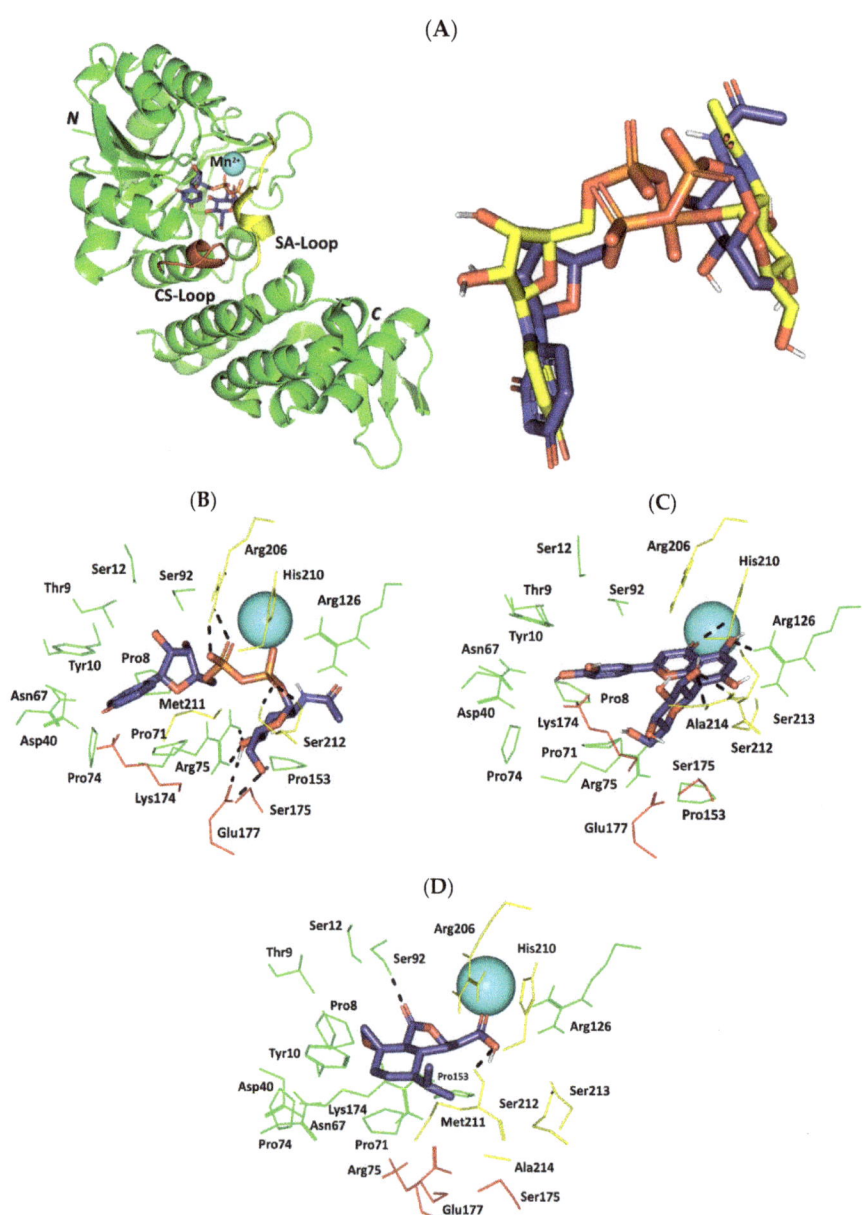

Figure 10. The architecture of *S. aureus* TarS and depicted molecular docking poses. (**A**) **Left Panel**: Cartoon 3D-representation of *S. aureus* TarS (PDB; 5tzj) catalytic domain in complex with co-crystalline substrate (UDP-GluNAc; blue sticks) with key structural loops; CS-loop (Glu171–Asp178; red) and SA-loop (Lys205–Tyr215; yellow). Bold letters C and N denote carboxy and amino terminals. **Right Panel**: Aligned redocked UDP-GluNAc (yellow) over its co-crystalline state (blue). Predicted binding mode of (**B**) UDP-GluNAc as a positive reference control, (**C**) orientin, and (**D**) heptelidic acid. Only surrounding residues within a 5 Å radius as lines are shown and colored as per constituting domains. Polar interactions are illustrated as black dash lines.

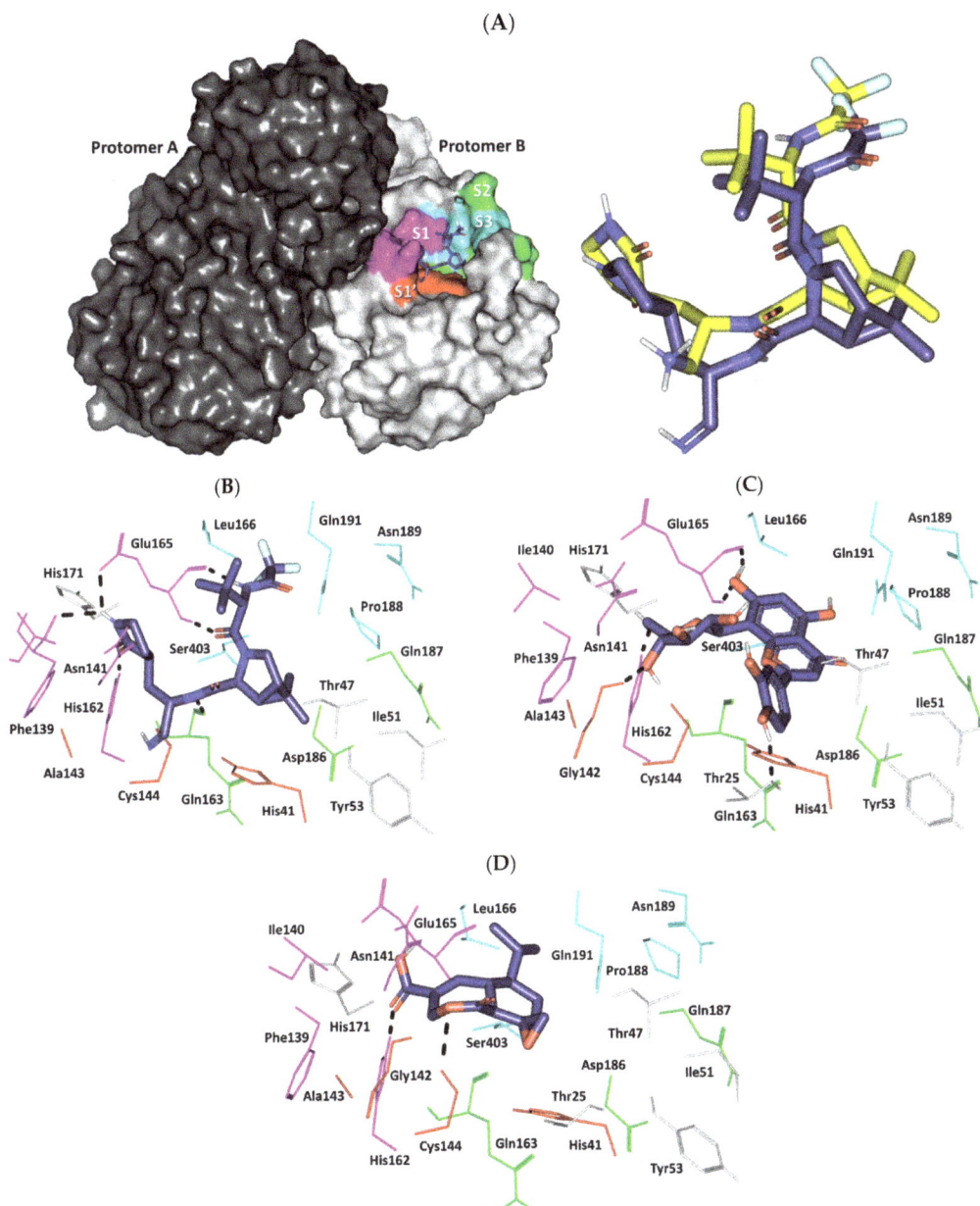

Figure 11. The architecture of HCoV-229E Mpro and depicted molecular docking poses. (**A**) **Left Panel**: Cartoon 3D-representation of HCoV-229E Mpro (PDB; 7yrz) in its dimeric state (dark/light grey surface colors for respective protomers A/B) with its canonical substrate binding site comprising important sub-sites (S1′ as red, S1 as magenta, S2 as green, and S3 as cyan). **Right Panel**: Aligned redocked nirmatrelvir (yellow) over its co-crystalline state (blue); predicted binding mode of (**B**) Nirmatrelvir as positive reference control, (**C**) orientin, and (**D**) heptelidic acid. Only surrounding residues within a 5 Å radius as lines are shown and colored as per constituting domains. Polar interactions are illustrated as black dash lines.

2.9. Molecular Dynamics Studies

Thermodynamic behaviors of the identified compound–target proteins complexes were monitored to positive reference compounds (MurE: T26, PBP2a: Ceftaroline, TarS: UDP-GluNAc, and Mpro: Nirmatrelvir) as well as the Apo protein states through explicit molecular dynamics simulation. This approach has provided valuable molecular insights regarding compound–target relative stabilities, conformational changes, and favored interactions under near-physiological conditions [103–105]. Regarding the initial structures, the RMSD trajectories were tracked for the simulated bound proteins. Altered conformations and compromised stability are typically correlated with high target RMSDs [106], whereas ligands with excellent pocket accommodation are related to steady and small-value RMSDs [107]. Simulated proteins showed typical thermodynamic behaviors as alpha-carbon RMSD trajectories showed low initial values that increased within the first few steps and then more or less leveled off around respective averages for more than half of the simulations (Figure 12A). Interestingly, monitored RMSDs for all compound-bound (holo) proteins were at lower average values and less fluctuating tones (4.54 ± 0.61 Å to 3.00 ± 0.49 Å for MurE, 3.40 ± 0.57 Å to 3.47 ± 0.63 Å for PBP2a, 2.30 ± 0.20 Å to 3.00 ± 0.41 Å for TarS, and 2.46 ± 0.33 Å to 2.58 ± 0.28 Å for Mpro) as compared to the apo state (without bound compound/unliganded) (5.92 ± 1.22 Å for MurE, 6.77 ± 0.96 Å for PBP2a, 3.92 ± 0.67 Å for TarS, and 3.46 ± 0.47 Å for Mpro). Higher fluctuation patterns (in terms of magnitudes and/or frequencies) were assigned for proteins inbound with heptelidic acid over those of orientin only at TarS, while being indistinguishable across the PBP2a and Mpro simulations. In terms of MurE protein RMSDs, the heptelidic acid-bound protein exhibited higher tones than that of orientin only across 30–70 ns, while kept at lower values for most of the simulation run. Differential RMSDs for simulated proteins showed higher fluctuations for MurE and PBP2a as compared to those of TarS and Mpro. These depicted dynamic behaviors can be partially correlated to the differential secondary structure and B-factor index of the bound proteins.

Concerning the sole simulated compound monitoring, the ligands' RMSDs were monitored through selecting the carbon-alpha center of their respective bound proteins for their least-square fit analysis. The grouped ligands atoms (based on the GROMACS index.file) were selected to perform the RMSD analysis. Applying such approaches have provided a relevant description to understand whether a specific ligand was retained within its binding pose, confined within the binding site, or not throughout the dynamics runs. Notably, RMSDs of respective compounds highlighted differential stability across the simulated times (Figure 12B). As a general observation, limited fluctuations with steady RMSD tones were assigned for orientin across all bound protein systems (3.25 ± 0.50 Å for MurE, 3.01 ± 0.42 Å for PBP2a, 3.17 ± 0.68 Å for TarS, and 3.29 ± 0.56 Å for Mpro) as compared to heptelidic acid (4.68 ± 1.78 Å for MurE, 64.43 ± 39.37 Å for PBP2a, 35.29 ± 11.97 Å for TarS, and 9.67 ± 4.09 Å for Mpro). The orientin's RMSD tones within the MurE complex were observed statistically indistinguishable from its own ones at the other three complexes. On the other hand, orientin's RMSD trajectories were depicted comparable to those of the TarS reference ligand (UDP-GluNAc; 2.48 ± 0.68 Å) and positive controls at the Mpro model (Nirmatrelvir; 2.48 ± 0.36 Å), yet even lower than references at the MurE and PBP2a models (T26; 9.17 ± 2.57 Å and Ceftaroline; 4.16 ± 1.74 Å, respectively). The latter thermodynamic behaviors would confer preferential orientin's dynamic stability and confinement within the different target binding sites. Despite higher fluctuations for heptelidic acid, its RMSDs leveled off around an average value starting from a 55-to-60 ns timeframe and till the end of the simulation runs only at the MurE and Mpro systems. Notably, T26 at the MurE complex across 30–70 ns showed high RMSDs before they descend and come to their initial tones. In cases of PBP2a and TarS complexes, heptelidic acid RMSDs were far beyond range (>16 Å) conferring significant drift at new protein sites much farther away from the initial location at the proteins' canonical binding site. Both large RMSDs for heptelidic acid in all complexes and T26 in the MurE complex confer that the ligands moved from their original binding sites to new ones but still remained in contact with the protein, rather than simply dissociating into the solvent. For T26 at MurE, the ligand's drift was quite transit as it managed to return back to the initial site for the last 30 ns of the simulation run. It is worth noting that the depicted RMSDs for orientin and reference ligands never exceeded 2.5-fold the RMSD values of their respective bound proteins, with an exception only for T26 at MurE (~ 4.0-folds). This has been confirmed relevant in the literature for the compound's existence within the binding site as well as successful protein convergence at the end of the simulation demanding no further time extensions [108,109]. Further compound-active site stability was investigated through the time evolution of the ligand–protein complex conformations and ligand orientation analysis. Overlaid timeframes at the beginning and end of the simulation run confirm orientin and reference ligand as relevant accommodations of the binding site (Figure 13).

Monitoring the RMS fluctuations for the holo/apo target proteins to their alpha-carbon references provided further stability analysis. Protein stability and flexibility/immobility profiles were dissected down to their constituent amino acids [110]. RMSFs allow the researchers to comprehend the residue-wise dynamic behaviors at the protein's binding pocket/vicinal loops in addition to pinpointing the key amino acids being significant for ligand binding [111,112]. Normalized RMSFs (ΔRMSF = apoRMSF − holoRMSF) were adopted as better representations of the protein's local flexibility in relation to its apo state. Adopting a ΔRMSF cut-off value of 0.30 Å has been reported as relevant for estimating the significant alterations within the protein's structural movements, meaning that residues depicting ΔRMSF greater than 0.30 Å indicated reduced backbone mobility upon binding [113]. In concordance with the RMSD findings, lower flexibility and mobility tones across almost all protein regions were assigned for the holo proteins in relation to their apo states where the earlier were shown with almost positive ΔRMSF values (Figure 14). This confers the impact of ligand binding on the stabilizing of target proteins' secondary structures. This further suggests that ligand binding would impact protein stability in a manner much extended beyond the canonical binding site affecting even the far protein regions. Additionally, typical protein dynamic behavior was illustrated since higher flexibility profiles were seen for the terminal residues as compared to the core regions, except for the carboxy terminals of *S. aureus* MurE proteins bound with orientin and T26 where binding sites are at proximity distances to the protein's C-terminus. Secondly, the stability-driven impacts of orientin and reference compound binding on the four protein targets were more profound than those of heptelidic acid where the latter was assigned with lower ΔRMSF values. This would further highlight the lower stability profiles of heptelidic acid–protein complexes in relation to orientin and reference compounds in good agreement with ligand drift away from the initial binding site.

Free-binding energy calculations using the trajectory-oriented Molecular Mechanics-Poisson Boltzmann Surface Area (MM-PBSA) approach were performed to understand the nature of top-stable ligand–protein binding, estimating affinity magnitude, as well as individual energy contribution of key binding residues [114]. MM-PBSA calculation possesses the advantage of being comparably accurate to free-energy perturbations, yet with lower computational expenditure [115]. Notably, the free binding energies of the simulated orientin were quite second to the simulated reference compounds at the complex targets: MurE (−66.00 ± 4.46 vs. −71.70 ± 13.08), PBP2a (−41.09 ± 6.87 vs. −51.66 ± 35.89), and Mpro (−115.41 ± 14.87 vs. −176.27 ± 16.42), except for TarS where the identified phytochemical was just superior (−43.76 ± 12.58 vs. −41.51 ± 46.35) (Figure 15). However, the uncertainties on the free binding energies for the reference compounds are so large that they encompass the orientin values. The latter would argue that orientin is second to the reference compound only at the Mpro complex. On the other hand, the provided total ΔG is a relevant translation for all previously presented data including the RMSD and ΔRMSF fluctuations, as well as ligand–target conformational analysis. Just because the Mpro–orientin complex was the one showing the largest difference in total energy from its reference compound does not confer that other protein systems are of negligible difference. In this regard, we would argue that the differential binding free energies for orientin and references incorporate contributions from RMSD and ΔRMSF so that these are already accounted for, due to the fact that the orientin's binding to a specific protein is quite different as compared to that of the reference compounds. To further confirm such an argument, the differential binding energy terms between orientin and reference compounds were investigated within the forthcoming text.

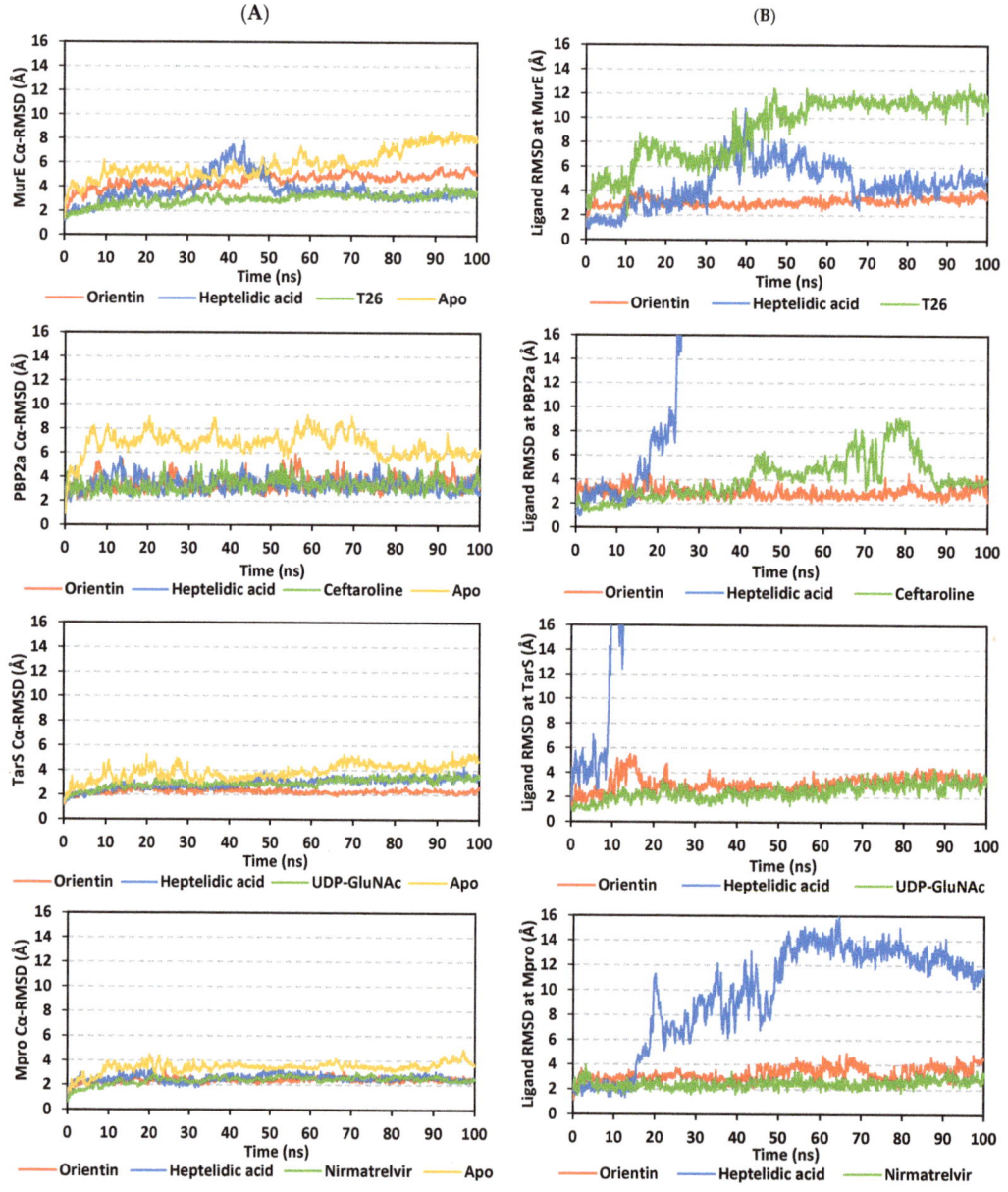

Figure 12. Thermodynamic stability analysis of the explicit molecular dynamics simulated compounds inbound to *S. aureus* and HCoV-229E biotargets. (**A**) Alpha-carbon atom RMSDs for protein (holo and apo states); (**B**) sole ligand RMSDs, in relation to simulation timeframes (ns).

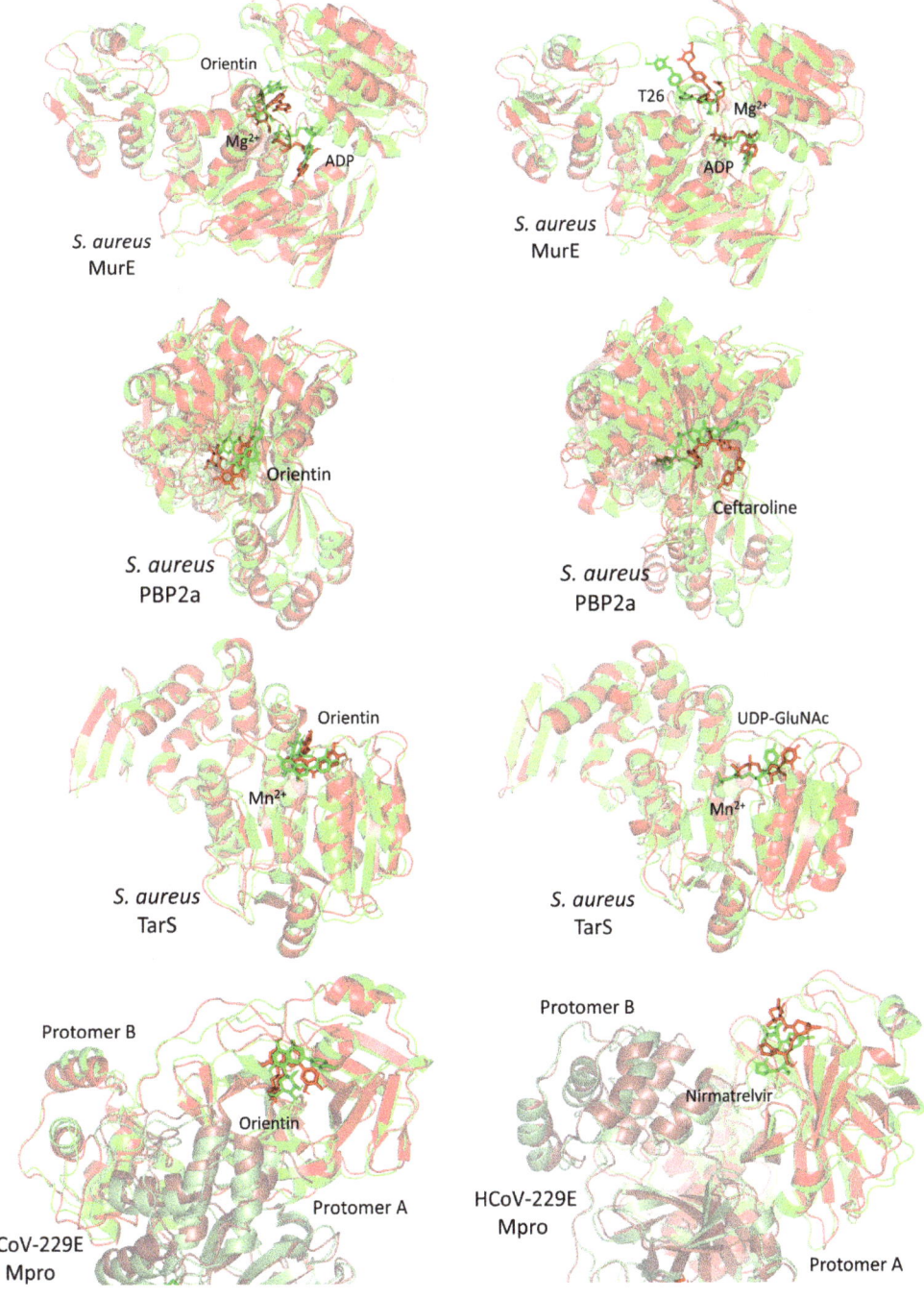

Figure 13. Conformational analysis for molecular dynamics simulated compounds inbound to *S. aureus* and HCoV-229E biotargets. Overlaid ligand–target snapshots at initial and final timeframes. Top-stable compounds (orientin and reference ligands—sticks) and bound proteins (cartoons) are colored green and red concerning 0 ns and 100 ns extracted frames.

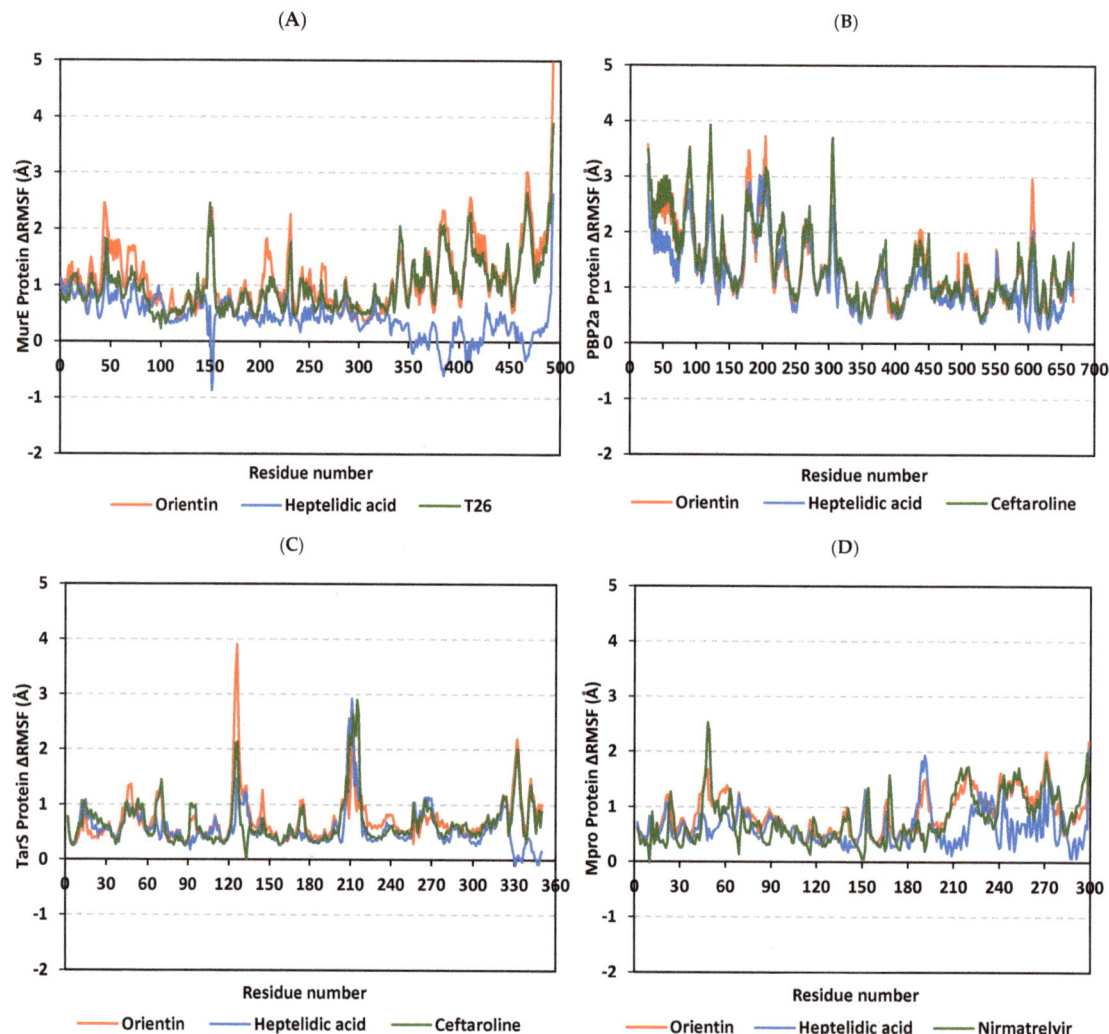

Figure 14. Global stability of simulated proteins down to their constituting residues. Difference RMSF analysis of inbound *S. aureus* (**A**) MurE, (**B**) PBP2a, (**C**) TarS, and (**D**) HCoV-229E Mpro proteins along the whole molecular dynamics runs highlighting the residue-wise flexible contributions of holoprotein in relation to the apo/unliganded states.

Dissection of the free binding energies showed that the electrostatic potential energies ($\Delta G_{electrostatic}$) were dominant over van der Waal hydrophobic energy contributions (ΔG_{vdW}) driving both the orientin and T26 stabilities at the MurE and TarS complex systems. On the other hand, ΔG_{vdW} showed predominant free-binding energy contributions for orientin's affinity towards the PBP2a and Mpro models. Dominant ΔG_{vdW} contribution fashions were also depicted with reference ligands only at the PBP2a and Mpro systems, while a profound $\Delta G_{electrostatic}$ contribution was seen for UDP-GluNAc at TarS. Interestingly, the high combined non-polar free binding interactions (sum of ΔG_{vdW} and non-polar solvation; ΔG_{SASA}) for the simulated ligand–target complexes might be directly related to the targets' large pocket surface area.

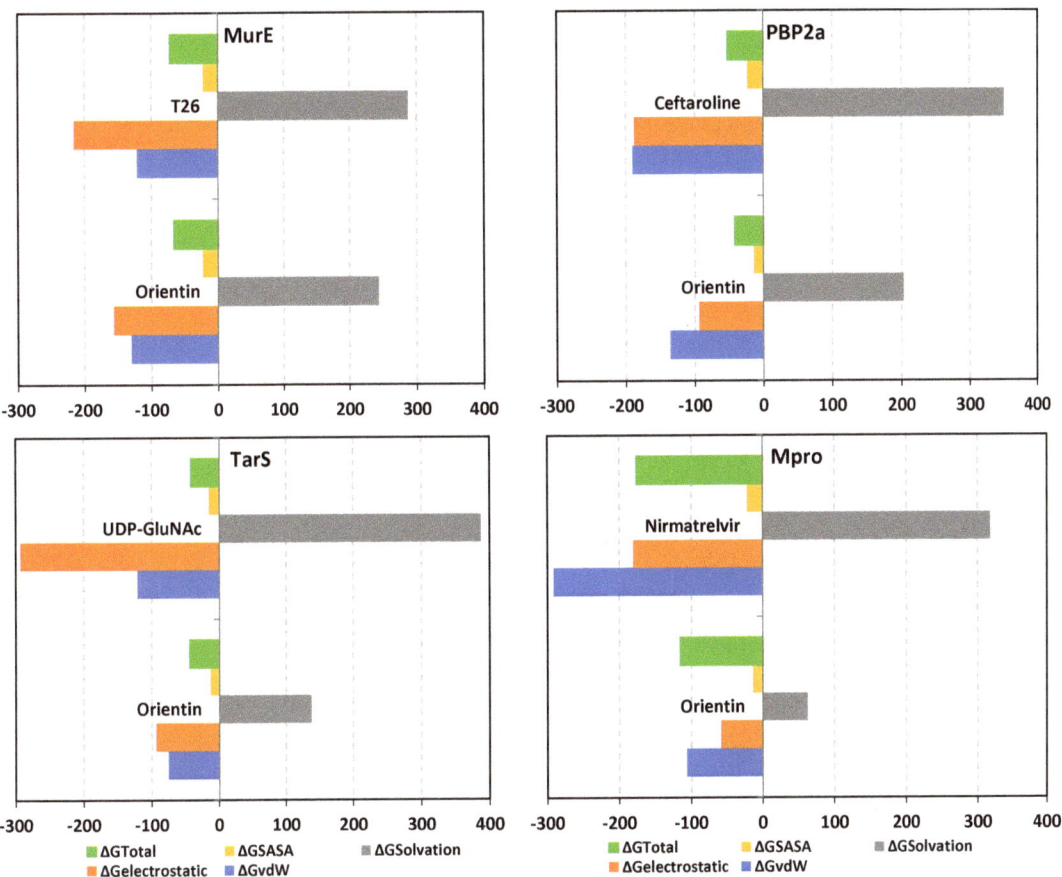

Figure 15. MM_PBSA free-binding energy calculations and constituting energy term contributions for the ligand–protein target complexes.

Concerning the polar solvation energy contributions, orientin was assigned with much lower polar solvation energies ($\Delta G_{polar\ solvation}$) across all simulated systems when being compared to reference ligands at corresponding target proteins. The latter was suggested to be in favor of orientin–target affinity since binding has been considered a solvent substitution process [116–120]. Harboring significant aromatic/heterocyclic structural features could allow reasonable compensation of solvation entropy and final relevant total free-binding energy profiles for orientin. On the other hand, higher solvation penalties for reference compounds could be related to the presence of several ionizable groups in contact with hydrophobic pocket sides that would compromise the totally free binding process. Based on the presented structural postulations, prospective structural optimization of orientin can be achieved through balanced hydrophobic/hydrophilic characters. Introducing ionizable scaffolds furnishing increased polarity while possessing relevant aromatic characteristics would be advantageous for minimizing the solvation penalty and maximizing the target affinity. Suggested scaffolds include tetrazole rings and other relevant cyclic carboxylate-related bioisosteres [121].

Finally, it is worth mentioning that significant differential patterns have been highlighted with distinctive energy term preferentiality between the reference compound and orientin at every target system. Thus, from the obtained MM-PBSA calculations, orientin's binding to a specific protein is quite different as compared to that of the reference compounds, the thing that again successfully reflects the findings obtained at the previous MD analysis parameters, including the RMSD and ΔRMSF analysis as well as the ligand–target conformational investigation.

For gaining more insights concerning ligand–residues interactions, the binding-free energy was dissected down to its residues' contribution to identifying key residues [115]. Residues of the active binding site depicted favored contributions (large negative values) within the ligand–protein binding energies of orientin and reference ligands (Figure 16). Adopting ≤ -5.00 kJ/mol cut-off for significant energy contributions [122], residues Lys62, Lys114, His205, His353, Arg383, Asp406, and Glu460 were illustrated as most important for compound binding at *S. aureus* MurE with the highest contributions being assigned for Lys114, Asp406, Glu460 (-15.37 to -17.24 kJ/mol), and His205 being the most (up to -28.35 kJ/mol). Concerning the PBP2a complex systems, top-favored contributing residues included Tyr446, Ser462, Asp463, Asp573, and Glu602 with the highest contributions for Asp463 (up to -12.03 kJ/mol) and Glu602 (up to -16.08 kJ/mol). Moving to the TarS models, residues like Tyr10, Arg75, Asp91, Asp94, Asp95, Arg126, Glu171, Glu172, Glu177, Asp178, Lys205, Arg206, Glu207, and Glu209 were significant for orientin and UDP-GluNAc binding stability. The dominant polar nature of top-contributing TarS residues further confirms the dominant impact of $\Delta G_{electrostatic}$ potentials on ligand binding. For the final target, HCoV-229E Mpro, several residues of the four sub-pocket and vicinal regions were involved in high-negative energy contributions (≤ -5.00 kJ/mol), including Ala49, Phe139, Cys144, His162, Gln163, Glu165, His171, and Phe184 with a dominant hydrophobic nature. It is worth mentioning that several other pocket residues showed significant positive energy contributions inferring repulsion forces and unfavored impact on the ligand's stability. Thus, the addition of balanced hydrophobic/ionizable scaffolds was further highlighted as significant for ligand anchoring. Finally, the above-depicted energy residue-wise findings were consistent with the above-described docking hydrophobic/polar contact preferentiality.

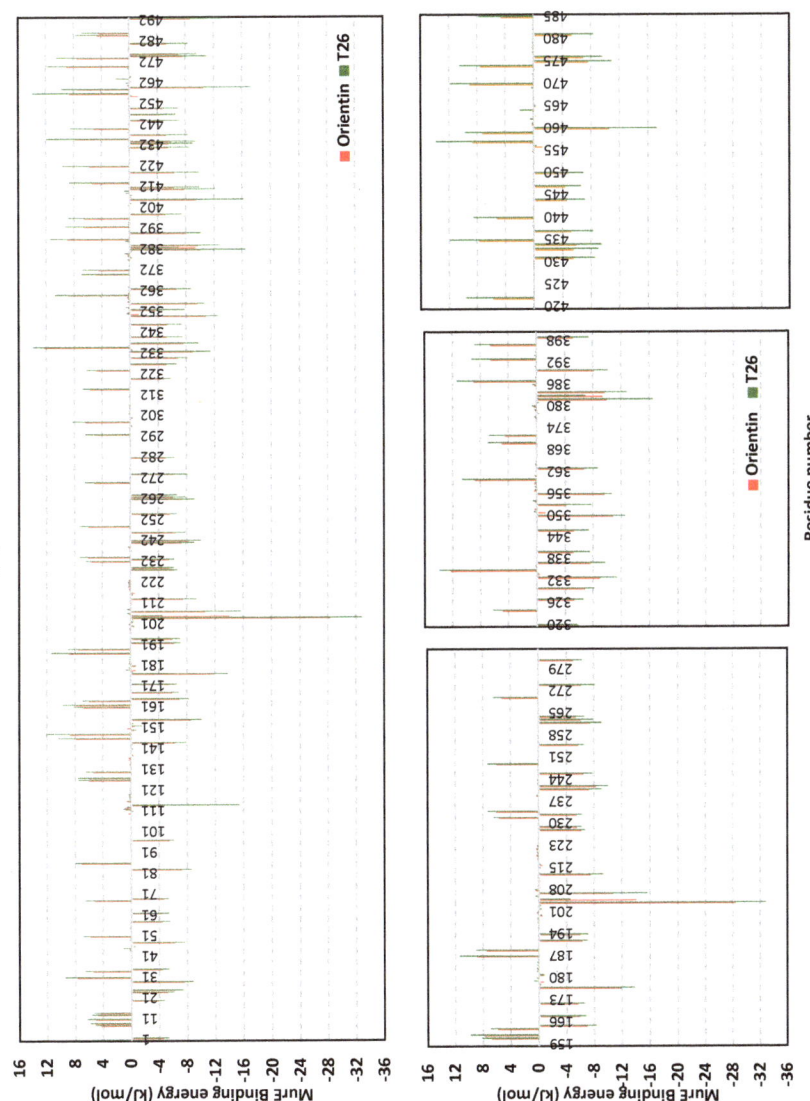

Figure 16. *Cont.*

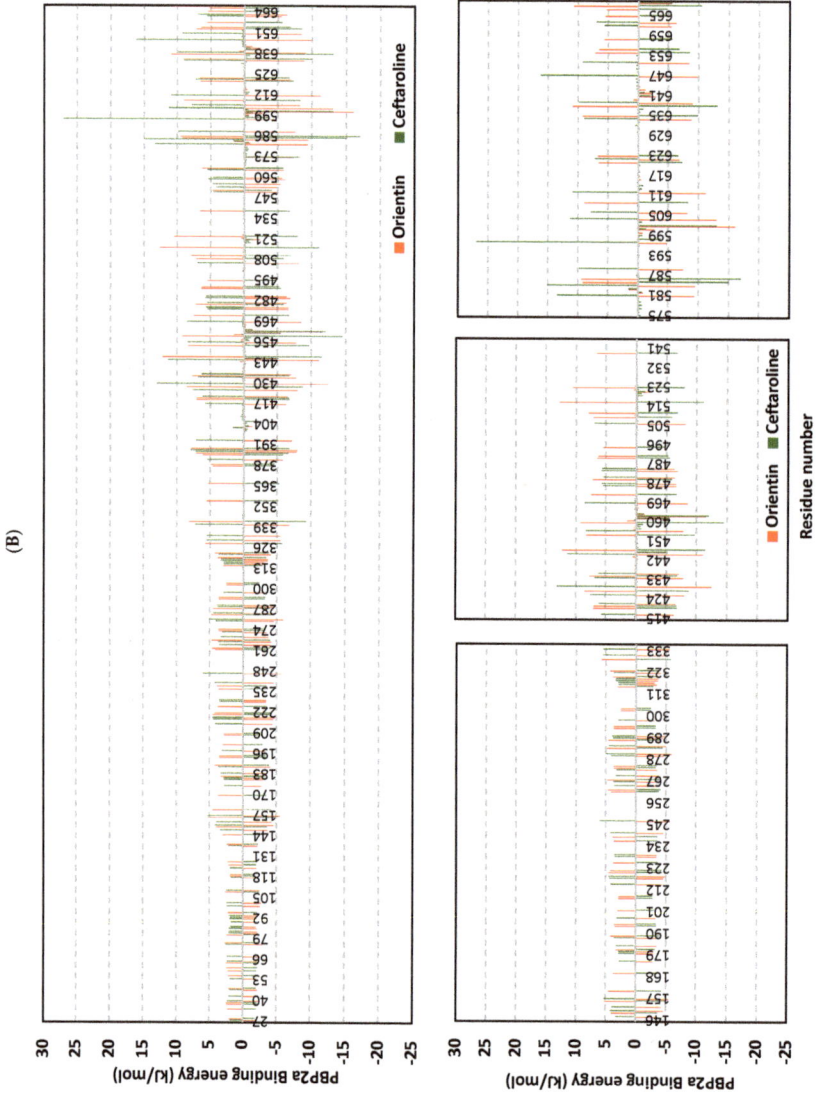

Figure 16. *Cont.*

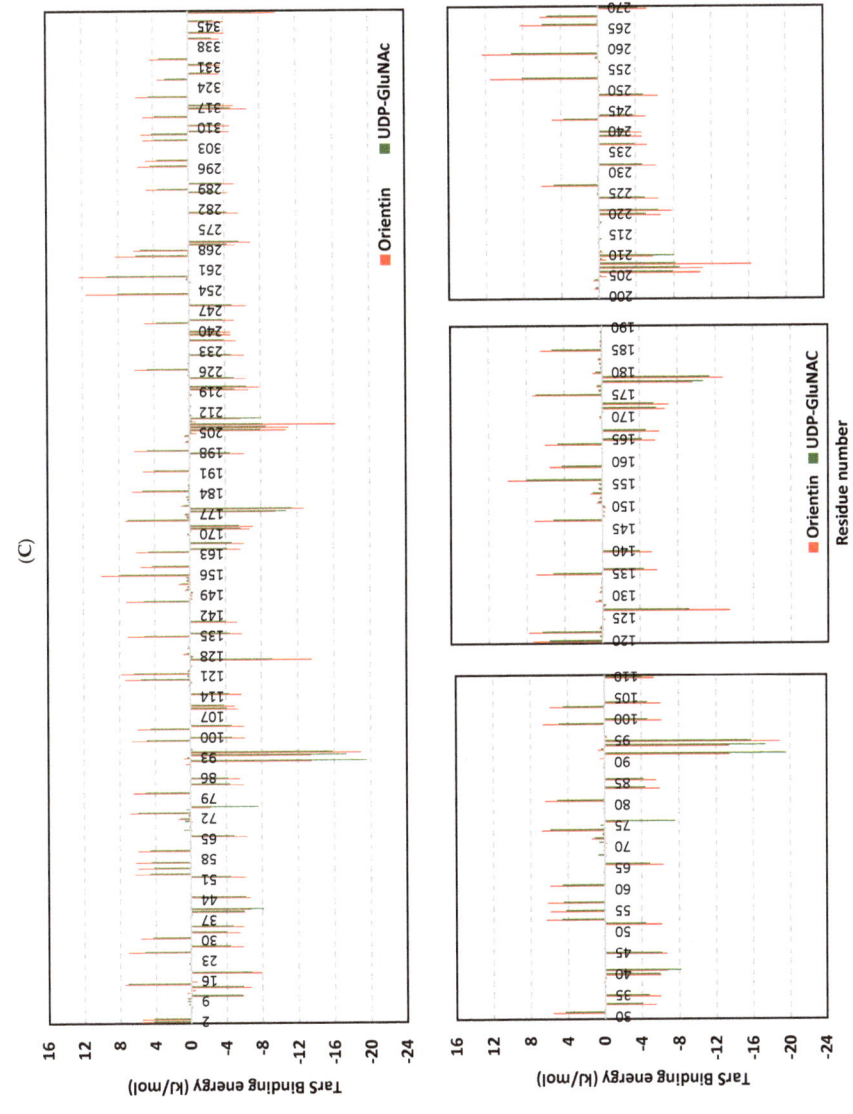

Figure 16. Cont.

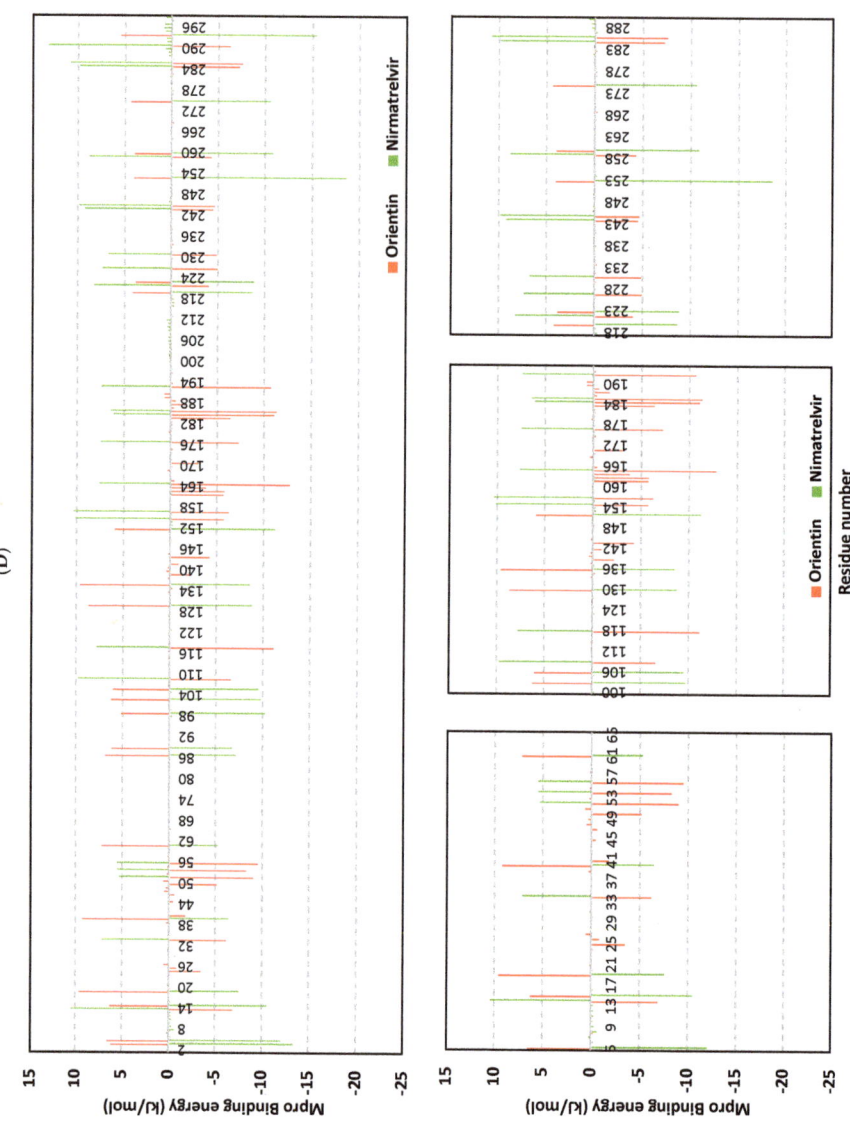

Figure 16. Residue-based energy contributions for ligand–protein target complexes. Lower panels are expanded residue ranges of the simulated protein complexes: (**A**) MurE, (**B**) PBP2a, (**C**) TarS, and (**D**) Mpro, showing their energy contributions within the MM_PBSA free-binding energy calculation.

3. Materials and Methods

3.1. Plant Material

In September 2021, fruits of *Annona squamosa* were acquired from a local Egyptian market. The plant sample was verified by Mrs. Therez Labib, a consultant in plant taxonomy for the Ministry of Agriculture and a former director of the El-Orman Botanical Garden. The Department of Pharmacognosy and Medicinal Plants at Future University's Faculty of Pharmacy (FUE) received a voucher specimen of the plant material (AS 101).

3.2. Isolation of the Endophytic Fungi

Endophytic fungi were separated using the procedure that Fathallah et al. and Hazalin et al. outlined [1,123]. In summary, *A. squamosa* fruits were sterilized for one minute using 70% ethanol, then rinsed twice with sterile water. To prevent bacterial development, the dried fruits (shade drying) were crushed and aseptically added to Potato Dextrose Agar (PDA) (Oxoid, Hampshire, UK) plates supplemented with 250 mg/L of gentamicin and streptomycin. In addition to non-inoculated PDA plates acting as a negative control, non-crushed, surface-sterilized fruits were also grown to rule out the presence of epiphytic fungus. The plates were incubated at 25 °C for seven to fourteen days. Different mycelia that emerged from the segments were grown, and PDA slants were used to preserve the isolated pure fungi.

3.3. Morphology of Fungi and Microscopic Analysis

As illustrated by [124], colony characteristics like texture, shape, and color as well as the conventional taxonomic key of the isolated fungus was morphologically identified. A prospective fungus was grown on PDA for seven days using the slide culture method [125]. After adding lactophenol cotton blue, the mycelia were found under a microscope. To identify fungi, hyphae and conidia's morphological characteristics were utilized.

3.4. Identification of Fungi Using a Molecular Approach

By Sigma Scientific Services Co., genomic DNA was extracted. To amplify the ribosomal ITS region, ITS 1 (5'-TCC GTA GGT GAA CCT GCG G-3') and ITS 4 (5'-TCC TCC GCT TAT TGA TAT GC-3') were utilized as forward and reverse primers, respectively. The following conditions applied to thermal cycling: ten minutes of initial denaturation at 95 °C, thirty seconds of denaturation at 95 °C, one minute of annealing at 57 °C, and one minute of extension at 72 °C were all included. One cycle of post-cycling expansion was performed for ten minutes at 72 °C. Following the manufacturer's instructions, the PCR yields were purified using the GeneJET PCR Purification Kit (Thermo K0701, Waltham, MA, USA), and the refined DNA was then stored. The PCR yields were purified using the GeneJET PCR Purification Kit (Thermo K0701, Waltham, MA, USA) under the manufacturer's instructions, and the refined DNA was thereafter stored at −20 °C. Ultimately, an ABI 3730xl DNA sequencer was used to sequence the improved PCR result. The final gene product sequence of the fungal isolate was aligned using NCBI BLAST (Basic Local Alignment Search Tool; http://blast.ncbi.nlm.nih.gov/; accessed on 14 November 2023) against sequences that were already available in the GenBank database. Using the MEGA 5 program, the phylogenetic tree was constructed using the neighbor-joining strategy. The isolate sequence that was found was entered into the GenBank database and given an entry number [126].

3.5. Fermentation in Solid-State Media and Extraction of the Fungi Metabolites

In 1L Erlenmeyer flasks sealed with cotton and autoclaved at 121 °C for 20 min, 100 g of rice combined with 120 mL of sterilized water was used to create a solid rice medium for mass manufacturing. Fifteen solid rice flasks were inoculated with plugs from PDA fungal cultures, and the cultures were left to develop for twenty-one days at room temperature.

3.6. Preparation of Ethyl Acetate Fungal Extract

Ethyl acetate (EtOAc) (4 × 300 mL) was used to extract fungal metabolites to exhaustion, as per the instructions [127]. Briefly, the fermentation process was terminated by the addition of ethyl acetate, the process was repeated 4 times and the pooled ethyl acetate extracts of the fungal material were evaporated under a vacuum resulting in a brown residue (FEA). In this investigation, the conventional procedure (solid-state fermentation) was utilized for large-scale fermentation and isolation of significant amounts of the chemicals of interest. VanderMolen et al. [128] suggested that solid media usually yield cultures that are one to two times higher in mass than those cultivated on liquid media.

3.7. Liquid Chromatography-Mass Spectrometry Analysis (LC/MS)

Using methanol HPLC grade, the ethyl acetate extract was dissolved and filtered by a membrane disc filter 0.2 µm, then into an RP C-18 column 5 µm, 125 mm × 4 mm, 10 µL of the sample was injected. Gradient elution was employed, with a flow rate of 0.2 mL/min. The total program time is 34 min. Mass spectra were detected in the ESI negative ion mode: source temperature 150 °C, cone voltage 30 eV, capillary voltage 3 KV, desolvation temperature 440 °C, cone gas flow 50 L/h, and desolvation gas flow 900 L/h.

3.8. Antimicrobial Screening

The screening of antibacterial and antifungal activities for FEA was conducted using the disk diffusion method, following the standard CLSI procedure [129]. Test microbes included two Gram-positive bacteria (*S. aureus* ATCC 25923; MSSA) and MRSA ATCC-700788), two Gram-negative bacteria (*E. coli* ATCC 25922 and *P. aeruginosa* ATCC9027), and one yeast strain (*C. albicans* ATCC 10231). FEA (20% w/v) was tested using bacterial and yeast suspensions adjusted to a turbidity equivalent to 0.5 McFarland (1.5×10^8 CFU/mL) in Trypticase Soy Broth (TSB). The prepared disks were placed on Muller–Hinton Agar, with DMSO disks as negative controls and disks containing antibiotics Vancomycin 10 µg, Gentamicin 10 µg, and Nystatin 10 µg were used as positive controls for Gram-positive, Gram-negative bacteria and yeast, respectively. After incubation at 37 °C for 24 h, zones of inhibition were determined according to [130,131].

3.9. Determination of Minimum Inhibitory Concentrations

FEA underwent further testing to determine its MIC against sensitive isolates employing the agar well diffusion method [131]. Various concentrations (10, 5, 2.5, 1.25, 0.6, and 0.3%) of the extract were meticulously prepared using a two-fold serial dilution. For each concentration, 1 mL of the prepared inoculum of sensitive isolates (log phase) was pipetted into sterile Petri dishes, followed by the addition of Trypticase Soy agar and thorough mixing. After solidification, wells were created using a sterile cork borer (6 mm in diameter) on agar plates containing the inoculums. Subsequently, 100 µL of the extract dilution was transferred to the respective wells, ensuring that each plate contained only four wells. Following a 30 min refrigeration period, the plates were incubated at 37 °C for 18 h. MIC was defined as the lowest concentration inhibiting the growth of the respective microorganisms. All assays were conducted in triplicate, and DMSO served as a control in these experiments.

3.10. Antibiofilm Screening

To assess the impact of extracts on biofilm formation, sublethal concentrations (75%, 50%, and 25% of MIC) were employed against biofilm-forming sensitive isolates, namely *S. aureus* ATCC 25923 (MSSA) and MRSA ATCC-700788 [132].

3.10.1. Inhibition of Biofilm Formation–Prevention of Initial Bacterial Cell Attachment

The potential of FEA to impede initial cell attachment was explored through the biofilm inhibition assay [133]. In brief, 100 µL of a standardized concentration of cultures with OD_{560} = 0.02 (1.0×10^6 CFU/mL) was added to individual flat-bottomed 96-well microtiter plates and incubated at 37 °C for 4 h without shaking. Subsequently, the plates were removed from the incubator, and 100 µL aliquots of the extract were added in triplicate to the wells, resulting in final sub-MIC concentrations (75%, 50%, and 25% of MIC). The plates were then further incubated at 37 °C for 24 h without agitation. DMSO served as the negative control. The biomass was quantified using the modified crystal violet staining method.

3.10.2. Inhibition of Development of Pre-formed Biofilms–Assessment of Destruction of Biofilm Mass

FEA was assessed for its ability to induce the destruction of pre-formed biofilms according to the method performed by Famuyide et al. (2019) [52]. A 100 µL aliquot of a standardized concentration of tested cultures with OD_{560} = 0.02 (1.0×10^6 CFU/mL) was added to individual flat-bottomed 96-well microtiter plates and incubated at 37 °C for 24 h without shaking to allow for the development of a multilayer biofilm. Subsequently, 100 µL aliquots of the extract or its fractions were added to the wells of a 96-well microtiter plate, achieving final sub-MIC concentrations (75%, 50%, and 25% of MIC), and the plates were further incubated at 37 °C for 24 h. The incubation was conducted without agitation. DMSO served as the negative control. The biomass was quantified using the modified crystal violet staining method [133].

3.10.3. The Crystal Violet Staining Assay

The assay followed the method outlined by Famuyide et al. (2019) [52]. Briefly, 96-well microtiter plates were washed five times with sterile distilled water, followed by air-drying and oven-drying at 60 °C for 45 min. Subsequently, wells were stained with 100 µL of 1% crystal violet and incubated at room temperature for 15 min. After three washes with sterile distilled water, a semi-quantitative assessment was conducted by destaining with 125 µL of 30% acetic acid solution for 10 min at room temperature. A 100 µL aliquot of the destaining solution was transferred to a new sterile plate, and absorbance at 590 nm was measured using a microplate reader (BioRad). The percentage inhibition of biofilm was calculated based on the mean absorbance of the samples using the equation below [53].

$$\text{Percentage Inhibition} = \frac{\text{OD negative control} - \text{OD experiment}}{\text{OD negativ econtrol}} * 100$$

3.11. Antiviral Activity

In this study, Nawah-Scientific, Egypt, provided the Low Pathogenic Corona Virus (229E) and Vero E6 cells. Vero E6 cells were cultured in DMEM medium supplemented with 10% fetal bovine serum and 0.1% antibiotic/antimycotic solution, with reagents sourced from Gibco BRL. Antiviral and cytotoxicity assays were conducted using the crystal violet method [134,135]. Briefly, Vero E6 cells were seeded one day before infection, and the infectivity of the Low Pathogenic Corona Virus (229E) was determined by monitoring cytopathic effects (CPE) and calculating cell viability percentages.

For the antiviral activity assessment, a 96-well culture plate was used, and 0.1 mL of diluted virus suspension was added to cells along with various concentrations of test compounds. The culture plates were then incubated, and the development of CPE was monitored. After staining and quantification, the antiviral activity was calculated using the Pauwels et al. (1988) equation [136], allowing for the determination of the 50% CPE inhibitory dose (IC_{50}).

To evaluate cytotoxicity, cells were seeded in a 96-well plate, treated with serially diluted samples, and incubated. After the incubation period, cells were processed similarly to the antiviral assay, and the 50% cytotoxic concentrations (CC_{50}) were determined. CC_{50} and IC_{50} were calculated using GraphPad PRISM Version 5.01 software.

3.12. Research on ADME (Absorption, Distribution, Metabolism, and Excretion) and Pharmacokinetics

The Absorption, Distribution, Metabolism, and Excretion (ADME) and Pharmacokinetic Studies were carried out using SWISSadme [137] (Swiss Institute of Bioinformatics online source), link: http://www.swissadme.ch accessed on 1 October 2023, to determine whether the compounds had the potential to be a promising pharmaceutical drug. By using the Swiss ADME (http://www.swissadme.ch/; accession date: 18 February 2024) molecules' bioavailability radar, the physicochemical properties of the identified compounds for oral bioavailability were determined. The pink area represents the optimal ranges for the represented compound's oral bioavailability based on six physicochemical characteristics: polarity, size, solubility, lipophilicity, flexibility, and saturation. To predict the compound's blood barrier and GIT absorption, the Boiled Egg approach was also used.

3.13. In Silico Studies (Molecular Docking-Coupled Dynamics Simulations)

The probable molecular binding mode between the identified compounds and different enzymes involved in the antimicrobial and antiviral activity was evaluated using the CDOCKER algorithm in Discovery Studio 4.5. (Accelrys Software, Inc., San Diego, CA, USA). The crystal structures of several different protein targets were obtained using the Protein Data Bank (http://www.rcsb.org/pdb/; accession date: 19 February 2024). The enzymes are *S. aureus* teichoic acid-associated β-glycosyltransferase enzyme (TarS; PDB = 5tzj) evaluating the antibiofilm activity; MurE ligase (PDB = 4c12), penicillin-binding proteins (PBP2a; PDB = 3zg0) for assessing the antibacterial activity; and finally HCoV-229E main protease (PDB = 7yrz) which demonstrates the antiviral activity. The protein was refined after the water molecules were eliminated. For each tested enzyme, the binding of the co-crystallized inhibitor and the target enzyme served as the basis for identifying the binding site. Rule-based docking was used to dock all identified compounds and the specific ligand for each enzyme into the protein-binding site, after the co-crystallized ligand was removed. The interaction energy was calculated to examine how the ligand molecules and receptors interacted. The best ligand-binding poses were chosen by sorting the top 10 ligand-binding poses for each ligand according to their CDOCKER interaction energies and looking at the predicted binding interactions.

Best docked complex model for each compound proceeded through molecular dynamics simulations using GROMACS-19 under CHARMM36m and CHARMM-General forcefields following solvation within the TIP3P-water model under periodic boundary conditions [138]. Models were ionized at physiological pH = 7.4 and neutralized using a sufficient number of chloride and potassium ions. System minimization was performed by steepest-descent algorithm-minimization steps (5 ps), then equilibrated at NVT followed by NPT ensembles for 500 ps each [104,138]. Systems were produced for 100 ns molecular dynamics simulations under the NPT ensemble and far-range electrostatic interactions were computed using Particle-Mesh/Ewald algorithm [139]. Root-mean-square deviations (RMSDs_Å) and RMS-fluctuations (RMSFs_Å) were monitored regarding the entire trajectories, while free-binding energies of compound-NCAPG-kleisin complexes were estimated via Molecular Mechanics/Poisson-Boltzmann (MM-PBSA_kJ/mol) single-trajectory approach [115]. Visualizing the simulated complexes at specified timeframes as well as conformational analysis were performed using PyMOL 2.0.6 software.

4. Conclusions

This study represents a sustainability approach for fruit peels, which are regarded as industrial waste. Peels can be used as a valuable source of endophytic fungi to enhance their economic value. The isolated *A. flavus* is an endophytic fungus that has significant secondary metabolites and owns a selective antibacterial and antibiofilm potential against Gram-positive microorganisms such as MSSA and MRSA; in addition, it exhibited a promising antiviral activity. The promising computational findings encourage deeper biological in vivo experiments for the identified metabolites which can be used singly, in combination, or in addition to presently prescribed antibiotics to increase their effectiveness and lessen the microbes' resistance. Additional research is recommended to assess the potential of this promising endeavor across diverse clinical bacterial strains, with a particular emphasis on further exploration concerning coronaviruses, particularly the recently emerged SARS-CoV-2. It is also advised that more research be conducted to identify the various endophytic fungal species that are concealed within fruit peels, as well as their secondary metabolites, modes of action, and biological activities.

Supplementary Materials: The following supporting information can be downloaded at: https://www.mdpi.com/article/10.3390/ph17050656/s1, Figure S1: Negative ionization mode chromatogram representing the major compounds identified from the fungal ethyl acetate extract numbered according to their relative abundance where (1) heptelidic acid, (2) ferulic acid, (3) oleic acid, (4) paxilline, (5) indole, (6) orientin, and (7) kojic acid.

Author Contributions: N.F. and Y.A.E., conceptualization and methodology; S.A.Z. isolated the endophytic fungus and assessed the antimicrobial and antibiofilm activities; Y.A.E., W.M.E., S.S.E. and N.F. designed the phytochemistry work and interpreted the UPLC; S.S.E. and K.M.D., validation, investigation, resources, and funding acquisition; S.S.E. and K.M.D., formal analysis, data curation, and in silico investigation. All authors have shared manuscript writing and reviewing. All authors have read and agreed to the published version of the manuscript.

Funding: This research work was funded by Institutional Fund Projects under grant no. (IFPIP:1625-166-1443). The authors gratefully acknowledge the technical and financial support provided by the Ministry of Education and King Abdulaziz University, DSR, Jeddah, Saudi Arabia.

Data Availability Statement: Data are available within the article.

Acknowledgments: The simulations in this work were performed at King Abdulaziz University's High Performance Computing Center (Aziz Supercomputer) (http://hpc.kau.edu.sa, accessed on 11 February 2024); the authors, therefore, acknowledge with thanks the center for technical support.

Conflicts of Interest: The authors declare no conflicts of interest.

Abbreviations

HCoV 229E: human coronavirus 229E; ITS-rDNA: internal transcribed space; UPLC/MS: ultra-high performance liquid chromatography-MS; CPE: cytopathic effects; MRSA: methicillin-resistant *Staphylococcus aureus*; MSSA: methicillin-susceptible *Staphylococcus aureus*; ADME: absorption, distribution, metabolism, and excretion; UV: ultra violet; IC_{50}: half-maximal inhibitory concentration; ARIs: acute respiratory infections; ATCC: American Type Culture Collection; CC_{50}/IC_{50}: cytotoxicity (CC_{50}) and inhibition concentration (IC_{50}); BBB: blood–brain barrier; POLAR, LIPO, INSOLU, and IN-SATU: polarity, lipophilicity, solubility, and saturation; TPSA: topological polar surface area; WLOGP: the atomic log *p*; HCoV-OC43: human coronavirus OC43; UDP-GluNAc: uridine diphosphate *N*-acetylglucosamine; RMSD: Root Mean Square Deviation; PBP2a: penicillin-binding protein.

References

1. Fathallah, N.; Raafat, M.M.; Issa, M.Y.; Abdel-Aziz, M.M.; Bishr, M.; Abdelkawy, M.A.; Salama, O. Bio-guided fractionation of prenylated benzaldehyde derivatives as potent antimicrobial and antibiofilm from *Ammi majus* L. fruits-associated *Aspergillus amstelodami*. *Molecules* **2019**, *24*, 4118. [CrossRef]
2. Abdel-Maksoud, M.; El-Shokry, M.; Ismail, G.; Hafez, S.; El-Kholy, A.; Attia, E.; Talaat, M. Methicillin-resistant *Staphylococcus aureus* recovered from healthcare-and community-associated infections in Egypt. *Int. J. Bacteriol.* **2016**, *2016*, 5751785. [CrossRef]
3. Kandeel, A.; Fahim, M.; Deghedy, O.; Roshdy, W.H.; Khalifa, M.K.; El Shesheny, R.; Kandeil, A.; Wagdy, S.; Naguib, A.; Afifi, S. Multicenter study to describe viral etiologies, clinical profiles, and outcomes of hospitalized children with severe acute respiratory infections, Egypt 2022. *Sci. Rep.* **2023**, *13*, 21860. [CrossRef] [PubMed]
4. Hasöksüz, M.; Kilic, S.; Saraç, F. Coronaviruses and SARS-CoV-2. *Turk. J. Med. Sci.* **2020**, *50*, 549–556. [CrossRef]
5. Santhoshkumar, R.; Kumar, N.S. Phytochemical analysis and antimicrobial activities of *Annona squamosa* (L) leaf extracts. *J. Pharmacogn. Phytochem.* **2016**, *5*, 128–131.

6. Elkady, W.M.; Raafat, M.M.; Abdel-Aziz, M.M.; Al-Huqail, A.A.; Ashour, M.L.; Fathallah, N. Endophytic Fungus from *Opuntia ficus-indica*: A Source of Potential Bioactive Antimicrobial Compounds against Multidrug-Resistant Bacteria. *Plants* **2022**, *11*, 1070. [CrossRef]
7. Singab, A.N.B.; Elkhawas, Y.A.; Al-Sayed, E.; Elissawy, A.M.; Fawzy, I.M.; Mostafa, N.M. Antimicrobial activities of metabolites isolated from endophytic *Aspergillus flavus* of *Sarcophyton ehrenbergi* supported by in-silico study and NMR spectroscopy. *Fungal Biol. Biotechnol.* **2023**, *10*, 16. [CrossRef] [PubMed]
8. Khattak, S.U.; Lutfullah, G.; Iqbal, Z.; Ahmad, J.; Rehman, I.U.; Shi, Y.; Ikram, S. Aspergillus flavus originated pure compound as a potential antibacterial. *BMC Microbiol.* **2021**, *21*, 322. [CrossRef]
9. Nair, D.N.; Padmavathy, S. Impact of endophytic microorganisms on plants, environment and humans. *Sci. World J.* **2014**, *2014*, 250693. [CrossRef]
10. Rodriguez, R.; White, J., Jr.; Arnold, A.; Redman, A.R.A. Fungal endophytes: Diversity and functional roles. *New Phytol.* **2009**, *182*, 314–330. [CrossRef]
11. Chang, Y.; Xia, X.; Sui, L.; Kang, Q.; Lu, Y.; Li, L.; Liu, W.; Li, Q.; Zhang, Z. Endophytic colonization of entomopathogenic fungi increases plant disease resistance by changing the endophytic bacterial community. *J. Basic Microbiol.* **2021**, *61*, 1098–1112. [CrossRef] [PubMed]
12. Mengistu, A.A. Endophytes: Colonization, behaviour, and their role in defense mechanism. *Int. J. Microbiol.* **2020**, *2020*, 6927219. [CrossRef]
13. Busby, P.E.; Ridout, M.; Newcombe, G. Fungal endophytes: Modifiers of plant disease. *Plant Mol. Biol.* **2016**, *90*, 645–655. [CrossRef] [PubMed]
14. Mohammed, M.A.; El-Gengaihi, S.; Enein, A.; Hassan, E.M.; Ahmed, O.; Asker, M. Chemical constituents and antimicrobial activity of different *Annona* species cultivated in Egypt. *J. Chem. Pharm. Res* **2016**, *8*, 261–271.
15. Nguyen, T.T.; Tran, P.N.T.; Phan, H.T. Evaluation of anti-inflammatory effect of fruit peel extracts of *Annona squamosa* L. on mouse models of rheumatoid arthritis. *J. Microbiol. Biotechnol. Food Sci.* **2021**, *11*, e2075.
16. Vikas, B.; Anil, S.; Remani, P. Cytotoxicity profiling of *Annona squamosa* in cancer cell lines. *Asian Pac. J. Cancer Prev. APJCP* **2019**, *20*, 2831. [CrossRef] [PubMed]
17. Zahid, M.; Arif, M.; Rahman, M.A.; Mujahid, M. Hepatoprotective and antioxidant activities of Annona squamosa seed extract against alcohol-induced liver injury in Sprague Dawley rats. *Drug Chem. Toxicol.* **2020**, *43*, 588–594. [CrossRef] [PubMed]
18. Tomar, R.S.; Sisodia, S.S. Antidiabetic activity of *Annona squamosa* L. in experimental induced diabetic rats. *Int. J. Pharm. Biol. Arch.* **2012**, *3*, 1492–1495.
19. Intaranongpai, J.; Chavasiri, W.; Gritsanapan, W. Anti-head lice effect of Annona squamosa seeds. *S. Asian J. Trop. Med. Public Health* **2006**, *37*, 532.
20. Pandey, N.; Barve, D. Phytochemical and pharmacological review on *Annona squamosa* Linn. *Int. J. Res. Pharm. Biomed. Sci.* **2011**, *2*, 1404–1412.
21. Chowdhury, S.S.; Tareq, A.M.; Tareq, S.M.; Farhad, S.; Sayeed, M.A. Screening of antidiabetic and antioxidant potential along with phytochemicals of *Annona* genus: A review. *Future J. Pharm. Sci.* **2021**, *7*, 144. [CrossRef]
22. Zaghlol, A.A.; Kandil, Z.A.; Yousif, M.F.; EL-Dine, R.S.; Elkady, W.M. Unveiling the anti-cancer potential of *Euphorbia greenwayi*: Cytotoxicity, cell migration, and identification of its chemical constituents. *Future J. Pharm. Sci.* **2024**, *10*, 24. [CrossRef]
23. El-Chaghaby, G.A.; Ahmad, A.F.; Ramis, E.S. Evaluation of the antioxidant and antibacterial properties of various solvents extracts of *Annona squamosa* L. leaves. *Arab. J. Chem.* **2014**, *7*, 227–233. [CrossRef]
24. Yunita, M.N.; Raharjo, A.P.; Wibawati, P.A.; Agustono, B. Antiviral activity of ethanolic extract of srikaya seeds (*Annona squamosa* L.) against avian influenza virus. *Indian Vet. J.* **2019**, *96*, 26–29.
25. Ola, A.R.; Soa, C.A.; Sugi, Y.; Cunha, T.; Belli, H.; Lalel, H. Antimicrobial metabolite from the endophytic fungi *Aspergillus flavus* isolated from Sonneratia alba, a mangrove plant of Timor-Indonesia. *Rasayan J. Chem.* **2020**, *13*, 377–381. [CrossRef]
26. Liu, Z.; Zhao, J.-Y.; Sun, S.-F.; Li, Y.; Qu, J.; Liu, H.-T.; Liu, Y.-b. Sesquiterpenes from an endophytic *Aspergillus flavus*. *J. Nat. Prod.* **2019**, *82*, 1063–1071. [CrossRef] [PubMed]
27. Patil, M.; Patil, R.; Maheshwari, V. Biological activities and identification of bioactive metabolite from endophytic *Aspergillus flavus* L7 isolated from *Aegle marmelos*. *Curr. Microbiol.* **2015**, *71*, 39–48. [CrossRef]
28. Lin, M.; Dianese, J. A coconut-agar medium for rapid detection of aflatoxin production by *Aspergillus* spp. *Phytopathology* **1976**, *66*, 1466–1469. [CrossRef]
29. Rao, K.R.; Vipin, A.; Venkateswaran, G. Molecular profile of non-aflatoxigenic phenotype in native strains of *Aspergillus flavus*. *Arch. Microbiol.* **2020**, *202*, 1143–1155. [CrossRef]
30. Mowaka, S.; Ayoub, B.M. Comparative study between UHPLC-UV and UPLC-MS/MS methods for determination of alogliptin and metformin in their pharmaceutical combination. *Die Pharm.-Int. J. Pharm. Sci.* **2017**, *72*, 67–72.
31. Tanaka, Y.; Shiomi, K.; Kamei, K.; Sugoh-Hagino, M.; Enomoto, Y.; Fang, F.; Yamaguchi, Y.; Masuma, R.; Zhang, C.G.; Zhang, X.W. Antimalarial activity of radicicol, heptelidic acid and other fungal metabolites. *J. Antibiot.* **1998**, *51*, 153–160. [CrossRef] [PubMed]
32. Itoh, Y.; Kodama, K.; Furuya, K.; Takahashi, S.; Haneishi, T.; Takiguchi, Y.; Arai, M. A new sesquiterpene antibiotic, heptelidic acid producing organisms, fermentation, isolation and characterization. *J. Antibiot.* **1980**, *33*, 468–473. [CrossRef] [PubMed]
33. Kim, J.-H.; Lee, C.-H. Heptelidic acid, a sesquiterpene lactone, inhibits etoposide-induced apoptosis in human leukemia U937 cells. *J. Microbiol. Biotechnol.* **2009**, *19*, 787–791. [CrossRef] [PubMed]

34. Pinheiro, P.G.; Santiago, G.M.P.; da Silva, F.E.F.; de Araújo, A.C.J.; de Oliveira, C.R.T.; Freitas, P.R.; Rocha, J.E.; de Araújo Neto, J.B.; da Silva, M.M.C.; Tintino, S.R. Ferulic acid derivatives inhibiting *Staphylococcus aureus* tetK and MsrA efflux pumps. *Biotechnol. Rep.* 2022, 34, e00717. [CrossRef] [PubMed]
35. Ibitoye, O.; Ajiboye, T. Ferulic acid potentiates the antibacterial activity of quinolone-based antibiotics against *Acinetobacter baumannii*. *Microb. Pathog.* 2019, 126, 393–398. [CrossRef] [PubMed]
36. Batista, R. Uses and potential applications of ferulic acid. In *Ferulic Acid: Antioxidant Properties, Uses and Potential Health Benefits*; Nova Science Publishers, Inc.: Hauppauge, NY, USA, 2014; pp. 39–70.
37. Deakin, N.O.; Turner, C.E. Distinct roles for paxillin and Hic-5 in regulating breast cancer cell morphology, invasion, and metastasis. *Mol. Biol. Cell* 2011, 22, 327–341. [CrossRef] [PubMed]
38. Lee, M.; Cho, J.-Y.; Lee, Y.G.; Lee, H.J.; Lim, S.-I.; Lee, S.-Y.; Nam, Y.-D.; Moon, J.-H. Furan, phenolic, and heptelidic acid derivatives produced by *Aspergillus oryzae*. *Food Sci. Biotechnol.* 2016, 25, 1259–1264. [CrossRef] [PubMed]
39. Kovač, T.; Borišev, I.; Kovač, M.; Lončarić, A.; Čačić Kenjerić, F.; Djordjevic, A.; Strelec, I.; Ezekiel, C.N.; Sulyok, M.; Krska, R. Impact of fullerol C60(OH)24 nanoparticles on the production of emerging toxins by *Aspergillus flavus*. *Sci. Rep.* 2020, 10, 725. [CrossRef]
40. Todokoro, T.; Negoro, H.; Kotaka, A.; Hata, Y.; Ishida, H. Aspergillus oryzae FaeA is responsible for the release of ferulic acid, a precursor of off-odor 4-vinylguaiacol in sake brewing. *J. Biosci. Bioeng.* 2022, 133, 140–145. [CrossRef]
41. Faulds, C.; Williamson, G. Release of ferulic acid from wheat bran by a ferulic acid esterase (FAE-III) from *Aspergillus niger*. *Appl. Microbiol. Biotechnol.* 1995, 43, 1082–1087. [CrossRef]
42. Amaike, S.; Keller, N.P. *Aspergillus flavus*. *Annu. Rev. Phytopathol.* 2011, 49, 107–133. [CrossRef] [PubMed]
43. Zhang, S.; Monahan, B.J.; Tkacz, J.S.; Scott, B. Indole-diterpene gene cluster from *Aspergillus flavus*. *Appl. Environ. Microbiol.* 2004, 70, 6875–6883. [CrossRef]
44. Nicholson, M.J.; Koulman, A.; Monahan, B.J.; Pritchard, B.L.; Payne, G.A.; Scott, B. Identification of two aflatrem biosynthesis gene loci in *Aspergillus flavus* and metabolic engineering of *Penicillium paxilli* to elucidate their function. *Appl. Environ. Microbiol.* 2009, 75, 7469–7481. [CrossRef] [PubMed]
45. Cole, R.J.; Dorner, J.W.; Springer, J.P.; Cox, R.H. Indole metabolites from a strain of *Aspergillus flavus*. *J. Agric. Food Chem.* 1981, 29, 293–295. [CrossRef]
46. Adione, N.M.; Onyeka, I.P.; Abba, C.C.; Okoye, N.N.; Okolo, C.C.; Eze, P.M.; Umeokoli, B.O.; Anyanwu, O.O.; Okoye, F.B.C. Detection, isolation and identification of more bioactive compounds from *Fusarium equiseti*, an endophytic fungus isolated from *Ocimum gratissimum*. *GSC Biolog. Pharm. Sci.* 2022, 20, 130–140. [CrossRef]
47. Ola, A.R.; Metboki, G.; Lay, C.S.; Sugi, Y.; Rozari, P.D.; Darmakusuma, D.; Hakim, E.H. Single production of kojic acid by *Aspergillus flavus* and the revision of flufuran. *Molecules* 2019, 24, 4200. [CrossRef]
48. de Caldas Felipe, M.T.; do Nascimento Barbosa, R.; Bezerra, J.D.P.; de Souza-Motta, C.M. Production of kojic acid by *Aspergillus* species: Trends and applications. *Fungal Biol. Rev.* 2023, 45, 100313.
49. Bhattacharya, A.; Chakraverty, R. The pharmacological properties of *Annona squamosa* Linn: A Review. *Int. J. Pharm. Eng.* 2016, 4, 692–699.
50. Clegg, J.; Soldaini, E.; McLoughlin, R.M.; Rittenhouse, S.; Bagnoli, F.; Phogat, S. *Staphylococcus aureus* vaccine research and development: The past, present and future, including novel therapeutic strategies. *Front. Immunol.* 2021, 12, 705360. [CrossRef]
51. Howden, B.P.; Giulieri, S.G.; Wong Fok Lung, T.; Baines, S.L.; Sharkey, L.K.; Lee, J.Y.; Hachani, A.; Monk, I.R.; Stinear, T.P. *Staphylococcus aureus* host interactions and adaptation. *Nat. Rev. Microbiol.* 2023, 21, 380–395. [CrossRef]
52. Famuyide, I.M.; Aro, A.O.; Fasina, F.O.; Eloff, J.N.; McGaw, L.J. Antibacterial and antibiofilm activity of acetone leaf extracts of nine under-investigated south African *Eugenia* and *Syzygium* (Myrtaceae) species and their selectivity indices. *BMC Complement. Altern. Med.* 2019, 19, 141. [CrossRef] [PubMed]
53. Sandasi, M.; Leonard, C.; Viljoen, A. The effect of five common essential oil components on *Listeria monocytogenes* biofilms. *Food Control* 2008, 19, 1070–1075. [CrossRef]
54. Pasquereau, S.; Nehme, Z.; Haidar Ahmad, S.; Daouad, F.; Van Assche, J.; Wallet, C.; Schwartz, C.; Rohr, O.; Morot-Bizot, S.; Herbein, G. Resveratrol inhibits HCoV-229E and SARS-CoV-2 coronavirus replication in vitro. *Viruses* 2021, 13, 354. [CrossRef] [PubMed]
55. Selick, H.E.; Beresford, A.P.; Tarbit, M.H. The emerging importance of predictive ADME simulation in drug discovery. *Drug Discov. Today* 2002, 7, 109–116. [CrossRef]
56. Attique, S.A.; Hassan, M.; Usman, M.; Atif, R.M.; Mahboob, S.; Al-Ghanim, K.A.; Bilal, M.; Nawaz, M.Z. A molecular docking approach to evaluate the pharmacological properties of natural and synthetic treatment candidates for use against hypertension. *Int. J. Environ. Res. Public Health* 2019, 16, 923. [CrossRef] [PubMed]
57. Daina, A.; Zoete, V. A boiled-egg to predict gastrointestinal absorption and brain penetration of small molecules. *ChemMedChem* 2016, 11, 1117–1121. [CrossRef] [PubMed]
58. Maddiboyina, B.; Roy, H.; Ramaiah, M.; Sarvesh, C.N.; Kosuru, S.H.; Nakkala, R.K.; Nayak, B.S. Methicillin-resistant *Staphylococcus aureus*: Novel treatment approach breakthroughs. *Bull. Natl. Res. Cent.* 2023, 47, 95. [CrossRef]
59. Monteiro, J.M.; Covas, G.; Rausch, D.; Filipe, S.R.; Schneider, T.; Sahl, H.G.; Pinho, M.G. The pentaglycine bridges of *Staphylococcus aureus* peptidoglycan are essential for cell integrity. *Sci. Rep.* 2019, 9, 5010. [CrossRef]

60. Ruane, K.M.; Lloyd, A.J.; Fülöp, V.; Dowson, C.G.; Barreteau, H.; Boniface, A.; Dementin, S.; Blanot, D.; Mengin-Lecreulx, D.; Gobec, S.; et al. Specificity determinants for lysine incorporation in *Staphylococcus aureus* peptidoglycan as revealed by the structure of a MurE enzyme ternary complex. *J. Biol. Chem.* **2013**, *288*, 33439–33448. [CrossRef]
61. Bouhss, A.; Trunkfield, A.E.; Bugg, T.D.; Mengin-Lecreulx, D. The biosynthesis of peptidoglycan lipid-linked intermediates. *FEMS Microbiol. Rev.* **2008**, *32*, 208–233. [CrossRef]
62. Nikolaidis, I.; Favini-Stabile, S.; Dessen, A. Resistance to antibiotics targeted to the bacterial cell wall. *Protein Sci.* **2014**, *23*, 243–259. [CrossRef] [PubMed]
63. Stelitano, G.; Sammartino, J.C.; Chiarelli, L.R. Multitargeting Compounds: A Promising Strategy to Overcome Multi-Drug Resistant Tuberculosis. *Molecules* **2020**, *25*, 1239. [CrossRef] [PubMed]
64. Igler, C.; Rolff, J.; Regoes, R. Multi-step vs. single-step resistance evolution under different drugs, pharmacokinetics, and treatment regimens. *eLife* **2021**, *10*, e64116. [CrossRef] [PubMed]
65. Sobhanifar, S.; Worrall, L.J.; King, D.T.; Wasney, G.A.; Baumann, L.; Gale, R.T.; Nosella, M.; Brown, E.D.; Withers, S.G.; Strynadka, N.C. Structure and Mechanism of *Staphylococcus aureus* TarS, the Wall Teichoic Acid β-glycosyltransferase Involved in Methicillin Resistance. *PLoS Pathog.* **2016**, *12*, e1006067. [CrossRef]
66. Oku, Y.; Kurokawa, K.; Matsuo, M.; Yamada, S.; Lee, B.L.; Sekimizu, K. Pleiotropic roles of polyglycerolphosphate synthase of lipoteichoic acid in growth of *Staphylococcus aureus* cells. *J. Bacteriol.* **2009**, *191*, 141–151. [CrossRef] [PubMed]
67. Draing, C.; Sigel, S.; Deininger, S.; Traub, S.; Munke, R.; Mayer, C.; Hareng, L.; Hartung, T.; von Aulock, S.; Hermann, C. Cytokine induction by Gram-positive bacteria. *Immunobiology* **2008**, *213*, 285–296. [CrossRef] [PubMed]
68. Morath, S.; Geyer, A.; Hartung, T. Structure-function relationship of cytokine induction by lipoteichoic acid from *Staphylococcus aureus*. *J. Exp. Med.* **2001**, *193*, 393–397. [CrossRef] [PubMed]
69. Gautam, S.; Kim, T.; Lester, E.; Deep, D.; Spiegel, D.A. Wall teichoic acids prevent antibody binding to epitopes within the cell wall of *Staphylococcus aureus*. *ACS Chem. Biol.* **2016**, *11*, 25–30. [CrossRef] [PubMed]
70. Gross, M.; Cramton, S.E.; Götz, F.; Peschel, A. Key role of teichoic acid net charge in *Staphylococcus aureus* colonization of artificial surfaces. *Infect. Immun.* **2001**, *69*, 3423–3426. [CrossRef]
71. Bera, A.; Biswas, R.; Herbert, S.; Kulauzovic, E.; Weidenmaier, C.; Peschel, A.; Götz, F. Influence of wall teichoic acid on lysozyme resistance in *Staphylococcus aureus*. *J. Bacteriol.* **2007**, *189*, 280–283. [CrossRef]
72. Lee, J.H.; Kim, N.H.; Winstel, V.; Kurokawa, K.; Larsen, J.; An, J.H.; Khan, A.; Seong, M.Y.; Lee, M.J.; Andersen, P.S.; et al. Surface Glycopolymers Are Crucial for In Vitro Anti-Wall Teichoic Acid IgG-Mediated Complement Activation and Opsonophagocytosis of *Staphylococcus aureus*. *Infect. Immun.* **2015**, *83*, 4247–4255. [CrossRef] [PubMed]
73. Brown, S.; Xia, G.; Luhachack, L.G.; Campbell, J.; Meredith, T.C.; Chen, C.; Winstel, V.; Gekeler, C.; Irazoqui, J.E.; Peschel, A.; et al. Methicillin resistance in Staphylococcus aureus requires glycosylated wall teichoic acids. *Proc. Natl. Acad. Sci. USA* **2012**, *109*, 18909–18914. [CrossRef] [PubMed]
74. Zhou, Y.; Wang, W.; Zeng, P.; Feng, J.; Li, D.; Jing, Y.; Zhang, J.; Yin, X.; Li, J.; Ye, H.; et al. Structural basis of main proteases of HCoV-229E bound to inhibitor PF-07304814 and PF-07321332. *Biochem. Biophys. Res. Commun.* **2023**, *657*, 16–23. [CrossRef] [PubMed]
75. Dai, W.; Zhang, B.; Jiang, X.M.; Su, H.; Li, J.; Zhao, Y.; Xie, X.; Jin, Z.; Peng, J.; Liu, F.; et al. Structure-based design of antiviral drug candidates targeting the SARS-CoV-2 main protease. *Science* **2020**, *368*, 1331–1335. [CrossRef] [PubMed]
76. Li, J.; Zhou, X.; Zhang, Y.; Zhong, F.; Lin, C.; McCormick, P.J.; Jiang, F.; Luo, J.; Zhou, H.; Wang, Q.; et al. Crystal structure of SARS-CoV-2 main protease in complex with the natural product inhibitor shikonin illuminates a unique binding mode. *Sci. Bull.* **2021**, *66*, 661–663. [CrossRef] [PubMed]
77. Pasquereau, S.; Galais, M.; Bellefroid, M.; Pachón Angona, I.; Morot-Bizot, S.; Ismaili, L.; Van Lint, C.; Herbein, G. Ferulic acid derivatives block coronaviruses HCoV-229E and SARS-CoV-2 replication in vitro. *Sci. Rep.* **2022**, *12*, 20309. [CrossRef] [PubMed]
78. V'Kovski, P.; Kratzel, A.; Steiner, S.; Stalder, H.; Thiel, V. Coronavirus biology and replication: Implications for SARS-CoV-2. *Nat. Rev. Microbiol.* **2021**, *19*, 155–170. [CrossRef] [PubMed]
79. van der Hoek, L. Human coronaviruses: What do they cause? *Antivir. Ther.* **2007**, *12 Pt B*, 651–658. [CrossRef]
80. Ren, Z.; Yan, L.; Zhang, N.; Guo, Y.; Yang, C.; Lou, Z.; Rao, Z. The newly emerged SARS-like coronavirus HCoV-EMC also has an "Achilles' heel": Current effective inhibitor targeting a 3C-like protease. *Protein Cell* **2013**, *4*, 248–250. [CrossRef]
81. Gupta, S.; Lynn, A.M.; Gupta, V. Standardization of virtual-screening and post-processing protocols relevant to in-silico drug discovery. *3 Biotech* **2018**, *8*, 504. [CrossRef]
82. Bender, B.J.; Gahbauer, S.; Luttens, A.; Lyu, J.; Webb, C.M.; Stein, R.M.; Fink, E.A.; Balius, T.E.; Carlsson, J.; Irwin, J.J.; et al. A practical guide to large-scale docking. *Nat. Protoc.* **2021**, *16*, 4799–4832. [CrossRef] [PubMed]
83. Tomašić, T.; Šink, R.; Zidar, N.; Fic, A.; Contreras-Martel, C.; Dessen, A.; Patin, D.; Blanot, D.; Müller-Premru, M.; Gobec, S.; et al. Dual Inhibitor of MurD and MurE Ligases from Escherichia coli and *Staphylococcus aureus*. *ACS Med. Chem. Lett.* **2012**, *3*, 626–630. [CrossRef] [PubMed]
84. Gordon, E.; Flouret, B.; Chantalat, L.; van Heijenoort, J.; Mengin-Lecreulx, D.; Dideberg, O. Crystal structure of UDP-N-acetylmuramoyl-L-alanyl-D-glutamate: Meso-diaminopimelate ligase from *Escherichia coli*. *J. Biol. Chem.* **2001**, *276*, 10999–11006. [CrossRef] [PubMed]

85. Otero, L.H.; Rojas-Altuve, A.; Llarrull, L.I.; Carrasco-López, C.; Kumarasiri, M.; Lastochkin, E.; Fishovitz, J.; Dawley, M.; Hesek, D.; Lee, M.; et al. How allosteric control of *Staphylococcus aureus* penicillin binding protein 2a enables methicillin resistance and physiological function. *Proc. Natl. Acad. Sci. USA* **2013**, *110*, 16808–16813. [CrossRef] [PubMed]
86. Duplessis, C.; Crum-Cianflone, N.F. Ceftaroline: A New Cephalosporin with Activity against Methicillin-Resistant *Staphylococcus aureus* (MRSA). *Clin. Med. Rev. Ther.* **2011**, *3*, a2466. [PubMed]
87. NCBI. PubChem Compound Summary for CID 155903259, Nirmatrelvir. Available online: https://pubchem.ncbi.nlm.nih.gov/compound/Nirmatrelvir (accessed on 9 February 2024).
88. Billones, J.; Bangalan, M. Structure-Based Discovery of Inhibitors Against MurE in Methicillin-Resistant *Staphylococcus aureus*. *Orient. J. Chem.* **2019**, *35*, 618–625. [CrossRef]
89. Zouhir, A.; Jemli, S.; Omrani, R.; Kthiri, A.; Jridi, T.; Sebei, K. In Silico Molecular Analysis and Docking of Potent Antimicrobial Peptides against MurE Enzyme of Methicillin Resistant *Staphylococcus aureus*. *Int. J. Pept. Res. Ther.* **2021**, *27*, 1253–1263. [CrossRef]
90. Hervin, V.; Roy, V.; Agrofoglio, L.A. Antibiotics and Antibiotic Resistance—Mur Ligases as an Antibacterial Target. *Molecules* **2023**, *28*, 8076. [CrossRef]
91. Oselusi, S.O.; Fadaka, A.O.; Wyckoff, G.J.; Egieyeh, S.A. Computational Target-Based Screening of Anti-MRSA Natural Products Reveals Potential Multitarget Mechanisms of Action through Peptidoglycan Synthesis Proteins. *ACS Omega* **2022**, *7*, 37896–37906. [CrossRef]
92. Kontoyianni, M.; McClellan, L.M.; Sokol, G.S. Evaluation of docking performance: Comparative data on docking algorithms. *J. Med. Chem.* **2004**, *47*, 558–565. [CrossRef]
93. Ghuysen, J.M. Serine beta-lactamases and penicillin-binding proteins. *Annu. Rev. Microbiol.* **1991**, *45*, 37–67. [CrossRef]
94. Frère, J.M.; Page, M.G. Penicillin-binding proteins: Evergreen drug targets. *Curr. Opin. Pharmacol.* **2014**, *18*, 112–119. [CrossRef]
95. Drawz, S.M.; Bonomo, R.A. Three decades of beta-lactamase inhibitors. *Clin. Microbiol. Rev.* **2010**, *23*, 160–201. [CrossRef] [PubMed]
96. Sobhanifar, S.; Worrall, L.J.; Gruninger, R.J.; Wasney, G.A.; Blaukopf, M.; Baumann, L.; Lameignere, E.; Solomonson, M.; Brown, E.D.; Withers, S.G.; et al. Structure and mechanism of *Staphylococcus aureus* TarM, the wall teichoic acid α-glycosyltransferase. *Proc. Natl. Acad. Sci. USA* **2015**, *112*, E576–E585. [CrossRef] [PubMed]
97. Swoboda, J.G.; Meredith, T.C.; Campbell, J.; Brown, S.; Suzuki, T.; Bollenbach, T.; Malhowski, A.J.; Kishony, R.; Gilmore, M.S.; Walker, S. Discovery of a small molecule that blocks wall teichoic acid biosynthesis in *Staphylococcus aureus*. *ACS Chem. Biol.* **2009**, *4*, 875–883. [CrossRef] [PubMed]
98. Wang, H.; Gill, C.J.; Lee, S.H.; Mann, P.; Zuck, P.; Meredith, T.C.; Murgolo, N.; She, X.; Kales, S.; Liang, L.; et al. Discovery of wall teichoic acid inhibitors as potential anti-MRSA β-lactam combination agents. *Chem. Biol.* **2013**, *20*, 272–284. [CrossRef] [PubMed]
99. Ragab, S.S.; Sweed, A.M.K.; Elrashedy, A.A.; Allayeh, A.K. Design, Synthesis, Antiviral Evaluation, and Molecular Dynamics Simulation Studies of New Spirocyclic Thiopyrimidinones as Anti HCoV-229E. *Chem. Biodivers.* **2022**, *19*, e202200632. [CrossRef]
100. Li, J.; Lin, C.; Zhou, X.; Zhong, F.; Zeng, P.; McCormick, P.J.; Jiang, H.; Zhang, J. Structural Basis of Main Proteases of Coronavirus Bound to Drug Candidate PF-07304814. *J. Mol. Biol.* **2022**, *434*, 167706. [CrossRef]
101. Li, J.; Lin, C.; Zhou, X.; Zhong, F.; Zeng, P.; Yang, Y.; Zhang, Y.; Yu, B.; Fan, X.; McCormick, P.J.; et al. Structural Basis of the Main Proteases of Coronavirus Bound to Drug Candidate PF-07321332. *J. Virol.* **2022**, *96*, e0201321. [CrossRef]
102. Anand, K.; Ziebuhr, J.; Wadhwani, P.; Mesters, J.R.; Hilgenfeld, R. Coronavirus main proteinase (3CLpro) structure: Basis for design of anti-SARS drugs. *Science* **2003**, *300*, 1763–1767. [CrossRef]
103. Zaki, A.A.; Ashour, A.; Elhady, S.S.; Darwish, K.M.; Al-Karmalawy, A.A. Calendulaglycoside A showing potential activity against SARS-CoV-2 main protease: Molecular docking, molecular dynamics, and SAR studies. *J. Tradit. Complement. Med.* **2022**, *12*, 16–34. [CrossRef] [PubMed]
104. Elhady, S.S.; Abdelhameed, R.F.A.; Malatani, R.T.; Alahdal, A.M.; Bogari, H.A.; Almalki, A.J.; Mohammad, K.A.; Ahmed, S.A.; Khedr, A.I.M.; Darwish, K.M. Molecular Docking and Dynamics Simulation Study of *Hyrtios erectus* Isolated *Scalarane* sesterterpenes as Potential SARS-CoV-2 Dual Target Inhibitors. *Biology* **2021**, *10*, 389. [CrossRef] [PubMed]
105. Soltan, M.A.; Eldeen, M.A.; Elbassiouny, N.; Kamel, H.L.; Abdelraheem, K.M.; El-Gayyed, H.A.; Gouda, A.M.; Sheha, M.F.; Fayad, E.; Ali, O.A.A.; et al. In Silico Designing of a Multitope Vaccine against *Rhizopus microsporus* with Potential Activity against Other Mucormycosis Causing Fungi. *Cells* **2021**, *10*, 3014. [CrossRef] [PubMed]
106. Arnittali, M.; Rissanou, A.N.; Harmandaris, V. Structure Of Biomolecules through Molecular Dynamics Simulations. *Procedia Comput. Sci.* **2019**, *156*, 69–78. [CrossRef]
107. Liu, K.; Watanabe, E.; Kokubo, H. Exploring the stability of ligand binding modes to proteins by molecular dynamics simulations. *J. Comput.-Aided Mol. Des.* **2017**, *31*, 201–211. [CrossRef]
108. Manandhar, A.; Blass, B.E.; Colussi, D.J.; Almi, I.; Abou-Gharbia, M.; Klein, M.L.; Elokely, K.M. Targeting SARS-CoV-2 M3CLpro by HCV NS3/4a Inhibitors: In Silico Modeling and In Vitro Screening. *J. Chem. Inf. Model.* **2021**, *61*, 1020–1032. [CrossRef]
109. Almalki, A.J.; Ibrahim, T.S.; Elhady, S.S.; Hegazy, W.A.H.; Darwish, K.M. Computational and Biological Evaluation of β-Adrenoreceptor Blockers as Promising Bacterial Anti-Virulence Agents. *Pharmaceuticals* **2022**, *15*, 110. [CrossRef]
110. Benson, N.C.; Daggett, V. A comparison of multiscale methods for the analysis of molecular dynamics simulations. *J. Phys. Chem. B* **2012**, *116*, 8722–8731. [CrossRef]

111. Singh, W.; Karabencheva-Christova, T.G.; Black, G.W.; Ainsley, J.; Dover, L.; Christov, C.Z. Conformational Dynamics, Ligand Binding and Effects of Mutations in NirE an S-Adenosyl-L-Methionine Dependent Methyltransferase. *Sci. Rep.* **2016**, *6*, 20107. [CrossRef]
112. Fatriansyah, J.F.; Rizqillah, R.K.; Yandi, M.Y.; Fadilah; Sahlan, M. Molecular docking and dynamics studies on propolis sulabiroin-A as a potential inhibitor of SARS-CoV-2. *J. King Saud Univ. Sci.* **2022**, *34*, 101707. [CrossRef]
113. de Souza, A.S.; Pacheco, B.D.C.; Pinheiro, S.; Muri, E.M.F.; Dias, L.R.S.; Lima, C.H.S.; Garrett, R.; de Moraes, M.B.M.; de Souza, B.E.G.; Puzer, L. 3-Acyltetramic acids as a novel class of inhibitors for human kallikreins 5 and 7. *Bioorg. Med. Chem. Lett.* **2019**, *29*, 1094–1098. [CrossRef] [PubMed]
114. Cavasotto, C.N. Binding Free Energy Calculation Using Quantum Mechanics Aimed for Drug Lead Optimization. *Methods Mol. Biol.* **2020**, *2114*, 257–268. [PubMed]
115. Kumari, R.; Kumar, R.; Lynn, A. g_mmpbsa—A GROMACS Tool for High-Throughput MM-PBSA Calculations. *J. Chem. Inf. Model.* **2014**, *54*, 1951–1962. [CrossRef] [PubMed]
116. Choi, J.M.; Serohijos, A.W.R.; Murphy, S.; Lucarelli, D.; Lofranco, L.L.; Feldman, A.; Shakhnovich, E.I. Minimalistic predictor of protein binding energy: Contribution of solvation factor to protein binding. *Biophys. J.* **2015**, *108*, 795–798. [CrossRef] [PubMed]
117. Genheden, S.; Kuhn, O.; Mikulskis, P.; Hoffmann, D.; Ryde, U. The normal-mode entropy in the MM/GBSA method: Effect of system truncation, buffer region, and dielectric constant. *J. Chem. Inf. Model.* **2012**, *52*, 2079–2088. [CrossRef] [PubMed]
118. Genheden, S.; Ryde, U. The MM/PBSA and MM/GBSA methods to estimate ligand-binding affinities. *Expert Opin. Drug Discov.* **2015**, *10*, 449–461. [CrossRef] [PubMed]
119. Schrader, A.M.; Donaldson, S.H.; Song, J.; Cheng, C.-Y.; Lee, D.W.; Han, S.; Israelachvili, J.N. Correlating steric hydration forces with water dynamics through surface force and diffusion NMR measurements in a lipid–DMSO–H_2O system. *Proc. Natl. Acad. Sci. USA* **2015**, *112*, 10708–10713. [CrossRef] [PubMed]
120. Shoichet, B.K.; Leach, A.R.; Kuntz, I.D. Ligand solvation in molecular docking. *Proteins* **1999**, *34*, 4–16. [CrossRef]
121. Zou, Y.; Liu, L.; Liu, J.; Liu, G. Bioisosteres in drug discovery: Focus on tetrazole. *Future Med. Chem.* **2020**, *12*, 91–93. [CrossRef]
122. Behairy, M.Y.; Soltan, M.A.; Adam, M.S.; Refaat, A.M.; Ezz, E.M.; Albogami, S.; Fayad, E.; Althobaiti, F.; Gouda, A.M.; Sileem, A.E.; et al. Computational Analysis of Deleterious SNPs in NRAS to Assess Their Potential Correlation With Carcinogenesis. *Front. Genet.* **2022**, *13*, 872845. [CrossRef]
123. Hazalin, N.A.; Ramasamy, K.; Lim, S.S.M.; Wahab, I.A.; Cole, A.L.; Abdul Majeed, A.B. Cytotoxic and antibacterial activities of endophytic fungi isolated from plants at the National Park, Pahang, Malaysia. *BMC Complement. Altern. Med.* **2009**, *9*, 46. [CrossRef] [PubMed]
124. Hubka, V.; Kolařík, M.; Kubátová, A.; Peterson, S.W. Taxonomic revision of Eurotium and transfer of species to *Aspergillus*. *Mycologia* **2013**, *105*, 912–937. [CrossRef]
125. Turan, C.; Nanni, I.M.; Brunelli, A.; Collina, M. New rapid DNA extraction method with Chelex from *Venturia inaequalis* spores. *J. Microbiol. Methods* **2015**, *115*, 139–143. [CrossRef] [PubMed]
126. Morgan, M.C.; Boyette, M.; Goforth, C.; Sperry, K.V.; Greene, S.R. Comparison of the Biolog OmniLog Identification System and 16S ribosomal RNA gene sequencing for accuracy in identification of atypical bacteria of clinical origin. *J. Microbiol. Methods* **2009**, *79*, 336–343. [CrossRef]
127. Elissawy, A.M.; Ebada, S.S.; Ashour, M.L.; El-Neketi, M.; Ebrahim, W.; Singab, A.B. New secondary metabolites from the mangrove-derived fungus *Aspergillus* sp. AV-2. *Phytochem. Lett.* **2019**, *29*, 1–5. [CrossRef]
128. VanderMolen, K.M.; Raja, H.A.; El-Elimat, T.; Oberlies, N.H. Evaluation of culture media for the production of secondary metabolites in a natural products screening program. *AMB Express* **2013**, *3*, 71. [CrossRef] [PubMed]
129. CLSI. *Performance Standards for Antimicrobial Susceptibility Testing: 20th Informational Supplement*; CLSI Doc. M100-S20; Clinical and Laboratory Standards Institute: Wayne, PA, USA, 2010.
130. Karamolah, K.S.; Mousavi, F.; Mahmoudi, H. Antimicrobial inhibitory activity of aqueous, hydroalcoholic and alcoholic extracts of leaves and stem of Daphne mucronata on growth of oral bacteria. *GMS Hyg. Infect. Control* **2017**, *12*, 301.
131. Gonelimali, F.D.; Lin, J.; Miao, W.; Xuan, J.; Charles, F.; Chen, M.; Hatab, S.R. Antimicrobial properties and mechanism of action of some plant extracts against food pathogens and spoilage microorganisms. *Front. Microbiol.* **2018**, *9*, 1639. [CrossRef] [PubMed]
132. Sánchez, E.; Morales, C.R.; Castillo, S.; Leos-Rivas, C.; García-Becerra, L.; Martínez, D.M.O. Antibacterial and antibiofilm activity of methanolic plant extracts against nosocomial microorganisms. *Evid.-Based Complement. Altern. Med. Ecam* **2016**, *2016*, 1572697. [CrossRef]
133. Djordjevic, D.; Wiedmann, M.; McLandsborough, L. Microtiter plate assay for assessment of Listeria monocytogenes biofilm formation. *Appl. Environ. Microbiol.* **2002**, *68*, 2950–2958. [CrossRef]
134. Choi, H.J.; Song, J.H.; Park, K.S.; Kwon, D.H. Inhibitory effects of quercetin 3-rhamnoside on influenza A virus replication. *Eur. J. Pharm. Sci.* **2009**, *37*, 329–333. [CrossRef] [PubMed]
135. Donalisio, M.; Nana, H.M.; Ngono Ngane, R.A.; Gatsing, D.; Tiabou Tchinda, A.; Rovito, R.; Cagno, V.; Cagliero, C.; Boyom, F.F.; Rubiolo, P. In vitro anti-Herpes simplex virus activity of crude extract of the roots of Nauclea latifolia Smith (Rubiaceae). *BMC Complement. Altern. Med.* **2013**, *13*, 266. [CrossRef] [PubMed]
136. Pauwels, R.; Balzarini, J.; Baba, M.; Snoeck, R.; Schols, D.; Herdewijn, P.; Desmyter, J.; De Clercq, E. Rapid and automated tetrazolium-based colorimetric assay for the detection of anti-HIV compounds. *J. Virol. Methods* **1988**, *20*, 309–321. [CrossRef] [PubMed]

137. Daina, A.; Michielin, O.; Zoete, V. SwissADME: A free web tool to evaluate pharmacokinetics, drug-likeness and medicinal chemistry friendliness of small molecules. *Sci. Rep.* **2017**, *7*, 42717. [CrossRef]
138. Páll, S.; Abraham, M.J.; Kutzner, C.; Hess, B.; Lindahl, E. Solving Software Challenges for Exascale. In *Tackling Exascale Software Challenges in Molecular Dynamics Simulations with GROMACS*; Markidis, S., Laure, E., Eds.; Springer International Publishing: Cham, Switzerland, 2015; pp. 3–27.
139. Darden, T.; York, D.; Pedersen, L. Particle mesh Ewald: An N·log(N) method for Ewald sums in large systems. *J. Chem. Phys.* **1993**, *98*, 10089–10092. [CrossRef]

Disclaimer/Publisher's Note: The statements, opinions and data contained in all publications are solely those of the individual author(s) and contributor(s) and not of MDPI and/or the editor(s). MDPI and/or the editor(s) disclaim responsibility for any injury to people or property resulting from any ideas, methods, instructions or products referred to in the content.

Article

Anticancer Evaluation of Novel Benzofuran–Indole Hybrids as Epidermal Growth Factor Receptor Inhibitors against Non-Small-Cell Lung Cancer Cells

Yechan Lee [1,†], Sunhee Lee [1,†], Younho Lee [1], Doona Song [2], So-Hyeon Park [1], Jieun Kim [1], Wan Namkung [1,*] and Ikyon Kim [1,*]

1 College of Pharmacy and Yonsei Institute of Pharmaceutical Sciences, Yonsei University, 85 Songdogwahak-ro, Yeonsu-gu, Incheon 21983, Republic of Korea; llyycc94@naver.com (Y.L.); lovelysh.94@hanmail.net (S.L.); zz265zz@hanmail.net (Y.L.); sohyeon0605@yonsei.ac.kr (S.-H.P.); jieunkimx@yonsei.ac.kr (J.K.)
2 Department of Biotechnology, College of Life Science and Biotechnology, Yonsei University, Seoul 03722, Republic of Korea; doona.s@yonsei.ac.kr
* Correspondence: wnamkung@yonsei.ac.kr (W.N.); ikyonkim@yonsei.ac.kr (I.K.); Tel.: +82-32-749-4519 (W.N.); +82-32-749-4515 (I.K.); Fax: +82-32-749-4105 (W.N. & I.K.)
† These authors contributed equally to this work.

Citation: Lee, Y.; Lee, S.; Lee, Y.; Song, D.; Park, S.-H.; Kim, J.; Namkung, W.; Kim, I. Anticancer Evaluation of Novel Benzofuran–Indole Hybrids as Epidermal Growth Factor Receptor Inhibitors against Non-Small-Cell Lung Cancer Cells. *Pharmaceuticals* **2024**, *17*, 231. https://doi.org/10.3390/ph17020231

Academic Editors: Valentina Noemi Madia and Davide Ialongo

Received: 11 January 2024
Revised: 2 February 2024
Accepted: 7 February 2024
Published: 9 February 2024

Copyright: © 2024 by the authors. Licensee MDPI, Basel, Switzerland. This article is an open access article distributed under the terms and conditions of the Creative Commons Attribution (CC BY) license (https://creativecommons.org/licenses/by/4.0/).

Abstract: The epidermal growth factor receptor (EGFR), also known as ErbB1 and HER1, belongs to the receptor tyrosine kinase family. EGFR serves as the primary driver in non-small-cell lung cancer (NSCLC) and is a promising therapeutic target for NSCLC. In this study, we synthesized a novel chemical library based on a benzofuran–indole hybrid scaffold and identified **8aa** as a potent and selective EGFR inhibitor. Interestingly, **8aa** not only showed selective anticancer effects against NSCLC cell lines, PC9, and A549, but it also showed significant inhibitory effects against the double mutant L858R/T790M EGFR, which frequently occurs in NSCLC. In addition, in PC9 and A549 cells, **8aa** potently blocked the EGFR signaling pathway, cell viability, and cell migration. These findings suggest that **8aa**, a benzofuran–indole hybrid derivative, is a novel EGFR inhibitor that may be a potential candidate for the treatment of NSCLC patients with EGFR mutations.

Keywords: benzofuran; indole; hybrid structure; NSCLC; EGFR

1. Introduction

The identification of new chemical entities with pharmacologically modulating properties is important in the early stages of drug discovery processes. Accordingly, the generation of new drug-like chemical scaffolds and their derivatives for biological screening is highly desired. Against this backdrop, we were able to find several chemical motifs (**I–VI**) with significant pharmacological functions through the synthesis and biological evaluation of novel heterocycles (Figure 1) [1–3].

Lung cancer is the most the common cause of cancer-related death worldwide, with a 5-year patient survival rate of less than 15%. Non-small-cell lung cancer (NSCLC), which is commonly found in lung cancer, accounts for 85% of lung cancer cases [4,5].

The epidermal growth factor receptor (EGFR) belongs to the receptor tyrosine kinase family, which is highly expressed in NSCLC patients [6]. Given the pivotal role of the EGFR signaling pathway in regulating tumorigenesis, cell growth, and proliferation in NSCLC, the EGFR emerges as an attractive therapeutic target [7]. For example, EGFR overexpression or mutation has been demonstrated in 43–89% of NSCLC patients [8]. In addition, it has been observed that 25% of NSCLC patients exhibited mutations in the EGFR tyrosine kinase domain, with 75% of these mutations being associated with overexpression of EGFR [9]. EGFR overexpression or abnormalities trigger sustained signal transduction, promoting cell survival, proliferation, relapse, tumorigenesis, and metastasis in NSCLC through the MAPK, PI3K/AKT, and signal transducer and activator of transcription (STAT) factors [10,11]. To date, clinically available EGFR inhibitors comprise EGFR tyrosine kinase inhibitors (TKIs) like erlotinib, gefitinib, afatinib, and osimertinib, and monoclonal antibodies (mAbs) such as panitumumab and cetuximab [12]. However, despite initial robust responses to first- and second-generation EGFR-TKIs, a considerable number of NSCLC patients develop acquired resistance during EGFR-TKI treatment within 9 to 14 months after starting treatment [13]. Therefore, there remains a necessity for the development of novel EGFR inhibitors to address drug resistance in the treatment of NSCLC.

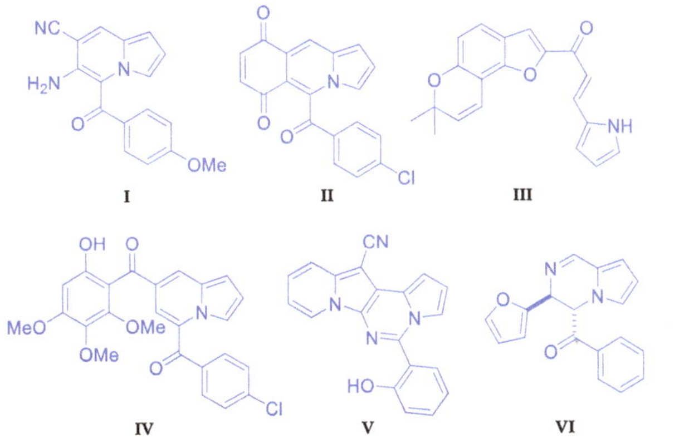

Figure 1. Bioactive heterocycles developed in our laboratory.

Erlotinib and gefitinib are two representative EGFR-TKIs with a 4-anilinoquinazoline skeleton (Figure 2). In 2008, Lüth and Löwe reported the synthesis of quinazoline–indole hybrids (A) which were found to exhibit EGFR inhibitory activity [14]. In connection with our continued interest in the design and synthesis of new anticancer agents [15,16], we hoped to find a new heterocyclic skeleton to replace the quinazoline moiety while retaining the indole group. Along this line, we wondered whether benzofuran could be used instead of quinazoline.

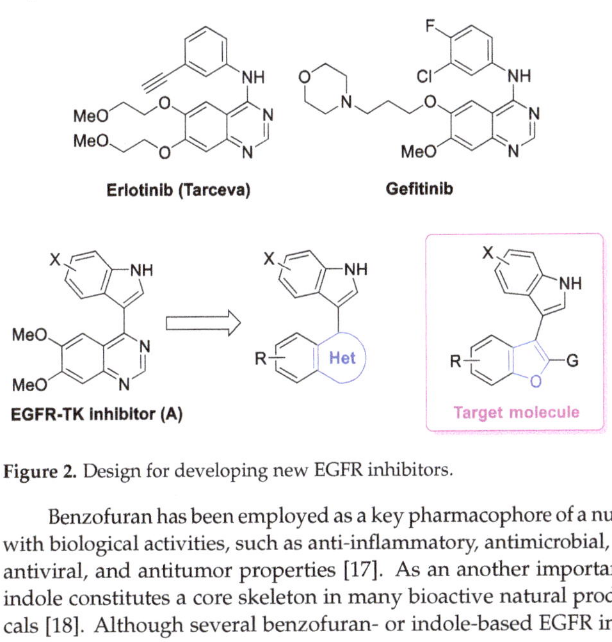

Figure 2. Design for developing new EGFR inhibitors.

Benzofuran has been employed as a key pharmacophore of a number of small molecules with biological activities, such as anti-inflammatory, antimicrobial, antifungal, antioxidant, antiviral, and antitumor properties [17]. As an another important privileged structure, indole constitutes a core skeleton in many bioactive natural products and pharmaceuticals [18]. Although several benzofuran- or indole-based EGFR inhibitors (VII–XI) have been discussed in the literature (Figure 3) [19–23], no chemical scaffolds consisting of both benzofuran and indole have been reported as EGFR inhibitors. Here, we wish to describe the modular synthesis and biological evaluation of benzofuran–indole hybrids as a new class of highly promising EGFR inhibitors.

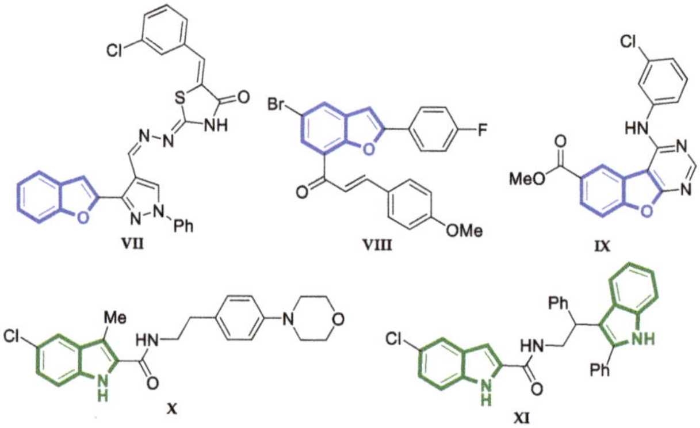

Figure 3. Benzofuran and indole EGFR inhibitors.

As part of our general plan to make poly-functionalized benzofurans, a domino nucleophilic substitution–dehydrative cyclization procedure with **1** was deemed to give 2,3-disubstituted benzofuran **2** (Scheme 1a). To validate our hypothesis, the requisite starting material **1** (when LG is OH) was envisioned to be easily prepared via a Friedel–Crafts-type reaction between phenol and arylglyoxal [24]. Inspired by our recent success in achieving the hexafluoroisopropanol (HFIP)-mediated hydroxyalkylation of indolizine **3** to arylglyoxal to afford **4** (Scheme 1b) [25], we expected that a HFIP-promoted Friedel–Crafts-type reaction between phenol **5** and arylglyoxal would give rise to **7**, which could be converted to benzofuran **8**, having an indole at the C3 position upon exposure to indole and *p*-toluenesulfonic acid (PTSA) (Scheme 1c). The biological investigation of benzofuran–indole hybrid **8** [26] revealed that this class of compounds exhibit anticancer activity against PC9 and A549 lung cancer cells via the inhibition of phosphorylated EGFR. Here, we wish to describe our findings along this line.

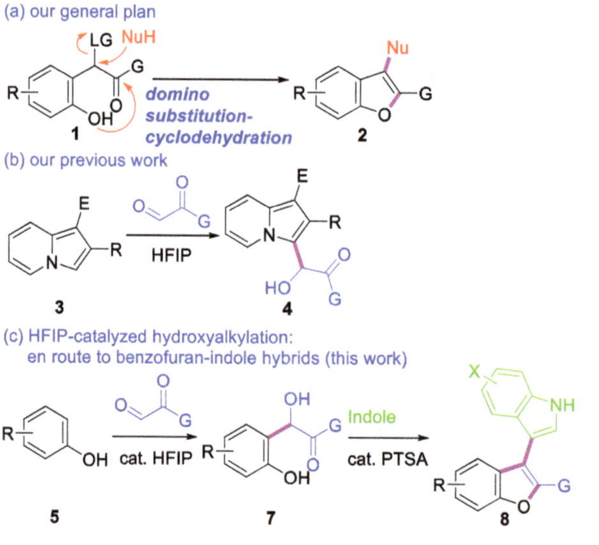

Scheme 1. Synthetic plans.

2. Results and Discussion
2.1. Design and Synthesis of Benzofuran–Indole Hybrids

When we reacted **5a** (2 equiv) with phenylglyoxal (1 equiv) in the presence of HFIP (0.5 equiv) in toluene at 70 °C, the desired product **7a** was isolated in 95% yield (Scheme 2). The subsequent treatment of **7a** with indole and PTSA (0.2 equiv) in CHCl₃ at 60 °C provided benzofuran possessing an indole at the C3 site in 97% yield.

Scheme 2. Synthesis of **8a** [a,b]. [a] A mixture of **5a** (2 equiv), phenylglyoxal (0.33 mmol, 1 equiv), and HFIP (0.5 equiv) in toluene (4 mL) was stirred at 70 °C for 36 h. A mixture of **7a** (0.08 mmol, 1 equiv), indole (1.5 equiv), and PTSA (0.2 equiv) in CHCl₃ (2 mL) was stirred at 60 °C for 16 h. [b] Isolated yield (%).

Having found the optimal conditions for the synthesis of **7** and **8**, we examined the reaction scope with several phenols, (hetero)arylglyoxals, and indoles (Table 1). In general, these two-step sequences allowed for a variety of 2-arylbenzofurans **8** bearing an indole at the C3 site via intermediates of **7** in good to excellent yields. Various functional groups, such as alkoxy, alkyl, and halogen, were well tolerated under these conditions.

Table 1. Synthesis of **8** [a].

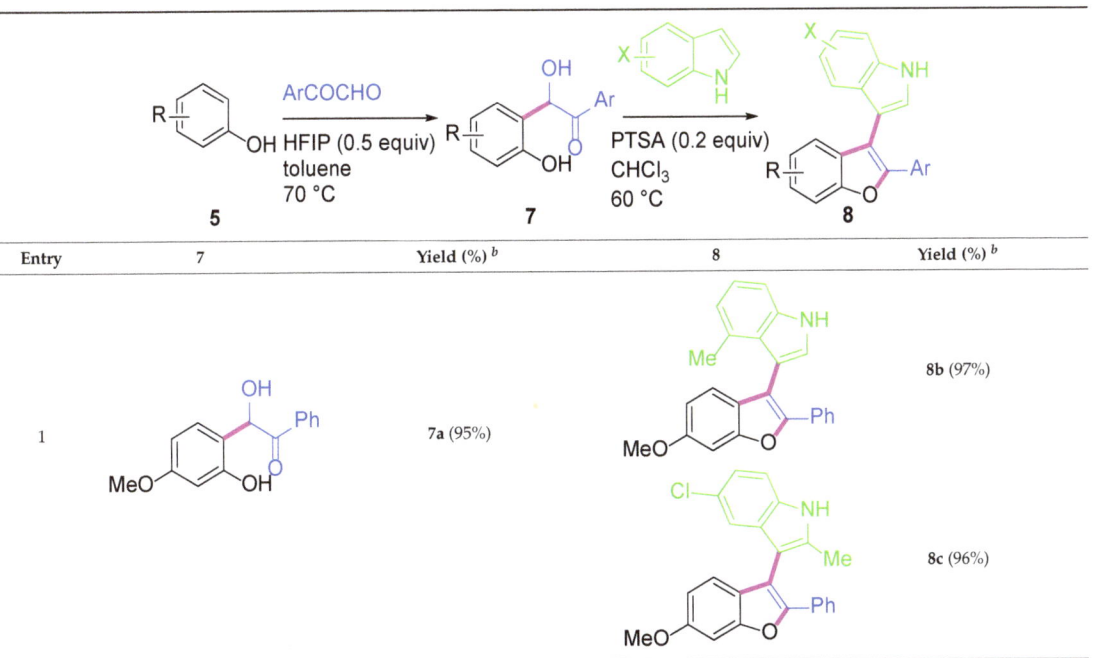

Table 1. Cont.

Entry	7	Yield (%) [b]	8	Yield (%) [b]
			8d (structure: 6-chloroindole-3-yl, 6-methoxybenzofuran, 2-Ph)	8d (98%)
			8e (structure: 5-bromoindole-3-yl, 6-methoxybenzofuran, 2-Ph)	8e (92%)
			8f (structure: 7-bromoindole-3-yl, 6-methoxybenzofuran, 2-Ph)	8f (69%) [d]
			8g (structure: 5-iodoindole-3-yl, 6-methoxybenzofuran, 2-Ph)	8g (94%)
			8h (structure: N-methylindole-3-yl, 6-methoxybenzofuran, 2-Ph)	8h (54%)
2	7b (structure: 4-MeO-2-hydroxyphenyl CH(OH)C(O)-3-MeO-C6H4)	7b (96%)	8i (structure: indole-3-yl, 6-methoxybenzofuran, 2-(3-MeO-C6H4))	8i (65%)

Table 1. *Cont.*

Entry	7	Yield (%) [b]	8	Yield (%) [b]
3	7c	64%	8j	65%
4	7d	89%	8k	87%
5	7e	65%	8l	75%
6	7f	81%	8m	54% [d]
7	7g	98%	8n	80%
8	7h	81%	8o	89%

Table 1. *Cont.*

Entry	7	Yield (%) [b]	8	Yield (%) [b]
9	7i	7i (95%)	8p	8p (81%)
10	7j	7j (94%)	8q	8q (99%)
11	7k	7k (62%) [c]	8r	8r (80%)
12	7l	7l (60%)	8s	8s (99%)
13	7m	7m (42%) [c]	8t	8t (65%)

Table 1. *Cont.*

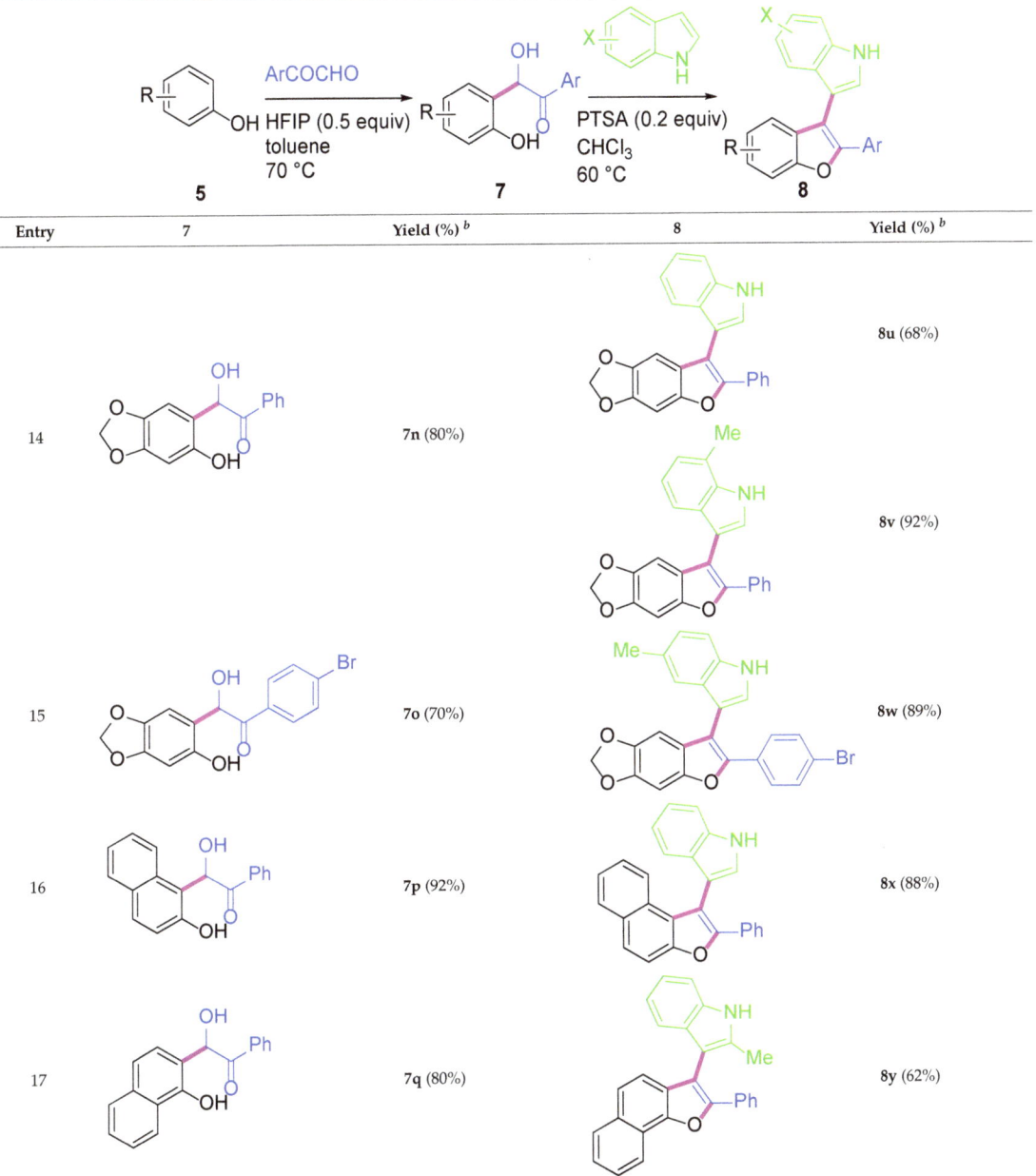

a A mixture of **5** (2 equiv), arylglyoxal (0.33 mmol, 1 equiv), and HFIP (0.5 equiv) in toluene (4 mL) was stirred at 70 °C for 36 h. A mixture of **7** (0.08 mmol, 1 equiv), indole (1.5 equiv), and PTSA (0.2 equiv) in CHCl₃ (2 mL) was stirred at 60 °C for 18 h. *b* Isolated yield (%). *c* Reaction at 60 °C. *d* Reaction at rt.

An immunoblot analysis of these compounds indicated that **8e** significantly inhibited the phosphorylation of the EGFR (Figure 4). In addition, **8g** showed a weak ability to reduce p-EGFR levels. The cytotoxicity of these benzofuran–indole hybrids **8** against lung cancer cell lines was evaluated in PC9 cells (Table 2). Consistent with the immunoblot analysis results, both **8e** and **8g** showed potent cytotoxicity in PC9 cells.

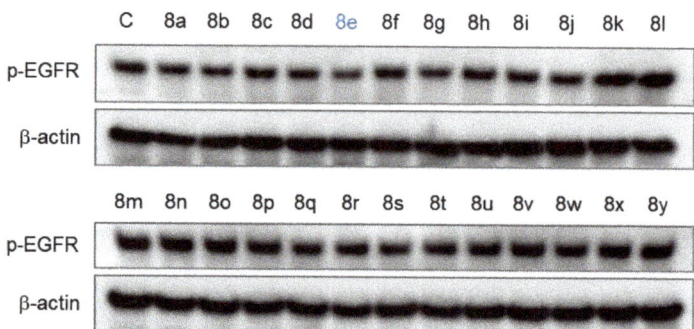

Figure 4. The inhibitory effects of 8a–y on the EGFR in PC9 cells. PC9 cells were pretreated with 10 µM of **8a–y** for 6 h, and then the cells were treated with EGF (20 ng/mL) for 30 min. The expression levels of p-EGFR were observed by immunoblot analysis.

Table 2. Inhibitory effects of **8a–y** on cell viability of PC9 cells (mean, n = 4).

Compound	IC$_{50}$ (µM)
8a	7.53
8b	24.67
8c	6.3
8d	13.38
8e	0.56
8f	31.98
8g	0.85
8h	18.59
8i	11.33
8j	6.33
8k	6.76
8l	14.99
8m	10.61
8n	7.38
8o	2.44
8p	4.04
8q	2.24
8r	5.59
8s	4.46
8t	9.58
8u	2.32
8v	2.11
8w	1.58
8x	27.86
8y	1.58

As **8e** showed promising anticancer activity, more close analogs (**8z–ad**) were synthesized for secondary screening. Our immunoblot analysis showed that **8aa** reduced p-EGFR more than **8e**, and **8aa** inhibited EGFR kinase activity with IC$_{50}$ values of 0.44 ± 0.02 µM (Figure 5). The cytotoxic activities of these derivatives against the lung cancer cell lines (PC9 and A549) indicated that **8aa** exhibited the most remarkable cytotoxicity among the derivatives, with IC$_{50}$ values of 0.32 ± 0.05 µM and 0.89 ± 0.10 µM, respectively (Table 3). To identify whether **8aa** is a potent EGFR inhibitor, further studies were conducted.

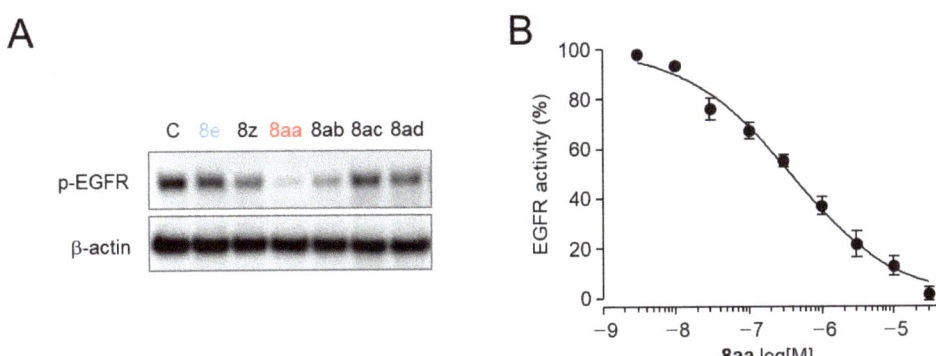

Figure 5. The inhibitory effects of **8e** and **8z–ad** on the phosphorylation of EGFR in PC9 cells. (**A**) PC9 cells were pretreated with 10 μM of **8e** and **8z–ad** for 6 h, and then the cells were treated with EGF (20 ng/mL) for 30 min. The expression levels of p-EGFR were observed by immunoblot analysis. (**B**) The kinase inhibitory activity of **8aa** on the EGFR was assessed using an EGFR kinase assay kit.

Table 3. Inhibitory effects of **8e** and **8z–ad** on cell viabilities of PC9 and A549 cells (mean, $n = 6$).

Compound	Structure	IC$_{50}$ (μM)	
		PC9	A549
8e		0.50	2.74
8z		0.58	1.36
8aa		0.32	0.89
8ab		0.45	6.51

Table 3. *Cont.*

Compound	Structure	IC$_{50}$ (μM)	
		PC9	A549
8ac		2.42	3.53
8ad		1.65	3.21

2.2. Inhibitory Effect of 8aa on EGFR Signaling Pathways in PC9 and A549 Cells

Previous studies have reported that the upregulation of the EGFR occurs frequently in NSCLC, and the EGFR plays an important role in the development and progression of NSCLC [27,28]. To investigate the effects of **8aa** on multiple EGFR-mediated signaling pathways, we performed an immunoblot analysis on the EGF-induced phosphorylation of the EGFR, AKT, and ERK1/2 in NSCLC cell lines, namely PC9 and A549 cells. As shown in Figure 6, EGF strongly increased the phosphorylation of EGFR, and **8aa** significantly reduced the EGF-induced phosphorylation of the EGFR in a dose-dependent manner. In addition, **8aa** also reduced the phosphorylation of AKT and ERK1/2, downstream signaling pathways of the EGFR. These results indicated that **8aa** can effectively block the signal transduction pathway through EGFR phosphorylation.

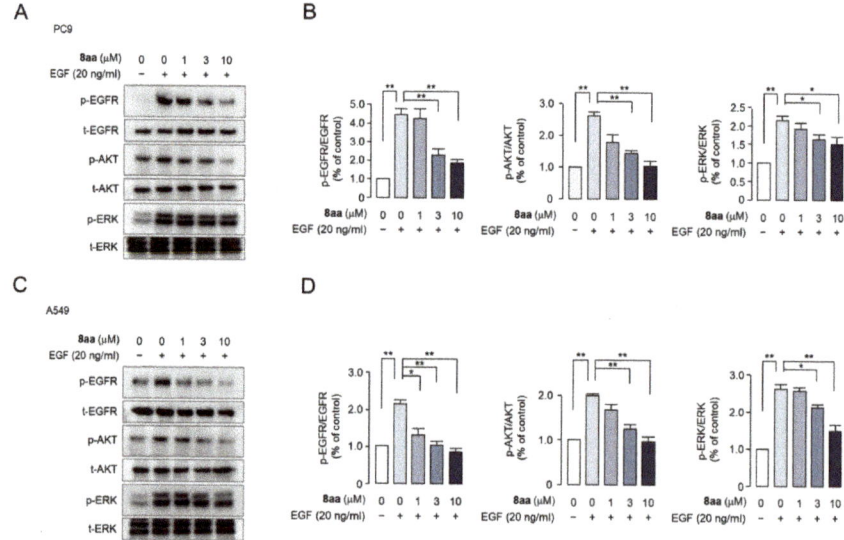

Figure 6. The inhibitory effects of **8aa** on the EGFR signaling pathways in PC9 and A549 cells. (**A,C**) PC9 and A549 cells were pretreated with **8aa** at the indicated concentrations for 6 h, and then

the cells were incubated with EGF (20 ng/mL) for 30 min. The expression levels of p-EGFR, t-EGFR, p-AKT, t-AKT, p-ERK1/2, and t-ERK1/2 were measured by immunoblotting. (**B**,**D**) p-EGFR, p-AKT, and p-ERK1/2 protein intensities were normalized to t-EGFR, t-AKT, and t-ERK1/2, respectively (mean ± S.E., $n = 3$). * $p < 0.05$; ** $p < 0.01$.

2.3. Molecular Modeling of 8aa

To elucidate the underlying mechanism driving the preferential binding of **8aa** to the active conformation of EGFR-TKD, molecular docking studies were carried out using the tyrosine kinase domain of the EGFR (PDB ID: 1M17), which provided the initial erlotinib conformation [29]. The binding mode of **8aa** to the EGFR is depicted in Figure 7, showing its possible molecular interactions. The methoxy oxygens at the C5 and C6 positions of **8aa** form hydrogen bond interactions with the kinase hinge that is an amide backbone of Met793 (Figure 7). The benzofuran moiety within **8aa**, situated at the core, maintains hydrophobic interactions with Val726 and Leu844. In addition, the phenyl group at the C1 site of benzofuran **8aa** is shown to have a π-π interaction with Phe723. Based on these results, compound **8aa** induced the intended mechanism of action by conserving the overall interaction with the tyrosine kinase domain of the EGFR.

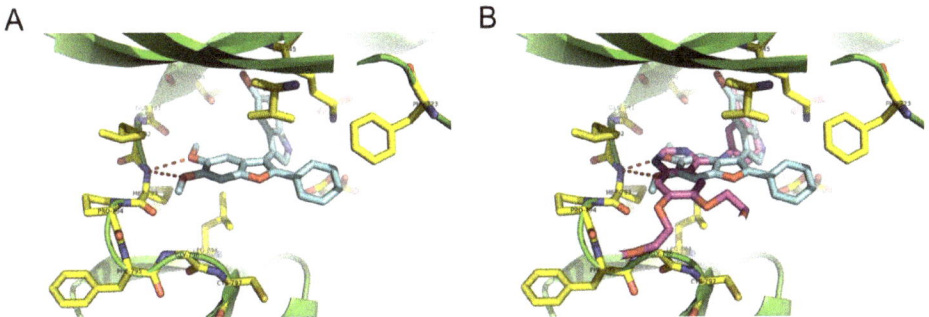

Figure 7. Binding mode of **8aa** with the EGFR. (**A**) A structural simulation of the **8aa**-EGFR complex showed that some residues (yellow stick) were involved in binding with **8aa** (cyan—carbon), including non-bonded interactions (Phe723, Val726, Met677, and Leu844) and red dot hydrogen bonds (Met793). (**B**) A super-imposed model of the co-crystal structure (1M17.pdb) of **8aa** and erlotinib (magenta—carbon).

2.4. Effect of 8aa on Cell Viability in PC9, A549, MCF7, HepG2, PC3, HT29, HaCaT, and HEK293T Cells

To investigate whether **8aa** shows selective cytotoxicity on lung cancer cells, we performed cell proliferation assays on PC9 and A549 non-small-cell lung adenocarcinoma, MCF7 breast adenocarcinoma, HepG2 hepatocellular carcinoma, PC3 prostate adenocarcinoma, HT29 colorectal adenocarcinoma, HaCaT human skin keratinocyte, and HEK293T human embryonic kidney cells. As expected, **8aa** significantly inhibited cell viability in both the PC9 and A549 cells with IC_{50} values of 0.32 ± 0.05 and 0.89 ± 0.10 μM, respectively (Figure 8A,B). Interestingly, **8aa** showed weak inhibitory effects on other cancer cell lines, namely MCF7, HepG2, PC3, and HT29 (Figure 8C–F). In addition, **8aa** weakly reduced cell viability in the non-tumorigenic cell lines, including HaCaT and HEK293T (Figure 8G,H). These findings indicate that **8aa** has the potential to serve as a potent and selective anticancer agent for NSCLC.

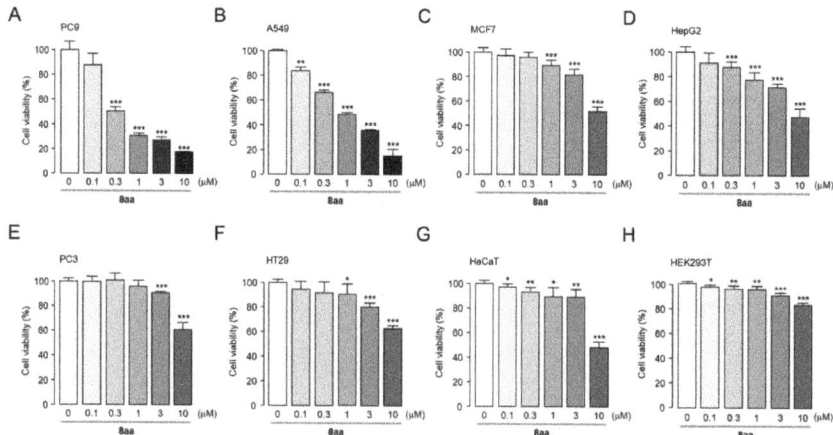

Figure 8. Effects of **8aa** on cell viability in PC9, A549, MCF7, HepG2, PC3, HT29, HaCaT, and HEK293T cells. (**A–H**) PC9, A549, MCF7, HepG2, PC3, HT29, HaCaT, and HEK293T cells were treated with **8aa** at the indicated concentrations for 72 h, and the medium was changed every 24 h with newly added **8aa**. Cell viability was estimated with the MTS assay (mean ± S.D., $n = 6$). * $p < 0.05$; ** $p < 0.01$; *** $p < 0.001$.

2.5. 8aa Inhibits Cell Migration in PC9 and A549 Cells

To assess the potential effect of **8aa** on NSCLC cell migration, an in vitro wound healing assay was performed using PC9 and A549 cells. Interestingly, **8aa** significantly inhibited cell migration in both the PC9 and A549 cells in a dose-dependent manner. In the PC9 cells, treatment with 0.1, 1, and 3 µM of **8aa** reduced cell migration by 21.6%, 42.0%, and 63.7%, respectively. Similarly, in the A549 cells, exposure to 0.1, 1, and 3 µM of **8aa** inhibited cell migration by 19.7%, 33.0%, and 59.6%, respectively (Figure 9).

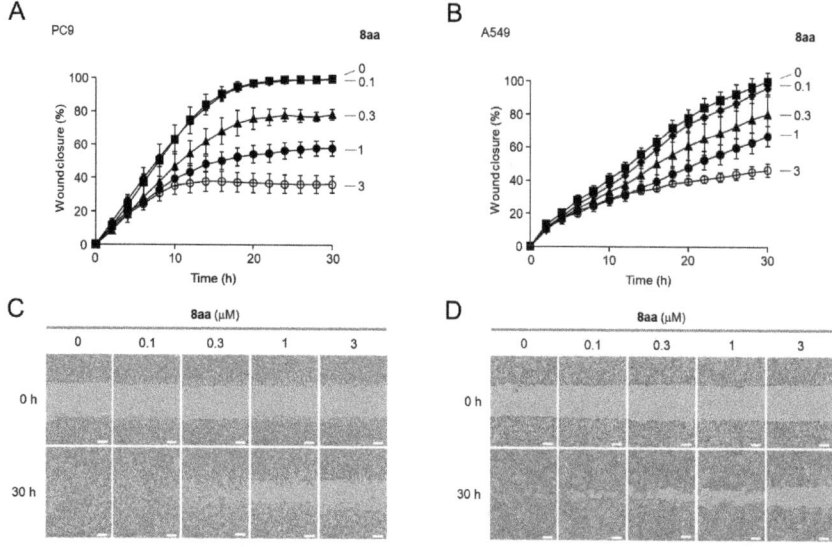

Figure 9. Effect of **8aa** on cell migration in PC9 and A549 cells. (**A,B**) An in vitro wound healing assay was performed on PC9 and A549 cells for 30 h (mean ± S.D., $n = 3$). The PC9 and A549 cells

were treated with the indicated concentrations of **8aa**, and time-lapse images were obtained every 2 h after wound infliction. (**C,D**) Representative wound images were taken at 0 h and 30 h following the administration of **8aa** at the indicated concentrations. The scale bars represent 300 µm.

2.6. *8aa Significantly Induces Apoptosis in PC9 and A549 Cells*

The pharmacological inhibition of the EGFR signaling pathway causes apoptosis in various solid tumors [30,31]. To investigate the apoptotic potential of **8aa** in PC9 and A549 cells, we evaluated its influence on caspase-3 activity and PARP cleavage, established markers of apoptotic signaling. Interestingly, caspase-3 activity was significantly increased by **8aa** in the PC9 and A549 cells in a dose-dependent manner, and the increased caspase-3 activity was completely inhibited by AC-DEVD-CHO, a potent caspase-3 inhibitor (Figure 10A–D). In addition, the expression levels of cleaved PARP were significantly increased by **8aa** in both the PC9 and A549 cells in a dose-dependent manner (Figure 10E–H). These results reveal that **8aa** exhibits potent anticancer effects by inducing apoptosis in NSCLC cells.

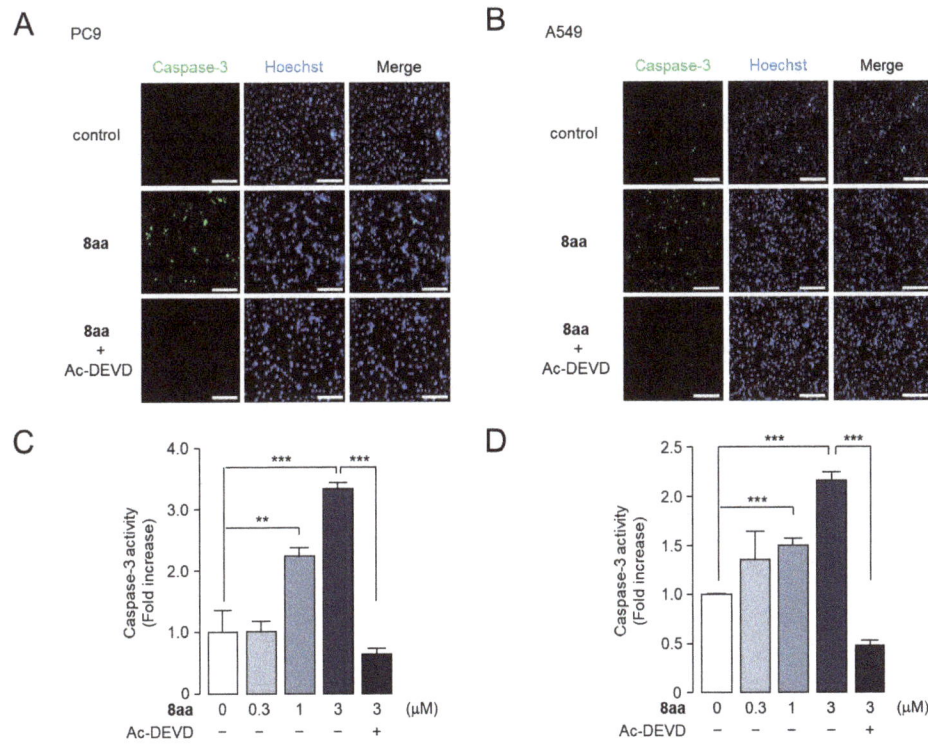

Figure 10. *Cont.*

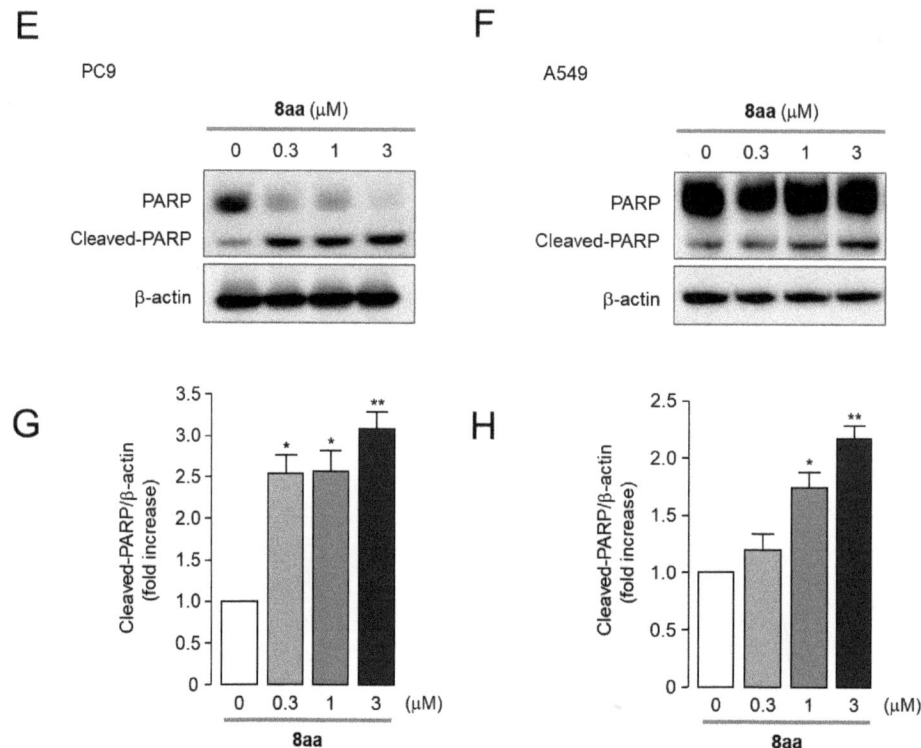

Figure 10. Effects of **8aa** on caspase-3 activity and PARP cleavage in PC9 and A549 cells. (**A,B**) PC9 and A549 cells were treated with 3 µM of **8aa** for 24 h and then incubated with 1 µM of caspase-3 substrate (green) and 1 µM of Hoechst 33342 (blue) for 30 min before image acquisition. The scale bars represent 200 µm. (**C,D**) The PC9 and A549 cells were treated with **8aa** at the indicated concentrations for 24 h, and then 1 µM of caspase-3 substrate was treated for 30 min. Caspase-3 activity was inhibited by 20 µM of Ac-DEVD-CHO (mean ± S.D., $n = 3$). (**E,F**) The cells were treated with **8aa** at the indicated concentrations for 24 h, and the expression levels of PARP, cleaved-PARP, and β-actin were measured by immunoblotting. (**G,H**) Cleaved-PARP protein intensities were normalized to β-actin (mean ± S.E., $n = 3$). * $p < 0.05$; ** $p < 0.01$; *** $p < 0.001$.

To further investigate the effect of **8aa** on the cell cycles of PC9 and A549 cells, we carried out cell cycle analysis using propidium (PI) staining. As shown in Figure 11, **8aa** significantly promoted the ratios in the Sub-G1 (apoptotic peak) phase compared to the control group, but the G2/M phase was not affected by **8aa**. In the case of the **8aa** treatment group, the G0/G1 phase reduced from 73.01% to 47.50% and from 81.64% to 64.24% in the PC9 and A549 cells, respectively. Also, the Sub-G1 phase increased from 8.52% to 35.23% and from 8.48% to 21.40% in the PC9 and A549 cells, respectively. These results suggest that **8aa** significantly induces apoptosis without exerting an effect on cell cycle arrest.

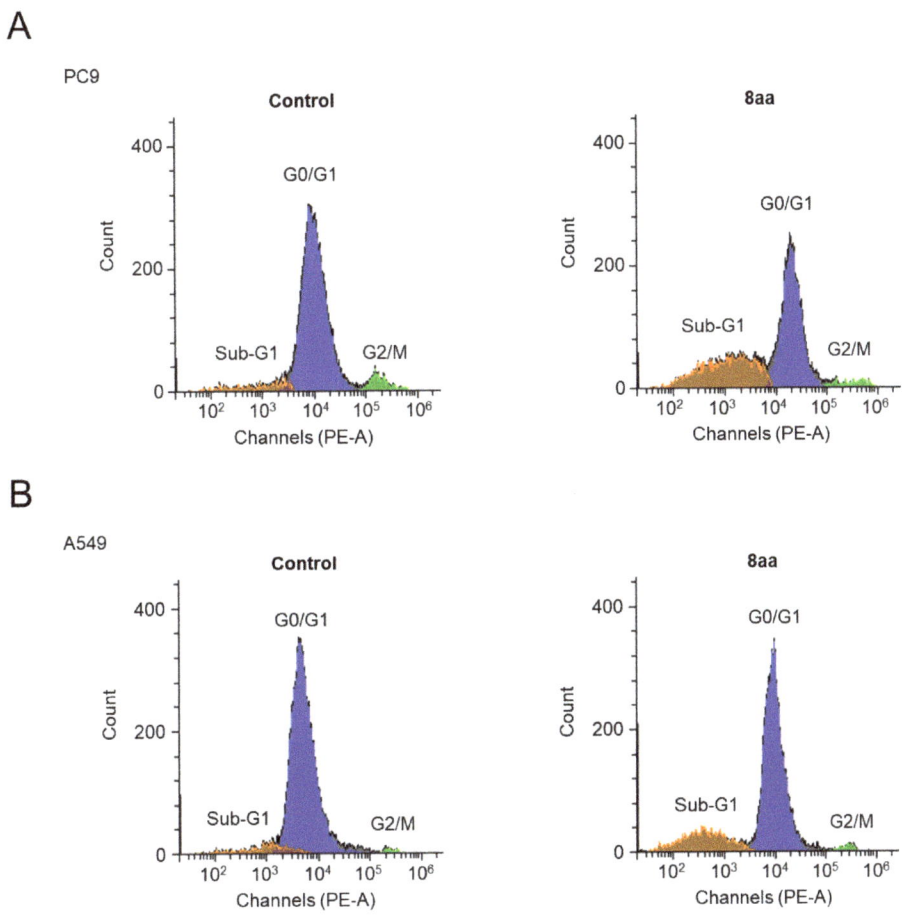

Figure 11. Effect of 8aa on cell cycles of PC9 and A549 cells. (**A**,**B**) PC9 and A549 cells were treated with 10 μM of **8aa** for 24 h and then cell cycle phases were estimated by using propidium iodide (PI) staining followed by cell cycle analysis.

2.7. 8aa Potently Inhibits EGFR$^{L858R/T790M}$ in H1975 Cells

Drug resistance in NSCLC patients is predominantly attributed to EGFR mutations, with the L858R and T790M mutations being the most prevalent EGFR mutations in NSCLC [32–34], and these mutations are associated with resistance to EGFR-TKIs in NSCLC [35,36]. To investigate whether **8aa** inhibits EGFR$^{L858R/T790M}$, we performed immunoblot analysis on the EGF-induced phosphorylation of EGFR$^{L858R/T790M}$ in H1975 cells expressing both EGFR mutations L858R and T790M. Notably, **8aa** potently inhibited the EGF-induced phosphorylation of EGFR$^{L858R/T790M}$ compared to erlotinib (Figure 12A,B). In addition, a structural simulation of the **8aa** and EGFR$^{L858R/T790M}$ complex revealed that **8aa** can interact with Asp855 of EGFR$^{L858R/T790M}$. The indole N-H bond seemed to form a hydrogen bond with the carboxylic acid of Asp855, whereas the same type of hydrogen bonding interaction was not observed in erlotinib (Figure 12C,D). These results suggest that **8aa** can potently inhibit EGFR$^{L858R/T790M}$ in NSCLC.

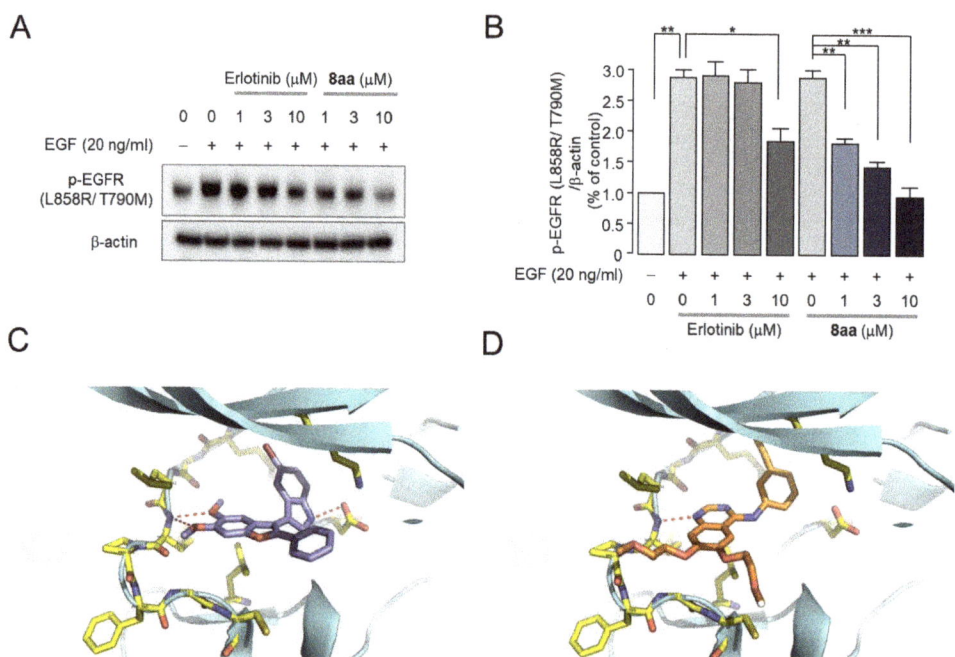

Figure 12. Effect of erlotinib and **8aa** on EGFR[L858R/T790M]. (**A**) H1975 cells were pretreated with erlotinib and **8aa** at the indicated concentrations for 6 h, and then EGF (20 ng/mL) was treated for 30 min. (**B**) p-EGFR (L858R/T790M) band intensities were normalized to β-actin (mean ± S.E., n = 3). (**C**) A structural simulation of the **8aa** complex showed that some residues (yellow stick) were involved in binding with **8aa** (purple—carbon) red dot hydrogen bonds (Met793 and Asp855). (**D**) A docking model of erlotinib (orange—carbon) established by using the co-crystal structures of EGFR[L858R/T790M] (4I22.pdb). * $p < 0.05$; ** $p < 0.01$; *** $p < 0.001$.

3. Experimental Section

3.1. General Methods

Unless specified, all reagents and starting materials were purchased from commercial sources and used as received without purification. "Concentrated" refers to the removal of volatile solvents via distillation using a rotary evaporator. "Dried" refers to pouring onto or passing through anhydrous magnesium sulfate followed by filtration. Flash chromatography was performed using silica gel (230–400 mesh) with hexanes, ethyl acetate, and dichloromethane as the eluents. All reactions were monitored by thin-layer chromatography on 0.25 mm silica plates (F-4) visualized with UV light. Melting points were measured by using a capillary melting point apparatus. ^{1}H and ^{13}C NMR spectra were recorded on a 400 MHz NMR spectrometer and were described as chemical shifts, multiplicity (s, singlet; d, doublet; t, triplet; q, quartet; m, multiplet), coupling constant in hertz (Hz), and number of protons. High-Resolution Mass Spectra (HRMS) were measured with an electrospray ionization (ESI) and Q-TOF mass analyzer.

3.1.1. General Procedure for the Synthesis of **7**

A reaction mixture of glyoxal (0.33 mmol, 1 equiv), **5** (2.0 equiv), and HFIP (0.5 equiv) in toluene (4.0 mL) was stirred at 70 °C for 36 h. The reaction mixture was concentrated in vacuo to give the crude residue, which was purified by silica gel column chromatography (hexane/ethyl acetate/dichloromethane = 8:1:2) to afford **7**.

2-Hydroxy-2-(2-hydroxy-4-methoxyphenyl)-1-phenylethan-1-one (7a). Ivory solid, mp: 108.1–108.8 °C (81 mg, 95%); ^{1}H NMR (400 MHz, (CD$_3$)$_2$CO) δ 8.04 (d, J = 8.0 Hz, 2H), 7.55 (t, J = 7.6 Hz, 1H), 7.44 (t, J = 7.6 Hz, 2H), 7.05 (d, J = 8.4 Hz, 1H), 6.42 (d, J = 2.4 Hz, 1H), 6.37 (dd, J = 8.4, 2.4 Hz, 1H), 6.32 (s, 1H), 3.68 (s, 3H); ^{13}C NMR (100 MHz, CDCl$_3$) δ 199.3, 161.0, 155.5, 133.9, 133.5, 129.6, 129.0, 128.6, 117.3, 106.9, 102.6, 71.7, 55.2; **HRMS** (ESI-QTOF) m/z [M+H]$^+$ calcd for C$_{15}$H$_{15}$O$_4$ 259.0965, found 259.0993.

2-Hydroxy-2-(2-hydroxy-4-methoxyphenyl)-1-(3-methoxyphenyl)ethan-1-one (7b). Ivory solid, mp: 105.9–106.2 °C (91 mg, 96%); ^{1}H NMR (400 MHz, (CD$_3$)$_2$CO) δ 7.70–7.55 (m, 2H), 7.34 (t, J = 8.0 Hz, 1H), 7.10 (d, J = 6.8 Hz, 1H), 7.05 (d, J = 8.0 Hz, 1H), 6.44 (s, 1H), 6.38 (d, J = 8.0 Hz, 1H), 6.31 (s, 1H), 3.80 (s, 3H), 3.68 (s, 3H); ^{13}C NMR (100 MHz, (CD$_3$)$_2$CO) δ 199.0, 160.8, 159.7, 155.4, 152.5, 135.7, 129.5, 120.8, 119.4, 118.8, 113.1, 105.5, 101.5, 70.1, 54.8, 54.5; **HRMS** (ESI-QTOF) m/z [M+Na]$^+$ calcd for C$_{16}$H$_{16}$NaO$_5$ 311.0890, found 311.0908.

2-Hydroxy-2-(2-hydroxy-4-methoxyphenyl)-1-(naphthalen-2-yl)ethan-1-one (7c). Ivory solid, mp: 107.2–107.9 °C (65 mg, 64%); ^{1}H NMR (400 MHz, CDCl$_3$) δ 8.53 (s, 1H), 8.00 (d, J = 8.4 Hz, 1H), 7.90 (d, J = 8.0 Hz, 1H), 7.86–7.82 (m, 2H), 7.59 (t, J = 6.8 Hz, 1H), 7.53 (t, J = 7.2 Hz, 1H), 7.02 (d, J = 8.8 Hz, 1H), 6.69 (s, 1H), 6.35 (s, 3H), 4.48 (s, 1H), 3.67 (s, 3H); ^{13}C NMR (100 MHz, CDCl$_3$) δ 199.2, 161.1, 155.8, 135.9, 132.3, 131.2, 131.0, 129.9, 129.8, 129.0, 128.6, 127.8, 126.9, 124.1, 117.1, 106.9, 102.9, 72.7, 55.3; **HRMS** (ESI-QTOF) m/z [M+Na]$^+$ calcd for C$_{19}$H$_{16}$NaO$_4$ 331.0941, found 331.0938.

1-(3-Chlorophenyl)-2-hydroxy-2-(2-hydroxy-4-methoxyphenyl)ethan-1-one (7d). Ivory solid, mp: 107.4–107.9 °C (86 mg, 89%); ^{1}H NMR (400 MHz, (CD$_3$)$_2$CO) δ 8.02 (s, 1H), 7.96 (d, J = 7.2 Hz, 1H), 7.58 (d, J = 8.4 Hz, 1H), 7.47 (t, J = 7.6 Hz, 1H), 7.09 (d, J = 8.4 Hz, 1H), 6.43 (s, 1H), 6.40 (d, J = 8.4 Hz, 1H), 6.28 (s, 1H), 3.69 (s, 3H); ^{13}C NMR (100 MHz, CDCl$_3$) δ 198.1, 161.2, 155.3, 135.1, 135.0, 133.8, 130.0, 129.8, 128.9, 127.0, 116.6, 107.0, 102.8, 72.1, 55.3; **HRMS** (ESI-QTOF) m/z [M+Na]$^+$ calcd for C$_{15}$H$_{13}$ClNaO$_4$ 315.0395, found 315.0411.

1-(4-Bromophenyl)-2-hydroxy-2-(2-hydroxy-4-methoxyphenyl)ethan-1-one (7e). Ivory solid, mp: 108.1–108.8 °C (72 mg, 65%); ^{1}H NMR (400 MHz, (CD$_3$)$_2$CO) δ 7.95 (d, J = 7.6 Hz, 2H), 7.62 (d, J = 8.4 Hz, 2H), 7.07 (d, J = 8.4 Hz, 1H), 6.43 (s, 1H), 6.39 (d, J = 7.2 Hz, 1H), 6.26 (s, 1H), 3.68 (s, 3H); ^{13}C NMR (100 MHz, (CD$_3$)$_2$CO) δ 160.9, 155.6, 155.5, 133.8, 131.7, 130.3, 129.6, 127.5, 118.3, 105.6, 101.5, 70.5, 54.5; **HRMS** (ESI-QTOF) m/z [M+Na]$^+$ calcd for C$_{15}$H$_{13}$BrNaO$_4$ 358.9889, found 358.9912.

1-(5-Bromothiophen-2-yl)-2-hydroxy-2-(2-hydroxy-4-methoxyphenyl)ethan-1-one (7f). Yellow solid, mp: 125.3–125.9 °C (92 mg, 81%); ^{1}H NMR (400 MHz, (CD$_3$)$_2$CO) δ 7.76 (d, J = 3.6 Hz, 1H), 7.26–7.17 (m, 2H), 6.47–6.42 (m, 2H), 6.05 (s, 1H), 3.71 (s, 3H); ^{13}C NMR (100 MHz, (CD$_3$)$_2$CO) δ 191.4, 161.0, 155.6, 142.6, 133.7, 131.8, 129.7, 121.8, 118.5, 105.6, 101.7, 71.3, 54.6; **HRMS** (ESI-QTOF) m/z [M+H]$^+$ calcd for C$_{13}$H$_{12}$BrO$_4$S 342.9634, found 342.9626.

2-Hydroxy-2-(2-hydroxy-4,5-dimethoxyphenyl)-1-phenylethan-1-one (7g). Ivory solid, mp: 145.9–146.3 °C (93 mg, 98%); ^{1}H NMR (400 MHz, CDCl$_3$) δ 7.97 (d, J = 7.6 Hz, 2H), 7.51 (t, J = 7.2 Hz, 1H), 7.37 (t, J = 7.6 Hz, 2H), 6.51 (s, 1H), 6.33 (s, 1H), 6.24 (s, 1H), 3.67 (s, 3H), 3.66 (s, 3H); ^{13}C NMR (100 MHz, CDCl$_3$) δ 199.3, 150.2, 148.4, 143.2, 134.0, 133.4, 128.9, 128.6, 115.5, 111.3, 101.8, 71.5, 56.4, 55.7; **HRMS** (ESI-QTOF) m/z [M+Na]$^+$ calcd for C$_{16}$H$_{17}$O$_5$ 311.0890, found 311.0908.

2-Hydroxy-2-(2-hydroxy-4,5-dimethoxyphenyl)-1-(3-methoxyphenyl)ethan-1-one (7h). Ivory solid, mp: 162.9–163.2 °C (85 mg, 81%); ^{1}H NMR (400 MHz, (CD$_3$)$_2$CO) δ 7.66 (d, J = 6.8 Hz, 1H), 7.62 (s, 1H), 7.34 (t, J = 8.0 Hz, 1H), 7.10 (d, J = 6.4 Hz, 1H), 6.73 (s, 1H), 6.51 (s, 1H), 6.32 (s, 1H), 3.81 (s, 3H), 3.70 (s, 3H), 3.64 (s, 3H); ^{13}C NMR (100 MHz, (CD$_3$)$_2$CO) δ 199.1, 159.7, 150.5, 148.6, 143.1, 135.8, 129.5, 120.8, 119.4, 116.8, 113.1, 112.8, 101.0, 70.2, 56.0, 55.0, 54.8; **HRMS** (ESI-QTOF) m/z [M+Na]$^+$ calcd for C$_{17}$H$_{18}$NaO$_6$ 341.0996, found 341.1006.

2-Hydroxy-2-(2-hydroxy-4,5-dimethoxyphenyl)-1-(4-methoxyphenyl)ethan-1-one (7i). Ivory solid, mp: 109.7–110.0 °C (100 mg, 95%); ^{1}H NMR (400 MHz, CDCl$_3$) δ 7.97 (d, J = 8.0 Hz, 2H), 6.86 (d, J = 8.0 Hz, 2H), 6.52 (s, 2H), 6.37 (s, 1H), 6.16 (s, 1H), 4.59 (s, 1H), 3.82 (s, 3H), 3.73 (s, 3H), 3.70 (s, 3H); ^{13}C NMR (100 MHz, CDCl$_3$) δ 197.6, 164.2, 150.2, 148.4, 143.3, 131.4, 126.2, 116.0, 113.9, 111.3, 101.9, 71.2, 56.5, 55.7, 55.5; **HRMS** (ESI-QTOF) m/z [M+H]$^+$ calcd for C$_{17}$H$_{19}$O$_6$ 319.1176, found 319.1253.

1-(3-Chlorophenyl)-2-hydroxy-2-(2-hydroxy-4,5-dimethoxyphenyl)ethan-1-one (7j).
Ivory solid, mp: 139.6–140.2 °C (100 mg, 94%); ^{1}H NMR (400 MHz, (CD$_3$)$_2$CO) δ 8.05 (s, 1H), 7.98 (d, J = 7.6 Hz, 1H), 7.57 (d, J = 8.0 Hz, 1H), 7.46 (t, J = 7.6 Hz, 1H), 6.76 (s, 1H), 6.49 (s, 1H), 6.29 (s, 1H), 3.68 (s, 3H), 3.64 (s, 3H); ^{13}C NMR (100 MHz, CDCl$_3$) δ 199.3, 161.2, 157.6, 142.1, 141.6, 135.0, 130.5, 129.0, 127.8, 125.7, 109.9, 108.0, 100.6, 74.6, 56.3, 56.2; **HRMS** (ESI-QTOF) m/z [M+Na]$^+$ calcd for C$_{16}$H$_{15}$ClNaO$_5$ 345.0500, found 340.0502.

2-Hydroxy-2-(2-hydroxy-4,6-dimethoxyphenyl)-1-phenylethan-1-one (7k). Ivory solid, mp: 145.9–146.3 °C (58 mg, 62%); ^{1}H NMR (400 MHz, (CD$_3$)$_2$CO) δ 7.91 (d, J = 5.2 Hz, 2H), 7.50 (t, J = 6.8 Hz, 1H), 7.39 (t, J = 7.6 Hz, 2H), 6.25 (s, 1H), 6.04 (d, J = 8.0 Hz, 2H), 3.72 (s, 3H), 3.68 (s, 3H); ^{13}C NMR (100 MHz, CDCl$_3$) δ 200.2, 161.6, 158.3, 156.9, 133.9, 133.5, 128.4, 128.3, 106.1, 94.6, 91.7, 68.7, 55.6, 55.2; **HRMS** (ESI-QTOF) m/z [M+Na]$^+$ calcd for C$_{16}$H$_{16}$NaO$_5$ 311.0890, found 311.0917.

2-(5-(Benzyloxy)-2-hydroxy-4-methoxyphenyl)-2-hydroxy-1-phenylethan-1-one (7l).
Ivory solid, mp: 150.1–150.6 °C (72 mg, 60%); ^{1}H NMR (400 MHz, CDCl$_3$) δ 7.86 (d, J = 7.2 Hz, 2H), 7.52 (t, J = 7.2 Hz, 1H), 7.37 (d, J = 7.6 Hz, 2H), 7.34–7.27 (m, 6H), 6.59 (s, 1H), 6.39 (s, 1H), 6.06 (s, 1H), 4.95 (s, 2H), 3.76 (s, 3H); ^{13}C NMR (100 MHz, CDCl$_3$) δ 199.3, 149.5, 148.4, 143.8, 136.5, 134.0, 133.5, 129.0, 128.7, 128.5, 127.9, 127.2, 116.0, 112.3, 103.9, 71.7, 70.7, 56.7; **HRMS** (ESI-QTOF) m/z [M+Na]$^+$ calcd for C$_{22}$H$_{20}$NaO$_5$ 387.1203, found 387.1227.

2-(5-(Benzyloxy)-2-hydroxy-4-methoxyphenyl)-2-hydroxy-1-(4-methoxyphenyl)ethan-1-one (7m). Ivory solid, mp: 151.2–151.8 °C (54 mg, 42%); ^{1}H NMR (400 MHz, CDCl$_3$) δ 7.97 (d, J = 8.4 Hz, 2H), 7.37–7.28 (m, 5H), 6.85 (d, J = 8.4 Hz, 2H), 6.55 (s, 1H), 6.50 (s, 1H), 6.38 (s, 1H), 6.14 (s, 1H), 4.96 (s, 2H), 4.58 (s, 1H), 3.81 (s, 3H), 3.68 (s, 3H); ^{13}C NMR (100 MHz, CDCl$_3$) δ 197.6, 164.2, 149.4, 148.4, 143.8, 136.6, 131.5, 128.5, 127.9, 127.2, 126.2, 116.6, 113.9, 112.2, 103.9, 71.2, 70.7, 56.8, 55.5; **HRMS** (ESI-QTOF) m/z [M+Na]$^+$ calcd for C$_{23}$H$_{22}$NaO$_6$ 417.1309, found 417.1324.

2-Hydroxy-2-(6-hydroxybenzo[d][1,3]dioxol-5-yl)-1-phenylethan-1-one (7n). Ivory solid, mp: 139.2–139.9 °C (72 mg, 80%); ^{1}H NMR (400 MHz, CDCl$_3$) δ 7.97 (d, J = 7.2 Hz, 2H), 7.54 (d, J = 7.6 Hz, 1H), 7.41 (t, J = 7.6 Hz, 2H), 6.48 (s, 1H), 6.44 (s, 1H), 6.31 (s, 1H), 6.21 (d, J = 4.4 Hz, 1H), 5.82 (d, J = 4.0 Hz, 2H), 4.55 (d, J = 4.4 Hz, 1H); ^{13}C NMR (100 MHz, CDCl$_3$) δ 199.0, 149.3, 148.6, 141.8, 134.1, 133.3, 129.0, 128.7, 116.7, 107.4, 101.3, 99.6, 71.7; **HRMS** (ESI-QTOF) m/z [M+Na]$^+$ calcd for C$_{15}$H$_{12}$NaO$_5$ 295.0577, found 295.0592.

1-(4-Bromophenyl)-2-hydroxy-2-(6-hydroxybenzo[d][1,3]dioxol-5-yl)ethan-1-one (7o).
Ivory solid, mp: 108.1–108.8 °C (81 mg, 70%); ^{1}H NMR (400 MHz, (CD$_3$)$_2$SO) δ 9.62 (s, 1H), 7.87 (d, J = 6.8 Hz, 2H), 7.67 (d, J = 8.0 Hz, 2H), 6.62 (s, 1H), 6.40 (s, 1H), 6.12 (s, 1H), 5.88 (s, 1H), 5.84 (s, 1H), 5.68 (s, 1H); ^{13}C NMR (100 MHz, (CD$_3$)$_2$SO) δ 198.6, 149.4, 147.7, 140.4, 134.4, 132.1, 130.6, 127.7, 118.0, 107.5, 101.3, 98.1, 69.7; **HRMS** (ESI-QTOF) m/z [M+Na]$^+$ calcd for C$_{15}$H$_{11}$BrNaO$_5$ 372.9682, found 372.9685.

2-Hydroxy-2-(2-hydroxynaphthalen-1-yl)-1-phenylethan-1-one (7p). White solid, mp: 114.8–115.5 °C (84 mg, 92%); ^{1}H NMR (400 MHz, (CD$_3$)$_2$CO) δ 8.04 (d, J = 6.8 Hz, 1H), 7.90–7.75 (m, 4H), 7.47 (t, J = 6.8 Hz, 1H), 7.42–7.29 (m, 4H), 7.21 (d, J = 8.0 Hz, 1H), 6.25 (s, 1H); ^{13}C NMR (100 MHz, (CD$_3$)$_2$CO) δ 155.3, 135.1, 132.5, 131.0, 129.4, 128.6, 128.5, 128.2, 127.6, 126.8, 126.1, 123.0, 122.8, 118.3, 117.7, 108.8; **HRMS** (ESI-QTOF) m/z [M+Na]$^+$ calcd for C$_{18}$H$_{14}$NaO$_3$ 301.0835, found 301.0854.

2-Hydroxy-2-(1-hydroxynaphthalen-2-yl)-1-phenylethan-1-one (7q). Ivory solid, mp: 118.8–119.2 °C (73 mg, 80%); ^{1}H NMR (400 MHz, (CD$_3$)$_2$CO) δ 8.23 (s, 1H), 8.06 (s, 2H), 7.76 (s, 1H), 7.56–7.51 (m, 1H), 7.49–7.41 (m, 4H), 7.36 (s, 2H), 6.54 (s, 1H); ^{13}C NMR (100 MHz, CDCl$_3$) δ 198.9, 151.2, 134.4, 134.1, 133.6, 129.0, 128.7, 127.5, 126.9, 125.7, 125.6, 125.5, 121.9, 120.7, 117.0, 74.2; **HRMS** (ESI-QTOF) m/z [M+Na]$^+$ calcd for C$_{18}$H$_{14}$NaO$_3$ 301.0835, found 301.0855.

3.1.2. General Procedure for the Synthesis of **8**

A reaction mixture of **7** (0.08 mmol, 1 equiv), indole (1.5 equiv), and PTSA (0.2 equiv) in CHCl$_3$ (2.0 mL) was stirred at 60 °C for 18 h. The reaction mixture was concentrated in vacuo to give the crude residue, which was purified by silica gel column chromatography (hexane/ethyl acetate/dichloromethane = 30:1:2) to afford 8.

3-(6-Methoxy-2-phenylbenzofuran-3-yl)-1H-indole (8a). Ivory solid, mp: 69.5–70.1 °C (26 mg, 97%); **^{1}H NMR** (400 MHz, CDCl$_3$) δ 8.36 (s, 1H), 7.72 (d, J = 7.2 Hz, 2H), 7.49 (d, J = 7.6 Hz, 1H), 7.37 (s, 1H), 7.32 (t, J = 8.8 Hz, 2H), 7.25–7.21 (m, 3H), 7.14 (s, 1H), 7.06 (t, J = 6.8 Hz, 1H), 6.86 (d, J = 8.8 Hz, 1H), 3.91 (s, 3H); **^{13}C NMR** (100 MHz, CDCl$_3$) δ 158.3, 155.0, 150.0, 136.4, 131.3, 128.3, 127.6, 126.6, 126.2, 124.7, 123.5, 122.5, 120.8, 120.7, 112.0, 111.6, 111.3, 110.2, 107.9, 95.7, 55.8; **HRMS** (ESI-QTOF) m/z [M+H]$^+$ calcd for C$_{23}$H$_{18}$NO$_2$ 340.1332, found 340.1316.

3-(6-Methoxy-2-phenylbenzofuran-3-yl)-4-methyl-1H-indole (8b). Brown solid, mp: 74.8–75.2 °C (27 mg, 97%); **^{1}H NMR** (400 MHz, CDCl$_3$) δ 8.34 (s, 1H), 7.67 (d, J = 8.0 Hz, 2H), 7.35 (d, J = 8.0 Hz, 1H), 7.25–7.21 (m, 2H), 7.20–7.14 (m, 4H), 7.12 (s, 1H), 6.88–6.81 (m, 2H), 3.90 (s, 3H), 2.18 (s, 3H); **^{13}C NMR** (100 MHz, CDCl$_3$) δ 158.3, 154.4, 150.7, 136.6, 131.7, 131.3, 128.4, 127.4, 126.6, 126.3, 125.6, 123.6, 122.6, 121.3, 120.8, 112.0, 111.8, 109.1, 107.6, 95.5, 55.8, 18.8; **HRMS** (ESI-QTOF) m/z [M+H]$^+$ calcd for C$_{24}$H$_{20}$NO$_2$ 354.1489, found 354.1472.

5-Chloro-3-(6-methoxy-2-phenylbenzofuran-3-yl)-2-methyl-1H-indole (8c). Yellow solid, mp: 94.2–94.8 °C (30 mg, 96%); **^{1}H NMR** (400 MHz, CDCl$_3$) δ 8.13 (s, 1H), 7.64 (d, J = 7.6 Hz, 2H), 7.32–7.27 (m, 3H), 7.25–7.21 (m, 2H), 7.18–7.12 (m, 3H), 6.85 (d, J = 8.4 Hz, 1H), 3.91 (s, 3H), 2.17 (s, 3H); **^{13}C NMR** (100 MHz, CDCl$_3$) δ 158.4, 155.0, 150.5, 134.6, 134.1, 131.4, 129.4, 128.4, 127.6, 125.6, 124.3, 121.8, 120.8, 118.9, 111.7, 111.4, 109.2, 107.5, 104.5, 95.8, 55.8, 12.6; **HRMS** (ESI-QTOF) m/z [M+H]$^+$ calcd for C$_{24}$H$_{19}$ClNO$_2$ 388.1099, found 388.1025.

6-Chloro-3-(6-methoxy-2-phenylbenzofuran-3-yl)-1H-indole (8d). Brown solid, mp: 128.5–128.9 °C (29 mg, 98%); **^{1}H NMR** (400 MHz, CDCl$_3$) δ 8.35 (s, 1H), 7.66 (d, J = 6.8 Hz, 2H), 7.48 (s, 1H), 7.37 (s, 1H), 7.29 (d, J = 8.8 Hz, 2H), 7.25–7.22 (m, 2H), 7.20 (d, J = 8.4 Hz, 1H), 7.13 (s, 1H), 7.01 (d, J = 8.8 Hz, 1H), 6.86 (d, J = 8.4 Hz, 1H), 3.90 (s, 3H); **^{13}C NMR** (100 MHz, CDCl$_3$) δ 158.4, 155.0, 150.2, 136.7, 131.1, 128.4, 128.4, 127.7, 126.2, 125.1, 124.4, 124.1, 121.6, 120.8, 120.5, 111.8, 111.2, 109.5, 108.2, 95.8, 55.8; **HRMS** (ESI-QTOF) m/z [M+H]$^+$ calcd for C$_{23}$H$_{17}$ClNO$_2$ 374.0942, found 374.0922.

5-Bromo-3-(6-methoxy-2-phenylbenzofuran-3-yl)-1H-indole (8e). Brown solid, mp: 183.5–184.6 °C (31 mg, 92%); **^{1}H NMR** (400 MHz, CDCl$_3$) δ 8.38 (s, 1H), 7.69–7.65 (m, 2H), 7.46 (s, 1H), 7.35 (s, 3H), 7.30–7.27 (m, 3H), 7.25–7.24 (m, 1H), 7.16–7.13 (m, 1H), 6.87 (d, J = 8.4 Hz, 1H), 3.91 (s, 3H); **^{13}C NMR** (100 MHz, CDCl$_3$) δ 158.4, 155.0, 150.4, 135.0, 131.1, 128.5, 128.4, 127.8, 126.2, 125.5, 124.6, 124.4, 123.1, 120.5, 113.4, 112.7, 111.8, 109.3, 107.8, 95.8, 55.8; **HRMS** (ESI-QTOF) m/z [M+H]$^+$ calcd for C$_{23}$H$_{17}$BrNO$_2$ 418.0437, found 418.0437.

7-Bromo-3-(6-methoxy-2-phenylbenzofuran-3-yl)-1H-indole (8f). Brown solid, mp: 107.2–107.9 °C (23 mg, 69%); **^{1}H NMR** (400 MHz, CDCl$_3$) δ 8.55 (s, 1H), 7.68 (d, J = 7.2 Hz, 2H), 7.43–7.39 (m, 2H), 7.32–7.27 (m, 2H), 7.25–7.22 (m, 3H), 7.14 (s, 1H), 6.93 (t, J = 8.0 Hz, 1H), 6.86 (d, J = 8.4 Hz, 1H), 3.90 (s, 3H); **^{13}C NMR** (100 MHz, CDCl$_3$) δ 158.4, 155.0, 150.2, 135.1, 131.1, 128.3, 127.7, 126.2, 125.9, 124.8, 124.4, 124.0, 121.2, 120.5, 120.0, 111.8, 109.6, 109.3, 104.8, 95.8, 55.8; **HRMS** (ESI-QTOF) m/z [M+H]$^+$ calcd for C$_{23}$H$_{17}$BrNO$_2$ 418.0437, found 418.0415.

5-Iodo-3-(6-methoxy-2-phenylbenzofuran-3-yl)-1H-indole (8g). Brown solid, mp: 135.1–135.9 °C (35 mg, 94%); **^{1}H NMR** (400 MHz, CDCl$_3$) δ 8.39 (s, 1H), 7.65 (s, 3H), 7.50 (d, J = 8.4 Hz, 1H), 7.32–7.27 (m, 4H), 7.25–7.23 (m, 2H), 7.13 (s, 1H), 6.87 (d, J = 7.6 Hz, 1H), 3.90 (s, 3H); **^{13}C NMR** (100 MHz, CDCl$_3$) δ 158.4, 155.0, 150.4, 135.4, 131.1, 130.9, 129.4, 129.2, 128.3, 127.8, 126.2, 124.4, 124.2, 120.5, 113.2, 111.8, 109.3, 107.4, 95.8, 83.5, 55.8; **HRMS** (ESI-QTOF) m/z [M+H]$^+$ calcd for C$_{23}$H$_{17}$INO$_2$ 466.0298, found 466.0244.

3-(6-Methoxy-2-phenylbenzofuran-3-yl)-1-methyl-1H-indole (8h). Ivory solid, mp: 148.9–149.5 °C (15 mg, 54%); **^{1}H NMR** (400 MHz, CDCl$_3$) δ 7.72 (d, J = 6.4 Hz, 2H), 7.42 (d, J = 7.6 Hz, 1H), 7.35–7.27 (m, 3H), 7.26–7.21 (m, 4H), 7.12 (s, 1H), 7.04 (t, J = 7.2 Hz, 1H), 6.85 (d, J = 8.4 Hz, 1H), 3.90–3.87 (m, 6H); **^{13}C NMR** (100 MHz, CDCl$_3$) δ 158.3, 155.0, 149.8, 137.2, 131.4, 128.2, 128.0, 127.5, 127.0, 126.2, 124.7, 121.9, 120.9, 120.8, 119.5, 111.6, 110.2, 109.4, 106.2, 95.7, 55.8, 33.0; **HRMS** (ESI-QTOF) m/z [M+H]$^+$ calcd for C$_{24}$H$_{20}$NO$_2$ 354.1489, found 354.1474.

3-(6-Methoxy-2-(3-methoxyphenyl)benzofuran-3-yl)-1H-indole (8i). White solid, mp: 136.5–136.9 °C (19 mg, 65%); **¹H NMR** (400 MHz, CDCl$_3$) δ 8.37 (s, 1H), 7.48 (d, J = 7.2 Hz, 1H), 7.40 (s, 1H), 7.32 (d, J = 8.4 Hz, 3H), 7.24 (s, 2H), 7.16 (d, J = 7.6 Hz, 1H), 7.13 (s, 1H), 7.06 (t, J = 7.6 Hz, 1H), 6.86 (d, J = 8.4 Hz, 1H), 6.76 (d, J = 7.6 Hz, 1H), 3.90 (s, 3H), 3.51 (s, 3H); **¹³C NMR** (100 MHz, CDCl$_3$) δ 159.3, 158.4, 154.9, 149.8, 136.3, 132.5, 129.3, 126.4, 124.6, 123.6, 122.5, 120.8, 120.7, 120.0, 118.6, 114.1, 111.7, 111.2, 110.7, 107.9, 106.4, 95.6, 55.8, 54.9; **HRMS** (ESI-QTOF) m/z [M+H]$^+$ calcd for C$_{24}$H$_{20}$NO$_3$ 370.1438, found 370.1427.

3-(6-Methoxy-2-(naphthalen-2-yl)benzofuran-3-yl)-1-methyl-1H-indole (8j). Ivory solid, mp: 110.5–110.9 °C (21 mg, 65%); **¹H NMR** (400 MHz, CDCl$_3$) δ 8.30 (s, 1H), 7.80–7.70 (m, 3H), 7.61 (d, J = 8.8 Hz, 1H), 7.47–7.42 (m, 3H), 7.36 (t, J = 8.4 Hz, 2H), 7.31–7.27 (m, 2H), 7.17 (s, 1H), 7.02 (t, J = 7.6 Hz, 1H), 6.87 (t, J = 8.4 Hz, 1H), 3.91 (s, 6H); **¹³C NMR** (100 MHz, CDCl$_3$) δ 158.4, 155.1, 149.7, 137.2, 133.3, 132.6, 128.9, 128.3, 128.2, 127.6, 127.6, 127.1, 126.2, 126.0, 125.0, 124.7, 124.2, 122.0, 121.0, 120.9, 119.6, 111.6, 110.9, 109.4, 106.2, 95.6, 55.8, 33.1; **HRMS** (ESI-QTOF) m/z [M+H]$^+$ calcd for C$_{28}$H$_{22}$NO$_2$ 404.1645, found 404.1626.

3-(2-(3-Chlorophenyl)-6-methoxybenzofuran-3-yl)-1H-indole (8k). Ivory solid, mp: 140.3–140.8 °C (26 mg, 87%); **¹H NMR** (400 MHz, CDCl$_3$) δ 8.38 (s, 1H), 7.78 (s, 1H), 7.49 (d, J = 8.0 Hz, 2H), 7.37 (s, 1H), 7.34–7.28 (m, 2H), 7.26–7.23 (m, 1H), 7.18–7.14 (m, 1H), 7.12–7.05 (m, 3H), 6.85 (d, J = 8.4 Hz, 1H), 3.89 (s, 3H); **¹³C NMR** (100 MHz, CDCl$_3$) δ 158.7, 155.1, 148.3, 136.3, 134.3, 133.0, 129.5, 128.7, 127.4, 126.3, 125.8, 124.3, 124.1, 123.5, 122.6, 121.0, 120.6, 120.1, 111.9, 111.3, 107.5, 95.6, 55.8; **HRMS** (ESI-QTOF) m/z [M+H]$^+$ calcd for C$_{23}$H$_{17}$ClNO$_2$ 374.0942, found 374.0919.

4-Bromo-3-(2-(4-bromophenyl)-6-methoxybenzofuran-3-yl)-1H-indole (8l). Yellow solid, mp: 207.5–207.9 °C (30 mg, 75%); **¹H NMR** (400 MHz, CDCl$_3$) δ 8.47 (s, 1H), 7.47 (t, J = 8.4 Hz, 3H), 7.34 (d, J = 8.4 Hz, 2H), 7.29 (d, J = 7.6 Hz, 1H), 7.25–7.23 (m, 1H), 7.16–7.08 (m, 3H), 6.84 (d, J = 8.4 Hz, 1H), 3.89 (s, 3H); **¹³C NMR** (100 MHz, CDCl$_3$) δ 158.4, 154.4, 150.4, 137.2, 131.5, 130.4, 127.2, 126.6, 125.9, 125.2, 124.7, 123.6, 121.4, 121.0, 114.5, 111.9, 110.9, 110.7, 108.0, 95.4, 55.8; **HRMS** (ESI-QTOF) m/z [M+H]$^+$ calcd for C$_{23}$H$_{16}$Br$_2$NO$_2$ 495.9542, found 495.9517.

3-(2-(5-Bromothiophen-2-yl)-6-methoxybenzofuran-3-yl)-1H-indole (8m). Brown solid, mp: 74.8–75.2 °C (18 mg, 54%); **¹H NMR** (400 MHz, CDCl$_3$) δ 8.40 (s, 1H), 7.50 (d, J = 8.4 Hz, 1H), 7.45 (d, J = 2.0 Hz, 1H), 7.39 (d, J = 8.0 Hz, 1H), 7.30–7.27 (m, 2H), 7.17 (d, J = 4.4 Hz, 1H), 7.13–7.07 (m, 2H), 6.94 (t, J = 4.8 Hz, 1H), 6.84 (dd, J = 8.4, 1.6 Hz, 1H), 3.89 (s, 3H); **¹³C NMR** (100 MHz, CDCl$_3$) δ 158.4, 154.8, 146.6, 136.3, 133.2, 127.2, 126.6, 125.0, 124.7, 124.4, 124.1, 122.5, 120.6, 120.6, 120.0, 111.7, 111.3, 109.4, 107.0, 95.7, 55.8; **HRMS** (ESI-QTOF) m/z [M+H]$^+$ calcd for C$_{21}$H$_{15}$BrNO$_2$S 424.0001, found 424.0004.

3-(5,6-Dimethoxy-2-phenylbenzofuran-3-yl)-1H-indole (8n). Brown solid, mp: 183.7–184.2 °C (24 mg, 80%); **¹H NMR** (400 MHz, CDCl$_3$) δ 8.39 (s, 1H), 7.67 (d, J = 8.0 Hz, 2H), 7.50 (d, J = 8.0 Hz, 1H), 7.38–7.34 (m, 2H), 7.27–7.19 (m, 4H), 7.15 (s, 1H), 7.07 (t, J = 8.0 Hz, 1H), 6.85 (s, 1H), 3.98 (s, 3H), 3.80 (s, 3H); **¹³C NMR** (100 MHz, CDCl$_3$) δ 150.2, 148.7, 148.3, 146.6, 136.4, 131.4, 128.2, 127.4, 126.6, 126.0, 123.4, 123.2, 122.5, 120.7, 120.0, 111.3, 110.4, 108.1, 101.7, 95.1, 56.4; **HRMS** (ESI-QTOF) m/z [M+H]$^+$ calcd for C$_{24}$H$_{20}$NO$_3$ 370.1438, found 370.1428.

3-(5,6-Dimethoxy-2-(3-methoxyphenyl)benzofuran-3-yl)-1H-indole (8o). Brown solid, mp: 128.9–129.4 °C (28 mg, 89%); **¹H NMR** (400 MHz, CDCl$_3$) δ 8.44 (s, 1H), 7.49 (d, J = 7.2 Hz, 1H), 7.42–7.34 (m, 2H), 7.32–7.27 (m, 1H), 7.22–7.12 (m, 4H), 7.08 (t, J = 7.2 Hz, 1H), 6.86 (s, 1H), 6.76 (d, J = 5.2 Hz, 1H), 3.99 (s, 3H), 3.81 (s, 3H), 3.51 (s, 3H); **¹³C NMR** (100 MHz, CDCl$_3$) δ 159.3, 150.0, 148.6, 148.4, 146.6, 136.3, 132.5, 129.3, 126.6, 123.5, 123.2, 122.5, 120.7, 120.1, 118.4, 113.9, 111.3, 110.7, 110.6, 108.0, 101.7, 95.1, 56.4, 54.9; **HRMS** (ESI-QTOF) m/z [M+H]$^+$ calcd for C$_{25}$H$_{22}$NO$_4$ 400.1543, found 400.1550.

6-Bromo-3-(5,6-dimethoxy-2-(4-methoxyphenyl)benzofuran-3-yl)-1H-indole (8p). Brown solid, mp: 249.9–250.4 °C (31 mg, 81%); **¹H NMR** (400 MHz, CDCl$_3$) δ 8.43 (s, 1H), 7.64 (s, 1H), 7.56 (d, J = 8.4 Hz, 2H), 7.34 (s, 1H), 7.21–7.13 (m, 3H), 6.79 (s, 2H), 6.77 (s, 1H), 3.97 (s, 3H), 3.81 (s, 3H), 3.78 (s, 3H); **¹³C NMR** (100 MHz, CDCl$_3$) δ 159.1, 150.5, 148.4, 147.9, 146.5, 137.1, 127.4, 125.5, 124.0, 123.9, 123.4, 123.1, 121.9, 116.1, 114.2, 113.8, 108.5,

108.0, 101.3, 95.1, 56.4, 55.2; **HRMS** (ESI-QTOF) m/z [M+Na]$^+$ calcd for C$_{25}$H$_{20}$BrNNaO$_4$ 500.0468, found 500.0473.

3-(2-(3-Chlorophenyl)-5,6-dimethoxybenzofuran-3-yl)-5-iodo-1H-indole (8q). Brown solid, mp: 230.7–231.0 °C (42 mg, 99%); **^{1}H NMR** (400 MHz, CDCl$_3$) δ 8.48 (s, 1H), 7.71 (s, 2H), 7.53 (d, J = 8.4 Hz, 1H), 7.45 (d, J = 7.6 Hz, 1H), 7.34–7.28 (m, 2H), 7.20–7.12 (m, 3H), 6.79 (s, 1H), 3.99 (s, 3H), 3.83 (s, 3H); **^{13}C NMR** (100 MHz, CDCl$_3$) δ 148.8, 148.7, 146.8, 135.4, 134.3, 132.8, 131.1, 129.6, 129.3, 129.0, 127.5, 125.7, 124.2, 123.9, 122.6, 113.4, 110.8, 107.0, 101.5, 95.1, 83.7, 56.4; **HRMS** (ESI-QTOF) m/z [M+Na]$^+$ calcd for C$_{24}$H$_{17}$ClINNaO$_3$ 551.9834, found 551.9838.

3-(4,6-Dimethoxy-2-phenylbenzofuran-3-yl)-6-methyl-1H-indole (8r). Brown solid, mp: 154.9–155.5 °C (25 mg, 80%); **^{1}H NMR** (400 MHz, CDCl$_3$) δ 8.10 (s, 1H), 7.58 (d, J = 8.0 Hz, 2H), 7.24–7.17 (m, 6H), 6.86 (d, J = 8.0 Hz, 1H), 6.76 (s, 1H), 6.30 (s, 1H), 3.89 (s, 3H), 3.58 (s, 3H), 2.47 (s, 3H); **^{13}C NMR** (100 MHz, CDCl$_3$) δ 159.0, 156.0, 155.0, 149.2, 136.4, 131.6, 131.4, 128.1, 127.2, 126.1, 125.3, 123.8, 121.4, 120.6, 114.0, 110.9, 109.5, 108.3, 94.5, 87.9, 55.8, 55.5, 21.8; **HRMS** (ESI-QTOF) m/z [M+H]$^+$ calcd for C$_{25}$H$_{22}$NO$_3$ 384.1594, found 384.1590.

3-(5-(Benzyloxy)-6-methoxy-2-phenylbenzofuran-3-yl)-1H-indole (8s). Brown solid, mp: 162.7–163.2 °C (35 mg, 99%); **^{1}H NMR** (400 MHz, CDCl$_3$) δ 8.38 (s, 1H), 7.68 (d, J = 6.8 Hz, 2H), 7.48 (d, J = 8.0 Hz, 1H), 7.43–7.39 (m, 2H), 7.37–7.22 (m, 9H), 7.18 (s, 1H), 7.05 (t, J = 7.2 Hz, 1H), 6.95 (s, 1H), 5.05 (s, 2H), 3.98 (s, 3H); **^{13}C NMR** (100 MHz, CDCl$_3$) δ 150.1, 149.2, 149.2, 145.6, 137.2, 136.3, 131.4, 128.4, 128.2, 127.8, 127.6, 127.4, 126.5, 126.1, 123.4, 123.2, 122.4, 120.7, 120.1, 111.2, 110.4, 108.0, 105.4, 95.6, 71.9, 56.5; **HRMS** (ESI-QTOF) m/z [M+H]$^+$ calcd for C$_{30}$H$_{24}$NO$_3$ 446.1751, found 446.1755.

2-(6-(Benzyloxy)-5-methoxy-2-(4-methoxyphenyl)benzofuran-3-yl)-3-methyl-1H-indole (8t). Brown solid, mp: 194.8–195.2 °C (25 mg, 65%); **^{1}H NMR** (400 MHz, CDCl$_3$) δ 7.98 (s, 1H), 7.67 (d, J = 7.6 Hz, 1H), 7.53 (d, J = 8.8 Hz, 2H), 7.49 (d, J = 7.6 Hz, 2H), 7.43–7.37 (m, 3H), 7.36–7.31 (m, 1H), 7.25–7.18 (m, 2H), 7.12 (s, 1H), 6.83 (d, J = 8.8 Hz, 2H), 6.80 (s, 1H), 5.24 (s, 2H), 3.85 (s, 3H), 3.79 (s, 3H), 2.20 (s, 3H); **^{13}C NMR** (100 MHz, CDCl$_3$) δ 159.5, 151.7, 148.1, 147.5, 147.0, 136.9, 136.3, 129.4, 128.6, 127.9, 127.3, 126.0, 123.3, 122.9, 122.1, 119.3, 118.9, 114.1, 110.9, 107.0, 101.7, 97.9, 71.5, 56.6, 55.2, 9.5; **HRMS** (ESI-QTOF) m/z [M+H]$^+$ calcd for C$_{32}$H$_{28}$NO$_4$ 490.2013, found 490.2007.

3-(6-Phenyl-[1,3]dioxolo[4,5-f]benzofuran-7-yl)-1H-indole (8u). Brown solid, mp: 193.9–194.4 °C (19 mg, 68%); **^{1}H NMR** (400 MHz, CDCl$_3$) δ 8.33 (s, 1H), 7.66 (d, J = 8.0 Hz, 2H), 7.47 (d, J = 8.0 Hz, 1H), 7.34–7.30 (m, 2H), 7.25–7.18 (m, 4H), 7.08–7.03 (m, 2H), 6.79 (s, 1H), 5.97 (s, 2H); **^{13}C NMR** (100 MHz, CDCl$_3$) δ 150.6, 149.2, 146.4, 144.5, 136.3, 131.3, 128.3, 127.5, 126.5, 126.0, 124.7, 123.4, 122.5, 120.6, 120.0, 111.3, 110.7, 107.9, 101.3, 99.1, 93.3; **HRMS** (ESI-QTOF) m/z [M+H]$^+$ calcd for C$_{23}$H$_{16}$NO$_3$ 354.1125, found 354.1118.

7-Methyl-3-(6-phenyl-[1,3]dioxolo[4,5-f]benzofuran-7-yl)-1H-indole (8v). Brown solid, mp: 185.8–186.4 °C (27 mg, 92%); **^{1}H NMR** (400 MHz, CDCl$_3$) δ 8.27 (s, 1H), 7.69 (d, J = 7.6 Hz, 2H), 7.31 (s, 1H), 7.25–7.20 (m, 4H), 7.11–7.05 (m, 2H), 7.00 (t, J = 6.4 Hz, 1H), 6.82–6.80 (m, 1H), 5.99 (s, 2H), 2.58 (s, 3H); **^{13}C NMR** (100 MHz, CDCl$_3$) δ 150.5, 149.2, 146.3, 144.5, 136.0, 131.3, 128.3, 127.5, 126.1, 126.0, 124.7, 123.2, 123.0, 120.5, 120.2, 118.4, 110.9, 108.3, 101.3, 99.1, 93.3, 16.7; **HRMS** (ESI-QTOF) m/z [M+H]$^+$ calcd for C$_{24}$H$_{18}$NO$_3$ 368.1281, found 368.1259.

3-(6-(4-Bromophenyl)-[1,3]dioxolo[4,5-f]benzofuran-7-yl)-5-methyl-1H-indole (8w). Brown solid, mp: 205.4–206.0 °C (32 mg, 89%); **^{1}H NMR** (400 MHz, CDCl$_3$) δ 8.28 (s, 1H), 7.53 (d, J = 8.4 Hz, 2H), 7.38 (d, J = 8.8 Hz, 1H), 7.35 (d, J = 8.8 Hz, 2H), 7.13–7.09 (m, 2H), 7.06 (s, 1H), 6.78 (s, 1H), 5.99 (s, 2H), 2.36 (s, 3H); **^{13}C NMR** (100 MHz, CDCl$_3$) δ 149.5, 149.2, 146.6, 144.6, 134.6, 131.4, 130.2, 129.6, 127.3, 126.7, 124.7, 124.3, 123.5, 121.3, 119.9, 111.5, 111.0, 106.9, 101.3, 99.1, 93.3, 21.5; **HRMS** (ESI-QTOF) m/z [M+H]$^+$ calcd for C$_{24}$H$_{16}$BrNO$_3$ 445.0314, found 445.0315.

3-(2-Phenylnaphtho[2,1-b]furan-1-yl)-1H-indole (8x). Brown solid, mp: 148.2–148.9 °C (25 mg, 88%); **^{1}H NMR** (400 MHz, CDCl$_3$) δ 8.41 (s, 1H), 7.92 (d, J = 8.0 Hz, 1H), 7.79 (s, 2H), 7.67–7.61 (m, 3H), 7.57 (d, J = 8.4 Hz, 2H), 7.43 (d, J = 7.6 Hz, 1H), 7.39–7.32 (m, 2H), 7.31–7.29 (m, 1H), 7.25–7.19 (m, 3H), 7.15 (t, J = 7.6 Hz, 1H), 7.09 (t, J = 7.2 Hz, 1H); **^{13}C**

NMR (100 MHz, CDCl$_3$) δ 151.69, 151.66, 136.5, 131.1, 130.8, 128.7, 128.6, 128.3, 127.8, 127.7, 126.1, 125.9, 125.7, 124.6, 124.2, 123.6, 123.2, 122.7, 120.5, 120.4, 112.3, 111.3, 111.2, 109.2; **HRMS** (ESI-QTOF) m/z [M+H]$^+$ calcd for C$_{26}$H$_{18}$NO 360.1383, found 360.1370.

2-Methyl-3-(2-phenylnaphtho[1,2-b]furan-3-yl)-1H-indole (8y). Brown solid, mp: 95.3–95.9 °C (19 mg, 62%); **^{1}H NMR** (400 MHz, CDCl$_3$) δ 8.51 (d, J = 8.4 Hz, 1H), 8.13 (s, 1H), 7.96 (d, J = 8.0 Hz, 1H), 7.83 (d, J = 8.0 Hz, 2H), 7.69–7.61 (m, 2H), 7.53 (t, J = 7.6 Hz, 1H), 7.43 (d, J = 7.6 Hz, 2H), 7.37 (d, J = 8.0 Hz, 1H), 7.32 (t, J = 7.2 Hz, 2H), 7.28–7.27 (m, 1H), 7.25–7.21 (m, 1H), 7.09 (t, J = 8.0 Hz, 1H), 2.23 (s, 3H); **^{13}C NMR** (100 MHz, CDCl$_3$) δ 150.6, 149.5, 135.8, 133.1, 131.7, 131.6, 130.1, 129.2, 128.5, 128.3, 127.6, 126.7, 126.5, 126.3, 125.8, 125.1, 123.2, 121.6, 121.4, 120.2, 119.9, 119.7, 119.6, 110.4, 12.7; **HRMS** (ESI-QTOF) m/z [M+H]$^+$ calcd for C$_{27}$H$_{20}$NO 374.1539, found 374.1544.

5-Chloro-3-(6-methoxy-2-phenylbenzofuran-3-yl)-1H-indole (8z). Ivory solid, mp: 88.2–88.9 °C (28 mg, 92%); **^{1}H NMR** (400 MHz, CDCl$_3$) δ 8.40 (s, 1H), 7.67 (d, J = 8.0 Hz, 2H), 7.42–7.36 (m, 2H), 7.31–7.27 (m, 3H), 7.25–7.19 (m, 3H), 7.14 (s, 1H), 6.87 (d, J = 8.4 Hz, 1H), 3.91 (s, 3H); **^{13}C NMR** (100 MHz, CDCl$_3$) δ 158.4, 155.0, 150.3, 134.7, 131.1, 128.4, 127.8, 126.1, 125.8, 124.8, 124.4, 122.9, 120.5, 120.0, 112.3, 111.8, 109.4, 107.8, 95.7, 55.8; **HRMS** (ESI-QTOF) m/z [M+H]$^+$ calcd for C$_{23}$H$_{17}$ClNO$_2$ 374.0942, found 374.0951.

5-Bromo-3-(5,6-dimethoxy-2-phenylbenzofuran-3-yl)-1H-indole (8aa). White solid, mp: 100.2–100.5 °C (29 mg, 81%); **^{1}H NMR** (400 MHz, CDCl$_3$) δ 8.46 (s, 1H), 7.63 (d, J = 7.6 Hz, 2H), 7.49 (s, 1H), 7.37–7.31 (m, 3H), 7.25–7.21 (m, 3H), 7.15 (s, 1H), 6.79 (s, 1H), 3.97 (s, 3H), 3.81 (s, 3H); **^{13}C NMR** (100 MHz, CDCl$_3$) δ 150.4, 148.6, 148.3, 146.6, 134.9, 131.1, 128.5, 128.3, 127.6, 125.9, 125.5, 124.6, 123.0, 122.9, 113.4, 112.8, 109.5, 107.8, 101.4, 95.1, 56.4, 56.3; **HRMS** (ESI-QTOF) m/z [M+Na]$^+$ calcd for C$_{24}$H$_{18}$BrNNaO$_3$ 470.0362, found 470.0356.

5-Bromo-3-(5,6-dimethoxy-2-(4-methoxyphenyl)benzofuran-3-yl)-1H-indole (8ab). Ivory solid, mp: 92.2–92.9 °C (34 mg, 89%); **^{1}H NMR** (400 MHz, (CD$_3$)$_2$CO) δ 7.67 (s, 1H), 7.59–7.55 (m, 2H), 7.52 (d, J = 8.4 Hz, 1H), 7.30–7.28 (m, 1H), 7.26–7.24 (m, 1H), 6.88 (s, 1H), 6.84 (d, J = 8.8 Hz, 3H), 3.90 (s, 3H), 3.77 (s, 3H), 3.73 (s, 3H); **^{13}C NMR** (100 MHz, (CD$_3$)$_2$CO) δ 159.4, 150.0, 148.8, 148.5, 147.2, 135.6, 128.3, 127.2, 126.2, 126.0, 124.4, 124.0, 122.9, 122.2, 113.8, 112.1, 108.5, 106.5, 102.0, 95.5, 55.68, 55.65, 54.7; **HRMS** (ESI-QTOF) m/z [M+Na]$^+$ calcd for C$_{25}$H$_{20}$BrNNaO$_4$ 500.0468, found 500.0454.

5-Bromo-3-(6-phenyl-[1,3]dioxolo[4,5-f]benzofuran-7-yl)-1H-indole (8ac). Ivory solid, mp: 110.2–110.9 °C (33 mg, 96%); **^{1}H NMR** (400 MHz, CDCl$_3$) δ 8.38 (s, 1H), 7.63–7.59 (m, 2H), 7.44 (s, 1H), 7.34 (s, 2H), 7.31–7.29 (m, 1H), 7.25–7.21 (m, 3H), 7.07 (s, 1H), 6.73 (s, 1H), 5.98 (s, 2H); **^{13}C NMR** (100 MHz, CDCl$_3$) δ 150.9, 149.2, 146.5, 144.6, 134.9, 131.0, 128.4, 128.3, 127.7, 125.9, 125.5, 124.6, 124.5, 122.9, 113.4, 112.8, 109.9, 107.6, 101.3, 98.8, 93.4; **HRMS** (ESI-QTOF) m/z [M+H]$^+$ calcd for C$_{23}$H$_{15}$BrNO$_3$ 432.0230, found 423.0211.

5-Iodo-3-(6-phenyl-[1,3]dioxolo[4,5-f]benzofuran-7-yl)-1H-indole (8ad). Ivory solid, mp: 113.9–114.2 °C (36 mg, 94%); **^{1}H NMR** (400 MHz, CDCl$_3$) δ 8.38 (s, 1H), 7.63 (s, 1H), 7.60 (d, J = 8.0 Hz, 2H), 7.49 (d, J = 8.8 Hz, 1H), 7.28–7.21 (m, 5H), 7.07 (s, 1H), 6.72 (s, 1H), 5.98 (s, 2H); **^{13}C NMR** (100 MHz, CDCl$_3$) δ 151.0, 149.2, 146.5, 144.6, 135.4, 131.0, 130.9, 129.2, 129.1, 128.3, 127.7, 125.9, 124.5, 124.2, 113.3, 109.8, 107.3, 101.3, 98.7, 93.4, 83.6; **HRMS** (ESI-QTOF) m/z [M+H]$^+$ calcd for C$_{23}$H$_{15}$INO$_3$ 480.0091, found 480.0061.

3.2. Bioassay

3.2.1. Cell Culture

A549, HT29, MCF7, HepG2, PC3, and HEK293T cells were purchased from the Korean Cell Line Bank (Seoul, Republic of Korea). HaCaT, H1975, and PC9 cells were obtained from Prof. Sohee Kwon (Yonsei University, Incheon, Republic of Korea), Prof. Dosik Min, and Daegu Gyeongbuk Medical Innovation Foundation, respectively. The hepatocellular carcinoma cells (HepG2), breast adenocarcinoma cells (MCF-7), human keratinocyte cells (HaCaT), and human embryonic kidney cells (HEK293T) were cultured in high-glucose DMEM (Welgene Inc., Gyeongsan, Republic of Korea), and the non-small-cell lung carcinoma cells (PC9, A549, H1975), colorectal adenocarcinoma cells (HT29), and prostate adenocarcinoma cells (PC3) were grown in RPMI 1640 medium (Welgene Inc., Gyeongsan,

Republic of Korea). In addition, 10% fetal bovine serum (FBS), 100 µg/mL streptomycin, and 100 Units/mL penicillin were supplemented in all media. All cells were grown at 5% CO_2, 37 °C, and 95% humidity.

3.2.2. Cell Proliferation Assay

Our cell proliferation assay was performed using the Cell Titer 96® AQueous One Solution Cell proliferation Assay kit (Promega, Madison, WI, USA). All cells were grown in 96-well microplates in RPMI1640 medium containing 10% fetal bovine serum (FBS) for 24 h. After 24 h incubation, the cells were treated with DMSO and test compounds. The medium and test compounds were changed every 24 h. To estimate cell viability, the cells were incubated with MTS solution for 50 min. The absorbance was quantified by using an Infinite M200 microplate reader (Tecan, Männedorf, Switzerland) at 490 nm.

3.2.3. In Vitro Wound Healing Assay

The inhibitory effect of **8aa** on cell migration was measured through an in vitro wound healing assay. PC9 and A549 cells were cultured to approximately 100% confluence in a 96-well microplate for 24 h to form a monolayer. After 24 h, wounds were formed through 96-well wound maker (Essen BioScience, Ann Arbor, MI, USA). Then, the growth medium was washed out three times with phosphate-buffered saline (PBS) and replaced with 200 µL of DMEM and RPMI 1640 medium containing **8aa** or DMSO. Images of the wound area were acquired by using IncuCyte ZOOM (Essen BioScience, Ann Arbor MI, USA), and wound closure rate was measured using IncuCyte software (2018A).

3.2.4. Caspase-3 Activity Assay

PC9 and A549 cells were grown in 96-black well plates to approximately 50% confluence, and then the cells were incubated with **8aa** for 24 h. To measure caspase-3 activity, the growth medium was replaced with 100 µL of PBS containing 1 µM of caspase-3 substrate and NucView 488 and incubated at room temperature for 30 min. Then, the cells were stained with 1 µM of Hoechst 33342. Caspase-3 activity was completely inhibited by Ac-DEVD-CHO, a potent caspase-3 inhibitor. The FLUOstar Omega microplate reader (BMG Labtech, Ortenberg, Germany) was used to measure the fluorescence of Hoechst 33342 and NucView 488, and a Lionheart FX Automated Microscope (BioTek, Winooski, VT, USA) was used to obtain the fluorescence microscopy images.

3.2.5. Immunoblot Analysis

The preparation of the protein sample was conducted as described previously [5]. The samples were centrifuged at 13,000 RPM for 20 min at 4 °C to eliminate cell debris, and the samples were separated using 4–12% Tris Glycine Precast Gel (KOMA BIOTECH, Seoul, Republic of Korea) for 60 min at 130 V. After 60 min, the samples were transferred onto a polyvinylidene Fluoride membrane (PVDF) (Millipore, Billerica, MA, USA) for 90 min at 30 V. Membrane blocking was conducted using Tris-buffered saline with 0.1% Tween 20 (TBST) containing 5% bovine serum albumin (BSA) at room temperature for 60 min. After membrane blocking, the membranes were incubated overnight at 4 °C with the following primary antibodies: anticleaved PARP (BD Biosciences, Franklin Lakes, NJ, USA), anti-β-actin (Santa Cruz Biotechnology, Dallas, TX, USA), anti-phospho-EGFR (Tyr1068) (Cell Signaling, Danvers, MA, USA), anti-EGFR (Santa Cruz Biotechnology, Dallas, TX, USA), anti-phospho-AKT (Santa Cruz Biotechnology, Dallas, TX, USA), anti-AKT (Santa Cruz Biotechnology, Dallas, TX, USA), anti-phospho-p42/44 (Cell Signaling, Danvers, MA, USA), and anti-p42/44 (Cell Signaling, Danvers, MA, USA). Then, the membranes were washed out three times every five minutes with 0.1% TBST and incubated with horseradish peroxidase (HRP) conjugated secondary IgG antibodies at room temperature for 60 min. After washing three times, the membranes were visualized using the EzWestLumi Plus (mid-femto-grade ECL) (ATTO, Amherst, NY, USA) immunoblot analysis detection system (GE Healthcare, Piscataway, NJ, USA). All experiments were repeated three times independently.

3.2.6. Cell Cycle Analysis

PC9 and A549 cells were grown to ~60% confluence in a 6-well plate; then, the cells were treated with 10 µM of **8aa** for 24 h. After 24 h, the PC9 and A549 cells were washed out twice with phosphate-buffered saline (PBS) and trypsinized using 0.5% trypsin-EDTA before the cells were centrifuged at 1000 RPM for 2 min at room temperature. Finally, the cells were stained with propidium iodide (PI) for 30 min, and then cell cycle phases were measured by using FACS (Beckman Coulter, Fullerton, CA, USA).

3.3. Molecular Docking Analysis

Molecular docking was studied using Maestro (Schrödinger Release 2022-1). The X-ray crystal structures of the EGFR (1M17.pdb) and EGFR$^{L858R/T790M}$ (4I22.pdb) were prepared by removing all water and hydrogen assignments at pH 7.0 using the Protein Preparation Wizard module. Compounds were minimized by using the conjugate gradient algorithm and the OPLS2005 force field with Minimization module in Maestro. The Glide module was used to generate the receptor grid and carry out ligand docking. The docking model figures were generated using PyMOL version 1.8.6.1. The amino acid numbers of 1M17.pdb were corrected based on other published X-ray cocrystal structures (7UKV, 7U99.pdb).

3.4. EGFR Kinase Activity Assay

The inhibitory effect of **8aa** on EGFR kinase activity was evaluated using an EGFR kinase assay kit (BPS Bioscience, San Diego, CA, USA) according to the manufacturer's instructions. Briefly, a mixture of 5X kinase buffer 1, ATP (500 µM), PTK substrate (10 mg/mL), and water was prepared. Subsequently, **8aa** was treated at various concentrations, and the reaction was initiated by adding the EGFR (1 ng/µL). After a 40-min incubation period at 30 °C, each well was treated with Kinase-Glo Max reagent (Promega, Madison, WI, USA) and incubated for 15 min at room temperature. Luminescence was measured using an Infinite M200 microplate reader (Tecan, Männedorf, Switzerland).

4. Conclusions

In summary, we established highly efficient modular access to a range of 2-arylbenzofurans with an indole at the C3 position via the HFIP-catalyzed hydroxyalkylation of phenols with (hetero) arylglyoxals, followed by PTSA-catalyzed substitution–cyclodehydration with indoles, enabling the installation of two distinct substituents at the C2 and C3 sites of benzofuran with the formation of two C-C bonds and one C-O bond. Biological evaluations and structure–activity relationship (SAR) studies of these products against the EGFR in NSCLC cells led us to identify **8aa** as a novel EGFR inhibitor. Notably, **8aa** potently inhibited the EGFR and EGFR-mediated signaling pathways such as AKT and ERK1/2 in a dose-dependent manner, and it also showed selective reductions in cell viability against human NSCLC cell lines PC9 and A549. **8aa** exhibited limited impact on cell viability in other cancer cell lines, including MCF7, HepG2, PC3, and HT29 cells, as well as non-tumorigenic cells such as HaCaT and HEK293T cells. Moreover, **8aa** significantly inhibited cell migration and induced apoptosis via increasing caspase-3 activity and PARP cleavage in PC9 and A549 cells. Of interest, **8aa** exhibited significant efficacy in suppressing the EGFR$^{L858R/T790M}$ resistance mutation, which frequently occurs in NSCLC. Molecular docking analysis suggests that this results from a hydrogen bonding interaction between **8aa** and Asp855 of EGFR$^{L858R/T790M}$. Overall, **8aa** has the potential to be developed as a novel EGFR inhibitor to treat NSCLC patients in general, as well as those with L858R and T790M mutations that are resistant to conventional EGFR-TKIs.

Supplementary Materials: The following supporting information can be downloaded at: https://www.mdpi.com/article/10.3390/ph17020231/s1, Supplementary Materials: NMR spectra (^{1}H and ^{13}C NMR) and HRMS of synthesized compounds and HPLC chromatogram of **8aa**.

Author Contributions: Conceptualization, W.N. and I.K.; methodology, W.N. and I.K.; software, Y.L. (Younho Lee) and D.S.; validation, Y.L. (Yechan Lee), S.L. and Y.L. (Younho Lee); formal analysis, Y.L. (Yechan Lee), S.L., S.-H.P. and J.K.; investigation, Y.L. (Yechan Lee), S.L., S.-H.P. and J.K.; resources, Y.L. (Yechan Lee), S.L., S.-H.P. and J.K.; data curation, Y.L. (Yechan Lee) and S.L.; writing—original draft preparation, Y.L. (Yechan Lee) and S.L.; writing—review and editing, Y.L. (Yechan Lee), S.L., W.N. and I.K.; visualization, Y.L. (Yechan Lee) and S.L.; supervision, W.N. and I.K.; project administration, W.N. and I.K.; funding acquisition, W.N. and I.K. All authors have read and agreed to the published version of the manuscript.

Funding: This research was funded by the National Research Foundation of Korea (NRF-2018R1A6A1-A03023718 and NRF-2020R1A2C2005961).

Institutional Review Board Statement: Not applicable.

Informed Consent Statement: Not applicable.

Data Availability Statement: Data is contained within the article and Supplementary Materials.

Conflicts of Interest: The authors declare no conflict of interest.

References

1. Lee, Y.; Joshi, D.R.; Namkung, W.; Kim, I. Generation of a poly-functionalized indolizine scaffold and its anticancer activity in pancreatic cancer cells. *Bioorg. Chem.* **2022**, *126*, 105877. [CrossRef] [PubMed]
2. Nam, S.; Lee, Y.; Park, S.-H.; Namkung, W.; Kim, I. Synthesis and Biological Evaluation of a Fused Structure of Indolizine and Pyrrolo[1,2-c]pyrimidine: Identification of Its Potent Anticancer Activity against Liver Cancer Cells. *Pharmaceuticals* **2022**, *15*, 1395. [CrossRef]
3. Lee, S.; Dagar, A.; Cho, I.; Kim, K.; Park, I.W.; Yoon, S.; Cha, M.; Shin, J.; Kim, H.Y.; Kim, I.; et al. 4-Acyl-3,4-dihydropyrrolo[1,2-a]pyrazine Derivative Rescued the Hippocampal-Dependent Cognitive Decline of 5XFAD Transgenic Mice by Dissociating Soluble and Insoluble Aβ Aggregates. *ACS Chem. Neurosci.* **2023**, *14*, 2016–2026. [CrossRef] [PubMed]
4. Gridelli, C.; Rossi, A.; Carbone, D.P.; Guarize, J.; Karachaliou, N.; Mok, T.; Petrella, F.; Spaggiari, L.; Rosell, R. Non-small-cell lung cancer. *Nat. Rev. Dis. Primers* **2015**, *1*, 15009. [CrossRef] [PubMed]
5. Cruz, C.S.D.; Tanoue, L.T.; Matthay, R.A. Lung cancer: Epidemiology, etiology, and prevention. *Clin. Chest Med.* **2011**, *32*, 605–644. [CrossRef] [PubMed]
6. Fontanini, G.; Vignati, S.; Chinè, S.; Lucchi, M.; Silvestri, V.; Mussi, A.; Placido, S.D.; Tortora, G.; Bianco, A.R.; Gullick, W.; et al. Evaluation of epidermal growth factor-related growth factors and receptors and of neoangiogenesis in completely resected stage I-IIIA non-small-cell lung cancer: Amphiregulin and microvessel count are independent prognostic indicators of survival. *Clin. Cancer Res.* **1998**, *4*, 241–249. [PubMed]
7. Bethune, G.; Bethune, D.; Ridgway, N.; Xu, Z. Epidermal growth factor receptor (EGFR) in lung cancer: An overview and update. *J. Thorac. Dis.* **2010**, *2*, 48.
8. Gupta, R.; Dastane, A.M.; Forozan, F.; Riley-Portuguez, A.; Chung, F.; Lopategui, J.; Marchevsky, A.M. Evaluation of EGFR abnormalities in patients with pulmonary adenocarcinoma: The need to test neoplasms with more than one method. *Mod. Pathol.* **2009**, *22*, 128–133. [CrossRef]
9. Shigematsu, H.; Gazdar, A.F. Somatic mutations of epidermal growth factor receptor signaling pathway in lung cancers. *Int. J. Cancer* **2006**, *118*, 257–262. [CrossRef]
10. Inamura, K.; Ninomiya, H.; Ishikawa, Y.; Matsubara, O. Is the epidermal growth factor receptor status in lung cancers reflected in clinicopathologic features? *Arch. Pathol. Lab. Med.* **2010**, *134*, 66–72. [CrossRef]
11. Massarelli, E.; Varella-Garcia, M.; Tang, X.; Xavier, A.C.; Ozburn, N.C.; Liu, D.D.; Bekele, B.N.; Herbst, R.S.; Wistuba, I.I. KRAS mutation is an important predictor of resistance to therapy with epidermal growth factor receptor tyrosine kinase inhibitors in non–small-cell lung cancer. *Clin. Cancer Res.* **2007**, *13*, 2890–2896. [CrossRef]
12. Gerber, D.E. EGFR inhibition in the treatment of non-small cell lung cancer. *Drug Dev. Res.* **2008**, *69*, 359–372. [CrossRef] [PubMed]
13. Westover, D.; Zugazagoitia, J.; Cho, B.; Lovly, C.; Paz-Ares, L. Mechanisms of acquired resistance to first-and second-generation EGFR tyrosine kinase inhibitors. *Ann. Oncol.* **2018**, *29*, i10–i19. [CrossRef] [PubMed]
14. Lüth, A.; Löwe, W. Syntheses of 4-(indole-3-yl)quinazolines—A new class of epidermal growth factor receptor tyrosine kinase inhibitors. *Eur. J. Med. Chem.* **2008**, *43*, 1478–1488. [CrossRef] [PubMed]
15. Park, S.; Kim, E.H.; Kim, J.; Kim, S.H.; Kim, I. Biological evaluation of indolizine-chalcone hybrids as new anticancer agents. *Eur. J. Med. Chem.* **2018**, *144*, 435–443. [CrossRef] [PubMed]
16. Seo, Y.; Lee, J.H.; Park, S.-H.; Namkung, W.; Kim, I. Expansion of chemical space based on a pyrrolo[1,2-a]pyrazine core: Synthesis and its anticancer activity in prostate cancer and breast cancer cells. *Eur. J. Med. Chem.* **2020**, *188*, 111988. [CrossRef] [PubMed]
17. Nevagi, R.J.; Dighe, S.N.; Dighe, S.N. Biological and medicinal significance of benzofuran. *Eur. J. Med. Chem.* **2015**, *97*, 561–581. [CrossRef] [PubMed]

18. Heravi, M.M.; Amiri, Z.; Kafshdarzadeh, K.; Zadsirjan, V. Synthesis of indole derivatives as prevalent moieties present in selected alkaloids. *RSC Adv.* **2021**, *11*, 33540–33612. [CrossRef]
19. Abbas, H.-A.S.; El-Karim, S.S.A. Design, synthesis and anticervical cancer activity of new benzofuran–pyrazol-hydrazono-thiazolidin-4-one hybrids as potential EGFR inhibitors and apoptosis inducing agents. *Bioorg. Chem.* **2019**, *89*, 103035. [CrossRef]
20. Mphahlele, M.J.; Maluleka, M.M.; Parbhoo, N.; Malindisa, S.T. Synthesis, evaluation for cytotoxicity and molecular docking studies of benzo[c]furan-chalcones for potential to inhibit tubulin polymerization and/or EGFR-tyrosine kinase phosphorylation. *Int. J. Mol. Sci.* **2018**, *19*, 2552. [CrossRef]
21. Sheng, J.; Liu, Z.; Yan, M.; Zhang, X.; Wang, D.; Xu, J.; Zhang, E.; Zou, Y. Biomass-involved synthesis of N-substituted benzofuro[2,3-d]pyrimidine-4-amines and biological evaluation as novel EGFR tyrosine kinase inhibitors. *Org. Biomol. Chem.* **2017**, *15*, 4971–4977. [CrossRef]
22. Al-Wahaibi, L.H.; Mostafa, Y.A.; Abdelrahman, M.H.; El-Bahrawy, A.H.; Trembleau, L.; Youssif, B.G. Synthesis and Biological Evaluation of Indole-2-Carboxamides with Potent Apoptotic Antiproliferative Activity as EGFR/CDK2 Dual Inhibitors. *Pharm. Res.* **2022**, *15*, 1006. [CrossRef] [PubMed]
23. Al-Wahaibi, L.H.; Mohammed, A.F.; Abdelrahman, M.H.; Trembleau, L.; Youssif, B.G. Design, Synthesis, and Biological Evaluation of Indole-2-carboxamides as Potential Multi-Target Antiproliferative Agents. *Pharm. Res.* **2023**, *16*, 1039. [CrossRef] [PubMed]
24. Eftekhari-Sis, B.; Zirak, M.; Akbari, A. Arylglyoxals in synthesis of heterocyclic compounds. *Chem. Rev.* **2013**, *113*, 2958–3043. [CrossRef] [PubMed]
25. Jung, E.; Jeong, Y.; Kim, H.; Kim, I. C3 Functionalization of Indolizines via HFIP-Promoted Friedel–Crafts Reactions with (Hetero)arylglyoxals. *ACS Omega* **2023**, *8*, 16131–16144. [CrossRef] [PubMed]
26. Cheng, C.; Liu, C.; Gu, Y. One-pot three-component reactions of methyl ketones, phenols and a nucleophile: An expedient way to synthesize densely substituted benzofurans. *Tetrahedron* **2015**, *71*, 8009–8017. [CrossRef]
27. Ramalingam, S.S.; Vansteenkiste, J.; Planchard, D.; Cho, B.C.; Gray, J.E.; Ohe, Y.; Zhou, C.; Reungwetwattana, T.; Cheng, Y.; Chewaskulyong, B.; et al. Overall survival with osimertinib in untreated, EGFR-mutated advanced NSCLC. *N. Engl. J. Med.* **2020**, *382*, 41–50. [CrossRef] [PubMed]
28. Rolfo, C.; Giovannetti, E.; Hong, D.S.; Bivona, T.; Raez, L.E.; Bronte, G.; Buffoni, L.; Reguart, N.; Santos, E.S.; Germonpre, P.; et al. Novel therapeutic strategies for patients with NSCLC that do not respond to treatment with EGFR inhibitors. *Cancer Treat. Rev.* **2014**, *40*, 990–1004. [CrossRef]
29. Phillips, J.C.; Braun, R.; Wang, W.; Gumbart, J.; Tajkhorshid, E.; Villa, E.; Chipot, C.; Skeel, R.D.; Kale, L.; Schulten, K. Scalable molecular dynamics with NAMD. *J. Comput. Chem.* **2005**, *26*, 1781–1802. [CrossRef]
30. Goel, S.; Hidalgo, M.; Perez-Soler, R. EGFR inhibitor-mediated apoptosis in solid tumors. *J. Exp. Ther. Oncol.* **2007**, *6*, 305–320.
31. Okamoto, K.; Okamoto, I.; Okamoto, W.; Tanaka, K.; Takezawa, K.; Kuwata, K.; Yamaguchi, H.; Nishio, K.; Nakagawa, K. Role of survivin in EGFR inhibitor–induced apoptosis in non–small cell lung cancers positive for EGFR mutations. *Cancer Res.* **2010**, *70*, 10402–10410. [CrossRef]
32. Johnson, B.E.; Jänne, P.A. Epidermal growth factor receptor mutations in patients with non–small cell lung cancer. *Cancer Res.* **2005**, *65*, 7525–7529. [CrossRef] [PubMed]
33. Paez, J.G.; Jänne, P.A.; Lee, J.C.; Tracy, S.; Greulich, H.; Gabriel, S.; Herman, P.; Kaye, F.J.; Lindeman, N.; Boggon, T.J.; et al. EGFR mutations in lung cancer: Correlation with clinical response to gefitinib therapy. *Science* **2004**, *304*, 1497–1500. [CrossRef] [PubMed]
34. Sharma, S.V.; Bell, D.W.; Settleman, J.; Haber, D.A. Epidermal growth factor receptor mutations in lung cancer. *Nat. Rev. Cancer* **2007**, *7*, 169–181. [CrossRef] [PubMed]
35. Hong, W.; Wu, Q.; Zhang, J.; Zhou, Y. Prognostic value of EGFR 19-del and 21-L858R mutations in patients with non-small cell lung cancer. *Oncol. Lett.* **2019**, *18*, 3887–3895. [CrossRef]
36. Yu, H.A.; Arcila, M.E.; Rekhtman, N.; Sima, C.S.; Zakowski, M.F.; Pao, W.; Kris, M.G.; Miller, V.A.; Ladanyi, M.; Riely, G.J. Analysis of tumor specimens at the time of acquired resistance to EGFR-TKI therapy in 155 patients with EGFR-mutant lung cancers. *Clin. Cancer Res.* **2013**, *19*, 2240–2247. [CrossRef] [PubMed]

Disclaimer/Publisher's Note: The statements, opinions and data contained in all publications are solely those of the individual author(s) and contributor(s) and not of MDPI and/or the editor(s). MDPI and/or the editor(s) disclaim responsibility for any injury to people or property resulting from any ideas, methods, instructions or products referred to in the content.

 *pharmaceuticals*

Review

Recent Advances in Pyrimidine-Based Drugs

Baskar Nammalwar [1] and Richard A. Bunce [2,*]

[1] Vividion Therapeutics, 5820 Nancy Ridge Drive, San Diego, CA 92121, USA; nbaskarphd@gmail.com
[2] Department of Chemistry, Oklahoma State University, Stillwater, OK 74078, USA
* Correspondence: rab@okstate.edu; Tel.: +1-405-744-5952

Abstract: Pyrimidines have become an increasingly important core structure in many drug molecules over the past 60 years. This article surveys recent areas in which pyrimidines have had a major impact in drug discovery therapeutics, including anti-infectives, anticancer, immunology, immuno-oncology, neurological disorders, chronic pain, and diabetes mellitus. The article presents the synthesis of the medicinal agents and highlights the role of the biological target with respect to the disease model. Additionally, the biological potency, ADME properties and pharmacokinetics/pharmacodynamics (if available) are discussed. This survey attempts to demonstrate the versatility of pyrimidine-based drugs, not only for their potency and affinity but also for the improved medicinal chemistry properties of pyrimidine as a bioisostere for phenyl and other aromatic π systems. It is hoped that this article will provide insight to researchers considering the pyrimidine scaffold as a chemotype in future drug candidates in order to counteract medical conditions previously deemed untreatable.

Keywords: pyrimidines; building block; antibiotic; antifungal; antiviral; anticancer; immunological treatment; neurological disorders; anti-inflammatory; chronic pain; diabetes mellitus

1. Introduction

Pyrimidine is an important electron-rich aromatic heterocycle, and, as a building block of DNA and RNA, is a critical endogenous component of the human body [1]. Due to its synthetic accessibility and structural diversity, the pyrimidine scaffold has found widespread therapeutic applications, including antimicrobial, antimalarial, antiviral, anticancer, antileishmanial, anti-inflammatory, analgesic, anticonvulsant, antihypertensive, and antioxidant applications [2–9]. Furthermore, pyrimidines are also reported to possess potential medicinal properties important to central nervous system (CNS)-active agents, calcium channel blockers and antidepressants [10,11]. Due to its broad biological activity, pyrimidines have piqued tremendous interest among organic and medicinal chemists. In addition to its ready availability, the pyrimidine skeleton can be easily modified for structural diversity at the 2, 4, 5 and 6 positions (Figure 1).

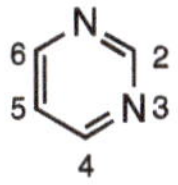

Figure 1. Structure and numbering scheme for pyrimidine.

Some of the known commercial pyrimidine-based drugs **1**–**16** are shown in Figure 2. Due to the pyrimidine ring's ability to interact with various targets by effectively forming hydrogen bonds and by acting as bioisosteres for phenyl and other aromatic π systems, they often improve the pharmacokinetic/pharmacodynamic properties of the drug. The pyrimidine ring has unique physiochemical attributes that have led to its widespread incorporation into drug candidates with a broad spectrum of activities. The chemical space portfolio of drugs relying on this privileged scaffold has increased at a rapid rate for a wide variety of biological targets with different therapeutic requirements.

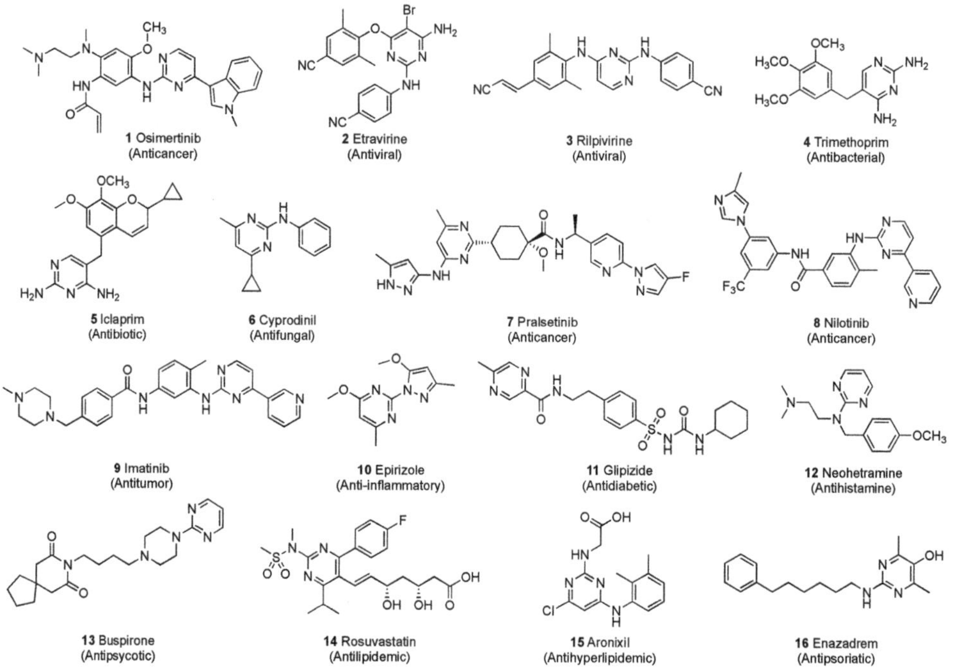

Figure 2. Marketed drugs containing the pyrimidine scaffold.

This review article attempts to comprehensively outline the synthetic strategies employed to prepare pyrimidine derivatives as well as the biological and clinical significance of these systems for various therapeutic needs. The review covers the literature of the past two years and drug candidates are organized according to the specific medical conditions they are designed to treat: bacterial, fungal and viral infections; cancer; immunological and neurological disorders; inflammation; chronic pain; and diabetes mellitus. The sheer volume of reports over this short period is a testament to the impact and potential of pyrimidine-based compounds in drug research. Throughout the manuscript, the activities of new drug prototypes are compared with numerous commercial and experimental drugs. Many of these compounds do not incorporate pyrimidine and are not shown in the text of this review. However, as readers may not be familiar with these drugs, structures for these standards are pictured in the SI along with some of the commercial names.

2. Pyrimidine-Based Drugs for Treatment of Infections

2.1. Pyrimidines as Antibacterials

Luo et al. [12] have focused on developing a lead molecule against tuberculosis (TB), a lethal infectious disease caused by *Mycobacterium tuberculosis* which is a prevalent problem in Asia and Europe [13]. Though the current treatment protocol for this infection involves a combination regimen which is very effective, resistance has started to emerge for this treatment option. Thus, a promising antitubercular compound with a novel mechanism of action must be developed against drug-resistant TB. In this work, the authors noted that Certinib (see Supplementary Materials), an approved antitumor drug for anaplastic lymphoma kinase, expressed antitubercular properties through a phenotypic screening approach. The compound exhibited a modest minimum inhibitory concentration (MIC) of

9.0 µM/mL against the H3Ra variant. The authors in this work were able to identify the lead pharmacophore and quickly develop a structure activity relationship (SAR) for this series of compounds.

The synthetic approach to model compounds for this study is depicted in Scheme 1. Nucleophilic aromatic substitution (S_NAr) reaction of 2,4,5-trichloropyrimidine (**17**) at C4 with commercial 2-isopropoxy-5-methyl-4-(piperdin-4-yl)aniline (**18**) in the presence of *N,N*-diisopropylethylamine (DIPEA) in isopropanol at 80 °C provided **19**. Compound **19** was further substituted by anilines at C2 under acidic conditions to generate compound **20**. Ammonolysis of **19** and subsequent removal of the *tert*-butoxycarbonyl (Boc)-protecting group from nitrogen with trifluoroacetic acid (TFA) in dichloromethane (DCM) afforded **21**. Finally, Suzuki–Miyaura coupling of **19** with an arylboronic acid using [1,1'-bis(diphenylphosphino)ferrocene]dichloropalladium(II) (Pd(dppf)Cl$_2$) with cesium carbonate (Cs$_2$CO$_3$) as the base in aqueous dioxane under reflux, followed by Boc deprotection, afforded targets **22**.

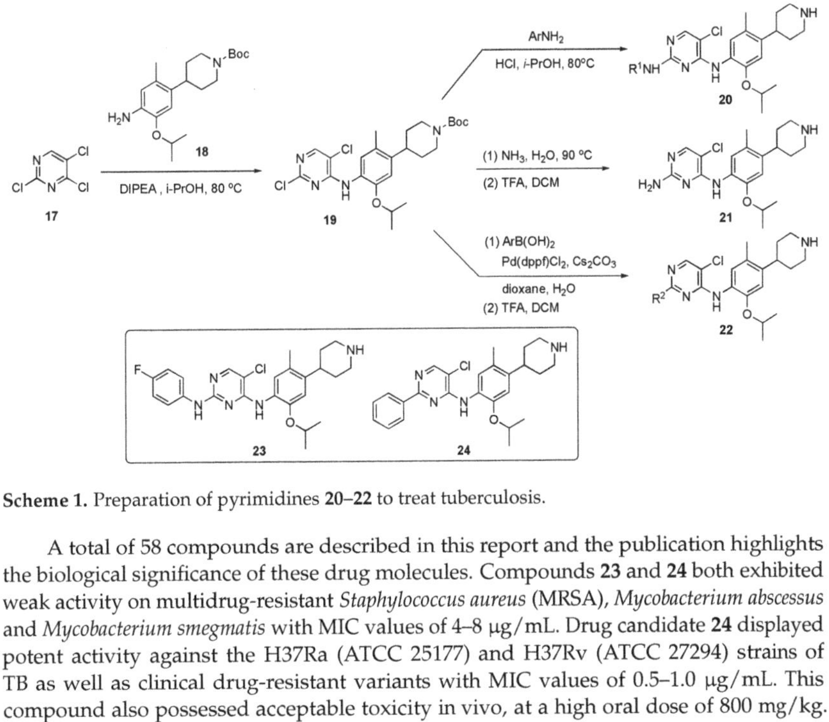

Scheme 1. Preparation of pyrimidines **20–22** to treat tuberculosis.

A total of 58 compounds are described in this report and the publication highlights the biological significance of these drug molecules. Compounds **23** and **24** both exhibited weak activity on multidrug-resistant *Staphylococcus aureus* (MRSA), *Mycobacterium abscessus* and *Mycobacterium smegmatis* with MIC values of 4–8 µg/mL. Drug candidate **24** displayed potent activity against the H37Ra (ATCC 25177) and H37Rv (ATCC 27294) strains of TB as well as clinical drug-resistant variants with MIC values of 0.5–1.0 µg/mL. This compound also possessed acceptable toxicity in vivo, at a high oral dose of 800 mg/kg. The pharmacokinetic (PK) properties were evaluated for **24** using Sprague–Dawley rats, and the results are summarized in Table 1. Compound **24** exhibited moderate exposure with a C_{max} = 592 ± 62 mg/mL, slow elimination with a $t_{1/2}$ = 26.2 ± 0.9 h, a low clearance (CL) value of 1.5 ± 0.3 L/h/kg following intravenous (i.v.) administration, and promising oral bioavailability (F) with a value of 40.7%.

Table 1. Pharmacokinetic properties of compound **24**.

Cpd No.	Administration	C_{max} (µg/mL)	$t_{\frac{1}{2}}$ (h)	CL (L/h/kg)	AUC$_{0-t}$ (µL/L·h)	F
24	i.v. (5 mg/kg)	592 ± 62	26.2 ± 0.9	1.5 ± 0.3	1694 ± 201	n.a.
	p.o. (15 mg/kg)	108 ± 18	—	—	2079 ± 274	40.7%

The Mohamady group also explored the use of pyrimidines as anti-infectives against *M. tuberculosis* [14]. TB is an airborne infectious disease which primarily targets the lungs and other organs, such as spine, kidney and brain, by alternating between the active and latent phases and challenges the immune system defense mechanism [15]. Multidrug-resistant (MDR) variants of this pathogen are on the rise for Isoniazid and Rifampicin (see Supplementary Materials) which are first line of defense drugs [16]. In this study, these researchers tried to inhibit *M. tuberculosis* by targeting the fatty acid biosynthesis pathway of the pathogen. They specifically targeted acyl carrier protein reductase which is an essential component of the mycobacterial survival pathway. Upon disruption of this pathway, the bacteria starve, resulting in cell death and eradication of TB. In this work, the concept of molecular hybridization was used to design the desired library of compounds.

Several model compounds were synthesized according to the generalized route outlined in Scheme 2. The synthesis began with a one-pot multi-component reaction of benzaldehydes **25**, ethyl cyanoacetate, and thiourea with potassium bicarbonate (KHCO$_3$) in ethanol to give pyrimidine derivatives **26**. Hydrazinolysis of thiol **26** with hydrazine hydrate in ethanol under reflux afforded the 2-hydrazinyl-6-oxo-4-aryl-1,6-dihydropyrimidine-5-carbonitriles **27**. These derivatives were subsequently condensed with isatin derivatives **28** in ethanol containing drops of acetic acid to produce the isatin–pyrimidine hybrids **29**.

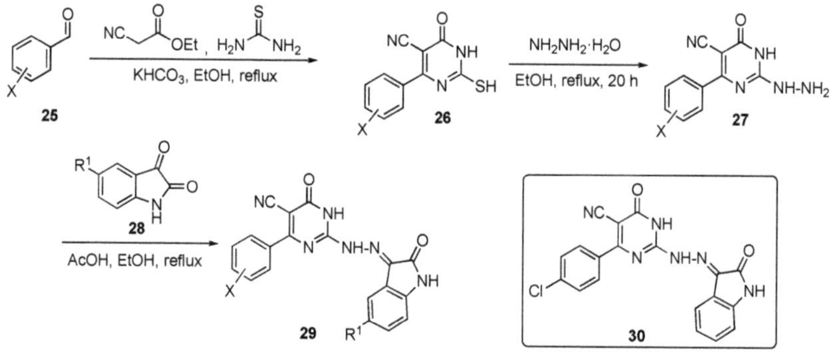

Scheme 2. Synthesis of the antitubercular isatin–pyrimidine hybrids **29**.

The inhibitory activity of compound **29** was screened against three strains of TB, including a sensitive strain (ATCC 25177 = H37Ra) as well as an isoniazid (see Supplementary Materials)-resistant strain (ATCC 35822). Some of the promising derivatives showed an MIC of <1 µg/mL on both strains, including the MDR and extremely drug-resistant (XDR) strains. Of all of the derivatives, compound **30** exhibited the maximum inhibition of MDR and XDR *M. tuberculosis* with MICs of 0.48 and 3.9 µg/mL, respectively. On the other hand, the same compound also proved the most potent against inhibin subunit alpha (InhA) with a half maximal inhibitory concentration (IC$_{50}$) of 0.6 ± 0.94 µM. In this paper, the authors were successfully able to co-crystallize compound **30** in the ligand-active site of InhA.

Recently, the Yang research group has worked to develop a broad-spectrum antibiotic by extending the scope of Linezolid (see Supplementary Materials) by appending a pyrimidine ring to its framework [17]. Oxazolidinone ring cores are known for their potent activity in the antibacterial space, especially toward Gram-positive pathogens [18]. Due to prevalent resistance by the pathogens, however, there needs to be continuous improvement over current medications to develop next-generation antibiotics. The authors sought antibiotic candidates that would also possess antibiofilm activity, specifically targeting urinary tract infections. The authors hypothesized that linking a pyrimidine moiety to the Linezolid structure would improve the drug's ability to form hydrogen bonds as well as improve the permeability of the compound.

The synthesis shown in Scheme 3 illustrates the preparation of a pyrimidine-linked prototype from piperazine–Linezolid precursor **31** which was synthesized by a known method [19]. Compound **31** underwent a regioselective reaction with various 2,4-dichloropyrimidine derivatives **32** in the presence of triethylamine (TEA) in ethanol to afford intermediates **33**. These individual intermediates were further reacted with various amines under mild acidic conditions in the presence of *p*-toluenesulfonic acid monohydrate (*p*-TsOH·H$_2$O) in ethanol, to eventuate C4 chloride displacement from the pyrimidine ring and afford the required pyrimidine–Linezolid structures **34**.

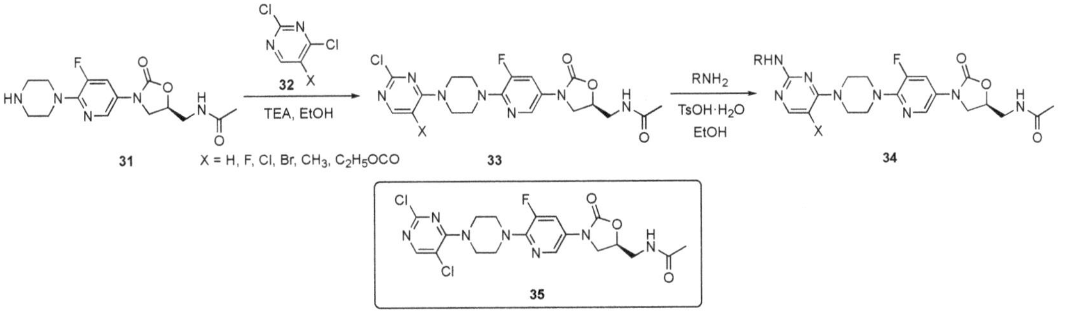

Scheme 3. Synthesis of pyrimidine-appended Linezolids **34**.

Compounds **34** were evaluated for their antibacterial activity against seven different strains of mostly Gram-positive bacteria including *S. aureus*, *Streptococcus pneumoniae*, *Enterococcus faecalis*, *Bacillus subtilis*, *Staphylococcus xylosus*, and *Listeria monocytogens*. Several derivatives exhibited very good activity against a subset of these organisms (MIC = 0.25–1 µg/mL) with **35** exhibiting an MIC of 0.25–1 µg/mL against all of these pathogens. Candidate **35** also displayed an MIC of 1 µg/mL against the Methicillin (see Supplementary Materials)-resistant *S. aureus* (MRSA) and Vancomycin-resistant *Enterococcus* (VRE) bacteria. Furthermore, **35** displayed a minimum biofilm inhibitory concentration (MBIC) ranging from 0.5–4 µg/mL against a series of four bacterial strains including MRSA, VRE, Linezolid-resistant *S. aureus* and Linezolid-resistant *S. pneumoniae*. These results have led the authors to conclude that compound **35** has high potential for further development as an antibacterial drug.

Kumari et al. sought to develop an antibacterial agent incorporating pyrimidines by targeting the enzyme DNA gyrase which is a bacterial topoisomerase II [20]. DNA gyrase is responsible for DNA replication, transcription, repair and decatenation in the bacteria. The gyrase is composed of two subunits, gyrase A and gyrase B, one of which controls ATPase activity while the other works by breaking and reassembling bacterial DNA. The role of DNA gyrase is to maintain the topology of the DNA present only in bacteria, which makes it an ideal target for developing antibiotics [21,22].

In this work, the Kumari group devised a route to a family of chrysin-substituted pyrimidine–piperazine hybrids and this is shown in Scheme 4. Initially, 4,6-dihydroxy-2-methylpyrimidine (**36**) was assembled from acetamidine hydrochloride and diethyl malonate in the presence of sodium acetate (NaOAc) in methanol. Conversion of the hydroxyl groups in **36** to chlorides using phosphorus oxychloride (POCl$_3$) gave dichloropyrimidine **37**, which was substituted by chrysin (**38**) in a S$_N$Ar reaction promoted by a potassium carbonate (K$_2$CO$_3$) base in *N*,*N*-dimethylformamide (DMF) to give **39**. The final nucleophilic substitution reaction of piperazinyl derivatives with **39** in the presence of DIPEA in ethanol afforded targets **40**.

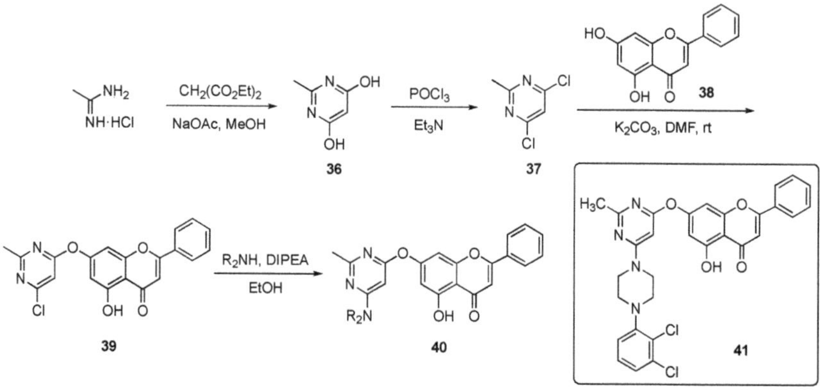

Scheme 4. Synthesis of chrysin-based pyrimidine–piperazine hybrid antibacterial **40**.

The synthesized compounds were evaluated against twelve different strains of bacteria and two fungi. The compounds generally exhibited modest antibacterial and antifungal activity, except for **41**, which achieved an MIC = 6.5 μg/mL against *Escherichia coli* and a respectable antifungal MIC = 250 μg/mL against *Candida albicans*.

2.2. Pyrimidines as Antifungals

A patent by Li et al. strived to develop a new class of safe and non-toxic anti-infective agents against a series of fungal infections, including *Candida albicans*, *Saccharomyces cerevisiae*, and *Candida parapsilosis* [23]. The patent described the preparation and evaluation of various pyrimidine derivatives from aryl sulfonamides **42** by reaction with ethyl chloroformate using K_2CO_3 in acetone to afford the arylsulfonylurethane **43** as shown in Scheme 5. Further reaction of **43** with various substituted 2-aminopyrimidines then produced sulfonylureas **44**.

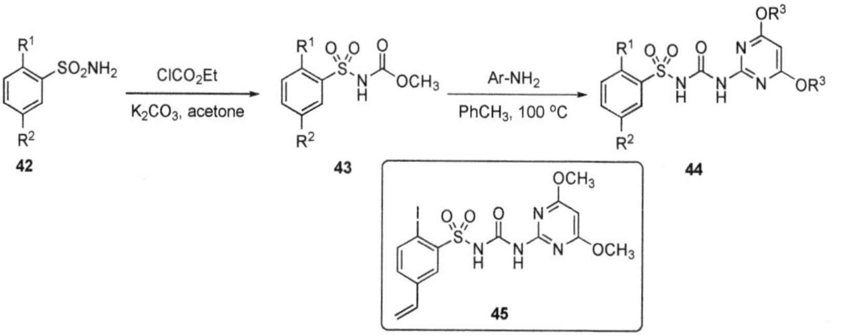

Scheme 5. Synthesis of pyrimidine antifungals **44**.

Results in the patent indicate that some of the model compounds had antifungal activity equivalent or up to 5-fold higher than the known drug Amphotericin B (see Supplementary Materials), and that the MIC_{90} of some of these compounds, including compound **45**, exhibited 3- to 30-fold better activity than Fluconazole (see Supplementary Materials). The drug prototypes exhibited very strong antifungal properties against many strains of *C. albicans*, *S. cerevisiae* and *C. parapsilosis*. Compound **45** exhibited potent activity toward *C. albicans* in RPMI 1640, YNB, and YPD media and had the most promising MIC_{90} = 0.05–0.3 μg/mL. It also showed encouraging activity with an MIC_{90} < 0.05–0.1 μg/mL against Fluconazole-resistant

C. albicans, *S. cerevisiae* and *C. parapsilosis* strains. Most of the compounds tested in this series had persistent antifungal activity which did not decrease even after 72 h.

2.3. Pyrimidines as Antivirals

Kang et al. advanced an interesting study on pyrimidines as drug scaffolds for anti-infectives which effectively studied acquired immunodeficiency syndrome (AIDS) caused by human immunodeficiency virus (HIV) [24]. With the advent of antiretroviral therapy, the disease classification of HIV has changed from being a mortal disease to a manageable chronic disorder. In this effort, the authors were focused on developing non-nucleoside reverse transcriptase inhibitors (NNRTIs) which are common among antiviral drugs. The prescribed first-generation and second-generation NNRTIs suffered from serious drug resistance due to mutant strains that rendered these drugs ineffective. The most common mutant strains, K103N and Y181C, arose against first-generation drugs while E138K developed toward second-generation drugs [25,26]. Apart from this, newer compounds had limited solubility profiles resulting in low bioavailability. Thus, it was imperative to create new structures with better adsorption, distribution, metabolism, and excretion (ADME) properties as leads for antiviral drugs.

In this work, the final target compounds were prepared in one step from previously reported compounds **46a–c** [27] as shown in Scheme 6. Suzuki–Miyaura coupling of **46a–c** with arylboronic acids in the presence of tetrakis(triphenylphosphine)palladium(0) (Pd(Ph$_3$P)$_4$) and K$_2$CO$_3$ in DMF at 100 °C gave candidates **47** for this study.

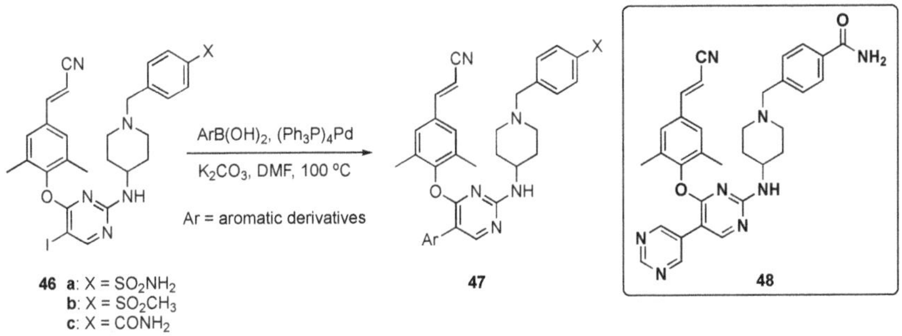

Scheme 6. Synthesis of pyrimidine NNRT inhibitors **47**.

The results were summarized for 39 compounds that were prepared and evaluated. Screening showed that most of the compounds exhibited antiviral activity (half maximal effective concentration, EC$_{50}$ < 10 nM) against the HIV-1-IIIB strain compared with Etravirine (**2**, EC$_{50}$ = 3.5 nM). Most compounds also exhibited activity on the double mutant strain RES056 (K103N/Y181C) with an EC$_{50}$ = 50 nM, and compound **48** displayed the most potent EC$_{50}$ values between 3.43–11.8 nM against a panel of wild type (WT) and resistant mutants. Notably, **48** displayed the highest potency against K103N and Y188L with EC$_{50}$ values of 4.77 nM and 15.3 nM, respectively. From an ADME and toxicological perspective, **48** demonstrated no cytochrome P450 (CYP450) inhibition (IC$_{50}$ > 10 µM) and had a favorable PK profile as shown in Table 2. The clearance of the compound was 82.7 ± 1.97 mL/h/kg (slightly high) after i.v. administration of 2 mg/kg with a sufficient oral bioavailability (F) of 31.8% following oral administration (p.o.) of 10 mg/kg. Finally, **48** did not show any acute toxicity in Kunming mice up to the maximum concentration of 2000 mg/kg.

Table 2. Pharmacokinetic properties of compound 48.

Cpd No.	Administration	C_{max} (ng/mL)	$t_{\frac{1}{2}}$ (h)	CL (mL/h/kg)	AUC_{0-t} (nL/L·h)	F
48	i.v. (2 mg/kg)	587 ± 63.1	1.33 ± 0.11	82.7 ± 1.97	400 ± 9.29	n.a.
	p.o. (10 mg/kg)	252 ± 22.0	1.20 ± 0.12	—	654 ± 125	31.8%

Liu et al. also sought to develop a new drug against the HIV-1 reverse transcriptase (RT) for the treatment of AIDS [28]. RT has an important biochemical function for viral replication of DNA/RNA-dependent DNA polymerase and ribonuclease H [29]. Currently, RT inhibitors are mainly divided into nucleoside RT inhibitors (NRTIs) and non-nucleoside RT inhibitors (NNRTIs) [27]. NNRTIs have been key components in highly active antiretroviral therapies due to their promising anti-HIV-1 properties, high specificity, and relatively low toxicity. In this work, the authors emphasize the need for novel NNRTIs with higher anti-HIV-1 activities against resistant mutant strains and improved drug properties.

The synthesis of candidate compounds for this project is shown in Scheme 7. Initially, nitroaryl halides **49a-c** were reacted with Boc-piperazine in the presence of K_2CO_3 in DMF at 120 °C to afford **50**. These adducts were converted to the first precursor by reduction to the aniline derivatives **51** with H_2 and Pd/C in methanol. To prepare the second precursors, 2,4-dichloropyrimidine (**52**) underwent reaction with cyanophenols **53** (a: 4-hydroxy-3,5-dimethylbenzonitrile or b: (E)-3-(4-hydroxy-3,5-dimethylphenyl)acrylonitrile) in the presence of K_2CO_3 in DMF to generate **54**. Ether **54b** was further subjected to Buchwald–Hartwig coupling with **51** in the presence of palladium acetate (Pd(OAc)$_2$), 3,5-bis(diphenylphosphino)-9,9-dimethylxanthene (xantphos), and Cs_2CO_3 in dioxane to produce **55b**. Boc deprotection of **55b** using TFA in DCM to afford **56b** which was acylated or sulfonated at the piperazine nitrogen to deliver the final targets **57b**.

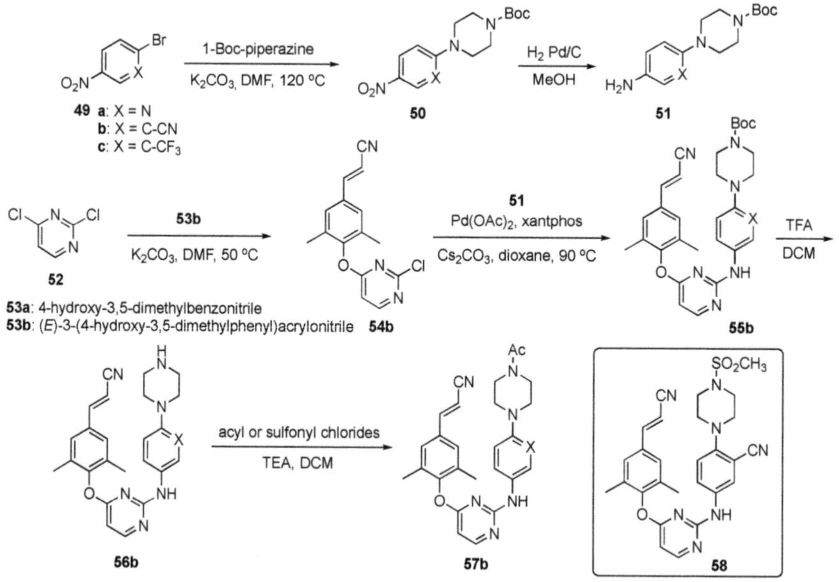

Scheme 7. Synthesis of pyrimidine RT inhibitors **57b**.

Interestingly, among the newly synthesized compounds, **58** demonstrated significantly improved antiretroviral activity compared with the known NNRTI BH-11c (see Supplementary Materials) against all tested HIV-1 strains. Furthermore, **58** possessed subnanomolar potency (0.1–2.6 nM) against WT and five mutant HIV-1 strains, including L100I, K103N, Y181C, E138K and F227L/V106A. Further molecular dynamics simulation studies were conducted to explain the differences between the inhibitory activity of **58** and Etravirine (**2**) against RT variants. Candidate **58** displayed improved water solubility (13.46 μg/mL at pH 7.0) compared with **2** (<1 μg/mL at pH 7.0), with an appropriate ligand efficiency (LE) value of 0.32. Moreover, **58** expressed significantly lower inhibitory activity than **2** and Rilpivirine (**3**) against CYP2C9, indicating that **58** was less likely to cause drug–drug interactions.

Compound **58** was evaluated for its PK profile in a Sprague–Dawley rat model as shown in Table 3 and demonstrated an acceptable $t_{1/2}$, moderate clearance and favorable distribution volume after an i.v. dose of 2.0 mg/kg. When administered p.o. at a dose of 10.0 mg/kg, **58** had a poor C_{max} and concentration–time curve (AUC). Nevertheless, the oral bioavailability (F) of **58** was determined to be 1.34%. Consequently, pending further optimization, the authors tagged **58** as a promising lead compound worthy of additional study.

Table 3. Pharmacokinetic properties of compound **58**.

Cpd No.	Administration	C_{max} (μg/mL)	$t_{\frac{1}{2}}$ (h)	CL (L/h/kg)	AUC_{0-t} (μL/L·h)	F
58	i.v. (2 mg/kg)	1233.33 ± 96.09	2.42 ± 0.34	2.54 ± 0.25	785.06 ± 74	n.a.
	p.o. (10 mg/kg)	16.05 ± 6.59	3.05 ± 1.44	—	45.1 ± 12.9	1.34%

Chen et al. have endeavored to optimize a pyrimidine-based drug scaffold to target the influenza virus [30]. Influenza is an infectious disease of the respiratory tract and results in >300,000 deaths/year worldwide, posing a huge social and economic burden to society [31]. One way to lessen this burden is to effectively use the influenza vaccine; however, due to antigenic drift and mismatch between the vaccine and circulating strains, the vaccine is not always effective. Although neuraminidase inhibitors (Oseltamivir, see Supplementary Materials) and RNA polymerase (RNAP) inhibitors (Baloxavir, see Supplementary Materials) are used as first-line-of-defense drugs, resistant mutations have evolved, leading to efforts to develop new direct-acting antivirals to combat drug-resistant mutations and to identify new drugs with novel mechanisms of action [32]. These researchers reported a medicinal chemistry strategy for anchoring aza-β^3- or $\beta^{2,3}$-amino acids on a 7-azaindole ring to the RNAP subunit PB2 which has proven to be an ideal target for antiviral drug development. Benefiting from facile structural elaboration, aza-β-amino acid motifs with diverse size, shape, steric hindrance, and configuration were linked to a pyrimidine and evaluated for their antiviral activities.

The preparation of experimental compounds is presented in Scheme 8 and started with 2,4-dichloro-5-fluoropyridine (**59**). After substitution of **59** at C4 by ethyl *N*-amino-*N*-(alkyl)glycinates **60** using DIPEA in THF, intermediates **61** were subjected to Suzuki–Miyaura coupling with azaindoleboronic acids **62** using $(Ph_3P)_4Pd$ and K_2CO_3 in aqueous acetonitrile (ACN) to afford **63**. This structure was hydrolyzed using lithium hydroxide (LiOH) in aqueous THF to afford the desired acid targets **64**.

Tests on these compounds revealed that **65** (HAA-09) targets the influenza PB2_cap binding domain with potent anti-influenza virus efficacies using both in vitro and in vivo models. This drug candidate possessed high inhibition against influenza A virus polymerase and was active in submicromolar concentrations, with an IC_{50} = 0.06 μM and an EC_{50} = 0.03 μM. Compound **65** also exhibited superior antiviral activity against the Oseltamivir-sensitive A/WSN/33 and Oseltamivir-resistant H275Y variants. It showed no inhibition of the human ether-á-go-go related gene (hERG) channel, demonstrating a low risk for hERG-related cardiac repolarization (manual patch, IC_{50} > 10 μM) and high plasma stability ($t_{1/2}$ > 12 h). Moreover, a subacute toxicity study was carried out in healthy mice to assess the safety profiles in vivo. Lead compound **65** demonstrated a favorable safety profile with oral administration in healthy mice at a high dose of 40 mg/kg once daily for three days. The PD read-out on oral administration for **65** indicated more than a 2-log viral load reduction and survival benefit in a mouse lethal infection model. The rapid reduction in the amount of influenza A virus in the lungs of infected mice confirmed that **65** had a direct effect on viral replication. Based on these findings, structure **65** was considered a potential PB2 inhibitor suitable for further anti-influenza drug development.

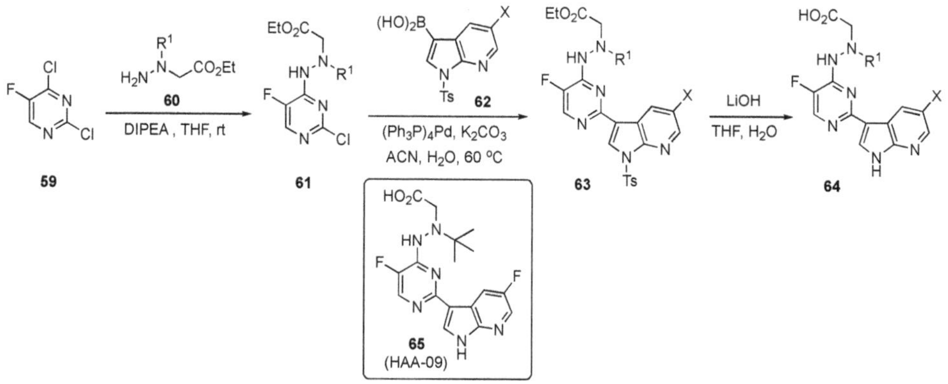

Scheme 8. Synthesis of azaindole-linked pyrimidine antivirals **64**.

3. Pyrimidine Based Drugs for the Treatment of Cancer

Pyrimidines have myriad biological activities, including as anticancer pharmacophores. Zhang et al. have explored the role of pyrimidine rings as anticancer agents in breast cancer cell lines [33]. Triple-negative breast cancer (TNBC) is a heterogenous aggressive breast cancer which leads to high mortality rates due to distant metastasis and lack of efficient targeted therapeutics [34].

Focal adhesion kinase (FAK) is a non-receptor tyrosine kinase which plays a significant role in integrin-activated signal transduction by initiating a cascade of biological functions [35]. These FAK-mediated signaling pathways lead to tumor progression and metastasis by regulating proliferation through invasion and cell survival strategies. Breast cancer cells overexpress FAK kinases, which in turn activate FAK signaling pathways for cell proliferation and metastasis. Thus, inhibition of the FAK kinase could potentially slow the signaling that leads to the spread of TNBC.

The synthesis of potential FAK inhibitors is shown in Scheme 9. Initial S_NAr reaction of 2,4,5-trichloropyrimidine (**17**) with 2-amino-N-methylbenzamide (**66**) using sodium bicarbonate ($NaHCO_3$) in ethanol afforded the monosubstituted pyrimidine **67**. A second S_NAr reaction on **67** by methyl 2-(4-aminophenyl)acetate (**68**) promoted by 12 N hydrochloric acid (HCl) in isopropanol yielded the 2,4-diamino-5-chloropyrimidine **69**. This compound was then condensed with 1,2,5-oxadiazole-2-oxide derivatives **70** [36] in the presence of N-(3-dimethylaminopropyl)-N′-ethylcarbodiimide hydrochloride (EDAC) and 4-(dimethylamino)pyridine (DMAP) to provide targets **71**.

The biological results in the paper pinpointed compound **72**, which exhibited FAK inhibition (IC_{50} = 27.4 nM) and displayed a strong inhibitory effect on cell proliferation with an IC_{50} = 0.126 µM. The compound further exhibited potent inhibitory effects on an MDA-MB-231 TNBC cell line but displayed a 19-fold lesser effect on non-cancer MCF10A, giving a nearly 20-fold window for cell differentiation. Importantly, treatment with **72** inhibited lung metastasis of TNBC more potently than known compound TAE226 (see Supplementary Materials) in mice. The compound also exhibited some off-target activity by showing significant inhibition activity against matrix metalloproteinase-2 (MMP-2) and MMP-9. A pharmacodynamic effect was also observed in a BALB/c nude mouse model by inoculation with MDA-MB-231 TNBC cells in the tail vein. Once the metastatic nodules were formed, the mice were randomly injected with **72** over a period of 30 days. It was found that **72** at 15 mg/kg significantly reduced the lung tumor nodules relative to the vehicle-treated control.

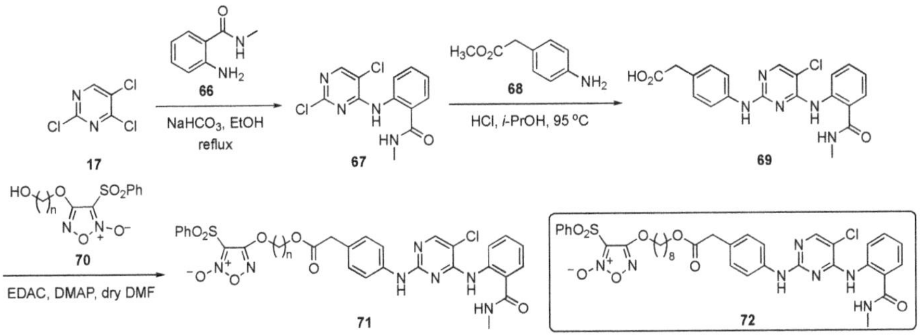

Scheme 9. Synthesis of pyrimidine-based FAK inhibitors **71**.

Concurrent with the work of Zhang, the Badawi group was also intent on developing a drug scaffold to challenge TNBC using *N*-pyrimidin-4-ylhydrazones [37]. Though the authors mentioned breast cancer as their focus, they did not clearly define their goal in this work. The report indicates they were leaning toward the epidermal growth factor receptor (EGFR) or the estrogen receptor as possible targets for inhibition.

The preparation of prospective drug compounds for this study are shown in Scheme 10. This involved reaction of cyano ester **73** and methyl carbamimidothioate (**74**) using NaOAc in DMF to prepare dihydropyrimidinone **75** by a known method [38]. Compound **75** was then treated with a series of cyclic aliphatic amines to provide **76**. Exposure of **76** to POCl$_3$ and *N,N*-dimethylaniline at 60 °C gave derivatives **77** which were reacted with hydrazine to deliver hydrazinyl derivatives **78**. The hydrazinyl function of these structures was finally condensed with benzaldehyde (**79**) or acetophenones **81** to generate the requisite pyrimidine–hydrazone conjugates **80** and **82**, respectively.

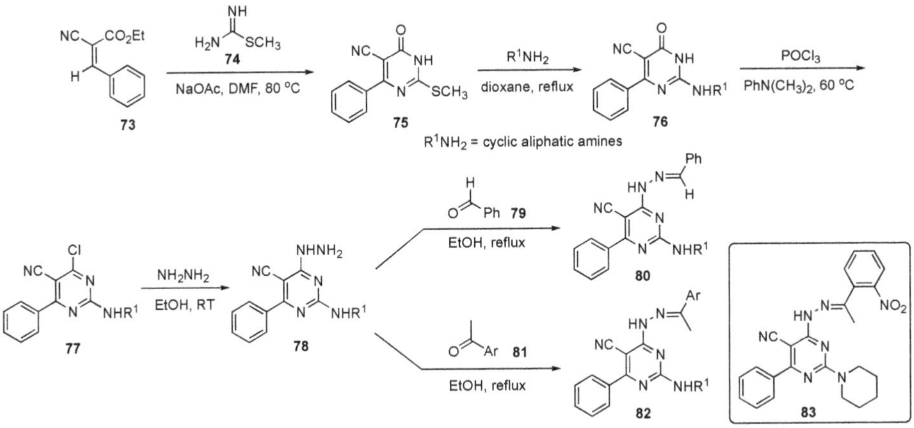

Scheme 10. Synthesis of hydrazonylpyrimidine EGFR inhibitors **80** and **82**.

Preliminary screening for antiproliferative activity revealed that some screened candidates exhibited nearly equal IC$_{50}$ values of 0.87–12.91 μM in MCF-7 and 1.75–9.46 μM in MDA-MB-231 cells, and better growth inhibition activities than those of the positive control

5-Fluorouracil (5-FU, see Supplementary Materials)) which showed IC$_{50}$ values of 17.02 µM and 11.73 µM, respectively. Compound **83** offered the best selectivity index with respect to both MCF-7 and MDA-MB-231 cancer cells in comparison with 5-FU and elicited the highest increase in caspase 9 levels in MCF-7 treated samples, attaining 27.13 ± 0.54 ng/mL compared with 19.011 ± 0.40 ng/mL observed from a Staurosporine (see Supplementary Materials) standard.

Research by El Hamd et al. sought to develop imidazole–pyrimidine–sulfonamide hybrids as inhibitors for the EGFR in mutant cancer cells [39]. Currently, there are many research groups across the world interested in developing inhibitors for EGFR, which plays a crucial role in many human cancers. EGFR, a member of the ErbB subfamily of tyrosine kinases, is overexpressed in many cancers, including those of the breast, colon, ovaries and prostrate [40]. Due to its impact on cancer progression, many therapies are currently approved for this target, including notables such as Cetuxiab, Pantitumumab and Necitumumab in antibody treatment (see Supplementary Materials) and Neratinib, Gefitinib, Lapatinib, Afatinib and Vandetinib in small molecule treatment (see Supplementary Materials).

Pyrimidine pharmacophores are well established as anti-EGFR lung cancer agents, and the authors in this work proposed to couple the pyrimidine ring with a sulfonamide core to bring dual activity against EGFR/human epidermal growth factor receptor 2 (HER2) breast cancer cell lines [41]. Additionally, the authors were also interested in developing inhibitors against drug-resistant mutant EGFR-L858R/T790M/C797S cell lines.

Scheme 11 outlines a multi-component synthesis route to access *N*-(pyrimidin-2-yl) 4-(2-aryl-4,5-diphenyl-1*H*-imizazol-1-yl)benzenesulfonamides **87**. The reactions were carried out using 4-amino-*N*-(pyrimidin-2-yl)benzenesulfonamide (**84**), aryl aldehydes **85**, and benzil (**86**) with ammonium acetate (NH$_4$OAc) and dimethylamine using diethyl ammonium hydrogen sulfate (ionic liquid) under reflux to afford the final targets **87**.

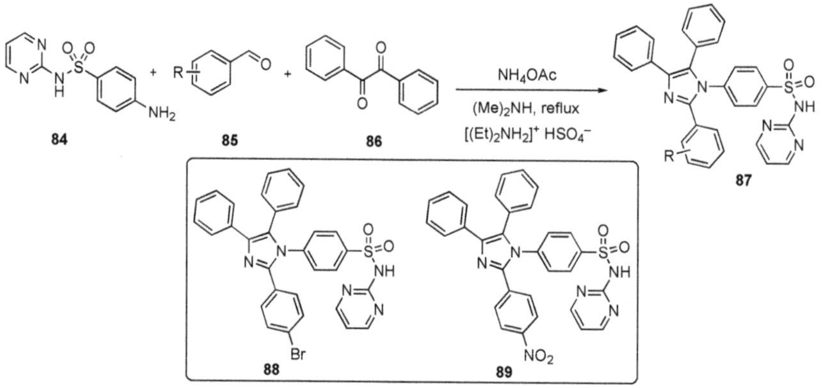

Scheme 11. Synthesis of imidazole–pyrimidine–sulfonamide hybrids **87**.

The compounds synthesized were screened against a panel of 60 cancer cell lines at a single dose of 10 µM at the National Cancer Institute. The results revealed 9 compounds that showed excellent cytotoxicity against all tested cell lines with growth inhibitions up to 95%. Two compounds, **88** and **89**, demonstrated inhibition against HER2 (IC$_{50}$ = 81 ± 40 ng/mL and 208 ± 110 ng/mL, respectively), against the EGFR-L858R mutant (IC$_{50}$ = 59 ± 30 ng/mL and 112 ± 60 ng/mL, respectively), and against the EGFR-T790M mutant (IC$_{50}$ = 49 ± 20 ng/mL and 152 ± 70 ng/mL, respectively). Both compounds induced MCF-7 cell death with a Bax/Bcl-2 expression ratio pointing to a mitochondrial apoptosis pathway. The authors are currently optimizing the active candidates to identify the most promising inhibitors for development.

The work of Zhang et al. expanded the scope of EGFR inhibitors, especially for targeting non-small cell lung cancer (NSCLC) cell lines [42]. As mutation or overexpression of EGFR is the main cause of NSCLC, it is considered the main target for treating this disease [43]. The authors focused on a fourth-generation reversible EGFR-tyrosine kinase inhibitor (TKI) by targeting the cysteine in the active binding site and focusing on the mutant EGFR cancer cell lines. Gefitinib and Erlotinib are first-generation EFGR-TKIs and have been shown to be very effective in NSCLC patients [44,45]. The gatekeeper mutation T790M in the ATP binding domain in EGFR is the primary mechanism of resistance which first develops in patients after 6–12 months of treatment. Considering this new mutant, second- and third-generation EGFR-TKIs were developed which display potent activity against EGFR-T790M while sparing WT cells.

A generalized route to the required compounds for this study is outlined in Scheme 12. S_NAr reaction of 5-bromo-2,4-dichloropyrimidine (**90**) with various anilines in the presence of DIPEA in isopropanol afforded intermediates **91**. The pyrimidine ring of **91** was subjected to a second nucleophilic substitution with various amines (mostly aniline derivatives) in the presence of p-TsOH·H$_2$O in butanol to afford bromopyrimidines **92**. Finally, these bromides were coupled under Suzuki–Miyaura conditions with various arylboronate esters **93** using Pd(dppf)Cl$_2$ and potassium acetate (KOAc) in dioxane to furnish drug candidates **94**.

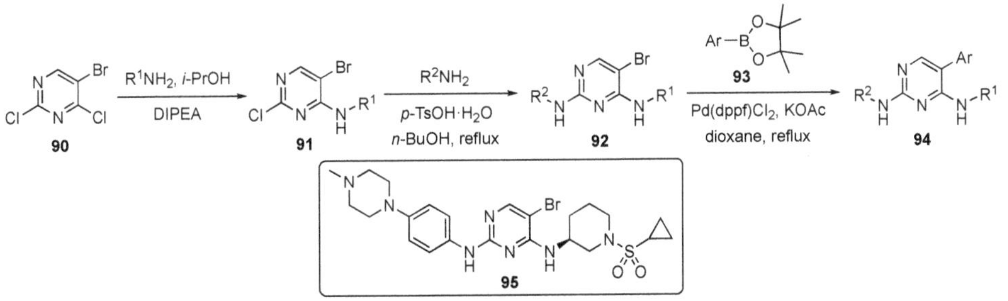

Scheme 12. Synthesis of 2-(phenylamino)pyrimidine EGFR inhibitors **95**.

The authors screened the newly designed and synthesized 2-(phenylamino)pyrimidine derivatives from this sequence for activity against EGFR triple mutant cell lines. One compound, **95**, showed a promising IC$_{50}$ value of 0.2 ± 0.01 µM against proliferation of the EFGR-Dell9/T790M/C797S and EGFR-L858R/T790M/C797S cell lines. The same compound exhibited a slightly higher antiproliferative activity than the commercial drug Brigatinib (see Supplementary Materials). Most of the compounds exhibited weak activities on EGFR-WT, which indicates that the compound was selective for mutant EGFR. Compound **95** also significantly inhibited EGFR phosphorylation, induced apoptosis in EGFR-Dell9/T790M/C797S, and arrested the cell cycle at the G2/M phase. The results indicate that **95** was a potent fourth-generation reversible EGFR-TKI which warranted further study.

As with most medicinal agents, drug resistance has become an issue for Osimetinib, and this has been a driving force behind the development of EFGR inhibitors. This drug, which is currently used for NSCLC, showed drug resistance after the median survival time of 9.6 months [46]. Thus, Xu et al. determined to solve this problem by developing fourth-generation inhibitors with additional interactions between the compound and the protein to compensate for the loss of the conventional covalent cysteine interaction [47]. The 2,4-di(arylamino)pyrimidine core is a key ring scaffold for maintenance of activity in these known inhibitors of mutant EGFR kinases [48]. All compounds synthesized in this work were designed following a molecular modelling analysis of the crystal structure of EGFR-L858R/T790M/C797S (PDB code: 6LUD) using Autodock 4.2 software. The synthesis of prototype molecules is shown in Scheme 13. S_NAr reaction between 2,4,5-trichloropyrimidine (**17**) and phenylenediamine (**96**) in ACN at −10 °C provided the C4-substituted pyrimidine derivative **97**. This compound subsequently underwent amide formation with acryloyl chloride and DIPEA in dioxane to afford **98**. Some of the pyrimidinamides were also prepared from acetic anhydride (Ac$_2$O) in the presence of TEA in ethyl acetate. Pyrimidinamides **98** were subsequently coupled with various substituted anilines using Pd(OAc)$_2$, xantphos, and Cs$_2$CO$_3$ in dioxane to furnish targets **99**.

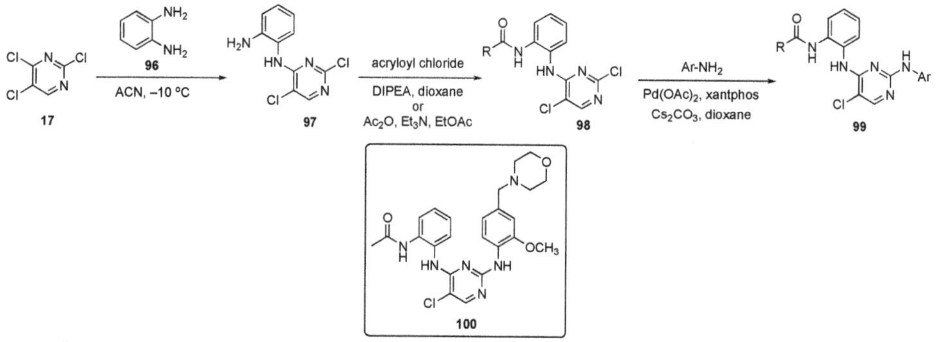

Scheme 13. Synthesis of 2,4-di(arylamino)pyrimidine EGFR kinase inhibitors **99**.

All derivatives were evaluated for their effect on the enzymatic activity of EGFR-WT and mutant EGFR-L858R/T790M/C797S and EGFR-L858R/T790M kinases using the ADP-Glo Kinase Kit. Osimertinib was employed as a positive control. One of the inhibitors, **100**, was identified as the most favorable compound and strongly inhibited EGFR-L858R/T790M/C797S and EGFR-L858R/T790M activity with IC$_{50}$ values of 5.51 nM and 33.35 nM, respectively. In addition, **100** exhibited stronger antiproliferative activity against NSCLC cells (H1975), expressing high levels of EGFR-L858R/T790M and Ba/F3-EGFR-L858R/T790M/C797S cells with IC$_{50}$ values of 0.442 µM and 0.433 µM, respectively. Proliferation was inhibited by arresting the H1975 cells at the G2/M phase, promoting apoptosis of the cells, and reducing phosphorylation of EGFR and extracellular signal-related kinase 1/2 in a dose-dependent manner. The wound-healing assay data showed that H1975 migration and invasion abilities were effectively inhibited by **100** in a concentration-dependent manner. Compound **100** also expressed a 27-fold lower toxicity against normal liver cells, indicating an improved dosage safety margin. The results further suggest that this compound could be used as a competitive ATP inhibitor, as well as an allosteric inhibitor of EGFR-L858R/T790M/C797S.

A patent developed by Lee et al. featured a lung cancer subtype which is an EGFR mutation positive for NSCLC [49]. More than 50% of NSCLC patients have EGFR activating mutations. Currently, third-generation EGFR-TKIs are being explored to overcome this resistance. Osimertinib is a powerful inhibitor that suppresses EGFR mutations and T790M resistant mutations, but it causes ineffective binding and subsequent C797S resistance in NSCLC patients. When Osimertinib was administered as a front-line therapy, the most common resistance mechanisms proved to be the C797S mutation (7%) and mesenchymal epithelial transition amplification (15%) [50]. The next-generation EGFR compounds would need to inhibit Del19/T790M/C797S, L858R/T790M/C797S, Del19/C797S, and L858R/C797S, and be highly selective versus EGFR-WT to avoid adverse effects. The work in this patent focused on an unmet need, to develop a next-generation TKI targeting both C797S triple and double mutants. It was imperative to create a selective, next-generation inhibitor for NSCLC patients with advanced or metastatic diseases carrying Del19/T790M/C797S, L858R/T790M/C797S, Del19/C797S, or L858R/C797S mutations following second-line or upfront use of third-generation EGFR-TKIs.

The synthetic approach to these next-generation anticancer agents is depicted in Scheme 14. Initially, amination of the commercial pyridine derivative **101** by S$_N$Ar displacement of the C4 chloride using DIPEA in DMF at 90 °C provided **102**. Subsequent exposure to Buchwald–Hartwig coupling conditions with pyrimidine derivative **103** using tris(dibenzylideneacetone)dipalladium(0) (Pd$_2$(dba)$_3$), xantphos, and Cs$_2$CO$_3$ in dioxane provided products **104**.

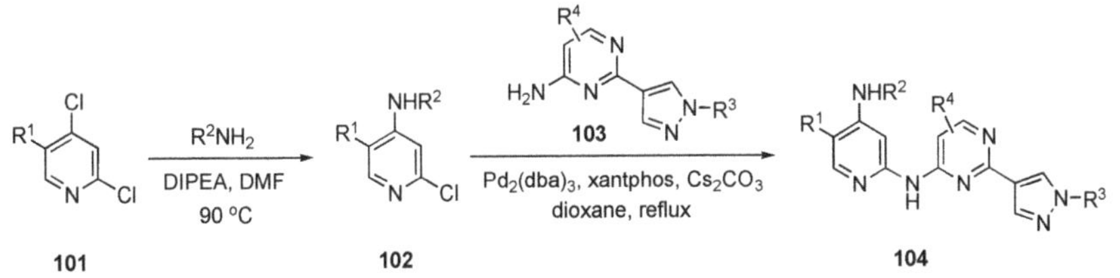

Scheme 14. Synthesis of pyrazole-appended pyrimidines as anticancer agents **104**.

The enzymatic biochemical assays for the EGFR kinases were reported in the patent. The assays were conducted and reported for the EGFR-WT, double mutants Del19/C797S and L858R/C797S, and triple mutants Del19/T790M/C797S and L858R/T790M/C797S. There were many promising compounds that had IC$_{50}$ values in the 0.1–100 nM range but no further biological data were reported.

A patent by the Dai group disclosed the use of pyrimidines with deuterated substituents to target cyclin-dependent kinases (CDKs) [51]. CDKs are part of a subfamily of serine/threonine protein kinases which play a significant role in regulating cell cycle progression [52]. They are essential cell cycle drivers, especially CDK2, which helps cells to transition from late G1 into S and G2 phases. CDK2 plays a prominent role in proliferative pathways, which are not important for normal cell proliferation but are essential for cancer cells [52]. Selective CDK2 inhibitors might target tumors which are highly cyclin E1 and E2 expressive. Cyclin E1 is always overexpressed in human cancer. Cyclin E1 amplified ovarian cancer cell lines are sensitive to reagents that inhibit CDK2 activity or decrease cellular CDK2 protein. Some of the pyrimidine-based drug candidates in this patent specifically targeted CDK2 and offered selectivity over other kinases in treating patients with tumors.

A strategy by which to synthesize potential CDK2 inhibitors is shown in Scheme 15. Compound **105** underwent a S$_N$Ar reaction with deuterated 4-aminobenzenesulfonamide **106** in the presence of zinc chloride and TEA in DCM–*t*-butanol to provide **107**. Intermediate **107** was further reacted with lithium hexamethyldisilazide (LiHMDS)-derived alkoxide from tetrahydro-2*H*-pyran-4-ol **108** in tetrahydrofuran (THF) to afford **109**. Finally, removal of the *p*-methoxybenzyl (PMB) protecting group was carried out in the presence of 2,3-dichloro-5,6-dicyano-*p*-benzoquinone (DDQ) in DCM and H$_2$O to furnish derivatives **110**. Some derivatives without the deuterium are also reported in this patent using the same sequence.

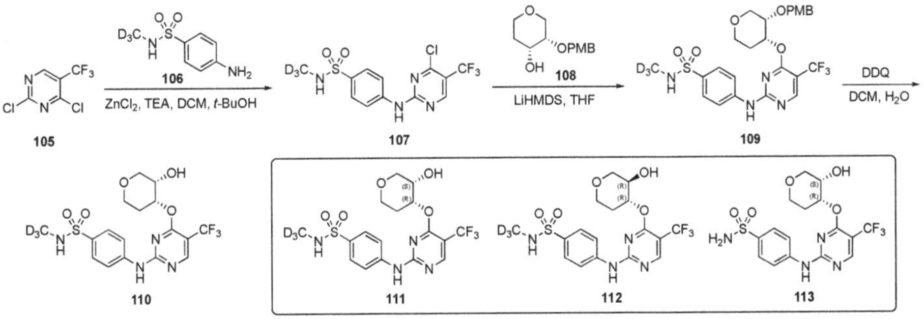

Scheme 15. Synthesis of deuterated sulfonamide-pyrimidine CDK inhibitors **110**–**113**.

The most promising compounds, **111–113**, had IC$_{50}$ = 1–10 nM for CDK2/cyclin E1 activity whereas for CDK1/cyclin B1 activity was nearly 10–20 times weaker offering only 20-fold selectivity. With respect to other isoforms of cyclin—CDK4, CDK6, CDK7, and CDK9—**111–113** offered 100–1000-fold greater selectivity. The patent asserts that these compounds were evaluated against breast, ovarian, bladder, uterine, prostate, lung (including NSCLC, SCLC, squamous cell carcinoma or adenocarcinoma), esophageal, head and neck, colorectal, kidney (including renal cell carcinoma), liver (including hepatocellular carcinoma), pancreatic, stomach, and thyroid cancers. The patent further claims that these derivatives were tested for estrogen receptor-positive/hormone receptor-positive, HER2-negative, HER2-positive, triple negative, and inflammatory breast cancer, but few results are reported from these experiments.

As in the previous entry, Zhou et al. were involved in developing a drug scaffold targeting the CDKs [53]. CDKs are important in many crucial processes, such as cell cycle and transcription, as well as communication, metabolism, and apoptosis. Deregulation of any stage of the cell cycle or transcription leads to apoptosis but, if uncorrected, can result in a series of diseases, such as cancer, neurodegenerative diseases (Alzheimer's or Parkinson's diseases), and stroke [50]. CDK4/6 is considered a potential anticancer drug target. To date, three CDK4/6 inhibitors have been approved; however, there is still a gap between the clinical requirements and the approved drugs [54]. Thus, selective and oral CDK4/6 inhibitors are urgently needed, particularly for monotherapy. This study investigated the interaction between Abemaciclib (see Supplementary Materials) and human CDK6 using molecular dynamics simulations. Based on these modelling studies, a candidate compound was designed that was predicted to show a significant inhibitory effect on a human breast cancer cell line.

The strategy to prepare the designed model compound is outlined in Scheme 16. Initially, Suzuki–Miyaura coupling of 2,4-dichloro-5-fluoropyridine (**59**) with pyrazolopyridineboronate ester **114** in the presence of Pd(dppf)Cl$_2$ and K$_2$CO$_3$ in aqueous dioxane afforded intermediate **115**. This was followed by Buchwald–Hartwig coupling with a 4-(4-isopropylpiperazin-1-yl)aniline (**116**) under standard conditions, to furnish **117**.

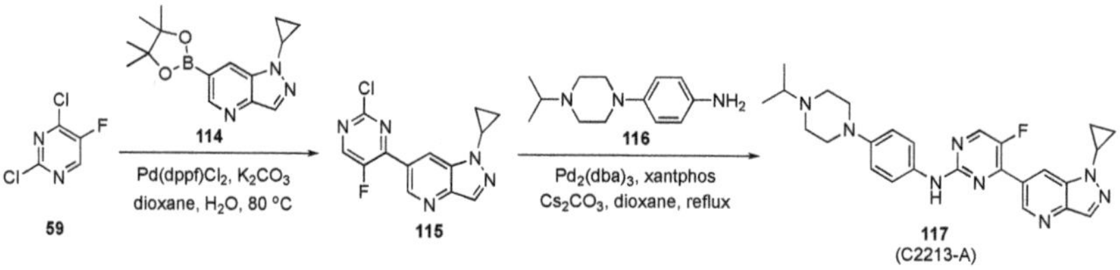

Scheme 16. Synthetic route to CDK inhibitor **117** (C2213-A).

The inhibitory activity of **117** (C2213-A) was validated against CDK6 using a kinase profiling radiometric protein kinase assay. The IC$_{50}$ value for **117** was 290 nM, comparable to the estimate of 238 nM for Abemaciclib targeting human CDK6/cyclin D3. The antiproliferative activity of **117** was significantly higher than Abemaciclib (positive control) with an IC$_{50}$ = 2.95 ± 0.15 µM. The inhibitory activity of **117** was tested against MCF-7 cells as well as other breast cancer cell lines such as T-47D, MDA-MB-452 and MDA-MB-468 and showed a better inhibitory effect than the control. The CDK4/6 inhibition by **117** and the phosphorylation of retinoblastoma tumor suppressor were assessed by a Western blot assay on MDA-MB-231 cells. Compound **117** was found to block the CDK4/6/Rb/E2F signaling pathway in a dose-dependent manner after 24 h of incubation.

Zhang et al. have identified a new anticancer drug incorporating the pyrimidine scaffold that targets microtubules [55]. Microtubules are essential structural components of the cytoskeleton and are composed of α- and β-tubulin heterodimers [56]. Due to the polymerization dynamics of tubulin, microtubules are important targets for anticancer drugs known as microtubule-targeting agents (MTAs). A total of 7 binding sites on tubulin have been found including Paclitaxel (see Supplementary Materials), Laulimalide, Colchicine (see Supplementary Materials), Vinblastine, Maytansine, and Pironetin, as well as a 7th binding site. To date, no tubulin inhibitors targeting the colchicine binding site (CBS) have been specifically approved for clinical application. On the other hand, one CBS inhibitor, namely ABI (**118**), has been reported to manifest nanomolar potency against multidrug resistant strains with significant in vivo antitumor efficacy [57]. Despite the excellent biological activities, ABI analogs contain a ketone group between the imidazole and the C-ring which is a metabolic soft spot susceptible to reduction by liver microsomes. Osimertinib (**1**), on the other hand, is an approved pyrimidine-containing anticancer drug for NSCLC [58]. In this work, the authors designed the series of Osimertinib–ABI hybrids shown in Figure 3 for use as MTAs to treat cancer.

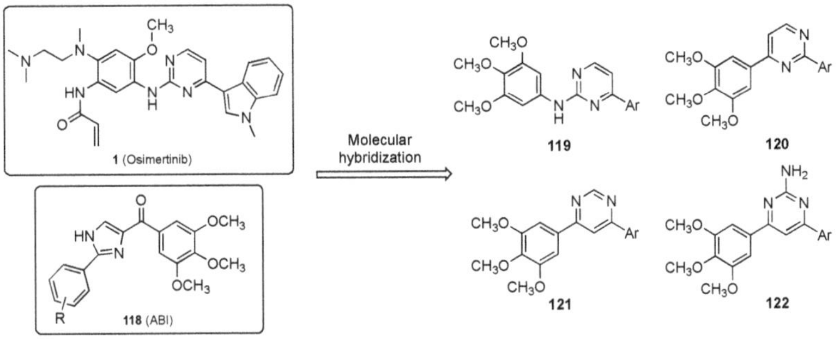

Figure 3. Molecular hybridization of Osimertinib (**1**) and ABI (**118**).

The general synthetic approach to hybrids **119–122** is shown in Scheme 17. In Scheme 17-I, construction of **119** involved Suzuki–Miyaura coupling between C4 of 2,4-dichloropyrimidine (**52**) and a wide range of commercially available arylboronic acids to generate intermediates **123**. S_NAr reaction of **123** with 3,4,5-trimethoxyaniline (**124**) provided drug candidates **119**. In Scheme 17-II, independent Suzuki–Miyaura couplings were carried out between 3,4,5-trimethoxyphenylboronic acid (**125**) and compounds **52** and 4,6-dichloropyrimidine (**127**) to furnish intermediates **126** and **128**. Each of these was further coupled with arylboronic acids to give 4-(3,4,5-trimethoxyphenyl)-2-arylpyrimidines **120** and 4-(3,4,5-trimethoxyphenyl)-6-arylpyrimidines **121**, respectively. In Scheme 17-III, aldol condensation of 1-(3,4,5-trimethoxyphenyl)ethan-1-one (**129**) and a series of benzaldehyde derivatives yielded 1,3-diaryl-2-propen-1-ones (chalcones) **130**, which were reacted with guanidine hydrochloride under basic conditions to produce a library of 2-amino-3,5-diarylpyrimidine derivatives **122**.

A total of 43 pyrimidine analogs were synthesized and evaluated for their antiproliferative activity. Among these, prototype **131**, bearing a fused 1,4-benzodioxane moiety, exhibited the best potency, inhibiting four cancer cell lines including A549–lung (IC_{50} = 0.80 ± 0.09 µM), HepG2–liver (IC_{50} = 0.11 ± 0.02 µM), U937–lymphoma (IC_{50} = 0.07 ± 0.01 µM), and Y79–retinoblastoma (IC_{50} = 0.10 ± 0.02 µM). Furthermore, **131** suppressed tubulin polymerization and disrupted the microtubule network of HepG2 cells. Molecular dynamics simulations suggested that **131** blocked the cell cycle at the G2/M phase and eventually induced HepG2 cell apoptosis by regulation of G2/M related protein expression of cyclin B1 and P21. Both scratch and transwell assays have indicated that this derivative inhibited migration and invasion of HepG2 cells in a dose-dependent manner. Overall, these results indicate that **131** has potential as a tubulin polymerization inhibitor targeting the CBS and merited further investigation.

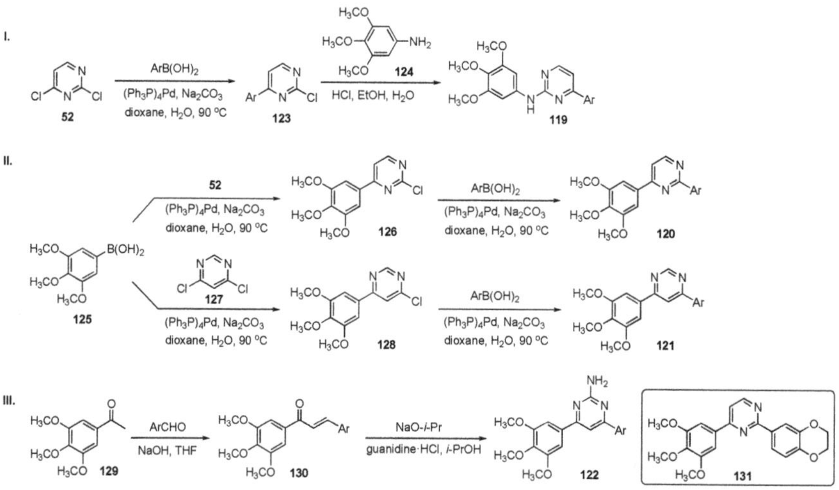

Scheme 17. Synthesis of pyrimidine MTAs 119–122.

MTAs are an important chemotherapeutic class of drugs that interfere with tubulin dynamics by disrupting the formation of the mitotic spindle, arresting cell cycles, and finally promoting apoptosis of tumor cells. An investigation by Wang et al. [59] has demonstrated that, while microtubule disruption can interfere with cancer development, it can also affect normal cells, leading to two major toxicities: neutropenia and peripheral neuropathy in postmitotic neurons [60]. Although achievements have been made in clinical treatment, toxicities still limit the utility of MTAs [61]. The CBSs, located at the interface between α- and β-tubulin heterodimers [62], effectively impact protein trafficking and could serve as a key entry point for anticancer agents. Evidence suggests that CBS inhibitors can overcome drug resistance mediated by P-glycoprotein (P-gp), multidrug resistance protein 1 (MRP1) and MRP2, and destroy the vascular networks that exist in tumor tissues serving as vascular damaging agents (VDAs). For these reasons, the CBS is an attractive target for the development of chemotherapeutic drugs, including those featured in their paper. The authors have identified a novel MTA skeleton which inhibits tubulin polymerization at 5 µM. The team further optimized this series of compounds by evaluating a structure–activity relationship (SAR) derived from an X-ray co-crystal with the target.

A concise synthesis of the required pyrimidine analogs is presented in Scheme 18. A series of aromatic amines was prepared and reacted with 4,6-dichloropyrimidine (127) using DIPEA in ethanol to produce 132. This was followed by a second nucleophilic substitution reaction with various alkyl and aromatic amines, which subsequently led to products 133.

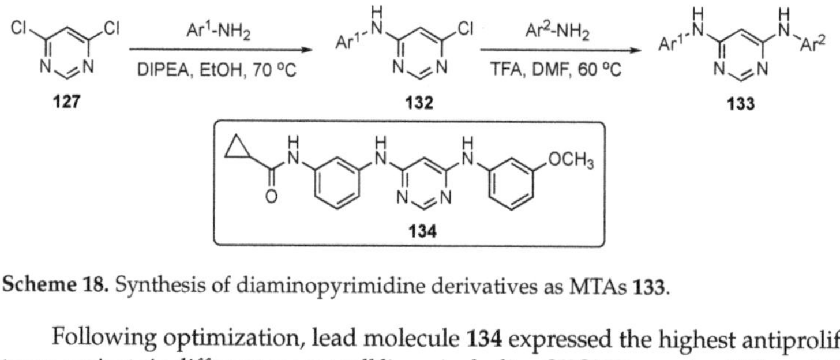

Scheme 18. Synthesis of diaminopyrimidine derivatives as MTAs 133.

Following optimization, lead molecule 134 expressed the highest antiproliferative potency against six different cancer cell lines, including SKOV-3–ovarian (EC$_{50}$ = 1.5 ± 0.2 nM),

HepG2–liver (EC$_{50}$ = 1.8 ± 0.6 nM), MDA-MB-231–breast (EC$_{50}$ = 4.4 ± 0.6 nM), HeLa–cervical (EC$_{50}$ = 3.6 ± 0.3 nM), B16-F10–melanoma (EC$_{50}$ = 3.3 ± 0.1 nM), and A549–lung (EC$_{50}$ = 1.1 ± 0.2 nM). This compound also exhibited more potent antiproliferative activities than Colchicine and Paclitaxel against the paclitaxel-resistant ovarian cancer cell line A2780/T and its parental cell line A2780, indicating that **134** could overcome P-gp-mediated paclitaxel resistance in vitro. The compound also showed equal activity against lung tumors A549-WT and low EGFR expression A549, proving that EGFR inhibition was not the major reason for the antitumor activity. The PK results show that **134** can be absorbed rapidly from the intestine with $t_{1/2}$ = 0.22 ± 0.02 h (see Table 4). The lead had a slightly high, but acceptable, CL of 69.84 ± 4.97 mL/min/kg and an AUC of 239.43 ± 16.39 ng/mL·h. These results establish that **134** has acceptable pharmacokinetic properties, and therefore, is suitable for further development.

Table 4. Pharmacokinetic properties of compound **134**.

Cpd No.	Administration	CL (mL/min/kg)	C$_{max}$ (ng/mL)	$t_{\frac{1}{2}}$ (h)	AUC$_{0-t}$ (ng/mL·h)
134	i.v. (1 mg/kg)	69.84 ± 4.97	87.86 ± 9.20	0.22 ± 0.02	239.23 ± 16.32

Abdel-Aal et al. have developed anticancer compounds specifically targeting the tubulins [62]. Though anticancer drugs are already in place for this target, the authors sought to address the poor oral bioavailability and the multidrug resistance of current tubulin drugs. This work focused on the modification of the Combretastatin and Phenstatin drug scaffolds. Chalcones can be simply viewed as keto stilbenes, mimicking both Combretastatin and Phenstatin. Various modifications of the 1,3-diaryl scaffold were developed without affecting their tubulin inhibitory activity, including phenoxy substitution and replacement with heterocyclic rings [63]. The authors developed lipidated 4,6-diarylpyrimidines as tubulin polymerization inhibitors (or antiproliferative agents) which improved the interaction in the hydrophobic pocket and enhanced their physiochemical properties and cell penetration. The pyrimidine moiety in this series of compounds offered extra hydrophilic interactions and rigidity relative to the propanone scaffold, which may enhance tubulin binding.

The syntheses of molecules for this study are depicted in Scheme 19. The plan targeted lipidated chalcones, which were prepared using known condensation chemistry [64]. The final lipidated 4,6-diarylpyrimidines were prepared by refluxing long-chain alkoxy-substituted chalcones **135** with urea, thiourea, or guanidine carbonate in alkaline medium to produce the required drug candidates **136–138**.

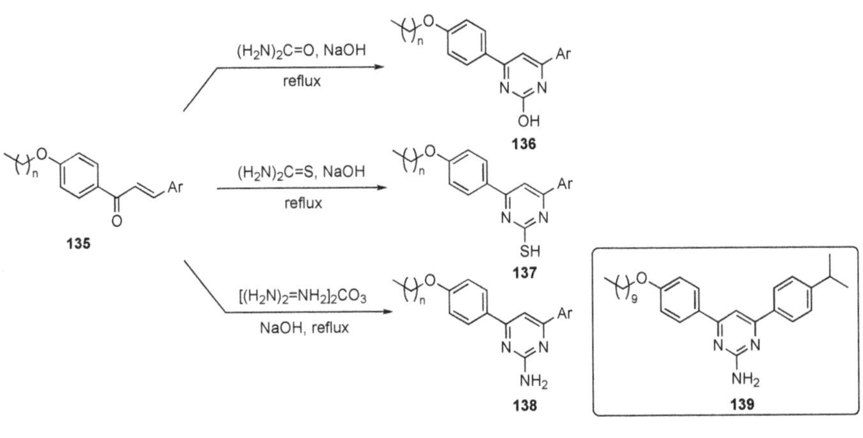

Scheme 19. Synthesis of 4,6-pyrimidine (lipidated chalcone) tubulin polymerization inhibitors **136–138**.

Eighteen chalcones and their lipidated pyrimidine derivatives were designed and synthesized as tubulin polymerization inhibitors. In general, the synthetic pyrimidine derivatives had improved antiproliferative activity over the corresponding chalcones against the MCF-7 cancer cell line. The pyrimidin-2-amine **139** showed dual antiproliferative activity against MCF-7–breast (IC_{50} = 10.95 µM) and HepG2–liver (IC_{50} = 11.93 µM) cell lines, induced apoptosis and cell cycle arrest, and displayed tubulin inhibitory activity against MCF-7 at low micromolar concentration. The compound also induced S-phase cell cycle arrest and apoptosis in MCF-7 cells with a tubulin IC_{50} = 9.7 µM. These findings established **139** as an anticancer lead worthy of further optimization and development.

A patent by Boeckman et al. describes inhibitors of histone H3K27 demethylase JMJD3 [65]. The Jumonji C (JMJC) domain, containing proteins which include histone H3K27, plays a significant role in tumorigenesis and has been identified as a key target for anti-cancer agents [66]. The patent highlights the critical role of Jumonji kinases and inhibitors of H3K27 to target diffuse intrinsic pontine glioma (DIPG), which is the most frequent brain stem tumor in pediatrics and has a survival rate of 9–12 months from diagnosis [67]. There are no surgical options for this brain stem tumor and conventional chemotherapy is used solely to alleviate pain. Due to this issue, an efficacious therapeutic agent is needed for these DIPG patients. DIPG is uniquely dependent on the H3K27 mutation for cancer initiation/maintenance and is detected in more than 80% of patients. However, more than 250 clinical trials have been executed on this target without much success.

The synthesis of prospective targets, shown in Scheme 20, started with 2-cyanopyridine (**140**), which was converted to picolinimidamide hydrochloride (**141**) with ammonium chloride (NH_4Cl) in the presence of HCl in ethanol and DCM. Hydrochloride **141** was cyclized with diethyl malonate using sodium ethoxide in ethanol to provide dihydroxypyrimidine **142**, which was transformed to the corresponding dichloro derivative **143** using $POCl_3$. Subsequent S_NAr reaction of **143** with amine **144**, promoted by DIPEA at reflux, provided **145** [68]. Finally, **145** was reacted with various aliphatic amino alcohols to generate the desired model compounds **146**, which were evaluated for biological activity. Some of the final alcohols employed prodrug approaches as part of this screening.

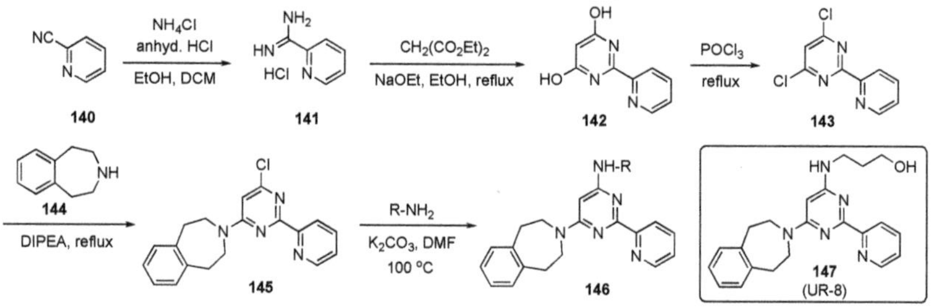

Scheme 20. Synthesis of pyrimidine JMJD3 kinase inhibitors **146**.

Synthesized derivative **147** (UR-8) demonstrated selective cytotoxic activity against human DIPG-K27M cells (IC_{50} = 4–6 µM) in vitro and was apparently transported to the brain due to its in vivo stability. In a mouse study, prototype **147** showed a favorable biodistribution in the brain stem compared with a competitor compound, GSK-J1. The concentration of **147** was found to be around 4455 ± 1576 ng/mL in serum and around 409.5 ± 243.9 ng/mL in the brain stem. Extraction of brain stem tissue from mice treated with **147** followed by HPLC assay revealed 8.77 ± 2.37% of **147** in this tissue. A similar experiment using GSK-J1 (see Supplementary Materials) detected no significant amount of the competitor compound in brain stem tissue. Further data indicate that **147** was likely active in its original form and therefore this subclass of inhibitors offers high potential for clinical application. The compound also inhibited tumor growth and prolonged survival rates in mice with human DIPG xenografts. To determine the in vivo antitumor activity of this analog, the mice were implanted with DIPG-SF8628 cells in the brain stem and treated with 100 mg/kg of **147** for 10 consecutive days. This experiment confirmed that **147** outperformed other current drugs for this tumor.

Ling et al. have developed an inhibitor for acute myeloid leukemia (AML), which is a life-threatening malignancy with a 5-year survival rate. This cancer is characterized by its disruption of hematopoietic progenitor cell differentiation and proliferation [68]. AML treatment, which includes chemotherapy, does not exhibit long-term efficacy, and 70% of people do not survive beyond 1 year [69]. Some of the known chemotherapeutic drugs, such as Cytarabine (see Supplementary Materials) and Daunorubicin (see Supplementary Materials), have already encountered drug resistance in patients. Bruton's tyrosine kinase (BTK), a member of the TEC kinase family, plays a critical role in multiple signaling pathways and significantly impacts proliferation, survival, and differentiation of B-lineage and myeloid cells [70]. BTK is highly expressed and activated in more than 90% of AML patients, and, thus, could offer a potential strategy for treatment. These researchers also specified interest in a second target, namely the FMS-like tyrosine kinase 3 (FLT3), which is expressed in most AML cell lines. The authors resolved to seek a dual inhibitor for these two kinase targets to address the issue of drug resistance.

The synthesis of possible BTK/FLT3 dual inhibitors is shown in Scheme 21. Two sequential $S_N Ar$ reactions of 2,4-dichloro-5-fluoropyrimidine (**59**), the first, at C4 by aniline esters **148** using DIPEA in isopropanol at 80 °C, gave **149**, while the second, at C2 by aniline **150** in the presence of TFA in butanol, produced **151**. Subsequently, the ester group in **151** was converted to hydroxamic acid with hydroxylamine hydrochloride and KOH in methanol to afford the final targets **152**.

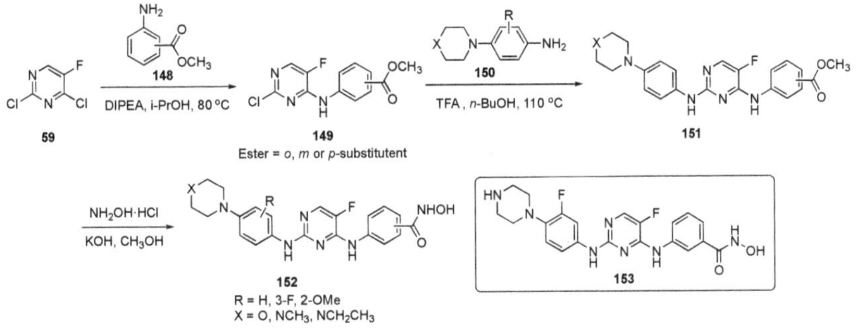

Scheme 21. Synthesis of pyrimidine analogs as BTK/FLT3 inhibitors **152**.

Some of the compounds synthesized as BTK/FLT3 dual inhibitors exhibited IC_{50} values at low nanomolar levels. Among these dual inhibitors, **153** exhibited activity against FLT3/D835Y mutant cells with single digit nanomolar potency (IC_{50} = 5.9 ± 0.1 nM). This inhibitor showed powerful antiproliferative activity against AML cells and inhibited the growth of other leukemia cells: MV-4-11 (IC_{50} = 0.29 ± 0.02 nM) Molm13 (IC_{50} = 0.45 ± 0.03 nM), K562 (IC_{50} = 73 ± 13 nM), Molt4 (IC_{50} = 1.4 ± 0.3 nM), and THP1 (IC_{50} = 37 ± 5 nM) which are all BTK and FLT3 positive. Additionally, compound **153** effectively induced apoptosis and upregulated proapoptotic protein levels in MV-4-11 cells in a dose-dependent manner. Finally, **153** effectively suppressed the growth of MV-4-11 cells in the xenograft tumor model with a 20 mg/kg intraperitoneal (i.p.) injection and showed an antitumor effect, like Sorafenib (20 mg/kg, see Supplementary Materials)), with no significant toxicity.

Yang et al. undertook a study to develop a drug for prostate cancer (PCa), which is a major threat to male health and results in a high mortality rate worldwide [71]. Hormonal therapies for PCa play a major role by decreasing androgen levels. However, once resistance develops in hormonal therapies, it renders this approach unusable, so there is an urgent need to develop alternative drugs for PCa [72]. In this work, the Yang group focused on dual-specificity tyrosine phosphorylation-regulated kinases (DYRKs), which belong to the CMGC kinase family, where DYRK2 plays an important role in cell proliferation, apoptosis, and migration. By downregulating DYRK2, PCa is suppressed which makes this a prominent target for inhibition [73].

The synthetic route to compounds needed for this program is shown in Scheme 22. Bromobenzothiazole **154** was coupled with bis(pinacolato)diboron (**155**) in the presence of Pd(dppf)Cl$_2$ and KOAc to generate boronate ester **156**, which was further coupled with pyrimidine **59** using bis(triphenylphosphine)palladium(II) chloride ((PPh$_3$)$_2$PdCl$_2$) and K$_2$CO$_3$ in aqueous 1,2-dimethoxyethane (DME) to afford **157**. Compound **157** underwent Buchwald–Hartwig coupling at C4 with various protected amines **158** to provide adducts that were deprotected with ethyl acetate-HCl in DCM to deliver targets **159** for biological evaluation.

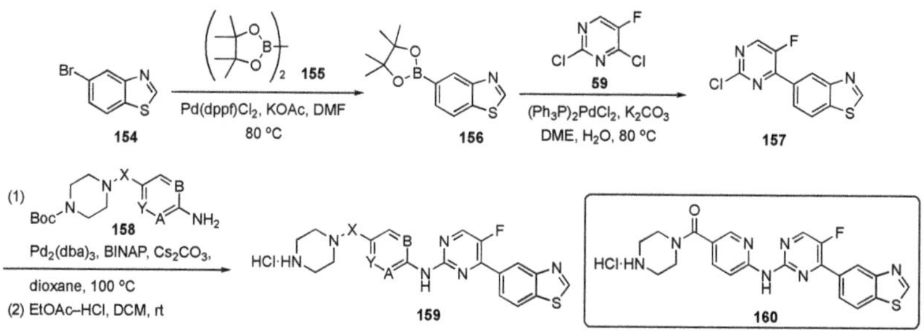

Scheme 22. Synthesis of pyrimidines targeting prostate cancer **159**.

The authors used structure-based virtual screening to develop these DYRK2 inhibitors of which the most potent was **160** with an IC$_{50}$ = 0.6 nM. This compound also elicited good inhibitory activity against proliferation and migration and promoted apoptosis on PCa cells. The ADME properties of **160** were presented along with a thermodynamic solubility of 29.5 mg/mL, a parallel artificial membrane permeation assay (PAMPA) value of log Pe = −5.98, and liver microsomal stability of ca. 16 mL/min/kg with t$_{1/2}$ = 78 min. There was no hERG inhibition with QPloghERG = −6.743 and the compound had an excellent LD$_{50}$ > 10,000 mg/kg. At a high concentration of 200 mg/kg, the compound displayed tumor growth inhibition better than Enzalutamide (see Supplementary Materials), which was the positive control (100 mg/kg) in the PCa xenograft models. The mice in this study did not undergo any significant weight loss, suggesting that these compounds likely have a good safety profile.

Xie et al. have developed adenosine A$_{2A}$ receptor (A$_{2A}$R) antagonists as a novel strategy for cancer immunotherapy [74]. Adenosine triphosphate (ATP) is an endogenous ligand that is widely distributed throughout the human body. ATP is involved in numerous functions, including cell growth, hearth rhythm, immune function, sleep regulation and angiogenesis. Though there are four subtypes of adenosine receptor, only A$_{2A}$R has been sufficiently investigated to attract much attention as a potential drug target for cancer and various inflammatory and neurodegenerative diseases [75]. These researchers sought to use this A$_{2A}$R strategy to develop a treatment for colon cancer.

The synthetic route to pyrimidine derivatives for this investigation is shown in Scheme 23. The synthesis leveraged two consecutive Suzuki–Miyaura couplings to 2-amino-4,6-dichloropyrimidines **161**, the first with arylboronate esters **162** to afford intermediate **163** and the second with various methyl protected pyridinones **164** to yield two sets of pyridine derivatives, **165** and **166**. Demethylation of the pyridine moieties on these structures using HBr, and subsequent N-alkylation afforded drug candidates **167** and **168**.

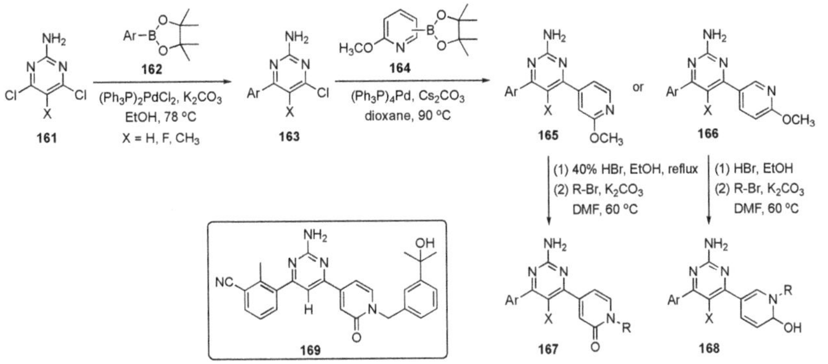

Scheme 23. Synthesis of pyrimidine–pyridinone $A_{2A}R$ antagonists 167 and 168.

Evaluations based on SAR and ADME properties led to compound 169 with improved potency (IC_{50} = 29 nM vs. $A_{2A}R$) and better mouse liver microsomal metabolic stability ($t_{1/2}$ = 86 min). The compound expressed preferential activity against $A_{2A}R$ over A_1R, $A_{2B}R$, and A_3R (>100-fold selectivity, IC_{50} > 3 μM), and the compound demonstrated good oral bioavailability in mice. Compound 169 showed excellent anticancer activity, with a total growth inhibition of 56.0% and good safety characteristics in the mouse MC38 colon cancer model at an oral dose of 100 mg/kg. The PK of this drug candidate was assessed in mice following i.v. (2 mg/kg) and p.o. (10 mg/kg) administration to C57BL/6 mice (n = 3 peer groups), and the results are shown in Table 5. The oral bioavailability (F) of compound 169 in mice was excellent (86.1%), and the compound had a plasma protein binding ratio of 98.6%. No significant body weight loss was observed in experimental mice, indicating that compound 169 was well tolerated at the given dosage. With these encouraging results, the anticancer agent 169 with an appended pyridinone moiety was deemed an excellent prospect for further refinement as an immunotherapeutic.

Table 5. Pharmacokinetic properties of compound 169.

Cpd No.	Administration	C_{max} (ng/mL)	$t_{\frac{1}{2}}$ (h)	AUC_{0-t} (ng/mL·h)	F
169	i.v. (2 mg/kg)	2584 ± 201	0.93 ± 0.05	5577 ± 667	n.a.
	p.o. (10 mg/kg)	8823 ± 1701	2.35 ± 0.20	24,008 ± 351	86.1%

Huang et al. actuated a study to develop a hematopoietic progenitor inhibitor (HPK1) as a cancer immunotherapy [76]. HPK1 is a mitogen-activated kinase 1 (MAP4K1), a cytosolic STE20 serine/threonine kinase from the germinal kinase family which is highly expressed in immune populations, including T cells, B cells, and dendritic cells [77]. Recent evidence in this field suggests that HPK1 activation can significantly limit the intensity and duration of T-cell receptor signaling, resulting in cell dysfunction. Their results demonstrate that loss of HPK1 kinase function can increase cytokine secretion and enhance T cell signaling, virus clearance, and tumor inhibition. Thus, HPK1 has potential as a novel and effective target for cancer immune response enhancement.

In this work, rational design, synthesis, and SAR exploration were carried out for novel 2,4-disubstituted pyrimidine derivatives as potent HPK1 inhibitors by a scaffold hopping (heterocycle replacement) approach. The design of this compound was based on a reverse indazole derivative discovered by Merck, and which demonstrated highly potent and selective inhibition of HPK1 [78]. The authors used this scaffold hopping strategy for drug design and diversification of chemotypes to identify pyrimidines as alternatives for indazole rings.

Scheme 24-I summarizes the preparation of one subset of target molecules. Initially, 2-chloro-4-aminopyrimidine (**170**) was condensed with 5-fluoro-2-morpholinobenzoic acid (**171**) in the presence of O-(7-azabenzotriazol-1-yl)-N,N,N′,N′-tetramethyluronium hexafluorophosphate (HATU) and DIPEA in THF to afford **172**. Amide **172** underwent Suzuki–Miyaura coupling with arylboronic acids/aryl(pinacolato)boronate esters in the presence of Pd(dppf)Cl$_2$ and K$_2$CO$_3$ in dioxane to provide **173**. Access to the second subset of compounds, outlined in Scheme 24-II, arose from Suzuki coupling of 2-chloropyrimidine-4-carboxylic acid (**174**) with 2-fluoro-6-methoxyphenylboronic acid (**175**) in the presence of Pd(dppf)Cl$_2$ and K$_2$CO$_3$ in aqueous dioxane to give **176**. Finally, linkage of various anilines to **176** using HATU and DIPEA in THF furnished amides **177**.

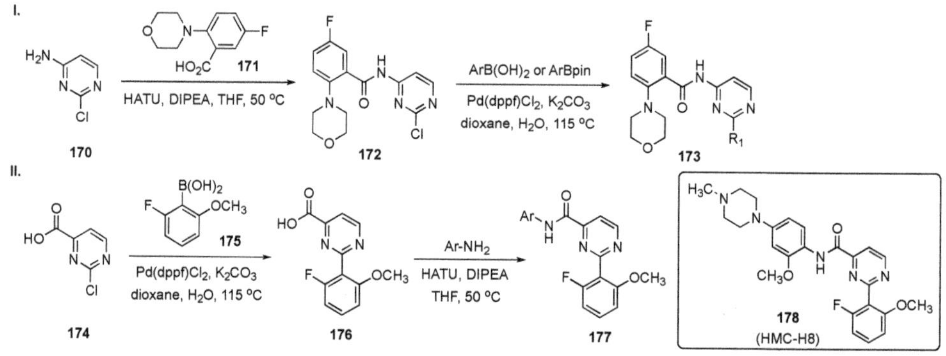

Scheme 24. Synthesis of pyrimidine HPK1 inhibitors **173** and **177**.

Upon screening, the synthetic 2,4-disubstituted pyrimidines proved to be powerful and selective HPK1 inhibitors. The most promising compound, **178** (HMC-H8), potently inhibited HPK1 with an IC$_{50}$ = 1.11 nM. The selectivity profile demonstrated that **178** exhibited good target differentiation and moderate preference against T-cell receptor-related targets such as lymphocyte-specific protein tyrosine kinase, germinal center kinase and protein kinase C-θ. In addition, the interleukin-2 (IL-2) and interferon-γ (IFN-γ) stimulation assay indicated that **178** actuated cytokine reproduction in a dose dependent manner. Notably, the reversal of immunosuppression evaluation revealed that **178** effectively restored IL-2 production, with up to 2.5 times greater increase in the IL-2 level over dimethyl sulfoxide (DMSO) treatment. The ADME properties for **178** demonstrated that the compound does not have significant CYP450 inhibition in human liver microsomes at 10 μM. The compound has low to moderate intrinsic clearance (CL$_{int}$) = 24.37 L/min/mg in a human liver microsomal stability assay. A single PK was conducted for compound **178** on Sprague–Dawley rats (190–200 g, n = 3 peer groups) with an i.v. of 1 mg/kg and a p.o. of 10 mg/kg, and the results are summarized in Table 6. Based on the data from the table, the compound appeared to have high clearance after both i.v. and p.o. administration. Finally, the compound has a very good C$_{max}$ and AUC with a bioavailability (F) of 15.05%.

Table 6. Pharmacokinetic properties of compound **178**.

Cpd No.	Administration	C$_{max}$ (ng/mL)	T$_{max}$ (h)	t$_{\frac{1}{2}}$ (h)	CL (L/h/kg)	AUC$_{0-t}$ (ng/mL·h)	F
178	i.v. (1 mg/kg)	634.68 ± 114.79	0.08 ± 0.01	0.52 ± 0.08	3922.19 ± 777.61	261.65 ± 51.94	n.a.
	p.o. (10 mg/kg)	141.69 ± 72.41	2.67 ± 1.15	0.93 ± 0.17	27,688.35 ± 8935.88	388.48 ± 149.14	15.05%

A patent by Ding et al. synthesized a class of kinesin family member 18A (KIF18A) inhibitors specifically to treat cancer [79]. Various kinases and kinesins are responsible for division in normal cells and cancer cells. The KIF18A gene belongs to the kinesin-8 subfamily and is a plus-end oriented motor. KIF18A is thought to affect the dynamics of the plus ends of centromere microtubules to control correct chromosome positioning and spindle tension. Depletion of human KIF18A in longer spindles increases chromosome oscillations in the metaphase of HeLa cervical cancer cells and activation of the mitotic spindle assembly checkpoint. KIF18A appears to be a viable target for cancer therapy. KIF18A has been overexpressed in various cancers, including, but not limited to, colon, breast, lung, pancreatic, prostate, bladder, head and neck, cervical, and ovarian cancers. Furthermore, in cancer cells, gene deletion, knockout, or KIF18A inhibition affects the mitotic spindle body device. Inhibition of KIF18A has been found to induce mitotic cell arrest, a known weakness that can be facilitated by apoptosis, mitotic catastrophe, or heterogeneously driven lethality following mitotic slippage in interphase mitotic cell death [80].

The preparation of drug candidates for this investigation is shown in Scheme 25. Initial S_NAr of 2-chloro-4-methyl-6-aminopyrimidine (**179**) with 4,4-difluoropiperidine (**180**) in the presence of DIPEA in 1-methyl-2-pyrrolidinone (NMP) at 140 °C produced adduct **181**. Derivative **181** was condensed with three different synthesized acids to give the final pyrimidinamide derivatives **182**. There were only three compounds reported in this patent.

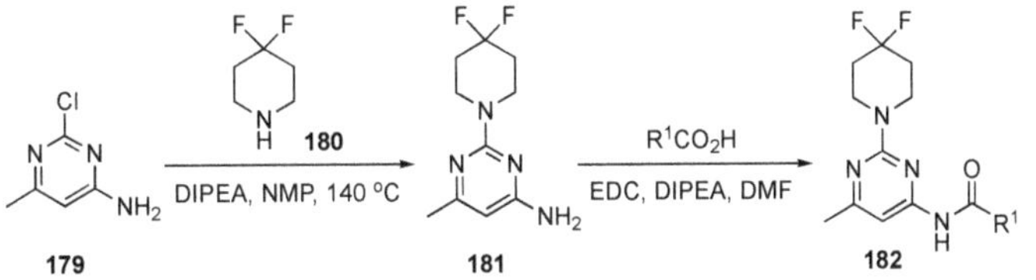

Scheme 25. Preparation of pyrimidine KIF18A inhibitors **182**.

The patent did not elaborate on the biological activity of the compounds but rather reported the IC_{50} values for the pyrimidinamide derivatives. The IC_{50} values for the enzymatic inhibition of KIF18A claimed for the three derivatives ranged from 27–120 nM. However, no specific data were presented.

A patent filed by Hergenrother and Kelly focused on metastatic melanoma, a cancer that readily spreads beyond its original location to other parts of the body [81]. This cancer results from genetic mutation and environmental factors. v-Raf murine sarcoma viral oncogene homolog B (BRAF) inhibitors are drugs that can shrink the growth of metastatic melanoma in patients whose tumors have a BRAF mutation. BRAF mutations are found in more than half of patients diagnosed with cutaneous melanoma. In BRAF-mutated melanoma, the BRAF kinase becomes hyperactivated, resulting in elevated cell proliferation and survival. The BRAF inhibitors Vemurafenib (**183**), Encorafenib (**184**) and Dabrafenib (**185**) are used in patients with BRAF-mutated melanoma. These inhibitors specifically target BRAF kinase and thus interfere with the mitogen-activated protein kinase signaling pathway that regulates the proliferation and survival of melanoma cells [82]. In this study, a new BRAF inhibitor, Everafenib-CO_2H (**187**), was envisioned by combining the structural features of **183–185** to reduce P-gp efflux propensity as well as to enhance brain penetration and activity in challenging intracranial mouse model melanoma (Figure 4).

The synthesis of **187** is illustrated in Scheme 26. In the first step, 3-amino-5-chloro-2-fluorobenzoic acid (**188**) was esterified to **189** using methanol and thionyl chloride. Ester **189** underwent reaction with propanesulfonyl chloride (**190**) using pyridine in DCM to afford the sulfonamide derivative **191** which was subsequently treated with the LiHMDS-derived anion of 2-chloro-4-methylpyrimidine (**192**) to provide **193**. Benzylic bromination of **193** and treatment with 2,2-dimethylpropanethioamide (**194**) resulted in cyclization to provide thiazole **195**. Compound **195** was then subjected to a S_NAr reaction with methyl 5-aminopentanoate hydrochloride (**196**) in the presence of DIPEA in N,N-dimethylacetamide

(DMA) under microwave irradiation to generate **197**. Finally, hydrolysis of **197** using LiOH furnished acid **187**.

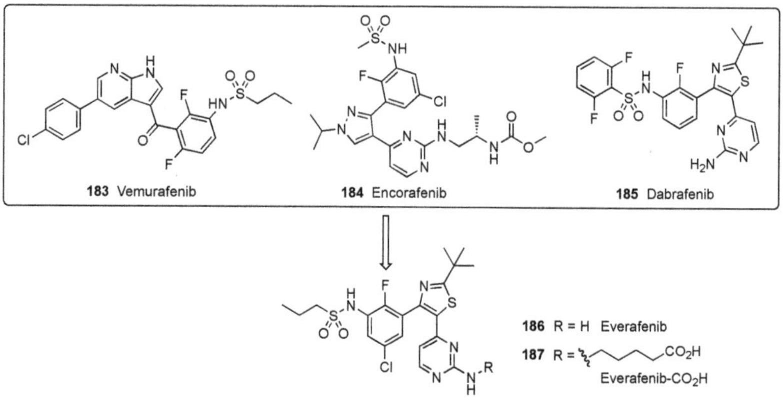

Figure 4. Design of Everafenib analogs.

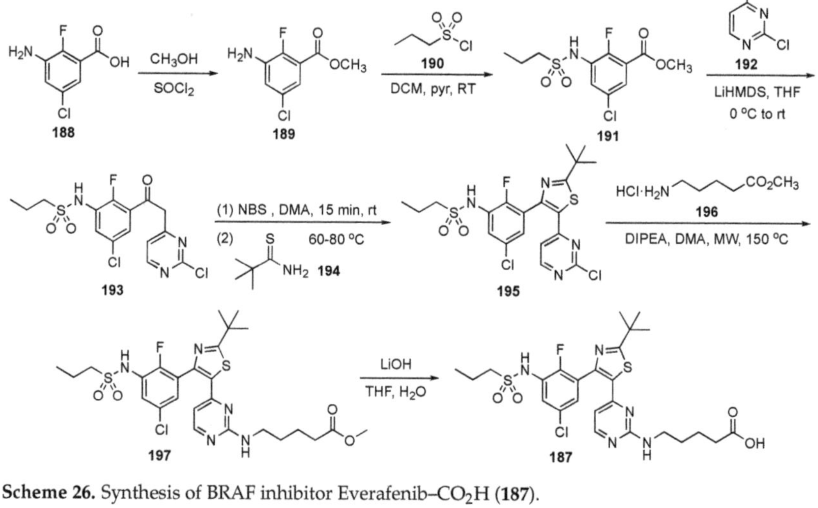

Scheme 26. Synthesis of BRAF inhibitor Everafenib–CO$_2$H (**187**).

The biological properties of Everafenib–CO$_2$H are summarized in Table 7. The compound displayed a similar potency in A375–human melanoma cells when compared with Dabrafenib. In cell permeability assays (Table 8), apparent permeability (P$_{app\ A-B}$) is like that of Dabrafenib, but P$_{app\ B-A}$ is lower, which leads to an improved efflux ratio of 1.17 ± 0.22.

Table 7. IC$_{50}$ values of Everafenib (**186**) and Everafenib–CO$_2$H (**187**) against BRAF-sensitive melanoma cell lines.

Melanoma Cell Line	Vemurafenib (183) IC$_{50}$ (nM)	Dabrafenib (185) IC$_{50}$ (nM)	Everafenib (186) IC$_{50}$ (nM)	Everafenib–CO$_2$H (187) IC$_{50}$ (nM)
A375	331 ± 93	7.46 ± 1.10	2.80 ± 0.53	27.3 ± 3.01
HT29	240 ± 61.7	4.74 ± 0.84	1.67 ± 0.36	21.4 ± 4.21
AM-38	344 ± 59.3	6.84 ± 0.84	8.50 ± 1.2	93.2 ± 14.1
3K-MEL-28	47% (10 μM)	4.76 ± 0.42	2.35 ± 0.13	22.5 ± 1.20

Table 8. Permeability and efflux ratios assessed in MDR1-MDCK transwell assay.

	Dabrafenib (185)	Everafenib (186)	Everafenib-CO$_2$H (187)
A375-melanoma cells	IC$_{50}$ (nM) = 7.46 ± 1.10	IC$_{50}$ (nM) = 2.80 ± 0.53	IC$_{50}$ (nM) = 27.3 ± 3.01
P$_{app\ A-B}$ (nm/s)	12.6 ± 6.72	49.2 ± 8.06	21.5 ± 0.70
P$_{app\ B-A}$ (nm/s)	132 ± 16.3	64.2 ± 24.5	24.9 ± 4.35
ER	17.9 ± 6.97	1.40 ± 0.65	1.17 ± 0.22

Work by De Vivo et al. highlighted the targeted cell division cycle GTPases (CDC42, RHOJ, and RHOQ), which are small guanosine triphosphate (GTP)-binding proteins that are known to regulate tumor growth, angiogenesis, metastasis, and cell resistance to targeted therapies [83]. CDC42 GTPases are essential molecular switches within the cell for which their active/inactive state depends on whether they are bound to GTP or guanosine diphosphate. When CDC42 GTPases are bound to GTP, the former change their structural conformation, allowing protein surface interactions that are complementary to their downstream effectors [84]. These include, but are not limited to, p21-activated protein kinases (PAKs). Notably, PAKs are known to be involved in invasion, migration, and oncogenic transformation. Many groups have sought to design small molecules that inhibit PAK kinases by targeting the large and flexible ATP binding pocket in the kinase domain or by targeting a large auto-inhibitory region that is observed in group I PAKs (PAK1, 2, and 3). However, the developed agents have failed to reach phase 2 due to their poor selectivity. For example, existing PAK inhibitors act on multiple isoforms of PAKs, including PAK2, which is thought to induce cardiotoxicity with a narrow therapeutic window. Potential modifications to GTPase inhibitors considered by the De Vivo team are summarized on the generalized structure in Figure 5.

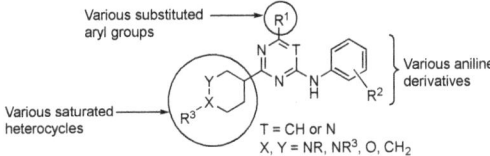

Figure 5. Possible structural modifications in the pyrimidine for GTP inhibition.

The synthetic plan for this work is delineated in Scheme 27. In Scheme 27-I, 2,4,6-trichloropyrimidine (**198**) underwent a Suzuki–Miyaura coupling with phenylboronic acid (**199**) under standard conditions to afford **200**. Subsequent S$_N$Ar reaction with various substituted anilines **201** in the presence of LiHMDS in THF at −60 °C provided intermediates **202**. Suzuki coupling of **202** with **203** gave cyclic alkene **204** which was hydrogenated in the presence of ammonium formate and Pd(OH)$_2$/C or triethylsilane with Pd/C to give **205**. When X = N-Boc, the amine **205** was deprotected with 4 M HCl in dioxane to yield **206**. In Scheme 27-II, amine **207** was condensed with isobutyric acid in the presence of HATU and DIPEA in DMF to afford amide **208**. Alkylation of **207** with 1-bromo-2-methoxyethane in the presence of DIPEA in ACN at 60 °C furnished ether **209**.

Based on the recent discovery of lead compound **210**, which showed anticancer activity in vivo, the authors expanded this new chemical class of CDC42/RHOJ inhibitors. Importantly, they identified and characterized two back-up compounds, namely, **211** and **212**, derived from a SAR study with ~30 close analogs bearing different substituents on the pyrimidine or triazine core. The most potent IC$_{50}$ values were observed from **211** against five different melanoma cell lines, including SKM28 (IC$_{50}$ = 6.1 µM), SKMeI3 (IC$_{50}$ = 4.6 µM), WM3248 (IC$_{50}$ = 9.3 µM), A375 (IC$_{50}$ = 5.1 µM), and SW480 (IC$_{50}$ = 5.9 µM). Compound **211** also had good kinetic solubility (168 µM), a t$_{1/2}$ > 120 min in plasma and acceptable microsomal stability (t$_{1/2}$ = 45 min). The PK profile for compound **211** is shown in Table 9.

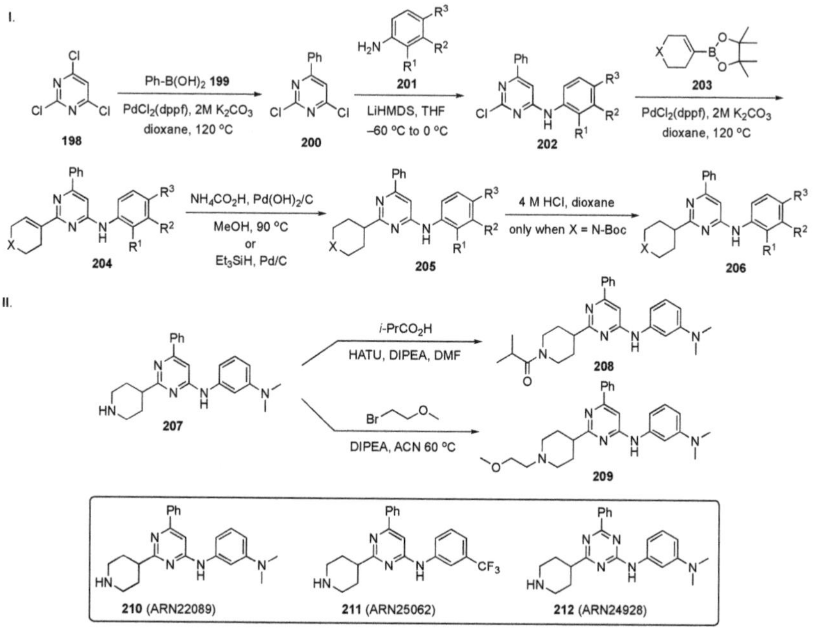

Scheme 27. Syntheses of common intermediate 206 and pyrimidine CDC42 GTPase inhibitors 208 and 209.

Table 9. Pharmacokinetic properties of compound 211.

Cpd No.	Administration	C_{max} (ng/mL)	$t_{\frac{1}{2}}$ (min)	V_D (L/kg)	CL (L/h/kg)	AUC_{0-t} (ng/mL·h)	F
211	i.v. (3 mg/kg)	1298	25	2.3	63	63	n.a.
	p.o. (10 mg/kg)	216	108	41.2	265	265	18%

Back-up compounds 211 and 212 have also displayed stable binding in the target pocket via molecular dynamics simulations and favorable PK profiles comparable to 210. Notably, the authors also measured the in vivo efficacy of the two lead compounds 211 and 212, with analog 211 exhibiting a significant ability to inhibit tumor growth in patient-derived xenografts in vivo, similar to lead compound 210.

Gray et al. have investigated inhibitors of the Hippo pathway, an important, evolutionarily conserved signaling cascade pathway with >30 components and which play a crucial role in organ size control, tissue homeostasis, stem cell renewal, cell proliferation, angiogenesis, and tumorigenesis [85]. Dysregulation of the Hippo pathway through merlin neurofibromin-2 loss, large tumor suppressor kinase 1 fusion, yes-associated protein (YAP) and transcriptional co-activator with PDZ-binding (TAZ) fusions, and YAP/TAZ amplification have been linked to the occurrence and progression of tumor malignancies in mesothelioma, meningioma, lung cancer, liver cancer, and other solid tumors [86]. Although the Hippo pathway has significant therapeutic potential, direct targeting of this cascade has been difficult. Thus, instead of directly targeting Hippo, the authors employed a reversible post-translational palmitoylation of the transcriptional enhanced associate domain (TEAD). Hyperactivation of TEAD–YAP/TAZ leads to human cancers and is associated with cancer cell proliferation, survival, and immune evasion. Therefore, targeting the TEAD–YAP/TAZ complex has emerged as an attractive therapeutic approach.

The synthesis of potential inhibitors for this work is shown in Scheme 28-I. S_NAr reaction of amino ether 213 with 2-chloropyrimidine (214) in the presence of DIPEA in butanol at 70 °C produced amino ethers **215**. These intermediates were subsequently Boc deprotected and reacted with acryloyl chloride and TEA or condensed with acrylic acid to give amides **216**. In Scheme 28-II, a 7-membered ring diamine **217** was reacted with **214** using DIPEA in DMSO at 90 °C to provide **218**. Intermediate **218** underwent the same amide formation with acryloyl chloride to afford target **219**.

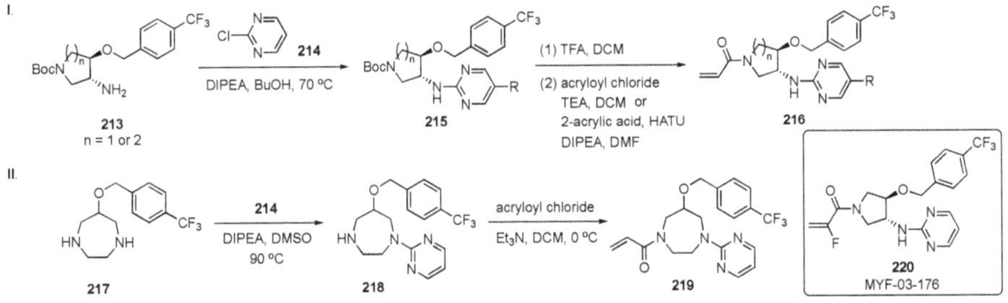

Scheme 28. Synthesis of pyrimidines **216** and **219** for inhibition of the Hippo pathway.

Time-resolved fluorescence energy transfer and TEAD reporter assays in this work demonstrated that the overall Y-shaped scaffold improved the potency of the compounds to an IC_{50} < 50 nM. The results suggest that selectivity could be achieved between TEAD isoforms due to modifications in different parts of the ring. Optimization of the chemistry on this series of compounds resulted in the development of a potent pan-TEAD inhibitor **220** (MYF-03-176). This structure exhibited potent inhibition of TEAD transcription with an IC_{50} = 17 ± 5 nM and significantly inhibited TEAD-regulated gene expression and proliferation of the cell lines with TEAD dependence, including those derived from mesothelioma and liposarcoma. Compound **220** also expressed the best antiproliferation activity on both the 94T778–liposarcoma (IC_{50} = 40 nM) and NCI-H226–squamous cell carcinoma (IC_{50} = 24 nM) cell lines.

A patent invention by Yu et al. sought to use small molecule pyrimidine derivatives to mitigate proliferative disorders caused by the expression of various kinases [87]. The required compounds would need to inhibit the growth of wild tumor beads with high kinase expression and tumor cell lines with corresponding kinase mutations. These proliferative disorders include rearranged transfection (RET), glial-derived neurotrophic factor (GDNF), platelet-derived growth factor receptor (PDGFR), and vascular endothelial growth factor receptor (VEGFR). Kinase-derived RET is a neuronal growth factor receptor tyrosine kinase and a transmembrane glycoprotein. The proto-oncogene, located on chromosome 10, is expressed during the embryonic stage, plays an important role in the development of the kidney and enteric nervous system, and is also critical in the homeostasis of neurons, neuroendocrine cells, hematopoietic tissue, and male germ cells [88]. These RET kinase inhibitors may find use in the treatment of cancer and gastrointestinal disorders. The growth of solid tumors is highly dependent on vascular proliferation, especially PDGFR and VEGFR. These are the main mediators of angiogenesis and act as two indicators of the angiogenic potential of human gliomas. Neurturin and persephin are ligands belonging to the GNDF family (GFLs). GFLs usually bind to the GDNF family receptor α (GFRa), and the formed GFL–GFRa complexes mediate the self-dimerization of RET proteins, causing trans-autophosphorylation of tyrosine in the intracellular domain. This complex also recruits related adapter proteins and activates the cascade reaction of signal transduction such as cell proliferation and related signaling pathways that include mitogen-activated protein kinase (MAPK), phosphoinositide 3-kinase (PI3K), Janus kinase signal transducer and activator of transcription (JAK-STAT), protein kinase A (PKA), and protein kinase C (PKC). Thus, the patent sought a small molecule inhibitor to block these kinases in order to restrict the proliferation of cancer cells.

A route to small molecule kinase inhibitors for this project is outlined in Scheme 29. Sandmeyer reaction of 2-amino-4-chloropyrimidine derivatives **221** using tert-butylnitrite and diiodomethane in ACN solvent afforded iodopyrimidines **222**. Intermediates **222** underwent a S$_N$Ar reaction with various amines and DIPEA to form **223**. Anilinic amines **223** were Boc-protected to give **224** and subjected to Heck coupling with methyl acrylate using Pd(OAc)$_2$ and TEA in ACN to give **225**. Acrylic esters **225** were hydrolyzed to acid **226** using LiOH and condensed with another set of amines using standard conditions to give amides **227**. Final Boc deprotection of **227** with TFA in DCM delivered the required drug candidates **228**.

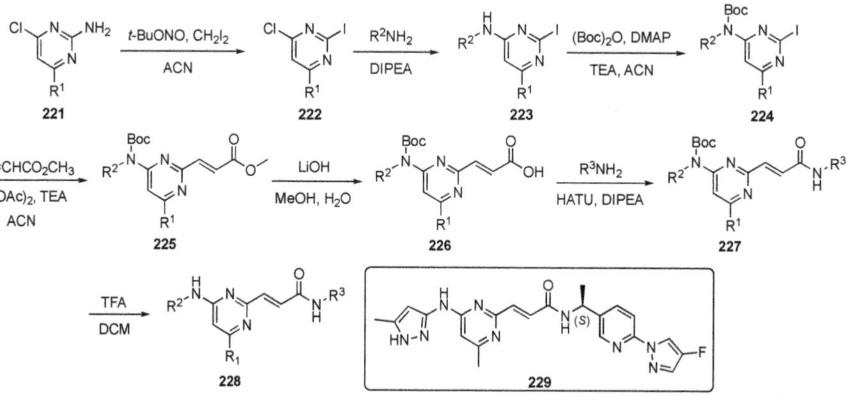

Scheme 29. Synthesis of pyrimidines **228** for proliferative disorders.

Compounds **228** were evaluated as inhibitors of WT and mutant RET kinases, namely RET-V804M and RET-V804L. Several compounds recorded potencies ranging from 1 nM–1 µM in these assays. Compound **229** exhibited the best activity on two cell lines, TT–human thyroid and KM12–human colon adenocarcinoma, with IC$_{50}$ values of 22 nM and 1.16 nM, respectively. Prototype **229** also showed inhibitory activity toward KIF5B-RET fusion (IC$_{50}$ = 22 nM) and CCDC6-RET fusion cell lines (no IC$_{50}$ given) which were developed to establish the antitumor activity. Most compounds tested on these cell lines showed IC$_{50}$ values between 20 nM and 1 µM. The antitumor activity was determined by a pharmacodynamic model of human cancer in BALB/c nude mice with a xenograft tumor derived from the TT cell line. Compound **229** gave tumor shrinkage of close to 80% at a dose of 40 mg/kg thereby exhibiting a robust anti-tumor effect. Finally, the mice showed no significant change in body weight which signified a good tolerance for the compound.

Chen et al. reported the synthesis of inhibitors toward salt-inducible kinases (SIKs), intracellular serine/threonine kinases which belong to the adenosine monophosphate activated (AMPK) superfamily [89]. The important role of SIKs is to act as molecular switches to regulate the transformation of macrophages (M1/M2) by phosphorylating CREB regulated transcription co-activator 3 (CRTC3) and to control its localization by activating the CRTC3 gene [90]. Some of the SIKs are involved in tumor cell resistance to cell-mediated immune responses and in resistance to tumor necrosis factor. In this work, the authors were focused on improving the ADME and pharmacokinetic properties of a known SIK inhibitor, HG-9-91-01 (see Supplementary Materials), which suffered from poor drug properties, including rapid clearance, low in vivo exposure, and high plasma protein binding [91]. To overcome these deficiencies, the authors sought to hybridize Dasatinib (see Supplementary Materials) and HG-9-91-01 to optimize the drug properties.

The syntheses of prospective drug molecules for this study are shown in Scheme 30. The starting 2,4-dichloropyrimidine ester **230** underwent C4 substitution with various alkyl amines in the presence of TEA in ACN to afford **231** which was subsequently hydrolyzed with LiOH in aqueous THF to form 232. Acids 232 were further transformed to their acid chlorides which reacted with substituted benzylamines and 2,6-dimethylaniline to provide amides 233 and 235, respectively. Finally, these compounds underwent C2 S_NAr reaction with a series of anilines in AcOH to furnish the desired derivatives 234 and 236.

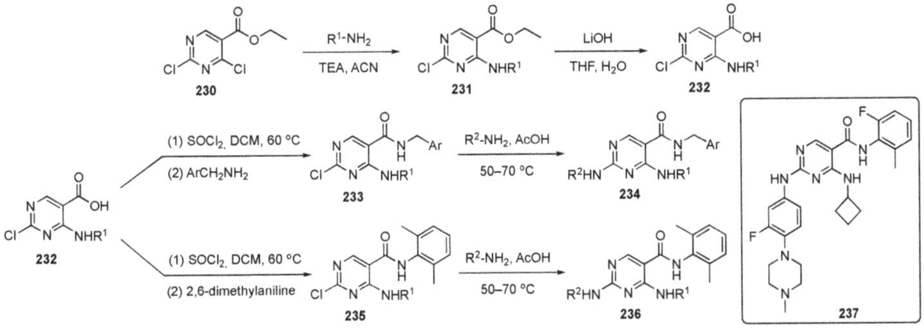

Scheme 30. Synthesis of pyrimidine SIK inhibitors 234 and 236.

Once these compounds were available, the pharmacokinetic profiles were evaluated. Each compound showed a modest improvement with a longer half-life, lower clearance and enhanced metabolic stability to human liver microsomes ($t_{1/2}$ = 120 min). The plasma protein binding for the most promising structure, 237, was ca. 79.4%, compared with >99% for HG-9-91-01. In addition to demonstrating good SIK inhibitory activity, 237 had medium selectivity among the subtypes of SIKs and exhibited excellent anti-inflammatory properties in a dextran sulfate sodium-induced colitis model. The in vitro anti-inflammation activity evaluation by cell-based phenotypes for 237 showed SIK inhibition via up-regulated IL-10 and reduced IL-12 at both the gene and protein level. The macrophage markers were also observed in LIGHT, SPHK1 and Arginase 1 proteins for the best compound synthesized. The PK profile of 237 is condensed in Table 10.

Table 10. Pharmacokinetic properties of compound 237.

Cpd No.	Administration	C_{max} (ng/mL)	$t_{\frac{1}{2}}$ (h)	CL (L/h/kg)	AUC_{0-t} (ng/mL·h)	F
237	i.v. (5 mg/kg)	11,959.8 ± 2015.9	0.1	0.2 ± 0.1	20,338.2 ± 8284.9	n.a.
	p.o. (10 mg/kg)	571.9 ± 101.1	8.0 ± 2.0	0.7 ± 0.3	9341.6 ± 1963.8	22.97%

A patent disclosure from Marseglia et al. also imagined inhibitors targeting SIK3 [92]. SIK3 is involved in tumor cell resistance to cell-mediated immune responses, specifically, tumor cell resistance to tumor necrosis factor. Recent reports have demonstrated that SIK3 expression regulates transforming growth factor-β mediated transcriptional activity and apoptosis [93].

The synthesis of a drug candidate for this study is shown in Scheme 31. S_NAr reaction of 4,6-dichloro-2-methylpyrimidine (17) by ethyl 2-aminothiazole-5-carboxylate (238) in the presence Cs_2CO_3 in DMF furnished aminothiazole-pyrimidine 239. A second S_NAr reaction of 239 with 4-methylpiperazine using DIPEA in butanol produced 240, which was saponified to acid 241. Amidification of 241 with 3-aminothiophene in the presence of chloro-N,N,N',N'-tetramethylformamidinium hexafluorophosphate (TCFH) and DIPEA in ACN generated amide 243.

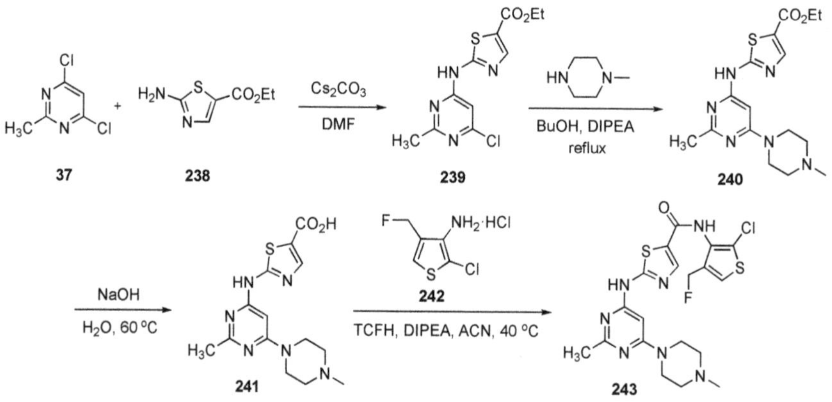

Scheme 31. Synthesis of pyrimidine inhibitor **243** targeting SIK3 and NF-κB.

Model compound **243** inhibited both SIK2 and SIK3 with IC$_{50}$ values of ca. 10 and 20 nM, respectively. The compound also exhibited significant inhibition of (1) nuclear factor kappa-light chain enhancer of activated B cells (NF-κB) in MC38–colon and EMT6–epithelial carcinoma cells and (2) histone deacetylase 4 phosphorylation. Agent **243** had a reasonable ADME profile, with a human liver microsomal stability of 131 µL/min/mg and a human hepatocyte stability of 58 µL/min/mg. Compound **243** also demonstrated good plasma exposure with 1% unbound and a recovery of 92% after 4 h. Moreover, **243** showed good permeability with Madin Darby canine kidney MDR1 cells with a P$_{app\ A>B}$ of ca. 5×10^{-6} cm/s. However, the efflux ratio was nearly 26, suggesting that the compound was an efflux substrate. Compound **243** presented a good PK profile, exhibiting low clearance with a good AUC following p.o. administration of 30 mg/kg (Table 11).

Table 11. Pharmacokinetic properties of compound **243**.

Cpd No. Administration	Plasma Conc.	C$_{max}$ (ng/mL)	T$_{max}$ (h)	t$_{\frac{1}{2}}$ (h)	CL (L/h/kg)	AUC$_{0-t}$ (ng/mL·h)
243 p.o. (30 mg/kg)	total free	2688.72 66.92	0.25 0.25	2.76 2.76	0.2 ± 0.1 0.7 ± 0.3	9584.81 238.75

Finally, treatment of established tumors (MC38–colon, EMT6–epithelial) in different syngeneic tumor mouse models with **243** resulted in significant tumor growth inhibition in a monotherapy protocol. Compound **243** showed tumor shrinkage of 74% with a 25 mg/kg twice daily dosing and a body weight increase of 16.8%, which was comparable or even superior to anti-programmed cell death-1 treatment alone. Furthermore, immune cell profiling of treated mice showed a significant infiltration of activated T cells, along with excellent reduction in immunosuppressive regulatory T cells and M2 tumor-associated microphages.

4. Pyrimidine-Based Drugs for Immunological Treatments

A patent by the Zhaoxing team investigated Janus kinase 2 (JAK2), which is an important pathogenic factor for various diseases [94]. Upon activation of the JAK kinases by receptor activators, these enzymes phosphorylate cytokine receptors and activate the signal transducers and activators of transcription (STAT) family [95]. In recent years, the therapeutic potential of JAK inhibitors has focused on diseases affecting various pathological conditions of the immune system, including atopy, cell-mediated hypersensitivity (allergic contact dermatitis, hypersensitivity pneumonitis), systemic lupus erythematosus, rheumatoid arthritis, psoriasis, transplantation (graft rejection, graft-versus-host disease), etc. [96]. Recently, erythropoietin-JAK2 signaling pathways have been implicated in myeloid proliferative disorders and proliferative diabetes mellitus, which are important in omental disease. Fedratinib (see Supplementary Materials), a JAK2 inhibitor, has been approved by the U.S. and is currently on the market. There are multiple deficiencies in this inhibitor, including its degree of selectivity for JAK2. The low selectivity, low bioavailability, and high toxicity limits its safe drug use in clinical practice. Therefore, it is necessary to develop a new more selective inhibitor of JAK2 that can overcome these shortcomings.

The synthesis of a prospective JAK2 inhibitor is shown in Scheme 32. Treatment of 2,4-dichloro-5-trifluoromethylpyrimidine (**105**) with aqueous ammonia in THF afforded the corresponding aminopyrimidine **244**. Intermediate **244** underwent Buchwald–Hartwig coupling with 4-bromobenzenesulfonamide **245** using Pd$_2$(dba)$_3$, xantphos, and Cs$_2$CO$_3$ in dioxane to afford the corresponding sulfonamide-pyrimidine derivative **246**. Derivative **246** was further subjected to a S$_N$Ar amination reaction with the 4-substituted aniline **247** to generate drug candidate **248**. Following conversion to various salts, this compound was evaluated for its biological significance against the JAK2 kinase.

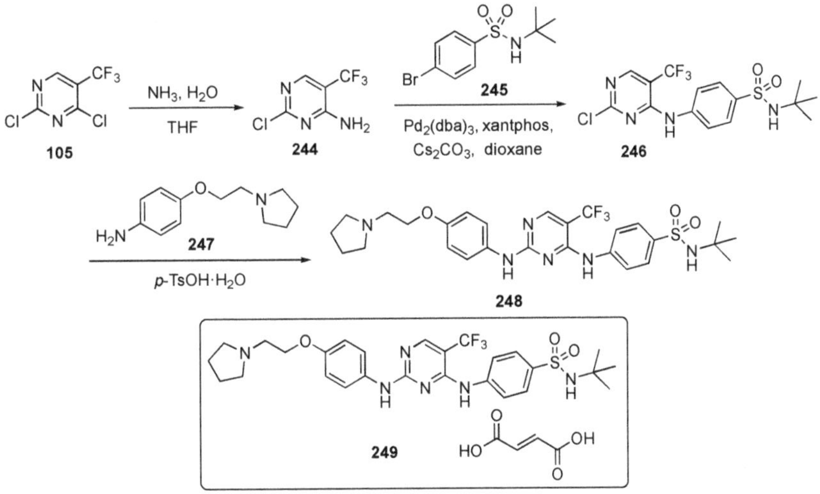

Scheme 32. Synthesis of pyrimidines as JAK2 inhibitor **248**.

Compound **248** or its salts had a 96-fold preference for JAK2 over JAK1, whereas the Fedratinib competitor showed only a 3-fold preference. Prototype **248** was also very selective for JAK2 kinase with an IC$_{50}$ = 5.86 nM and was much less active toward JAK3 (IC$_{50}$ = 538.5 nM) and tyrosine kinase 2 (TYK2, IC$_{50}$ = 700.4 nM).

A PK study was conducted for several of the compounds prepared, and the results for the most potent, **249** (the fumarate salt of **248**), when administered p.o. (10 mg/kg), are shown in Table 12. The compound showed a very low clearance, and the oral bioavailability (F) of the salt was 13.4%, whereas the Fedratinib competitor was 7.24%. In a guinea pig allergic conjunctivitis model, however, it was found that **249**, even at a 10-fold lower dose, elicited a better therapeutic effect than Fedratinib.

Table 12. Pharmacokinetic properties of compound **249**.

Cpd No. Administration	C$_{max}$ (ng/mL)	T$_{max}$ (h)	t$_{\frac{1}{2}}$ (h)	CL (L/h/kg)	AUC$_{0-t}$ (ng/mL·h)	F
249 p.o. (10 mg/kg)	66	1.00	3.08	1.3	513	13.4%

An investigation by Ellis and co-workers focused on developing a drug scaffold aimed at dry eye disease (DED) which affects more than 39 million adults in the US [97]. To date, only four therapies for this affliction have been approved by the FDA. Though the target and mechanism are not well defined, it is understood that there are underlying cytokine and receptor-mediated pathogenic inflammatory states conspiring to break T cell and B cell tolerance against self-antigens, resulting in an undesirable autoimmune response that leads to ocular surface inflammation and loss of tear film homeostasis [98]. The work in this article focused on the Janus family of intracellular tyrosine kinases. Since many of the cytokines signaling through the JAKs (IL-2 (JAK1/JAK3), IL-6 (JAK1/JAK2/TYK2), IL-12 (JAK2/TYK2), IL-23 (JAK2/TYK2), and IFN-γ (JAK1/JAK2)) are implicated in the immunoinflammatory pathogenesis and pathophysiology of DED [99], a new small molecule JAK inhibitor, which is potent and water-soluble, could represent an ideal drug for pharmacological intervention against this condition.

The synthesis of several potential anti-inflammatory agents is summarized in Scheme 33. In Scheme 33-I, 2,4-dichloro-5-methylpyrimidine (**250**) underwent a regioselective Suzuki–Miyaura cross-coupling with (*R*)-3-cyclopentyl-3-(4-(4,4,5,5-tetramethyl-1,3,2-dioxaborolan-2-yl)-1*H*-pyrazol-1-yl)propanenitrile (**251**) to provide **252** using $(Ph_3P)_4Pd$ and Na_2CO_3 in aqueous dioxane at 100 °C. Subsequent linking of **252** with 1-methyl-1*H*-pyrazol-4-amine (**253**) under Buchwald–Hartwig conditions using $(Ph_3P)_4Pd$, 2,2′-bis(diphenylphosphino)-1,1′-binaphthalene (BINAP) and K_2CO_3 in dioxane-*t*-butanol at 100 °C completed the synthesis of target **254**. In Scheme 33-II, a similar Suzuki–Miyaura coupling was carried out between **255** and **256** to afford **257**, which underwent *m*-chloroperoxybenzoic acid (*m*-CPBA) oxidation with sulfur to give sulfone **258**. Independent S_NAr displacement of the sulfone in **258** with two amines yielded products **259a** and **259b**. In Scheme 33-III, S_NAr displacement of the C4 chloride of 2,4-dichloro-5-methylpyrimidine (**250**) with tert-butyl 3,9-diazabicyclo[3.3.1]nonane-9-carboxylate (**260**) using DIPEA in DMF gave **261**. Intermediate **261** was then coupled with amine **253** under Buchwald–Hartwig conditions and Boc deprotected with ethanolic HCl to generate **262**. Final *N*-alkylation of **262** with either bromoacetonitrile or acrylonitrile in the presence of DIPEA delivered target **263**. In Scheme 33-IV, regioselective S_NAr displacement of the C4 chloride from **250** with 2-(azetidine-3-yl)acetonitrile hydrochloride (**264**) using DIPEA in DMF afforded **265**. Subsequent conjugate addition of 1-methylpiperazine to this intermediate in the presence of DBU yielded **266**. Finally, Buchwald–Hartwig amination of **266** with 2-methylisothiazol-3-amine (**267**) completed the synthesis of prototype **268**.

A pharmacophore-based SAR helped to identify a lead JAK inhibitor as an immunomodulating anti-inflammatory agent for topical ocular disposition. Compound **268**, which had a unique 3-aminoazetidine bridging scaffold, offered good JAK-STAT potency and excellent aqueous solubility. Overall, **268** displayed suitable, low single-digit nanomolar potency toward JAK2 (IC_{50} = 3.9 nM) and good on-target cellular potency (STAT3 (IL-6), IC_{50} = 162 nM), excellent aqueous solubility (24,904 µM), minimal off-target kinase activity (S(35) = 0.055), and no observable genotoxicity in the micronucleus (a biomarker for genotoxicity) assay at the highest concentration tested (3 × micronucleus concentration, ≥50 µM). The pharmacodynamics of **268** were evaluated on a murine model of allergic eye disease in vivo against inflammation-driven Meibomian gland dysfunction. A three-day study using 0.1% and 0.3% of **268** demonstrated marked improvement in clinical scores relative to the vehicle control. Notably, no statistical differences were observed at any timepoint between 0.3% **268** and 0.1% Dexamethasone (positive control, see Supplementary Materials)). This striking result strongly indicates the potential of **268** to affect a rapid, sustained, and robust anti-inflammatory response by inhibition of the JAK-STAT pathway. Thus, compound **268** was deemed a safe, fast-acting and well-tolerated noncorticosteroid eye drop to treat DED as well as other inflammatory ocular surface diseases.

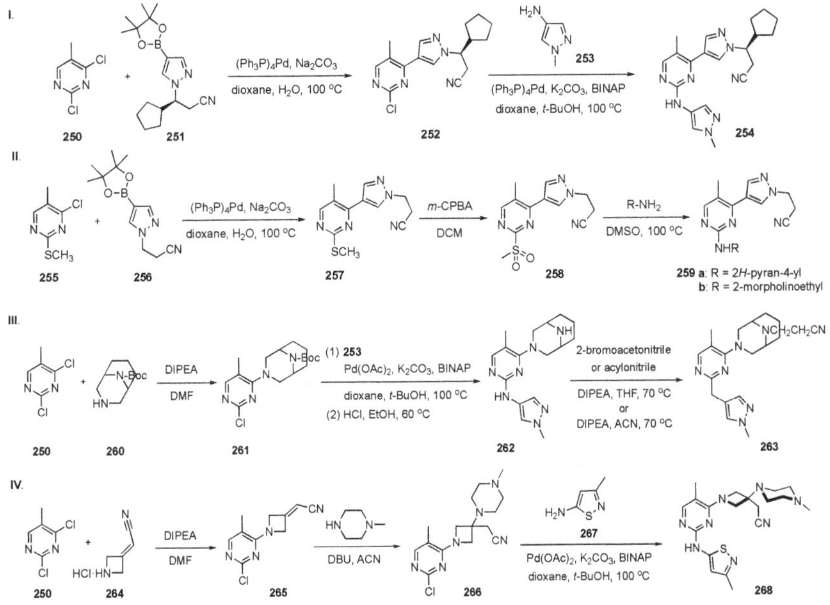

Scheme 33. Synthesis of agents **254**, **259**, **263** and **268** to inhibit JAKs.

A patent by Zhang et al. pursued the development of 2,4-diarylaminopyrimidine derivatives for treatment of inflammation [100]. The authors did not specify the mode of action or the target, though the work emphasized the use of lipopolysaccharides (LPS) to stimulate the inhibitory activity of human airway epithelial cells to release inflammatory cytokines. The LPS-induced human airway epithelial cell inflammation model was used to evaluate the anti-inflammatory activity of the synthesized derivatives [101].

The synthesis of potential anti-inflammatory candidates is summarized in Scheme 34. In this work, various aliphatic and aryl amines participated in a S_NAr reaction to displace the C4 chloride in 2,4-dichloropyrimidine (**52**) in the presence of TEA in ethanol or DIPEA in *t*-butanol to provide **269**. Subsequent reaction of derivatives **269** with various nitroaryl amines **270**, promoted by *p*-TsOH·H$_2$O in dioxane, gave nitroaryl pyrimidine-diamines **271**, which upon reduction, gave aminoaryl pyrimidinediamines **272**.

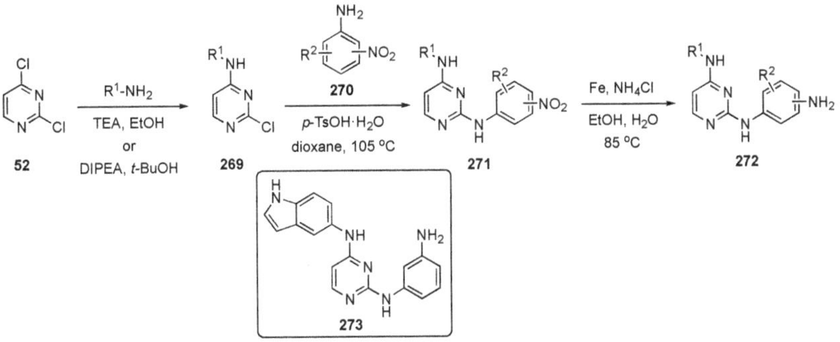

Scheme 34. Synthesis of pyrimidine anti-inflammatory agents **272**.

All of the prepared derivatives expressed good inhibitory effects on inflammatory cytokines IL-6 and IL-8. The effective compounds also had inhibitory effects on the release of IL-6 and IL-8 stimulated by LPS. The authors have asserted that all of the compounds in this series demonstrated excellent inflammatory response and showed high bioavailability. Among the series evaluated, the patent claimed that compound **273** exhibited the best anti-inflammatory activity at 5 µM concentration with inhibition of IL-6 and IL-8 reaching 66% and 71%, respectively. Several of the candidate compounds were endorsed as having high potential as future drugs.

5. Pyrimidine-Based Drugs for the Treatment of Neurological Disorders

A patent by the Defossa team focused on inflammatory responses to harmful stimuli, such as the invasion of pathogens and tissue damage [102]. Chronic inflammation is an important underlying factor in many human diseases, such as neurodegeneration, rheumatoid arthritis, autoimmune and inflammatory diseases, and cancer. Receptor-interacting protein kinase 1 (RIPK1, UniProtKB Q13546) is a key regulator of inflammation, apoptosis, and necroptosis. Receptor-interacting protein kinase 1 has an important role in modulating inflammatory responses mediated by NF-κB [103]. Dysregulation of RIPK1 signaling can lead to excessive inflammation or cell death, and, conversely, inhibition of RIPK1 can be an effective therapy for chronic neurodegenerative diseases involving inflammation or cell death. RIPK1 inhibition has been identified as a promising target for different diseases, like rheumatoid arthritis, psoriasis, multiple sclerosis, and Alzheimer's disease, and of inflammatory bowel diseases, such as Crohn's disease or ulcerative colitis [104]. Dihydropyrazoles and isoxazolidines as RIPK1 inhibitors are well known in phase II clinical trials for ulcerative colitis [105].

The syntheses of prospective agents to treat these conditions are shown in Scheme 35. Reaction of 4,6-dichloropyrimidine (**127**) with the 4-piperidinecarboxylic ester hydrochloride **274** in the presence of DIPEA in butanol afforded C4-aminated intermediate **275**. S_NAr reaction of **275** at C6 with imidazoles **276** and triazoles **278** using K_2CO_3 and Cs_2CO_3 in various solvents afforded **277** and **279**. Compounds **279** were further transformed by saponification of the 4-piperidinyl ester to give **280** and condensation with the 3-arylisoxazoline derivative **281** to give **282**. Incorporation of a pyrazole on the pyrimidine ring of **275** was achieved by Suzuki–Miyaura coupling with **283** to generate **284**. On the other hand, a methyltriazole was added by reaction of **275** with hydrazine hydrate in isopropanol, followed by condensation with N-(E)-[(dimethylamino)methylidene]acetamide (**285**) using catalytic p-TsOH·H_2O in ethanol to afford **286**.

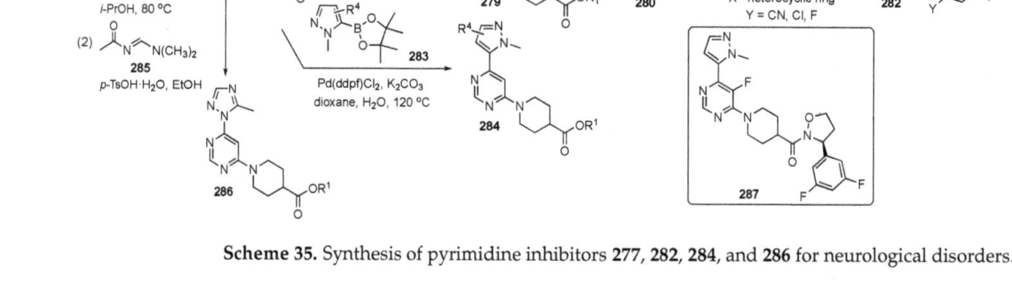

Scheme 35. Synthesis of pyrimidine inhibitors **277**, **282**, **284**, and **286** for neurological disorders.

The compounds were evaluated in the receptor-interacting protein kinase 1 (RIPK1) inhibition assay. The catalytic activity of RIPK1 was measured using an ADP-Glo Kinase Kit and the cellular assay was measured in U937–lymphoma cells for RIPK1 inhibition activity causing cell death. The patent spotlighted 20 compounds in both assays and the potency for most compounds ranged from 4–200 nM. Overall, compound **287** performed the best, having an IC_{50} = 9 nM in the RIPK1 kinase inhibition assay and an IC_{50} = 4 nM in the U937 cellular assay.

Hartz et al. sought to develop a drug for Alzheimer's disease, which is a neurodegenerative disorder characterized by memory loss and cognitive impairment [106]. As the disease progresses, it also causes deterioration in behavioral functions leading to communication problems, spatial disorientation, and changes in personality. More than 5.8 million Americans over the age of 65 currently live with Alzheimer's disease [107]. In this work, the authors targeted glucogen synthase kinase-3 (GSK-3) which is a proline-directed serine/threonine kinase that is widely distributed in the human body. GSK-3β is an important isoform of GSK-3 found in most areas of the brain. The most recent studies suggest that this kinase offers a therapeutic window wherein the GSK-3β inhibition that modulates key neuronal molecular targets can be achieved while avoiding mechanism-based β-catenin-driven effects [108]. As a result of its multifaceted role, GSK-3 has been linked to numerous conditions, including Alzheimer's disease, mood disorders and type 2 diabetes as well as cancer and myocardial disease.

The syntheses of candidate structures for this research are shown in Scheme 36. In Scheme 36-I, 2-aminopyrimidine-4-carboxylic acid (**288**) was reacted with a set of 3- aminopyridines **289** in the presence of HATU with DIPEA in DCM or DMF to afford amides **290**. One of the compounds prepared, **291**, was further subjected to a Buchwald–Hartwig coupling with bromobenzene (**292**) in the presence of $Pd_2(dba)_3$, xantphos, and Cs_2CO_3 in dioxane to provide targets **293**. Scheme 36-II employed the same reaction conditions as above, starting with 2-chloropyrimidine-4-carboxylic acid (**294**) along with a different subset of 3-aminopyridines **295**, to provide amides **296**. Compounds **296** underwent amination either by a Buchwald–Hartwig coupling or by S_NAr reaction in nMP at 150 °C to furnish prototypes **297**.

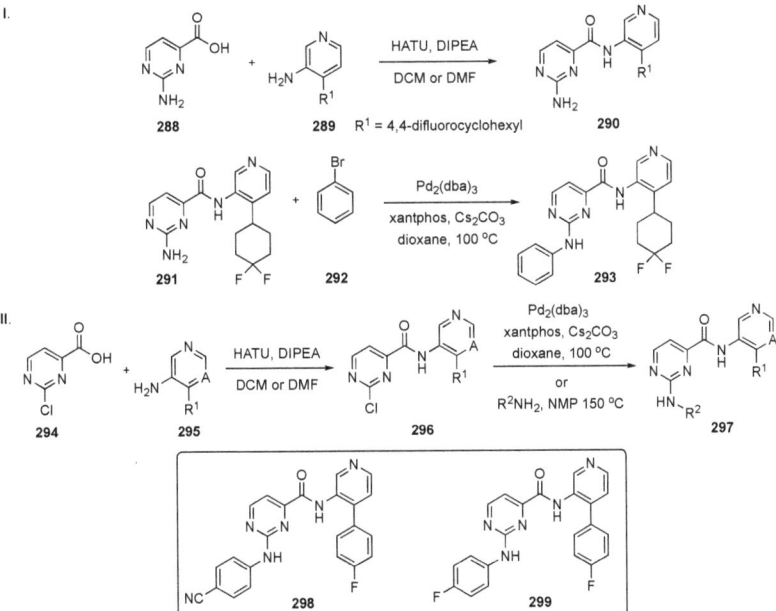

Scheme 36. Synthesis of pyrimidine analogs **293** and **297** targeting Alzheimer's disease.

In this investigation, two highly potent pyrimidine-based GSK-3 inhibitors were discovered. Amides **298** and **299** displayed potent IC_{50} values of 0.35 nM and 0.56 nM on GSK-3β and inhibition of p-tau with IC_{50} values of 10 nM and 34 nM, respectively. However, both compounds also exhibited respective IC_{50} values of 0.25 nM and 0.45 nM on GSK-3α, as this isoform has a 90% similarity to GSK-3β. Some of the ADME properties of **298** and **299** are summarized in Table 13. PAMPA and Caco-2 assay results show that these compounds are highly permeable to membranes. Kinase selectivity assessment in the Ambit panel of 412 kinases indicated that analogs with an aromatic group at the 4-position of the pyridine exhibited excellent kinase discrimination. Both **298** and **299** showed high selectivity against CDK2 and CDK5 in a standard panel of kinases.

Table 13. ADME properties of compound **298** and **299**.

Cpd No.	PAMPA P_c pH 5.5/7.4 (10^{-6} cm/s)	Caco-2 P_c A-B/B-A (10^{-6} cm/s)	PPB (% Unbound, H/R/M)	Brain Tissue (% Unbound, M)	cLogP	tPSA	CNS MPO Score
298	121/84	24.4/9.8	1.1/1.9/3.0	1.2	3.4	104	3.5
299	62.0/59.0	11.8/4.4	0.2/0.7/0.8	0.3	4.0	80	3.7

The PK properties of **298** and **299** are summarized in Table 14. Both compounds had excellent oral bioavailability in a triple-transgenic mouse Alzheimer's disease model with a good volume of distribution. The compound had low clearance and an average $t_{1/2}$. In vivo studies have demonstrated that these compounds were brain-penetrant GSK-3 inhibitors that significantly lowered tau phosphorylation. The results described herein may encourage further investigation of this class of GSK-3 inhibitors as a potential treatment for Alzheimer's disease.

Table 14. Pharmacokinetic properties of compound **298** and **299**.

Cpd No.	Administration	C_{max} (μg/mL)	V_{ss} (L/kg)	$t_{1/2}$ (h)	CL (mL/h/kg)	AUC$_{0-t}$ (μM·h)	F
298	i.v. (2 mg/kg)	—	5.4	2.5	23.3	—	n.a.
	p.o. (10 mg/kg)	3.5	—	—	—	24.3	100%
299	i.v. (2 mg/kg)	—	3.6	2.8	18.9	—	n.a.
	p.o. (10 mg/kg)	1.5	—	—	—	11.7	23%

A recent patent disclosure by Kumaravel et al. describes the utilization of pyrimidine-based compounds to treat a wide variety of neurological disorders attributable to transactive response DNA binding protein-43 (TDP-43), such as immunoreactive pathology, chronic traumatic encephalopathy, amyotrophic lateral sclerosis (ALS), Parkinson's disease, Alzheimer's disease, myofibrillar myopathy, sporadic inclusion body myositis, dementia pugilistica, chronic traumatic encephalopathy, Alexander disease, progressive supranuclear palsy, corticobasal degeneration, and frontotemporal lobar degeneration [109]. TDP-43 is an important nuclear DNA/RNA binding protein involved in RNA splicing. With pathological stress, TDP-43 translocates to the cytoplasm and aggregates into stress granules and related protein inclusions [110]. These phenotypes are known to degrade motor neurons and are found in 97% of all ALS cases. These TDP-43 mutations promote aggregation and are linked to a higher risk of developing ALS, suggesting that protein misfolding and aggregation act as drivers of toxicity [111]. In this invention, inhibitors of a Fyve-type zinc finger containing phosphoinositide kinase (PIKfyve) were prepared and studied in order to treat or prevent neurological disorders, such as the conditions enumerated above. The disclosure was based, in part, on the discovery that PIKfyve inhibition modulates TDP-43 aggregation in cells. Suppressing this aggregation exerts beneficial effects in patients suffering from neurological decline.

The synthetic work from this patent is outlined in Scheme 37 and involves the construction of the pyrimidine core appended with a morpholine. Reaction of morpholine-4-carboximidamide hydrochloride (**300**) with 2-methoxymalonate ester **301** in the presence of methanolic NaOMe provided **302**. Dihydroxypyrimidine **302** was reacted with POCl$_3$ to give the dichloride **303** which served as the starting material for a variety of analogs. Initially, **303** was subjected to Suzuki coupling with boronate ester **304** with Pd(dppf)Cl$_2$ and Cs$_2$CO$_3$ to generate **305**. Subsequently, **305** underwent methoxycarbonylation with CO and methanol in the presence of Pd(OAc)$_2$, Pd(dppf)Cl$_2$, and TEA to give ester **306**. This ester was converted to an acid which was condensed with amines using HATU to yield **307**. Chloropyrimidine **305** was also subjected to Buchwald–Hartwig coupling with a series of amines to afford targets **308**. A similar approach, in which intermediate **303** was reacted with various alkyl alcohols in the presence of NaH gave **309**. Final S$_N$Ar reaction with a series of amines resulted in **310**.

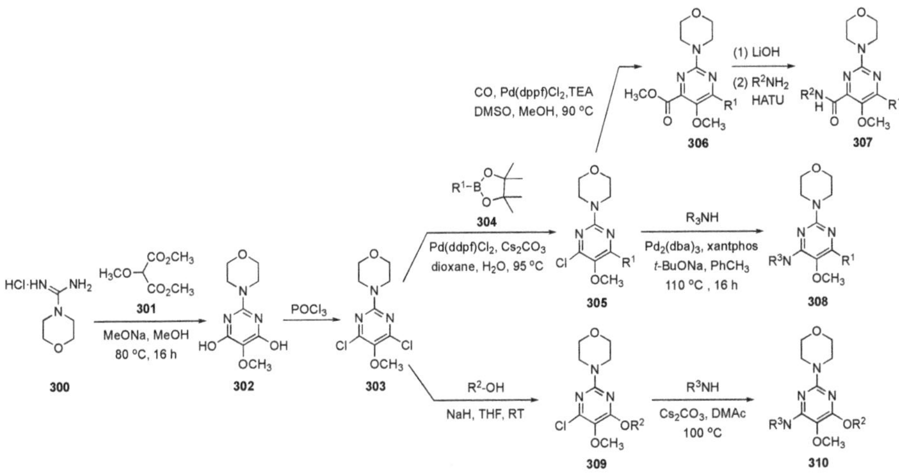

Scheme 37. Synthesis of pyrimidines **307**, **308** and **310** targeting TDP-43 binding protein.

Some of the analogs synthesized in this patent were evaluated in a biochemical inhibition assay for PIKfyve, and the results reveal more than 10 compounds exhibiting single-digit nM potency. The compounds were further evaluated in a PIKfyve early endosome antigen 1 assay in which the best compounds demonstrated activity in the 10–100 nM range. The compounds were further evaluated with ferrous amyloid buthionine 1 (FAB1) mouse and PIKfyve TDP-43 yeast models. Compounds that demonstrated low nanomolar potency in the biochemical assay were also active in the PIKfyve TDP-43 yeast model while structures that showed weak activity in the biochemical assay proved ineffective in the PIKfyve TDP-43 models.

An invention publication by Lei et al. explored two small molecule inhibitors of Parkinson's disease [112]. Previous research by others [113] has revealed that leucine-rich repeat kinase 2 (LRRK2) might prove to be a key link to understanding the etiology of Parkinson's disease. The LRRK2 mechanism functions by blocking the molecular chaperone mediation which induces autophagy and leads to the degradation of α-synuclei that results in toxicity. This chemical mechanism also induces mitochondrial damage and endolysosomal dysfunction, which paves the way for disease progression. LRRK2 kinase inhibitors can reduce damage in a Parkinson's disease model and can improve motor function in patients. Nitrogen-containing heterocycles, especially the pyrimidine core skeleton, are known to play an important role as LRRK2 inhibitors. This invention aimed to create a drug that could effectively inhibit LRRK2 as a potential treatment for neurodegenerative diseases. The patent identifies several drug scaffolds that are brain penetrating and incorporates these in effective inhibitors of LRRK2 as well as mutant LRRK2 that could be used as a treatment option for chronic neurodegenerative diseases.

In this patent, the requisite pyrimidine derivatives were assembled in two steps shown in Scheme 38. Reaction of 2,4-dichloro-5-(trifluoromethyl)pyrimidine (**105**) with methylamine hydrochloride in methanol afforded **311**. Intermediate **311** was then independently subjected to S_NAr reaction with aminopyrazole-piperidine derivatives **312** and **314** in the presence of 0.5 N HCl in *t*-butanol at 80 °C to produce candidate LRRK2 inhibitors **313** and **315**.

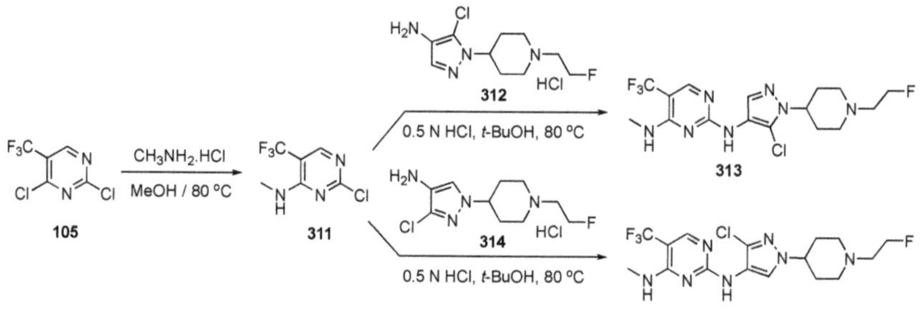

Scheme 38. Synthesis of compounds **313** and **315** as LRRK2 inhibitors.

While little about the biological properties of these compounds is revealed in this patent, the compounds synthesized are claimed to have IC_{50} values of ca. 13.0 nM compared with a known LRRK2 inhibitor which had an IC_{50} = 90 nM. Though the patent reports that **311** and **313** have better pharmacological profiles than those of known LRRK2 inhibitors, no data are included to support this claim.

A paper by Gelin et al. aimed to develop a potent brain penetrant and a selective glutamate receptor N2B (GluN2B) inhibitor [114]. Glutamate receptors serve an important function in neuronal activity, by regulating the brain's predominant excitatory neurotransmitter. Dysfunction of the glutamate receptor *N*-methyl-D-aspartate (NMDA) leads to many neurological and psychiatric disorders, including Alzheimer's disease, Parkinson's disease, neuropathic pain, stroke, brain trauma, schizophrenia, and depression [115]. It has been shown that modulation of nMDA with a ketamine antagonist results in very robust antidepressant activity.

The preparations of several pyrimidine–triazoles for this investigation are shown in Scheme 39. A flow chemistry method was used to convert various aniline derivatives to aryl azides that were cyclized with propargyl alcohol to produce 1,2,3-triazole alcohols **316**. Deprotonation of these alcohols with NaH in DMF and S_NAr etherification of 2-chloropyrimidine **317** delivered the target derivatives **318**.

The pyrimidine–triazole ethers were evaluated and shown to exhibit very favorable profiles, especially with respect to cardiovascular safety issues. Optimization of both the potency and metabolic characteristics of these model compounds was achieved by the introduction of a metabolic soft spot (a C4-methoxymethyl on the pyrimidine) to trigger metabolic switching. This design feature in the most active compound **319** precluded the formation of metabolites M1 and M2, a result of the loss of the pyrimidine moiety's ability to afford the triazole alcohol and acid, which in turn manifested in a saturable, nonlinear PK. Some of the PK parameters are condensed for **319** in Table 15.

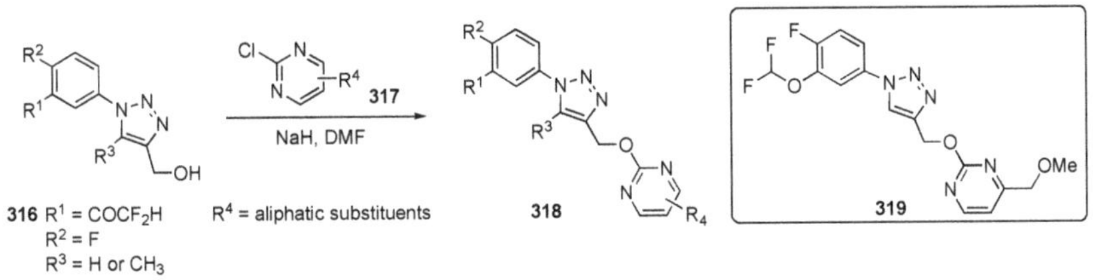

Scheme 39. Synthesis of pyrimidines 318 for glutamate receptor N2B.

Table 15. Pharmacokinetic profile for compound 319.

Cpd No.	Administration	C_{max} (µg/mL)	V_{ss} (L/kg)	CL (mL/h/kg)	F
319	i.v. (1 mg/kg)	—	0.4 ± 0.1	23 ± 3	n.a.
	p.o. (5 mg/kg)	1630 ± 1060	—	—	42 ± 43%

Diether 319 also showed very high aqueous solubility, and the compound was highly selective for GluN2B negative allosteric modulator (hGluN2A/C/D, $IC_{50} > 10$ µM) over other isoforms. Compound 319 did not have any hERG drawbacks or drug–drug interactions (over 10 µM) and the compound was not found to be a P-glycoprotein (P-gp) substrate. Candidate 319 also achieved 77% GluN2B receptor occupancy 0.5 h after a p.o. dose of 10 mg/kg, with excellent brain permeation (unbound partitioning coefficient, $K_{p,uu}$ = 0.65). Finally, the compound also had an efficacious plasma EC_{50} = 541 ng/mL and brain EC_{50} = 121 ng/mL.

A patent invention by Wagner, et al. sought to develop novel small molecule splicing modulators (SMSMs) for use in treating a variety of diseases, including neurodegenerative and repeat expansion diseases [116]. The disclosure primarily focused on the neurodegenerative disorder known as Huntington's disease by targeting the spliceosome. Currently, there is no cure for Huntington's disease or any way to mitigate its progression. Splicing is carried out by spliceosomes and is an essential process for generating distinct transcripts in different cells and tissue types during the developmental process [117]. Most cases of the disease are caused by mutation in the spliceosome, while others arise from mutations at the splicing sites, branchpoints, or by various splicing enhancers and silencers. Small molecules, such as RNA splicing modulators, are a recent area of exploration for identifying small molecule modulators with limited chemical series. Thus, there is a great need in this area for the discovery of SMSMs, due to the ability of small molecules to be effective delivery options with good bioavailability.

Access to potential SMSMs for this research is outlined in Scheme 40. Starting with thioether-substituted 4-chloropyrimidine ester 320, etherification by S_NAr displacement of the C4 chloride afforded 321. Compounds 321 were converted to 322 by oxidation of the thioethers to the sulfones with m-CPBA. The sulfone groups were displaced by various amines in the presence of K_2CO_3 in ACN to provide the 2-aminopyrimidine esters 323. Hydrolysis of the ester function in 321 with LiOH gave acids 324 which were converted to amide products 325.

The patent did not divulge any ADME or PK properties of the drug candidates synthesized but only claimed IC_{50} values < 500 nM for some of the promising compounds on the minigene reporter assay PMS1.

Another patent disclosure by Burli and Doyle promoted the invention of N-(4-aminocyclohexyl)pyrimidine-4-carboxamides as brain permeable cluster of differentiation 38 (CD38) inhibitors for treating disorders associated with CD38 activity [118]. Nicotinamide adenine dinucleotide (NAD+) is an essential cellular component in most living organisms and is responsible for redox functions. Though this is the primary role of NAD+ in most organisms, there are other functions for which NAD+ is important. An example of this is the necessity for NAD+ to be maintained to ensure long-term tissue homeostasis.

Due to aging, there is a decrease in NAD+ levels, which lowers metabolic function [119] and leads to debilitating conditions, such as Alzheimer's and Parkinson's disease. One way to stop the consumption of NAD+ is by inhibiting CD38, which has emerged as a valuable therapeutic approach for age-related disorders. CD38 is a multifunctional protein involved in (1) cellular NAD+ homeostasis via its hydrolase function and (2) the generation of second messengers such as adenosine diphosphate ribose (ADPR) and cyclic-ADPR. Several experiments using CD38 knockout mice have demonstrated the positive effects of CD38 deletion in models of neurodegeneration.

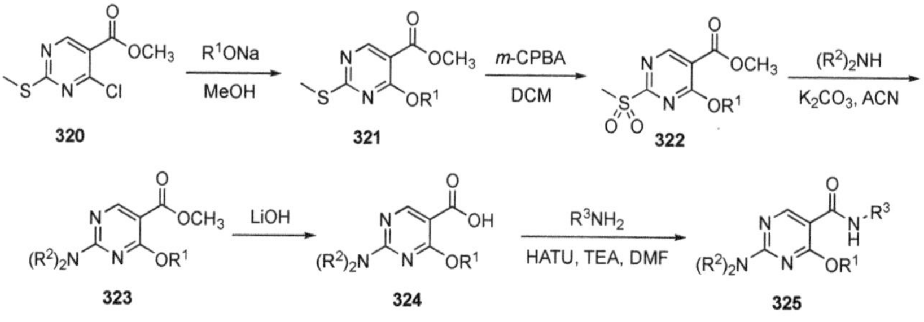

Scheme 40. Synthesis of pyrimidine-based SMSMs **325** to target Huntington's disease.

The synthesis of several potential pyrimidine CD38 inhibitors was accomplished in concise fashion as depicted in Scheme 41. Compound assembly was initiated by reacting 2-chloropyrimidine ester **326** with various five-membered nitrogen heterocycles **327** in the presence of DIPEA or Cs_2CO_3 in DMF at 100 °C to afford **328**. Hydrolysis of these heterocyclic pyrimidine esters **328** with LiOH in THF afforded acids **329**, which were reacted with various amines in the presence of TEA and propylphosphoric anhydride (T_3P) to deliver the required amides **330**.

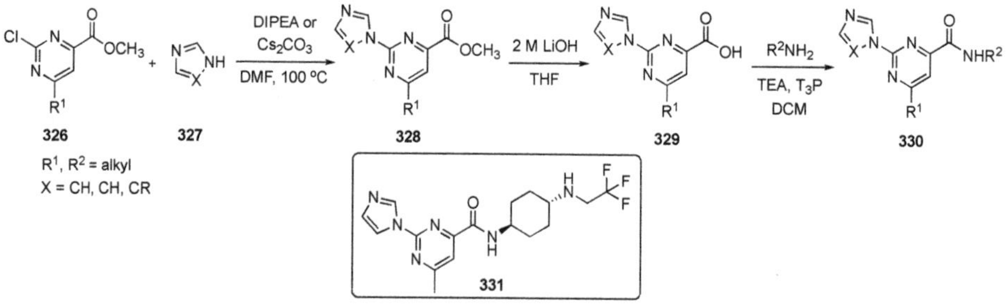

Scheme 41. Synthesis of heterocycle-linked pyrimidines ring as CD38 inhibitors **330**.

The analogs synthesized were evaluated for their CD38 hydrolase activity. Many of the compounds showed strong CD38 inhibition at ca. 40 nM or lower, including candidate **331**. In terms of pharmacokinetics, the tissue binding assay revealed a wide percentage range of unbound compound in mouse brain. Several compounds showed between 63–68% of unbound compound in mouse brain, with 12–22% of unbound compound in mouse plasma. A single PK study using 10 mg/kg p.o. was carried out to access the PK in brain permeability and revealed that **331** was the most promising derivative (see Table 16). The results demonstrate a robust 5512 nM concentration of **331** in the brain with a free brain concentration around 3801 nM and an unbound partitioning coefficient ($K_{p,uu}$) of around 1.33. The N-(4-aminocyclohexyl)pyrimidine-4-carboxamides displayed excellent brain permeability, whereas the corresponding cyclohexyl ethers or alcohols showed very low brain concentrations and lower $K_{p,uu}$ values.

Table 16. Brain PK study for compound 331.

Example	Total Plasma Conc. (nM)	Free Plasma Conc. (nM)	Total Brain Conc. (nM)	Free Brain Conc. (nM)	$K_{p,uu}$
331	21,941	2852	5512	3801	1.33

6. Pyrimidine-Based Drugs for the Treatment of Chronic Pain

A patent from Eli-Lilly focused on developing a potentiator for the human mas-related G-protein coupled receptor member X1 (hMRGXI) to address the issue of chronic pain [120]. This condition is often associated with older adults due to restricted mobility in daily activities. The major problem when treating chronic pain is due to dose-limiting adverse reactions, such as addiction, which is a problem for many analgesics currently available on the market. In this patent, several 2-aryloxy- and 2-arylthio-substituted pyrimidines that act as antagonists against the corticotropin releasing factor receptor were advanced to treat conditions such as depression, anxiety, drug addiction, and inflammatory disorders [121]. In this disclosure, certain (trifluoromethyl)pyrimidine-2-amines were identified as potentiators of hMRGXI that might prove to be a viable means to solve the issue of chronic pain.

The synthetic work from this patent is summarized in Scheme 42. Ethyl 4,4,4-trifluoro-3-oxobutanoate (332) reacted with sodium hydride and iodomethane-d_3 in methyl tert-butyl ether (MTBE) under reflux to afford 333. Derivative 333 further underwent cyclization with guanidine hydrochloride and sodium methoxide in methanol to provide pyrimidinol 334. Treatment of 335 with POCl$_3$ produced chloride 335 which reacted with phenol derivatives 336a-b using potassium phosphate (K$_3$PO$_4$) in DMA to deliver the desired ethers 337a-b.

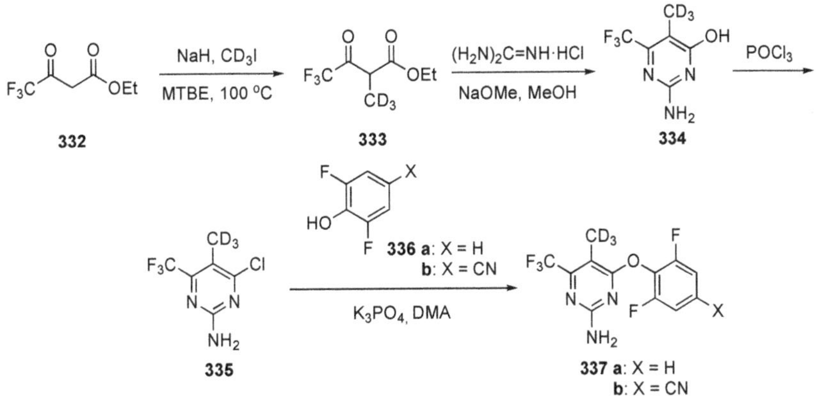

Scheme 42. Synthesis of deuterated pyrimidine derivatives 337a-b for chronic pain.

Some of the biological results for the most potent compounds are included in the patent [120]. Though the most potent compound was not disclosed, several derivatives showed an EC$_{50}$ between 20–30 nM against hMRGX1 inositol monophosphate. The most promising compound, 337b, had a very low CL$_{int}$ of ca. <1.80 and 6.47 μg/mL/min in mouse and human, respectively. Compound 337b exhibited a low clearance through IV in a mouse PK of around 7.1 ± 1.3 mL/min/kg with a volume distribution of 9.5 ± 3.1 L/kg and oral bioavailability (F) of 60 ± 2.6%. The low intrinsic clearance with high oral exposure of 337b would allow for lower dose quantity/frequency while achieving therapeutic levels of target engagement. Compound 337b also exhibited a very high total brain concentration of C$_{total\ brain}$ = 64,100 ± 22,700 nM with the K$_{p,uu}$ around 0.568 ± 0.165 for a 100 mg/kg single oral dose. The K$_{p,uu}$ was indicative of good penetration into the CNS, suggesting that an active transport mechanism was not operative in mouse brain tissue.

7. Pyrimidine-Based Drugs for the Treatment of Diabetes Mellitus

A publication by Alam et al. evaluated pyrimidine as a core ring for agents to treat diabetes mellitus [122]. Diabetes mellitus is a metabolic disorder which is caused by hyperglycemia due to insufficient secretion of insulin, or resistance to insulin, or both. The number of people affected by diabetes will reach around 643 million by 2030 and 783 million by 2045, an increase of over 8% per year [123]. In this work, researchers tried to couple thiazolidinedione rings with a pyrimidine derivative for insulin resistance in peroxisome proliferator-activated receptor-γ (PPAR-γ) peripheral tissues. PPAR-γ tissue enhances insulin formation and displays prominent antihyperglycemic activity without causing hypoglycemia [124]. The team resorted to an in-silico modelling method for docking the core, and then designed, synthesized, and evaluated the biological activity of each compound.

Assembly of the pyrimidines of interest is outlined in Scheme 43. The first step involved a one-pot Biginelli reaction using a series of *p*-substituted aryl aldehydes **338**, ethyl cyanoacetate and thiourea in the presence of K_2CO_3 in ethanol to form 1,6-dihydro-2-mercapto-6-oxopyrimidine-5-carbonitriles **339**. Intermediates **339** underwent S-alkylation with isopropyl bromide and NaOH in methanol to provide **340**. Chlorination of **340** with $POCl_3$ produced chloropyrimidine **341** which underwent S_NAr etherification with phenol-substituted thiazolidinone derivatives **342** in the presence Cs_2CO_3 to form drug candidates **343**.

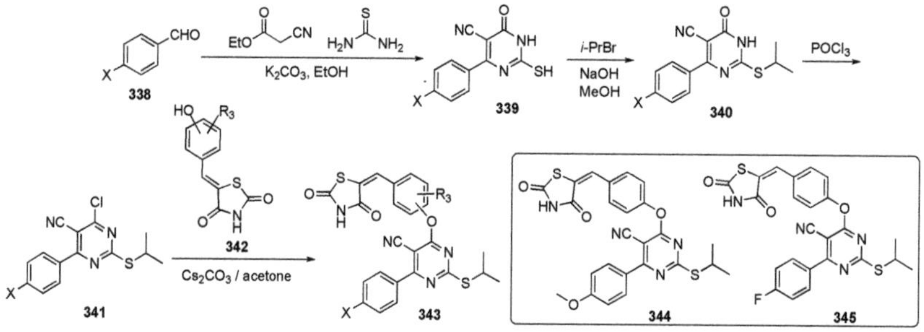

Scheme 43. Preparation pyrimidines **343** to treat diabetes mellitus.

The synthesis yielded 13 derivatives using different benzaldehydes and all were evaluated for biological activity. Screening procedures identified two compounds, **344** and **345**, which demonstrated very good oral glucose tolerance test results in vivo using streptozotocin-induced diabetic rats for 28 days, and they both reduced blood glucose levels significantly. The compounds caused a significant ($p < 0.0001$) decrease in blood glucose levels compared with the standard drug Pioglitazone (see Supplementary Materials). Compounds **344** and **345** decreased the blood glucose levels to 145.2 ± 1.35 and 146.6 ± 0.81, respectively, compared with Pioglitazone (150.2 ± 1.06). The compounds also showed a significant ($p < 0.0001$) decrease in triglycerides, total cholesterol and low-density lipoprotein cholesterol and an increase of high-density lipoprotein cholesterol. The biochemical estimations of hepatotoxicity using alanine transaminase, aspartate transaminase, and alkaline phosphatase, along with urea, creatinine, blood urea nitrogen, total protein, and lactate dehydrogenase, indicated that the levels were restored to normal by **344** and **345** in treatment groups compared with a diabetic control group. Histopathological investigations revealed a normal architecture of the pancreas, liver, heart, and kidneys following administration of **344**. Finally, compounds **344** and **345** did not show any toxicity to mice or cause an increase in body mass.

8. Addendum

While this manuscript was being written, several reviews appeared with content overlapping the material in this paper. The first, by the Farghaly team [125], presented an excellent survey of the patent literature from 1980–2021 as it pertains to pyrimidines as antiviral compounds. A second review, by the Roh group [126] outlined recent pyrimidine derivatives developed as antitubercular agents. Finally, Saleem et al. [127], provided a summary on pyrimidine-based drugs as antibacterials. These contributions are more detailed and comprehensive treatments of three of the topics covered in the current review which is limited to compounds studied during the past 2–3 years of developmental work.

9. Conclusions

Molecular dynamics modelling and a growing body of knowledge have revealed numerous new structures that can interact with key enzymes important to the etiology of many debilitating conditions. With the plethora of precursors available, pyrimidine drugs predicted to bind with these enzymes should be readily accessible for screening. Many of the pyrimidines cited have shown IC_{50} values in the nM range, exhibited favorable ADME properties, and demonstrated compelling pharmacokinetic/pharmacodynamic readouts to become successful new drug candidates for various health conditions. The greater potency of these agents should translate to lower doses, causing fewer side effects relative to current pharmaceuticals. In addition to their high potency, pyrimidine-based drugs often avoid off-target toxicity, including hERG, ion-channels, CYP450 inhibition and induction, and cytotoxicity toward normal cell lines, thus establishing a safety window for their use. Indeed, many of the most potent prototype compounds showed low toxicity in test animals and this would hopefully extend to humans. Additionally, several of the candidate molecules discussed in this article presented evidence that pyrimidines hybridized with other ring systems were often competent at overcoming the increasing problems associated with drug resistance. Though pyrimidines were used as anti-infective and anticancer agents in the past, the recent discoveries with this ring have broken boundaries and extended its functional role against conditions which are difficult to treat, including neurological disorders such as Alzheimer's disease, Parkinson's disease and chronic pain.

The research summarized in this review should convince the reader of the high potential of targets built around the pyrimidine core ring structure. Overall, pyrimidine-based drugs and hybrid structures appear to be some of the most promising drug candidates among new medicinal agents on the horizon. While pyrimidines are already an important substructure within many current therapeutic agents, it is likely they will only gain increasing importance as new medications become necessary to maintain a healthy global society.

Supplementary Materials: The following supporting information can be downloaded at: https://www.mdpi.com/article/10.3390/ph17010104/s1, Comparison drug structures mentioned in the text are pictured in the Supplementary Materials. Many of these compounds are not pyrimidines.

Author Contributions: Conceptualization, B.N.; formal analysis, B.N.; writing—original draft preparation, writing—review and editing, B.N. and R.A.B. All authors have read and agreed to the published version of the manuscript.

Funding: This research received no external funding.

Institutional Review Board Statement: Not applicable.

Informed Consent Statement: Not applicable.

Data Availability Statement: No new data were created or analyzed in this review article. Data sharing is not applicable to this review.

Acknowledgments: The authors wish to thank the National Institutes of Health for their support over the years.

Conflicts of Interest: Baskar Nammalwar is employed by the company Vividion Therapeutics. The remaining author declares that the research was conducted in the absence of any commercial or financial relationships that could be construed as a potential conflict of interest.

References

1. Kumar, S.; Narasimhan, B. Therapeutic potential of heterocyclic pyrimidine scaffolds. *Chem. Cent. J.* **2018**, *12*, 1–29. [CrossRef] [PubMed]
2. Nadar, S.; Khan, T. Pyrimidine: An elite heterocyclic leitmotif in drug discovery and biological activity. *Chem. Biol. Drug. Des.* **2022**, *100*, 818–842. [CrossRef] [PubMed]
3. Rani, J.; Kumar, S.; Saini, M.; Mundlia, J.; Verma, P.K. Biological potential of pyrimidine derivatives in a new era. *Res. Chem. Intermed.* **2016**, *42*, 6777–6804. [CrossRef]
4. Cocco, M.T.; Congiu, C.; Onnis, V.; Piras, R. Synthesis and antitumor evaluation of 6-thioxo, 6-oxo- and 2,4-dioxopyrimidine derivatives. *Farmaco* **2001**, *56*, 741–748. [CrossRef] [PubMed]
5. Meneghesso, S.; Vanderlinden, E.; Stevaert, A.; McGuigan, C.; Balzarini, J.; Naesens, L. Synthesis and biological evaluation of pyrimidine nucleoside monophosphate prodrugs targeted against influenza virus. *Antivir Res.* **2012**, *94*, 35–43. [CrossRef] [PubMed]
6. Anupama, B.; Dinda, S.C.; Prasad, Y.R.; Rao, A.V. Synthesis and antimicrobial activity of some new 2,4,6-trisubstituted pyrimidines. *Int. J. Res. Pharm. Chem.* **2012**, *2*, 231–236.
7. Bhalgat, C.M.; Ali, M.I.; Ramesh, B.; Ramu, G. Novel pyrimidine and its triazole fused derivatives: Synthesis and investigation of antioxidant and anti-inflammatory activity. *Arab. J. Chem.* **2014**, *7*, 986–993. [CrossRef]
8. Kumar, D.; Khan, S.I.; Tekwani, B.L.; Diwan, P.P.; Rawat, S. 4-Aminoquinoline-pyrimidine hybrids: Synthesis, antimalarial activity, heme binding and docking studies. *Eur. J. Med. Chem.* **2015**, *89*, 490–502. [CrossRef]
9. Mallikarjunaswamy, C.; Mallesha, L.; Bhadregowda, D.G.; Pinto, P. Studies on synthesis of pyrimidine derivatives and their antimicrobial activity. *Arab. J. Chem.* **2017**, *10*, S484–S490. [CrossRef]
10. Rodrigues, A.L.S.; Rosa, J.M.; Gadotti, V.M.; Goulart, E.C.; Santos, M.M.; Silva, A.V.; Sehnem, B.; Rosa, L.S.; Goncalves, R.M.; Correa, R.; et al. Antidepressant-like and antinociceptive-like actions of 4-(4′-chlorophenyl)-6-(4″-methylphenyl)-2-hydrazinepyrimidine Mannich base in mice. *Pharmacol. Biochem. Behav.* **2005**, *82*, 156–162. [CrossRef]
11. Tani, J.; Yamada, Y.; Oine, T.; Ochiai, T.; Ishida, R.; Inoue, I. Studies on biologically active halogenated compounds. 1. Synthesis and central nervous system depressant activity of 2-(fluoromethyl)-3-aryl-4(3H)-quinazolinone derivatives. *J. Med. Chem.* **1979**, *22*, 95–99. [CrossRef]
12. Li, C.; Tian, X.; Huang, Z.; Gou, X.; Yusuf, B.; Li, C.; Gao, Y.; Liu, S.; Wang, Y.; Yang, T.; et al. Structure activity relationship of novel pyrimidine derivatives with potent inhibitory activities against Mycobacterium tuberculosis. *J. Med. Chem.* **2023**, *66*, 2699–2716. [CrossRef] [PubMed]
13. Li, C.; Tang, Y.; Sang, Z.; Yang, Y.; Gao, Y.; Yang, T.; Fang, C.; Zhang, T.; Luo, Y. Discovery of napabucasin derivatives for the treatment of tuberculosis. *MedChemComm* **2019**, *10*, 1635–1640. [CrossRef]
14. Khalifa, A.; Khalil, A.; Abdel-Aziz, M.M.; Albohy, A.; Mohamady, S. Isatin-pyrimidine hybrid derivatives as enoyl acyl carrier protein reductase (InhA) inhibitors against Mycobacterium tuberculosis. *Bioorg. Chem.* **2023**, *138*, 106591–106602. [CrossRef] [PubMed]
15. Companico, A.; Moreira, R.; Lopes, F. Drug discovery in tuberculosis. New drug targets and antimycobacterial agents. *Eur. J. Med. Chem.* **2018**, *150*, 525–545. [CrossRef]
16. Seung, K.J.; Keshavjee, S.; Rich, M.L. Multidurg-resistant tuberculosis and extensively drug-resistant tuberculosis. *Cold Spring Harb. Perspect. Med.* **2015**, *5*, a017863-a. [CrossRef] [PubMed]
17. Wang, X.; Jin, B.; Han, Y.; Wang, T.; Sheng, Z.; Tao, Y.; Yang, H. Optimization and antibacterial evaluation of novel 3-(5-fluoropyridine-3-yl)-2-oxazolidinone derivatives containing a pyrimidine substituted piperazine. *Molecules* **2023**, *28*, 4267. [CrossRef]
18. Swaney, S.M.; Aoki, H.; Ganoza, M.C.; Shinabarger, D.L. The oxazolidinone linezolid inhibits initation of protein synthesis in bacteria. *Antimicrob. Agents Chemother.* **1998**, *42*, 3251–3255. [CrossRef]
19. Tao, Y.; Chen, J.X.; Fu, Y.; Chen, K.; Luo, Y. Exploratory process development and kilogram-scale synthesis of a novel oxazolidinone antibacterial candidate. *Organ. Process Res. Dev.* **2014**, *18*, 511–519. [CrossRef]
20. Patel, K.B.; Rajani, D.; Ahmad, I.; Patel, H.; Kumari, P. Chrysin based pyrimidine-piperazine hybrids: Design, synthesis, in vitro antimicrobial and in silico E. coli topoisomerase II DNA gyrase efficacy. *Mol. Div.* **2023**, 1–6. [CrossRef]
21. Champoux, J.J. DNA topoisomerases: Structure, function and mechanism. *Annu. Rev. Biochem.* **2001**, *70*, 369–413. [CrossRef] [PubMed]
22. Mayer, C.; Janin, Y.L. Non-quinolone inhibitors of bacterial type IIA topoisomerases: A feat of bioisoterism. *Chem. Rev.* **2014**, *114*, 2313–2342. [CrossRef] [PubMed]
23. Li, Z.; Meng, F.; Yu, Z.; Wei, W.; Ren, J. Preparation of Pyrimidine Containing Sulfonylurea Compounds with Antibacterial Activity. Patent CN116023338A, 28 April 2023.
24. Zhao, F.; Zhang, H.; Xie, M.; Meng, B.; Liu, N.; Dun, C.; Qin, Y.; Gao, S.; Clercq, E.D.; Pannecouque, C.; et al. Potent HIV-1 non-nucleoside reverse transcriptase inhibitors: Exploiting the tolerant regions of the non-nucleoside reverse transcriptase inhibitors binding pocket. *J. Med. Chem.* **2023**, *66*, 2102–2115. [CrossRef] [PubMed]
25. Cilento, M.E.; Kirby, K.A.; Sarafianos, S.G. Avoiding drug resistance in HIV reverse transcriptase. *Chem. Rev.* **2021**, *121*, 3271–3296. [CrossRef]

26. Xu, H.T.; Asahchop, E.L.; Oliveira, M.; Quashie, P.K.; Quan, Y.; Brenner, B.G.; Wainberg, M.A. Compensation by the E138K mutation in HIV-1 reverse transcriptase for deficits in viral replication capacity and enzyme processivity associated with the M184I/V mutations. *J. Virol.* **2011**, *85*, 11300–11308. [CrossRef] [PubMed]
27. Kang, D.; Ruiz, F.X.; Sun, Y.; Feng, D.; Jing, L.; Wang, Z.; Zhang, T.; Gao, S.; Sun, L.; De Clercq, E.; et al. 2,4,5-Trisubsituted pyrimidines as potent HIV-1 NNRTIs: Rational design, synthesis, activity evaluation, and crystallographic studies. *J. Med. Chem.* **2021**, *64*, 4239–4256. [CrossRef]
28. Jiang, X.; Huang, B.; Rumrill, S.; Pople, D.; Zalloum, W.A.; Kang, D.; Zhao, F.; Ji, X.; Gao, Z.; Hu, L.; et al. Discovery of diarylpyrimidine derivatives bearing piperazine sulfonyl as potent HIV-1 nonucleoside reverse transcriptase inhibitors. *Commun. Chem.* **2023**, *6*, 83. [CrossRef]
29. Bec, G.; Meyer, B.; Gerard, M.-A.; Steger, J.; Fauster, K.; Wolff, P.; Bernouf, D.; Micura, R.; Dumas, P.; Ennifar, E. Thermodynamics of HIV-reverse transcriptase in action elucidates the mechanism of action of non-nucleoside inhibitors. *J. Am. Chem. Soc.* **2013**, *135*, 9743–9752. [CrossRef]
30. Wang, S.; Ying, Z.; Huang, Y.; Li, Y.; Hu, M.; Kang, K.; Want, H.; Shao, J.; Wu, G.; Yu, Y.; et al. Synthesis and structure-activity optimization of 7-azaindoles containing aza-β-amino acids targeting the influenza PB2 subunit. *Eur. J. Med. Chem.* **2023**, *250*, 115185. [CrossRef]
31. Carrat, F.; Flahault, A. Influenza vaccine: The challenge of antigenic drift. *Vaccine* **2007**, *25*, 6852–6862. [CrossRef]
32. Samson, M.; Pizzorno, A.; Abed, Y.; Boivin, G. Influenza virus resistance to neuraminidase inhibitors. *Antivir. Res.* **2013**, *98*, 174–185. [CrossRef] [PubMed]
33. Zhang, J.; Xu, K.; Yang, F.; Qiu, Y.; Li, J.; Wang, W.; Tan, G.; Zou, Z.; Kang, F. Design, synthesis and evaluation of nitric oxide releasing derivatives of 2,4-diaminopyrimidine as novel FAK inhibitors for intervention of metastatic triple-negative breast cancer. *Eur. J. Med. Chem.* **2023**, *250*, 115192. [CrossRef] [PubMed]
34. Won, K.A.; Spruck, C. Triple negative breast cancer therapy: Current and future perspectives. *Int. J. Oncol.* **2020**, *57*, 1245–1261. [CrossRef] [PubMed]
35. Sulzmaier, F.J.; Jean, C.; Schlaepfer, D.D. FAK in cancer: Mechanistic findings and clinical applications. *Nat. Rev. Cancer* **2014**, *14*, 598–610. [CrossRef]
36. Carragher, N.O.; Frame, M.C. Focal adhesion and actin dynamics: A place where kinases and proteases meet to promote invasion. *Trends Cell Biol.* **2004**, *13*, 241–249. [CrossRef]
37. Badawi, W.A.; Samir, M.; Fathy, H.M.; Okda, T.M.; Noureldin, M.H.; Atwa, G.M.K.; Aboulwafa, O.M. Design, synthesis and molecular docking study of new pyrimidine-based hydrazones and selective anti-proliferative activity against MCF-7 and MDA-MB-231 human breast cancer cell lines. *Bioorg. Chem.* **2023**, *138*, 106610. [CrossRef] [PubMed]
38. AboulWafa, O.M.; Daabees, H.M.; Badawi, W.A. 2-Anilinopyrimidine derivatives: Design, synthesis, in vitro anti-proliferative activity, EGFR and ARO inhibitory activity, cell cycle analysis and molecular docking study. *Bioorg. Chem.* **2020**, *99*, 103798. [CrossRef]
39. Alghamdi, E.M.; Alamshany, Z.M.; El Hamd, M.A.; Taher, E.S.; El-Behairy, M.F.; Norcott, P.L.; Marzouk, A.A. Anticancer activities of tetrasubstituted imidazole-pyrimidine-sulfonamide hybrids as inhibitors of EGFR mutant. *ChemMedChem* **2023**, *18*, e202200641. [CrossRef]
40. Sigismund, S.; Avanzato, D.; Lanzetti, L. Emerging functions of the EGFR in cancer. *Mol. Oncol.* **2017**, *12*, 3–20. [CrossRef]
41. Kamel, M.S.; Belal, A.; Aboelez, M.O.; Shokr, E.K.; Abdel-Ghany, H.; Mansour, H.S.; Shwaky, A.M.; Abd El Aleem Ali Ali El-Remaily, M. Microwave-assisted synthesis, biological activity evaluation, molecular docking, and ADMET studies of some novel pyrrolo[2,3-b]pyrrole derivatives. *Molecules* **2022**, *27*, 2061. [CrossRef]
42. Mao, Y.-Z.; Xi, X.-X.; Zhao, H.-Y.; Zhang, Y.-L.; Zhang, S.-Q. Design, synthesis and evaluation of new pyrimidine derivatives as EGFR[C797S] tyrosine kinase inhibitors. *Bioorg. Med. Chem. Lett.* **2023**, *91*, 129381. [CrossRef] [PubMed]
43. Harrison, P.T.; Vyse, S.; Huang, P.H. Rare epidermal growth factor receptor (EGFR) mutations in non-small cell lung cancer. *Semin. Cancer Biol.* **2020**, *61*, 167–179. [CrossRef] [PubMed]
44. Giaccone, G. The role of gefitinib in lung cancer treatment. *Clin. Cancer Res.* **2004**, *10*, 4233s–4237s. [CrossRef] [PubMed]
45. Li, T.; Ling, Y.H.; Goldman, I.D.; Perez-Soler, R. Schedule-dependent cytotoxic synergism of pemetrexed and erlotinib in human non-small cell lung cancer cells. *Clin. Cancer Res.* **2007**, *13*, 3413–3422. [CrossRef] [PubMed]
46. Wang, L.; Ding, X.; Wang, K.; Sun, R.; Li, M.; Wang, F.; Xu, Y. Structure-based modification of ortho-amidophenylaminopyrimidines as a novel mutant EGFR inhibitor against resistant non-small cell lung cancer. *J. Mol. Struct.* **2023**, *1274*, 134499. [CrossRef]
47. Janne, P.A.; Wang, J.C.; Kim, D.W.; Planchard, D.; Ohe, Y.; Ramalingam, S.S.; Ahn, M.J.; Ki, S.W.; Su, W.C.; Horn, L.; et al. AZD9291 in EGFR inhibitor-resistant non-small lung cancer. *N. Engl. J. Med.* **2015**, *372*, 1689–1699. [CrossRef]
48. Walter, A.O.; Sijin, R.T.; Haringsma, H.J.; Ohashi, K.; Sun, J.; Lee, K.; Dubrovskiy, A.; Labenski, M.; Zhu, Z.; Wang, Z.; et al. Discovery of mutant-selective covalent inhibitor of EGFR that overcomes T790M-mediated resistance in NSCLC. *Cancer Discov.* **2013**, *3*, 1404–1415. [CrossRef] [PubMed]
49. Lee, H.; Choi, S.-B.; Yoon, Y.A.; Hyun, K.H.; Sim, J.Y.; Bryan, M.C.; Kuduk, S.; Robertson, J.C.; Lee, J.; Salgaonkar, P.D.; et al. Substituted Aminopyrimidine Compounds as EGFR Inhibitors and Their Preparation. Patent WO2023027515A1, 2 March 2023.
50. Gomatou, G.; Syrigos, N.; Kotteas, E. Osimertinib resistance: Molecular mechanisms and emerging treatment options. *Cancers* **2023**, *15*, 841. [CrossRef]
51. Dai, C.; Qiang, D.; Tao, Z. Aminoheteroaryl Kinase Inhibitors. Patent WO2023093769A1, 1 June 2023.

52. Ding, L.; Cao, J.; Lin, W.; Chen, H.; Xiong, X.; Ao, H.; Yu, M.; Lin, J.; Cui, Q. The roles of cyclin-dependent kinases in cell-cycle progression and therapeutic strategies in human breast cancer. *Int. J. Mol. Sci.* **2020**, *21*, 1960. [CrossRef]
53. Zhou, Y.; Li, X.; Luo, P.; Chen, H.; Zhou, Y.; Zheng, X.; Yin, Y.; Wei, H.; Liu, H.; Xia, W.; et al. Identification of abemaciclib derivatives targeting cyclin-dependent kinase 4 and 6 using molecular dynamics, binding free energy calculation, synthesis and pharmacological evaluation. *Front. Pharmacol.* **2023**, *14*, 1154654. [CrossRef]
54. Kim, E.S. Abemaciclib: First global approval. *Drugs* **2017**, *77*, 2063–2070. [CrossRef]
55. Kang, Y.; Pei, Y.; Qin, J.; Zhang, Y.; Duan, Y.; Yang, H.; Yao, Y.; Sun, M. Design, synthesis and biological activity evaluation of novel tubulin polymerization inhibitors based on pyrimidine ring skeletons. *Bioorg. Med. Chem. Lett.* **2023**, *84*, 129195. [CrossRef]
56. Knossow, M.; Campanacci, V.; Khodja, L.A.; Gigant, B. The mechanism of tubulin assembly into microtubules: Insights from structural studies. *iScience* **2020**, *23*, 101511. [CrossRef]
57. Yang, J.; Yu, Y.; Li, Y.; Yan, W.; Ye, H.; Niu, L.; Tang, M.; Wang, Z.; Yang, Z.; Pei, H.; et al. Cevipabulin-tubulin complex reveals a novel agent binding site on α-tubulin with tubulin degradation effect. *Sci. Adv.* **2021**, *7*, eabg4168. [CrossRef] [PubMed]
58. Ayati, A.; Moghimi, S.; Toolabi, M.; Foroumadi, A. Pyrimidine-based EGFR TK inhibitors in targeted cancer therapy. *Eur. J. Med. Chem.* **2021**, *221*, 113523. [CrossRef] [PubMed]
59. Zhang, J.; Tan, L.; Wu, C.; Li, Y.; Chen, H.; Liu, Y.; Wang, Y. Discovery and biological evaluation of 4,6-pyrimidine analogues with potential anticancer agents as novel colchicine binding site inhibitors. *Eur. J. Med. Chem.* **2023**, *248*, 115085. [CrossRef] [PubMed]
60. Kavallaris, M. Microtubules and resistance to tubulin-binding agents. *Nat. Rev. Cancer* **2010**, *10*, 194–204. [CrossRef] [PubMed]
61. Wu, X.; Wang, Q.; Li, W. Recent advances in heterocyclic tubulin inhibitors targeting the colchicine binding site. *Anti-Cancer Agents Med. Chem.* **2016**, *16*, 1325–2338. [CrossRef]
62. Ali, F.E.; Salem, O.I.A.; El-Mokhtar, M.A.; Aboraia, A.S.; Abdel-Moty, S.G.; Abdel-Aal, A.-B.M. Design, synthesis and antiproliferative evaluation of lipidated 1,3-diaryl propenones and their cyclized pyrimidine derivatives as tubulin polymerization inhibitors. *Results Chem.* **2023**, *6*, 101016. [CrossRef]
63. Liu, W.; He, Y.; Li, Z.; Peng, Z.; Wang, G. A review on synthetic chalcone derivatives as tubulin polymerization inhibitors. *J. Enzym. Inhib. Med. Chem.* **2022**, *37*, 9–38. [CrossRef]
64. Ngaini, Z.; Fadzillah, S.M.H.; Hussain, H. Synthesis and antimicrobial studies of hydroxylated chalcone derivatives with variable chain length. *Nat. Prod. Res.* **2012**, *26*, 892–902. [CrossRef] [PubMed]
65. Srinivasan, V.; Ebetino, F.H.; Hashizume, R.; Boeckman, R.K., Jr. 4-Amino Pyrimidine Compounds for the Treatment of Cancer. Patent US20230150976A1, 18 May 2023.
66. Rotili, D.; Mai, A. Targeting histone demethylases: A new avenue for the fight against cancer. *Genes Cancer* **2011**, *2*, 663–679. [CrossRef] [PubMed]
67. Vitanza, N.A.; Monje, M. Diffuse intrinsic pontine glioma: From diagnosis to next-generation clinical trials. *Curr. Treat. Options Neurol.* **2019**, *21*, 37. [CrossRef] [PubMed]
68. Ran, F.; Liu, Y.; Zhu, J.; Deng, X.; Wu, H.; Tao, W.; Xie, X.; Hu, Y.; Zhang, Y.; Ling, Y. Design, synthesis and pharmacological characterization of aminopyrimidine derivatives as BTK/FLT3 dual-target inhibitors against acute myeloid leukemia. *Bioorg. Chem.* **2023**, *134*, 106479. [CrossRef]
69. Shallis, R.M.; Wang, R.; Davidoff, A.; Ma, X.; Zeidan, A.M. Epidemiology of acute myeloid leukemia: Recent progress and enduring challenges. *Blood Rev.* **2019**, *36*, 70–87. [CrossRef] [PubMed]
70. Rushworth, S.A.; Murray, M.A.; Zaitseva, L.; Bowles, K.M.; MacEwan, D.J. Identification of Bruton's tyrosine kinase as a therapeutic target in acute myeloid leukemia. *Blood* **2014**, *123*, 1229–1238. [CrossRef] [PubMed]
71. Yuan, K.; Shen, H.; Zheng, M.; Xia, F.; Li, Q.; Chen, W.; Ji, M.; Yang, H.; Zhuang, X.; Cai, Z.; et al. Discovery of potent DYRK2 inhibitors with high selectivity, great solubility, and excellent safety properties for the treatment of prostate cancer. *J. Med. Chem.* **2023**, *66*, 4215–4230. [CrossRef]
72. Teo, M.Y.; Rathkopf, D.E.; Kantoff, P. Treatment of advanced prostate cancer. *Annu. Rev. Med.* **2019**, *70*, 479–499. [CrossRef]
73. Yoshida, S.; Yoshida, K. Multiple functions of DYRK2 in cancer and tissue development. *FEBS Lett.* **2019**, *593*, 2953–2965. [CrossRef]
74. Zhu, C.; Ze, S.; Zhour, R.; Yang, X.; Wang, H.; Chai, X.; Fang, M.; Liu, M.; Wang, Y.; Lu, W.; et al. Discovery of pyridinone derivatives as potent, selective, and orally bioavailable adenosine A_{2A} receptor antagonists for cancer immunotherapy. *J. Med. Chem.* **2023**, *66*, 4734–4754. [CrossRef]
75. Hasko, G.; Linden, J.; Cronstein, B.; Pacher, P. Adenosine receptors: Therapeutic aspects for inflammatory and immune diseases. *Nat. Rev. Drug Discov.* **2008**, *7*, 759–770. [CrossRef] [PubMed]
76. Zeng, S.; Zeng, M.; Yuan, S.; He, L.; Jin, Y.; Huang, J.; Zhang, M.; Yang, M.; Pan, Y.; Wang, Z.; et al. Discovery of potent and selective HPK1 inhibitors based on the 2,4-disubstituted pyrimidine scaffold with immune modulatory properties for ameliorating T cell exhaustion. *Bioorg. Chem.* **2023**, *139*, 106728. [CrossRef] [PubMed]
77. Hernandez, S.; Qing, J.; Thibodeau, R.H.; Du, X.; Park, S.; Lee, H.M.; Xu, M.; Oh, S.; Navarro, A.; Roose-Girma, M.; et al. The kinase activity of hemotopoietic progenitor kinase 1 is essential for the regulation of T cell function. *Cell Rep.* **2018**, *25*, 80–94. [CrossRef] [PubMed]
78. Yu, E.C.; Methot, J.L.; Fradera, X.; Lesburg, C.A.; Lacey, B.M.; Siliphaivanh, P.; Liu, P.; Smith, D.M.; Xu, Z.; Piesvaux, J.A.; et al. Identification of potent reverse indazole inhibitors for HPK1. *ACS Med. Chem. Lett.* **2021**, *12*, 459–466. [CrossRef] [PubMed]

79. Ding, X.; Ren, F.; Wang, H.; Zheng, M.; Zhu, W. Preparation of Pyrimidine Compounds as KiF18a Inhibitors for Treatment of Cancer. Patent CN115925684A, 7 April 2023.
80. Czechanski, A.; Kim, H.; Byers, C.; Greenstein, I.; Stumpff, J.; Reinholdt, L.G. KiF18a is specifically required for mitotic progression during germ line development. *Dev. Biol.* **2015**, *402*, 253–262. [CrossRef] [PubMed]
81. Hergenrother, P.J.; Kelly, A.M. Compounds for Cancers Driven by BRAF Mutation. Patent WO2023070076A1, 27 April 2023.
82. Kelly, A.M.; Berry, M.R.; Tasker, S.Z.; Mckee, S.A.; Fan, T.M.; Hergenrother, P.J. Target-agnostic p-glycoprotein assessment yields strategies to evade efflux, leading to a BRAF inhibitor with intracranial efficacy. *J. Am. Chem. Soc.* **2022**, *144*, 12367–12380. [CrossRef] [PubMed]
83. Bridani, N.; Vuong, L.M.; Acquistapace, I.M.; La Serra, M.A.; Ortega, J.A.; Veronesi, M.; Bertozzi, S.M.; Summa, M.; Girotto, S.; Bertorelli, R.; et al. Design, synthesis, in vitro and in vivo characterization of CDC42 GTPase interaction inhibitors for the treatment of cancer. *J. Med. Chem.* **2023**, *66*, 5981–6001. [CrossRef] [PubMed]
84. Kumar, R.; Gururaj, A.E.; Barnes, C.J. p21-Activated kinases in cancer. *Nat. Rev. Cancer* **2006**, *6*, 459–471. [CrossRef]
85. Lu, W.; Fan, M.; Ji, W.; Tse, J.; You, I.; Ficarro, S.B.; Tavares, I.; Che, J.; Kim, A.Y.; Zhu, X.; et al. Structure-based design of Y-shaped covalent TEAD inhibitors. *J. Med. Chem.* **2023**, *66*, 4617–4632. [CrossRef]
86. Moroishi, T.; Hansen, C.G.; Guan, K.-L. The emerging roles of YAP and TAZ in cancer. *Nat. Rev. Cancer* **2015**, *15*, 73–79. [CrossRef]
87. Yu, N.; Dong, J.; Xia, W.; Wu, W. Preparation of Pyrimidine Derivatives for the Treatment of Proliferative Diseases. Patent CN115925684A, 7 April 2023.
88. Drilon, A.; Hu, Z.I.; Lai, G.G.Y.; Tan, D.S.W. Targeting RET-driven cancers: Lessons from evolving preclinical and clinical landscapes. *Nat. Rev. Clin. Oncol.* **2018**, *15*, 151–167. [CrossRef]
89. Cai, X.; Wang, L.; Yi, Y.; Deng, D.; Shi, M.; Tang, M.; Li, N.; Wei, H.; Zhang, R.; Su, K.; et al. Discovery of pyrimidine-5-carboxamide derivatives as novel salt-inducible kinases (SIKs) inhibitors for inflammatory bowel disease (IBD) treatment. *Eur. J. Med. Chem.* **2023**, *256*, 115469. [CrossRef]
90. Tesch, R.; Rak, M.; Raab, M.; Berger, L.M.; Kronenberger, T.; Joerger, A.C.; Berger, B.T.; Abdi, I.; Hanke, R.; Poso, A.; et al. Structure-based design of selective salt-inducible kinase inhibitors. *J. Med. Chem.* **2021**, *64*, 8142–8160. [CrossRef] [PubMed]
91. Hua, Y.; Yin, N.; Liu, X.; Xie, J.; Zhan, W.; Liang, G.; Shen, Y. Salt-inducible kinase 2-triggered release of its inhibitor from hydrogel to suppress ovarian cancer metastasis. *Adv. Sci.* **2022**, *9*, e2202260. [CrossRef] [PubMed]
92. Marseglia, G.; Caruana, L.; Canelli, T.; Zhao, X.; Wang, W.; Zhao, Z. A Synthesis Scheme and Procedures for Preparing SIK3 Inhibitor and Intermediates Thereof. Patent WO2023067021A1, 27 April 2023.
93. Zicheng, S.; Jiang, Q.; Li, J.; Guo, J. The potent roles of salt-inducible kinases (SIKs) in metabolic homeostasis and tumorigenesis. *Signal Transduct. Target. Ther.* **2020**, *5*, 150. [CrossRef]
94. Zhaoxing, C.; Li, S.; Yan, Z.; Jiajia, M.; Qinlong, X.; Gaofeng, L.; Guangwei, H. Preparation Method of JAK2 Kinase Selective Inhibitor, and Application Thereof in Preparation of Drugs for Preventing and/or Treating JAK2 Kinase-Mediated Diseases. Patent CN115745896A, 7 March 2023.
95. Seif, F.; Khoshmirsafa, M.; Aazami, H.; Mohsenzadegan, M.; Sedighi, G.; Bahar, M. The role of JAK-STAT signaling pathway and its regulators in the fate of T helper cells. *Cell Commun. Signal.* **2017**, *15*, 23. [CrossRef]
96. Banerjee, S.; Biehl, A.; Gadina, M.; Hasni, S.; Schwartz, D.M. JAK-STAT signaling as a target for inflammatory and autoimmune diseases. *Curr. Future Prospect.* **2017**, *77*, 521–546. [CrossRef]
97. Gordhan, H.M.; Miller, S.T.; Clancy, D.C.; Ina, M.; McDougal, A.V.; Cutno, D.K.; Brown, R.V.; Lichorowic, C.L.; Sturdivant, J.M.; Vick, K.A.; et al. Eyes on topical ocular disposition: The considered design of a lead Janus kinase (JAK) inhibitor that utilizes a unique azetidin-3-amino bridging scaffold to attenuate off-target kinase activity, while driving potency and aqueous solubility. *J. Med. Chem.* **2023**, *66*, 8929–8950. [CrossRef]
98. Baudouin, C. A new approach for better comprehension of diseases of the ocular surface. *J. Fr. Ophthalmol.* **2007**, *30*, 239–246. [CrossRef]
99. Pflugfelder, S.C.; de Paiva, C.S. The pathophysiology of dry eye disease: What we know and future directions for research. *Ophthalmology* **2017**, *12*, S4–S13. [CrossRef]
100. Zhang, X.; Shen, C.; Zhu, G. Preparation of 2,4-Diarylaminopyrimidine Derivatives as IL-6 and IL-8 Inhibitors for Treatment of Inflammatory Diseases. Patent CN116003331A, 25 April 2023.
101. Liu, X.; Yin, S.; Chen, Y.; Wu, Y.; Zheng, W.; Dong, H.; Bai, Y.; Qin, Y.; Li, J.; Feng, S.; et al. LPS-induced proinflammatory cytokine expression in human airway epithelial cells and macrophages via NF-κB, STAT3 or AP-1 activation. *Mol. Med. Rep.* **2018**, *17*, 5484–5491. [CrossRef] [PubMed]
102. Defossa, E.; Heinelt, U.; Matter, H.; Mendez-Perez, M.; Rackelmann, N.; Ritter, K.; Szillat, H.; Zech, G. Isoxazolidinones as RIPK1 Inhibitors and Use Thereof. Patent WO2023083847A1, 19 May 2023.
103. Liu, T.; Zhang, L.; Joo, D.; Sun, S.-C. NF-κB signaling in inflammation. *Signal Transduct. Target. Ther.* **2017**, *2*, e17023. [CrossRef] [PubMed]
104. Bandyopadhyay, D.; Eidam, P.M.; Gough, P.J.; Harris, P.A.; Jeong, J.U.; Kang, J.; King, B.W.; Lakdawala Shah, A.; Marquis, R.W., Jr.; Leister, L.K. Preparation of Heterocyclic Amides as RIP1 Kinase Inhibitors for Therapy. Patent WO2014125444A1, 21 August 2014.
105. Fox, R.M.; Harris, P.A.; Holenz, J.; Seefeld, M.A.; Zhou, D. Preparation of Heterocyclic Amides as Kinase Inhibitors. Patent WO2019130230A1, 4 July 2019.

106. Hartz, R.A.; Ahuja, V.T.; Luo, G.; Chen, L.; Sivaprakasam, P.; Xiao, H.; Krause, C.M.; Clarke, W.J.; Xu, S.; Tokarski, J.S.; et al. Discovery of 2-(anilino)pyrimidine-4-carboxamides as highly potent, selective, and orally active glycogen synthase kinase-3 (GSK-3) inhibitors. *J. Med. Chem.* **2023**, *66*, 7534–7552. [CrossRef] [PubMed]
107. Alzheimer's Association. 2020 Alzheimer's disease facts and figures. *Alzheimer's Dement.* **2020**, *16*, 391–460. [CrossRef]
108. Bernard-Gauthier, V.; Mossine, A.V.; Knight, A.; Patnaik, D.; Zhao, W.-N.; Cheng, C.; Krishnan, H.S.; Xuan, L.L.; Chindavong, P.S.; Reis, S.A.; et al. Structural basis for achieving GSK-3β inhibition with high potency, selectivity, and brain exposure for positron emission tomography imaging and drug discovery. *J. Med. Chem.* **2019**, *62*, 9600–9617. [CrossRef] [PubMed]
109. Kumaravel, G.; Macdonell, M.; Peng, H. Pyrimidines and Methods of Their Use. Patent WO2023107603A1, 15 June 2023.
110. Bright, F.; Chan, G.; van Hummel, A.; Ittner, L.; Ke, Y.D. TDP-43 and inflammation: Implications for amyotrophic lateral sclerosis and frontotemporal dementia. *Int. J. Mol. Sci.* **2021**, *22*, 7781. [CrossRef]
111. Johnson, B.S.; Snead, D.; Lee, J.J.; McCaffery, M.; Shorter, J.; Gitler, A.D. TDP-43 Is Intrinsically aggregation-prone, and amyotrophic lateral sclerosis-linked mutations accelerate aggregation and increase toxicity. *J. Biol. Chem.* **2009**, *284*, 20329–20339. [CrossRef] [PubMed]
112. Lei, H.; Hong, L.; Liu, C.; Liu, H.; Ke, S. Preparation of Pyrimidine Aminopyrazole Derivative as Leucine-Rich Repeat Kinase-2-Inhibitor. Patent CN115819405A, 21 March 2023.
113. Li, J.-Q.; Tan, L.; Yu, J.-T. The role of the LRRK2 gene in Parkinsonism. *Mol. Neurodegener.* **2014**, *9*, 47–63. [CrossRef]
114. Gelin, C.F.; Stenne, B.; Coate, H.; Hiscox, A.; Soyode-Johnson, A.; Wall, J.L.; Lord, B.; Schoellerman, J.; Coe, K.J.; Wang, K.; et al. Discovery of a series of substituted 1H-((1,2,3-triazol-4-yl)methoxy)pyrimidines as brain penetrant and potent GluN2B selective negative allosteric modulators. *J. Med. Chem.* **2023**, *66*, 2877–2892. [CrossRef]
115. Traynelis, S.F.; Wollmuth, L.P.; McBain, C.J.; Menniti, F.S.; Vance, K.M.; Ogden, K.K.; Hansen, K.B.; Yuan, H.; Myers, S.J.; Dingledine, R. Glutamate receptor ion channels: Structure, regulation, and function. *Pharmacol. Rev.* **2010**, *62*, 405–496. [CrossRef]
116. Wagner, T.T.; Weng, Z.; Xi, H.S. 5-Pyrimidinecarboxamide Derivatives and Methods of Using the Same. Patent WO202310223A1, 8 June 2023.
117. Elorza, A.; Marquez, Y.; Cabrera, J.R.; Sanchez-Trincado, J.L.; Santos-Galindo, M.; Hernandez, I.H.; Pico, S.; Diaz-Hernandez, J.I.; Garcia-Escudero, R.; Irimia, M.; et al. Huntington's disease-specific mis-splicing unveils key effector genes and altered splicing factors. *Brain* **2021**, *144*, 2009–2023. [CrossRef] [PubMed]
118. Burli, R.; Doyle, K. N-(4-Aminocyclohexyl)pyrimidine-4-carboxamide Derivatives as CD38 Inhibitors. Patent WO2023084206A1, 19 May 2023.
119. Covarrubia, A.J.; Perrone, R.; Grozio, A.; Verdin, E. NAD+ metabolism and its roles in cellular processes during aging. *Nat. Rev. Mol. Cell Biol.* **2021**, *22*, 119–141. [CrossRef] [PubMed]
120. Ruble, J.C.; Winneroski, L.L. Deuterated (Trifluoromethyl)pyrimidine-2-amine Compounds as Potentiators of the Hmrgxl Receptor. Patent WO2023081463A1, 11 May 2023.
121. Logrip, M.L.; Koob, G.F.; Zorrilla, E.P. Role of corticotropin-releasing factor in drug addiction: Potential for pharmacological intervention. *CNS Drugs* **2011**, *25*, 271–287. [CrossRef] [PubMed]
122. Amin, A.; Sheikh, K.A.; Iqubal, A.; Khan, M.A.; Shaquiquzzaman, M.; Tasneem, S.; Khanna, S.; Najmi, A.K.; Akhter, M.; Haque, A.; et al. Synthesis, in-silico studies and biological evaluation of pyrimidine-based thiazolidinedione derivatives as potential anti-diabetic agent. *Bioorg. Chem.* **2023**, *134*, 106449. [CrossRef] [PubMed]
123. Sun, H.; Saeedi, P.; Karuranga, S.; Pinkepank, M.; Ogurtsova, K.; Duncan, B.B.; Stein, C.; Basit, A.; Chan, J.C.; Mbanya, J.C. IDF Diabetes Atlas: Global, regional and country level diabetes prevalence estimates for 2021 and projections for 2045. *Diabetes Res. Clin. Pract.* **2022**, *183*, 109119. [CrossRef]
124. Thangavel, N.; Al Bratty, M.; Javed, S.A.; Ahsan, W.; Alhazmi, H.A. Targeting peroxisome proliferator-activated receptors using thiazolidinediones: Strategy for design of novel antidiabetic drugs. *Int. J. Med. Chem.* **2017**, *2017*, 1069718. [CrossRef]
125. Farghaly, T.A.; Harras, M.F.; Alsaeda, A.M.R.; Thakir, H.A.; Mahmoud, H.K.; Katowah, D.F. Antiviral activity of pyrimidine containing compounds: Patent review. *Mini-Rev. Med. Chem.* **2023**, *23*, 821–851. [CrossRef]
126. Finger, V.; Kufa, M.; Soukup, O.; Castagnolo, D.; Roh, J. Pyrimidine derivatives with antitubercular activity. *Eur. J. Med. Chem.* **2023**, *246*, 114946. [CrossRef] [PubMed]
127. Ahmed, K.; Choudhary, M.I.; Saleem, R.S.Z. Heterocyclic pyrimidine derivatives as promising antibacterial agents. *Eur. J. Med. Chem.* **2023**, *259*, 115701. [CrossRef]

Disclaimer/Publisher's Note: The statements, opinions and data contained in all publications are solely those of the individual author(s) and contributor(s) and not of MDPI and/or the editor(s). MDPI and/or the editor(s) disclaim responsibility for any injury to people or property resulting from any ideas, methods, instructions or products referred to in the content.

pharmaceuticals

Article

Discovery of Potent Indolyl-Hydrazones as Kinase Inhibitors for Breast Cancer: Synthesis, X-ray Single-Crystal Analysis, and In Vitro and In Vivo Anti-Cancer Activity Evaluation

Eid E. Salama [1,*], Mohamed F. Youssef [1], Ahmed Aboelmagd [1], Ahmed T. A. Boraei [1], Mohamed S. Nafie [1,2], Matti Haukka [3], Assem Barakat [4,*] and Ahmed A. M. Sarhan [5]

1 Department of Chemistry, Faculty of Science, Suez Canal University, Ismailia 41522, Egypt; mohamed_gomaa@science.suez.edu.eg (M.F.Y.); ahmed_mohamed@science.suez.edu.eg (A.A.); ahmed_tawfeek83@yahoo.com or ahmed_boraei@science.suez.edu.eg (A.T.A.B.); mohamed_nafie@science.suez.edu.eg (M.S.N.)
2 Department of Chemistry, College of Sciences, University of Sharjah, Sharjah P.O. Box 27272, United Arab Emirates
3 Department of Chemistry, University of Jyväskylä, P.O. Box 35, FI-40014 Jyväskylä, Finland; matti.o.haukka@jyu.fi
4 Chemistry Department, College of Science, King Saud University, P.O. Box 2455, Riyadh 11451, Saudi Arabia
5 Chemistry Department, Faculty of Science, Arish University, Al-Arish 45511, Egypt; ahmed_sarhan252@yahoo.com or asarhan@aru.edu.eg
* Correspondence: eid_mohamed@science.suez.edu.eg (E.E.S.); ambarakat@ksu.edu.sa (A.B.); Tel.: +966-11467-5901 (A.B.); Fax: +966-11467-5992 (A.B.)

Abstract: According to data provided by the World Health Organization (WHO), a total of 2.3 million women across the globe received a diagnosis of breast cancer in the year 2020, and among these cases, 685,000 resulted in fatalities. As the incidence of breast cancer statistics continues to rise, it is imperative to explore new avenues in the ongoing battle against this disease. Therefore, a number of new indolyl-hydrazones were synthesized by reacting the ethyl 3-formyl-1*H*-indole-2-carboxylate **1** with thiosemicarbazide, semicarbazide.HCl, 4-nitrophenyl hydrazine, 2,4-dinitrophenyl hydrazine, and 4-amino-5-(1*H*-indol-2-yl)-1,2,4-triazole-3-thione to afford the new hit compounds, which were assigned chemical structures as thiosemicarbazone **3**, *bis*(hydrazine derivative) **5**, semicarbzone **6**, Schiff base **8**, and the corresponding hydrazones **10** and **12** by NMR, elemental analysis, and X-ray single-crystal analysis. The MTT assay was employed to investigate the compounds' cytotoxicity against breast cancer cells (MCF-7). Cytotoxicity results disclosed potent IC_{50} values against MCF-7, especially compounds **5**, **8**, and **12**, with IC_{50} values of 2.73 ± 0.14, 4.38 ± 0.23, and 7.03 ± 0.37 µM, respectively, compared to staurosproine (IC_{50} = 8.32 ± 0.43 µM). Consequently, the activities of compounds **5**, **8**, and **12** in relation to cell migration were investigated using the wound-healing test. The findings revealed notable wound-healing efficacy, with respective percentages of wound closure measured at 48.8%, 60.7%, and 51.8%. The impact of the hit compounds on cell proliferation was assessed by examining their apoptosis-inducing properties. Intriguingly, compound **5** exhibited a significant enhancement in cell death within MCF-7 cells, registering a notable increase of 39.26% in comparison to the untreated control group, which demonstrated only 1.27% cell death. Furthermore, the mechanism of action of compound **5** was scrutinized through testing against kinase receptors. The results revealed significant kinase inhibition, particularly against PI3K-α, PI3K-β, PI3K-δ, CDK2, AKT-1, and EGFR, showcasing promising activity, compared to standard drugs targeting these receptors. In the conclusive phase, through in vivo assay, compound **5** demonstrated a substantial reduction in tumor volume, decreasing from 106 mm³ in the untreated control to 56.4 mm³. Moreover, it significantly attenuated tumor proliferation by 46.9%. In view of these findings, the identified leads exhibit promises for potential development into future medications for the treatment of breast cancer, as they effectively hinder both cell migration and proliferation.

Keywords: indole; hydrazone; MCF7; anticancer; apoptosis; kinase inhibition

1. Introduction

Globally, cancer stands as the foremost contributor to mortality, with breast cancer ranking among the leading causes of death in women. The complexity of this ailment poses a significant challenge to medical therapy. Despite their capacity to eliminate cancer cells, conventional anticancer treatments often induce a multitude of adverse effects, proving detrimental to healthy tissues. Traditional antineoplastic drugs, aimed at inhibiting specific molecules fostering tumor growth, commonly result in side effects. In response, scientists are actively exploring novel anticancer drugs, engaging in the design and discovery of new compounds tailored for treating various cancer types. This pursuit aims to identify targeted therapies that can potentially offer more effective and less harmful treatment options [1,2]. Kinases, constituting the sixth largest class of proteins in the human body [3–5], play a crucial role in cellular function. Inhibitors of kinases are indispensable for maintaining proper cellular activities, as they modulate kinase dysregulation associated with various diseases and disorders, including cancer, inflammatory conditions, and responses to external stimuli. Through the regulation of protein kinases, these inhibitors effectively impede the growth of their substrates, thereby exerting control over the viability and proliferation of cells [6–9].

Various therapeutic targets for kinase inhibitors exist, encompassing EGFR, CDK, AKT, PI3K, and other specific targets [10–12]. Indole-containing compounds have garnered prolonged attention from researchers and evolved into a dynamic field of study. The indole moiety demonstrates a high affinity for binding to several receptors, paving the way for the development of new bioactive medications. Its widespread utilization in target-based discovery and the design of anticancer drugs is well-documented [13–18].

N'-Methylene-5-((4-(pyridin-3-yl)pyrimidin-2-yl)amino)1H-indole-2-carbohydrazide moiety showed CDK9 inhibitory activity for cancer therapy treatment, according to Hu et al. [19]. Some representative examples discussed, such as compound **I**, were assessed, and it was revealed that they showed potential against CDK9 inhibition. Al-warhi et al. reported that oxindole **II** displayed anti-tumor activity targeting the CDK4 inhibitor [20]. N-substituted iso-indigo compounds were designed, synthesized, and biologically evaluated by Zhao et al. as inhibitors of cyclin-dependent kinase 2 (CDK2). Iso-indigo compound **III** was found to stop the S-phase of the cell cycle [21] (Figure 1).

Sunitinib is used for the treatment of gastrointestinal stromal tumors (GIST) and advanced renal cell carcinoma (RCC) [22–25]. Semaxanib is a tyrosine kinase inhibitor drug that is used in cancer therapeutics [26]. Indole-triazole alkylated system (Figure 1) displayed significant anti-cancer activity. For example: 3-benzylsulfanyl-5-(1H-indol-2-yl)-2H-1,2,4-triazole **IV** showed promising antiproliferative activity against HEPG-2 and MCF-7 cancer cell lines [27]; 3-(allylsulfanyl)-4-phenyl-5-(1H-indol-2-yl)-1,2,4-triazole **V** and its analogues showed interesting anti-proliferative potential against breast cancer [28]; substituted 3-(triazolo-thiadiazin-3-yl)-indolin-2-one derivatives **VI** displayed dual inhibition activity for c-Met (a receptor tyrosine kinase) and VEGFR2 enzymes, with an effective anti-proliferative potency against different subpanels of the most NCI 58 tumor cell lines [29]; and indole-triazole hybrid **VII** and its analogues revealed a potent inhibition against vascular endothelial growth factor receptor-2 (VEGFR-2), with potential anti-renal cancer activity [30]. Alkylated indolyl-triazole Schiff bases **VIII** targeted breast cancer via VEGFR2 tyrosine kinase inhibition [31].

Building upon the findings from the aforementioned studies, the conceptualization of synthesizing novel compounds that incorporate ester and azomethine groups, along with an indole scaffold in a single compound, is expected to yield potent anticancer medicines. This anticipation stems from the potential of these compounds to function as both hydrogen bond donors and/or acceptors upon interaction with receptors.

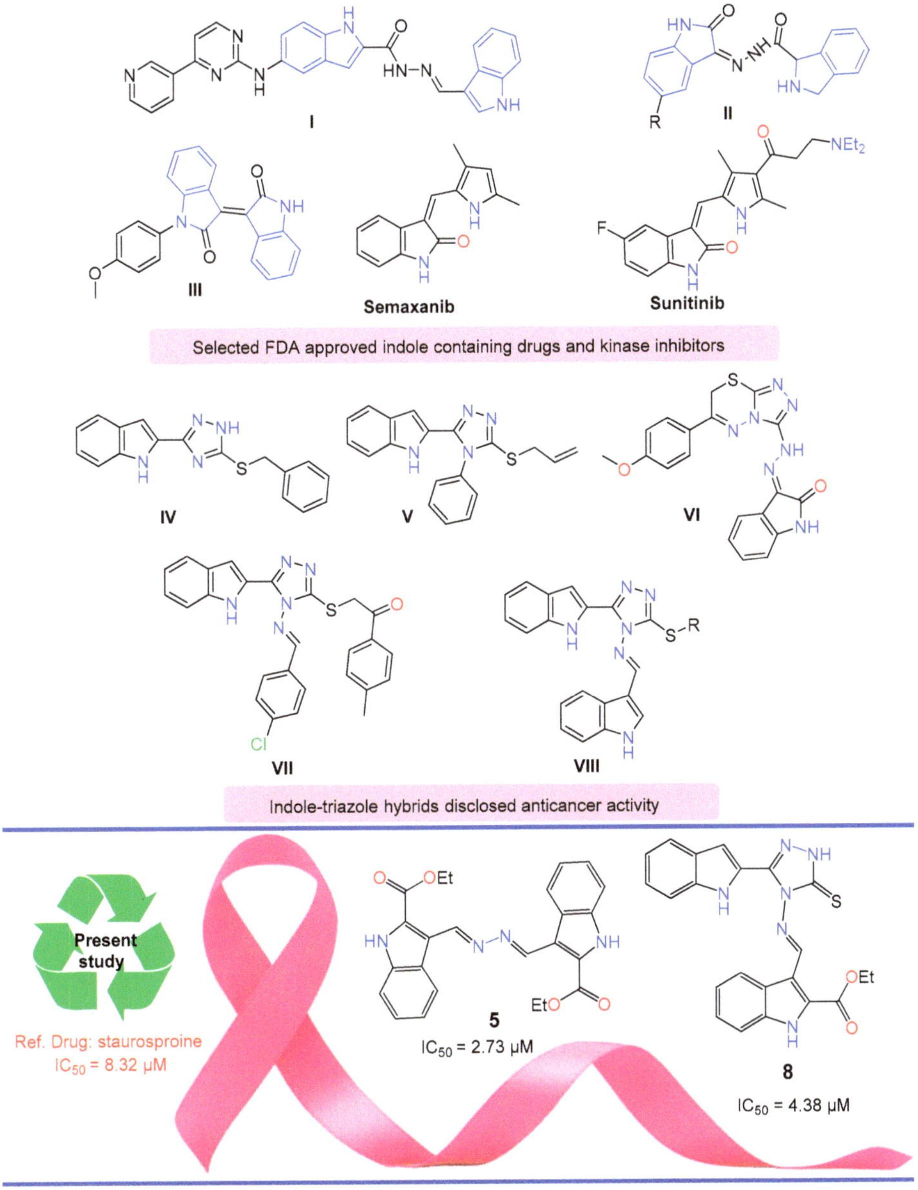

Figure 1. Selected indole structures and drugs that possess anticancer activity.

2. Results and Discussion

2.1. Synthesis

Condensation of thiosemicarbazide **2** with ethyl 3-formyl-1*H*-indole-2-carboxylate **1** by fusion for 5 min led to the formation of the thiosemicarbazone derivative **3**. Under the same fusion condition, the condensation of **1** with semicarbazide.HCl **4** interestingly afforded *bis*-ester derivative **5**. Through the reaction of **1** with semicarbazide.HCl **4** under reflux in AcOH/MeOH, semicarbazone derivative **6** was obtained (Scheme 1).

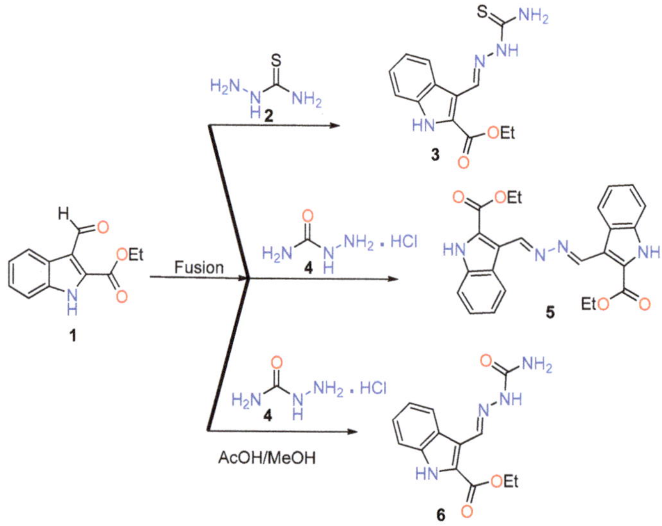

Scheme 1. Reaction of thiosemicarbazide **2** and semicarbazide.HCl **4** with ethyl 3-formyl-1*H*-indole-2-carboxylate **1**.

Under fusion conditions, the reaction between ethyl 3-formyl-1*H*-indole-2-carboxylate **1** with amine functionality derivatives, such as 4-amino-5-(1*H*-indol-2-yl)-1,2,4-triazole-3-thione **7**, 4-nitrophenyl hydrazine **9**, and 2,4-dinitrophenyl hydrazine **11**, resulted in the formation of condensed hydrazones **8**, **10**, and **12**, respectively (Scheme 2). The structural assignment for the newly synthesized hits was established through a comprehensive set of spectroscopic tools (see Section 3), which included nuclear magnetic resonance (NMR), mass spectrometry (MS), CHN analysis, and single-crystal X-ray diffraction analysis.

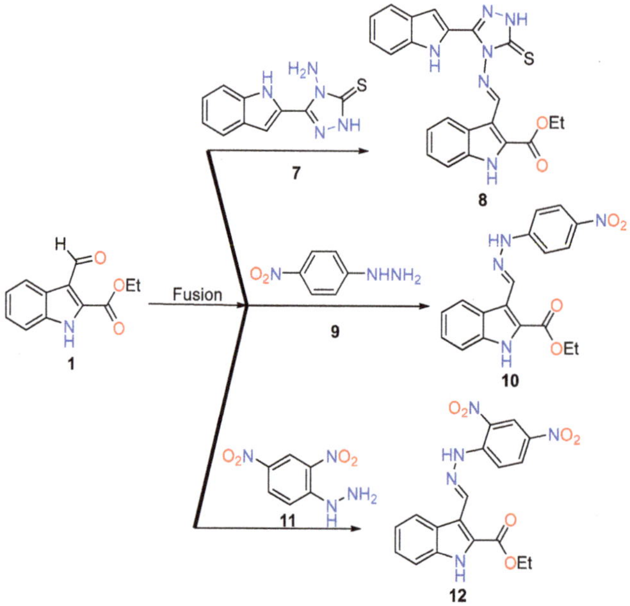

Scheme 2. Synthesis of hydrazones **8**, **10**, and **12**.

2.2. X-ray Single-Crystal Analysis for Compounds 3 and 5

Using single-crystal X-ray analysis, structures of compounds **3** and **5** were conclusively confirmed. The unit cell parameters of compound **3** (a = 9.23490(10) Å, b = 19.4168(2) Å, c = 7.89280(10) Å, and V = 1374.36(3) Å^3) that crystallized in monoclinic space group $P2_1/c$ and compound **5** (a = 5.4774(3) Å, b = 9.2031(5) Å, c = 10.6234(8) (10) Å, and V = 531.26(6) show the Å^3 that crystallized in the triclinic space group $P\bar{1}$ (Table 1). The crystal structure (Figure 2) revealed that compound **3** was a thiosemicarbazone structure, while compound **5** was a *bis*-hydrazino derivative.

Table 1. Crystals data of compounds **3** and **5**.

	3	5
CCDC	2293777	2293778
empirical formula	$C_{13}H_{14}N_4O_2S$	$C_{24}H_{22}N_4O_4$
fw	290.34	430.45
temp (K)	120(2)	120(2)
Λ (Å)	1.54184	0.71073
cryst syst	Monoclinic	Triclinic
space group	$P2_1/c$	$P\bar{1}$
a (Å)	9.23490(10)	5.4774(3)
b (Å)	19.4168(2)	9.2031(5)
c (Å)	7.89280(10)	10.6234(8)
A (deg)	90	93.995(5)
β (deg)	103.8110(10)	94.008(6)
γ (deg)	90	94.201(5)
V (Å^3)	1374.36(3)	531.26(6)
Z	4	1
ρ_{calc} (Mg/m^3)	1.403	1.345
μ(Mo Kα) (mm^{-1})	2.168	0.094
No. reflns.	16497	8485
Unique reflns.	2884	2637
Completeness to θ = 67.684°	99.8%	
Completeness to θ = 25.242°		99.9%
GOOF (F^2)	1.058	1.046
R_{int}	0.0187	0.0300
R_1 [a] ($I \geq 2\sigma$)	0.0279	0.0469
wR_2 [b] ($I \geq 2\sigma$)	0.0780	0.1123

[a] $R_1 = \Sigma||F_o| - |F_c||/\Sigma|F_o|$. [b] $wR_2 = \{\Sigma[w(F_o^2 - F_c^2)^2]/\Sigma[w(F_o^2)^2]\}^{1/2}$.

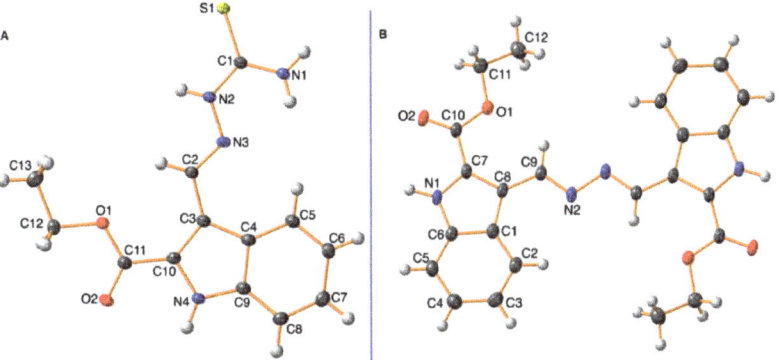

Figure 2. Ortep representation of **3** (**A**) and **5** (**B**).

2.3. Biology

2.3.1. MTT Assay for the Synthesized Compounds

The produced compounds were examined for their cytotoxicity against MCF-7 breast cancer cells using the MTT assay. As summarized in Table 2, they demonstrated potent IC_{50} values against MCF-7, especially compounds **5**, **8**, and **12**, with IC_{50} values of 2.73 ± 0.14, 4.38 ± 0.23, and 7.03 ± 0.37 µM, compared to staurosproine, with an IC_{50} value of 8.32 ± 0.43 µM, while compounds **1**, **3**, and **6** exhibited promising cytotoxicity, with IC_{50} values of 19.7 ± 2.31, 10.2 ± 0.53, and 9.42 ± 0.57 µM, respectively. Compound **10** exhibited moderate cytotoxicity, with a high concentration of IC_{50} (25.4 ± 1.54 µM).

Table 2. Cytotoxicities of the investigated compounds against MCF-7 cells using the MTT assay.

Compounds	$IC_{50} \pm SD$ [µM]
1	19.7 ± 2.31
3	10.2 ± 0.53
5	2.73 ± 0.14
6	9.42 ± 0.57
8	4.38 ± 0.23
10	25.4 ± 1.54
12	7.03 ± 0.37
Staurosporine	8.32 ± 0.43

IC_{50} values were calculated using "Mean ± SD" of three independent values.

2.3.2. Wound-Healing Activity

As shown in Table 3 and Figure 3, the wounded area between cell layers following a scratch was partially filled by migrating MCF-7 control cells (94.07% wound closure), while treatments of compounds **5**, **8**, and **12** significantly inhibited wound-healing activity, with percentages of wound closure of 48.88, 60.74, and 51.85%, respectively, compared to control.

Table 3. The percentage of wound healing (% closure) for untreated and **5**-treated MCF-7 cells.

Compound	%Closure *, MCF-7
5	48.88 [#] ± 2.7
8	60.74 [#] ± 3.43
12	51.85 [#] ± 2.92
Untreated control	94.07 ± 5.5

* Values are expressed as "Mean ± SD". [#] Significance level ($p < 0.05$) indicates a significant difference (unpaired Student's t-test) from the untreated control group. Data for length of migration (mm) and area are supported in the Supplementary Materials.

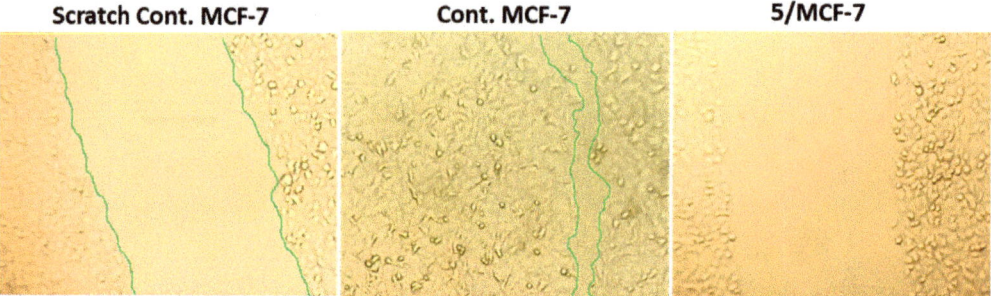

Figure 3. Migration of MCF-7 cells treated with compound **5** for 72 h observed under a light microscope as detected by the wound-healing assay.

2.3.3. Apoptotic Induction Activity

To investigate the apoptotic activity of compounds **5**, **8**, and **12**, flow cytometric evaluation of Annexin V/PI staining was utilized to examine apoptotic cell death in untreated and treated MCF-7 cells. Table 4 shows that compound **5** dramatically increased cell death in MCF-7 cells by 39.26% (29.35% for apoptosis and 9.91% for necrosis), compared to the untreated control group, which increased it by 1.27% (0.4% for apoptosis and 0.87% for necrosis). Additionally, compounds **8** and **12** caused total cell death by 24.4% and 37%, with apoptosis ratios of 15.72% and 21.0%, respectively.

Table 4. Flow cytometry results of the three promising cytotoxic agents using Annexin V/PI and DNA-aided flow cytometry.

Compound	Annexin V/PI Staining				DNA Content			
	Total	Early	Late	Necrosis	%G0-G1	%S	%G2/M	%Pre-G1
Cont.MCF7	1.27	0.29	0.11	0.87	52.91	41.33	5.76	1.27
5	39.26	7.11	22.24	9.91	39.07	56.19	4.74	39.26
8	24.38	2.27	13.45	8.66	47.10	46.31	6.57	24.38
12	37.05	4.59	17.44	15.02	59.33	38.10	2.55	37.05

After being treated with a cytotoxic chemical, the cell population in each cell phase was then ascertained by DNA flow cytometry. Compound **5** increased the S-phase cell population by 56.2%, compared to control, which increased it by 41.33%, as Figure 4 illustrates, whereas cells in other phases decreased negligibly. Consequently, compound **5** stopped MCF-7 cells from proliferating at the S-phase by inducing apoptosis.

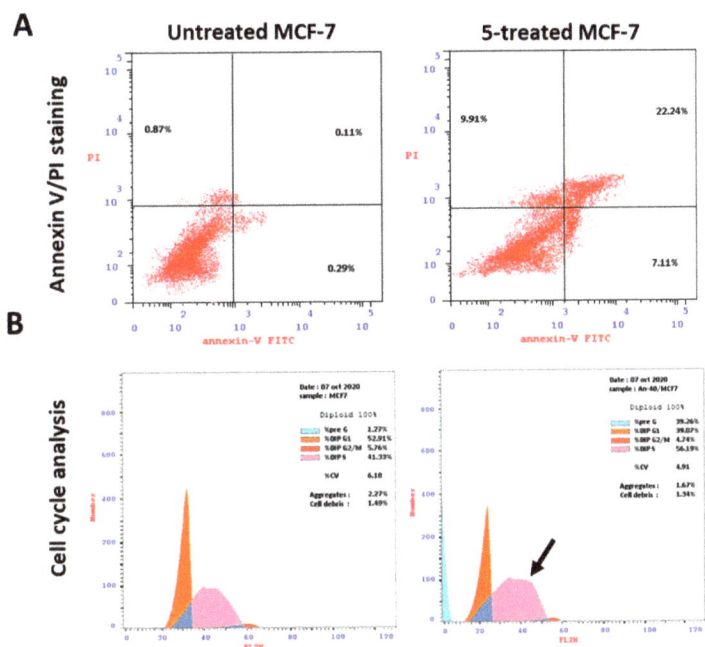

Figure 4. Analysis using flow cytometry. Upper panel (**A**): Annexin V/PI staining for evaluating necrosis and apoptosis, Q1;eEarly apoptosis is Q4, and late apoptosis is Q2. The DNA content histograms of untreated and **5**-treated MCF-7 cells at each phase, "Pre-G, G1, G2/M, S" phases, with an IC_{50} value of 2.73 µM, 48 h, are displayed in the lower panel (**B**).

2.3.4. Kinase-Inhibition Activity

To highlight their effective molecular target, the most cytotoxic and apoptotic compound **5** was screened for its activity towards a panel of kinase activities, including PI3K-α, PI3K-β, PI3K-δ, CDK2, AKT-1, and EGFR, compared to their standard drugs. It caused promising kinase inhibitory activities, as summarized in Table 5. Interestingly, compound **5** exhibited significant inhibitory potential against PI3K-α and showed selectivity, with a 4.92-fold higher potency than LY294002. However, in the cases of PI3K-β and PI3K-δ, compound **5** demonstrated lower activity compared to LY294002. Moreover, compound **5** (IC$_{50}$ = 0.156 ± 0.01 µM) showed a reactivity profile against CDK2 closer to the standard drug erlotinib (IC$_{50}$ of 0.173 ± 0.01 µM). On the other hand, compound **5** (IC$_{50}$ = 0.602 ± 0.03 µM) demonstrated lower reactivity towards AKT-1, compared to the standard drug A-674563 (IC$_{50}$ of 0.26 ± 0.01 µM). Finally, compound **5** (IC$_{50}$ = 0.058 ± 0.029 µM) demonstrated lower reactivity against EGFR, compared to the standard drug erlotinib (IC$_{50}$ of 0.038 ± 0.019 µM). Compound **5** was found to possess the potential for inhibiting multiple kinases.

Table 5. IC$_{50}$ values of kinase activities of the tested compounds.

Compound	IC$_{50}$ [µM] ± SD *					
	PI3K-α	PI3K-β	PI3K-δ	CDK2	AKT-1	EGFR
5	1.73 ± 0.1	2.27 ± 0.11	2.68 ± 0.15	0.156 ± 0.01	0.602 ± 0.03	0.058 ± 0.029
LY294002	8.52 ± 0.48	0.44 ± 0.02	0.85 ± 0.05	NT	NT	NT
erlotinib	NT	NT	NT	0.173 ± 0.01	NT	0.038 ± 0.019
A-674563	NT	NT	NT	NT	0.26 ± 0.01	NT

* "Values are expressed as an average of three independent replicates". "IC$_{50}$ values were calculated using a sigmoidal non-linear regression curve fit of percentage inhibition against five concentrations of each compound". NT = Not tested.

2.3.5. In Vivo (SEC-Bearing Mice)

A solid Ehrlich carcinoma cell was implanted, and **5** was injected intraperitoneally (IP) throughout the experiment to confirm its anticancer efficacy, as shown in Figure 5, which summarizes the main findings of the antitumor activity experiments. As a result, tumor proliferation revealed an increase in solid tumor mass of approximately 398.1 mg, which is related to tumor proliferation. Following treatment with **5**, the solid tumor mass decreased to 126.5 mg, compared to 110 mg in the 5-FU treatment. As a result, treatments with **5** considerably reduced tumor volume from 10^6 mm^3 in the untreated control to 56.4 mm^3 and significantly decreased tumor proliferation by 46.9%, while 5-FU reduced tumor volume to 43.7 mm^3 and inhibited tumor development by 58.8%.

2.3.6. Molecular Docking

To illustrate the virtual mechanism of binding towards the EGFR, PI3K, and CDK2 binding sites, molecular docking research was carried out. As seen in Figure 6, compound **5** was properly docked inside the protein active sites of EGFR (A), PI3K (B), and CDK2 (C), with binding energies of −23.15, −21.32, and −23.44 Kcal/mol, and it formed good binding interactions with their active sites. Compound **5** exhibited strong binding interactions with the amino acids Lys721, Cys773, and Leu694 inside EGFR. It formed two H-bond interactions with Val882 inside the PI3K active site, and it formed two arene–cation interactions with Lys89 inside the CDK2 active site like the co-crystallized ligands. These outcomes corroborated the kinase inhibition experiment findings. Previous literature reported the downstream inhibition pathway of EGFR/PI3K/AKT, which is linked to CDK2 inhibition, as a promising target for inducing apoptosis in cancer cells [32].

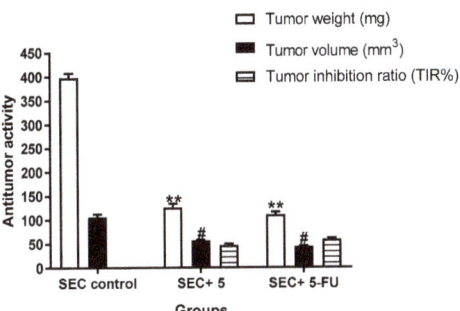

Figure 5. Measurements of antitumor potentiality in the SEC-bearing mice treated with compound **5** and 5-FU. "Mean ± SD values of mice in each group (n = **6**)". "** Values are highly significantly different ($p \leq 0.01$) between treated and SEC control", while "# values are significantly different ($p \leq 0.05$) between treated SEC and SEC control mice using the un-paired test in GraphPad prism". TIR% = C − T/C × 100.

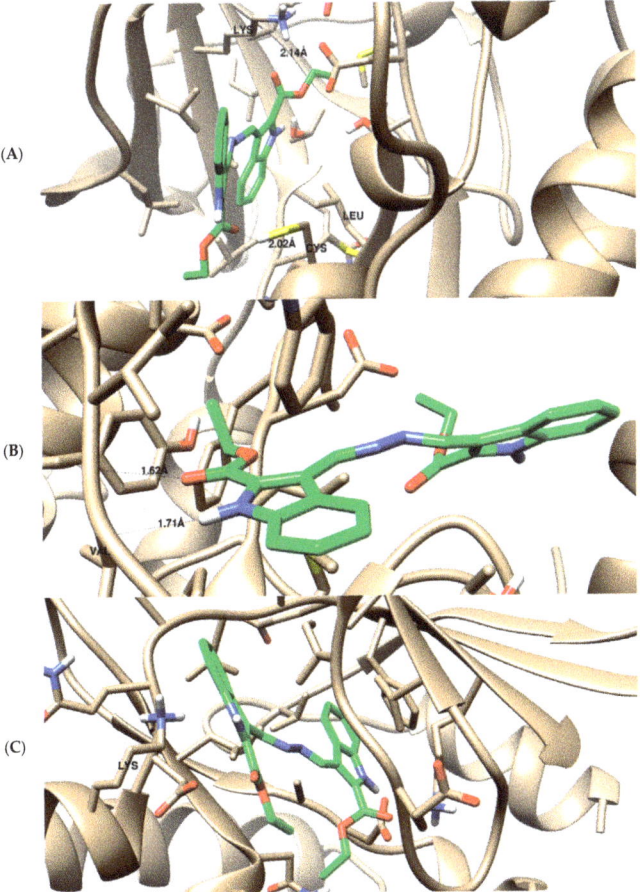

Figure 6. The binding disposition of docked compound **5** with ligand–receptor interactions inside the EGFR receptor (**A**), phosphoinositide 3-kinase (**B**), and cyclin-dependent kinase (**C**). Visualization was carried out using Chimara-UCSF.

As summarized in Figure 7, compound **5**, as an indolyl-hydrazone derivative, induced potent cytotoxicity against MCF-7 as an apoptosis inducer through the downstreaming pathway of EGFR/PI3K/AKT and CDK2 inhibition. The effective pathway induced cell cycle arrest at the S-phase, and it led to apoptosis in the MCF-7 cells. EGFR, and its downstreaming pathway is considered one of the promising effective pathways for cancer treatment, and our results agreed with previous reported studies for the same compounds' scaffold affecting cytotoxic activities through apoptosis [33–35].

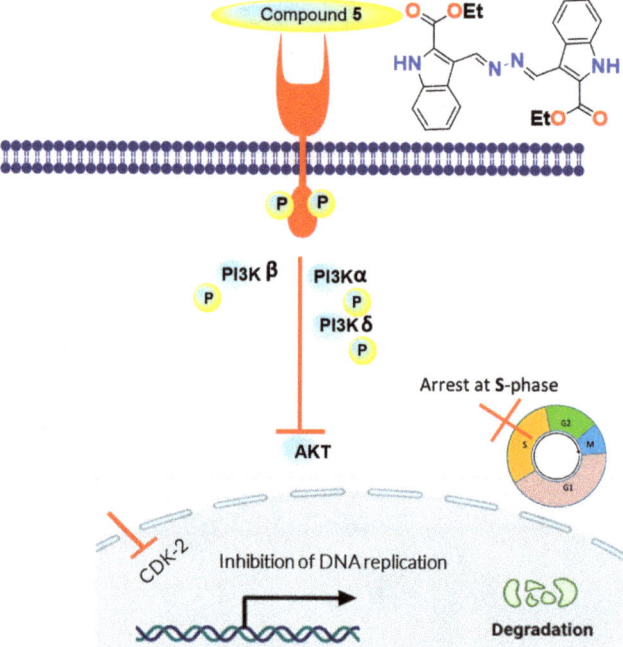

Figure 7. Schematic diagram for the mechanistic pathway for EGFR/PI3K/CDK2 inhibition as apoptosis inducers.

2.3.7. SAR

The structure–reactivity relationship of the synthesized compounds is summarized as follows in Figure 8: The hydrazone derivative **10**, featuring a *p*-nitro group-substituted benzene ring, exhibited the lowest reactivity, with an IC_{50} value of 25.4 ± 1.54 µM. In contrast, the aldehyde-based indole derivative **1**, the starting material, demonstrated better reactivity, with an IC_{50} value of 19.7 ± 2.31 µM. The thiosemicarbazide **3** and its isosteric semicarbazide **6** enhanced reactivity, with IC_{50} values of 9.42 ± 0.57 and 10.2 ± 0.53 µM, respectively. The presence of two nitro groups on the substituted benzene ring of hydrazone **12** significantly improved reactivity, with an IC_{50} value of 7.03 ± 0.37 µM, due to the high electron-withdrawing group effect. The introduction of the thio-triazole indole-based Schiff base **8** significantly increased activity (IC_{50} = 4.38 ± 0.23 µM) up to 1.9-fold higher than the reference drug, while the symmetrical *bis*-esters azine **5** emerged as the most potent compound in inhibiting breast cancer cells, with an IC_{50} of 2.73 ± 0.14 µM, 3-fold more potent than the standard drug staurosporine (IC_{50} = 8.32 ± 0.43 µM).

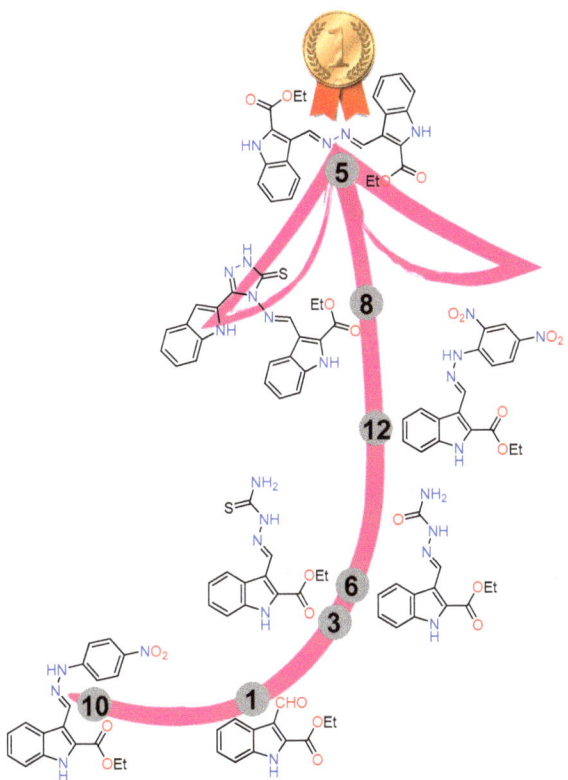

Figure 8. SAR for the synthesized compounds.

3. Materials and Methods

3.1. Chemistry

3.1.1. General

The values for the melting points were uncorrected and were determined in open capillaries using a Temp-melt II melting point equipment. On silica gel 60 (230–400 mesh ASTM), flash chromatography was carried out. On silica gel 60 F254 aluminum plates (E. Merck, layer thickness 0.2 mm), thin-layer chromatography (TLC) was performed. The spots were found using a UV lamp. Using DMSO-d_6 and CDCl$_3$ as solvents, the ^{1}H and ^{13}C-NMR spectra were captured on Bruker instruments at 400 MHz for ^{1}H NMR and 101 MHz for ^{13}C NMR, respectively. Using KBr and a PerkinElmer 1430 ratio-recording infrared spectrophotometer, Bruker's Fourier-transform infrared (FT-IR) spectrophotometry was used to record the IR spectra.

3.1.2. Synthesis

A mixture of **1** (1.0 mmol, 0.22 g), thiosemicarbazide, and semicarbazide·HCl (1.1 mmol, 0.1 g, and 0.12 g respectively) was grinded and fused on a hotplate for 5 min until all reactants turned to products. The products were purified by recrystallization from DMF/EtOH to **3** and **5**, respectively.

Ethyl (*E*)-3-((2-carbamothioylhydrazineylidene)methyl)-1*H*-indole-2-carboxylate **3**.

There was 81% yield, 0.23 g, and m.p. 229–230 °C. ^{1}H NMR (400 MHz, DMSO-d_6): δ 1.41 (t, 3 H, *J* = 6.8 Hz, CH$_3$), 4.42 (q, 2 H, *J* = 6.8 Hz, OCH$_2$), 7.21 (dd, 1H, *J* = 7.2, 7.6 Hz), 7.35 (dd, 1 H, *J* = 7.2, 7.6 Hz), 7.51 (d, 1 H, *J* = 8.0 Hz), 8.17 (brs, 2 H), 8.40 (d, 1 H, *J* = 8.4 Hz), 9.01 (s, 1H, CH=N-), 11.58 (s, 1H, NH), and 12.15 (s, 1 H, NH indole); ^{13}C NMR

(100 MHz, DMSO-d_6): δ 14.69 (CH$_3$), 61.60 (OCH$_2$), 113.06, 115.69, 122.32, 124.57, 124.69, 126.12, 127.90, 137.04, 141.10 (9 C), 161.41 (C=O), and 177.97 (C=S); and elemental analysis calculated for [C$_{13}$H$_{14}$N$_4$O$_2$S]: C, 53.78; H, 4.86; N, 19.30; S, 11.04; found C, 53.89; H, 4.93; N, 19.23; and S, 11.09

Diethyl 3,3'-((1E,1'E)-hydrazine-1,2-diylidene*bis*(methaneylylidene))*bis*(1H-indole-2-carboxylate) **5**.

There was 78% yield, 0.34 g, and m.p. 302–303 °C. ^{1}H NMR (400 MHz, DMSO-d_6): δ 1.45 (t, 3 H, *J* = 6.4 Hz, CH$_3$), 4.48 (q, 2H, *J* = 6.4 Hz, OCH$_2$), 7.29 (dd, 1 H, *J* = 6.8, 7.2 Hz), 7.41 (dd, 1 H, *J* = 7.2, 6.8 Hz), 7.58 (d, 1 H, *J* = 7.6 Hz), 8.57 (d, 1 H, *J* = 8.0 Hz), 9.59 (s, 1 H, CH=N), and 12.37 (s, 1 H, NH indole); ^{13}C NMR (100 MHz, DMSO-d_6): δ 14.67 (CH$_3$), 61.70 (OCH$_2$), 113.39, 116.09, 122.52, 124.53, 125.34, 126.25, 128.92, 137.07 (8 C), 157.27 (C=O), and 161.24 (C=O); elemental analysis computed for [C$_{24}$H$_{22}$N$_4$O$_4$]: C, 66.97; H, 5.15; N, 13.02; found C, 66.99; H, 5.21; and N, 13.13.

Ethyl (*E*)-3-((2-carbamoylhydrazineylidene)methyl)-1*H*-indole-2-carboxylate **6**.

A mixture of **1** (1.0 mmol) and semicarbazide.HCl (1.1 mmol) was refluxed in equal volumes of MeOH/AcOH 10 mL for 8 h until all reactants formed products. A precipitate was formed upon cooling, which was filtered, dried, and recrystallized from MeOH to obtain **6**.

There was 88% yield, 0.25 g, and m.p. 285–286 °C. ^{1}H NMR (400 MHz, DMSO-d_6): δ 1.36 (t, *J* 7.1 Hz, 3 H), 4.37 (q, *J* = 7.1 Hz, 2 H), 6.46 (s, 2 H), 7.19 (t, *J* = 7.5 Hz, 1 H), 7.47–7.32 (m, 2H), 7.55 (d, *J* = 8.3 Hz, 1 H), 7.66 (s, 1 H), 8.67 (s, 1 H), and 12.31 (s, 1 H, NH indole); ^{13}C NMR (101 MHz, DMSO-d_6): δ 14.63, 61.45, 112.10, 113.63, 121.50, 121.95, 124.78, 125.65, 126.23, 134.06, 136.85, 156.88, and 161.24; elemental analysis calculated for [C$_{13}$H$_{14}$N$_4$O$_3$]: C, 56.93; H, 5.15; N, 20.43; found C, 57.01; H, 5.13; and N, 20.52.

Ethyl (*E*)-3-(((3-(1*H*-indol-2-yl)-5-thioxo-1,5-dihydro-4*H*-1,2,4-triazol-4-yl)imino)methyl)-1*H*-indole-2-carboxylate **8**.

There was 81% yield, 0.36 g, and m.p. 248–249 °C. ^{1}H NMR (400 MHz, DMSO-d_6): δ 1.41 (t, 3 H, *J* = 6.8 Hz,CH$_3$), 4.46 (q, 2 H, *J* = 6.8 Hz,OCH$_2$), 7.02 (dd, 1H, *J* = 7.2, 7.6 Hz), 7.20–7.23 (m, 2H), 7.31 (dd, 1H, and *J* = 7.6 Hz), 7.44–7.52 (m, 3 H), 7.65 (d, 1 H, *J* = 8 Hz), 8.48 (d, 1 H, *J* = 8 Hz), 10.41 (s, 1 H), 11.88 (s, 1 H), 12.79 (s, 1 H, NH indole), and 14.20 (brs, 1 H, NH$_{trz}$); ^{13}C NMR(100 MHz, DMSO-d_6): δ 14.60 (CH$_3$), 62.17 (CH$_2$), 105.59, 112.41, 113.11, 113.86, 120.45, 121.55, 123.10, 123.53, 123.86, 124.19, 125.09, 126.61, 127.73, 131.19, 137.12, 137.41, 144.01, 160.84, 162.73, and 163.31. Calculated elemental analysis for [C$_{22}$H$_{18}$N$_6$O$_2$S]: found C, 61.44; H, 4.37; N, 19.43; S, 7.39; C, 61.38; H, 4.21; N, 19.52; O, 7.43; and S, 7.45

Ethyl (*E*)-3-((2-(4-nitrophenyl)hydrazineylidene)methyl)-1*H*-indole-2-carboxylate **10**.

There was 89% yield, 0.32 g, and m.p. 270–271 °C. ^{1}H NMR (400 MHz, DMSO-d_6): δ 1.43 (t, *J* = 7.0 Hz, 3 H), 4.44 (q, *J* = 6.9 Hz, 2 H), 7.28–7.11 (m, 2 H), 7.30 (d, *J* = 7.4 Hz, 1 H), 7.40 (t, *J* = 7.6 Hz, 1 H), 7.54 (d, *J* = 8.2 Hz, 1 H), 8.18 (d, *J* = 8.6 Hz, 2 H), 8.45 (d, *J* = 8.1 Hz, 1 H), 8.99 (s, 1 H), 11.39 (s, 1 H), and 12.13 (s, 1 H, NH indole); ^{13}C NMR (101 MHz, DMSO-d_6): δ 14.76, 61.44, 111.43, 113.29, 117.04, 122.20, 124.15, 124.56, 126.23, 126.81, 137.10, 138.40, 139.67, 151.01, and 161.46; elemental analysis calculated for [C$_{18}$H$_{16}$N$_4$O$_4$]: C, 61.36; H, 4.58; N, 15.90; found C, 61.47; H, 4.43; and N, 15.82.

Ethyl (*E*)-3-((2-(2,4-dinitrophenyl)hydrazineylidene)methyl)-1*H*-indole-2-carboxylate **12**.

There was 89% yield, 0.36 g, and m.p. 292–293 °C. ^{1}H NMR (400 MHz, DMSO-d_6): δ 1.44 (t, 3 H, CH$_3$), 4.46 (q, 2 H, OCH$_2$), 7.30–7.54 (m, 3 H), 8.00 (brs, 1 H), 8.38 (brs, 2 H), 8.83 (brs, 1 H), 9.36 (s, 1 H, CH=N), 11.73 (s, 1 H, NH), and 12.33 (s, 1 H, NH indole); ^{13}C NMR (100 MHz, DMSO-d_6): δ 14.72 (CH$_3$), 61.67 (OCH$_2$), 113.49, 115.73, 122.76, 123.89, 126.33, 130.43, 137.08, 146.62, and 161.20; elemental analysis calculated for [C$_{18}$H$_{15}$N$_5$O$_6$]: C, 54.41; H, 3.81; N, 17.63; found C, 54.53; H, 3.88; and N, 17.49.

3.1.3. X-ray Structure Determination

The general protocol for the collection of crystalline compounds **3** and **5** is provided in the supporting materials [36–38].

4. Cytotoxicity

The National Research Institute in Egypt donated the breast cancer (MCF-7) cells, which were collected and cultured in RPMI-1640 medium L-glutamine (Lonza Verviers SPRL, Verviers, Belgium, cat#12-604F). The 10% fetal bovine serum (FBS; Sigma-Aldrich, St. Louis, MO, USA) and 1% penicillin-streptomycin (Lonza, Belgium) were given to each of the two cell lines.

All cells were cultured, following routine tissue culture work, in 5% CO_2 humidified at 37 °C. Cells were exposed to compounds at concentrations of 0.01, 0.1, 1, 10, and 100 µM on the second day of culturing. After 48 h, cell viability was evaluated using the MTT solution (Promega, Madison, WI, USA) [38]. MTT dye (20 µL) was placed into each well, and the plate was then incubated for three hours. Absorbance was subsequently measured at 570 nm using the ELISA microplate reader (BIO-RAD, model iMark, Tokyo, Japan), and the percentage of cell viability was calculated, compared to control, as (mean absorbance of tested compound)/(mean absorbance in control) × 100. Finally, IC_{50} values were found using the nonlinear dose-response sigmoidal curve in GraphPad Prism 7 [39].

4.1. Investigation of Apoptosis

Annexin V/PI staining and cell cycle analysis 3-10^5 MCF-7 cells were added to 6-well culture plates, which were then placed in the incubator for the night. Following that, cells were treated for 48 h to compound **5** at its IC_{50} levels. Following that, PBS was rinsed with ice-cold water before cells and media supernatants were gathered. The cells were then treated with "Annexin V-FITC solution (1:100) and propidium iodide (PI)" at a concentration of 10 g/mL for 30 min in the dark after being suspended in 100 L of annexin-binding buffer solution, which was composed of 25 mM $CaCl_2$, 1.4 M NaCl, and 0.1 M Hepes/NaOH, pH 7.4. Then, labeled cells were collected using the Cytoflex FACS system. The data were assessed using the cytExpert program [39].

4.2. Wound-Healing Assay (Scratch Assay)

The wound-healing test was mentioned in previous research [40,41]. Six-well plates containing starvation media were filled with four 10^5 MCF-7 cells per well, and the plates were subsequently incubated at 37 °C for the whole night. A sterile 1 mL pipette tip was used to generate a scratch of the cell monolayer once it was established the following day that the cells had adhered to the well and that cell confluence had reached 90%. Starvation media was used to clean the cells before they were removed from the plates. For 48 h, the cells were cultivated in a CO_2 incubator with the IC_{50} of compounds **5**, **8**, and **12** in the full medium. After 48 h, the medium was immediately changed to PBS, the wound gap was examined, and cells—both control and treated—were captured on camera with a digital camera attached to an Olympus microscope. The region where the wound closes was measured [42,43].

4.3. Kinase Inhibitory Assays

EGFR (catalog #40321), CDK2 (catalog #79599), AKT (catalog #78038), PI3K-α (catalog #40639), β (catalog #79802), and δ (catalog #40628) kinase inhibitions were conducted using an ELISA kit in accordance with the manufacturer instructions from Bioscience, USA. To assess the inhibitory potency of compound **5** against the kinase activity, kinase inhibitory tests were carried out. The following calculation was used to compute the proportion that chemicals inhibited autophosphorylation: $100 - [(A_{Control})/(A_{Treated}) - A_{Control}]$. Using the GraphPad prism7 program, the IC_{50} was calculated using the curves of percentage inhibition of five concentrations of each chemical [44].

4.4. In Vivo (SEC-Bearing Model)

The Suez Canal University Research Ethics Committee gave the experimental procedure their seal of approval (approval number REC219/2023, Faculty of Science, Suez Canal University) [45,46]. The full, detailed methodology is supported in the Supplementary Materials.

4.5. Molecular Docking

Maestro was used to construct, optimize, and energetically favor ligand structures. The X-ray crystallographic structures of EGFR kinase (PDB ID: 1M17), PI3K (PDB = 1E7V), and CDK2 (PDB = 2A4L) [47,48] were subjected to a molecular docking investigation using the AutoDock Vina 1.2.0 software following routine work, which was followed by the Chimera-UCSF 1.17.3 software.

5. Conclusions

Ethyl 3-formyl-1*H*-indole-2-carboxylate **1** serves as a precursor in a series of condensation reactions under fusion conditions, aiming to discover novel bioactive lead compounds. Structure assignments were accomplished through NMR and X-ray single-crystal analysis. The majority of the synthesized compounds exhibited noteworthy anti-breast cancer activity. Compounds **5**, **8**, and **12** exhibited superior activity, compared to the standard drug. Compound **5** demonstrated potent cytotoxicity, with IC_{50} values of 2.73 ± 0.14 µM, surpassing staurosporine ($IC_{50} = 8.32 \pm 0.43$ µM) by three-fold. Additionally, it exhibited significant wound-healing activity, with a wound-closure percentage of 48.8%. Notably, in terms of apoptosis induction, compound **5** markedly increased cell death through apoptosis by inhibiting PI3K-α, PI3K-β, PI3K-δ, CDK2, AKT-1, and EGFR kinases, compared to their respective standard drugs. Consequently, the newly identified lead compound is recommended for further development as a kinase-targeted anti-breast cancer agent.

Supplementary Materials: The following documents are available for download at https://www.mdpi.com/article/10.3390/ph16121724/s1, where you can find the titles of Figures S1–S12: NMR spectrum of the synthesized compounds, Figure S13: Wound-healing assay for the most active compounds. In vivo assay protocol. References [49–52] are cited in the supplementary materials.

Author Contributions: Conceptualization, A.T.A.B., A.B., E.E.S., M.S.N., M.F.Y. and A.A.M.S.; methodology, A.T.A.B., M.S.N., E.E.S., A.A.M.S. and A.A.; software, M.H. and M.S.N.; validation, A.T.A.B., A.B. and E.E.S.; formal analysis, A.T.A.B.; investigation, M.S.N. and E.E.S.; resources, E.E.S.; data curation, M.H. and M.S.N.; writing—original draft preparation, A.T.A.B., M.F.Y., M.S.N. and A.A.; writing—review and editing, A.T.A.B., A.A. and M.S.N.; visualization, A.B., M.H. and M.S.N.; supervision, M.F.Y.; project administration, A.B.; funding acquisition, A.B. All authors have read and agreed to the published version of the manuscript.

Funding: The authors would like to extend their sincere appreciation to the Researchers Supporting Project (RSP2023R64), King Saud University, Riyadh, Saudi Arabia.

Institutional Review Board Statement: All procedures related to care and maintenance of the animals were performed according to the international guiding principles for animal research and approved by Faculty of Science, Suez Canal University bioethics and animal ethics committee (Approval number REC219/2023).

Informed Consent Statement: Not applicable.

Data Availability Statement: Data is contained within the article and supplementary material.

Acknowledgments: The authors would like to extend their sincere appreciation to the Researchers Supporting Project (RSP2023R64), King Saud University, Riyadh, Saudi Arabia.

Conflicts of Interest: The authors declare no conflict of interest.

References

1. Ye, F.; Dewanjee, S.; Li, Y.; Jha, N.K.; Chen, Z.-S.; Kumar, A.; Vishakha; Behl, T.; Jha, S.K.; Tang, H. Advancements in clinical aspects of targeted therapy and immunotherapy in breast cancer. *Mol. Cancer* **2023**, *22*, 105. [CrossRef] [PubMed]
2. Senapati, S.; Mahanta, A.K.; Kumar, S.; Maiti, P. Controlled drug delivery vehicles for cancer treatment and their performance. *Signal Transduct. Target. Ther.* **2018**, *3*, 7. [CrossRef] [PubMed]
3. Ward, R.A.; Goldberg, F.W. *Introduction to Kinase Drug Discovery: Modern Approaches*; Royal Society of Chemistry: London, UK, 2018; Volume 67.
4. Zhao, Z.; Bourne, P.E. Progress with covalent small-molecule kinase inhibitors. *Drug Discov. Today* **2018**, *23*, 727–735. [CrossRef]

5. Bhanumathy, K.K.; Balagopal, A.; Vizeacoumar, F.S.; Vizeacoumar, F.J.; Freywald, A.; Giambra, V. Protein Tyrosine Kinases: Their roles and their targeting in leukemia. *Cancers* **2021**, *13*, 184. [CrossRef]
6. Patterson, H.; Nibbs, R.; McInnes, I.; Siebert, S. Protein kinase inhibitors in the treatment of inflammatory and autoimmune diseases. *Clin. Exp. Immunol.* **2014**, *176*, 1–10. [CrossRef]
7. Bhullar, K.S.; Lagarón, N.O.; McGowan, E.M.; Parmar, I.; Jha, A.; Hubbard, B.P.; Rupasinghe, H.P.V. Kinase-targeted cancer therapies: Progress, challenges and future directions. *Mol. Cancer* **2018**, *17*, 48. [CrossRef]
8. Bononi, A.; Agnoletto, C.; De Marchi, E.; Marchi, S.; Patergnani, S.; Bonora, M.; Giorgi, C.; Missiroli, S.; Poletti, F.; Rimessi, A.; et al. Protein kinases and phosphatases in the control of cell fate. *Enzym. Res.* **2011**, *2011*, 329098. [CrossRef]
9. Valdespino-Gómez, V.M.; Valdespino-Castillo, P.M.; Valdespino-Castillo, V.E. Cell signaling pathways interaction in cellular proliferation: Potential target for therapeutic interventionism. *Cirugía Cir.* **2015**, *83*, 165–174. [CrossRef] [PubMed]
10. Zhong, L.; Li, Y.; Xiong, L.; Wang, W.; Wu, M.; Yuan, T.; Yang, W.; Tian, C.; Miao, Z.; Wang, T.; et al. Small molecules in targeted cancer therapy: Advances, challenges, and future perspectives. *Signal Transduct. Target. Ther.* **2021**, *6*, 201. [CrossRef] [PubMed]
11. You, K.S.; Yi, Y.W.; Cho, J.; Park, J.S.; Seong, Y.S. Potentiating therapeutic effects of epidermal growth factor receptor inhibition in triple-negative breast cancer. *Pharmaceuticals* **2021**, *14*, 589. [CrossRef] [PubMed]
12. Roskoski, R., Jr. Properties of FDA-approved small molecule protein kinase inhibitors: A 2021 update. *Pharmacol. Res.* **2021**, *165*, 105463. [CrossRef]
13. Spanò, V.; Barreca, M.; Rocca, R.; Bortolozzi, R.; Bai, R.; Carbone, A.; Raimondi, M.V.; Piccionello, A.P.; Montalbano, A.; Alcaro, S.; et al. Insight on [1,3]thiazolo[4,5-e]isoindoles as tubulin polymerization inhibitors. *Eur. J. Med. Chem.* **2021**, *212*, 113122. [CrossRef] [PubMed]
14. Ostacolo, C.; Di Sarno, V.; Lauro, G.; Pepe, G.; Musella, S.; Ciaglia, T.; Vestuto, V.; Autore, G.; Bifulco, G.; Marzocco, S.; et al. Identification of an indol-based multi-target kinase inhibitor through phenotype screening and tar-get fishing using inverse virtual screening approach. *Eur. J. Med. Chem.* **2019**, *167*, 61–75. [CrossRef] [PubMed]
15. Kryshchyshyn-Dylevych, A.; Radko, L.; Finiuk, N.; Garazd, M.; Kashchak, N.; Posyniak, A.; Niemczuk, K.; Stoika, R.; Lesyk, R. Synthesis of novel indole-thiazolidinone hybrid structures as promising scaffold with anticancer potential. *Bioorganic Med. Chem.* **2021**, *50*, 116453. [CrossRef] [PubMed]
16. Wang, G.; He, M.; Liu, W.; Fan, M.; Li, Y.; Peng, Z. Design, synthesis and biological evaluation of novel 2-phenyl-4,5,6,7-tetrahydro-1H- indole derivatives as potential anticancer agents and tubulin polymerization inhibitors. *Arab. J. Chem.* **2022**, *15*, 103504. [CrossRef]
17. Ponnam, D.; Arigari, N.K.; Kalvagunta Venkata Naga, S.S.; Jonnala, K.K.; Singh, S.; Meena, A.; Misra, P.; Luqman, S. Synthesis of non-toxic anticancer active forskolin-indole-triazole conjugates along with their in silico succinate dehydrogenase inhibition studies. *J. Heterocycl. Chem.* **2021**, *58*, 2090–2101. [CrossRef]
18. Saruengkhanphasit, R.; Butkinaree, C.; Ornnork, N.; Lirdprapamongkol, K.; Niwetmarin, W.; Svasti, J.; Ruchirawat, S.; Eurtivong, C. Identification of new 3-phenyl-1H-indole-2-carbohydrazide derivatives and their structure-activity relationships as potent tubulin inhibitors and anticancer agents: A combined in silico, in vitro and synthetic study. *Bioorganic Chem.* **2021**, *110*, 104795. [CrossRef]
19. Hu, H.; Wu, J.; Ao, M.; Zhou, X.; Li, B.; Cui, Z.; Wu, T.; Wang, L.; Xue, Y.; Wu, Z.; et al. Design, synthesis and biological evaluation of methylenehydrazine-1-carboxamide derivatives with (5-((4-(pyridin-3-yl)pyrimidin-2-yl)amino)-1H-indole scaffold: Novel potential CDK9 inhibitors. *Bioorganic Chem.* **2020**, *102*, 104064. [CrossRef]
20. Al-Warhi, T.; El Kerdawy, A.M.; Aljaeed, N.; Ismael, O.E.; Ayyad, R.R.; Eldehna, W.M.; Abdel-Aziz, H.A.; Al-Ansary, G.H. Synthesis, biological evaluation and in silico studies of certain oxindole–indole conjugates as anticancer CDK inhibitors. *Molecules* **2020**, *25*, 2031. [CrossRef]
21. Zhao, P.; Li, Y.; Gao, G.; Wang, S.; Yan, Y.; Zhan, X.; Liu, Z.; Mao, Z.; Chen, S.; Wang, L. Design, synthesis and biological evaluation of N-alkyl or aryl substituted isoindigo derivatives as potential dual Cyclin-Dependent Kinase 2 (CDK2)/Glycogen Synthase Kinase 3β (GSK-3β) phosphorylation inhibitors. *Eur. J. Med. Chem.* **2014**, *86*, 165–174. [CrossRef]
22. Oudard, S.; Beuselinck, B.; Decoene, J.; Albers, P. Sunitinib for the treatment of metastatic renal cell carcinoma. *Cancer Treat. Rev.* **2011**, *37*, 178–184. [CrossRef]
23. Xu, D.; Wang, T.-L.; Sun, L.-P.; You, Q.-D. Recent progress of small molecular VEGFR inhibitors as anticancer agents. *Mini-Rev. Med. Chem.* **2011**, *11*, 18–31. [CrossRef]
24. Gridelli, C.; Maione, P.; Del Gaizo, F.; Colantuoni, G.; Guerriero, C.; Ferrara, C.; Nicolella, D.; Comunale, D.; De Vita, A.; Rossi, A. Sorafenib and Sunitinib in the Treatment of Advanced Non-Small Cell Lung Cancer. *Oncologist* **2007**, *12*, 191–200. [CrossRef] [PubMed]
25. Imming, P.; Sinning, C.; Meyer, A. Drugs, their targets and the nature and number of drug targets. *Nat. Rev. Drug Discov.* **2006**, *5*, 821–834. [CrossRef]
26. Lockhart, A.C.; Cropp, G.F.; Berlin, J.D.; Donnelly, E.; Schumaker, R.D.; Schaaf, L.J.; Hande, K.R.; Fleischer, A.C.; Hannah, A.L.; Rothenberg, M.L. Phase I/pilot study of SU5416 (semaxinib) in combination with irinotecan/bolus 5-FU/LV (IFL) in patients with metastatic colorectal cancer. *Am. J. Clin. Oncol.* **2006**, *29*, 109–115. [CrossRef] [PubMed]
27. Boraei, A.T.A.; Gomaa, M.S.; El Ashry, E.S.H.; Duerkop, A. Design, selective alkylation and X-ray crystal structure determination of dihydro-indolyl-1,2,4-triazole-3-thione and its 3-benzylsulfanyl analogue as potent anticancer agents. *Eur. J. Med. Chem.* **2017**, *125*, 360–371. [CrossRef] [PubMed]

28. Boraei, A.T.A.; Singh, P.K.; Sechi, M.; Satta, S. Discovery of novel functionalized 1,2,4-triazoles as PARP-1 inhibitors in breast cancer: Design, synthesis and antitumor activity evaluation. *Eur. J. Med. Chem.* **2019**, *182*, 111621. [CrossRef] [PubMed]
29. Mohamady, S.; Galal, M.; Eldehna, W.M.; Gutierrez, D.C.; Ibrahim, H.S.; Elmazar, M.M.; Ali, H.I. Dual Targeting of VEGFR2 and C-Met Kinases via the Design and Synthesis of Substituted 3-(Triazolo-thiadiazin-3-yl)indolin-2-one Derivatives as Angiogenesis Inhibitors. *ACS Omega* **2020**, *5*, 18872–18886. [CrossRef]
30. Al-Hussain, S.A.; Farghaly, T.A.; Zaki, M.E.A.; Abdulwahab, H.G.; Al-Qurashi, N.T.; Muhammad, Z.A. Discovery of novel indolyl-1,2,4-triazole hybrids as potent vascular endothelial growth factor receptor-2 (VEGFR-2) inhibitors with potential anti-renal cancer activity. *Bioorganic Chem.* **2020**, *105*, 104330. [CrossRef]
31. Nafie, M.S.; Boraei, A.T.A. Exploration of novel VEGFR2 tyrosine kinase inhibitors via design and synthesis of new alkylated indolyl-triazole Schiff bases for targeting breast cancer. *Bioorg. Chem.* **2022**, *122*, 105708. [CrossRef]
32. Lee, D.H.; Szczepanski, M.J.; Lee, Y.J. Magnolol induces apoptosis via inhibiting the EGFR/PI3K/Akt signaling pathway in human prostate cancer cells. *J. Cell. Biochem.* **2009**, *106*, 1113–1122. [CrossRef] [PubMed]
33. Sreenivasulu, R.; Reddy, K.T.; Sujitha, P.; Kumar, C.G.; Raju, R.R. Synthesis, antiproliferative and apoptosis induction potential activities of novel bis (indolyl) hydrazide-hydrazone derivatives. *Bioorg. Med. Chem.* **2019**, *27*, 1043–1055. [CrossRef] [PubMed]
34. Das Mukherjee, D.; Kumar, N.M.; Tantak, M.P.; Das, A.; Ganguli, A.; Datta, S.; Kumar, D.; Chakrabarti, G. Development of novel bis(indolyl)-hydrazide–hydrazone derivatives as potent microtubule-targeting cytotoxic agents against A549 lung cancer cells. *Biochemistry* **2016**, *55*, 3020–3035. [CrossRef] [PubMed]
35. Kilic-Kurt, Z.; Acar, C.; Ergul, M.; Bakar-Ates, F.; Altuntas, T.G. Novel indole hydrazide derivatives: Synthesis and their antiproliferative activities through inducing apoptosis and DNA damage. *Arch. Pharm.* **2020**, *353*, 2000059. [CrossRef]
36. Rikagu Oxford Diffraction. *CrysAlisPro*; Rikagu Oxford Diffraction Inc.: Yarnton, UK, 2020.
37. Sheldrick, G.M. SHELXT–Integrated space-group and crystal-structure determination. *Acta Cryst.* **2015**, *A71*, 3–8. [CrossRef]
38. Hübschle, C.B.; Sheldrick, G.M.; Dittrich, B. ShelXle: A Qt graphical user interface for SHELXL. *J. Appl. Crystallogr.* **2011**, *44*, 1281–1284. [CrossRef]
39. Van Meerloo, J.; Kaspers, G.J.L.; Cloos, J. Cell Sensitivity Assays: The MTT Assay. *Methods Mol. Biol.* **2011**, *731*, 237–245.
40. Mohamed, F.A.M.; Gomaa, H.A.M.; Hendawy, O.M.; Ali, A.T.; Farghaly, H.S.; Gouda, A.M.; Abdelazeem, A.H.; Abdelrahman, M.H.; Trembleau, L.; Youssif, B.G.M. Design, Synthesis, and Biological Evaluation of Novel EGFR Inhibitors Containing 5-Chloro-3-Hydroxymethyl-Indole-2-Carboxamide Scaffold with Apoptotic Antiproliferative Activity. *Bioorganic Chem.* **2021**, *112*, 104960. [CrossRef]
41. Ali, G.M.E.; Ibrahim, D.A.; Elmetwali, A.M.; Ismail, N.S.M. Design, Synthesis and Biological Evaluation of Certain CDK2 Inhibitors Based on Pyrazole and Pyrazolo[1,5-a] Pyrimidine Scaffold with Apoptotic Activity. *Bioorganic Chem.* **2019**, *86*, 1–14. [CrossRef]
42. Qin, X.; Han, X.; Hu, L.; Li, Z.; Geng, Z.; Wang, Z.; Zeng, C.; Xiao, X. Design, Synthesis and Biological Evaluation of Quinoxalin-2(1H)-One Derivatives as EGFR Tyrosine Kinase Inhibitors. *Anticancer. Agents Med. Chem.* **2015**, *15*, 267–273. [CrossRef]
43. AkgÜl, Ö.; ErdoĞan, M.A.; Bİrİm, D.; KayabaŞi, Ç.; GÜndÜz, C.; ArmaĞan, G. Design, Synthesis, Cytotoxic Activity, and Apoptosis Inducing Effects of 4- and N-Substituted Benzoyltaurinamide Derivatives. *Turk. J. Chem.* **2020**, *44*, 1674–1693. [CrossRef]
44. Turner, D.P.; Moussa, O.; Sauane, M.; Fisher, P.B.; Watson, D.K. Prostate-derived ETS factor is a mediator of metastatic potential through the inhibition of migration and invasion in breast cancer. *Cancer Res.* **2007**, *67*, 1618–1625. [CrossRef]
45. Chen, Y.; Lu, B.; Yang, Q.; Fearns, C.; Yates, J.R.; Lee, J.-D. Combined Integrin Phosphoproteomic Analyses and siRNA-based Functional Screening Identified Key Regulators for Cancer Cell Adhesion and Migration. *Cancer Res.* **2009**, *69*, 3713–3720. [CrossRef] [PubMed]
46. Barghash, R.F.; Eldehna, W.M.; Kovalová, M.; Vojáčková, V.; Kryštof, V.; Abdel-Aziz, H.A. One-Pot Three-Component Synthesis of Novel Pyrazolo[3,4-b]Pyridines as Potent Antileukemic Agents. *Eur. J. Med. Chem.* **2022**, *227*, 113952. [CrossRef]
47. Noser, A.A.; Abdelmonsef, A.H.; Salem, M.M. Design, Synthesis and Molecular Docking of Novel Substituted Azepines as Inhibitors of PI3K/Akt/TSC2/mTOR Signaling Pathway in Colorectal Carcinoma. *Bioorganic Chem.* **2023**, *131*, 106299. [CrossRef]
48. Lapillo, M.; Tuccinardi, T.; Martinelli, A.; Macchia, M.; Giordano, A.; Poli, G. Extensive Reliability Evaluation of Docking-Based Target-Fishing Strategies. *Int. J. Mol. Sci.* **2019**, *20*, 1023. [CrossRef] [PubMed]
49. Dhanalakshmi, B.; Anil Kumar, B.M.; Srinivasa Murthy, V.; Srinivasa, S.M.; Vivek, H.K.; Sennappan, M.; Rangappa, S. Design, Synthesis and Docking Studies of Novel 4-Aminophenol-1,2,4-Oxadiazole Hybrids as Apoptosis Inducers against Triple Negative Breast Cancer Cells Targeting MAP Kinase. *J. Biomol. Struct. Dyn.* **2023**, 1–17. [CrossRef] [PubMed]
50. Dicato, M.; Plawny, L.; Diederich, M. Anemia in cancer. *Ann. Oncol.* **2010**, *21*, vii167–vii172. [CrossRef]
51. HEl Zahabi, S.A.; Nafie, M.S.; Osman, D.; Elghazawy, N.H.; Soliman, D.H.; EL-Helby, A.A.H.; Arafa, R.K. Design, synthesis and evaluation of new quinazolin-4-one derivatives as apoptotic enhancers and autophagy inhibitors with potent antitumor activity. *Eur. J. Med. Chem.* **2021**, *222*, 113609.
52. Boraei, A.T.A.; Eltamany, E.H.; Ali, I.A.I.; Gebriel, S.M.; Nafie, M.S. Synthesis of new substituted pyridine derivatives as potent anti-liver cancer agents through apoptosis induction: In vitro, in vivo, and in silico integrated approaches. *Bioorganic Chem.* **2021**, *111*, 104877. [CrossRef]

Disclaimer/Publisher's Note: The statements, opinions and data contained in all publications are solely those of the individual author(s) and contributor(s) and not of MDPI and/or the editor(s). MDPI and/or the editor(s) disclaim responsibility for any injury to people or property resulting from any ideas, methods, instructions or products referred to in the content.

Article

Novel Coumarin Derivatives as Potential Urease Inhibitors for Kidney Stone Prevention and Antiulcer Therapy: From Synthesis to In Vivo Evaluation

Kiran Shahzadi [1,2], Syed Majid Bukhari [1], Asma Zaidi [1,*], Tanveer A. Wani [3], Muhammad Saeed Jan [4], Seema Zargar [5], Umer Rashid [1], Umar Farooq [1,6], Aneela Khushal [1] and Sara Khan [1,*]

1. Department of Chemistry, COMSATS University Islamabad, Abbottabad Campus, Abbottabad 22060, KPK, Pakistan; kiranshahzadi12@gmail.com (K.S.); majidbukhari@cuiatd.edu.pk (S.M.B.); umerrashid@cuiatd.edu.pk (U.R.); umarf@cuiatd.edu.pk (U.F.); aneelakhan460@gmail.com (A.K.)
2. School of Chemistry and Chemical Engineering Technology, Beijing Institute of Technology, Beijing 100811, China
3. Department of Pharmaceutical Chemistry, College of Pharmacy, King Saud University, P.O. Box 2457, Riyadh 11451, Saudi Arabia; twani@ksu.edu.sa
4. Department of Pharmacy, The Professional Institute of Health Sciences, Mardan 23200, KPK, Pakistan
5. Department of Biochemistry, College of Science, King Saud University, P.O. Box 22452, Riyadh 11451, Saudi Arabia; szargar@ksu.edu.sa
6. Beijing National Laboratory for Molecular Sciences, State Key Laboratory of Molecular Reaction Dynamics, Institute of Chemistry, Chinese Academy of Sciences, Beijing 100190, China

* Correspondence: asmazaidi@cuiatd.edu.pk (A.Z.); sarakhan@cuiatd.edu.pk (S.K.)

Citation: Shahzadi, K.; Bukhari, S.M.; Zaidi, A.; Wani, T.A.; Jan, M.S.; Zargar, S.; Rashid, U.; Farooq, U.; Khushal, A.; Khan, S. Novel Coumarin Derivatives as Potential Urease Inhibitors for Kidney Stone Prevention and Antiulcer Therapy: From Synthesis to In Vivo Evaluation. *Pharmaceuticals* **2023**, *16*, 1552. https://doi.org/10.3390/ph16111552

Academic Editors: Roberta Rocca and Valentina Noemi Madia

Received: 7 September 2023
Revised: 25 October 2023
Accepted: 27 October 2023
Published: 2 November 2023

Copyright: © 2023 by the authors. Licensee MDPI, Basel, Switzerland. This article is an open access article distributed under the terms and conditions of the Creative Commons Attribution (CC BY) license (https://creativecommons.org/licenses/by/4.0/).

Abstract: The presence of ammonium ions in urine, along with basic pH in the presence of urease-producing bacteria, promotes the production of struvite stones. This causes renal malfunction, which is manifested by symptoms such as fever, nausea, vomiting, and blood in the urine. The involvement of urease in stone formation makes it a good target for finding urease enzyme inhibitors, which have the potential to be developed as lead drugs against kidney stones in the future. The documented ethnopharmacology of coumarin 2-one against bacterial, fungal and viral strains encouraged us to synthesize new derivatives of coumarins by reacting aromatic aldehydes with 4-aminocoumarin. The synthesized compounds (**2a** to **11a**) were evaluated for their antimicrobial, in vitro, and in silico properties against the urease enzyme. The study also covers in vivo determination of the synthesized compounds with respect to different types of induced ulcers. The molecular docking study along with extended MD simulations (100 ns each) and MMPBSA study confirmed the potential inhibitory candidates as evident from computed ΔG_{bind} (**3a** = −11.62 and **5a** = −12.08 Kcal/mol) against the urease enzyme. The in silico analyses were augmented by an enzymatic assay, which revealed that compounds **3a** and **5a** had strong inhibitory action, with IC_{50} of 0.412 µM (64.0% inhibition) and 0.322 µM (77.7% inhibition), respectively, compared to standard (Thiourea) with 82% inhibition at 0.14 µM. Moreover, the most active compound, **5a**, was further tested in vivo for antiulcer activity by different types of induced ulcers, including pyloric ligation-, ethanol-, aspirin-, and histamine-induced ulcers. Compound **5a** effectively reduced gastric acidity, lipid peroxidation, and ulceration in a rat model while also inhibiting gastric ATPase activity, which makes it a promising candidate for ulcer treatment. As a result of the current research, **3a** and **5a** may be used as new molecules for developing potent urease inhibitors. Additionally, the compound **3a** showed antibacterial activity against *Staphylococcus aureus* and *Salmonella typhimurium*, with zones of inhibition of 41 ± 0.9 mm and 35 ± 0.9 mm, respectively. Compound **7a** showed antibacterial activity against *Staphylococcus aureus* and *Salmonella typhimurium*, with zones of inhibition of 30 ± 0.8 mm and 42 ± 0.8 mm, respectively. These results prove that the synthesized compounds also possess good antibacterial potential against Gram-positive and Gram-negative bacterial strains.

Keywords: 4-aminocoumarin; Schiff bases; urease inhibition; molecular docking; antibacterial activity; MD simulation

1. Introduction

The development of kidney stones is one of the major health hazards faced by several human beings. The elements for this clinical condition are basic urinary pH with the existence of ammonium ions in the urinary tract and urease-producing bacteria present in the vicinity. Urine formation is known to involve bacterial strains (Gram-positive and Gram-negative), as well as yeast and mycoplasma. Urease present in certain bacterial strains may act as a virulence factor in this scenario. Hence, the combination of phosphate with urine containing magnesium, carbonate apatite, and ammonium leads to the formation of kidney stones. Certain heterocyclic compounds are well known for their antibacterial, antifungal, and antiviral potential. Therefore, the synthesis of new derivatives of such heterocyclic compounds may lead medicinal chemistry towards the discovery of good and cost-effective antimicrobial agents against the drawbacks caused by excessive urease enzyme production [1].

One of the classes of such heterocyclic compounds is coumarin, which belongs to benzopyrans, the leading members of heterocyclic organic compounds. Currently, there is a great emphasis on the synthesis of coumarin-based compounds due to their pharmacological and biological activities. Derivatives of coumarin can be synthesized by using different methods, including Knoevenagel, Kostanecki–Robinson, Perkin, Wittig, Reformatsky, solid-state synthesis of coumarin, and Pechmann reactions [2–4].

The 4-hydroxycoumarin anticoagulant is one of the most important derivatives of coumarin. It belongs to the class of anticoagulant (vitamin K antagonist) drug molecules. Said molecule is considerably important, as it can inhibit vitamin K epoxide reductase. It is also of notable interest as being an important reagent for heterocyclization, which will become one of the primary approaches in the forthcoming synthesis of heterocyclic derivatives in the area of combinatorial chemistry [5]. The 4-hydroxycoumarin and its derivatives namely coumatetralyl, acenocoumarol, warfarin, phenprocoumon, and coumachlor show a large biological potential which includes analgesic, anticoagulant, cytotoxic, antiproliferative, hypnotic and sedative, anti-tumor, anti-inflammatory, antibacterial, antifungal, anticancer, antioxidant and anti-HIV [6,7].

Anti-inflammatory and antioxidant properties of coumarins are known for minimizing the risk of diabetes, cancer, and cardiovascular diseases [8–11]. Furthermore, the derivatives of coumarin are used for the treatment of renal stones and burns. The inhibition of 5α-reductase as well as platelet accumulation inhibition has also been reported.

The Schiff base synthesis of coumarins further enhances their biological activities. Schiff bases are obtained by the reaction of amines with a stoichiometric quantity of aldehydes. Schiff bases that contain the azomethine group play an integral role in the field of drug discovery and development and present a large number of pharmacological applications, including anticoagulant, anticonvulsant, antibacterial, antifungal, anti-inflammatory [12–15], sedative, and antioxidant.

The presence of ammonium ions in urine, along with a basic pH in the presence of urease-producing bacteria, promotes the production of struvite stones. This causes renal malfunction, which is manifested by symptoms such as fever, nausea, vomiting, and blood in the urine. The involvement of urease in stone formation makes it a good target for finding urease enzyme inhibitors, which have the potential to be developed as lead drugs against kidney stones in the future. The documented ethnopharmacology of coumarin 2-one against bacterial, fungal, and viral strains encouraged us to synthesize new derivatives of coumarins by reacting aromatic aldehydes with 4-aminocoumarin. Along with the antibacterial and anti-urease potential of these compounds experimentally, urease inhibition has also been evaluated by an in silico approach followed by the determination of antiulcer properties of the synthesized compounds in different ulcer-induced models in rats.

2. Results and Discussion

The synthesis of coumarin derivatives was initially optimized using a range of different conditions, as indicated in Table 1. Following the successful optimization of the initial reaction, we proceeded to synthesize various coumarin derivatives. Additionally, we carried out further optimization for the second step of this process, as outlined in Table 2.

Table 1. Optimization conditions for the first step of the reaction.

1st Reactant (Equiv.)	2nd Reactant (Equiv.)	Solvents	Temp. (°C)	Time (h)	Results
1	1	Ethoxyethanol	170	48	Too many side products are present
1	2	Glacial acetic acid	180	72	Too many side products are present
1	2	Solvent-free	130	05	No reaction
1	5	Solvent-free	170	14–15	70% yield

Table 2. Optimization conditions for the second step of reaction.

1st Reactant (Equiv.)	2nd Reactant (Equiv.)	Solvents	Temp. (°C)	Time (h)	Results
1	1	Ethanol	170	48	10% yield
1	1.5	Ethanol + glacial acetic acid (3–4 drops)	170	72	10% yield
1	1.5	Methanol + glacial acetic acid (3–4 drops)	170	72	15% yield
1	1	Analytical grade methanol	170	36–82 Continuous reflux	65–75% yield

4-aminocoumarin

Light yellowish powder; % age yield: 59%; melting point range = 226–227 °C; R_f = 0.51 ($CH_3COOC_2H_5:CH_3OH::$ 1.5:1); UV-vis (THF) λ_{max} (nm) = 305; FT-IR: ν (cm^{-1}) = 3370 (N-H stretching), 1596 (C=O vibration), 3196 (=C-H stretching vibration), 1455 (C=C vibration), 1321 (C-N stretching vibration); ^{1}HNMR: (400 MHz, DMSO-d_6); δ: 2.51 (s, 2H, H-11), 5.22 (s, 1H, H-9), 7.98 (d, J = 8 Hz, 1H, H-6), 7.59 (m, 1H, H-1), 7.31 (t, J = 8 Hz, 2H, H-2, 3). ^{13}C NMR δ:159.58, 153.47, 152.63, 132.53, 125.74, 122.64, 114.66, 114.45, 86.62.

Schiff bases (**2a–11a**)

The structures (Figure 1) and spectral data of synthesized derivatives of 4-aminocoumarin are given below.

(E)-4-(5-fluoro-2-hydroxybenzylideneamino)-2H-chromen-2-one (**2a**):

White solid; yield = 69%; R_f = 0.5 ($CH_3COOC_2H_5:C_6H_{14}::$ 1.5:1); melting point range = 295–298 °C; UV-vis (THF) λ_{max} (nm) = 307; FT-IR: ν (cm^{-1}) = 1568 (C=N vibrations), 1701 (C=O stretch), 3048 (=C-H stretching), 1383 (C-F vibration), 3248 (OH vibration); ^{1}H-NMR: (400 MHz, CDCl$_3$); δ: 10.43 (s, 1H, H-13), 5.37 (s, 1H, H-9), 8.16 (d, J = 8 Hz, 1H, H-6), 7.53 (m, 1H, H-1), 7.67 (m, 1H, H-2), 8.06 (d, J = 8 Hz, 1H, H-3), 7.49 (s, 1H, H-15), 7.03 (m, 1H, H-17), 6.87 (dd, J = 8 Hz and 4 Hz, 1H, H-18), 3.51 (s, 1H, H-20). ^{13}C NMR δ: 164.44, 164.41, 163.28, 156.23, 156.16, 156.14, 154.45, 154.22, 151.59, 133.22, 128.76, 123.92, 121.28, 121.21, 119.59, 119.43, 118.50, 118.25, 118.18, 117.69, 116.66, 116.50, 102.80

(E)-4-(3-nitrobenzylideneamino)-2H-chromen-2-one (**3a**):

White powder; yield = 64%; R_f = 0.6 ($CH_3COOC_2H_5:C_6H_{14}::$ 1.5:1); melting point range = 200–202 °C; UV-vis (THF) λ_{max} (nm) = 304; FT-IR: ν (cm^{-1}) = 1560 (C=N stretching), 1650 (C=O stretching vibration), 3047 (=C-H vibrations), 1524 and 1341 (NO$_2$ stretching), 1443 (C=C, Ar stretching vibrations); ^{1}H-NMR: (400 MHz, CDCl$_3$); δ: 8.10 (s, 1H, H-13), 6.15 (s, 1H, H-9), 7.58 (d, J = 8 Hz, 1H, H-6), 7.47 (m, 2H, H-1, 3), 7.54 (t, J = 8 Hz, 1H, H-2), 8.02 (d, J = 8 Hz, 1H, H-15), 8.12 (d, J = 8 Hz, 1H, H-16), 8.17 (d, J = 8 Hz, 1H, H-17), 7.69 (s,

1H, H-19). ^{13}C NMR δ: 163.28, 163.22, 154.56, 151.39, 146.84, 136.79, 133.55, 130.37, 129.03, 128.60, 123.69, 122.65, 122.13, 118.52, 117.61, 103.03.

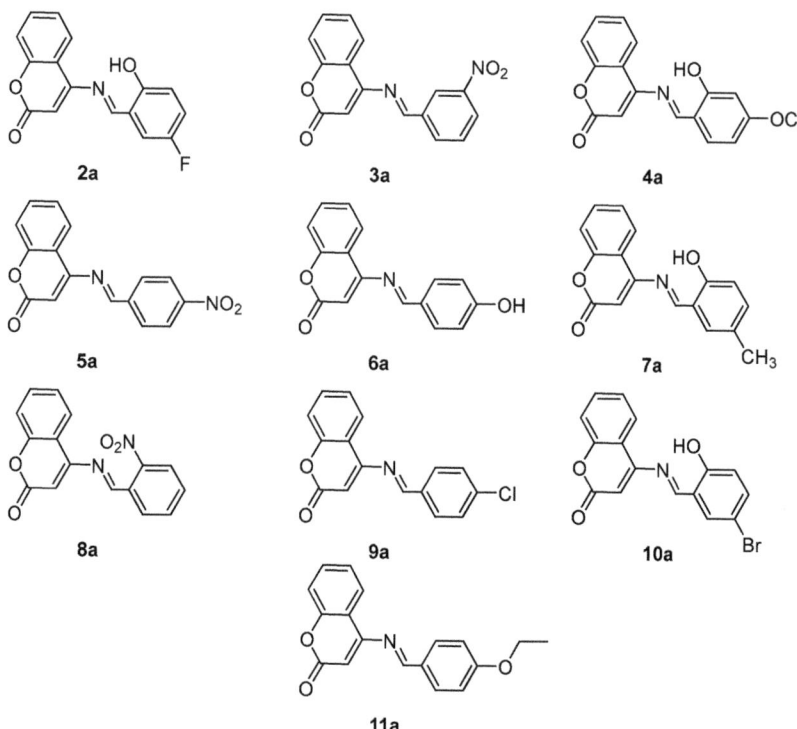

Figure 1. The structures of synthesized derivatives (2a–11a).

(*E*)-4-(2-hydroxy-4-methoxybenzylideneamino)-2*H*-chromen-2-one (**4a**):

Pale-yellow solid; yield = 57%; R_f = 0.6 (CH$_3$COOC$_2$H$_5$:C$_6$H$_{14}$:: 1.5:1); melting point range = 185–188 °C; UV-vis (THF) $λ_{max}$ (nm) = 368; FT-IR: ν (cm^{-1}) = 1543 (C=N stretching), 1596 (C=O vibration), 3061 (=C-H stretching vibration), 3218 (OH stretch), 1453 (C=C vibrations); ^{1}H-NMR: (400 MHz, DMSO-d_6); δ: 7.33 (s, 1H, H-13), 5.22 (s, 1H, H-9), 7.99 (d, *J* = 8 Hz, 2H, H-6, 15), 7.59 (m, 2H, H-1, 2), 7.30 (d, *J* = 8 Hz, 2H, H-2, 16), 7.38 (s, 1H, H-18), 5.76 (s, 1H, H-20), 3.34 (s, 3H, H-21).

(*E*)-4-(4-nitrobenzylideneamino)-2*H*-chromen-2-one (**5a**):

Off-white solid; yield = 63%; R_f = 0.6 (CH$_3$COOC$_2$H$_5$:C$_6$H$_{14}$:: 1.5:1); melting point range = 255–257 °C; UV-vis (THF) $λ_{max}$ (nm) = 374; FT-IR: ν (cm^{-1}) = 1577 (C=N stretching vibration), 1670 (C=O stretch), 3050 (=C-H vibrations), 1509 and 1340 (NO$_2$ stretching vibrations), 1436 (C=C vibrations); ^{1}H-NMR: (400 MHz, CDCl$_3$); δ: 12.54 (s, 1H, H-13), 6.12 (s, 1H, H-9), 8.03 (d, *J* = 8 Hz, 1H, H-6), 7.64 (m, 1H, H-1), 7.69 (t, *J* = 8 Hz, 1H, H-2), 7.73 (d, *J* = 8 Hz, 1H, H-3), 8.17 (d, *J* = 8 Hz, 2H, H-15, 19), 8.19 (d, *J* = 8 Hz, 2H, H-16, 18). ^{13}C NMR δ: 164.73, 163.76, 163.28, 158.34, 154.56, 151.48, 133.55, 130.31, 128.60, 123.67, 118.50, 117.55, 114.58, 107.24, 102.80, 101.85, 55.57.

(*E*)-4-(4-hydroxybenzylideneamino)-2*H*-chromen-2-one (**6a**):

White solid; yield = 61%; R_f = 0.7 (CH$_3$COOC$_2$H$_5$:C$_6$H$_{14}$:: 1.5:1); melting point range = 280–282 °C; UV-vis (THF) $λ_{max}$ (nm) = 305; FT-IR: ν (cm^{-1})= 1582 (C=N stretching vibrations), 1517 (C=O stretching), 3191 (=C-H vibration), 3337 (OH stretching), 1436 (C=C, Ar stretch); ^{1}H-NMR: (400 MHz, DMSO-d_6); δ: 9.28 (s, 1H, H-13), 4.98 (s, 1H, H-9), 8.51 (d,

J = 8 Hz, 1H, H-6), 7.46 (t, J = 8 Hz, 1H, H-1), 7.68 (t, J = 8 Hz, 1H, H-2), 7.41 (d, J = 8 Hz, 1H, H-3), 7.11 (d, J = 0 Hz, 2H, H-15, 19), 6.64 (d, J = 8 Hz, 2H, H-16, 18), 9.71 (br.s, 1H, H-20, -OH). ^{13}C NMR δ: 164.05, 163.28, 154.56, 151.58, 148.24, 139.10, 133.55, 128.60, 128.41, 124.30, 123.69, 118.52, 117.61, 103.03.

(*E*)-4-(2-hydroxy-5-methylbenzylideneamino)-2*H*-chromen-2-one (**7a**):

Yellow solid; yield = 79%; R_f = 0.6 (CH$_3$COOC$_2$H$_5$:C$_6$H$_{14}$:: 1.5:1); melting point range = 253–255 °C; UV-vis (THF) λ_{max} (nm) = 368; FT-IR: ν (cm^{-1}) = 1545 (C=N stretch), 1608 (C=O stretching vibration), 3150 (=C-H stretching), 3374 (OH stretching vibrations), 2959 (C-H vibration), 1454 (C=C, Ar stretching vibration); ^{1}H-NMR: (400 MHz, DMSO-d_6); δ: 10.22 (s, 1H, H-13), 5.22 (s, 1H, H-9), 7.99 (d, J = 8 Hz, 1H, H-6), 7.33 (m, 1H, H-1), 7.59 (t, J = 8 Hz, 1H, H-2), 6.90 (d, J = 8 Hz, 1H, H-3), 7.31 (s, 1H, H-15), 7.45 (d, J = 8 Hz, 1H, H-17), 7.35 (d, J = 8 Hz, 1H, H-18), 7.38 (br.s, 1H, H-20, -OH), 2.24 (s, 3H, H-21, -CH$_3$). ^{13}C NMR δ: 164.31, 163.28, 157.14, 154.45, 151.59, 133.27, 132.49, 131.98, 129.44, 128.76, 123.92, 120.58, 118.50, 117.69, 116.51, 102.80, 20.61.

(*E*)-4-(2-nitrobenzylideneamino)-2*H*-chromen-2-one (**8a**):

White solid; yield = 60%; R_f = 0.6 (CH$_3$COOC$_2$H$_5$:C$_6$H$_{14}$:: 1.5:1); melting point range = 265–268 °C; UV-vis (THF) λ_{max} (nm) = 304; FT-IR: ν (cm^{-1}) = 1571 (C=N stretching vibrations), 1667 (C=O vibration), 3249 (=C-H stretching), 1517 and 1348 (NO$_2$ stretching vibrations), 1439 (C=C, Ar stretching), ^{1}H-NMR: (400 MHz, DMSO-d_6); δ: 8.88 (s, 1H, H-13), 6.49 (s, 1H, H-9), 7.75 (d, J = 8 Hz, 1H, H-6), 7.40–7.46 (m, 2H, H-1,3), 7.50 (t, J = 8 Hz, 1H, H-2), 7.92 (d, J = 8 Hz, 1H, H-15), 7.66 (t, J = 8 Hz, 1H, H-16), 7.62 (m, 1H, H-17), 8.12 (d, J = 8 Hz, 1H, H-18). ^{13}C NMR δ: 163.28, 159.16, 154.56, 151.06, 147.42, 133.55, 131.35, 130.56, 130.00, 128.60, 128.03, 125.27, 123.69, 118.53, 117.61, 102.80.

(*E*)-4-(4-chlorobenzylideneamino)-2*H*-chromen-2-one (**9a**):

Yellowish powder; yield = 57%; R_f = 0.6 (CH$_3$COOC$_2$H$_5$:C$_6$H$_{14}$:: 1.5:1); melting point range = 255–257 °C; UV-vis (THF) λ_{max} (nm) = 301; FT-IR: ν (cm^{-1}) = 1524 (C=N stretch), 1669 (C=O vibrations), 3047 (=C-H stretching vibration), 1437 (C=C vibrations); ^{1}H-NMR: (400 MHz, CDCl$_3$); δ: 11.34 (s, 1H, H-13), 6.07 (s, 1H, H-9), 8.08 (m, 1H, H-6), 7.44 (d, J = 8 Hz, 2H, H-1, 3), 7.66 (t, J = 8 Hz, 1H, H-2), 7.32 (d, J = 8 Hz, 2H, H-15, 19), 7.18 (d, J = 8 Hz, 2H, H-16, 18). ^{13}C NMR δ: 164.10, 163.28, 154.30, 151.72, 135.21, 133.39, 133.26, 129.32, 128.92, 128.76, 123.92, 118.56, 117.69, 103.16.

(*E*)-4-(5-bromo-2-hydroxybenzylideneamino)-2*H*-chromen-2-one (**10a**):

White solid; yield = 75%; R_f = 0.6 (CH$_3$COOC$_2$H$_5$:C$_6$H$_{14}$:: 1.5:1); melting point range = 276–278 °C; UV-vis (THF) λ_{max} (nm) = 321; FT-IR: ν (cm^{-1}) = 1524 (C=N stretching vibration), 3342 (O-H stretching vibration), 1670 (C=O stretching vibration), 3208 (=C-H stretching vibration), 1509 (C=C, Ar stretching vibration); ^{1}H-NMR: (400 MHz, DMSO-d_6); δ: 8.18 (s, 1H, H-13), 5.54 (s, 1H, H-9), 8.08 (d, J = 8 Hz, 1H, H-6), 7.56 (t, J = 8 Hz, 1H, H-1), 7.69 (t, J = 8 Hz, 1H, H-2), 7.36 (t, J = 8 Hz, 1H, H-3), 7.43 (s, 1H, H-15), 7.31 (d, J = 8 Hz, 1H, H-17), 7.22 (d, J = 8 Hz, 1H, H-18), 7.85 (br.s, 1H, H-20, -OH). ^{13}C NMR δ: 164.29, 163.28, 157.95, 154.56, 151.59, 136.03, 133.25, 133.20, 128.61, 123.83, 120.65, 118.50, 117.93, 117.80, 112.47, 102.94.

(*E*)-4-(4-ethoxybenzylideneamino)-2*H*-chromen-2-one (**11a**):

White powder; yield = 69%; R_f = 0.6 (ethyl acetate:*n*-hexane:: 1.5:1); m.p. = 264–266 °C; UV-vis (DCM) λ_{max} (nm) = 305 nm; FT-IR: ν = 1509 cm^{-1} (C=N stretching vibration), 1650 cm^{-1} (C=O stretching vibration), 3125 cm^{-1} (=C-H stretching vibration), 1443 cm^{-1} (C=C, Ar stretching vibration), 1099 cm^{-1} (C-O stretching vibration); ^{1}H-NMR: (400 MHz, DMSO-d_6); δ: 9.7 (s, 1H, H-13), 5.04 (s, 1H, H-9), 8.54 (d, J = 8 Hz, 1H, H-6), 7.50 (t, J = 8 Hz, 1H, H-1), 7.67 (t, J = 8 Hz, 1H, H-2), 7.45 (d, J = 8 Hz, 1H, H-3), 7.21 (d, J = 8 Hz, 2H, H-15, 19), 6.79 (d, J = 8 Hz, 2H, H-16, 18), 3.92 (q, J = 8 Hz, 2H, H-20), 1.26 (t, J = 8 Hz, 3H, H-21). ^{13}C NMR δ: 164.14, 163.28, 160.88, 154.56, 151.58, 133.55, 130.24, 129.16, 128.60, 123.69, 118.52, 117.61, 114.73, 103.03, 63.56, 14.67.

2.1. UV-Vis Analysis

Absorbance spectra of all the newly synthesized Schiff bases under investigation were recorded from 200 nm to 800 nm using a quartz cuvette (10 mm), and THF was used as solvent. All the newly synthesized compounds possess a coumarin ring in their chemical structure. The maximum absorbance of compound **2a** was recorded at 305 nm, whereas spectra for compounds **3a–11a** showed maximum absorbance at 307, 304, 368, 305, 374, 368, 304, 310, 301, 321, and 305 nm, respectively. The compounds **4a, 7a**, and **8a** showed a blue shift (hypsochromic), whereas the compounds **3a, 5a, 6a, 9a, 10a**, and **11a** showed a red shift (bathochromic) compared to the starting material (compound **2**). The shift in maximum absorbance can be attributed to extended conjugation and provides evidence in favor of the successful synthesis of the anticipated products.

2.2. FT-IR Analysis

The FT-IR spectrum of newly synthesized compound **2a** shows a characteristic absorption peak at 3370 cm^{-1} with a shoulder band at 3484 cm^{-1}, which corresponds to the primary amine group (N-H). These characteristic bands disappear in the FT-IR spectra of synthesized Schiff bases. Also, a characteristic absorption peak at around 1517–1575 cm^{-1} appears in each FT-IR spectrum of Schiff bases, which corresponds to strong C=N vibrations [16]. This analysis further confirms that the amino group of compound **2a** has successfully been converted to an imine group in each succeeding Schiff base compound. Furthermore, the FT-IR spectra of newly synthesized compounds exhibit a strong absorption peak at 3047–3214 cm^{-1}, which can be attributed to the presence of a methylene group (=C-H) [17]. The appearance of this particular signal indicates that this moiety did not get involved in the reaction with substituted aromatic aldehydes (Figure S1).

2.3. NMR Spectral Analysis

The NMR spectra of newly synthesized compounds were obtained using a Bruker AM-400 MHz NMR spectrometer. All of the compounds were either soluble in DMSO-d_6 or CDCl$_3$. The ^{1}H-NMR data of all the newly synthesized compounds confirmed their proposed structures.

In the case of compound **2a**, a singlet appearing at δ 2.51 ppm accounted for the two protons of the amino group (NH$_2$). The active methylene proton in compound **2a** at position 9 has no neighboring proton; therefore, it showed a singlet at δ 5.22 ppm. The proton at position 6 shows strong coupling with its ortho proton, does not show long-range coupling, and gives a doublet signal at δ 7.98 ppm. The proton at position 1 shows a multiplet signal at δ 7.59 ppm, which is due to a long-range coupling. Protons at positions 2 and 3 are neighboring protons and show a triplet signal at δ 7.31 ppm. All of these signals confirm the successful synthesis of compound **2a** (4-aminocoumarin).

The singlet peak appearing at δ 2.51 ppm due to two protons of the amino group (NH$_2$) in 4-aminocoumarin disappears in the ^{1}H-NMR spectrum of Schiff bases (**2a–11a**); instead, a singlet due to azomethine proton (N=CH) appears at 7.43–12.5 ppm. The shielding and deshielding of the imine proton are due to the mesomeric effect of pi-electrons of aromatic and aldehydic moiety. The appearance of a characteristic singlet peak due to methylene proton at δ 4.98–6.49 ppm in the ^{1}H-NMR spectra is strong evidence for the formation of Schiff bases (Figure S1).

2.4. Antibacterial Activity

All the newly synthesized Schiff bases of 4-aminocoumarin (**2a–11a**) were tested against two bacterial strains: *Salmonella typhimurium* (Gram-negative) and *Staphylococcus aureus* (Gram-positive). Zones of inhibition (mm) of all the synthesized compounds against these two strains are shown in Figure 2.

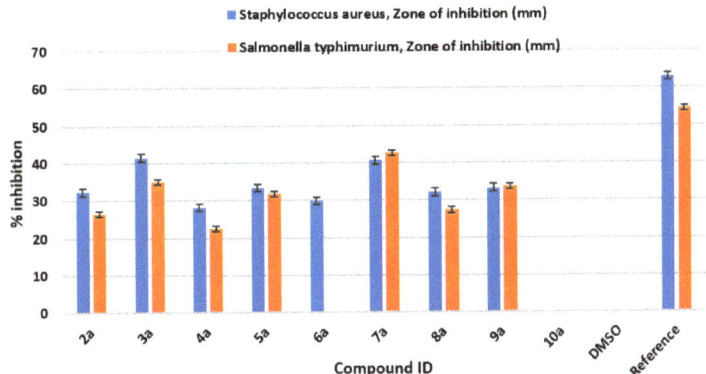

Figure 2. Antibacterial bioassay screening of newly synthesized derivatives (**2a–11a**).

The zones of inhibition were measured in millimeters. The 50 µg/mL concentration of ciprofloxacin (reference drug) showed 62.9 ± 0.94 and 54.4 ± 0.50 mm zones of inhibition against *Staphylococcus aureus* and *Salmonella typhimurium*, respectively. The negative control (DMSO) did not show any activity against these two strains. The synthesized compounds were also tested against the two bacterial strains (concentration: 200 µg/mL). The compounds **3a** and **7a** showed higher zones of inhibition against these two bacterial strains compared to other synthesized derivatives. Compound **3a** showed antimicrobial activity against *Staphylococcus aureus* with a zone of inhibition of 41.5 ± 0.96 mm and against *Salmonella typhimurium* with a zone of inhibition 35 ± 0.90. Also, compound **7a** showed antibacterial activity against *Staphylococcus aureus* with a zone of inhibition of 40.6 ± 0.76 mm and against *Salmonella typhimurium* with a zone of inhibition of 42.6 ± 0.76 mm. Hence, it can be deduced that compound **3a** is highly active against *Staphylococcus aureus* compared to compound **7a**, whereas compound **7a** is comparatively more effective against *Salmonella typhimurium* compared to compound **3a**. The antibacterial bioassay proves that compound **3a** can be used as a lead for the development of antibacterial drugs against *Staphylococcus aureus* and compound **7a** can be taken as an antibiotic against the Gram-negative bacterial strain *Salmonella typhimurium*.

2.5. In Vitro Urease Inhibition

In continuation of the in silico studies, compounds **3a** and **5a** were subjected to in vitro urease inhibition assay. The results of this assay further confirm the findings obtained as a result of in silico studies. The inhibition potential of compounds **3a** and **5a** is comparable to that of the standard used (thiourea). Further optimization of **3a** and **5a** may lead to the development of new standards for future use in urease inhibition assays (Table 3).

Table 3. Inhibition of synthesized compounds and their IC$_{50}$ value along with standard deviation/error of the mean.

Compound No.	% Inhibition ± STD	IC$_{50}$ ± SEM (µM)
3a	64.0 ± 0.42	0.412 ± 0.10
5a	77.7 ± 0.64	0.322 ± 0.13
7a	60.2 ± 0.32	0.112 ± 0.26
Thiourea	82.0 ± 0.15	0.140 ± 0.22

2.6. In Silico Studies

Urease has the tendency to catalyze the hydrolysis of urea into ammonia and carbon dioxide. It is a key virulence factor for a wide range of human infections: 10 of the 12 antibiotic-resistant priority pathogens designated by the World Health Organization (WHO) in 2017 are ureolytic bacteria that colonize and thrive in the host organism using urease activity [1,18].

All of the synthesized compounds were screened against a minimized crystal structure and the active site was identified by using the site finder tool embedded in MOE; the rest of the parameters were kept as default, with each molecule sampled in 10 conformations. All of the compounds docked into the urease's active pocket showed a strong binding affinity with the active site residues which is evident from ΔG_{bind} score (Table 4). All of the compounds have molecular weights within 300–350 and calculated logP values of 0.3–3.2, which places them within the lead-like range. Although all reported compounds are topologically similar to one another, binding affinity and docked poses suggest that compounds **3a** (−8.67 Kcal/mol), **5a** (−7.44 Kcal/mol) and **9a** (−6.96 Kcal/mol) fall into the best candidates for activity against the urease enzyme. In the docked poses, all the compounds showed van der Waal interaction as well as H–arene interaction through an unprecedented coumarin functional group.

Table 4. Binding energies and metal-oxygen binding of the compounds in the active site of urease enzyme.

Compounds ID	Binding Kcal/mol	Ni Binding
2a	−3.547	Absent
3a	−8.677	Present (Ni-O 2.3 Å)
4a	−5.180	Present (Ni-O 2.5 Å)
5a	−7.443	Present (Ni-O 2.4 Å)
6a	−5.892	Present (Ni-O 2.7 Å)
7a	−6.834	Present (Ni-O 2.3 Å)
8a	−4.131	Absent
9a	−6.967	Present (Ni-O 2.6 Å)

The lead compounds identified as a result of the molecular docking study were subjected to MD simulations in order to understand the structural dynamics, which is essential for identifying potential inhibitors associated with protein inhibition mechanisms. For all trajectories, RMSD calculations were performed to investigate the time-evolved behavior of the protein. Amplitude perturbations of about >2Å were detected in free protein, thereby illustrating the complicated structural changes involved with protein expansion over time. The bound protein has reduced dynamics relative to the free protein, as evidenced by the RMSD result. The average RMSD for all **3a** and **5a** complexed proteins revealed dynamics ranging from 1.2 to 1.4 Å. The attachment of **3a** and **5a** in the active site shifts the protein to a more rigid texture, whereas the free protein is more flexible in its motions. The RMSF is used to analyze the flexibility of individual residues in bound and free proteins. The orientation of the helix-turn-helix motif has a significant impact on urease enzyme binding to compounds (Figure 3A).

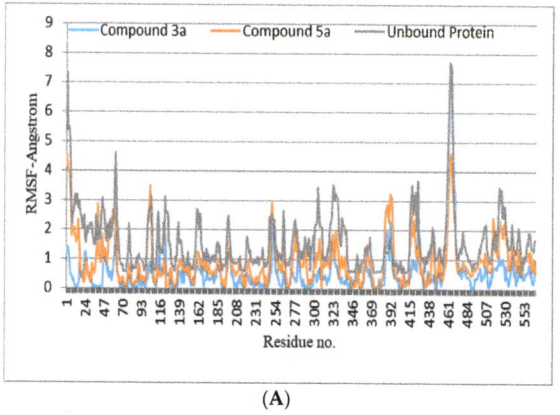

(**A**)

Figure 3. *Cont.*

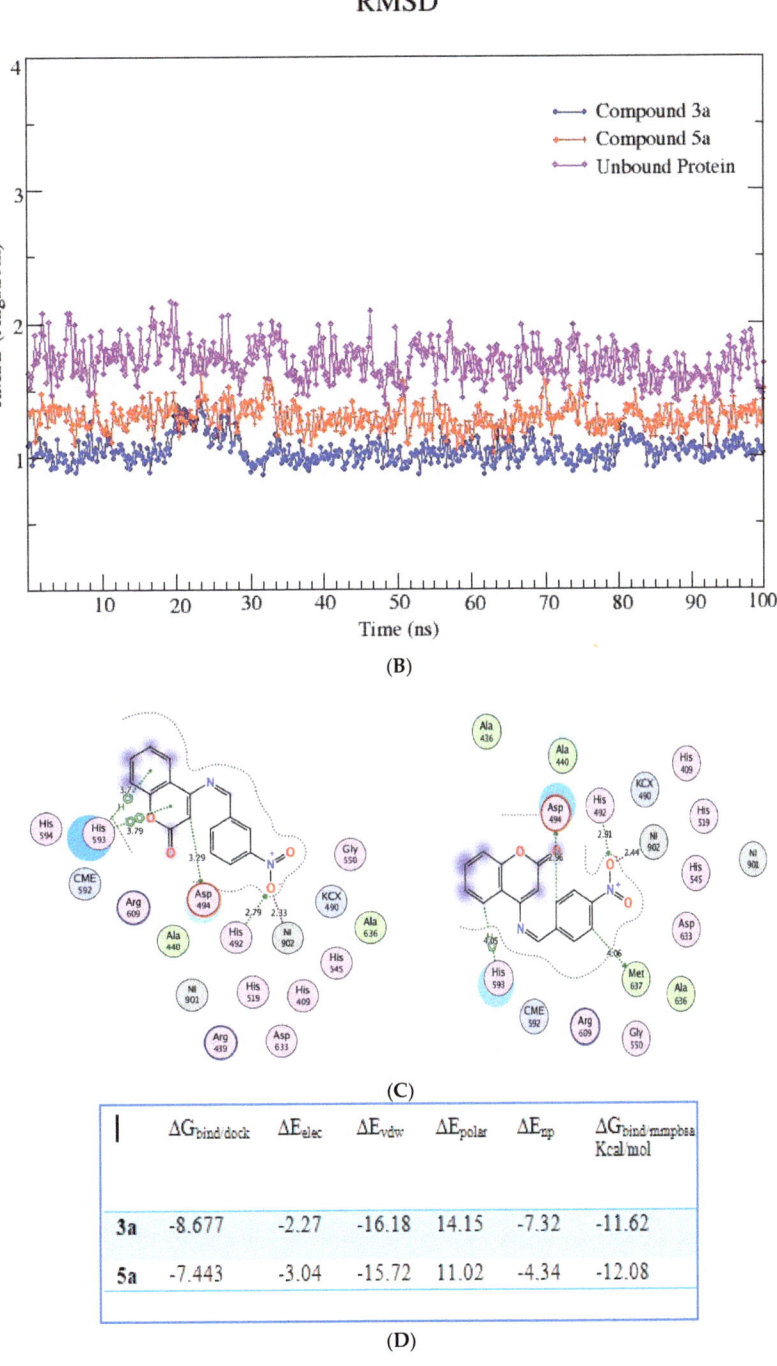

Figure 3. Cont.

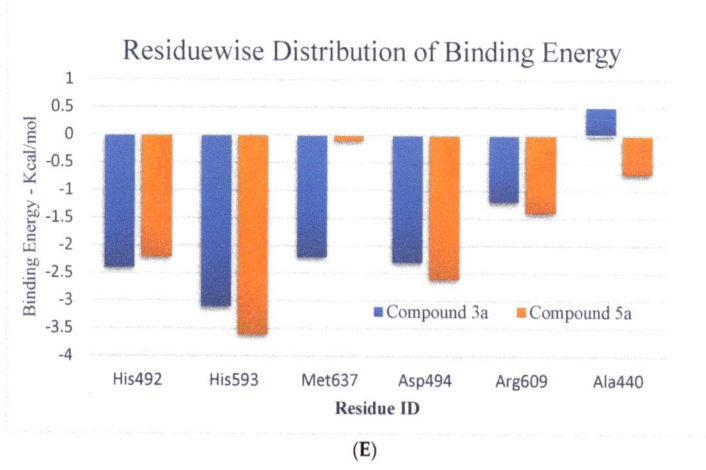

(E)

Figure 3. (**A**,**B**) RMSD (all-atom) and RMSF (per residue) plot of urease enzyme in free and bound form with respect to initial minimized structure. Protein—ligand complexes are color—coded as **3a** (blue), **5a** (red), and free protein (violet). (**C**) 2D interaction plots of **3a** and **5a** with urease protein using LigX tools of MOE software (https://www.chemcomp.com/Products.htm). The three-letter amino acid code is assigned to each residue. (**D**) Free energy of binding ΔG_{bind} for **3a** and **5a** compounds using molecular docking as well as an MMPBSA approach. (**E**) Major residues contributing toward protein—ligand interactions. (All values are given in Kcal/mol).

The RMSFs of ligand-bound urease are lower than those of free protein, and they are more prominent in the active site. An MMPBSA study was carried out to determine the thermodynamic parameters of the protein–ligand complexes, such as binding free energy and van der Waals, electrostatic, and polar solvation energies for the last 50 ns of the MD simulation trajectories. The binding energy of **3a** and **5a** bound urease complexes remained stable during the explored time scale, as evidenced by ΔG_{bind}.

The MMPBSA data analysis aids in calculating the contribution of individual amino acid residues to the total binding energy. The binding of both compounds is adjacent to the nickel ion, in a similar fashion to that of either urea or thiourea fragment which were being complexed by nickel (II) ion. The determination of trajectories that were gained also exposes that binding of **3a** and **5a** to the target protein occurs. It also shows effectively that the compounds under consideration efficiently take up the active sites of the target protein. It has also been observed that tight attachment of the helix-turn-helix motif occurs as a cover on the active site space. This results in obstruction of the flap closure of the urease active site and consequently leads towards inhibition of the urease enzyme. The studies of interaction showed that binding of the sample compounds within the active pocket leads towards the generation of Ni-O electrostatic bonds, whereby nickel-bound oxygen of the compounds **3a** and **5a** forms H-bonds with His492. Apart from this—the key residues (Arg609, Asp494, Arg439, Met697) facilitating the binding of the compounds **3a** and **5a** to the active site residues—a group of histidines (His593, His594, His492) encapsulated the active site, forming a highly charged atmosphere. Ni atom facilitates the binding of the substrate (urea) to the active site; however, the compounds **3a** and **5a** block the entry of the substrate via strongly coordinating with metal at one end and histidine at the other end. The formation of H-bonds between compounds **3a** and **5a** to that of active-site charged residues act as an additional force confirming the strong binding of compounds inside the active pocket.

2.7. In Vivo Pharmacology

2.7.1. Pylorus Ligation Activity

The pyloric ligation ulcer was induced in the rat's stomach, and the stomach was ligated for approximately 6 h. The insertion of compound **5a** resulted in a reduction in the ulcer index (14.30 ± 0.34 vs. 6.80 ± 0.24, Group I vs. Group V; see Figure 4). In terms of pH, fewer changes were observed (Figure 5), while the gastric content volume after the administration of **5a** was 2.75 ± 0.10 vs. 1.40 ± 0.14 (Group I vs. Group V; Figure 4). The total acidity was 43.10 ± 0.52 meq/L/100 g vs. 25.10 ± 0.16 meq per liter per 100 g (Group I vs. Group V; Figure 6), while the free acidity was 23.34 ± 0.42 meq per liter per 100 g vs. 9.50 ± 0.46 meq per liter per 100 g (Group I vs. Group V; Figure 6), and both were reduced upon the insertion of compound **5a**. Similarly, lipid peroxidation (0.70 ± 0.02; Group I) was reduced (0.42 ± 0.046; Group V, *** $p < 0.001$; Figure 7) upon the administration of compound **5a**.

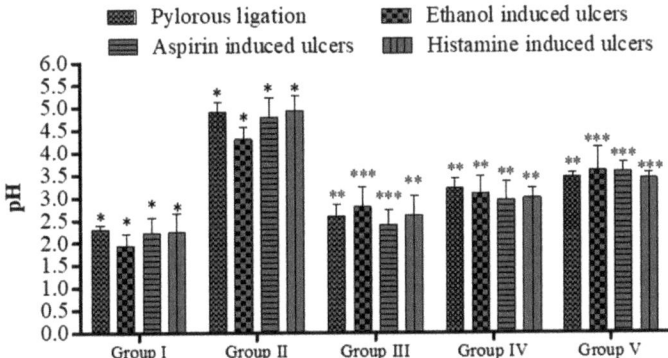

Figure 4. Effect of **5a** on pH of gastric contents in rats. Data are shown as mean ± SEM ($n = 6$). Group II was compared with groups III–V; * $p < 0.05$, ** $p < 0.01$; *** $p < 0.001$.

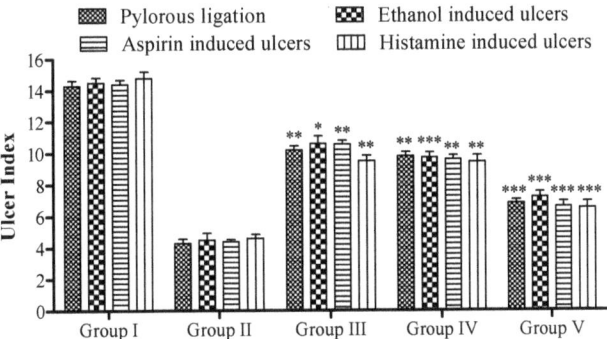

Figure 5. Effect of **5a** on rat's ulcer index. Data are shown as mean ± SEM ($n = 6$). Group II was compared with groups III–V; * $p < 0.05$, ** $p < 0.01$; *** $p < 0.001$.

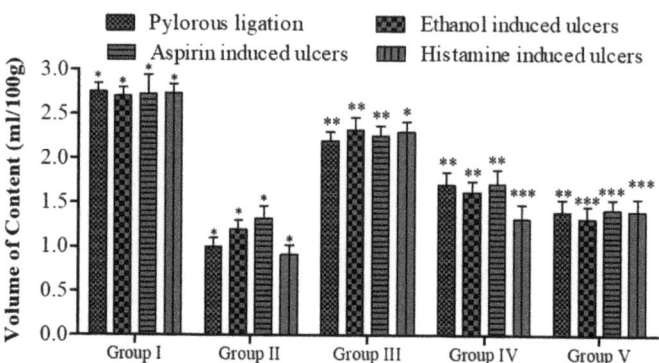

Figure 6. Effect of **5a** on gastric contents volume in rats. Data are shown as mean ± SEM ($n = 6$). Group II was compared with groups III–V; * $p < 0.05$, ** $p < 0.01$; *** $p < 0.001$.

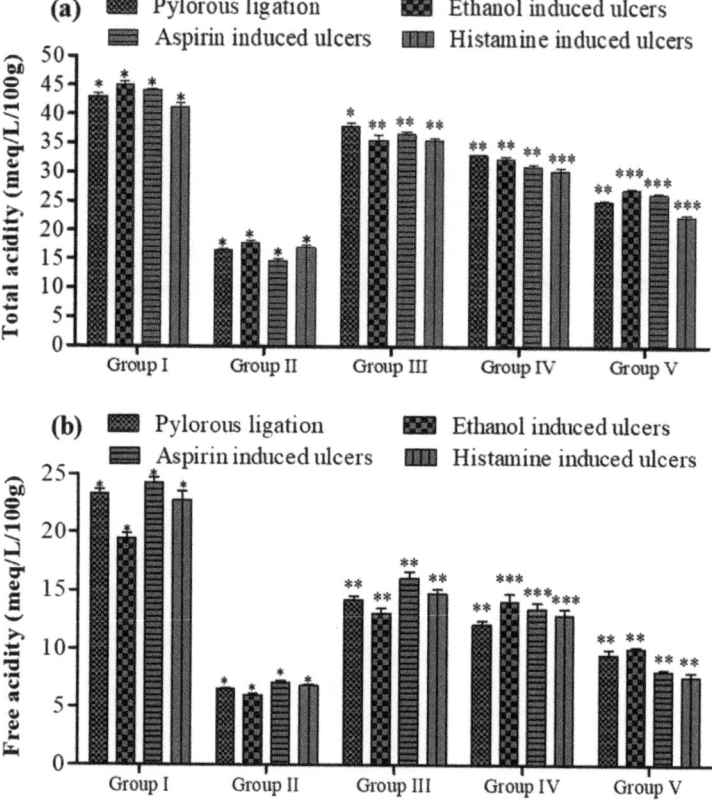

Figure 7. Effect of **5a** on (**a**) total and (**b**) free acidity of gastric contents in rats. Data are shown as mean ± SEM ($n = 6$). Group II was compared with groups III–V; * $p < 0.05$, ** $p < 0.01$; *** $p < 0.001$.

2.7.2. Ethanol-Induced Ulcer

To investigate ethanol-induced ulcers in the rat model, we administered compound **5a**, which exhibited promising results by significantly decreasing the ulcer index (*** $p < 0.001$; 14.50 ± 0.32 vs. 7.20 ± 0.36, Group I vs. Group V; see Figure 5). Gastric content pH is shown in Figure 4. Similarly, the gastric content volume (** $p < 0.01$) decreased after the insertion of the potent compound **5a** (2.70 ± 0.10 vs. 1.32 ± 0.14, Group I vs. Group V; Figure 6). The pH of the gastric contents (** $p < 0.01$, Group V) also increased (1.94 ± 0.26 in Group I, 3.60 ± 0.52 in Group V). The total acidity, initially at 45.10 ± 0.52 meq per liter per 100 g (Group I), was reduced to 27.10 ± 0.28 meq per liter per 100 g (Group V, *** $p < 0.001$; Figure 7), while the free acidity levels (19.45 ± 0.43 meq per liter per 100 g; Group I; Figure 8) decreased to 10.01 ± 0.16 meq per liter per 100 g (Group V, ** $p < 0.01$). Similarly, the rate of malondialdehyde formation (0.65 ± 0.040; Group I) was reduced to 0.38 ± 0.048 (Group V, *** $p < 0.001$; Figure 7).

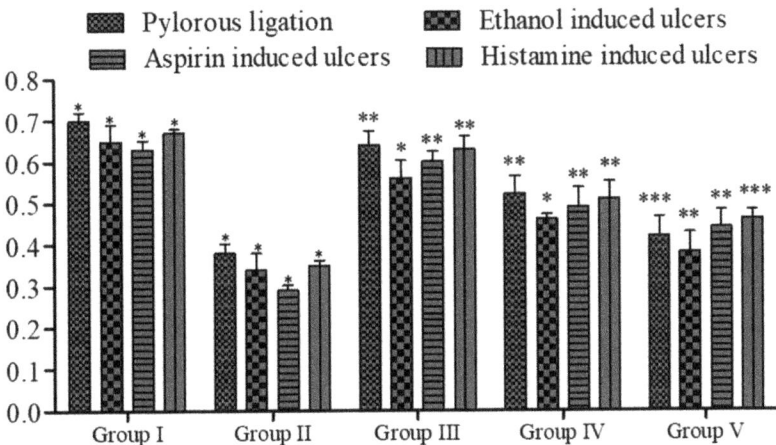

Figure 8. Effect of **5a** on lipid peroxidation of gastric contents in rats. Data are shown as mean ± SEM ($n = 6$). Group II was compared with groups III–V; * $p < 0.05$, ** $p < 0.01$; *** $p < 0.001$.

Conclusively, the administration of compound **5a** demonstrated promising results in reducing ethanol-induced ulcers in a rat model, as indicated by a significant decrease in the ulcer index and alterations in gastric pH, content volume, total acidity, free acidity, and malondialdehyde formation. These findings suggest the potential therapeutic value of compound **5a** for ulcer treatment.

2.7.3. Aspirin-Induced Ulcer

In the dose range of 20 and 40 mg/kg, compound **5a** demonstrated a significant reduction in the ulcer index (*** $p < 0.001$; 9.60 ± 0.26 vs. 14.40 ± 0.28, Group IV vs. Group I; 6.60 ± 0.34 vs. 14.40 ± 0.28, Group V vs. Group I; see Figure 5) in rats treated with aspirin, along with pH alterations (Figure 4). Similarly, compound **5a** elevated gastric pH from 2.22 ± 0.34 (Group I) to 2.94 ± 0.42 (Group IV, ** $p < 0.01$) and 3.58 ± 0.22 (Group V, *** $p < 0.001$). A significant decrease in gastric content volume was observed with the treatment of compound **5a** (** $p < 0.01$; Group IV and Group V; Figure 6). Total acidity, initially at 44.23 ± 0.17 meq per liter per 100 g (Group I), decreased to 26.34 ± 0.34 meq per liter per 100 g (Group V, *** $p < 0.001$; Figure 7), while free acidity values decreased from 24.23 ± 0.53 meq per liter per 100 g (Group I) to 8.10 ± 1.16 meq per liter per 100 g (Group V, ** $p < 0.01$; Figure 6). Correspondingly, the rate of malondialdehyde formation decreased from 0.63 ± 0.022 (Group I) to 0.44 ± 0.042 (Group V, *** $p < 0.001$; Figure 8).

2.7.4. Histamine-Induced Ulcer

Administration of histamine was shown to be ulcerogenic in the rats. The ulcer index in the group I animals was (14.78 ± 0.44; Figure 5). Compound **5a** was administered, and displayed a decline in ulcer severity and ulcer index (9.44 ± 0.42 in Group IV, ** $p < 0.01$; 6.52 ± 0.42 in Group V, *** $p < 0.001$; Figure 3). The pH of the gastric content (2.24 ± 0.42 in Group I; Figure 4) was elevated (2.62 ± 0.42 in Group III, ** $p < 0.01$; 2.98 ± 0.22 in Group

IV, ** $p < 0.01$; 3.42 ± 0.12 in Group V, *** $p < 0.001$; Figure 4). A noteworthy decline in the gastric contents volume was observed (2.73 ± 0.11 versus 1.41 ± 0.14., Group I versus Group V; *** $p < 0.001$ in Group V; Figure 6). The total acidity was 41.24 ± 0.66 meq per liter per 100 g (Group I) and reduced to 22.46 ± 0.42 meq per liter per 100 g (Group V, *** $p < 0.001$; Figure 5), while the free acidity was 22.78 ± 0.78 meq per liter per 100 g (Group I; Figure 7) decreased to 7.60 ± 0.44 meq per liter per 100 g (Group V, ** $p < 0.01$). Likewise, the rate of the formation of the malondialdehyde (0.67 ± 0.010; Group I) was reduced (0.46 ± 0.022; Group V, *** $p < 0.001$; Figure 7).

2.7.5. H^+–K^+ ATPase Assay

Compound **5a** was also demonstrated to significantly (* $p < 0.05$) affect gastric mucosal homogenate in rats. Its inhibitory potential was concentration-dependent, and the tested drug showed a comparable effect to omeprazole. Specifically, compound **5a** significantly reduced the hydrolysis of ATP (see Figure 9) through gastric ATPase, with an IC_{50} of 30 µg/mL, which was very similar to the IC_{50} of omeprazole, used as a positive control (IC_{50} of 22 µg/mL).

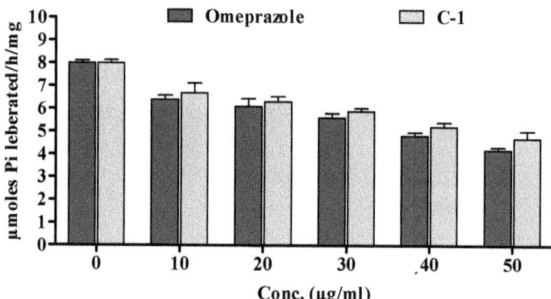

Figure 9. Effect of in vitro enzymatic studies of omeprazole and **5a**.

In the rat model with pyloric ligation-induced ulcers, the animal stomachs were ligated for approximately 6 h. This resulted in a notable increase in total acidity (43.10 ± 0.52 meq/L/100 g vs. 25.10 ± 0.16 meq per liter per 100 g) and free acidity (23.34 ± 0.42 meq per liter per 100 g vs. 9.50 ± 0.46 meq per liter per 100 g) in Group I compared to Group V (as shown in Figure 7b). However, upon the introduction of compound **5a**, these acidity levels were significantly reduced.

Likewise, lipid peroxidation, which was at 0.70 ± 0.02 in Group I, showed a significant reduction (0.42 ± 0.046) in Group V upon the administration of compound **5a**. This suggests the potential effectiveness of compound **5a** in reducing ulceration induced by ethanol in the rat model. Furthermore, the gastric content pH, as depicted in Figure 4, was notably affected. The insertion of compound **5a** resulted in a significant increase in gastric pH from 2.22 ± 0.34 (Group I) to 2.94 ± 0.42 (Group IV, ** $p < 0.01$) and 3.58 ± 0.22 (Group V, *** $p < 0.001$). Additionally, the volume of gastric content decreased significantly with the treatment of compound **5a**. Moreover, compound **5a** demonstrated a significant reduction in the hydrolysis of ATP, as shown in Figure 9, with an IC_{50} of 30 µg/mL, which is comparable to the positive control omeprazole (IC_{50} of 22 µg/mL). This suggests the potential of compound **5a** as a gastric ATPase inhibitor.

Purity of the synthesized compounds

The purity of two best-acting derivatives was confirmed using HPLC (Figure S1, Compound **3a** and **5a**).

3. Materials and Methods

3.1. Chemicals and Reagents

The ammonium acetate, 4-hydroxycoumarin, substituted aldehydes, methanol, and ethanol were acquired from Sigma Aldrich/Fluka. All of the purchased chemicals were used as acquired, as they were of analytical grade; no purifying methods were applied to the purchased chemicals. TLC plates (silica gel 60 F^{254}) used in this study were manufactured by MERCK®. Melting point determination of synthesized compounds was carried out by using the Stuart digital melting point apparatus SMP_{10}. The UV-vis analysis was performed on a PG Instruments T80 + UV-vis spectrometer (solvent: THF) using a 10 mm quartz cuvette. Fourier-transform infrared spectral analysis was carried out on a PerkinElmer spectrum 100 FT-IR spectrometer. The proton nuclear magnetic resonance spectroscopic analysis δ (ppm) was conducted in chloroform–DMSO-d_6 on a Bruker AM-400 MHz NMR spectrometer. The antimicrobial potential of synthesized derivatives was evaluated by means of the agar well diffusion method. The urease inhibition potential of these newly synthesized compounds was also evaluated on SpectraMax M2 by the indophenol method.

3.2. Experimental Synthesis of 4-Aminocoumarin and Its Schiff Base Derivatives

The analogue synthesis of 4-hydroxycoumarin was performed by following a two-step mechanism. In the 1st step, the synthesis of 4-aminocoumarin was carried out under solvent-free conditions as per reported literature [19]. With this method, 4-hydroxycoumarin (1.07 g; 6.6 mmol) and ammonium acetate (7.87 g; 0.1 mol) were melted at 170 °C for 12 h, followed by stirring at constant speed for 3 h at ambient temperature [20]. The mixture was left to cool down at room temperature. The reaction completion was checked on thin-layer chromatography (n-hexane: ethyl acetate: 1:1.5). Water was added to the reaction mixture on reaction completion, followed by filtration.

In the next step, Schiff bases of 4-aminocoumarin were synthesized by condensing a stoichiometric amount of 4-aminocoumarin (1.0 mmol) with substituted aromatic aldehydes (1.0 mmol) in the presence of dry distilled methanol. Refluxing of the mixture was carried out for 48–72 h depending upon the aldehyde used. The reaction completion was checked on thin-layer chromatography (n-hexane: ethyl acetate: 1:1.5). Acetic acid was added as a catalyst in small quantities. After the reaction, the product was precipitated and subsequently filtered, followed by its washing using chilled methanol to obtain pure product with moderate-to-good yield. The reactions take a long time to complete, though. The first step takes 14 to 15 h to complete, whereas, depending upon the aldehyde chosen, the second step may complete from 48 to 82 h, thereby giving 65% to 75% yields of resultant products (Scheme 1).

3.3. Antibacterial Activity

Antibacterial activity of the synthesized compounds was assessed using the agar well diffusion method. Two bacterial strains selected for this experiment were *Salmonella typhimurium* and *Staphylococcus aureus*. The former is a round-shaped bacterial strain that is Gram-positive and has the tendency to cause food poisoning, skin infections and respiratory tract infections. The latter is a rod-shaped Gram-negative bacterial strain capable of causing typhoid fever, weakness, stomach pain, headache and loss of appetite. Some compounds show best inhibition potential, as their zone of inhibition (mm) is less than the standard drug used.

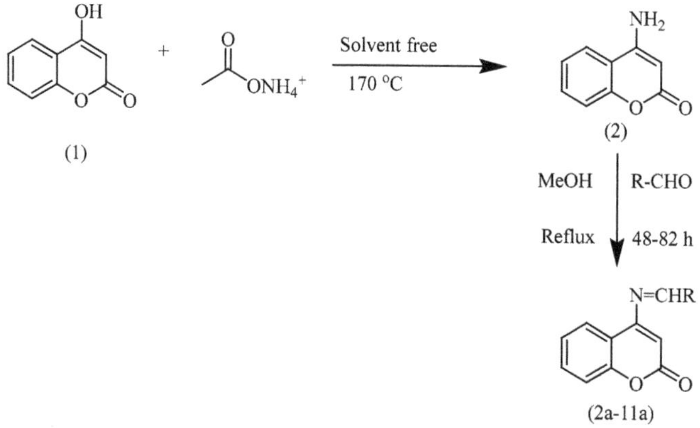

R=4-nitrobenzaldehyde, 3-nitrobenzaldehyde, 2-nitrobenzaldehyde
2-hydroxy-4-methoxybenzaldehyde, p-dimethylaminobenzaldehyde,
2-hydroxy-5-methylbenzaldehyde, 4-chlorobenzaldehyde,
5-fluoro-2-hydroxybenzaldehyde

Scheme 1. Synthetic scheme presenting two-step synthesis of 4—aminocoumarin and its derivatives. Step (1) requires 170 °C and is a solvent-free synthetic step. Step (2) is the addition of selected aldehyde under reflux conditions using methanol as solvent. This step requires a temperature ranging from 48 to 82 °C, depending upon the aldehyde used.

3.4. Urease Inhibition Assay

The presence of *H. pylori* in the stomach has been associated with a range of gastric disorders, including peptic ulcers, gastritis, and an increased risk of gastric cancer. The bacterium's ability to manipulate the gastric environment through urease activity underscores its role in the development and progression of these conditions. Different studies have targeted *H. pylori* and its urease activity as potential avenues for therapeutic intervention and disease management. The evaluation of urease inhibition potential of the compounds under discussion was carried out by the indophenol method. In this method, the production of NH_3 occurs in situ and is measured as a factor of urease inhibition/activation potential of the samples. In this bioassay, 25 µL of jack bean urease solution was mixed with phosphate buffer (pH 6.8), the volume of which was 55 µL. Addition of urea (100 mM) followed by incubation at 30 °C with 5 µL of test samples (0.5 mM each) was carried out. This final incubation was done for 15 min in 96-well plates. Next, the addition of phenol reagent (0.005% *w*/*v* sodium nitroprusside and 1% *w*/*v* phenol) was carried out. The volume of phenol reagent taken was 45 µL. In the next step, 70 µL of alkali reagent (0.1% NaOCl and 0.5% *w*/*v* NaOH) was added to each well. After 50 min, the absorbance was noted at 630 nm. Analysis in triplicate for each test sample was carried out in a final volume of 200 µL. The standard inhibitor (positive control) for this urease activity was thiourea [21].

3.5. In Silico Studies

Selection and Refinement of Protein Structure

The amino acid sequence of the target protein urease was obtained from UniProt (www.uniprot.org/uniprot/Q2G2K5 on 7 March 2022), the tertiary structure was designed and the ultimate model was prepared with Modeller 9. This obtained model was further utilized to produce the final full atomic model, and this final model was optimized by MD simulation using AMBER18 (Figure 10).

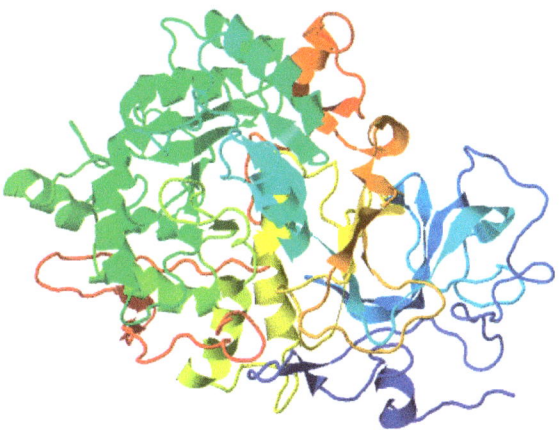

Figure 10. Modeled protein structure optimized using MD simulation.

3.6. Validation of Model, Active Site Determination, and Molecular Docking Studies

The model was authorized by examining the phi/psi distributions in the Ramachandran plot attained by the PROCHECK server (https://www.ebi.ac.uk/thornton-srv/software/PROCHECK/ on 7 March 2022). Further, structural quality was examined using ProSA web servers (https://prosa.services.came.sbg.ac.at/prosa.php on 7 March 2022). The prediction of the active site(s) was carried out by means of a site finder tool embedded in MOE-2016. The cavity prediction in a certain protein occurs by means of an active site prediction server that has the residues of the binding site surrounding the cavity at a distance of ~10 Å. Compounds were drawn using ChemDraw software (https://revvitysignals.com/products/research/chemdraw) and were optimized using Gaussian 09 software at 6–31 G basis set employing B3LYP functional prior to docking. Molecular docking was performed using MOE [22]. Site finder tools from MOE were employed to analyze potential protein binding residues and to create electrostatic surface maps around these residues in order to define docking regions. The MOE tools were utilized to dock synthesized compounds within the target proteins' specified docking sites. We utilized a triangle algorithm to determine the best-docked molecule poses that were later minimized using the force field refinement technique, and binding energies were taken into consideration while maintaining receptor residues rigid using GB solvation models. The receptor molecule preparation requires the addition of polar hydrogens as well as the merger of nonpolar hydrogens, in accordance with the standard procedure. The top ten ligands were chosen based on their binding energy, together with root-mean-square deviation (RMSD). Finally, all compounds were docked with protein. Drug-like characteristics of the compounds in question were retrieved by the Lipinski filter server.

3.7. MD Simulation Studies of Protein–Ligand Complexes as Well as Rescoring of Binding Energies

MD simulation analysis was carried out for the top two (**3a** and **5a**) receptor–ligand complexes. These complexes were selected on the basis of molecular docking studies for confirmation of not only the mode of binding but also the dynamic system stability. Molecular dynamic (MD) simulations were conducted using PMEMD.CUDA from the AMBER 18 suite of programs. To expedite simulation times, an NVIDIA Geforce GTX-1070 Ti graphics card was utilized. The General AMBER Force Field (GAFF) parameters were employed to generate the atomic parameters for each ligand, and Gasteiger charges were assigned to all ligands in the MD simulations.

For each complex structure, periodic boundary conditions were applied, and the system was solvated in a cubic box with TIP3P water molecules extending 12 Å in each direction from the complex model. Additionally, neutralizing Na+ ions were introduced.

The cutoff distance for computing unbonded interactions was set to 12 Å. All experimental simulations were conducted under periodic boundary conditions. The AMBER ff14SB force field parameters were employed to characterize the complex. Long-range electrostatic interactions were handled using the particle mesh Ewald (PME) method. To constrain bonds involving hydrogen atoms and maintain temperature control, the SHAKE algorithm and Langevin dynamics were applied. A time step of 2 fs was utilized, and trajectory data were recorded every 0.2 ps. The system's temperature was gradually raised from 0 to 310.15 K over 100 ps of NVT dynamics, followed by 10 ns of NPT equilibration at 310.15 K and 1 atm pressure. Finally, a total of 100 ns of production phase MD simulations were performed to collect properties [23]. Trajectory analyses, including root-mean-square deviation and fluctuation, were conducted using the CPPTRAJ module within the Amber 18 program. To calculate the binding free energy for each simulation complex, the Amber molecular mechanics Poisson–Boltzmann surface area (MM-PBSA) method was employed. A total of 300 structural frames were selected from the 80 ns trajectory data. Subsequently, 1000 snapshots were extracted from the trajectory data for binding free energy calculations using the MM-PBSA method. The grid size utilized in the Poisson–Boltzmann calculations within MM-PBSA was set to 0.5 Å [24].

3.8. In Vivo Pharmacology

Experimental Animals

The rats used in this study were Swiss albino mice (male and female; weight range: 30–35 g) The animals were acquired from the National Institute of Health, Islamabad, Pakistan. Written approval was obtained from the departmental ethics committee (DREC/20). Also, the approved animal house was used to preserve the animals [25]. All of the standard ethical guidelines were followed during the course of this study [26].

Pylorus ligation activity [27], ethanol-induced ulcer [28], aspirin-induced ulcer [29], histamine-induced ulcer [30] and H^+–K^+ ATPase assay [31] studies were conducted as per standard operating procedures with respective standards.

4. Conclusions

A total of 1 reported and 10 new derivatives of chromene were synthesized with enhanced/modified biological activity. The antibacterial activity hints at the antibacterial potential of synthesized 4-aminocoumarin derivatives. The present work also presents the in silico inhibitory potential of synthesized compounds against urease enzyme. Out of ten substituted chromenone Schiff bases, **3a** and **5a** showed significantly increased binding potential at the urease active site, as inferred from molecular docking and molecular dynamics simulation study. Compounds **3a** and **5a** interacted with Ni ion via Ni-O bonding with three key amino acids (His492, His 593, and Asp494) facilitated by strong H-bonds/hydrophobic interactions that are critical for the activity of urease receptors. These newly synthesized compounds were also tested for in vitro urease inhibition potential. This investigation yields a few crucial findings. First, two compounds were identified that represent scaffolds that are different from those previously recognized; they are all significantly smaller than most of the known urease inhibitors, giving them comparatively good ligand efficiencies. Furthermore, compounds **3a** and **5a** IC_{50} values of 0.412 µM (64.0% inhibition) and 0.322 µM (77.7% inhibition), respectively, together with their good physical qualities, place them in the lead-like range of compounds that might be optimized as bioactive molecules for the urease enzyme. Additionally, the rat model with pyloric ligation-induced ulcers exhibited elevated acidity levels, including total acidity and free acidity, in Group I compared to Group V, which indicated the severity of ulceration. However, the introduction of compound **5a** significantly reduced these acidity levels, suggesting its potential effectiveness in mitigating ulceration induced by ethanol. Furthermore, compound **5a** demonstrated a substantial reduction in lipid peroxidation and a significant increase in gastric pH. It also led to a remarkable decrease in the volume of gastric content. These findings collectively support the potential therapeutic role of compound **5a** in reducing ethanol-induced ulcers in the rat model. Moreover, compound **5a** displayed a noteworthy inhibitory effect on gastric ATPase, with an IC_{50} value of 30 µg/mL, comparable to the positive control omeprazole (IC_{50} of 22 µg/mL). This suggests that compound **5a** may act as a gastric ATPase inhibitor, further highlighting its potential as a treatment option for gastric ulcers. Overall, the comprehensive results presented in the study suggest that compound **5a** holds promise as a therapeutic agent for the management of ethanol-induced ulcers, with potential benefits in reducing acidity, lipid peroxidation, and gastric ATPase activity. Further investigations and clinical studies are warranted to fully explore its therapeutic efficacy and safety for clinical application.

Supplementary Materials: The following supporting information can be downloaded at: https://www.mdpi.com/article/10.3390/ph16111552/s1, Figure S1. Spectral Data of all compounds.

Author Contributions: Conceptualization, S.K.; Methodology, A.Z., A.K. and S.K.; Software, K.S., A.K. and S.K.; Validation, K.S., T.A.W. and M.S.J.; Formal analysis, K.S., S.M.B., S.Z., U.R., U.F. and S.K.; Investigation, S.M.B., A.Z. and U.R.; Resources, A.Z., T.A.W., M.S.J. and U.F.; Data curation, M.S.J.; Writing—original draft, T.A.W. and S.K.; Writing—review & editing, S.Z.; Funding acquisition, S.Z. All authors have read and agreed to the published version of the manuscript.

Funding: This research was funded by the Researchers Supporting Project (RSP2023R357), King Saud University, Riyadh, Saudi Arabia for funding this research.

Institutional Review Board Statement: The animal study protocol was approved by the Institutional Review Board of the Department of Pharmacy, The Professional Institute of Health Sciences, Mardan 23200, KPK, Pakistan (DAEC/Dec/2022/88 and date of approval was 15 December 2022).

Data Availability Statement: Data is contained within the article and supplementary material.

Acknowledgments: The authors extended their appreciation to the research supporting project number (RSP2023R357), King Saud University, Riyadh, Saudi Arabia, for funding this research. The authors also extend thanks to the COMSATS University Islamabad, Abbottabad campus for providing lab placement and NMR analysis facility.

Conflicts of Interest: The authors declare no conflict of interest.

References

1. Konieczna, I.; Żarnowiec, P.; Kwinkowski, M.; Kolesińska, B.; Frączyk, J.; Kamiński, Z.; Kaca, W. Bacterial urease and its role in long-lasting human diseases. *Curr. Protein Pept. Sci.* **2012**, *13*, 789–806. [CrossRef]
2. Vekariya, R.H.; Patel, H.D. Recent advances in the synthesis of coumarin derivatives via knoevenagel condensation: A review. *Synth. Commun.* **2014**, *44*, 2756–2788. [CrossRef]
3. Stefanachi, A.; Leonetti, F.; Pisani, L.; Catto, M.; Carotti, A. Coumarin: A natural, privileged and versatile scaffold for bioactive compounds. *Molecules* **2018**, *23*, 250. [CrossRef] [PubMed]
4. Valizadeh, H.; Shockravi, A. An efficient procedure for the synthesis of coumarin derivatives using $TiCl_4$ as catalyst under solvent-free conditions. *Tetrahedron Lett.* **2005**, *46*, 3501–3503. [CrossRef]
5. Abdou, M.M. 3-acetyl-4-hydroxycoumarin: Synthesis, reactions and applications. *Arab. J. Chem.* **2017**, *10*, S3664–S3695. [CrossRef]
6. Kostova, I.; Bhatia, S.; Grigorov, P.; Balkansky, S.; Parmar, S.V.; Prasad, K.A.; Saso, l. Coumarins as antioxidants. *Curr. Med. Chem.* **2011**, *18*, 3929–3951. [CrossRef]
7. Chohan, Z.H.; Shaikh, A.U.; Rauf, A.; Supuran, C.T. Antibacterial, antifungal and cytotoxic properties of novel N-substituted sulfonamides from 4-hydroxycoumarin. *J. Enzyme Inhib. Med. Chem.* **2006**, *21*, 741–748. [CrossRef]
8. Mitra, A.K.; Misra, S.K.; Patra, A. A New synthesis of 3-alkyl coumarins. *Synth. Commun.* **1980**, *10*, 915–919. [CrossRef]
9. Bose, D.S.; Rudradas, A.; Babu, M.H. The indium (III) chloride-catalyzed von Pechmann reaction: A simple and effective procedure for the synthesis of 4-substituted coumarins. *Tetrahedron Lett.* **2002**, *43*, 9195–9197. [CrossRef]
10. Ghosh, R.; Singha, P.S.; Das, L.K.; Ghosh, D.; Firdaus, S.B. Antiinflammatory activirty of natural coumarin compounds from plants of the Indo-Gangetic plain. *AIMS Mol. Sci.* **2010**, *10*, 79–98. [CrossRef]
11. Satyanarayana, V.; Sreevani, P.; Sivakumar, A.; Vijayakumar, V. Synthesis and antimicrobial activity of new Schiff bases containing coumarin moiety and their spectral characterization. *Arkovic* **2008**, *17*, 221–233. [CrossRef]
12. Paul, M.K.; Singh, Y.D.; Dey, A.; Saha, S.K.; Anwar, S.; Chattopadhyay, A.P. Coumarin based emissive rod shaped new schiff base mesogens and their zinc (II) complexes: Synthesis, photophysical, mesomorphism, gelation and DFT studies. *Liq. Cryst.* **2016**, *43*, 343–360. [CrossRef]
13. Creaven, B.S.; Devereux, M.; Karcz, D.; Kellett, A.; McCann, M.; Noble, A.; Walsh, M. Copper (II) complexes of coumarin-derived Schiff bases and their anti-candida activity. *J. Inorg. Biochem.* **2009**, *103*, 1196–1203. [CrossRef] [PubMed]

14. Kulkarni, A.; Avaji, P.G.; Bagihalli, G.B.; Patil, S.A.; Badami, P.S. Synthesis, spectral, electrochemical and biological studies of Co (II), Ni (II) and Cu (II) complexes with Schiff bases of 8-formyl-7-hydroxy-4-methyl coumarin. *J. Coord. Chem.* **2009**, *62*, 481–492. [CrossRef]
15. Al-Majedy, Y.K.; Kadhum, A.A.H.; Al-Amiery, A.A.; Mohamad, A.B. Coumarins: The antimicrobial agents. *Syst. Rev. Pharm.* **2017**, *8*, 62–70. [CrossRef]
16. Baboukani, A.R.; Sharifi, E.; Akhavan, S.; Saatchi, A. Co complexes as a corrosion inhibitor for 316 l stainless steel in H_2SO_4 solution. *J. Mater. Sci. Chem. Eng.* **2016**, *4*, 28–35.
17. Kadhum, A.A.H.; Al-Amiery, A.A.; Musa, A.Y.; Mohamad, A.B. The antioxidant activity of new coumarin derivatives. *Int. J. Mol. Sci.* **2011**, *12*, 5747–5761. [CrossRef] [PubMed]
18. Hassan, S.T.; Žemlička, M. Plant-derived urease inhibitors as alternative chemotherapeutic agents. *Arch. Pharm.* **2016**, *349*, 507–522. [CrossRef]
19. Stamboliyska, B.; Janevska, V.; Shivachev, B.; Nikolova, R.P.; Stojkovic, G.; Mikhova, B.; Popovski, E. Experimental and theoretical investigation of the structure and nucleophilic properties of 4-aminocoumarin. *Arkivoc* **2010**, *10*, 62–76.
20. Arshad, N.; Perveen, F.; Saeed, A.; Channar, P.A.; Farooqi, S.I.; Larik, F.A.; Ismail, H.; Mirza, B. Spectroscopic, molecular docking and structural activity studies of (E)-N′-(substituted benzylidene/methylene) isonicotinohydrazide derivatives for DNA binding and their biological screening. *J. Mol. Struct.* **2017**, *1139*, 371–380. [CrossRef]
21. Ali, M.; Bukhari, S.M.; Zaidi, A.; Khan, F.A.; Rashid, U.; Tahir, N.; Rabbani, B.; Farooq, U. Inhibition profiling of urease and carbonic anhydrase II by high throughput screening and molecular docking studies of structurally diverse organic compounds. *Lett. Drug Des. Discov.* **2021**, *18*, 299–312. [CrossRef]
22. Vilar, S.; Cozza, G.; Moro, S. Medicinal chemistry and the molecular operating environment (MOE): Application of QSAR and molecular docking to drug discovery. *Curr. Top. Med. Chem.* **2008**, *8*, 1555–1572. [CrossRef] [PubMed]
23. Case, D.A.; Aktulga, H.M.; Belfon, K.; Ben-Shalom, I.; Brozell, S.R.; Cerutti, D.; Cheatham, T., III; Cisneros, G.; Cruzeiro, V.; Darden, T. *Amber 2021*; University of California Press: San Francisco, CA, USA, 2021.
24. Zaman, Z.; Khan, S.; Nouroz, F.; Farooq, U.; Urooj, A. Targeting protein tyrosine phosphatase to unravel possible inhibitors for Streptococcus pneumoniae using molecular docking, molecular dynamics simulations coupled with free energy calculations. *Life Sci.* **2021**, *264*, 118621. [CrossRef] [PubMed]
25. Mahmood, F.; Khan, J.A.; Mahnashi, M.H.; Jan, M.S.; Javed, M.A.; Rashid, U.; Sadiq, A.; Hussan, S.S.; Bungau, S. Anti-Inflammatory, Analgesic and Antioxidant Potential of New (2S,3S)-2-(4-isopropylbenzyl)-2-methyl-4-nitro-3-phenylbutanals and Their Corresponding Carboxylic Acids through In Vitro, In Silico and In Vivo Studies. *Molecules* **2022**, *27*, 4068. [CrossRef]
26. Shah, S.M.M.; Ullah, F.; Shah, S.M.H.; Zahoor, M.; Sadiq, A. Analysis of chemical constituents and ntinociceptive potential of essential oil of Teucrium *Stocksianum* bioss collected from the North West of Pakistan. *BMC Complement. Med.* **2012**, *12*, 244.
27. Wang, X.-Y.; Yin, J.-Y.; Zhao, M.-M.; Liu, S.-Y.; Nie, S.-P.; Xie, M.-Y. Gastroprotective activity of polysaccharide from *Hericium erinaceus* against ethanol-induced gastric mucosal lesion and pylorus ligation-induced gastric ulcer, and its antioxidant activities. *Carbohydr. Polym.* **2018**, *186*, 100–109. [CrossRef]
28. Al-Qarawi, A.A.; Abdel-Rahman, H.; Ali, B.H. The ameliorative effect of dates (*Phoenix dactylifera* L.) on ethanol-induced gastric ulcer in rats. *J. Ethnopharmacol.* **2005**, *98*, 313–317. [CrossRef]
29. Choi, J.-I.; Raghavendran, H.R.B.; Sung, N.-Y.; Kim, J.-H.; Chun, B.S.; Ahn, D.H.; Choi, H.-S.; Kang, K.-W.; Lee, J.-W. Effect of fucoidan on asprin-induced stomachulceration in rats. *Chem. Biol. Interact.* **2010**, *183*, 249–254. [CrossRef]
30. Bodhankar, S.L.; Jain, B.B.; Ahire, B.P.; Daude, R.B.; Shitole, P.P. The effect of rabeprazole and its isomers on asprin and histamine-induced ulcers in rats. *Indian J. Pharmacol.* **2006**, *38*, 357–358. [CrossRef]
31. Razzaq, S.; Minhas, A.M.; Qazi, N.G.; Nadeem, H.; Khan, A.U.; Ali, F.; ul Hassan, S.S.; Bangau, S. Novel Isoxazole Derivative Attenuates Ethanol-Induced Gastric Mucosal Injury through Inhibition of H+/K+-ATPase Pump, Oxidative Stress and Inflammatory Pathways. *Molecules* **2022**, *27*, 5065. [CrossRef]

Disclaimer/Publisher's Note: The statements, opinions and data contained in all publications are solely those of the individual author(s) and contributor(s) and not of MDPI and/or the editor(s). MDPI and/or the editor(s) disclaim responsibility for any injury to people or property resulting from any ideas, methods, instructions or products referred to in the content.

Review

Synthesis and Biological Studies of Benzo[b]furan Derivatives: A Review from 2011 to 2022

Lizeth Arce-Ramos, Juan-Carlos Castillo * and Diana Becerra *

Escuela de Ciencias Químicas, Universidad Pedagógica y Tecnológica de Colombia, Avenida Central del Norte 39-115, Tunja 150003, Colombia; lizeth.arce@uptc.edu.co
* Correspondence: juan.castillo06@uptc.edu.co (J.-C.C.); diana.becerra08@uptc.edu.co (D.B.); Tel.: +57-8740-5626 (ext. 2425) (J.-C.C. & D.B.)

Abstract: The importance of the benzo[b]furan motif becomes evident in the remarkable results of numerous biological investigations, establishing its potential as a robust therapeutic option. This review presents an overview of the synthesis of and exhaustive biological studies conducted on benzo[b]furan derivatives from 2011 to 2022, accentuating their exceptional promise as anticancer, antibacterial, and antifungal agents. Initially, the discussion focuses on chemical synthesis, molecular docking simulations, and both in vitro and in vivo studies. Additionally, we provide an analysis of the intricate interplay between structure and activity, thereby facilitating comparisons and profoundly emphasizing the applications of the benzo[b]furan motif within the realms of drug discovery and medicinal chemistry.

Keywords: benzo[b]furans; benzofurans; anticancer; antibacterial; antifungal; medicinal chemistry; drug discovery

Citation: Arce-Ramos, L.; Castillo, J.-C.; Becerra, D. Synthesis and Biological Studies of Benzo[b]furan Derivatives: A Review from 2011 to 2022. *Pharmaceuticals* **2023**, *16*, 1265. https://doi.org/10.3390/ph16091265

Academic Editor: Valentina Noemi Madia

Received: 12 August 2023
Revised: 1 September 2023
Accepted: 1 September 2023
Published: 6 September 2023

Copyright: © 2023 by the authors. Licensee MDPI, Basel, Switzerland. This article is an open access article distributed under the terms and conditions of the Creative Commons Attribution (CC BY) license (https://creativecommons.org/licenses/by/4.0/).

1. Introduction

Heterocyclic chemistry plays a pivotal role in both chemical and life sciences, serving as a focal point for extensive global research. This branch of organic chemistry is dedicated to developing innovative molecules through the application of numerous synthetic protocols [1–3]. Heterocyclic compounds are widely distributed in naturally occurring and synthetic molecules, showcasing a broad range of physiological and pharmacological properties [4–7]. This diversity makes them particularly intriguing for applications in medicinal chemistry, material sciences, drug discovery, and the agrochemical and pharmaceutical industries [8–11]. Specifically, an analysis of the U.S. FDA-approved drug database revealed that around 60% of the top-selling drugs contain at least one heterocyclic nucleus [4]. These outstanding results can be attributed to the diverse intermolecular interactions between heterocycles and enzymes, involving hydrogen bonding, π-stacking, metal coordination bonds, and van der Waals and hydrophobic forces, along with their varied ring sizes, which enable a wide range of shapes to match the diverse enzyme binding pockets [12]. Moreover, drugs incorporating heterocycles exhibit improved solubility and the ability to facilitate salt formation, which are crucial factors in enhancing oral absorption and overall bioavailability [12]. Additionally, recent developments in synthetic methodologies targeting functionalized heterocycles play a key role in medicinal chemistry and drug discovery, effectively expanding the drug-like chemical space [1,13–15]. The establishment of reliable synthetic pathways for large-scale production further expedites drug development. Notably, the introduction of inventive heterocyclic syntheses, incorporating diverse bond-forming strategies, profoundly influences the pharmaceutical industry [1,13–15].

In this context, oxygen-containing heterocycles stand out due to their broad spectrum of biological and pharmacological activities. Their significance stems from their structural similarities with a variety of well-established natural and synthetic compounds [16]. The inherent importance of oxygen-containing heterocycles as therapeutic agents resides in their distinctive structural characteristics, closely resembling those present in biologically active compounds, like ribose derivatives [17,18]. In these compounds, the prevalence presence of oxygen atoms contributes to polar interactions that foster stabilization within the active site [17,18]. Among these compounds, the benzo[*b*]furan scaffold, positioned within the domain of oxa-heterocycles has recently garnered significant attention. Its distinct physiological and chemotherapeutic properties are accentuated, further accompanied by its widespread prevalence in the natural realm. Some of the most prominent benzo[*b*]furan derivatives exhibit remarkable pharmaceutical applications, such as Amiodarone, used to treat life-threatening ventricular arrhythmias [19], and Bufuralol, employed as a nonselective β-adrenoceptor antagonist, which can lead to potential complications like hepatic toxicity or adverse drug interactions (Figure 1) [20]. Recent studies have shown that Ailanthoidol exhibits antitumor potential by suppressing TGF-β1-promoted HepG2 hepatoblastoma cell progression [21]. On the other hand, benzo[*b*]furan derivatives are widely distributed in various plant families, including Asteraceae, Fabaceae, and Moraceae [22,23]. For instance, Moracin D, isolated from *Morus alba*, exhibited anti-inflammatory and antioxidant activities, as well as induced apoptotic effects in prostate and breast cancer cells (Figure 1) [24,25]. Furthermore, Cicerfuran, isolated from the roots of chickpea (*Cicer* spp.), demonstrated antibacterial and antifungal activities [26]. The myriad applications in medicinal chemistry, biomedical science, and drug discovery have spurred both academia and the pharmaceutical industry to develop novel, efficient, and straightforward synthetic protocols for preparing a wide array of structurally diverse benzo[*b*]furan derivatives [23,27–31].

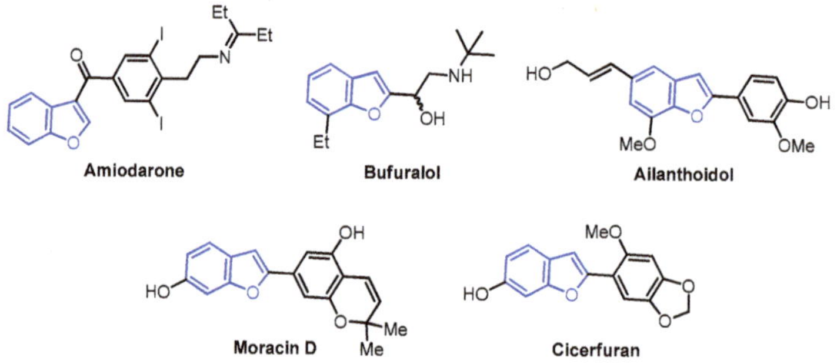

Figure 1. Bioactive benzo[*b*]furan derivatives.

Surprisingly, an exploration of the Scopus database covering the years from 2011 to 2022, encompassing all fields with the keywords "benzo[*b*]furan" and "biological activity," has revealed a total of 7482 documents. Among these, 5431 are categorized as articles, while 1451 are identified as reviews. Notably, the array of published articles showcases a diverse spectrum of keywords, spanning a wide range of activities, including anticancer, antitumor, antiproliferative, antibacterial, antifungal, anti-inflammatory, antiviral, antitubercular, antidepressant, antipsychotic, α7 nAChR agonist, and antiosteoporosis activities (Figure 2). The data related to the last two activities are somewhat limited and may not be readily distinguishable on a graphic. Within this broad spectrum of activities, it is important to emphasize that articles related to cancer (anticancer, antitumor, and antiproliferative), as well as those focused on antimicrobial activity (antibacterial and antifungal), collectively constitute approximately 70% of the observed activities (Figure 2).

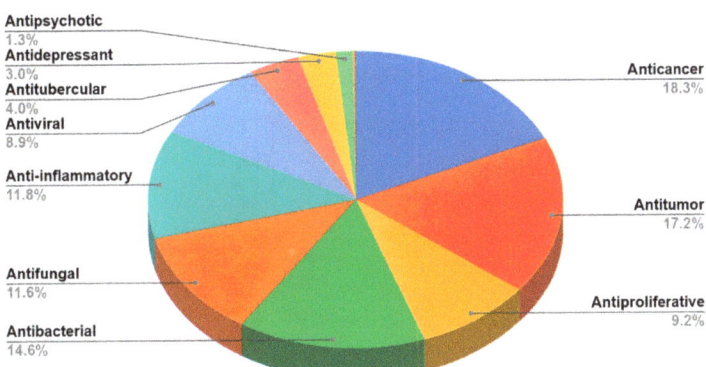

Figure 2. Bibliometric analysis: percentage distribution of articles across various biological activities of benzo[*b*]furan derivatives from 2011 to 2022 (data obtained from the Scopus database using the keywords: "benzo[*b*]furan" and each one activity).

The most representative articles from these activities were selected and grouped under the headings of anticancer, antibacterial, and antifungal activities. In doing so, a focus was primarily placed on aromatic benzo[*b*]furans, excluding related search terms like furans and furanones. This approach resulted in a collection of 36 articles for the current review, of which 27 pertain to cancer (anticancer, antitumor, and antiproliferative) and 9 involve antimicrobial activity (antibacterial and antifungal). The articles chosen for this review offer insights into various aspects, including chemical synthesis, molecular docking discussions involving enzyme–substrate complexes, in vitro and in vivo studies, and diverse analytical techniques, such as flow cytometry, Western blotting, and confocal microscopy. These methodologies elucidate how the compounds impact the inhibition of the cell cycle, tumor regression, reduction in colony-forming units, and microbial growth. This broadens the scope for future investigations in the field of benzo[*b*]furan derivatives, encompassing chemical synthesis and the evaluation of anticancer and antimicrobial activities.

2. Synthesis of Bioactive Benzo[*b*]furan Derivatives

2.1. Anticancer Activity

Benzo[b]furan derivatives have demonstrated a fascinating array of biological and pharmaceutical activities, including antitumor properties. For instance, Flynn et al. described the discovery of 7-hydroxy-6-methoxy-2-methyl-3-(3,4,5-trimethoxybenzoyl)benzo[b]furan (BNC105, **6a**, R^1 = OH, R^2 = Me), a potent and selective antiproliferative agent. They achieved the synthesis of various derivatives, many of which were obtained through a modified Larock-type coupling between o-iodophenol **1a,b** and 3-silyl-1-arylpropinone **2**, yielding 2-silylbenzo[b]furans **3a,b** in 59% and 69% yields, respectively (Scheme 1) [32,33]. Subsequently, silanes **3a** (R^1 = H) and **3b** (R^1 = Oi-Pr) underwent treatment with TBAF in methanol to remove the silyl groups. For compound **3b**, an additional reaction with AlCl$_3$ was performed to eliminate the isopropyl group, resulting in the formation of compounds **4a** (R^1 = H) and **4b** (R^1 = OAc) with yields of 83% and 86%, respectively. On the other hand, compound **3a** underwent bromodesilylation with a 59% yield, producing 2-bromobenzo[b]furan **5a** (R^1 = H). To avoid competitive bromination of the C–4 position of benzo[b]furan during the bromodesilylation process of **3b**, the isopropyl group was first exchanged for an acetyl group, yielding compound **3c**. Subsequent bromodesilylation of **3c** resulted in the formation of compound **5b** with a 69% yield. Additionally, the brominated derivatives **5a** and **5b** exhibited versatile functionality allowing bromine replacement through palladium coupling or nucleophilic displacement, leading to the formation of analog series **6**, which includes heterocyclic, carbocyclic, and alicyclic analogs at C–2 of benzo[b]furan (Table 1). One striking example of the versatility of the brominated derivatives is observed in the synthesis of the biologically significant compound **6a** (R^1 = OH, R^2 = Me), which was obtained via a Negishi reaction, coupling the derivative **5b** with methylzinc bromide using palladium, achieving an impressive 93% yield.

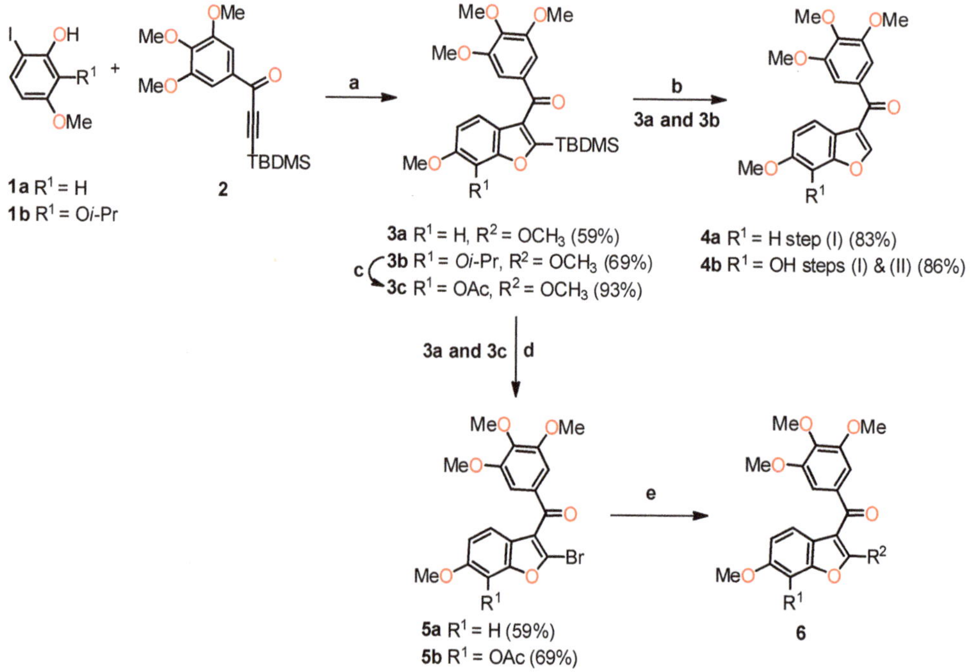

Scheme 1. Modified Larock coupling synthesis. Reagents and conditions: (**a**) Pd(OAc)$_2$, Na$_2$CO$_3$, DMF; (**b**) TBAF, THF, AlCl$_3$, CH$_2$Cl$_2$; (**c**) AlCl$_3$, CH$_2$Cl$_2$, Ac$_2$O, Pyridine; (**d**) Br$_2$, 1,2-dichloroethane; (**e**) coupling with palladium or nucleophilic displacement.

A molecular docking simulation was performed to investigate the interactions of compounds **4b** and **6a** with the α,β-tubulin dimer complexed with podophyllotoxin (PDB ID code: 1SA1). During the docking study, the structures of compounds **4b** and **6a** were oriented, considering the structural similarities between these synthesized compounds and colchicine **7**. Specifically, the study focused on the interaction of colchicine with the β-tubulin subunit, where the 3,4,5-trimethoxyphenyl rings overlapped with similar rings in colchicine. Additionally, the C6–OMe and C7–OH substituents were examined for their interactions with the methoxy and carbonyl groups on the tropone ring of colchicine (Figure 3). The study showed the formation of a hydrogen bond between the C7–OH group of the benzo[*b*]furan (**4b** and **6a**) with the Asn β258 side chain, as well as the formation of a hydrogen bond with the amide nitrogen Val 181 in the adjacent subunit of α-tubulin. Furthermore, the orientation of the C–2 position of benzo[*b*]furan toward a gap between the α- and β-tubulin subunits allows it to harbor large substituents.

Table 1. Anticancer evaluation of benzo[b]furans 4–6.

Compound	R¹	R²	Tubulin [a] IC$_{50}$ (μM)	MCF-7 [b] IC$_{50}$ (μM)	Activated HUVEC [c] IC$_{50}$ (μM)	Quiescent HUVEC [d] IC$_{50}$ (μM)	Selectivity Ratio [e]
CA-4 [f]	-	-	1.8 ± 0.2	2.9	3.6	3.9	1.1
4a	H	H	1.6 ± 0.2	55 ± 5	45	63	1.4
4b	OH	H	ND	45 ± 15	83	322	3.9
5a	H	Br	ND	495 ± 25	510	309	0.6
6a	OH	Me	3.0 ± 0.6	2.4 ± 2	0.31	25	81
6b	H	pyridyl-OMe	ND	71 ± 1	45	27	0.6
6c	H		ND	215 ± 15	485	405	0.8
6d	OH	2-Furanyl	8.7 ± 0.8	0.6 ± 0.2	3.6	3.4	0.9
6e	OH	2-Thiophenyl	ND	0.5 ± 0.2	0.31	0.36	1.2
6f	OH	Pyrrole	ND	ND	2.6	3.3	1.3
6g	OH	CH=CH-CO$_2$Me	ND	ND	0.38	2.3	6.0
6h	OH	CN	ND	ND	1.1	3.0	2.7
6i	OH	CH=CH-C(O)NH$_2$	ND	ND	2.6	2.3	0.9
6j	OH	NH-CH$_2$CH$_2$-N(Me)$_2$	ND	ND	29	36	1.2
6k	OH	NH-CH$_2$CH$_2$-OH	ND	ND	3.1	2.9	0.9
6l	OH	NH$_2$	ND	ND	1.9	8.8	4.6

[a] The tubulin concentration used was 10 μM. Inhibition of extent of assembly was the parameter measured (n = 2). [b] Cells were cultured for 48 h at 37 °C in a humidified atmosphere containing 5% CO$_2$. The data presented are the mean ± SD of two independent experiments. [c] HUVEC cells were seeded at 2500 and 500 cells/well and cultured in EGM-2 (Lonza) or F12K medium containing 0.03 mg/mL endothelial cell growth supplement. [d] HUVEC and HAAE-1 cells were seeded at 15,000 and 5000 cells/well, respectively, in basal medium (EBM-2 or F12K) containing 0.5% fetal calf serum and antibiotics. [e] Selectivity ratio (IC$_{50}$ quiescent)/(IC$_{50}$ activated). [f] Combretastatin-A4 (CA-4) is the standard drug for the study. ND means not detected. Reprinted (adapted) with permission from ref. [32]. Copyright American Chemical Society, 2023.

Another conducted study aimed to evaluate the effectiveness of benzo[b]furan **6a** on various human cancer cell lines using Combretastatin-A4 (CA-4) as a standard drug (Table 2). Interestingly, the results revealed that benzo[b]furan **6a** exhibited excellent selectivity against human aortic arterial endothelial cells (HAAECs), a characteristic not observed with CA-4. Furthermore, compound **6a** demonstrated significantly higher antiproliferative activity than CA-4, with up to a 10-fold increase in potency observed across many of the tested cell lines.

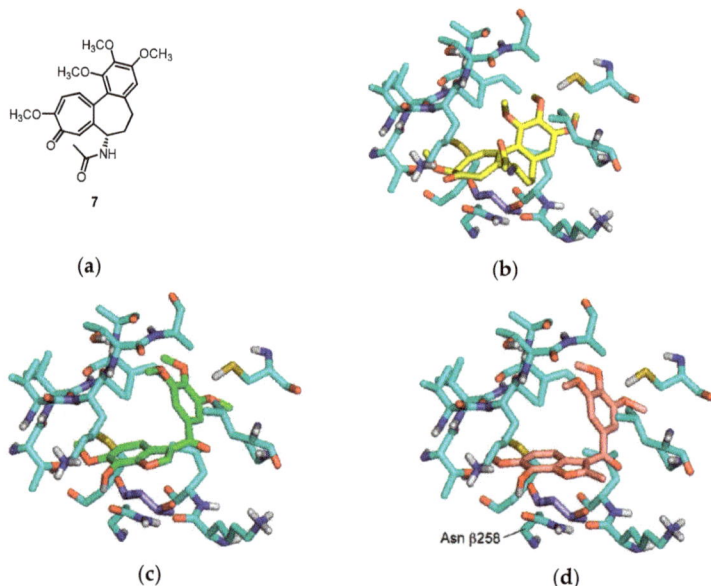

Figure 3. The molecular docking study utilized the X-ray crystal structure of the bovine α,β-tubulin dimer complexed with podophyllotoxin (PDB ID code: 1SA1). The benzo[b]furans were docked at the colchicine site using Glide. (**a**) The structure of colchicine **7**; (**b**) comparison with the orientation of colchicine **7** bound to the crystal; and (**c**,**d**) the docked orientations of **4b** and **6a**, respectively. The β-subunit of tubulin is represented by carbon atoms in light blue, and the Thr residue R179 is depicted in dark blue. Reprinted (adapted) with permission from ref. [32]. Copyright American Chemical Society, 2023.

Table 2. In vitro inhibition of cell proliferation in various cancer cell lines by **6a** and CA-4.

Entry	Cell Line [a]	Cell Type	6a IC$_{50}$ (nM)	CA-4 IC$_{50}$ (nM)
1	Activated HAAE-1	Endothelial cell	0.1 (120) [b]	2.2 (1.7) [b]
2	Quiescent HAAE-1	Endothelial cell	12	1.3
3	U87-MG	Brain glioblastoma	0.41	2.6
4	DU145	Prostate carcinoma	0.36	4
5	Calu-6	Lung anaplastic carcinoma	0.16	0.94
6	MDA-MB-231	Breast adenocarcinoma	0.63	3.2
7	A431	Epidermoid carcinoma	18.6	188
8	A375	Malignant skin melanoma	1.5	2.1
9	SKOV-3	Ovary adenocarcinoma	0.59	5.5
10	LoVo	Colorectal adenocarcinoma	0.24	2.9
11	AU565	Breast adenocarcinoma	5.8	4.4
12	BT549	Breast carcinoma	0.34	3.5

[a] HAAE-1 cells were seeded at 500 cells/well in medium containing 0.03 mg/mL endothelial cell growth supplement (activated) or in basal medium containing 0.5% fetal calf serum and antibiotics (quiescent). Cancer cell lines were seeded at an average of 500–2000 cells/well and cultured as recommended by the ATCC. [b] Selectivity ratio (IC$_{50}$ quiescent)/(IC$_{50}$ activated) in parentheses. Reprinted (adapted) with permission from ref. [32]. Copyright American Chemical Society, 2023.

On the other hand, Romagnoli et al. highlighted the significance of incorporating a 3,4,5-trimethoxybenzoyl group at the C–2 position of benzo[b]furan in determining the antiproliferative activity of benzofuran derivatives [34]. As depicted in Scheme 2, they synthesized a series of amino 2-(3′,4′,5′-trimethoxybenzoyl)-benzo[b]furan with good yields through three reaction steps: (1) a one-step cyclization reaction of nitrosalicylaldehydes or 2-hydroxyacetophenone **8** with 2-bromo-1-(3′,4′,5′-trimethoxyphenyl)ethanone and anhydrous potassium carbonate in acetone at reflux, yielding the nitro derivatives of 2-(3′,4′,5′-trimethoxybenzoyl)benzo[*b*]furanone **9a–j**; (2) the subsequent reduction of the nitro group using iron in a mixture of 37% HCl in water and ethanol at reflux, leading to the formation of the amino derivatives **10a–j**; and (3) the preparation of analogs **11a–i** in good yields through a substitution reaction between α-bromoacrylic acid and the amino benzo[*b*]furanone derivatives **10a–f** and **10h–j**, using an excess of two equivalents of EDCI and BtOH in dry DMF as a solvent. The glycine prodrug **13** was obtained in a 95% yield through the reaction between the amino derivative **10h** and *N*-Boc-glycine using EDCI and BtOH as coupling agents, with subsequent scission of the *N*-Boc protecting group with a solution of 3M HCl in ethyl acetate.

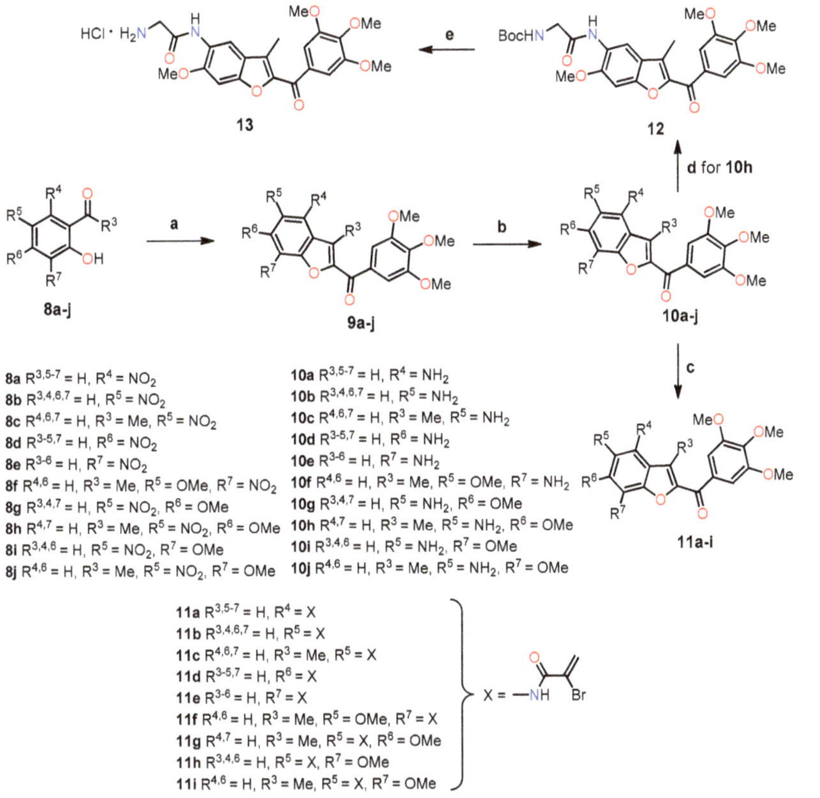

Scheme 2. Synthesis of benzo[*b*]furans **9–13**. Reagents and conditions: (**a**) 2-bromo-1-(3′,4′,5′-trimethoxyphenyl)ethanone, K$_2$CO$_3$, acetone, reflux, 18 h; (**b**) Fe, HCl (37% in H$_2$O), EtOH, reflux, 3 h, 42–79%; (**c**) α-bromoacrylic acid, EDCI, BtOH, DMF, r.t, 18 h, 44–66%; (**d**) *N*-Boc-glycine, EDCI, BtOH, DMF, r.t, 12 h, 83%; (**e**) 3M HCl in EtOAc, r.t, 3 h, 95%. Yields for compounds **9a–j** have not been reported by the authors.

The antiproliferative activity of this series of derivatives, including amino 2-(3′,4′,5′-trimethoxybenzoyl)-benzo[*b*]furans **10a–j** and **11a–I**, was evaluated against various cancer cell lines, along with CA-4 as the standard drug (Table 3). Compound **10h** ($R^{4,7}$ = H, R^3 = Me, R^5 = NH$_2$, R^6 = OMe) demonstrated the most promising results in the series, exhibiting significant growth inhibition against cancer cell lines L1210, FM3A/0, Molt4/C8, CEM/0, and HeLa, with IC$_{50}$ values ranging from 16 to 24 nM. Notably, compound **10h** exhibited higher activity in the FM3A/0 cell line, with an IC$_{50}$ value of 24 nM, compared to the standard drug CA-4 (IC$_{50}$ = 42 nM). SAR information derived from the comparison of unsubstituted compounds **10b** ($R^{3,4,6,7}$ = H, R^5 = NH$_2$), **10g** ($R^{3,4,7}$ = H, R^5 = NH$_2$, R^6 = OMe), and **10i** ($R^{3,4,6}$ = H, R^5 = NH$_2$, R^7 = OMe) vs. methyl derivatives **10c** ($R^{4,6,7}$ = H, R^3 = Me, R^5 = NH$_2$), **10h** ($R^{4,7}$ = H, R^3 = Me, R^5 = NH$_2$, R^6 = OMe), and **10j** ($R^{4,6}$ = H, R^3 = Me, R^5 = NH$_2$, R^7 = OMe) showed a significant increase in antiproliferative activity against cell lines by introducing the methyl group at the C–3 position of the benzofuran ring. Also, an increase in activity can be observed when comparing the methyl derivative **10c** with **10b**. When comparing the activities of compounds **10g** and **10h** with those of **10i** and **10j**, higher activity can be observed in compounds with methoxy groups at position C–6 rather than at position C–7 of the benzofuran ring. According to the results, compound **10h**, with a methyl group at the C–3 position and a methoxy group at the C–6 position, exhibited 2–4 times greater potency than the unsubstituted compound **10g** and 3–10 times higher activity than compound **10j**, which features a methoxy group at the C–7 position of the benzofuran ring. Changing the positions of the amino and methoxy groups from **10j** ($R^{4,6}$ = H, R^3 = Me, R^5 = NH$_2$, R^7 = OMe) to **10f** ($R^{4,6}$ = H, R^3 = Me, R^5 = OMe, R^7 = NH$_2$) resulted in a reduction in activity.

Table 3. In vitro inhibitory effects of benzo[*b*]furans **10**, **11**, and CA-4 against the proliferation of murine leukemia (L1210), murine mammary carcinoma (FM3A), human T-lymphocyte (Molt/4 and CEM), and human cervix carcinoma (HeLa) cells.

Compound	IC$_{50}$ (nM) [b]				
	L1210	FM3A/0	Molt4/C8	CEM/0	HeLa
10a	6700 ± 600	9200 ± 800	ND [c]	5400 ± 1200	2200 ± 600
10b	>10,000	>10,000	>10,000	>10,000	>10,000
10c	2500 ± 100	3600 ± 150	ND	2400 ± 400	5500 ± 530
10d	>10,000	>10,000	>10,000	>10,000	7500 ± 200
10e	>10,000	>10,000	>10,000	>10,000	>10,000
10f	>10,000	>10,000	>10,000	>10,000	>10,000
10g	73 ± 24	73 ± 2	66 ± 6.7	59 ± 4.1	42 ± 3.8
10h	19 ± 2	24 ± 6	22 ± 4	22 ± 5	16 ± 1
10i	600 ± 90	560 ± 90	870 ± 20	450 ± 10	370 ± 30
10j	67 ± 4	140 ± 13	360 ± 20	120 ± 10	490 ± 54
11a	78 ± 2.8	200 ± 90	ND	110 ± 20	1000 ± 80
11b	150 ± 8	320 ± 30	230 ± 1	390 ± 30	1100 ± 90
11c	320 ± 17	520 ± 29	ND	310 ± 80	430 ± 90
11d	420 ± 90	1200 ± 90	ND	600 ± 37	1200 ± 80
11e	800 ± 51	780 ± 62	1100 ± 90	970 ± 55	1000 ± 100
11f	1200 ± 40	5900 ± 380	ND	3200 ± 240	5900 ± 900
11g	430 ± 23	960 ± 78	ND	340 ± 20	1000 ± 100
11h	120 ± 14	250 ± 15	300 ± 40	240 ± 38	780 ± 54
11i	390 ± 20	1100 ± 100	920 ± 91	770 ± 15	1400 ± 120
13	14 ± 1	10 ± 8	18 ± 1	20 ± 2	14 ± 1
CA-4 [a]	2.8 ± 1.1	42 ± 6.0	16 ± 1.4	1.9 ± 1.6	1.9 ± 1.6

[a] Combretastatin-A4 (CA-4) is the standard drug for the study. [b] The data presented are the mean ± SD of three independent experiments. [c] ND means not detected. Reproduced with permission from ref. [34]. Copyright John Wiley & Sons Inc., 2023.

The unsubstituted α-bromoacryloylamide derivatives **11a** ($R^{3,5-7}$ = H, R^4 = X), **11b** ($R^{3,4,6,7}$ = H, R^5 = X), **11d** ($R^{3-5,7}$ = H, R^6 = X), and **11e** (R^{3-6} = H, R^7 = X) exhibited antiproliferative potency that was 10–100 times greater than their amino counterparts, demonstrating the direct relationship between the presence of α-bromoacryloylamides and increased activity (Table 3). Shifting the α-bromoacryloylamide group from the C–4 at the 5 position to the C–6 at the 7 position led to decreased activity. Finally, there were no significant differences in antiproliferative activity on all cell lines observed between compound **10h** and its glycine hydrochloride prodrug **13**.

To explore the potential correlation between antiproliferative activity and tubulin inhibition, the most active compounds, **10g**, **10h**, **10i**, **10j**, and **11a**, were evaluated in the inhibition of tubulin polymerization. Among these compounds, **10h** demonstrated the highest potency, displaying an IC$_{50}$ value of 0.56 µM, which was two times higher than that of CA-4 (IC$_{50}$ = 1.0 µM). Compounds **10g** and **10j** exhibited IC$_{50}$ values of 1.4 and 1.6 µM, respectively, for tubulin polymerization, showing a marginal decrease in potency compared to CA-4. Compound **10i** exhibited approximately half the activity of **10j**. In contrast, compound **11a** did not induce any alteration in tubulin assembly, even at a concentration as high as 40 µM, suggesting that the mechanism of action of this α-bromoacryloylamide derivative does not involve interaction with tubulin. Subsequently, the effect of the selected compounds, **10g, 10h, 10j**, and **11a**, was assessed on the cell cycle of human myeloid leukemia cell lines HL-60 and U937 using flow cytometry. The cells were cultured for 24 h with a concentration of 100 nM for each compound, and the two most active compounds (**10g** and **10h**) were examined at a lower concentration of 10 nM. Figure 4 shows the fractions of hypodiploid cells in the sub-G1 peak of each compound studied, representing apoptotic cells. Compound **10j** showed a modest increase in apoptotic cells at 100 nM, while compounds **10g** and **10h** presented different effects on the cell cycle in the two cell lines. It was observed that compound **11a** had no effect on cell cycle distribution at 100 nM, in agreement with the previous results observed in tubulin inhibition. The significant increase in the sub-G1 peak in both cell lines with increasing concentrations of compounds **10g** and **10h** suggests that these compounds exert their growth inhibitory effect by inducing apoptosis.

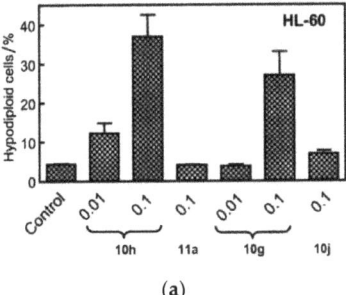
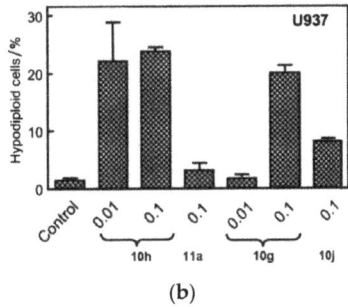

Figure 4. Compounds **10g, 10h, 10j**, and **11a** were put on approval against (**a**) HL-60 and (**b**) U937 cells for 24 h at indicated concentrations (in µM). Apoptosis was determined using flow cytometry. Values represent the mean ± SE of three independent experiments, each conducted in triplicate. Reproduced with permission from ref. [34]. Copyright John Wiley & Sons Inc., 2023.

To further examine the apoptotic effects of compounds **10g, 10h, 10j**, and **11a**, proteolytic processing of caspases in HL-60 and U937 cells was observed by Western blot analysis (Figure 5). Compounds **10g, 10h, 10j**, and **11a** allowed the cleavage of inactive procaspase-9 to the active 37 kDa fragment, while lower concentrations of compounds **10g** and **10h** (0.3 µM) significantly promoted procaspase-8 hydrolysis. Furthermore, compounds **10g** and **10h** significantly led to the cleavage of inactive procaspases-3 and -6 in both cell lines (Figure 5). Additionally, the induction of poly(ADP-ribose)polymerase (PARP) cleavage showed that compounds **10j** and **11a** exhibited lower potency in inducing PARP cleavage compared to compounds **10g** and **10h**. The appearance of the fragment at 85 kDa coincided with caspase-3 activation, observed by a decrease in the proenzyme at 36 kDa and an increase in cleaved procaspase-3 levels at 20 and 18 kDa (Figure 5). Dose–response studies were performed, and cytosolic preparations were analyzed by immunoblotting to investigate whether apoptosis induced by compounds **10g, 10h, 10j**, and **11a** in HL-60 and U937 cells involved the release of cytochrome c from mitochondria into the cytosol during the apoptotic event. The results revealed a significant increase in the amount of cytochrome c in the cytosol of both cell lines at 15 kDa (Figure 5).

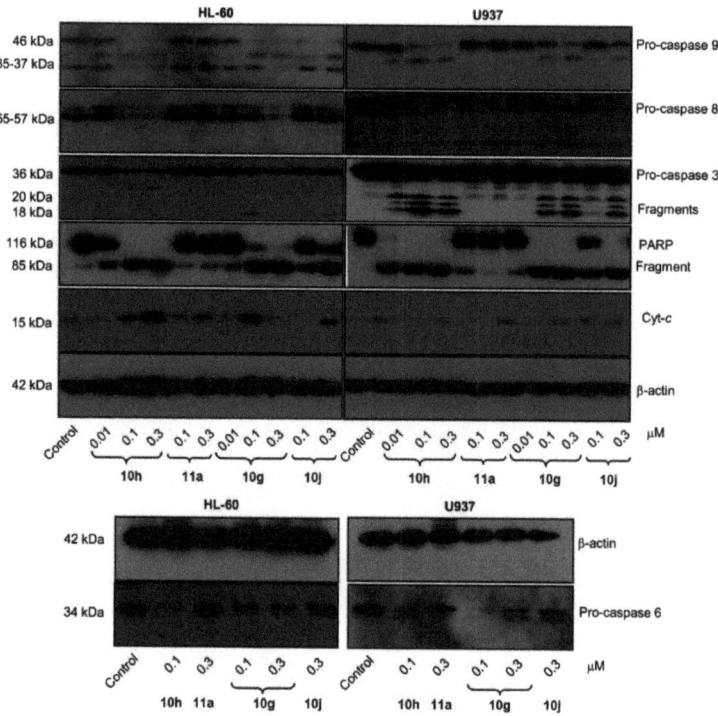

Figure 5. Relationship between caspases and the induction of apoptosis in two leukemia cell lines. Cells were incubated at different concentrations of compounds **10g**, **10h**, **10j**, or **11a**, and the cell lysates were analyzed by immunoblotting for cleavage of pro-caspases-9, -8, -6, and -3; poly(ADP-ribose) polymerase (PARP); and cytochrome c release. β-Actin served as the loading control. Reproduced with permission from ref. [34]. Copyright John Wiley & Sons Inc., 2023.

CA-4 and its analogs have been clinically termed vascular disruptors agents (VDAs) [35], so glycine prodrug **13** was tested to see its ability as a VDA in an in vivo model in rat breast cancer tumors, using the spatial frequency optical technique (SDFI). With 10 min of administration of compound **13** (30 mg kg^{-1}), there is a rapid decrease in oxygen saturation in tumor tissues similar to that observed with CA-4, confirming that prodrug **13** causes vascular disruption in vivo.

Interestingly, Wellington et al. conducted the synthesis of a variety of dihydroxylated 5,6-benzo[b]furans **16** with catechol derivative **14** using a commercial laccase, Suberase®, under different reaction conditions [36]. All the synthesized compounds were evaluated for their anticancer properties. The synthetic protocol consisted of reacting an equivalent of catechol **14** with an equivalent of the 1,3-dicarbonyl compound **15** at room temperature using Suberase® in an air-open vessel at pH 7.15 (Scheme 3). In method A, the reaction of catechol derivatives **14a–c** with 1,3-dicarbonyls compounds **15a–e** was performed at room temperature at pH 7.15 for 24 h. In method B, the reaction was conducted under similar conditions with an extended time of 44 h to investigate the potential enhancement in product yield with prolonged reaction time. In method C, a mixture of the 1,3-dicarbonyl compound and catechol, combined in a 4:1 ratio, was dissolved in DMF. Subsequently, the resulting mixture was left to react for 42 h. The results obtained from methods A–C are shown in Table 4. In particular, method A demonstrated the most favorable outcome of the three methods, exhibiting the highest yield of 98% for compound **16j** (Entry 15, Table 4). On the other hand, method B afforded compound **16k** in a 77% yield (Entry 18, Table 4), while method C yielded 71% for compound **16g** (Entry 11, Table 4). It is worth noting that the reaction time in Method B had minimal impact on the yield, whereas the presence of DMF in method C may have potentially deactivated the laccase, Suberase®, leading to lower yields.

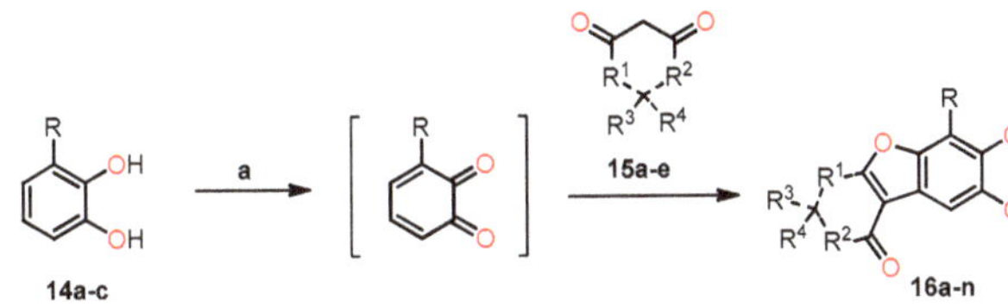

Scheme 3. Synthesis of dihydroxylated 5,6-benzo[b]furans **16a–n**. Reagents and conditions: (a) Suberase, phosphate buffer, rt, pH 7.15.

14a R = H
14b R = Me
14c R = OMe

15a $R^1 = R^2$ = Me
15b $R^1 = R^2 = CH_2$, $R^3 = R^4$ = H
15c $R^1 = R^2 = CH_2$, R^3 = H, R^4 = Me
15d $R^1 = R^2 = CH_2$, $R^3 = R^4$ = Me
15e $R^1 = R^2 = CH_2$, R^3 = H, R^4 = Ph

Table 4. Reaction conditions for the synthesis of dihydroxylated 5,6-benzo[b]furans **16a–n**.

Entry	Catechol Derivative	1,3-Dicarbonyl Compound	Reaction Time (h)	Method	Yield 16 (%)
1	14a	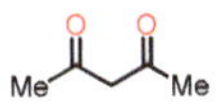	24	A	16a (48)
2	14a		44	B	16a (49)
3	14b		24	A	16b (50)
4	14a	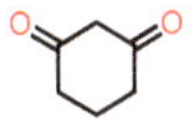	24	A	16c (65)
5	14b		24	A	16d (62)
6	14b		44	B	16d (67)
7	14c		24	A	16e (70)
8	14a	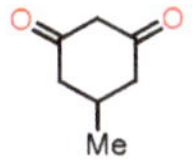	24	A	16f (59)
9	14a		42	C	16f (50)
10	14b		24	A	16g (78)
11	14b		42	C	16g (71)
12	14c		24	A	16h (37)
13	14a	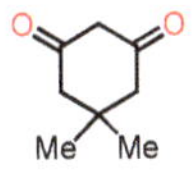	24	A	16i (58)
14	14a		42	C	16i (40)
15	14b		24	A	16j (98)
16	14b		42	C	16j (59)
17	14c		24	A	16k (73)
18	14c		44	B	16k (77)
19	14a	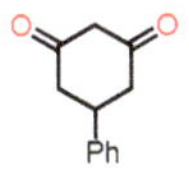	24	A	16l (76)
20	14a		42	C	16l (50)
21	14b		24	A	16m (80)
22	14c		24	A	16n (43)
23	14c		42	C	16n (15)

Anticancer studies were performed on various types of cancer, including renal (TK10), melanoma (UACC62), breast (MCF7), and cervical (HeLa), using a sulforhodamine B (SRB) assay to determine the growth inhibitory effects of these compounds. Notably, the 5,6-dihydroxylated benzo[b]furans **16e**, **16g**, **16h**, **16k**, **16m** and **16n** exhibited potent cytotoxic effects against the melanoma cell line (UAC62), with GI_{50} values ranging from 0.77 to 9.76 µM. Among these compounds, **16h** (R = OMe, $R^1 = R^2 = CH_2$, R^3 = H, R^4 = Me) and **16n** (R = OMe, $R^1 = R^2 = CH_2$, R^3 = H, R^4 = Ph) showed better activity than the standard drug Etoposide (GI_{50} = 0.89 µM). Moreover, compound **16n** showed potent activity (GI$_{50}$ = 9.73 µM) against the renal cancer cell line (TK10), while both **16h** and **16n** demonstrated strong activity against the breast cancer cell line (MCF7), with GI$_{50}$ values of 8.79 and 9.30 µM, respectively.

In 2013, Kamal et al. synthesized a series of benzo[b]furans with a modification at position 5 of the benzene ring by introducing C-linked substituents to generate 2-(3′,4′,5′-trimethoxybenzoyl)benzo[b]furan derivatives [37]. The most biologically interesting benzo[b]furan derivatives, **22** and **25**, were synthesized through a sequence of reactions depicted in Scheme 4, which included (a) the acylation of **17** to yield product **18**; (b) methylation using methyl iodide and potassium carbonate, resulting in **19**; (c) iodination with iodine and silver nitrate in a catalytic amount to produce **20**; (d) cyclization with 2-bromo-1-(3,4,5-trimethoxyphenyl)ethanone and potassium carbonate in acetone to furnish the benzofuran derivative **21**; (e) Sonogashira coupling reaction to obtain **22**; (f) Wittig reaction with the ylide generated from methyltriphenylphosphonium bromide in the presence of LiHMDS, leading to **23**; (g) Heck coupling reaction with ethyl acrylate to yield the ethyl cinnamate derivative **24**; and (h) ester reduction with DIBAL resulting in the formation of (E)-allyl alcohol **25**.

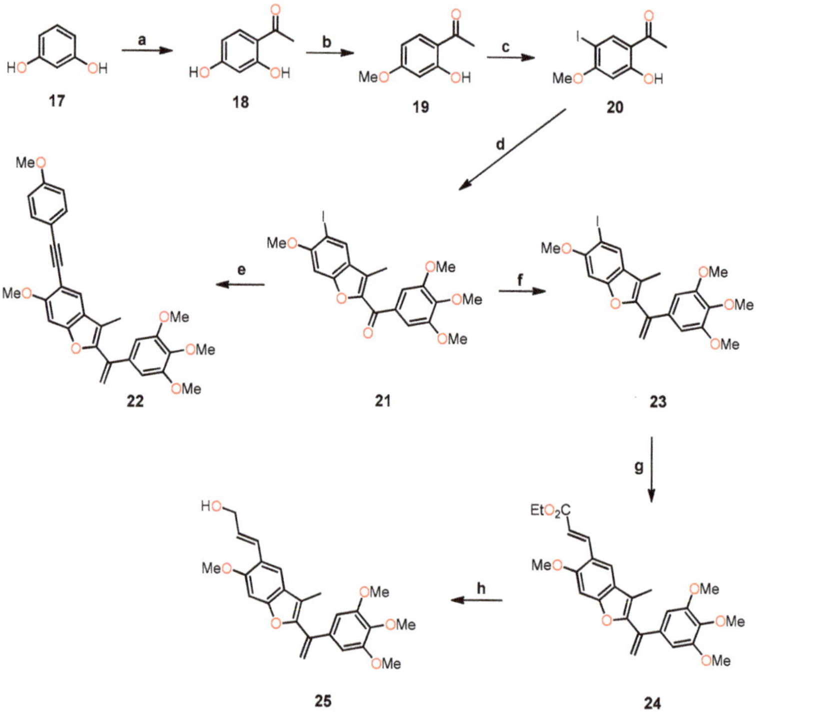

Scheme 4. Synthesis of benzo[b]furan derivatives **22** and **25**, and the reaction conditions: (**a**) anhydrous ZnCl$_2$, CH$_3$COOH, reflux, 12 h, 73%; (**b**) K$_2$CO$_3$, CH$_3$I, acetone, reflux, 18 h, 85%; (**c**) I$_2$, AgNO$_3$ (cat.), Ph$_3$PCH$_3$Br, THF, 0 °C, 50–70%; (**d**) 2-bromo-1-(3,4,5-trimethoxyphenyl)ethenone, K$_2$CO$_3$, acetone, reflux, 12 h, 75%; (**e**) substitution alkynes, PdCl$_2$(PPh$_3$)$_3$, CuI, Et$_3$N, THF, 60 °C, 14 h, 62–65%; (**f**) LiHMDS, Ph$_3$PCH$_3$Br, THF, 0 °C, 50–70%; (**g**) ethyl acrylate, PdCl$_2$(Ph$_3$P)$_2$, Et$_3$N, DMF, 60 °C, 24 h, 65%; (**h**) DIBAL, CH$_2$Cl$_2$, 0 °C, 1 h, 60%.

A study was performed to assess the cytotoxicity of benzofuran analogs against ME-180, A549, ACHNs, HT-29, and B-16 cell lines using a 3-(4,5-dimethylthiazol-2-yl)-2,5-diphenyltetrazolium bromide (MTT) assay and CA-4 as the standard drug (Table 5). The results from compound **22**, which contains a 4-MeO-phenylacetylene group, showed good activity, with IC$_{50}$ values in the range of 0.08–1.14 μM against all evaluated cell lines, as shown in Table 5. However, derivative **22** was two times less active against A549 and significantly less active against all four cell lines compared to CA-4. On the other hand, analog **25** exhibited the effect of the alkenyl substituent at position 5 of benzofuran on cytotoxicity, showing higher potency compared to compound **22** against ME-180, A549, ACHN, and B-16 cancer cell lines, with IC$_{50}$ values ranging from 0.06 to 0.17 μM. These values were comparable to those obtained with CA-4 against A549 and ACHN cancer lines, with IC$_{50}$ values of 0.05 and 0.09 μM, respectively. Furthermore, compounds **22** and **25** inhibited tubulin polymerization by 37.9 and 65.4%, respectively, which is comparable to the 70.5% tubulin inhibition observed with CA-4 (Table 5).

Table 5. The cytotoxicity against a panel of cancer cell lines and the inhibition of tubulin polymerization of compounds **22**, **25**, and CA-4.

Compound	IC$_{50}$ (μM) [b]					Tubulin	
	ME-180	A549	ACHN	HT-29	B-16	IC$_{50}$ (μM) [c]	%
22	0.19 ± 0.02	0.08 ± 0.16	0.97 ± 0.49	0.08 ± 0.10	1.14 ± 0.50	3.81 ± 0.4	37.9
25	0.09 ± 0.02	0.06 ± 0.11	0.09 ± 0.14	0.08 ± 0.01	0.17 ± 0.09	1.95 ± 0.1	65.4
CA-4 [a]	0.007 ± 0.06	0.05 ± 0.06	0.09 ± 0.08	0.008 ± 0.01	0.06 ± 0.01	1.86 ± 0.1	70.5

[a] Combretastatin-A4 (CA-4) is the standard drug for the study. [b] Values are the mean ± SD of three independent experiments determined after 48 h of treatment. [c] Values are the mean ± SD of two independent experiments performed in triplicate.

Apoptotic studies were performed using various assays, including Hoechst staining assay, caspase-3 activation, DNA fragmentation analysis, and Western blot analysis. Specifically, a Hoechst staining assay was utilized to study the effects of compounds **22** and **25** on nuclear condensation. Remarkably, cells treated with these compounds showed a pronounced increase in nuclear condensation compared to untreated cells, strongly suggesting their potent ability to induce cell apoptosis (Figure 6). In addition, caspase-3 activation analysis was conducted on A549 cells, treating them with concentrations of 50 and 100 nM of compounds **22** and **25** and comparing them to CA-4 at 100 nM. The results indicated a significant increase in caspase-3 activation, ranging from 1.5- to 3-fold compared to the control experiment (50 and 100 nM), demonstrating the programmed apoptotic activity induced by compounds **22** and **25** in the A549 cells.

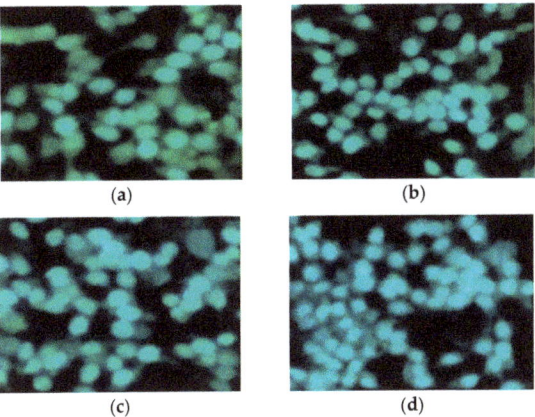

Figure 6. Hoechst staining in the A549 cell line. (**a**) A549 control cells; (**b**) CA-4 (50 nm); (**c**) compound **22** (50 nm); and (**d**) compound **25** (50 nm). Reproduced with permission from ref. [37]. Copyright John Wiley & Sons Inc., 2023.

DNA fragmentation analysis was conducted by incubating A549 cells with a 50 nM concentration of compounds **22** and **25**. The results revealed a discrete staircase pattern after 48 h of treatment, indicative of significant fragmentation associated with cell death (Figure 7a). Additionally, Western blot analysis was realized on the same cancer cell line, treating it with the same concentration of compounds **22** and **25** as used in the DNA fragmentation analysis (Figure 7b). After 48 h of treatment, it was observed that the anti-apoptotic protein Bcl-2 was down-regulated, while the pro-apoptotic protein Bax was up-regulated. The results provide evidence that the induction of apoptosis by compounds **22** and **25** is associated with Bcl-2 down-regulation.

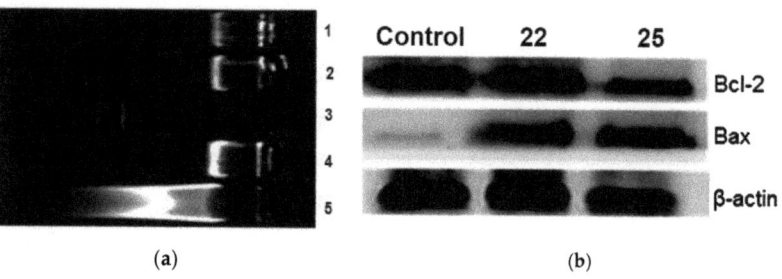

Figure 7. (a) DNA fragmentation analysis was performed on A549 lung cancer cells treated with compounds **22** and **25**. The gel electrophoresis results show lane 1: control (A549 cells), lane 2: CA-4 (50 nM), lane 3: marker (100 bp), lane 4: **22** (50 nM), and lane 5: **25** (50 nM). (b) The effect of compounds on Bcl-2 and Bax levels. A549 cells were treated with compounds **22** and **25** at 50 nM for 48 h. Cell lysates were subjected to Western blot analysis to determine the expression levels of Bcl-2 and Bax, using β-actin as the loading control. Reproduced with permission from ref. [37]. Copyright John Wiley & Sons Inc., 2023.

In a study conducted by Frías et al., an asymmetric synthesis of diheteroarylalkanes was presented. This synthesis involved a one-pot reaction using dienamine and Friedel–Crafts reactions between aldehyde **26** and indole **27**, catalyzed by the Hayashi–Jørgensen catalyst **28** (20 mol %). Various substituents at different positions on the aldehyde and indole were utilized during the reaction [38]. When starting, materials with electron-withdrawing groups (EWGs) or electron-donating groups (EDGs) were located at the *para* position to the oxygen atom, and products **29a,g–l** showed good yields and enantioselectivity ranging from 93% to 97% toward the (*S*)-enantiomer (Table 6). However, substrates substituted at the *ortho* and *meta* positions also enabled the synthesis of products **29j** and **29k** without a decrease in the final enantioselectivity (*ee* = 93% and 94%, respectively). Indoles with bromo (**29b**) or methoxy (**29c**) substituents exhibited satisfactory yields and enantioselectivity (>95%). Methyl groups displayed good yield and enantioselectivity in products **29d** and **29e** (*ee* = 99% and 96%, respectively). Additionally, the introduction of the 1*H*-benzo[*g*]indole group resulted in the desired aldehyde **29f** with excellent enantioselectivity (*ee* = 98%).

The enantioselectivity of compound **29** is explained through the proposed reaction mechanism depicted in Scheme 5. It begins with the condensation reaction between aldehyde **26** and organocatalyst **28** to form iminium ion **I**. Then, the isomerization of **I** gives dienamine intermediate **II**, which undergoes intramolecular condensation/dehydration sequence to afford iminium ion **III**. At this pivotal stage, indole **27** undergoes an attack on intermediate **III**, leading to the formation of product **29** with remarkable enantioselectivity. This precise enantiocontrol is facilitated by the steric shielding offered by the bulkier group (CPhPhOTMS) present in the organocatalyst.

Table 6. Synthesis of diheteroarylalkanal **29** through a one-pot Friedel–Crafts reaction.

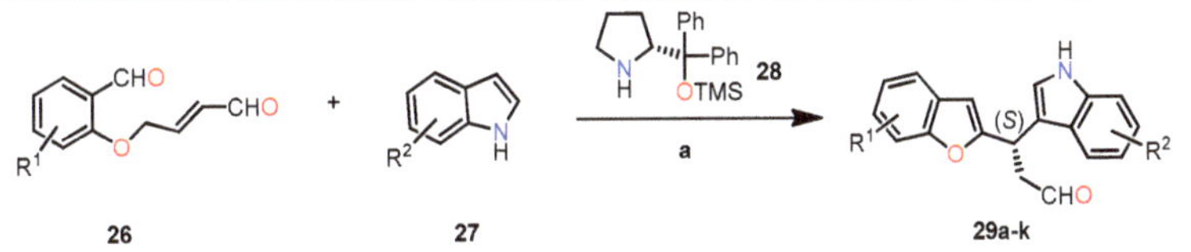

Compound	R^1	R^2	Yield (S)-29 (%)	% ee
29a	H	H	79	94
29b	H	5-Br	71	95
29c	H	5-Ome	75	96
29d	H	2-Me	79	99
29e	H	7-Me	78	96
29f	H	benzo[g]	85	98
29g	5-Cl	H	76	97
29h	5-Me	H	72	94
29i	5-NO_2	H	75	93
29j	7-Ome	H	71	93
29k	6-Ome	H	69	94

Reagent and conditions: (**a**) 26 (0.1 mmol), 27 (0.12 mmol), 28 (0.02 mmol), DABCO (0.02 mmol), and toluene (0.3 mL).

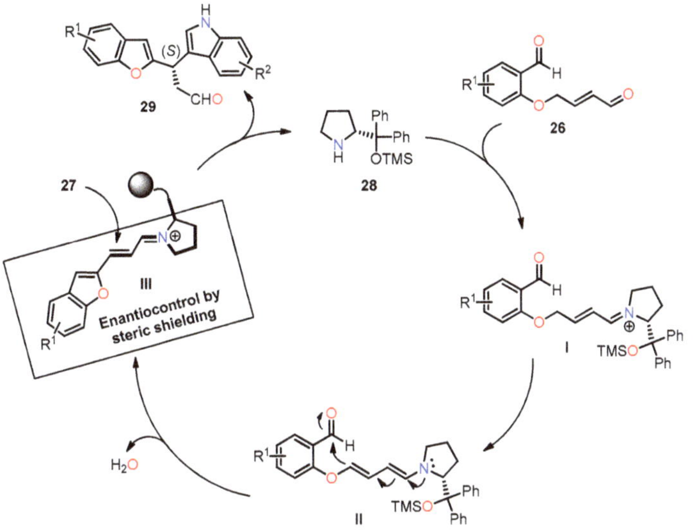

Scheme 5. Plausible reaction mechanism for the synthesis of diheteroalkane **29**.

The synthesized products **29a–k** were evaluated for their antiproliferative activity against a panel of tumor cell lines, including HBL-100 (breast), HeLa (cervix), SW1573 (non-small-cell lung), and WiDr (colon), using the SRB assay. Figure 8 presents the GI_{50} values, comparing them with *cis*-platin as the standard drug. The results highlight the significance of the substituent on the aryl moiety of the benzofuran ring in influencing the antiproliferative activity of product **29**. The introduction of strong electron-withdrawing groups (EWGs) and electron-donating groups (EDGs) resulted in a decrease in activity for compounds **29h**, **29i**, and **29j** in WiDr cell lines. Notably, compounds **29a** ($R^1 = R^2 = $ H) and **29g** ($R^1 = $ 5-Cl, $R^2 = $ H) exhibited GI_{50} values comparable to *cis*-platin in the WiDr cell line, with values of 28 and 16 μM for compounds **29a** and **29g**, respectively. However, the product of highest biological interest was **29f** ($R^1 = $ H, $R^2 = $ benzo[g]), which showed the most significant activity across all cell lines, achieving similar or even better potency than *cis*-platin (GI$_{50}$ = 2–18 µM).

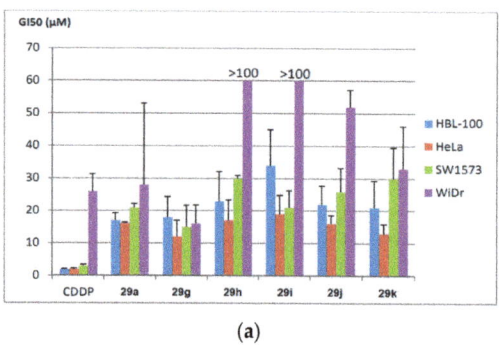
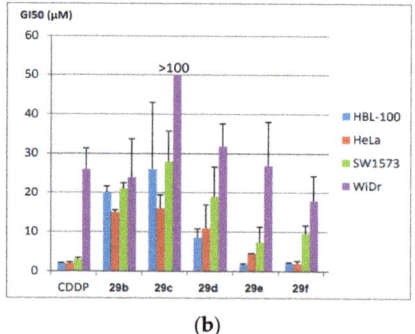

Figure 8. Antiproliferative activity (GI$_{50}$, µM) in various human solid tumor cell lines, including HBL-100 (breast), HeLa (cervix), SW1573 (non-small-cell lung), and WiDr (colorectal). The compounds evaluated are presented in two groups: (**a**) **29a**, **29g–k**, and (**b**) **29b–f**. Reproduced with permission from ref. [38]. Copyright John Wiley & Sons Inc., 2023.

In a separate study, Penthala et al. synthesized a series of heterocyclic analogs, including indoles, benzofurans, and benzothiophenes, based on Combretastatin, and assessed their anticancer activity against a panel of 60 human cancer cell lines [39]. The benzo[*b*]furans were synthesized by condensing benzo[*b*]furanocarbaldehyde **30** (1.0 mol) with phenylacetonitrile **31** (1.1 mol) in a 5% sodium methoxide/methanol solution for 3–6 h, resulting in the successful formation of the desired product **32** (Table 7). The evaluation of anticancer studies focused on compounds **32a**, **32b**, and **32d** against 60 cancer cell lines. Compound **32a** exhibited the most favorable results, displaying GI$_{50}$ values ranging from <0.01 to 73.4 µM across all 60 cell lines. It effectively inhibited the growth of 70% of the evaluated cancer cell lines, with a remarkable GI$_{50}$ value < 0.01 µM in almost all cases. On the other hand, substituting the 3,4,5-trimethoxyphenyl group in **32a** (R = 2-CHO, R^1 = OMe, R^2 = OMe) with the 3,4-dimethoxyphenyl group in **32b** (R = 2-CHO, R^1 = H, R^2 = OMe) resulted in reduced growth inhibition against 54% of the cancer cells, exhibiting GI$_{50}$ values ranging from 0.229 to 0.996 µM. Furthermore, compound **32b** exhibited potent anti-proliferative activity in MDA-MB-435 melanoma cells, exhibiting a remarkable GI$_{50}$ value of 0.229 µM. Similarly, compound **32d** (R = 3-CHO, R^1 = OMe, R^2 = OMe) displayed significant antiproliferative inhibition, with GI$_{50}$ values ranging from 0.237 to 19.1 µM, and effectively inhibited 52% of the evaluated cell lines with a GI$_{50}$ value < 1 µM. In addition, the evaluation of anti-leukemia activity against the MV4–11 cell line was performed for compounds **32a–d** (Table 7), demonstrating that **32a** emerged as the most active compound among the evaluated benzo[*b*]furans in the leukemia cell line. Later, a molecular docking simulation was performed to investigate the interactions of compound **32a** with α/β tubulin in complex with colchicine-DAMA (PDB ID code: 1SA0). The simulation revealed a hydrophobic interaction at the α–β interface, where colchicine binds, and stability was observed through van der Waals interactions with Asn101, Ser178, Thr179, and Val181 in α-tubulin, as well as Asn258 and Lys352 in β-tubulin (Figure 9). The calculated free energy value for these interactions was −7.74 kcal/mol.

Table 7. Synthesis of (Z)-benzo[b]furan-2-yl and (Z)-benzo[b]furan-3-yl cyanocombretastatins **32a–f** and the evaluation of their anti-leukemic activity (LD$_{50}$) against the MV4–11 AML cell line.

Compound	R	R^1	R^2	LD$_{50}$ (μM)
32a	2-CHO	OMe	OMe	0.047
32b	2-CHO	H	OMe	>20
32c	2-CHO	OMe	H	1.169
32d	3-CHO	OMe	OMe	4.529
32e	3-CHO	H	OMe	ND
32f	3-CHO	OMe	H	ND

Reagent and conditions: (**a**) 30 (1.0 mol), 31 (1.1 mol), NaOMe (5%), CH3OH, reflux, 3–6 h. ND means not detected.

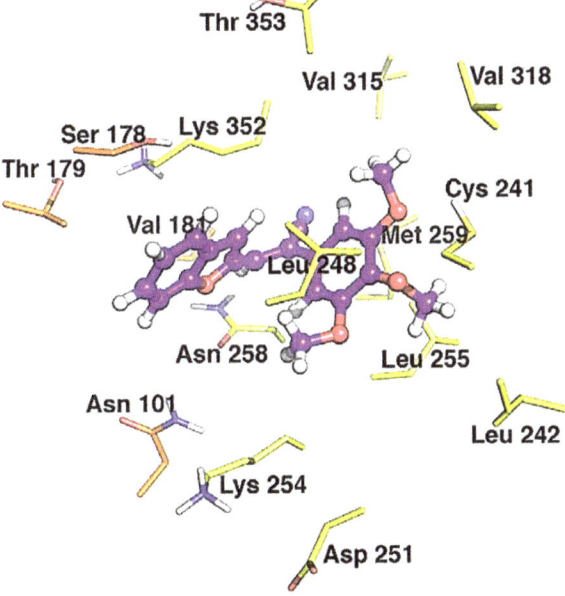

Figure 9. Binding modes between **32a** and α/β tubulin in complex with colchicine-DAMA (PDB ID code: 1SA0). The inhibitors are represented as purple ball-and-stick models, while the tubulin residues are shown as orange (α-tubulin) and yellow (β-tubulin) sticks. Reproduced with permission from ref. [39]. Copyright Elsevier Inc., 2023.

In the same year, Romagnoli et al. synthesized a series of compounds known as 3-(3′,4′,5′-trimethoxyanilino)benzo[b]furan, wherein a 2-methoxy/ethoxycarbonyl group was combined with either no substituent or a methoxy group at each position of the benzene ring [40]. The synthesis of compounds **35a–l** involved a two-step reaction process (Scheme 6). In the first step, 2-hydroxybenzonitrile derivatives **33a–f** were condensed with methyl or ethyl bromoacetate and K$_2$CO$_3$ in DMF, leading to the formation of 3-aminobenzo[b]furan analogs **34a–l** in high yields through a one-pot tandem cyclization method. Subsequently, compounds **35a–l** were synthesized via palladium-catalyzed C-N

Buchwald–Hartwig cross-coupling between the deactivated 3-aminobenzo[b]furans **34a–l** and 1-bromo-3,4,5-trimethoxybenzene in toluene at 100 °C, utilizing Pd(OAc)$_2$, *rac*-BINAP, and Cs$_2$CO$_3$ as the catalyst, ligand, and base, respectively.

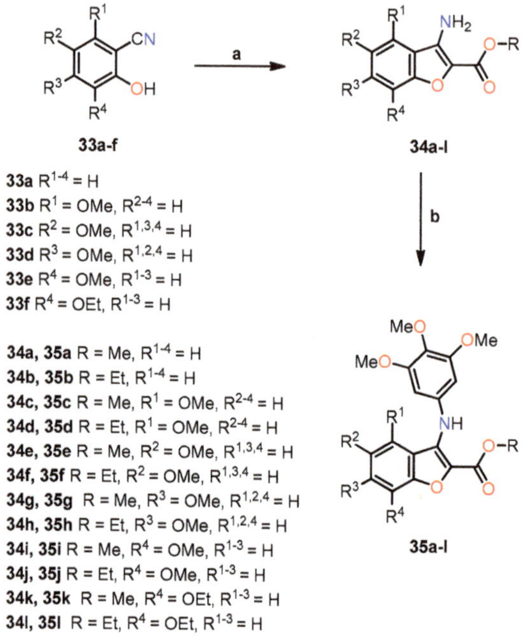

33a-f

33a R$^{1-4}$ = H
33b R^1 = OMe, R$^{2-4}$ = H
33c R^2 = OMe, R^{1,3,4} = H
33d R^3 = OMe, R^{1,2,4} = H
33e R^4 = OMe, R$^{1-3}$ = H
33f R^4 = OEt, R$^{1-3}$ = H

34a-l

35a-l

34a, 35a R = Me, R$^{1-4}$ = H
34b, 35b R = Et, R$^{1-4}$ = H
34c, 35c R = Me, R^1 = OMe, R$^{2-4}$ = H
34d, 35d R = Et, R^1 = OMe, R$^{2-4}$ = H
34e, 35e R = Me, R^2 = OMe, R^{1,3,4} = H
34f, 35f R = Et, R^2 = OMe, R^{1,3,4} = H
34g, 35g R = Me, R^3 = OMe, R^{1,2,4} = H
34h, 35h R = Et, R^3 = OMe, R^{1,2,4} = H
34i, 35i R = Me, R^4 = OMe, R$^{1-3}$ = H
34j, 35j R = Et, R^4 = OMe, R$^{1-3}$ = H
34k, 35k R = Me, R^4 = OEt, R$^{1-3}$ = H
34l, 35l R = Et, R^4 = OEt, R$^{1-3}$ = H

Scheme 6. Synthesis of benzo[b]furan analogs **35a–l** and the reactions conditions: (**a**) BrCH$_2$CO$_2$CH$_3$ or BrCH$_2$CO$_2$C$_2$H$_5$, K$_2$CO$_3$, DMF, 60 °C for 4 h, then reflux for 8 h, 52–78%; (**b**) 1-bromo-3,4,5-trimethoxybenzene, Pd(OAc)$_2$, BINAP, Cs$_2$CO$_3$, PhMe, 100 °C, 16 h, 52–78%. Reprinted (adapted) with permission from ref. [40]. Copyright American Chemical Society, 2023.

The in vitro antiproliferative activity was evaluated against seven cell lines, and the corresponding results are presented in Table 8. The findings revealed a notable correlation between the presence and position of the methoxy substituent on the benzene moiety of the benzo[b]furan system. Among the series of 2-alkoxycarbonyl derivatives, the highest activity was observed when the methoxy group was located at the C–6 position, as exemplified by compounds **35g** (R = Me, R^3 = OMe, R^{1,2,4} = H) and **35h** (R = Et, R^3 = OMe, R^{1,2,4} = H), exhibiting IC$_{50}$ values ranging from 0.3 to 27 nM for **35g** and from 13 to 100 nM for **35h**. On the contrary, compounds **35c** (R = Me, R^1 = OMe, R$^{2-4}$ = H) and **35d** (R = Et, R^1 = OMe, R$^{2-4}$ = H) displayed the lowest activity when the methoxy group was situated at the C–4 position, with IC$_{50}$ values exceeding 10 μM. Furthermore, the methoxycarbonyl group demonstrated superior efficacy compared to the ethoxycarbonyl substituent in all cell lines, except for MCF-7 cells, which exhibited equal sensitivity to both compounds. Notably, compounds **35i** (R = Me, R^4 = OMe, R$^{1-3}$ = H) (average IC$_{50}$ = 370 nM) and **35j** (R = Et, R^4 = OMe, R$^{1-3}$ = H) (average IC$_{50}$ = 670 nM), featuring a methoxy C–7 substituent, displayed higher activity compared to **35e** (R = Me, R^2 = OMe, R^{1,3,4} = H) (average IC$_{50}$ = 1.500 nM) and **35f** (R = Et, R^2 = OMe, R^{1,3,4} = H) (average IC$_{50}$ = 2.900 nM), which possessed a methoxy C–5 substituent. These compounds also demonstrated remarkable activity against RS 4;11 cells, with IC$_{50}$ values of 39 nM for **35e** and 1 nM for **35i** (R = Me, R^4 = OMe, R$^{1-3}$ = H). Additionally, in Jurkat cells, they displayed an IC$_{50}$ value of 30 nM for **35i**. In contrast, compounds **35a** (R = Me, R$^{1-4}$ = H) and **35b** (R = Et, R$^{1-4}$ = H) exhibited IC$_{50}$ values of 3.300 and 2.600 nM, respectively. The absence of a methoxy substituent led to lower activity, highlighting the significant enhancement achieved by including a methoxy substituent at C–5. Furthermore, the substitution with ethoxycarbonyl at C–7 resulted in notably lower potency when compared to the substitution with C–7-methoxy (35i–l).

Table 8. In vitro cell growth inhibitory effects of compounds 35a–l and CA-4 on various cancer cell lines.

Compound	IC$_{50}$ (nM) [b]						
	HeLa	A549	HT-29	Jurkat	RS 4; 11	MCF-7	HL-60
35a	260 ± 50	5280 ± 800	930 ± 35	4100 ± 200	430 ± 97	7800 ± 900	4400 ± 200
35b	1330 ± 580	5470 ± 700	1600 ± 120	180 ± 38	300 ± 80	6600 ± 310	2500 ± 130
35c	>10.000	>10.000	>10.000	>10.000	>10.000	>10.000	>10.000
35d	>10.000	>10.000	>10.000	>10.000	>10.000	>10.000	>10.000
35e	250 ± 88	1570 ± 430	240 ± 60	210 ± 20	39 ± 9	7900 ± 1300	470 ± 30
35f	1260 ± 510	8900 ± 1460	1400 ± 540	2300 ± 700	190 ± 15	2900 ± 400	3400 ± 120
35g	2 ± 0.1	9 ± 1.4	3 ± 0.9	8 ± 0.6	0.3 ± 0.1	27 ± 2	5 ± 1
35h	13 ± 8	36 ± 11	17 ± 8	22 ± 6	100 ± 10	25 ± 3	24 ± 7
35i	130 ± 60	1270 ± 400	290 ± 30	30 ± 5	1 ± 0.1	520 ± 40	320 ± 17
35j	270 ± 80	1100 ± 300	110 ± 50	290 ± 50	230 ± 10	2100 ± 90	590 ± 50
35k	2530 ± 280	8900 ± 1300	3200 ± 210	3700 ± 450	400 ± 100	>10.000	4200 ± 200
35l	3280 ± 370	7250 ± 237	5300 ± 290	9100 ± 820	3000 ± 400	>10.000	5500 ± 540
CA-4 [a]	4 ± 1	180 ± 30	3100 ± 100	5 ± 0.6	0.8 ± 0.2	370 ± 100	1 ± 0.2

[a] Combretastatin-A4 (CA-4) is the standard drug for the study. [b] Data are expressed as the mean ± SE from the dose–response curves of at least three independent experiments.

Compounds 35e and 35g–j, along with CA-4, were investigated to determine their inhibitory effects on tubulin polymerization and colchicine binding to tubulin. The aim was to gain insights into their mechanisms of antiproliferative action, particularly their interaction with tubulin microtubules (Table 9). The results revealed that compound 35g exhibited the highest potency among the tested compounds, with an IC$_{50}$ of 1.1 µM, comparable to that of CA-4. Meanwhile, compound 35h demonstrated slightly lower activity compared to CA-4. Compounds 35e, 35i, and 35j showed 6–7 times lower potency than CA-4, with IC$_{50}$ values of 7.5, 7.6, and 6.4 µM, respectively. Regarding colchicine binding studies, outcomes were observed exclusively for compounds 35g and 35h, exhibiting inhibition percentages of 83% and 74%, respectively, which are comparable to the 99% inhibition observed with CA-A. The findings underscore the intricate interplay between the inhibition of tubulin polymerization and the hindrance of colchicine binding, shedding light on their potential synergistic effects in influencing antiproliferative pathways.

Table 9. Antitubular and colchicine binding evaluation by compounds 35e, 35g–j, and CA-4.

Compound	Tubulin Assembly IC$_{50}$ (µM) [a]	Colchicine Binding % Inhibition [b]
35e	7.5 ± 0.5	ND
35g	1.1 ± 0.1	83 ± 0.5
35h	1.5 ± 0.2	74 ± 4.1
35i	7.6 ± 1.0	ND
35j	6.4 ± 0.9	ND
CA-4 [c]	1.1 ± 0.1	99 ± 0.1

[a] Inhibition of tubulin polymerization: tubulin was used at 10 µM. [b] Inhibition of [3H] colchicine binding: tubulin, colchicine, and the tested compound were used at 1, 5, and 5 µM, respectively. [c] Combretastatin-A4 (CA-4) is the standard drug for the study. Values represent the mean ± SE from three independent experiments. ND means not detected.

A molecular docking simulation was performed to investigate the interactions of compound 35g with the colchicine site of tubulin (PDB ID code: 3HKC) (Figure 10). This revealed that the trimethoxyphenyl ring of 35g resides near Cys241. Moreover, a potential hydrogen bond interaction was observed between the ester moiety and Ala250, consistent with other colchicine site agents. These findings underscore the potential impact of substitutions at C–4, C–5, and C–7 on the antiproliferative activity of the compounds.

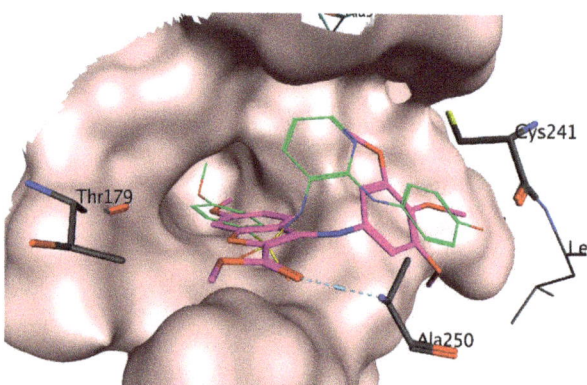

Figure 10. Binding modes of compound **35g** (represented in magenta) in the colchicine site of tubulin (PDB ID code: 3HKC). The co-crystallized ligand *N*-[2-[(4-hydroxyphenyl)amino]-3-pyridinyl]-4-methoxybenzenesulfonamide (PDB ID code: ABT751) is represented in green. Reprinted (adapted) with permission from ref. [40]. Copyright American Chemical Society, 2023.

Conducting Western blot studies, we aimed to explore the potential of compounds **35h** and **35g** in triggering apoptosis via the activation of caspase-3 and caspase-9, crucial components of the mitochondrial apoptotic pathway. Upon exposing HeLa cells to these compounds, we observed a concentration- and time-dependent activation of caspases, as depicted in Figure 11. Moreover, both in vitro and in vivo revealed the activation of poly(ADP-ribose) polymerase (PARP), a major substrate targeted by caspase-3. In addition to these findings, we carefully examined the role of Bcl-2 and Mcl-1 proteins, well known for their capacity to counteract pro-apoptotic proteins and preserve mitochondrial membrane potential. After 48 h of treatment with concentrations of 100 and 250 nM for both compounds, a decrease in Bcl-2 protein expression was observed, while Mcl-1 showed strong down-regulation. Interestingly, at 24 h, Mcl-1 expression increased for **35g** but not for **35h**. These results suggest that compounds **35g** and **35h** effectively down-regulate anti-apoptotic proteins.

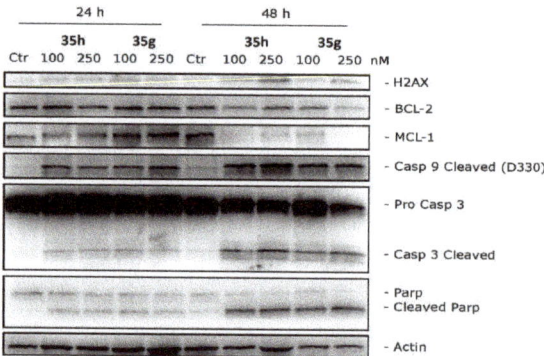

Figure 11. Western blot analysis was conducted on HeLa cells treated with different concentrations of compounds **35g** or **35h** to investigate their effects on H2AX, Bcl-2, Mcl-1, caspase-3, cleaved caspase-9, and PARP. β-Actin was used as the loading control. Reprinted (adapted) with permission from ref. [40]. Copyright American Chemical Society, 2023.

To assess angiogenesis, the vascular properties of **35g** (the most potent within the series) were investigated in vitro using HUVEC endothelial cells. The endothelial cell motility and the ability of **35g** to disrupt tubular structures formed by HUVECs on Matrigel were investigated. As shown in Figure 12a,b, at a concentration of 25 nM, compound **35g** exhibited significant inhibition of cell motility within just 6 h of incubation. This inhibitory effect remained highly significant at all concentrations after 24 h of incubation. Moreover, in Figure 12c, it was observed that compound **35g** disrupted the network of HUVECs compared to the control after 1 h of incubation. Remarkably, after 3 h, all tested concentrations demonstrated significant disruption of the tubular-like structures.

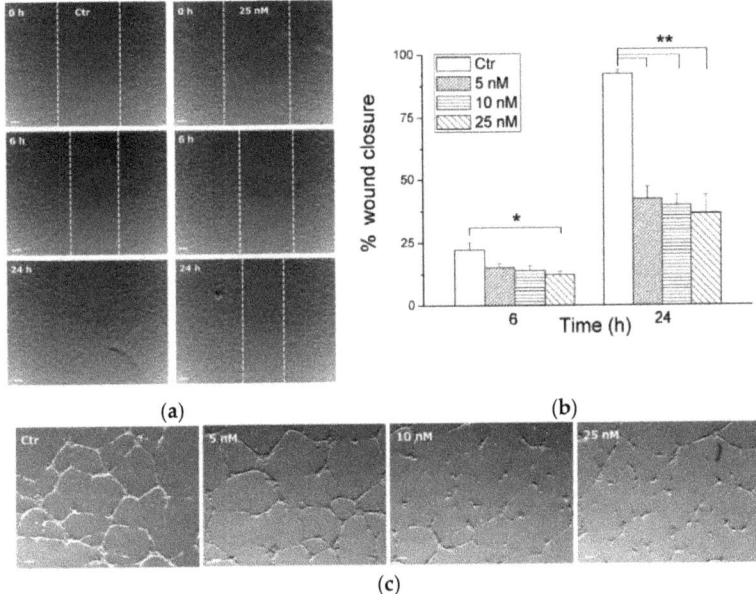

Figure 12. In vitro antivascular evaluation of compound **35g**. (**a**) Confluent HUVEC samples were prepared in a monolayer, and cells were treated with different concentrations of **35g**. The cells were photographed at various time intervals (magnification, 7×; bar, 100 µm). Dotted lines define areas lacking cells. (**b**) The graph shows the quantitative effect of compound **35g**, where gap closure was measured at specified time intervals to assess migration. Data are represented as mean of three independent experiments. * $p < 0.05$, ** $p < 0.01$ vs. control. (**c**) Inhibition of endothelial cell capillary tubule formation by compound **35g**. Representative images show preformed capillary tubules treated with increasing concentrations of **35g** for 1 or 3 h. Reprinted (adapted) with permission from ref. [40]. Copyright American Chemical Society, 2023.

Expanding on the encouraging results regarding the antiproliferative and anticancer activity [37], Kamal et al. conducted more extensive investigations on benzo[b]furans **22** and **25** to explore their potential efficacy against breast cancer cell lines, specifically MCF-7 and MDA MB-231. These studies involved assessments of the cell cycle and the PI3K/Akt/mTOR signaling pathway, along with other complementary studies [41]. Table 10 presents the results obtained for the antiproliferative activity of compounds **22** and **25** against the mentioned cancer cell lines. Remarkably, **22** and **25** displayed significant activity, especially in the MCF-7 cell line, exhibiting IC$_{50}$ values of 0.057 and 0.051 µM, respectively. Due to the MCF-7 cell line showing the highest anticancer activity among the tested cell lines, it was chosen for further analysis to investigate the correlation between cell growth inhibition and cell cycle arrest. In this study, MCF-7 cells were treated with the compounds **22** and **25** at concentrations of 25 nM and 50 nM for 48 h. The results revealed that these compounds induced G2/M cell cycle arrest compared to the untreated control cells. Specifically, at a concentration of 25 nM, compounds **22** and **25** caused a cell accumulation of 36.4% and 37.1%, respectively, in the G2/M phase. Moreover, at 50 nM, these percentages increased to 47.6% and 50.5%, respectively, in the same cell phase (Figure 13).

Table 10. Antiproliferative activity of compounds **22** and **25**.

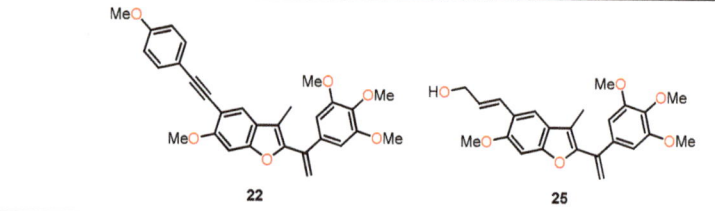

Compound	IC$_{50}$ (µM)	
	MCF-7	MDA MB-231
22	0.057 ± 0.008	0.168 ± 0.008
25	0.051 ± 0.002	0.093 ± 0.006

Data are presented as the mean ± SE from at least three independent experiments.

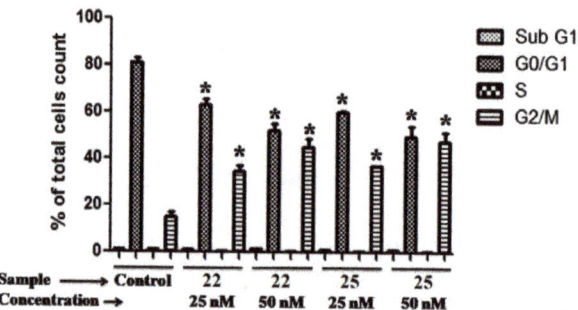

Figure 13. Cell cycle evaluation of compounds **22** and **25** in the MCF-7 cell line. The data are presented as the percentage of cell count at each induced cell cycle phase for each compound (* $p < 0.05$ vs. control). Reproduced with permission from ref. [41]. Copyright Elsevier Inc., 2023.

The PI3K/Akt/mTOR signaling pathway plays a crucial role in breast tumor cell growth. Thus, the impact of compounds **22** and **25** on this signaling pathway was investigated in MCF-7 cells. The results demonstrated effective suppression of p-Akt, p-mTOR, p-p70S6K, and p-4E-BP1 expression levels after 48 h of treatment with a concentration of 50 nM (Figure 14a). The findings strongly support the potent inhibitory activity of both compounds against the PI3K/Akt/mTOR pathway. Notably, given the involvement of this pathway in apoptosis regulation, the studies further unveiled that its inhibition resulted in the up-regulation of key apoptotic markers. These markers included the release of cytochrome c, up-regulation of p53, down-regulation of procaspase-9, cleavage of poly(ADP-ribose)polymerase (PARP), up-regulation of Bax, and down-regulation of Bcl-2 (Figure 14b). Collectively, these results firmly establish the inhibition of the PI3K/Akt/mTOR pathway as the primary mechanism underlying the induction of apoptosis in breast cancer cells by compounds **22** and **25**.

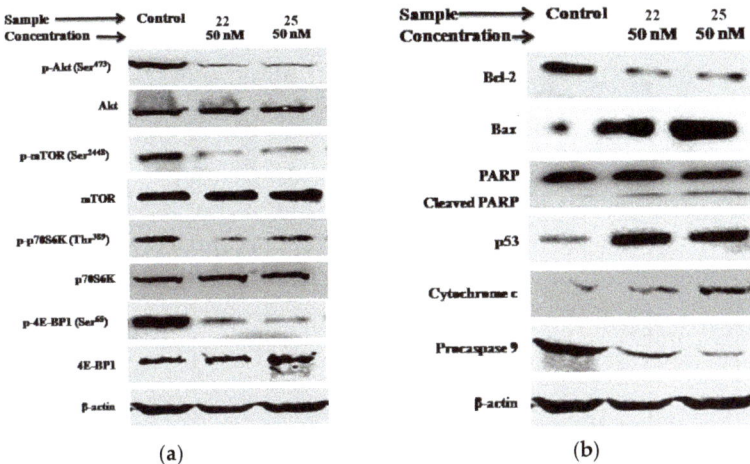

Figure 14. Analysis of the PI3K/Akt/mTOR pathway and apoptotic markers in MCF-7 cells treated with compounds **22** and **25** for 48 h at a concentration of 50 nM. The expression levels of (**a**) Akt, mTOR, p70S6K, and 4E-BP1, including their phosphorylated forms, as well as (**b**) cytochrome c, Bax, Bcl-2, p53, procaspase-9, and PARP, were measured. Protein expression levels were analyzed using Western blot analysis, with β-actin serving as the loading control in both cases. Reproduced with permission from ref. [41]. Copyright Elsevier Inc., 2023.

In their study, Yin et al. achieved the successful synthesis of 2,3-dihydrobenzo[*b*]furan **37** in a 32% yield through the dimerization of methyl caffeate **36** using silver oxide in the presence of anhydrous benzene and acetone (Scheme 7). The primary aim of this study was to explore the potential correlation between IL-25, an endogenous factor secreted by tumor-associated fibroblasts (TAFs), and the inhibition of metastasis in 4T1 mammary tumors in mice [42].

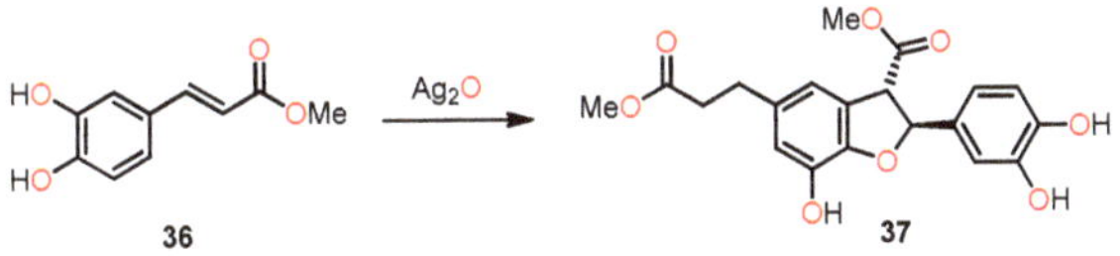

Scheme 7. Synthesis of 2,3-dihydrobenzo[*b*]furan **37** by the dimerization of methyl caffeate **36**.

The investigation into the antimetastatic effects of compound **37** involved the injection of luciferase-expressing 4T1-Luc2 transgenic mouse cells into the mammary fat pad of the experimental mice [42]. At 15 days after tumor cell implantation, the 4T1 tumors were surgically resected in situ. Over the following 8 weeks, a comparative analysis of tumor metastatic activity and survival was conducted between the control group and the mice treated with compound **37** (Figure 15a–c). By detection of luminescent activity of 4T1-Luc2 cells as an indicator of tumor metastasis, it was observed that treatment with **37** ($\geq$20 µg kg^{-1}) significantly suppressed 4T1 cell metastasis to the lung (Figure 15a). In addition, treatment with **37** at a relatively low dose (>20 µg kg^{-1}) had considerable antimetastatic activity in comparison with the treatment with Doxorubicin (2 mg kg^{-1}) (Figure 15b), used as a control drug for breast cancer. Treatment with **37** also significantly increased the survival rate of mice with tumor resection (Figure 15c). These results demonstrate that the in vivo administration of compound **37** effectively prevents breast tumor metastasis following tumor resection.

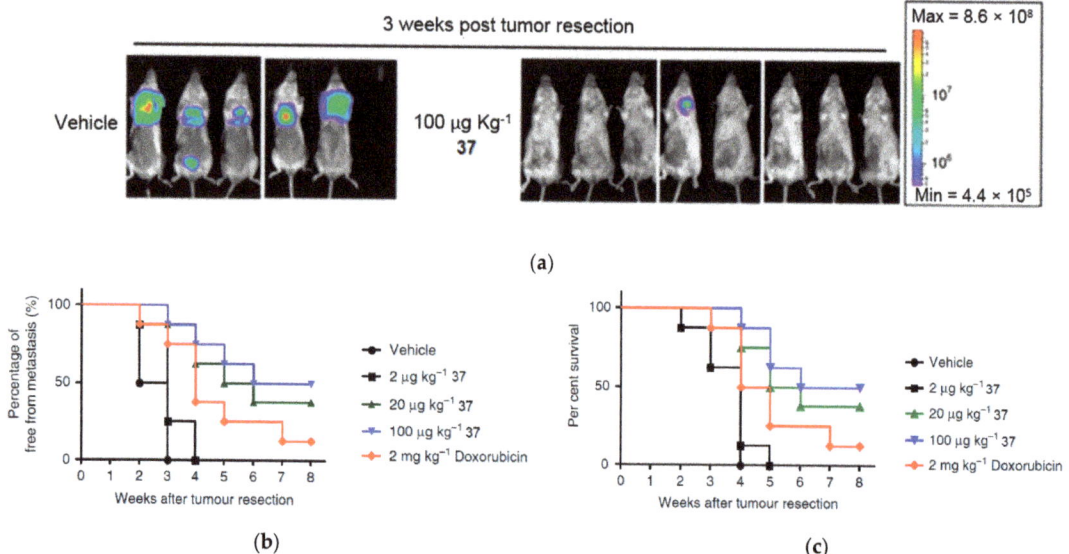

Figure 15. In vivo studies of compound **37** in suppressing metastasis in mouse mammary cells (4T1). (**a**) Bioluminescent images of mouse tumors ($n = 8$ per group) after treatment with PBS and at different doses of compound **37** after tumor resection. Three of the mice died after treatment with PBS before the end of treatment. (**b**) Percentage of metastasis by measurement of luciferase activity in photons over time ($n = 8$ per group). (**c**) Percentage survival of mice after treatment with compound **37**. This is an open-access article distributed under the terms of the Creative Commons CC BY license [42].

Following the previously mentioned findings, the study showed the potential regulatory effects of administering compound **37** in vivo on metastatic tissues [42]. To assess the physiological significance of compound **37** in the modulatory activity, the researchers evaluated the expression of various cytokines secreted in vivo in the lung tissue of the test mice. Remarkably, the findings demonstrated that the administration of compound **37** (at a dosage of 100 μg kg^{-1}) had a stimulatory effect on IL-25 activity in pulmonary fibroblasts surrounding the pulmonary artery and vein (Figure 16b). In contrast, little to no IL-25 expression was observed in lung fibroblasts from control and Docetaxel-treated mice, which suggests that the induction of IL-25 expression in fibroblasts of the lung tissue microenvironment specifically resulted from the administration of compound **37**. This finding is intriguing as IL-25 expression is not typically considered a conventional drug target for anticancer medications.

To quantify the change in the cell population of IL-25-expressing lung fibroblasts in response to treatment with compound **37**, the researchers assessed FSP-1 + ER-TR7 + IL-25 + cells from lung tissues of test mice and compared their IL-25 expression levels at 3 weeks after tumor resection. The results unveiled a remarkable increase in the IL-25 fibroblast cell population from 16.7% to 79.5% in **37**-treated mice compared to those treated with PBS (Figure 16a). Furthermore, the population of fibroblasts in FSP-1 + ER-TR7 + cells in mice treated with **37** exhibited a significant dose-dependent increase from 5.2% to 7.3% (Figure 16a), in contrast to docetaxel treatment, which showed no augmentation in the number or level of evaluated fibroblasts. These findings solidify the evidence that compound **37** effectively stimulates lung fibroblasts in an in vivo setting.

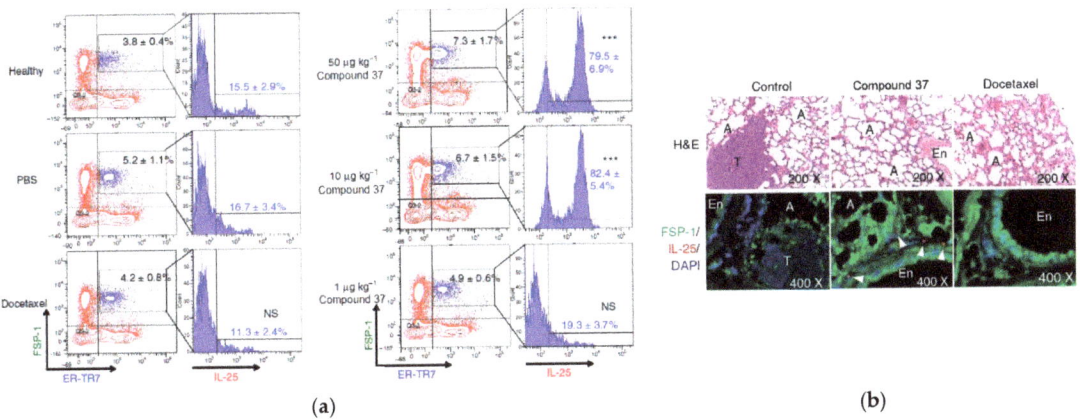

Figure 16. Effect of compound **37** on the regulation of IL-25 expression in tumor-associated fibroblasts. (**a**) Flow cytometry of the population change in FSP-1 + ER-TR7+ cells and their IL-25 expression level in the lungs of mice after treatment with compound **37**, PBS, and docetaxel. Data are reported as mean ± SD ($n = 3$). *** $p < 0.001$. NS—not significant (two-tailed t-test). (**b**) IF staining in mouse lung tissues after treatment with compound **37** and docetaxel 21 days after tumor resection. T—tumor, A—alveoli. This is an open-access article distributed under the terms of the Creative Commons CC BY license [42].

Additionally, the researchers conducted complementary studies to assess the potential suppressive effect of IL-25 secreted by fibroblasts on the growth activity of mammary tumor cells [42]. In this regard, they compared the levels of IL-25 in mouse (4T1) and human (MDA-MB-231) tumor cells treated with compound **37** using an immunoprecipitation method mediated by anti-IL-25 antibodies. To ensure accuracy, the samples were first immunodepleted with 3T3 (3T3-CM) and WI38 (WI38-CM) fibroblasts for IL-25, utilizing anti-rabbit IgG antibody (isotype control) as the negative control for immunodepletion (Figure 17a). The results indicated that the levels of IL-25 secreted in media conditioned with compound **37** treated fibroblasts were significantly higher compared to untreated conditioned media (Figure 17a). Moreover, the researchers detected only a minor fraction of unspecific binding to the protein in the IgG antibody, confirming the high specificity and efficiency of the anti-IL-25 antibody employed. On the other hand, when 4T1 and MDA-MB-231 cells were cultured with 3T3-CM, their growth was notably higher compared to cells cultured solely with fresh medium. This observation implies that in both cases, fibroblasts released critical cellular and molecular factors that contribute to the expansion of tumor cells, potentially influencing the suppression of metastatic mammary tumor cell growth (Figure 17b,c).

In their investigation, Yin et al. conducted a comparative analysis of the in vivo treatment effects of IL-25, examining its additive vs. overlapping impact [42]. They discovered that co-administration of compound **37** (at a dosage of 100 µg kg^{-1}) and IL-25 (at a dosage of 10 µg kg^{-1}) resulted in a similar antimetastatic activity compared to the group of mice treated solely with compound **37** (Figure 18a). The survival rate of mice in the co-treatment group (**37** + Anti-IL-25) was also comparable to the group treated with compound **37** alone, in contrast to the untreated group (Figure 18b). These findings suggest that the in vivo antimetastatic effect of **37** can be effectively substituted by the administration of exogenous IL-25.

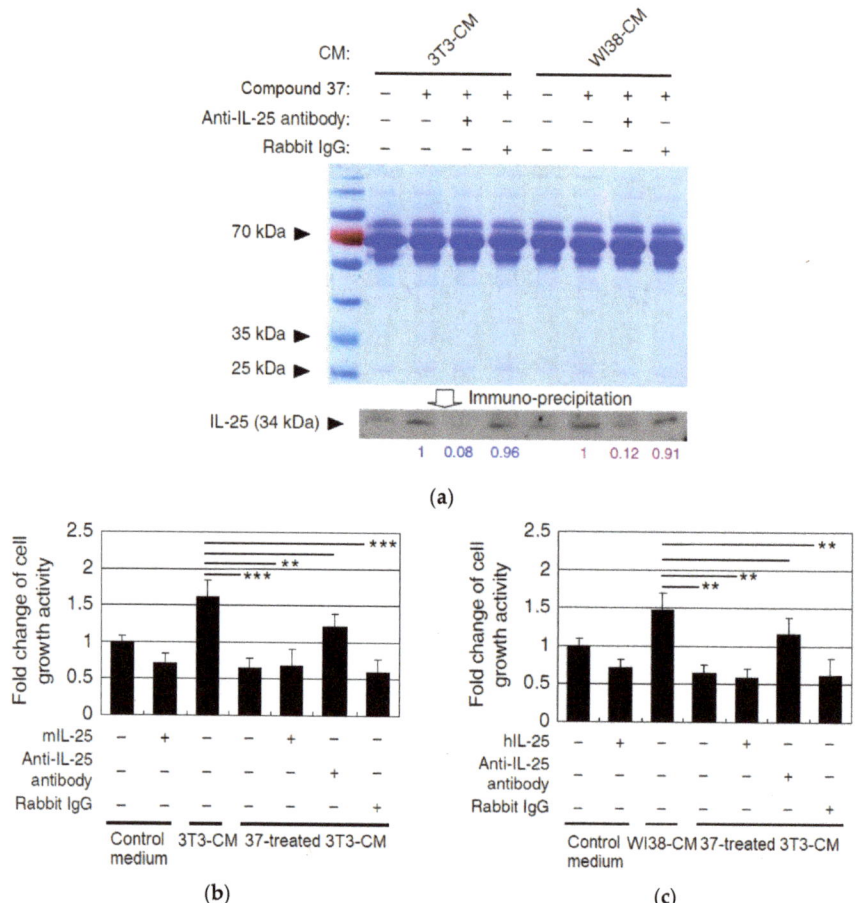

Figure 17. (a) Western blot of IL-25 secretion activity in mouse (3T3) and human (WI38) fibroblasts after treatment with compound **37** using Rabbit IgG as the negative control. (b) Evaluation of 3T3-CM cytotoxicity levels in 4T1 cells after IL-25 immunodepletion. (c) Evaluation of WI38-CM cytotoxicity levels in MDA-MB-231 cells after IL-25 immunodepletion. Data are reported as mean ± SD (n = 3). ** $p < 0.01$; *** $p < 0.001$ (two-tailed Student's t-test). This is an open-access article distributed under the terms of the Creative Commons CC BY license [42].

Moreover, the study evaluated the combined effect of compound **37** and docetaxel in suppressing the metastatic activities of human MDA-MB-231-Luc2 cells in mice through bioluminescent studies following the resection of mammary tumor tissues in the experimental mice (Figure 19a). The results demonstrated that the treatment with compound **37** (100 μg kg^{-1}) and docetaxel (5 mg kg^{-1}) showed significantly higher antimetastatic activity than the treatment with docetaxel alone (Figure 19b). As a result, the test mice receiving the combination of compound **37** and docetaxel exhibited a higher survival rate compared to those receiving single treatments (Figure 19c). These findings suggest a complementary effect on the anticancer activity of docetaxel when combined with compound **37**, effectively suppressing the metastatic activities of tumor cells by modulating the tumor-associated microenvironment.

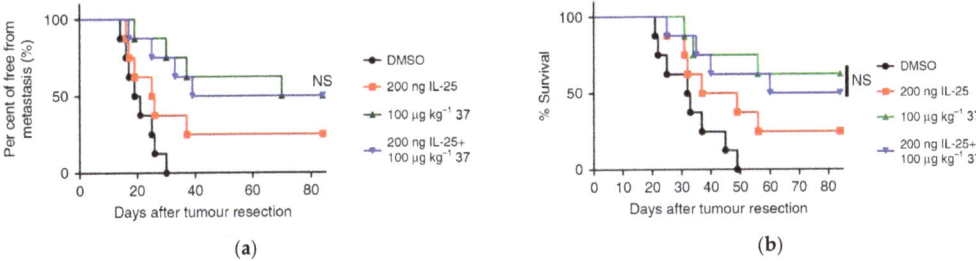

Figure 18. In vivo treatment of compound **37** exhibits comparable antimetastatic activity to IL-25 administration. (**a**) Mice with tumor resection ($n = 8$ per group) were treated with PBS, IL-25, compound **37**, or co-treated with IL-25 and compound **37** after three weeks. Tumor metastases were quantified by measuring luciferase activity over the indicated time course. (**b**) Percentage of survival of mice treated with PBS, compound **37**, or co-treated with IL-25 and compound **37**. The results were analyzed using the log-rank test. NS means not significant. This is an open-access article distributed under the terms of the Creative Commons CC BY license [42].

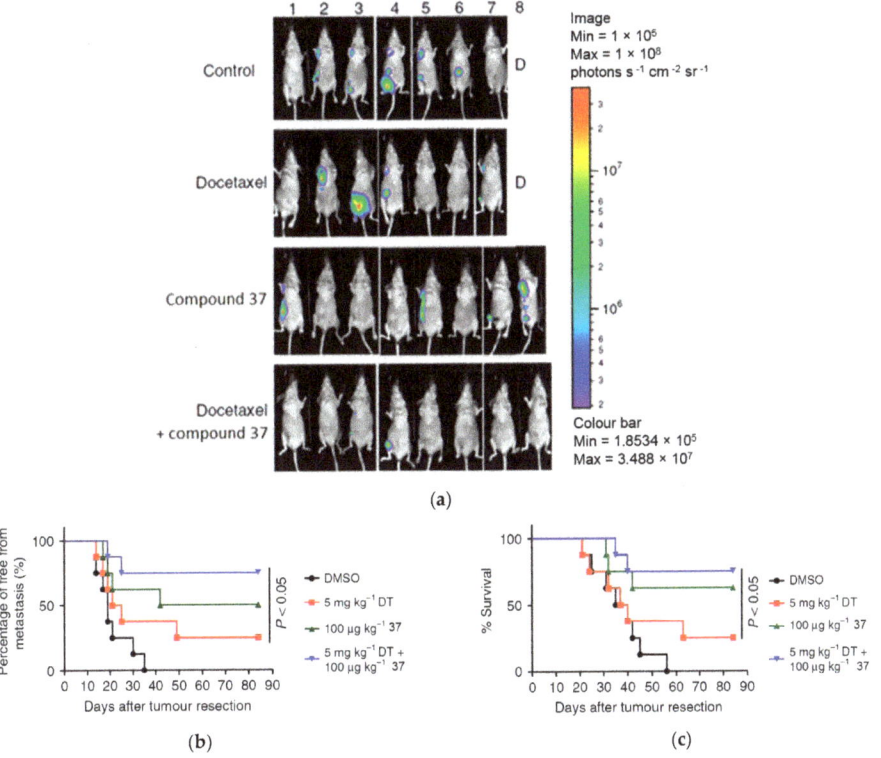

Figure 19. The antimetastatic effect after administration of compound **37** in MDA-MB-231 cells in combination with docetaxel. (**a**) Bioluminescent images of mice with tumor portions ($n = 8$ per group) after treatment with PBS, compound **37**, docetaxel, and co-treatment with docetaxel and compound **37** for three weeks and after tumor resection. (**b**) Levels of tumor metastasis by measurement of luciferase activity in photons in the treated mice over time. (**c**) Percentage survival of mice treated with docetaxel and those co-treated with docetaxel and compound **37**. This is an open-access article distributed under the terms of the Creative Commons CC BY license [42].

On a different note, Quan et al. conducted a molecular modeling study involving 64 Combretastatin A-4 analogs based on five-membered heterocycles. Their objective was to explore the development of novel anticancer agents by using 3D-QSAR, molecular docking, and molecular dynamic (MD) simulation [43]. Within the 3D-QSAR approach, both CoMFA and CoMSIA models were prepared for both the training and test sets. The CoMFA model incorporated steric and electrostatic fields, while the CoMSIA model included steric, electrostatic, hydrophobic, hydrogen bond donor, and hydrogen bond acceptor fields. Three-dimensional contour maps for both models were performed using the "Stdev*Coeff" field type. By analyzing the results obtained from the study, the researchers identified essential structure–activity relationships, which highlighted substitutions that could enhance biological activity. This was summarized in A–D regions, as shown in Figure 20. Building on this information, they designed five novel benzo[b]furan derivatives. The structures and predicted pIC$_{50}$c values of compounds **43a–e** are provided in Table 11. Although these data indicate the presence of inhibitory activity for the designed compounds, they were not comparable to CA-4.

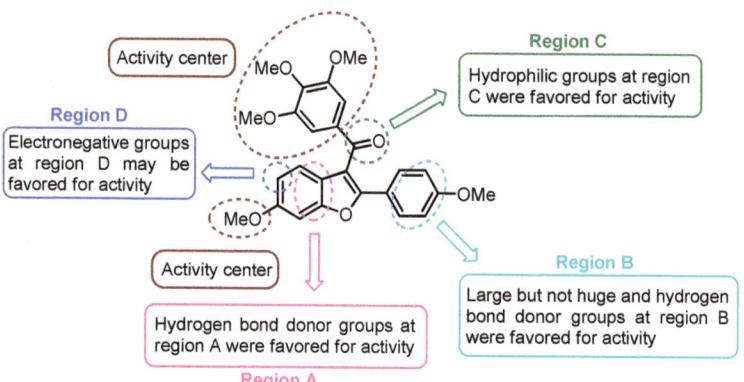

Figure 20. 3D-QSAR studies in benzo[b]furan derivatives **43a–e**.

Table 11. Prediction of pIC$_{50}$c values for compounds **43a–e**.

Compound	R^1	R^2	R^3	R^4	pIC$_{50}$c	
					CoMFA	CoMSIA
43a	H	H	OEt	O	0.351	0.291
43b	F	H	Ome	O	0.285	0.308
43c	H	Nme$_2$	Ome	O	0.348	0.282
43d	H	Me	Ome	O	0.425	0.346
43e	H	NH$_2$	Ome	CH$_2$	0.503	0.206
CA-4 [a]	-	-	-	-	0.000	0.037

[a] Combretastatin-A4 (CA-4) is the standard drug for the study.

Additionally, the researchers conducted 20 ns molecular dynamics (MD) simulations and binding free energy calculations using the Amber 12.0 package [43]. The stability of the tubulin–inhibitor complex in the designed compounds **43a–e** was evaluated, employing the general Amber force field (gaff) for ligands and the ff99SB force field for proteins. In this study, binding free energy calculations were performed using both MM/GBSA and MM/PBSA methods. The results showed that the calculated binding free energies using MM/GBSA for CA-4 and the inhibitors **43a–e** were as follows: −34.32, −57.52, −54.41, −55.78, −50.77, and −55.56 kcal mol^{-1}, respectively. Meanwhile, the MM/PBSA results revealed the binding free energies of −21.54, −43.45, −42.03, −40.99, −39.79, and −36.23 kcal mol^{-1} for CA-4 and the inhibitors **43a–e**, respectively. Among the five newly designed compounds, **43a** exhibited the most negative binding free energy, suggesting it has the potential for the best inhibitory activity within the series. These computational findings provide valuable insights into the potential efficacy of the designed compounds and can guide further experimental investigations to validate their inhibitory activity against tubulin.

Using the binding free energy calculations, two of the designed compounds, **43a** and **43b**, were selected for synthesis. The synthetic process involved several sequential steps to obtain the desired compounds. Firstly, the synthesis began with the iodination of methoxyphenol **38** using a catalytic amount of AgOTFA in chloroform at room temperature for 24 h to afford the iodinated compound **39** (Scheme 8). Subsequently, the acetylation of the iodine-phenol **39** was carried out in the presence of acetic anhydride in pyridine at room temperature for 4 h to yield the acetate derivative **40**. Next, the Sonogashira coupling of compound **40** with either 1-ethynyl-4-ethoxybenzene or 1-ethynyl-4-methoxybenzene was conducted in the presence of catalytic PdCl$_2$(PPh$_3$)$_3$, leading to the formation of alkyne **41**. The intermediates underwent an intramolecular cyclization reaction mediated by K$_2$CO$_3$ in methanol at 60 °C for 16 h to produce the benzo[*b*]furan derivative **42**. Finally, a Friedel–Crafts reaction was performed in the presence of 3,4,5-trimethoxybenzoyl chloride, followed by the addition of compound **42** and tin (IV) chloride to deliver compound **43**. By employing this multi-step synthetic approach, the researchers successfully synthesized compounds **43a** and **43b**, paving the way for further evaluation of their potential as benzo[*b*]furan-based anticancer agents.

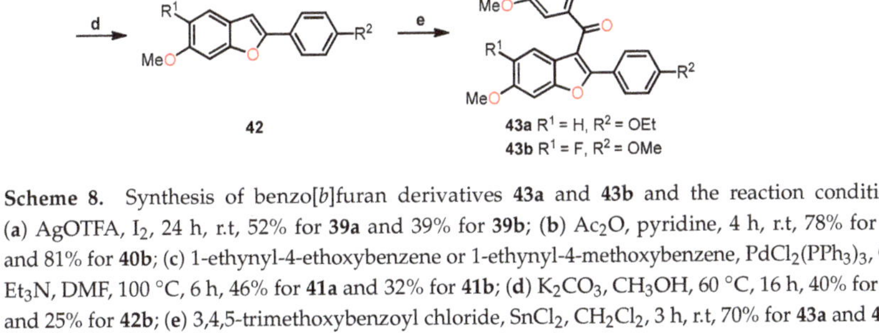

Scheme 8. Synthesis of benzo[*b*]furan derivatives **43a** and **43b** and the reaction conditions: (**a**) AgOTFA, I$_2$, 24 h, r.t, 52% for **39a** and 39% for **39b**; (**b**) Ac$_2$O, pyridine, 4 h, r.t, 78% for **40a** and 81% for **40b**; (**c**) 1-ethynyl-4-ethoxybenzene or 1-ethynyl-4-methoxybenzene, PdCl$_2$(PPh$_3$)$_3$, CuI, Et$_3$N, DMF, 100 °C, 6 h, 46% for **41a** and 32% for **41b**; (**d**) K$_2$CO$_3$, CH$_3$OH, 60 °C, 16 h, 40% for **42a** and 25% for **42b**; (**e**) 3,4,5-trimethoxybenzoyl chloride, SnCl$_2$, CH$_2$Cl$_2$, 3 h, r.t, 70% for **43a** and **43b**.

The in vitro antiproliferative activity of compounds **43a–b** was evaluated against six human cancer cell lines, and their tubulin inhibition was assessed, using CA-4 and CA-4P as standard drugs. As shown in Table 12, the compound **43a** (R^1 = H, R^2 = Oet) exhibits the highest activity with IC$_{50}$ values of 1.37, 8.99, 1.31, and 0.91 µM against A549, HeLa, HepG2, and MCF-7 cell lines, respectively [43]. Compound **43a** exhibits comparable and even superior activity than CA-4 in the mentioned cell lines. Specifically, it is 4.4-

fold more active against A549 cells and 12.2-fold more active against HepG2 cells than CA-4. In addition, compound **43a** showed only comparable activity with compound CA-4P in the MCF-7 cell line. On the other hand, compound **43b** (R^1 = H, R^2 = Ome) exhibited remarkable activity against A549 and HepG2 cells, with IC_{50} values of 6.87 and 4.75 μM, respectively, which is comparable to CA-4 (IC_{50} = 5.99 and 16.04 μM, respectively). Finally, in a tubulin polymerization assay, compound **43a** demonstrated potent inhibition of tubulin polymerization, with an IC_{50} value of 0.86 μM, which is comparable to CA-4 (IC_{50} = 0.88 μM) and superior to CA-4P (IC_{50} = 4.79 μM).

Table 12. Antiproliferative activities and inhibition of tubulin polymerization of compounds **43a–b**, CA-4, and CA-4P.

Compound	IC_{50} (μM) [a]						Tubulin
	A549	HCT-116	HeLa	HepG2	MGC-803	MCF-7	IC_{50} (μM) [a]
43a	1.37 ± 0.14	12.55 ± 0.60	8.99 ± 1.11	1.31 ± 0.17	9.97 ± 0.81	0.91 ± 0.16	0.86 ± 0.07
43b	6.87 ± 0.34	78.23 ± 3.83	19.1 ± 1.59	4.75 ± 0.20	47.26 ± 1.17	6.49 ± 0.55	>200
CA-4 [b]	5.99 ± 0.46	9.00 ± 0.85	11.33 ± 1.03	16.04 ± 0.50	1.76 ± 0.85	0.67 ± 0.13	0.88 ± 0.22
CA-4P [b]	0.40 ± 0.17	0.24 ± 0.05	0.56 ± 0.12	0.27 ± 0.04	0.47 ± 0.14	0.96 ± 0.22	4.79 ± 0.40

[a] Data are expressed as the mean ± SE of at least three independent experiments. [b] Combretastatin-A4 (CA-4) and its disodium phosphate form (CA-4P) serve as the standard drugs for this study.

Similarly, Lauria et al. conducted the synthesis of a novel series of 3-benzoylamino-5-(1H-imidazol-4-yl)methylaminobenzo[b]furans **51–53** and subsequently evaluated their potential as antitumor agents [44]. The synthetic route involved a sequence of steps (Scheme 9). Initially, 2-fluorobenzonitrile **44** underwent a nitration reaction in the presence of a mixture of concentrated nitric and sulfuric acids under a nitrogen atmosphere at 0 °C for 2 h to afford 2-fluoro-5-nitrobenzonitrile **45**. Afterward, ethyl glycolate was utilized for nucleophilic displacement in the presence of K_2CO_3 and anhydrous DMF at 100 °C for 12 h, facilitating in situ intramolecular cyclization and giving rise to the 3-amino-benzo[b]furan derivative **46**. Following this, intermediate **46** underwent acyl substitution with benzoyl chloride **47** in pyridine, serving as both base and solvent, at room temperature for 12 h, yielding compound **48**, containing amide functionality at the C–3 position. The subsequent reduction of the nitro group in compound **48** was performed through hydrogenation using a Parr hydrogenation apparatus at 500 psi in the presence of Pd/C (10%) as a catalyst in ethanol at room temperature for 2 h, yielding 5-amino-benzo[b]furan derivative **49**. Finally, the compounds **51a–f** and **52a–f** were obtained through a reductive amination with imidazole-4-carbaldehyde **50** using sodium cyanoborohydride as a selective reducing agent in a mixture of ethanol and acetic acid at room temperature for 6–24 h. In this final step, compound **53** was also isolated using carbaldehyde **50b**, wherein evidence of the hydrolysis of the amide functionality was observed. These novel compounds hold promise as potential antitumor agents and warrant further investigation to assess their efficacy in cancer treatment.

The biological studies focused on the antiproliferative activity in HeLa and MCF-7 cell lines using the MTT assay, cell cycle analysis, and in silico assessment [44]. Table 13 shows the GI_{50} values of compounds **51a–f** and **52a–f**. Notably, the insertion of a methyl group in the imidazole ring increased activity for compounds **52a** (R^1 = R^2 = R^3 = R^4 = H), **52e** (R^1 = H, R^2 = CF_3, R^3 = R^4 = H), and **52f** (R^1 = Cl, R^2 = F, R^3 = R^4 = H) without substitution on the benzoyl moiety (**52a** vs. **51a**), or functionalized with 4-trifluoromethyl- (**52e** vs. **51e**) and 3-chloro-4-fluoro- (**52f** vs. **51f**) substituents. Compound **52f** exhibited the highest activity against HeLa and MCF-7 cell lines, with GI_{50} values of 2.14 μM and 1.55 μM, respectively. In cell cycle analysis, compounds **52b**, **52c**, **52d**, and **52f** showed significant suppression of the G0/G1 phase and an accumulation of cells in G2/M at 1xGI_{50} concentrations (Figure 21). However, at 2xGI_{50} concentrations, changes in the distribution profile were observed. Albeit compounds **52a** and **52e** did not show a significant impact on the cell cycle at the evaluated concentrations, they induced G0/G1 arrest at 2.5xGI$_{50}$ and 5xGI$_{50}$ concentrations, which correlated with their antiproliferative effects in other phases.

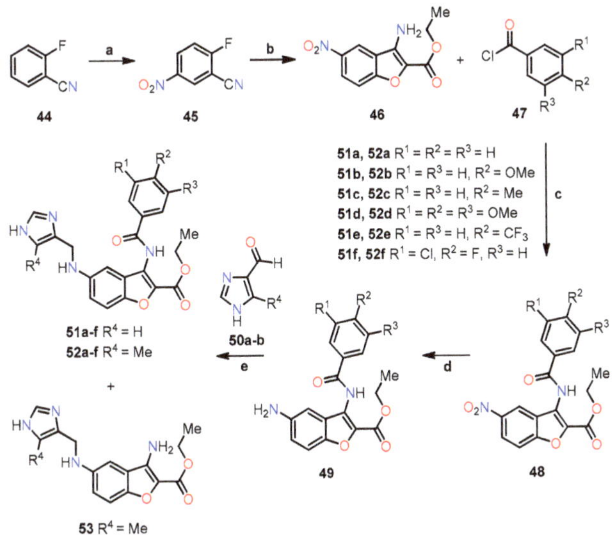

Scheme 9. Synthesis of 3-benzoylamino-5-(1H-imidazol-4-yl)methylaminobenzo[b]furan derivatives **51a–f**, **52a–f**, and **53**, and the reaction conditions: (**a**) HNO$_3$/H$_2$SO$_4$, N$_2$, 0 °C, 2 h; (**b**) ethyl glycolate, K$_2$CO$_3$, anhydrous DMF, 100 °C, 12 h, 63%; (**c**) pyridine, r.t, 12 h, 62–86%; (**d**) 10% Pd/C, H$_2$, ethanol, r.t, 2 h, 43–93%; (**e**) NaBH$_3$CN, ethanol, AcOH, r.t, 6–24 h, 23–65% for **51a–f**, 15–90% for **52a–f**, and 10–23% for **53**.

Table 13. Antiproliferative evaluation of compounds **51** and **52**.

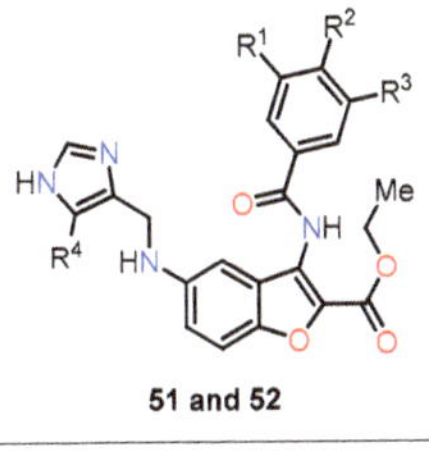

Compound	R^1	R^2	R^3	R^4	GI$_{50}$ (µM) [a]	
					HeLa	MCF-7
51a	H	H	H	H	41.59 ± 3.08	49.65 ± 4.91
51b	H	Ome	H	H	7.98 ± 0.67	13.36 ± 1.94
51c	H	Me	H	H	13.65 ± 1.02	18.22 ± 1.28
51d	Ome	Ome	Ome	H	12.18 ± 1.19	3.88 ± 0.40
51e	H	CF$_3$	H	H	>50	>50
51f	Cl	F	H	H	13.61 ± 0.86	11.25 ± 1.11
52a	H	H	H	Me	7.69 ± 0.46	12.27 ± 1.46
52b	H	Ome	H	Me	14.31 ± 1.97	17.35 ± 1.78
52c	H	Me	H	Me	9.20 ± 0.34	15.53 ± 0.98
52d	Ome	Ome	Ome	Me	12.19 ± 1.59	9.21 ± 1.02
52e	H	CF$_3$	H	Me	2.01 ± 0.13	14.16 ± 2.02
52f	Cl	F	H	Me	2.14 ± 0.35	1.55 ± 0.14

[a] Data are expressed as the mean ± SD of at least three independent experiments.

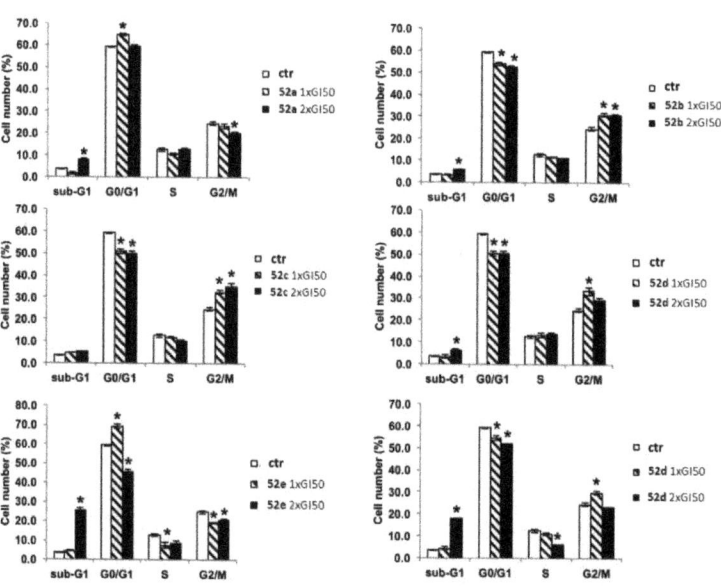

Figure 21. Analysis of the influence of the concentrations of compounds **52a–f** on the cell cycle of the HeLa cell line after 24 h using flow cytometry. Data are expressed as the mean of two independent experiments. Statistical analyses were performed using the Student's t test to determine the differences between the datasets. * $p < 0.05$ denote significant differences from untreated control cells. Reproduced with permission from ref. [44]. Copyright John Wiley & Sons Inc., 2023.

In the in silico studies, the researchers assessed the binding modes of each derivative at the colchicine binding site on tubulin, considering the involved amino acid residues [44]. To facilitate this analysis, they obtained the crystal structure of tubulin bound to colchicine from the PDB database (PDB ID code: 4O2B) and extracted the tubulin dimer with colchicine (chains A and B) from the protein model. Table 14 shows the favorable induced docking protocol (IFD) scores for all ligand–tubulin complexes, with compounds **51** and **52** showing affinity similar to colchicine and higher than CA-4. Notably, significant differences in the binding complexes were observed among the evaluated amino acids. These benzo[b]furans displayed strong interactions with amino acids Alaα180, Serα178, Ileβ318, Alaβ316, Leuβ255, Leuβ248, and Lysβ254, as indicated by the IFD scores. Interestingly, compounds **52b–d,f** presented an aromatic ring in contact with Cysβ241, a crucial and distinctive feature for identifying new antitubulin molecules. A more detailed illustration of these interactions is found in Figure 22, depicting the ligand interaction maps of compounds **52b** and **52e** with colchicine as a reference.

Afterward, Pervaram et al. synthesized 1,2,4-oxadiazole-fused benzo[b]furan derivatives **62a–j** and assessed their antiproliferative activity against four human cancer cell lines, including A549 (lung), MCF-7 (breast), A375 (melanoma), and HT-29 (colon), using the MTT method [45]. The synthetic route for compounds **62a–j** is shown in Scheme 10. The synthesis began with the reaction of 5-methoxybenzofuran-3-carbaldehyde **54** with 2-aminophenol **55** in refluxing ethanol for 4 h, leading to the 2,3-dihydrobenzo[d]oxazole intermediate, which was oxidized to 2-(5-methoxybenzofuran-3-yl)benzo[d]oxazole **56** by adding Pb(Oac)$_4$ and acetic acid at room temperature for 1 h. Next, compound **56** reacted with BBr$_3$ in anhydrous CH$_2$Cl$_2$ at room temperature for 5 h to give 3-(benzo[d]oxazol-2-yl)benzofuran-5-ol **57**, which is O-alkylated with 2-bromoacetonitrile **58** and K$_2$CO$_3$ in refluxing acetone for 5 h to furnish compound **59**. Subsequently, a nucleophilic addition reaction between **59** and hydroxylamine hydrochloride in the presence of K$_2$CO$_3$ in refluxing ethanol for 3 h yielded acetamide **60**. Finally, compound **60** was reacted with the different benzoyl chloride **61** using pyridine at room temperature for 4 h to obtain 1,2,4-oxadiazole-fused benzo[*b*]furan derivatives **62a–j** in yields ranging from 63% to 93%.

Table 14. IFD scores for compounds **51a–f**, **52a–f**, colchicine, and CA-4.

Compound	Prime Energy	Glide Score	IFD Score [a]
51a	−34,437	−7.17	−1729
51b	−34,462	−10.51	−1734
51c	−34,487	−11.96	−1736
51d	−34,496	−12.23	−1737
51e	−34,480	−10.45	−1734
51f	−34,395	−11.78	−1732
52a	−34,463	−8.16	−1731
52b	−34481	−11.04	−1735
52c	−34,478	−10.38	−1734
52d	−34,545	−10.11	−1737
52e	−34,371	−10.53	−1729
52f	−34,475	−11.28	−1735
Colchicine	−34,392	−10.38	−1730
CA-4	−34,316	−9.17	−1725

[a] Defined as: glide score + 0.05 prime energy.

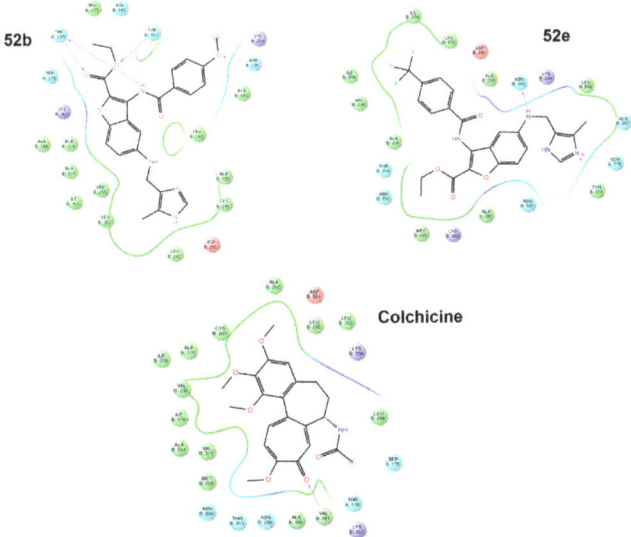

Figure 22. Ligand interaction maps of compounds **52b**, **52e**, and colchicine. Reproduced with permission from ref. [44]. Copyright John Wiley & Sons Inc., 2023.

The antiproliferative activity of all synthesized compounds **62a–j** was assessed against four cancer cell lines using the MTT method and CA-4 as the standard drug (Table 15) [45]. Notably, compounds **62b** (R^1 = 3,4,5-*tri*-(Ome)$_3$), **62g** (R^1 = 4-NO$_2$), **62h** (R^1 = 4-CN), and **62j** (R^1 = 4-CF$_3$) exhibited comparable and, in some cases, even higher potency than CA-4, with IC$_{50}$ values ranging from 0.012 to 1.45 µM for these compounds, while CA-4 had IC$_{50}$ values ranging from 0.11 to 0.93 µM.

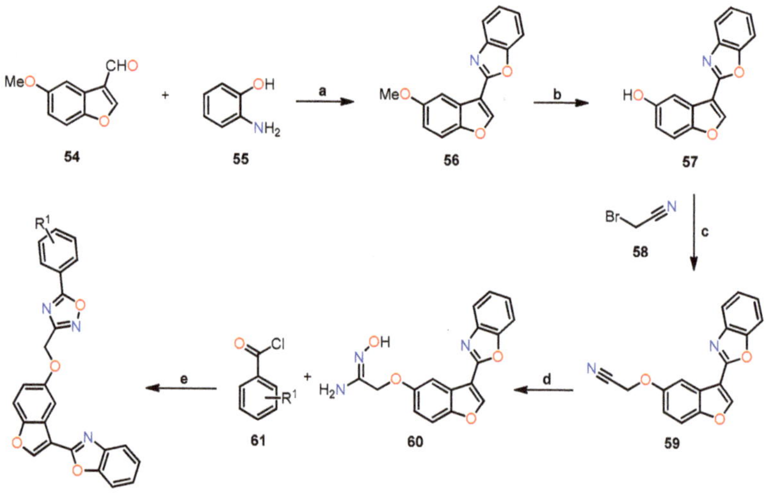

R¹ = H (**62a**), 3,4,5-*tri*-(OMe)₃ (**62b**), 4-OMe (**62c**), 4-Cl (**62d**), 4-Br (**62e**), 4-F (**62f**), 4-NO₂ (**62g**), 4-CN (**62h**), 4-Me (**62i**), 4-CF₃ (**62j**)

Scheme 10. Synthesis of benzo[*b*]furan-3-yl-1,3-benzoxazole derivatives **62a–j**, and reaction conditions: (**a**) EtOH, reflux, 4 h, then AcOH, Pb(Oac)₄, r.t., 1 h, 79%; (**b**) DCM, BBr₃, r.t., 5 h, 88%; (**c**) acetone, K₂CO₃, reflux, 5 h, 94%; (**d**) EtOH, NH₂OH. HCl, K₂CO₃, reflux, 3 h, 92%; (**e**) pyridine, r.t., 4 h, 63–93%.

Table 15. Antiproliferative evaluation of 1,2,4-oxadiazole fused benzo[*b*]furan derivatives **62a–j**.

Compound	IC$_{50}$ (µM)			
	A549	MCF-7	A375	HT-29
62a	2.900	10.560	ND	3.78
62b	0.030	0.011	1.45	0.33
62c	1.900	0.400	2.01	0.18
62d	1.030	1.760	ND	1.22
62e	3.200	ND	ND	9.70
62f	10.400	2.440	5.70	ND
62g	0.100	0.014	ND	1.89
62h	0.012	0.015	1.09	0.17
62i	7.100	13.600	ND	20.50
62j	0.100	0.020	0.19	1.23
CA-4 [a]	0.110	0.180	0.21	0.93

[a] Combretastatin-A4 (CA-4) is the standard drug for the study. ND means not detected.

Similarly, Kwiecień et al. conducted a study on the synthesis and evaluation of functionalization at position 3 of 2-phenyl- and 2-alkylbenzo[*b*]furans as potential antitumor agents [46]. The synthesis of 2-phenylbenzo[*b*]furan **67** involved a three-step reaction process (Scheme 11). Firstly, 2-hydroxybenzaldehyde **63** was *O*-alkylated with methyl 2-bromo-2-phenylacetate **64** in the presence of K₂CO₃ and DMF at 92–94 °C for 4 h to afford methyl 2-(2-formylphenoxy)-2-phenylacetate derivative **65** with yields in the range of 61–76%. Secondly, basic hydrolysis of ester **65** in methanol refluxing for 2 h, and then protonation yielded 2-(2-formylphenoxy)-2-phenylacetic acid **66** in acceptable yields (70–72%). Lastly, an intramolecular cyclization of compound **66** using a mixture of Ac₂O and AcONa at 125–130 °C for 4.5 h produced the benzo[*b*]furan **67** in excellent yields (90–97%).

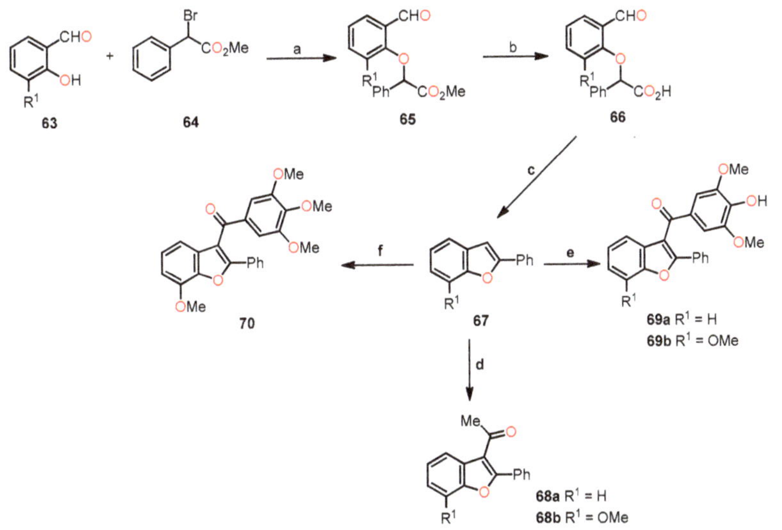

Scheme 11. Synthesis of 2-phenylbenzo[*b*]furan derivatives **68a–b**, **69a–b**, and **70**, and the reaction conditions: (**a**) K_2CO_3, DMF, 92–94 °C, 4 h, 61–76%; (**b**) KOH (10%), MeOH, 75–80 °C, 2 h, then HCl (10%), r.t., 70–72%; (**c**) Ac_2O, AcONa, 125–130 °C, 4.5 h, 90–97%; (**d**) Ac_2O, Amberlyst-15, 1,2-dichloroethane, reflux, 6 h, 69–72%; (**e**) 3,4,5-trimethoxybenzoyl chloride, $AlCl_3$, 1,2-dichloroethane, 45 °C, 3.5 h, 68–76%; (**f**) 3,4,5-trimethoxybenzoyl chloride, Amberlyst-15, 1,2-dichloroethane, reflux, 6.5 h, 69%.

Later, the researchers focused on the acylation of 2-phenylbenzo[*b*]furan **67**. Initially, the acetylation of **67** with Ac_2O in the presence of Amberlyst-15 in 1,2-dichloroethane refluxing for 6 h afforded 2-phenylbenzo[*b*]furan-3-yl)ethan-1-ones **68a–b** in acceptable yields (69–72%). Then, (4-hydroxy-3,5-dimethoxyphenyl)-(2-phenylbenzo[*b*]furan-3-yl)methanones **69a–b** were synthesized in 68–76% yields through an $AlCl_3$-mediated acylation with 3,4,5-trimethoxybenzoyl chloride in 1,2-dichloroethane at 45 °C for 3.5 h. Finally, the phenylbenzo[*b*]furan-3-yl-(3,4,5-trimethoxyphenyl)methanone **70** was obtained in a 69% yield through an Amberlyst-mediated acylation with 3,4,5-trimethoxybenzoyl chloride in 1,2-dichloroethane refluxing for 6.5 h.

A synthetic route for obtaining 3-phenyl-functionalized 2-alkylbenzo[*b*]furans **72** and **73** was developed (Scheme 12). Firstly, 1-(5-bromo-2-ethylbenzo[*b*]furan-3-yl)-2-(4-hydroxyphenyl)ethanone **72** was synthesized from 5-bromo-2-ethylbenzo[*b*]furan **71a** through an acylation with 4-methoxyphenylacetyl chloride, followed by demethylation to convert the methoxy group into a hydroxy group. Secondly, 2-(3,5-dibromo-4-hydroxyphenyl)-1-(2-butylbenzo[*b*]furan-3-yl)ethanone **73** was prepared from 2-butylbenzo[*b*]furan **71b** by carrying out three sequential reactions: acylation with 4-methoxyphenylacetyl chloride, demethylation of the methoxy group, and bromination.

The biological evaluation encompassed the examination of compounds **68b**, **69a–b**, **70**, **72**, and **73**, focusing primarily on antiproliferative studies, flow cytometry, confocal microscopy imaging, and the tubulin polymerization assay, among other complementary analyses [46]. For the in vitro antiproliferative assessment, the benzo[*b*]furan derivatives were tested against the A375 cancer cell line using a cell proliferation reagent WST-1 assay (Table 16). The results revealed that compounds **69a** (R^1 = H), **69b** (R^1 = Ome), and **70** exhibited the most potent antiproliferative activity, displaying IC_{50} values of 2.85, 0.86, and 0.09 µM, respectively. Conversely, compounds **68b** (R^1 = Ome), **72**, and **73** demonstrated low activity, with IC_{50} values exceeding 100 µM.

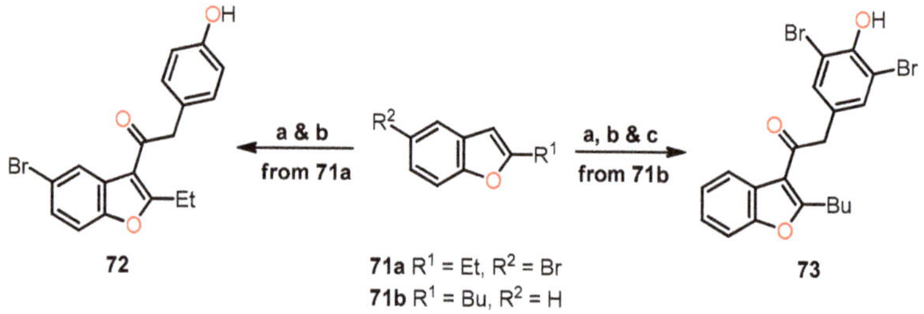

Scheme 12. Synthesis of 2-alkylbenzo[*b*]furan derivatives **72** and **73**, and the reaction conditions: (**a**) 2-(4-methoxyphenyl)acetyl chloride, AlCl$_3$, 1,2-dichloroethane, 0–15 °C, 6 h; (**b**) pyridine hydrochloride, reflux, 10–12 min; (**c**) HBr (20%), NaClO$_3$, r.t., 2 h. The authors have not provided information regarding the yields.

Table 16. Antiproliferative activity of benzo[*b*]furan derivatives against the A375 cancer cell line.

Compound	IC$_{50}$ (µM) [a]
68b	>100
69a	2.85 ± 0.82
69b	0.86 ± 0.37
70	0.09 ± 0.03
72	>100
73	>100

[a] The data indicate the mean ± SD of at least three independent experiments.

Flow cytometry analysis employed an apoptosis detection kit FITC Annexin V to evaluate apoptosis and necrosis in cells after 48 h of incubation with benzo[*b*]furan derivatives [46]. The results from the flow cytometry demonstrated minimal cytotoxicity for compounds **68b**, **72**, and **73**, with approximately 90% of live cells, which is similar to the control group. In contrast, compounds **69a**, **69b**, and **70** exhibited a significant increase in late apoptotic cells, with percentages reaching 57.63%, 71.21%, and 58.52%, respectively. These findings highlight the potent anticancer activity of 3-aryl-2-phenylbenzo[*b*]furan derivatives.

In addition, a complementary study on cell cycle distribution was performed, revealing that the 3-aryl-2-phenylbenzo[*b*]furan derivatives induced the accumulation of A375 cells in a tetraploid state (4N), resulting in a decrease in the percentage of cells in the G0/G1 phase. Specifically, compounds **69a**, **69b**, and **70** led to 66.34%, 58.86%, and 63.62% of A375 cells in the G2/M phase, respectively. In contrast, compounds **68b**, **72**, and **73** did not show significant differences in the percentages of A375 cells in the G0/G1 and G2/M phases compared to untreated control cells. These findings provide valuable insights into the mode of action of the 3-aryl-2-phenylbenzo[*b*]furan derivatives and highlight their potential as effective anticancer agents.

After analyzing the cell cycle distribution results, Kwiencień et al. conducted a confocal microscopy analysis [46]. Cells were incubated with the tested compounds for 7 h, fixed, and then stained for α-tubulin and chromosomes. The control cells displayed well-defined bipolar spindle formation with chromosome alignment at the metaphase central plate or anaphase distribution (Figure 23A,B). However, cells treated with 3-aryl-2-phenylbenzo[*b*]furans exhibited diverse phenotypes, characterized by enlarged nuclei and the absence of a visible mitotic spindle (Figure 23C–H). Notably, compound **70** showed a distinct phenotype with binuclear or enlarged nuclei. These observations suggest that the disparity between the tested compounds and control cells was specifically evident during mitosis, indicating a specific mitotic activity of the compounds. This finding sheds significant light on the potential of these compounds as specific mitosis-targeting agents.

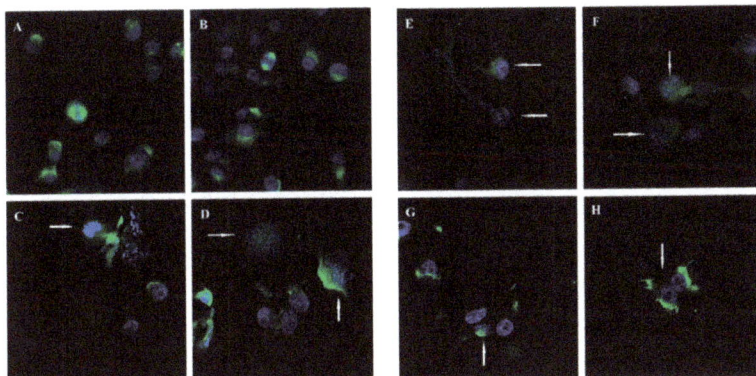

Figure 23. Confocal microscopy images in A375 cells: (**A,B**) control cells; (**C,D**) compound **69a**; (**E,F**) compound **69b** after 7 h of incubation and at a concentration of 100 µM; (**G,H**) compound **70** at 10 µM with α-tubulin stained (green) and chromosomes (blue). Arrows indicate abnormal spindles. Reproduced with permission from ref. [46]. Copyright Elsevier Inc., 2023.

Finally, the effects of benzo[*b*]furan derivatives on tubulin polymerization were evaluated based on fluorescence [46]. Paclitaxel (PTX), vinblastine (VBL), and DMSO (0.2%) were used as control and reference compounds. As depicted in Figure 24, DMSO had no direct effect on tubulin polymerization. In contrast, the reference compounds (PTX and VBL) interacted with tubulin, resulting in alterations to the normal polymerization curve. Upon comparing the curves of VBL, **69a**, **69b**, and **70**, it became evident that these compounds were the most effective in inhibiting tubulin polymerization, as indicated by a decrease in V_{max} (maximum slope values for the growth phase) and a reduction in the final mass of the protein polymer. These findings align with the observations from confocal microscopy and flow cytometry analyses, where the inhibition of tubulin polymerization led to the prevention of mitotic spindle formation, resulting in the presence of polyploid nuclei and cell cycle arrest at 4N. These collective data provide compelling evidence for the significant impact of the tested compounds on tubulin polymerization and their potential as potent agents affecting cell division and proliferation.

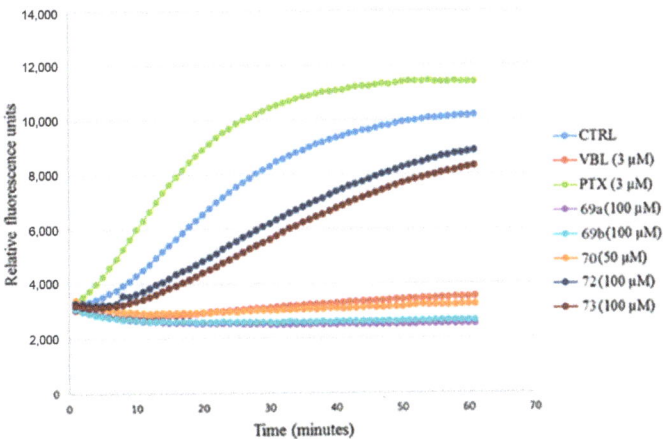

Figure 24. Analysis of the inhibition of tubulin polymerization in benzo[*b*]furan derivatives **69a**, **69b**, **70**, **72**, and **73**. Reproduced with permission from ref. [46]. Copyright Elsevier Inc., 2023.

Anwar et al. pursued a unique approach and conducted preliminary studies, which revealed a promising anticancer activity in a benzofuran–pyrazole hybrid **77** [47]. Encouraged by these favorable findings, the researchers explored the potential benefits of its nanorange form, aiming to investigate the influence of the nanorange and its effect on the cytotoxic potency of the hybrid **77** [48]. The synthesis of hybrid **77** involved a three-step process: (1) the conversion of 1-(benzofuran-2-yl)ethanone **74** into the pyrazole-4-carbaldehyde **75** through the Vilsmeier–Haack reaction; (2) Claisen–Schmidt condensation of compound **75** with 2-acetylpyrrole to give chalcone **76** in an 88% yield; and, finally, (3) the cyclocondensation of compound **76** with hydrazine hydrate in acetic acid to afford benzofuran–pyrazole hybrid **77** in an 85% yield (Scheme 13).

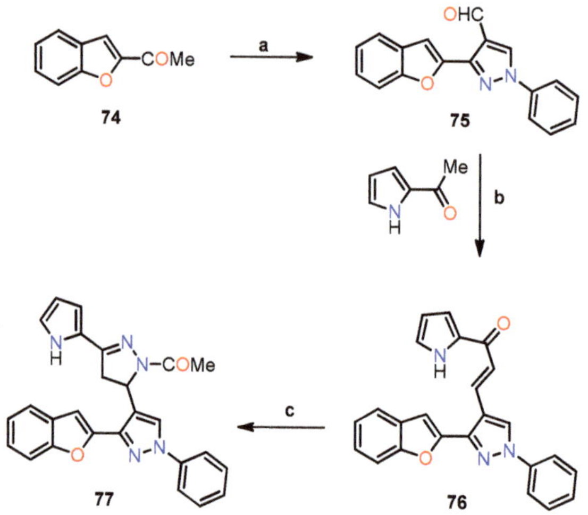

Scheme 13. Synthesis of benzofuran–pyrazole hybrid **77**, and reaction conditions: (**a**) PhNHNH$_2$, then POCl$_3$, DMF; (**b**) EtOH, NaOH, 88%; (**c**) N$_2$H$_4$/AcOH, 85%.

On the other hand, nanoparticles of the benzofuran–pyrazole hybrid **77** were synthesized using the nanoprecipitation method and exhibited sizes ranging from 3.8 to 5.7 nm. The characterization of these nanoparticles involved transmission microscopy (TEM) to confirm spherical shape and average size. Additionally, the surface charge and stability of the nanoparticles were analyzed utilizing the Malvern Zetasizer nano Zs instrument. These results indicated that the nanoparticles had an average size of 3.8–5.7 nm and a zeta potential of −27.3 mV, with a polydispersity index (PDI) of 0.77, confirming their uniformity and stability.

The anticancer activity of the benzofuran–pyrazole hybrid **77** and its nanoparticles was assessed against two breast cancer cell lines, MCF-7 and MDA-MB-231, using an MTT assay and doxorubicin as a standard drug (Table 17) [48]. The results showed that the nanoparticles of hybrid **77** exhibited the highest cytotoxic activity against both MCF-7 and MDA-MB-231 cell lines with IC$_{50}$ values of 1 and 0.6 nM, respectively, outperforming doxorubicin (IC$_{50}$ = 620 nM in both cases). In contrast, hybrid **77** showed lower activity than its nanoparticles against both cell lines, with IC$_{50}$ values of 7 and 10 nM. This difference in activity can be attributed to the high surface area/volume ratio of nanoparticles, which allows for selective targeting of cells and tissues, and more effective interactions compared to hybrid **77** (>100-fold). Finally, IC$_{50}$ values of hybrid **77** and its nanoparticles showed over a 1000-fold difference when targeting normal breast cells MCF-12A compared to cancer cells, indicating their safety profiles in normal cells.

Table 17. Antiproliferative activity of benzofuran–pyrazole hybrid **77** and its nanoparticles.

Compound	IC$_{50}$ (nM) [a]		
	MCF-7	MDA-MB-231	MCF-12A
77	7 ± 1	10 ± 1	87,600 ± 335
Nanoparticles—**77**	1 ± 0.4	0.6 ± 0.1	21,540 ± 66
Doxorubicin [b]	620 ± 31	620 ± 31	ND

[a] The data indicate the mean ± SD of at least three independent experiments. [b] Doxorubicin is the standard drug for the study. ND means not detected.

The researchers performed complementary analyses to explore the cell cycle, the effect on caspase-3/p53/Bax/Bcl-2 levels, and PARP-1 cleavage [48]. For the cell cycle analysis, the effects of hybrid **77** and its nanoparticles were evaluated against MCF-7 and MDA-MB-231 cell lines using flow cytometry, comparing the results with a DMSO control. As shown in Figure 25a, both the hybrid **77** and its nanoparticles induced apoptotic cells, resulting in percentages of 9.18% and 21.54% for the MCF-7 line, and 11.09% and 23.17%, for the MDA-MB-231 line, respectively. These findings underscore the significantly greater potency of the nanoparticles of **77**, being twice as effective as the hybrid **77** against both cell lines tested. Moreover, exposure to the hybrid **77** and its nanoparticles caused a noticeable disruption in cell cycle distribution, with percentages of 11.26% and 17.52% observed in MCF-7, and 12.11% and 19.24% in MDA-MB-231, respectively (Figure 25b). These results suggest that the inhibitory potency was predominantly associated with the nanoparticles of **77**, particularly in the G2/M phase.

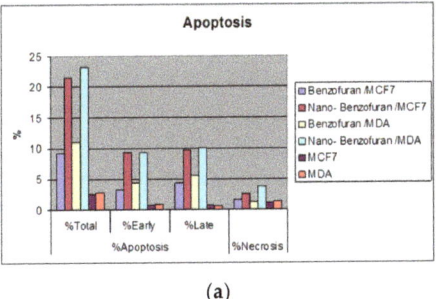
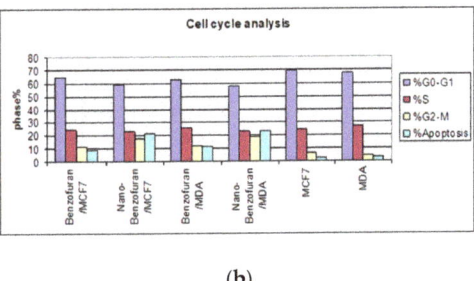

(a) (b)

Figure 25. Flow cytometry analysis, showing (a) apoptosis and (b) cell cycle analysis of the benzofuran–pyrazole hybrid **77** and its nanoparticles in MCF-7 and MDA-MB-231 cell lines. This is an open-access article distributed under the terms of the Creative Commons CC BY license [48].

In the study focusing on caspase-3/p53/Bax/Bcl-2 levels, the researchers employed an enzyme-linked immunosorbent assay (ELISA) to investigate two cancer cell lines [48]. As shown in Table 18, hybrid **77** exhibited a notable increase in caspase-3 levels (>5-fold) compared to untreated cells. However, the nanoparticles of **77** demonstrated an even more remarkable effect, surpassing the impact of hybrid **77**, with caspase-3 levels elevated by over 17-fold vs. untreated cells. Moreover, p53 levels showed an approximately 7-fold increase for hybrid **77** and a remarkable 14-fold increase for nanoparticles of **77** against both cell lines tested. Additionally, Bax levels displayed a significant increase of 5- to 13-fold, while Bcl-2 levels decreased by 4- to 7-fold in both cell lines compared to the untreated cells.

Table 18. Caspase-3, p53, Bax, and Bcl-2 levels of benzofuran–pyrazole hybrid 77 and its nanoparticles.

Compound	Cell Line	Caspasase-3	p53	Bax	Bcl-2
77	MCF-7	6.383836	7.453852	5.745321	0.272695
	MDA-MB-231	5.399087	7.792609	7.553853	0.181989
Nanoparticles—77	MCF-7	14.56524	12.51432	9.149760	0.131011
	MDA-MB-231	17.915	14.60536	13.19230	0.134738
Untreated control	MCF-7	1	1	1	1
	MDA-MB-231	1	1	1	1

In the final phase of the study, the researchers conducted a PARP-1 cleavage assay, using staurosporine as the standard drug [48]. As depicted in Table 19, hybrid 77 did not exhibit a considerable inhibitory effect on both cancer cell lines compared to staurosporine. However, the nanoparticles of 77 showed a remarkable inhibitory effect, proving to be 4 and 10 times more potent than the hybrid 77 for the MCF-7 and MDA-MB-231 cell lines, respectively. Notably, the nanoparticles of 77 exhibited an IC_{50} value of 6 nM, whereas staurosporine had an IC_{50} value of 8 nM for the MDA-MB-231 line.

Table 19. PARP-1 inhibitory assay of benzofuran–pyrazole hybrid 77 and its nanoparticles.

Compound	IC_{50} (nM) [a]	
	MCF-7	MDA-MB-231
77	40 ± 1	60 ± 1
Nanoparticles—77	10 ± 4	6 ± 3
Staurosporine [b]	10 ± 1	8 ± 1

[a] The data indicate the mean ± SD of at least three independent experiments. [b] Staurosporine is the standard drug for the study.

Shikonin–benzo[b]furan hybrids have shown potential as inhibitors of tubulin polymerization. To further explore this, Shao et al. synthesized the compound 79 using the Finkelstein reaction of salicylaldehyde 78 with ethyl bromoacetate in the presence of KI as a catalyst under mild reaction conditions (Scheme 14) [49]. Subsequently, compound 79 underwent condensation with K_2CO_3 in DMF at 120 °C for 3 h to afford ethyl benzofuran-2-carboxylate 80, which was then hydrolyzed with sodium hydroxide in DMF at 60 °C for 30 min, resulting in benzo[b]furan-2-carboxylic acid 81. Finally, the esterification reaction of carboxylic acid 81 with shikonin 82 using a mixture of DCC and DMAP in dichloromethane at 0 °C for 8 h afforded shikonin–benzo[b]furan hybrids 83a–q in low yields (20–36%).

The antiproliferative activity of the shikonin–benzo[b]furan hybrids 83a–q was evaluated against five human cancer cell lines (HepG2, HT29, HCT116, MDA-MB-231, and A549) and two non-cancerous cells (293T and LO2) using the MTT assay, with colchicine, shikonin, and CA-4 as the standard drugs (Table 20). The results demonstrated significant antiproliferative activity for the majority of the hybrids against the tested cancer cell lines. Remarkably, compounds 83c (R^1 = 3-OMe), 83o (R^1 = 3-OEt), and 83i (R^1 = 3-tert-butyl) exhibited outstanding activity in the HT29 cell line, with IC_{50} values of 0.18, 0.73, and 0.82 μM, respectively, compared to shikonin and colchicine (IC_{50} = 2.80 and 1.81 μM, respectively). Surprisingly, compound 83c (IC_{50} = 0.18 μM) demonstrated even higher activity than CA-4 (IC_{50} = 0.31 μM) against the HT29 cell line. The impact of substitutions on antiproliferative activity was evident, as the most active shikonin–benzo[b]furan hybrids 83c, 83i, and 83o with a substitution at position 3 of the salicylaldehyde 78 displayed greater potency than shikonin in four of the cancer cell lines tested. However, this trend was observed only in mono-substituted compounds because di-substituted compounds (83i and 83p) exhibited lower activity than them at the same position. On the other hand, compound 83f (R^1 = 2-Br) at position 2 demonstrated significant activity comparable to that of shikonin. Furthermore, the cytotoxicity of all compounds was evaluated in two non-cancerous cells (293T and LO2) (Table 20). The results indicated that all shikonin–benzo[b]furan hybrids exhibited low cytotoxicity (CC$_{50}$ > 100 µM). These findings suggest their potential as promising candidates for further exploration in cancer treatment research.

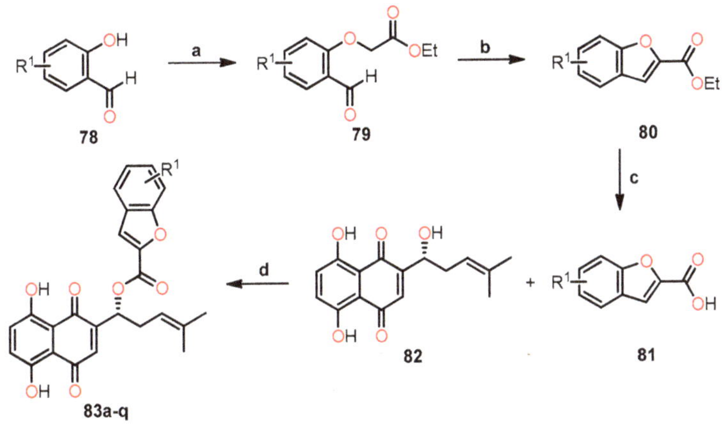

R^1 = H (**83a**), 4-Me (**83b**), 3-OMe (**83c**), 5-Br (**83d**), 2-Cl (**83e**), 2-Br (**83f**), 5-F (**83g**), 3,5-diCl (**83h**), 3-*tert*-butyl (**83i**), 6-F (**83j**), 5-Cl (**83k**), 3-Br-5-Cl (**83l**), 5-Me (**83m**), 4-Br (**83n**), 3-OEt (**83o**), 3,5-*tert*-butyl (**83p**), 5-OMe (**83q**)

Scheme 14. Synthesis of shikonin–benzo[*b*]furan hybrids **83a–q**, and reaction conditions: (**a**) KI, DMF, r.t, 30 min, then BrCH$_2$CO$_2$Et, 2 h, r.t; (**b**) K$_2$CO$_3$, DMF, 120 °C, 3 h; (**c**) NaOH (20%), DMF, 60 °C, 30 min; (**d**) DCC, DMAP, CH$_2$Cl$_2$, 0 °C, 8 h, 20–36%. Yields for compounds **79**, **80**, and **81** have not been reported by the authors.

Table 20. Antiproliferative activity of shikonin–benzo[*b*]furan hybrids **83a–q**, shikonin, colchicine, and CA-4.

Compound	IC$_{50}$ (µM) [a]					CC$_{50}$ (µM) [a]	
	HepG2	HT29	HCT116	MDA-MB-231	A549	293T [b]	LO2 [b]
83a	46.75 ± 6.96	7.16 ± 0.81	10.27 ± 1.60	6.00 ± 0.51	24.91 ± 2.29	169.08 ± 17.58	155.31 ± 9.20
83b	75.33 ± 3.23	1.22 ± 0.09	2.28 ± 0.32	1.91 ± 0.34	19.05 ± 1.83	ND	ND
83c	73.20 ± 4.03	0.18 ± 0.04	0.58 ± 0.11	0.81 ± 0.13	0.57 ± 0.79	184.86 ± 9.88	154.76 ± 9.98
83d	71.83 ± 5.19	7.62 ± 0.85	10.68 ± 0.18	8.22 ± 0.38	18.53 ± 1.43	158.90 ± 19.48	151.52 ± 8.23
83e	ND	6.06 ± 0.16	3.86 ± 0.47	3.59 ± 0.33	47.20 ± 2.24	ND	ND
83f	52.38 ± 5.27	1.03 ± 0.17	2.21 ± 0.40	2.63 ± 0.13	27.33 ± 1.34	162.38 ± 17.77	197.07 ± 9.88
83g	ND	10.37 ± 0.44	20.36 ± 1.82	10.42 ± 0.61	42.43 ± 5.35	170.07 ± 18.85	ND
83h	44.60 ± 4.88	5.89 ± 0.77	9.16 ± 0.44	4.09 ± 0.75	30.77 ± 2.11	158.61 ± 9.77	143.36 ± 9.88
83i	ND	0.82 ± 0.08	1.21 ± 0.07	1.51 ± 0.10	48.81 ± 5.66	ND	ND
83j	70.68 ± 6.96	4.04 ± 0.29	9.30 ± 0.76	10.74 ± 0.60	28.19 ± 2.53	194.98 ± 10.79	107.54 ± 9.93
83k	82.60 ± 9.68	9.14 ± 0.65	11.61 ± 0.91	4.88 ± 0.12	23.27 ± 2.65	121.92 ± 9.86	165.62 ± 11.98
83l	74.81 ± 8.50	7.55 ± 0.44	11.99 ± 1.21	15.25 ± 0.22	21.55 ± 3.26	148.46 ± 18.78	178.54 ± 8.50
83m	80.92 ± 9.13	16.35 ± 1.06	23.01 ± 0.97	18.33 ± 0.03	22.24 ± 3.83	149.38 ± 7.86	104.74 ± 9.85
83n	77.71 ± 5.81	1.12 ± 0.11	3.57 ± 0.50	2.01 ± 0.06	16.99 ± 0.66	ND	ND
83o	81.32 ± 8.76	0.73 ± 0.09	1.27 ± 0.34	1.33 ± 0.44	9.11 ± 2.21	ND	ND
83p	ND	ND	ND	24.68 ± 0.56	ND	ND	ND
83q	90.78 ± 9.73	10.94 ± 0.90	11.28 ± 0.94	6.21 ± 0.03	19.13 ± 1.89	108.40 ± 9.14	137.4 ± 8.81
Shikonin [c]	2.92 ± 0.09	2.80 ± 0.26	2.47 ± 0.19	2.77 ± 0.29	10.25 ± 1.37	7.00 ± 0.89	11.77 ± 0.14
Colchicine [c]	1.47 ± 0.07	1.81 ± 0.06	2.13 ± 0.13	2.94 ± 0.12	1.17 ± 0.09	8.43 ± 0.71	9.18 ± 0.62
CA-4 [c]	0.27 ± 0.007	0.31 ± 0.02	0.11 ± 0.006	0.09 ± 0.008	0.23 ± 0.006	1.44 ± 0.011	0.36 ± 0.003

[a] The data indicate the mean ± SD of at least three independent experiments. [b] Cytotoxicity in human normal cells. [c] Standard drugs for the study. ND means not detected.

Subsequently, the effect of hybrid **83c** on tubulin microtubule dynamics was evaluated in the HT29 cell line, using colchicine, CA-4, and paclitaxel as the standard drugs [49]. The hybrid **83c** exhibited a similar action to colchicine, indicating its potential as a microtubulin destabilizing agent with a more potent inhibitory effect on tubulin polymerization than both colchicine and CA-4, as evidenced by their respective IC$_{50}$ values of 0.98, 2.11, and 1.12 μM (Table 21). In addition, compound **83c** displayed a competitive trend at the tubulin binding site, demonstrating a remarkable 92.42% inhibition at 4 μM, comparable to colchicine.

Table 21. Evaluation of the impact of hybrids **83c**, **83f**, and **83i** on the inhibition of tubulin polymerization and colchicine binding to tubulin.

Compound	Inhibition of Tubulin Polymerization IC$_{50}$ (μM) [a]	Inhibition of Colchicine Binding (% Inhibition) [a]	
		2 μM	4 μM
83c	0.98 ± 0.014	82.36 ± 0.88	92.42 ± 0.79
83f	4.13 ± 0.26	ND	ND
83i	2.37 ± 0.21	ND	ND
Colchicine [b]	2.11 ± 0.32	ND	ND
CA-4 [b]	1.12 ± 0.13	81.46 ± 0.94	92.96 ± 0.87

[a] The data indicate the mean ± SD of at least three independent experiments. [b] Standard references for the study. ND means not detected.

The results led to additional biological investigations on hybrid **83c**, including cell cycle and apoptosis assays, tubulin antiangiogenesis, and antivascular assays [49]. Flow cytometry was employed to analyze the cell cycle using the HT29 cell line at varying concentrations of hybrid **83c**. The results revealed cell cycle arrest at the G2/M phase persisting over time (Figure 26a,b). Moreover, a Western blot study was performed to assess the impact of hybrid **83c** on cell-cycle-related proteins. The study revealed that hybrid **83c** increased the expression of P21 and cyclin B1, alongside reducing the expression of Cdc2, p-Cdc2, and p-Cdc25c, corroborating the findings from the cell cycle analysis (Figure 26c).

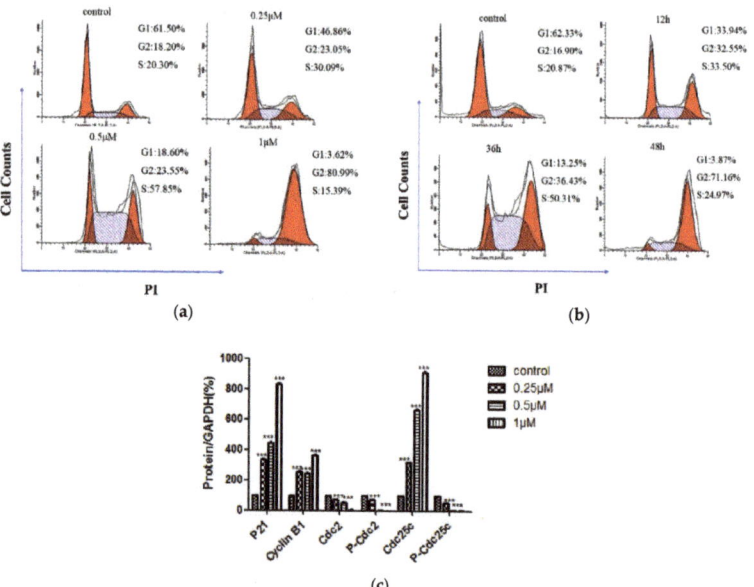

Figure 26. The cell cycle analysis of hybrid **83c**: (**a**) using HT29 cells at concentrations of 0.25, 0.5, and 1 μM; (**b**) with HT29 cells at time points of 12, 36, and 48 h; and (**c**) Western blot showing the expression of cell-cycle-related proteins. Data are means ± SD of three independent experiments. *** $p < 0.05$ vs. control. Reproduced with permission from ref. [49]. Copyright Elsevier Inc., 2023.

Cell apoptosis analysis was performed using the Annexin V-FITC/PI assay and confirmed by the Western blot in HT29 cells. Hybrid **83c** induced cell apoptosis in a concentration- and time-dependent manner (Figure 27a,b) [49]. Moreover, the expression of proteins associated with cell apoptosis, including Bax, PARP, caspase-3, and caspase-9, increased, while Bcl-2 expression decreased, aligning with the observed trend in cell apoptosis (Figure 27c).

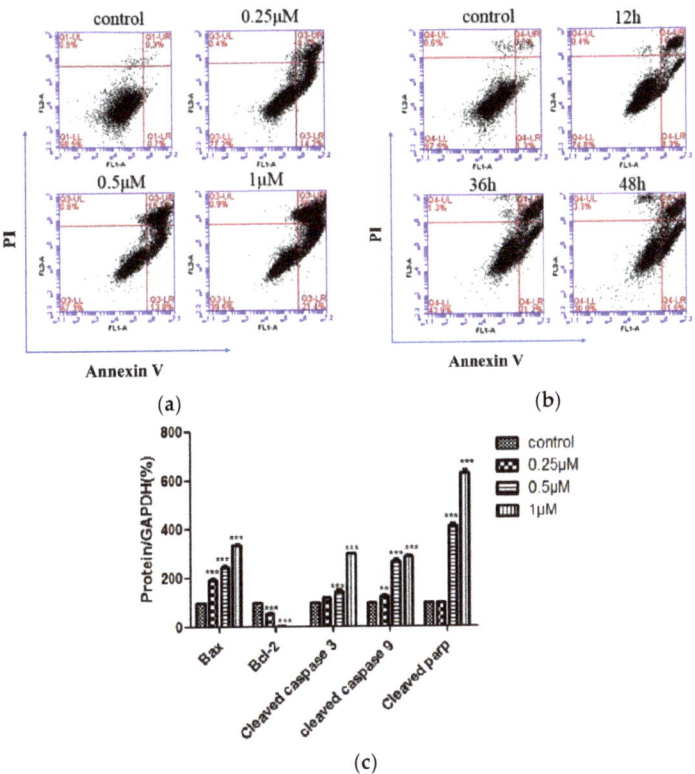

Figure 27. Apoptosis analysis of hybrid **83c**: (**a**) using HT29 cells at concentrations of 0.25, 0.5, and 1 µM; (**b**) with HT29 cells at time points of 12, 36, and 48 h; (**c**) Western blot showing the expression of apoptosis-related proteins. Data are means ± SD of three independent experiments. ** $p < 0.01$; *** $p < 0.05$ vs. control. Reproduced with permission from ref. [49]. Copyright Elsevier Inc., 2023.

A complementary study was conducted to explore the effect of hybrid **83c** on inhibiting tubulin polymerization and determine its potential to regulate microtubule dynamics in living cells [49]. The immunofluorescence assay was performed using the HT29 cell line, with colchicine and paclitaxel as positive control drugs (Figure 28). The results demonstrated that cells treated with paclitaxel showed more stable microtubules, whereas those treated with colchicine disintegrated and became soluble. Similarly, cells treated with hybrid **83c** showed a similar response to colchicine, inducing the collapse of microtubules in a dose-dependent manner. On the other hand, the potential antivascular activity of hybrid **83c** was investigated using endothelial cells (HUVECs), which were treated with different concentrations of **83c** and then seeded in Matrigel. After 6 h of treatment, cells treated with hybrid **83c** exhibited a dose-dependent inhibition in the formation of HUVEC cords, suggesting that hybrid **83c** can impede the formation of HUVEC tubes.

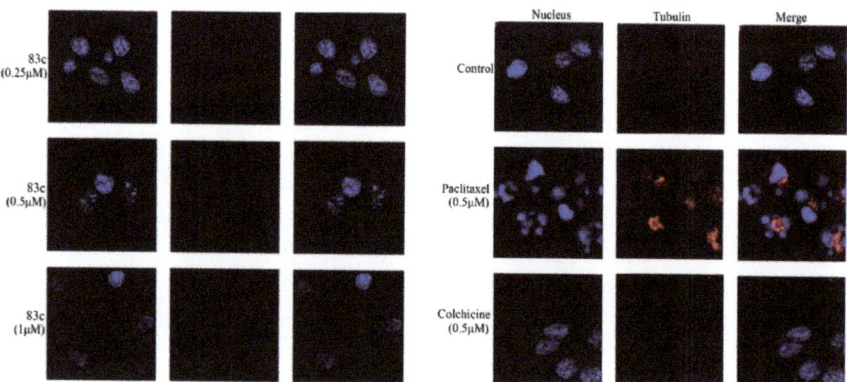

Figure 28. Evaluation of antimicrotubule effects against HT29 cells using confocal fluorescence microscopy for hybrid **83c** (0.25, 0.5, and 1 mM), colchicine (0.5 mM), paclitaxel (0.5 mM), and DMSO. The nucleus was stained with DAPI (blue), and tubulins were labeled with Alexa Fluor 594 (red), captured with a confocal fluorescence microscope. Reproduced with permission from ref. [49]. Copyright Elsevier Inc., 2023.

As mentioned earlier, aryl- and alkylbenzo[*b*]furan groups have exhibited significant biological activity against various types of human cancer. Building upon these promising findings, Sivaraman et al. undertook the synthesis of a series of 2-aryl[*b*]benzofurans, classified as lignane and neolignane, through a one-pot reaction using 2-bromobenzo[*b*]furans as crucial intermediates [50]. *Lavandula agustifola* served as the source for both natural (**89a**, **99**, and **100**) and non-natural (**88a–b**, **89b**, **94**, and **95**) compounds in the process (Schemes 15–17).

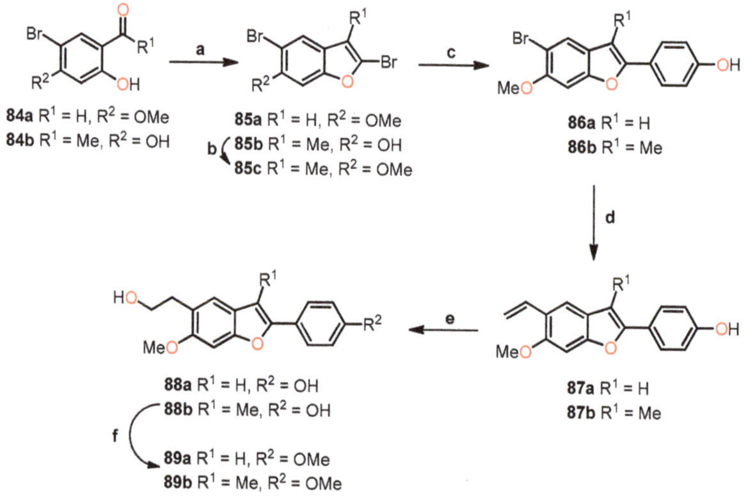

Scheme 15. Synthesis of 2-arylbenzo[*b*]furan derivatives **88a–b** and **89a–b**, and their reaction conditions: (**a**) CBr_4, TPP, Zn, NH_4Cl, Cs_2CO_3, CuI, CH_3CN/DCM (4:1), 85 °C, 2 h, 40–60%; (**b**) K_2CO_3, MeI, acetone, 60 °C, 1 h, 64%; (**c**) K_2CO_3, 4-(4,4,5,5-tetramethyl-1,3,2-dioxaborolan-2-yl)phenol, Pd(PPh$_3$)$_4$, THF/EtOH/H_2O (20:1:1), 80 °C, 16 h, 50–60%; (**d**) K_3PO_4, vinylboronic acid, Pd(dppf)Cl$_2$.CH$_2$Cl$_2$, DMF/H_2O (4:1), 90 °C, 2 h, 81–88%; (**e**) BH_3/THF, 2M NaOH–H_2O_2, 0 °C, r.t, 2 h, 53–68%; (**f**) K_2CO_3, MeI, acetone, 60 °C, 1 h, 72%.

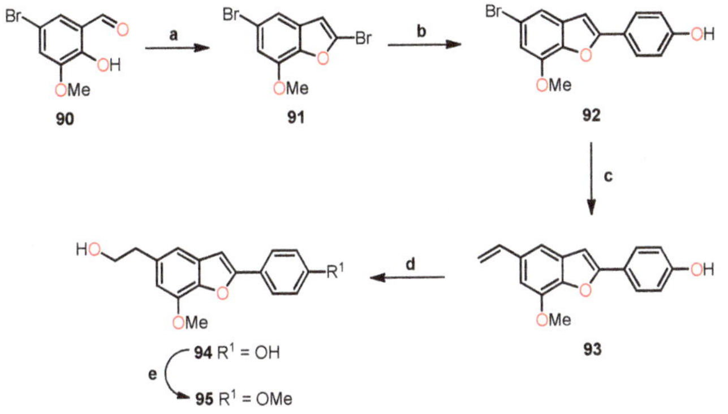

Scheme 16. Synthesis of 2-arylbenzo[*b*]furan derivatives **94** and **95** and their reaction conditions: (**a**) CBr$_4$, TPP, Zn, NH$_4$Cl, Cs$_2$CO$_3$, CuI, CH$_3$CN/DCM (4:1), 85 °C, 72%; (**b**) K$_2$CO$_3$, 4-(4,4,5,5-tetramethyl-1,3,2-dioxaborolan-2-yl)phenol, Pd(PPh$_3$)$_4$, THF/EtOH/H$_2$O (20:1:1), 80 °C, 16 h, 57%; (**c**) K$_3$PO$_4$, vinylboronic acid, Pd(dppf)Cl$_2$·CH$_2$Cl$_2$, DMF/H$_2$O (4:1), 90 °C, 2 h, 90%; (**d**) BH$_3$/THF, 2 M NaOH–H$_2$O$_2$, 0 °C, r.t, 2 h, 60%; (**e**) K$_2$CO$_3$, MeI, acetone, 60 °C, 1 h, 70%.

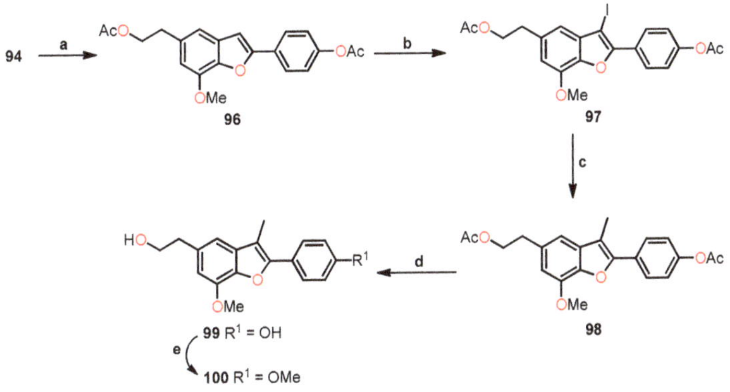

Scheme 17. Synthesis of 2-arylbenzo[*b*]furan derivatives **99** and **100** and their reaction conditions: (**a**) AcCl, DMAP, Et$_3$N, DCM, 0 °C, r.t, 2 h, 88%; (**b**) NIS, PTSA, CH$_3$CN, 0 °C, r.t, 2 h, 80%; (**c**) K$_3$PO$_4$, boronic acid, Pd(dppf)Cl$_2$·CH$_2$Cl$_2$, 1,4-dioxane/H$_2$O (100:1), 110 °C, 2 h, MWI, 73%; (**d**) LiOH/H$_2$O, THF/EtO/H$_2$O (4:4:1), 0 °C, r.t, 2 h, 89%; (**e**) K$_2$CO$_3$, MeI, acetone, 60 °C, 1 h, 76%.

In the initial synthesis, 2,5-dibromo-6-methoxybenzo[*b*]furan **85a** was obtained in a 60% yield through a *gem*-dibromo olefination/cyclization sequence (Scheme 15). Subsequently, a selective C–2 acylation of the Suzuki-type reaction resulted in the formation of 4-(5-bromo-6-methoxybenzofuran-2-yl)phenol **86a**. Further vinylation of **86a**, followed by a hydroboration/oxidation sequence, led to the compound **88a**, which was then subjected to methylation to afford compound **89a**. For the tetrasubstituted benzo[*b*]furans **88b** and **89b**, a similar reaction pathway was employed, utilizing 1-(5-bromo-2,4-dihydroxyphenyl)ethanone **84b**, leading to the formation of 2,5-dibromo-3-methylbenzo[*b*]furan-6-ol **85b**. Subsequent methylation of **85b** gave 2,5-dibromo-6-methoxy-3-methylbenzo[*b*]furan **85c**. By employing the identical reaction steps as previously described, compounds **88b** and **89b** were obtained with overall yields of 6% and 5%, respectively.

The second reaction was performed following a similar procedure as described in Scheme 15. Initially, a one-pot cyclization of 5-bromo-2-hydroxy-3-methoxybenzaldehyde **90** yielded benzo[*b*]furan **91**. Subsequently, a selective C–2 arylation gave **92**, and vinylation led to the formation of 4-(7-methoxy-5-vinyl-benzofuran-2-yl)phenol **93** (Scheme 16). Afterward, a hydroboration/oxidation sequence afforded compound **94**, and its methylation yielded compound **95**, with overall yields of 22% and 16%, respectively.

Benzo[*b*]furan derivatives **99** and **100** were synthesized by a multi-step process starting from **94**. Initially, compound **94** underwent diacetylation to afford **96** (Scheme 17). Subsequently, a regioselective iodination reaction of **96** with NIS afforded iodinated compound **97**, which underwent a Suzuki reaction to give compound **98**. Finally, benzofuran **98** suffered deprotection and methylation to obtain compounds **99** and **100** with overall yields of 10% and 8%, respectively.

The previously synthesized compounds were evaluated for their cytotoxic effects on five human cancer cell lines, including MCF-7, A549, PC3, HepG2, and Hep3B, utilizing the MTT assay [50]. Notably, compounds **88a**, **94**, and **99** exhibited significant reductions in cell counts, indicating promising cytotoxic activity. Subsequently, comprehensive biological studies were conducted using Western blot analysis to assess the effects of compounds **88a**, **94**, and **99** on proteins associated in cellular processes (Figure 29). As expected, all compounds induced the cleavage of PARP, indicative of activation of the apoptosis pathway. Additionally, MCF-7, A549, and HepG2 cells showed phosphorylation and stabilization of p53 in response to the compounds, while PC3 and Hep3B cells did not exhibit this response. Furthermore, the compounds similarly induced the p21 target gene across all cancer cell lines, regardless of the sensitivity of each cell line. This suggests that these compounds affect cancer cell survival through a combination of both p53-dependent and p53-independent mechanisms.

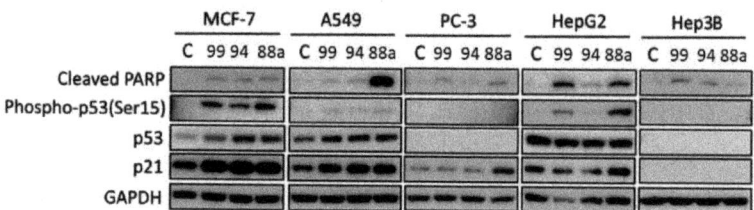

Figure 29. Western blot analysis for PARP, Phospho-p53, p53, p21, and GAPDH as the loading control for compounds **88a**, **94**, and **99**. Reprinted (adapted) with permission from ref. [50]. Copyright American Chemical Society, 2023.

Similarly, Oliva et al. synthesized a novel series of 2-amino-3-(3′,4′,5′-trimethoxybenzoyl)benzo[*b*]furan derivatives **103a–o** and evaluated their in vivo and in vitro anticancer activity [51]. The synthesis involved two distinct reaction steps, as illustrated in Scheme 18. Initially, the Knoevenagel condensation reaction of salicylaldehyde **101** with 2-azido-1-(3,4,5-trimethoxyphenyl)ethanone was conducted in methanol at room temperature for 24 h, utilizing piperidinium acetate to afford α-azido chalcone **102**. Subsequently, the chalcones were treated with PTSA (20 mol%) in refluxing acetonitrile for 12 h, resulting in the formation of 2-amino-3-(3′,4′,5′-trimethoxybenzoyl)benzo[*b*]furan derivatives **103a–n** in 60–80% yields. An alternative photochemical process was also employed, where thermal heating was replaced with irradiation at room temperature using a 25 W compact fluorescent lamp (CFL) to obtain compounds **103a–n**, albeit in reduced yields. Moreover, compound **103n** was obtained from **103i** through hydrogenation using Pd/C (10%) as the catalyst, while compound **103o** was derived from **103m** through reduction using iron and ammonium chloride.

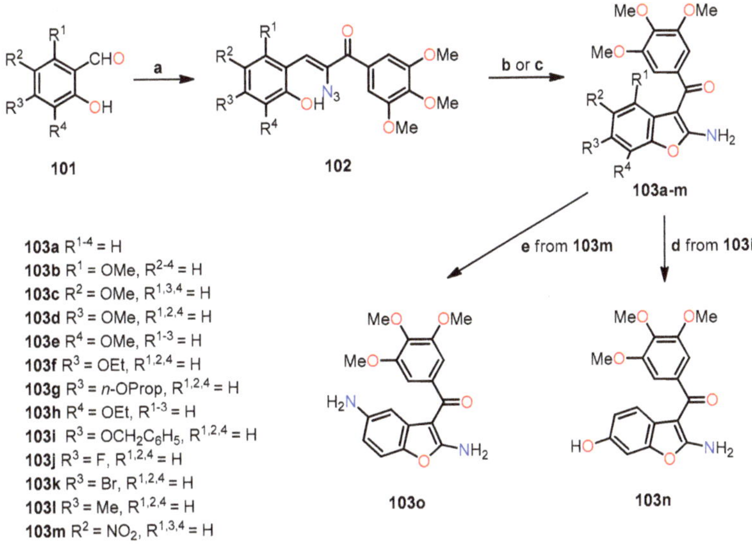

Scheme 18. Synthesis of 2-amino-3-(3′,4′,5′-trimethoxybenzoyl)benzo[b]furan derivatives **103a–o**, and the reaction conditions: (**a**) 2-azido-1-(3,4,5-trimethoxyphenyl)ethanone, piperidinium acetate, MeOH, r.t, 24 h; (**b**) PTSA (20 mol%), CH$_3$CN, reflux, 12 h; (**c**) white CFL (25 W), PTSA (20 mol%), CH$_3$CN, r.t, 24 h; (**d**) Pd/C (10%), EtOH, 40 bar, 60 °C, 30 min; (**e**) Fe, NH$_4$Cl, EtOH/H$_2$O, reflux, 3 h.

The antiproliferative activity of the compounds **103a–l** and **103n–o** was evaluated against six human cancer cell lines (HeLa, HT-29, Daoy, HL-60, SEM, and Jurkat) using the MTT assay and CA-4 as the standard drug (Table 22). The results revealed the remarkable activity of four of the compounds (**103d, 103f, 103k,** and **103l**), with IC$_{50}$ values below 5 nM. Specifically, compound **103f** (R^{1,2,4} = H, R^3 = OEt) displayed the highest potency with an IC$_{50}$ value of 5 pM against the Daoy cell line, outperforming the other cell lines evaluated. Both compounds **103f** and **103l** (R^{1,2,4} = H, R^3 = Me) exhibited greater potency than CA-4 in all cancer lines and demonstrated significantly higher sensitivity in Daoy, HL-60, and Jurkat cell lines, with IC$_{50}$ values ranging from 0.005 to 0.38 nM. Furthermore, a clear and consistent trend in the antiproliferative activity was observed among the methoxy-substituted compounds (**103b–e**). The compounds bearing a methoxy group at either the C–5 or C–6 position exhibited the highest activity, while those with the methoxy group at the C–4 or C–7 position displayed the lowest activity. Specifically, the order of potency was as follows: 6-OMe (**103d**) > 5-OMe (**103c**) > 7-OMe (**103e**) > 4-OMe (**103b**). In particular, compound **103b** exhibited a remarkable IC$_{50}$ value ranging from 2.8 to 8.5 nM. Moreover, the substitution of the methoxy group with a methyl group at C–6 (**103d** and **103l**, respectively) resulted in a 3–14-fold increase in activity against HeLa, HT-9, SEM, and Jurkat cell lines for **103l** compared to **103d**. However, the substitution of the methoxy group with a hydroxy group at C–6 (**103n**) did not enhance its activity in any significant manner. Finally, in the halide compounds, increasing the size from fluorine to bromine (**103k**) led to a 71–338-fold increase in activity across all six cell lines, with particularly pronounced effects observed in the Daoy line.

Table 22. Antiproliferative activity of compounds 103a–l, 103n–o, and CA-4.

Compound	IC$_{50}$ (nM) [a]					
	HeLa	HT-29	Daoy	HL-60	SEM	Jurkat
103a	413 ± 40.3	591 ± 29	371 ± 19	339 ± 21	250 ± 28	385 ± 52
103b	5225 ± 312	6512 ± 263	3950 ± 236	4891 ± 445	4557 ± 369	2760 ± 156
103c	56.0 ± 19	62.9 ± 12.9	36.2 ± 6.9	36.2 ± 11	38.9 ± 12	29.2 ± 9.3
103d	2.9 ± 0.3	5.6 ± 0.9	0.30 ± 0.08	3.0 ± 0.2	4.1 ± 0.2	2.7 ± 0.6
103e	496 ± 39	477 ± 42	345 ± 36	233 ± 31	285 ± 26	333 ± 42
103f	2.8 ± 0.3	2.1 ± 0.2	0.005 ± 0.001	2.7 ± 0.2	0.31 ± 0.05	0.28 ± 0.08
103g	22.7 ± 1.8	17.7 ± 1.2	4.4 ± 0.9	41.7 ± 15.9	29.7 ± 8.9	27.2 ± 1.9
103h	1250 ± 98	910 ± 58	1670 ± 87	428 ± 36	385 ± 27	390 ± 39
103i	9670 ± 125	8680 ± 458	>10,000	7056 ± 659	3675 ± 298	4390 ± 154
103j	591 ± 26	682 ± 45	846 ± 28	413 ± 56	333 ± 45	299 ± 39
103k	4.5 ± 0.5	5.4 ± 0.3	2.5 ± 0.2	5.6 ± 0.9	3.5 ± 0.1	2.8 ± 0.6
103l	1.1 ± 0.2	0.6 ± 0.02	0.34 ± 0.1	3.5 ± 0.4	0.30 ± 0.03	0.33 ± 0.05
103n	591 ± 63	3420 ± 368	428 ± 45	285 ± 2.5	371 ± 1.5	299 ± 37
103o	635 ± 1.5	790 ± 98	494 ± 56	413 ± 3.8	259 ± 1.8	399 ± 58
CA-4 [b]	4 ± 1	3100 ± 100	12.3 ± 0.09	1 ± 0.2	5 ± 0.1	0.8 ± 0.2

[a] The data indicate the mean ± SD of at least three independent experiments. [b] The standard drug for the study.

The antiproliferative studies facilitated the identification of the most promising compounds from the synthesized series of 2-amino-3-(3′,4′,5′-trimethoxybenzoyl)benzo[b]furans [51]. Subsequently, compounds 103c–d, 103f–g, and 103k–l were selected for further evaluation to assess their inhibitory effects on tubulin polymerization and [^{3}H]colchicine binding to tubulin (Table 23). The results demonstrated that all compounds exhibited comparable tubulin polymerization inhibition as compared to CA-4. Specifically, compounds 103f and 103l displayed greater potency than CA-4 (IC$_{50}$ = 0.54 nM), with IC$_{50}$ values of 0.37 and 0.39 nM, respectively. Moreover, in the colchicine binding studies to tubulin, compounds 103d, 103f, and 103l showed results similar to CA-4 at concentrations of 5 and 0.5 μM. These findings suggest that the compounds in this assay are robust antiproliferative and antitubulin agents.

Table 23. Evaluation of the impact of compounds 103c–d, 103f–g, 103k–l, and CA-4 on the inhibition of tubulin polymerization and colchicine binding to tubulin.

Compound	Inhibition of Tubulin Polymerization IC$_{50}$ (μM) [a]	Inhibition of Colchicine Binding (% Inhibition) [a]	
		5 μM	0.5 μM
103c	0.50 ± 0.09	73 ± 0.5	ND
103d	0.48 ± 0.05	99 ± 0.2	79 ± 1
103f	0.37 ± 0.02	99 ± 0.3	86 ± 0.8
103g	0.51 ± 0.1	91 ± 0.3	42 ± 0.4
103k	0.57 ± 0.04	92 ± 1	56 ± 2
103l	0.39 ± 0.04	97 ± 0.4	74 ± 0.1
CA-4 [b]	0.54 ± 0.06	97 ± 0.8	82 ± 2

[a] The data indicate the mean ± SD of at least three independent experiments. [b] The standard drug for the study. ND means not detected.

The potent inhibition of tubulin polymerization displayed by compounds 103f and 103l prompted the evaluation of their effects on cell cycle progression using flow cytometry with the HeLa cell line. As shown in Figure 30a,b, both compounds caused a remarkable cell cycle arrest at the G2/M phase after 24 h of treatment at a concentration of 10 nM. Furthermore, there was a notable reduction in the number of cells in the G1 phase, while no significant effect was observed on the S phase for both compounds. These findings indicate that compounds 103f and 103l exhibited a strong influence on cell cycle dynamics, which may play a crucial role in their antiproliferative activity.

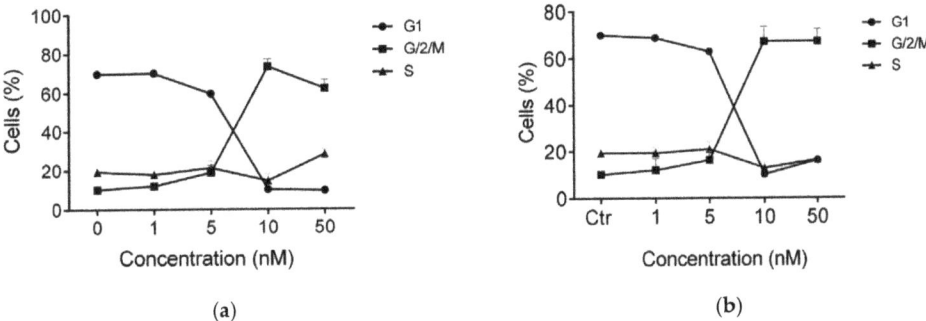

Figure 30. The cell cycle distribution in the HeLa cell line using flow cytometry analysis of compounds (a) **103f** and (b) **103l**. Reproduced with permission from ref. [51]. Copyright Elsevier Inc., 2023.

Furthermore, a study was conducted to investigate the effects of compound **103f** on two proteins, Bubr1 and Mad-2, which play essential roles in the spindle assembly checkpoint (SAC), and are associated with apoptotic cell death (Figure 31). The treatment with compound **103f** resulted in a significant reduction in the expression of both Bubr1 and Mad-2 proteins, even at low concentrations as low as 10 nM, indicating a potential arrest of the mitotic checkpoint. Moreover, the study examined cyclin B, a key regulator in the G2/M phase of the cell cycle, which showed a dose-dependent increase in expression in response to compound **103f**. This result aligns with the rapid accumulation of cells in the G2/M phase induced by the compound.

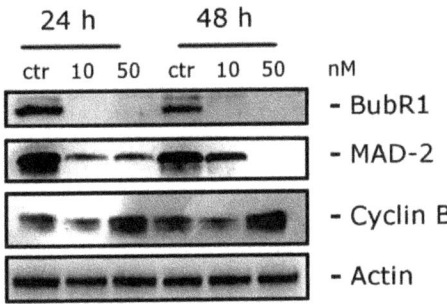

Figure 31. Analysis of the effect of compound **103f** on the regulation of cell-cycle-related protein levels. Reproduced with permission from ref. [51]. Copyright Elsevier Inc., 2023.

Considering the well-established association of tubulin-targeting agents with antivascular effects against tumor endothelium, the response of compound **103f** was evaluated to assess its antivascular activity in angiogenesis in vivo using HUVEC endothelial cells [51]. For this purpose, HUVECs were seeded on Matrigel to analyze the impact of compound **103f** on the formation of "tubule-like" structures in these cells. The results indicated that after 1 h of treatment, compound **103f** effectively disrupted the HUVEC network at both concentrations tested (10 and 100 nM) compared to the control cells. To quantitatively evaluate these effects, image analysis was conducted to measure parameters such as the tubule segment length, meshwork area, and number of branches. Notably, the results showed a statistically significant effect at a concentration of 10 nM on segment length and mesh area, underscoring the strong potential of compound **103f** to induce vascular disruption. These findings strongly suggest that compound **103f** holds promise as a potential antivascular agent for inhibiting angiogenesis.

Finally, compound **103f** underwent in vivo tests to evaluate its antitumor and cytotoxic effects in syngeneic mice. The method involved injecting E0771 murine breast cancer cells into the mammary fat pads of female C57BL/6 mice. Simultaneously, compound **103f** was administered intraperitoneally on alternate days at two doses (5 or 15 mg/kg). The results demonstrated a dose-dependent reduction in tumor growth upon treatment with compound **103f**, achieving a decrease of 45.7% and 16.9% at 15 and 5 mg/kg, respectively (Figure 32a). Compound **103f** exhibited higher potency than the reference drug (CA-4P), which reduced tumor growth by 26.5% at 30 mg/kg. Importantly, cytotoxicity tests revealed no apparent signs of toxicity at the 15 mg/kg doses of compound **103f** (Figure 32b). These findings highlight the potential of compound **103f** as a promising antitumor agent with limited toxicity in vivo.

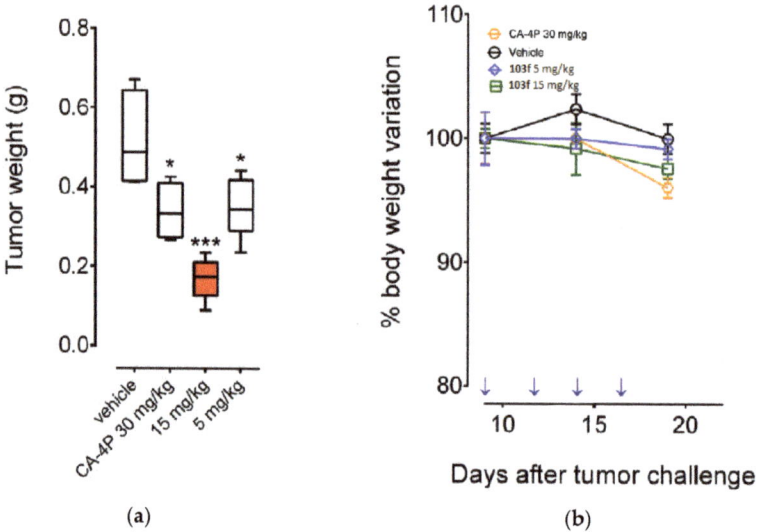

Figure 32. Suppression of mammary tumor growth in female C57BL/6 mice by compound **103f**: (**a**) tumor weight measurement at the end of the procedure ($n = 5$ mice/group) and (**b**) changes in mouse body weight during the procedure. Asterisks indicate significant difference between the treated and the control group. * $p < 0.05$, *** $p < 0.0001$. Reproduced with permission from ref. [51]. Copyright Elsevier Inc., 2023.

In 2021, Xu et al. accomplished the successful synthesis of novel polycyclic heterocycles derived from Evodiamine, a quinazolinocarboline alkaloid naturally occurring in the *Evodia rutaecarpa* plant native to China [52]. This research primarily aimed to evaluate these compounds as potential inhibitors of topoisomerase I (Top 1) for treating triple-negative breast cancer (TNBC), an aggressive subtype of breast cancer. The key focus of their synthetic efforts was the preparation of the 2-(5-methoxybenzofuran-3-yl)ethanamine **110**, which involved a five-step reaction sequence using 1,4-dimethoxybenzene **104** as the starting reagent (Scheme 19). Subsequently, intermediate **110** underwent an amidation reaction with ethyl formate in reflux conditions for 12 h to afford amide **111** in a 72% yield. The following step involved an intramolecular cyclization of compound **111** using POCl$_3$ in dichloromethane at room temperature for 12 h to give compound **112**, which was then reacted with substituted salicylic acid chlorides **114a–f** in dichloromethane at room temperature for 12 h to obtain the compounds **115a–f** in good yields. Finally, the intermediates **115a–f** underwent O-demethylation using BBr$_3$ in dichloromethane at −78 °C for 6 h to furnish evodiamine derivatives **116a–f** in good yields.

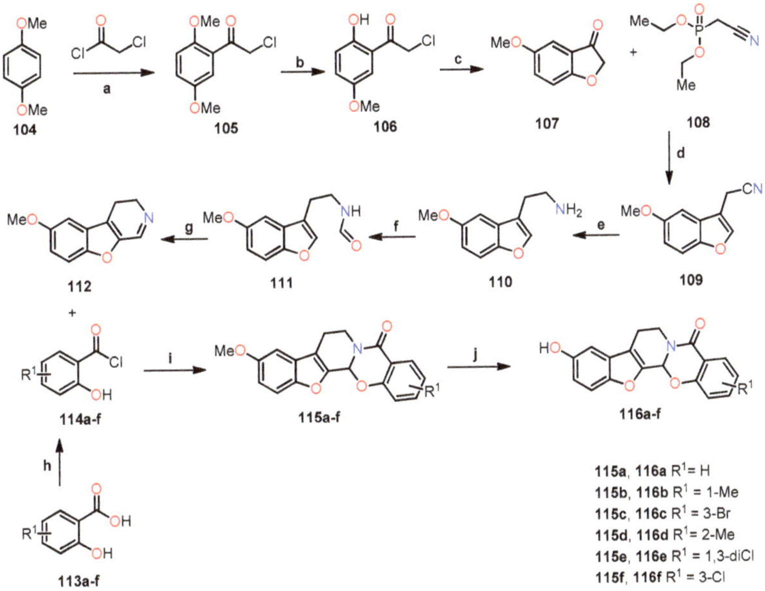

Scheme 19. Synthesis of evodiamine derivatives **115a–f** and **116a–f**, and reagents and reaction conditions: (**a**) AlCl$_3$, DCM, r.t, 24 h, 90%; (**b**) AlCl$_3$, DCM, r.t, 12 h, 88%; (**c**) AcONa, EtOH, reflux, 2 h, 91%; (**d**) THF, NaH, r.t, 12 h, 75%; (**e**) BH$_3$·THF, reflux, 3 h; (**f**) Ethyl formate, reflux, 12 h, 72%; (**g**) POCl$_3$, DCM, r.t, 12 h, 76%; (**h**) SOCl$_2$, reflux, 2 h; (**i**) DCM, r.t, 12 h, 78–82%; (**j**) BBr$_3$, DCM, −78 °C, 6 h, 82–91%.

The antiproliferative evaluation of evodiamine analogs **115a–f** and **116a–f** was performed on MDA-MB-435 human breast carcinoma cells using the MTT method, with evodiamine as the standard drug (Table 24) [52]. Notably, the results significantly favored compounds **116a–f**, exhibiting a higher percentage of inhibition at 10 μM compared to evodiamine. Building upon these promising findings, compounds **116a–f** underwent further evaluation against four human cancer cell lines: MDA-MB-435, MDA-MB-231, HCT116, and A549, using the MTT method in the presence of evodiamine and camptothecin as the standard drugs (Table 25). The breast cancer lines showed the highest sensitivity to these compounds. For instance, compounds **116a** (R^1 = H) and **116f** (R^1 = 3-Cl) showed the best activity against the MDA-MB-435 cell line, with IC$_{50}$ values of 0.47 and 0.42 μM, respectively, which were comparable to the camptothecin (IC$_{50}$ = 0.31 μM). The introduction of a halogen at position 3 significantly enhanced the antiproliferative activity of the analogs, as demonstrated by **116f**, which was the most active in the series with IC$_{50}$ values of 0.36, 0.42, and 0.76 μM against the MDA-MB-231, MDA-MB-435, and HCT116 cell lines, respectively.

The data presented above shed light on the main objective of this study, which involves evaluating the topoisomerase inhibitory activity of compounds **116a–f** [52]. To achieve this, a Top-1-mediated DNA cleavage assay was performed using purified Top 1 on compounds **116a–f**. During the assay, DNA and Top 1 were incubated with or without these compounds to observe their effect on the appearance of relaxed DNA fragments. Remarkably, among the tested compounds, only **116f** and camptothecin (CPT) exhibited significant inhibition of Top-1-mediated relaxation of supercoiled DNA at a concentration of 50 μM (Figure 33a). Furthermore, in a Top 1 inhibition study, compound **116f** showed activity at 20 μM, while evodiamine only displayed moderate activity up to 500 μM (Figure 33b,c). These findings strongly suggest that **116f** specifically targets Top 1 and holds great potential as a promising candidate for further investigation in cancer drug development.

Table 24. Percentage inhibition of evodiamine derivatives **115** and **116** on MDA-MB-435 cell line.

Compound	% Inhibition [a]		
	100 µM	10 µM	1 µM
115a	36.6 ± 2.7	10.6 ± 0.5	6.0 ± 0.4
115b	15.0 ± 1.9	11.9 ± 1.5	7.1 ± 1.2
115c	13.0 ± 0.7	9.2 ± 0.5	6.8 ± 2.4
115d	41.5 ± 2.6	19.2 ± 2.8	7.5 ± 1.0
115e	12.1 ± 0.7	5.5 ± 0.2	8.6 ± 0.7
115f	15.9 ± 0.9	7.3 ± 0.6	2.3 ± 0.1
116a	96.5 ± 9.6	61.2 ± 4.9	55.1 ± 3.0
116b	97.4 ± 8.3	66.5 ± 8.5	60.6 ± 12.8
116c	94.5 ± 6.2	83.1 ± 6.2	75.7 ± 8.2
116d	94.0 ± 6.1	64.8 ± 7.0	62.7 ± 9.3
116e	96.9 ± 9.8	59.5 ± 2.1	32.7 ± 4.6
116f	95.8 ± 4.1	84.3 ± 5.8	82.4 ± 7.4
Evodiamine [b]	73.5 ± 5.5	53.3 ± 3.1	38.9 ± 3.1

[a] The data indicate the mean ± SD of at least three independent experiments. [b] The standard drug for the study.

Table 25. Antiproliferative activity of evodiamine derivatives **116a–f**.

Compound	IC$_{50}$ (µM) [a]			
	MDA-MB-435	HCT116	A549	MDA-MB-231
116a	0.47 ± 0.03	1.26 ± 0.10	4.28 ± 0.30	0.86 ± 0.07
116b	1.95 ± 0.23	3.45 ± 0.24	6.79 ± 0.64	2.52 ± 0.33
116c	0.81 ± 0.06	1.71 ± 0.13	4.14 ± 0.33	1.16 ± 0.14
116d	1.14 ± 0.09	4.71 ± 0.32	10.49 ± 0.97	1.05 ± 0.09
116e	2.35 ± 0.18	6.09 ± 0.46	7.58 ± 0.67	3.73 ± 0.28
116f	0.42 ± 0.03	0.76 ± 0.07	5.24 ± 0.34	0.36 ± 0.02
Evodiamine [b]	6.30 ± 0.54	32.67 ± 2.30	77.24 ± 6.33	13.30 ± 1.84
Camptothecin [b]	0.31 ± 0.07	0.09 ± 0.008	0.10 ± 0.001	0.42 ± 0.03

[a] The data indicate the mean ± SD of at least three independent experiments. [b] The standard drugs for the study.

Complementary biochemical studies were performed to explore the ability of the compound **116f** to effectively trap Top 1–DNA cleavable complexes within cancer cells, potentially leading to cell death. To assess this, the researchers quantified the number of trapped cleavable complexes using [^{3}H]thymidine incorporation and SDS-K$^+$ precipitation methods. As shown in Figure 34a, there was a significant increase in the formation of the cleavable complex over prolonged periods in cells treated with both camptothecin and the compound **116f**. Furthermore, to confirm the formation of the cleavable complex, an immunological band depletion assay was performed to verify the presence of Top 1 in the precipitated complex, demonstrating that the complex could not migrate through the gel. Conversely, in the absence of the complex, Top 1 would have migrated through the gel. The results revealed a proportional decrease in the amount of free Top 1 with increasing time in cells treated with 15 µM of CPT and 30 µM of **116f**, particularly evident after 9 h, where almost no free Top 1 was detected (Figure 34b). These compelling findings strongly support the notion that the **116f** analog effectively interacts with Top 1, leading to the formation of the cleavable complex, underscoring its potential as a promising candidate for anti-cancer drug development.

Furthermore, a comprehensive study was performed to investigate the potential of stabilizing Top 1–DNA covalent complexes through an indirect process involving ROS generation and subsequent oxidative DNA damage, as observed in other studies with staurosporine. This investigation employed a fluorogenic ROS probe in combination with flow cytometry analysis. As demonstrated in Figure 35a, the treatment of MDA-MB-231 cells with the compound **116f** resulted in a compelling, dose-dependent increase in ROS levels. Notably, even in the absence of Top 1 in MDA-MB-231 cells (siRNA-Top 1), the **116f** analog induced ROS generation, indicating that the compound has the ability to generate ROS independently of Top 1 (Figure 35b).

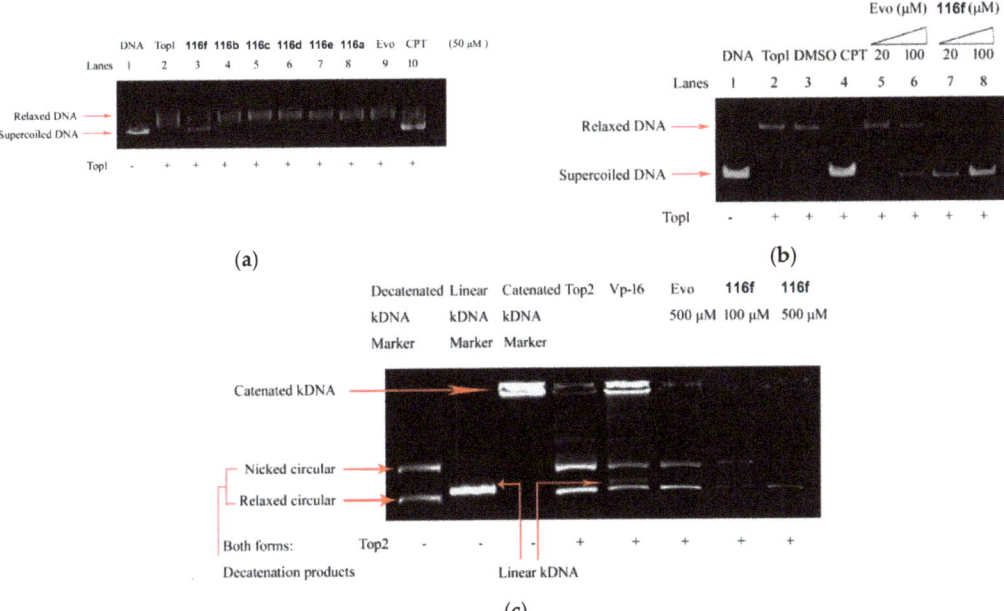

Figure 33. (a) Inhibition of Top 1 DNA relaxation activity in compounds **116a–f**, (b) inhibition of Top 1 relaxation in evodiamine (evo) and compound **116f**, (c) catalytic inhibition of Top 2 or a Top 2 poison using a kDNA substrate in compound **116f**. Reprinted (adapted) with permission from ref. [52]. Copyright American Chemical Society, 2023.

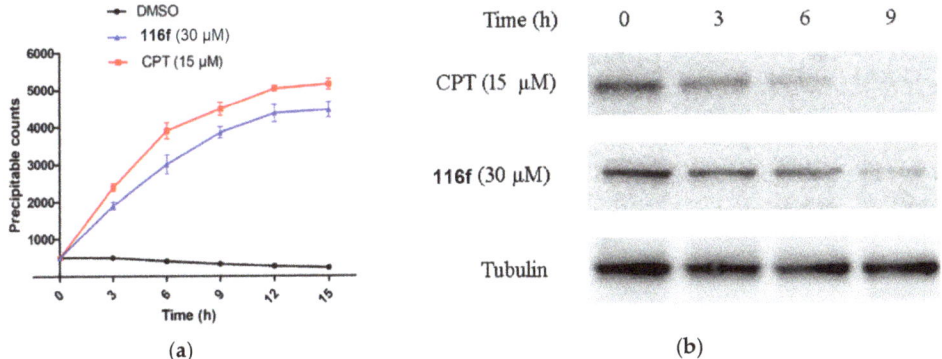

Figure 34. Formation of cleavable complexes of Top 1–DNA induced by compound **116f**: (a) [^{3}H]thymidine labeling of the MDA-MB-231 cell line over different time intervals, (b) immunological band assay for Top 1 in MDA-MB-231 across varied time periods. Reprinted (adapted) with permission from ref. [52]. Copyright American Chemical Society, 2023.

The cell cycle study focused on evaluating the mitochondrial dysfunction involved in the apoptosis process. To achieve this, a JC-1 fluorescent probe was employed to measure the mitochondrial membrane potential, incubated with MDA-MB-231 cells at different concentrations, and quantified by flow cytometry analysis. Mitochondrial dysfunction was found at 3.0% with 0 μM, 12.2% with 0.1 μM, 18.5% with 0.2 μM, and 25.0% with 0.4 μM of cells after treatment with the compound **116f** (Figure 36a). These data suggest a direct relationship with the mitochondrial pathway. Therefore, it was also necessary to evaluate the expression of apoptotic proteins, including Bax, Bcl-2, cytochrome C, and caspase-3, using a Western blot assay. Figure 36b shows that following 48 h of treatment with **116f**, the levels of Bax, cytochrome C, and caspase-3 proteins noticeably increased, while Bcl-2 expression significantly decreased.

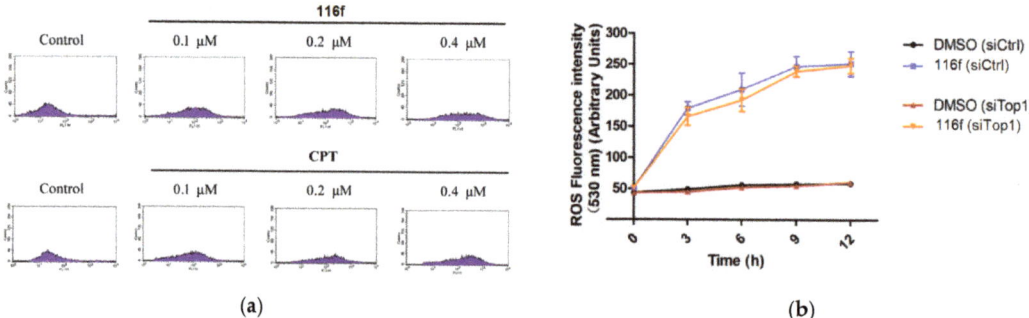

Figure 35. Generation of ROS by compound **116f** independent of Top 1: (**a**) flow cytometry analysis of ROS generation in MDA-MB-231 cells treated with **116f**, and (**b**) comparison of ROS generation in parental MDA-MB-231 cells, MDA-MB-231 cells with control siRNA (siRNA-Ctrl), and Top-1-deficient MDA-MB-231 cells (siRNA-Top 1). Reprinted (adapted) with permission from ref. [52]. Copyright American Chemical Society, 2023.

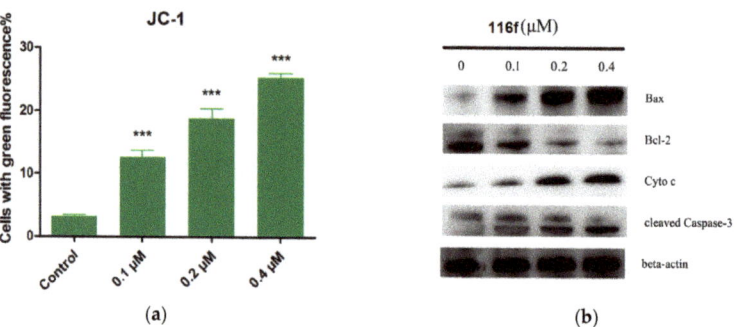

Figure 36. Effect of compound **116f** on (**a**) mitochondrial dysfunction using JC-1 staining and (**b**) expression levels of related proteins analyzed by Western blot in the MDA-MB-231 cell line. β-Actin was used as an internal control. *** $p < 0.001$. Reprinted (adapted) with permission from ref. [52]. Copyright American Chemical Society, 2023.

Finally, in vivo studies were conducted using a mouse xenograft model with surgical residual tumor samples from a patient with TNBC to assess the anticancer potency of compound **116f** in TNBC, with paclitaxel (PTX) used as the positive control. Different concentrations of **116f** were administered to the mice, followed by histological analysis of tumor tissue sections using H&E staining (Figure 37a). The results showed that the tumors from mice treated with **116f** exhibited reduced cell density and increased necrosis rates compared to the untreated mice. Furthermore, immunofluorescence labeling of cell proliferation marker Ki67 showed a significant decrease in proliferating Ki67 cells in tumors treated with **116f** compared to untreated tumors. This result convincingly demonstrates the potent inhibitory effect of **116f** on tumor growth. Additionally, the body weight of the evaluated mice treated with compound **116f** showed no significant changes, indicating the absence of apparent toxicity (Figure 37b). These promising findings highlight the potential of compound **116f** for future TNBC treatments, even at doses as low as 20 mg/kg.

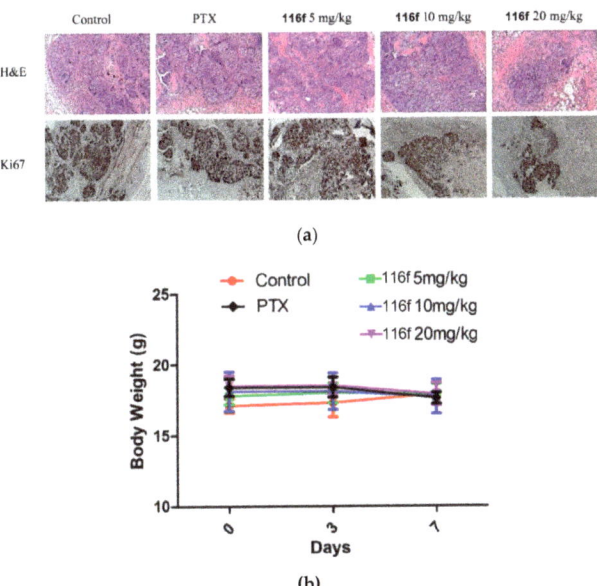

Figure 37. In vivo study of **116f** in a PDTX model: (**a**) immunohistochemical analysis with H&E staining and Ki67 expression in TNBC cells, (**b**) effects of **116f** on mice body weight during treatment. Reprinted (adapted) with permission from ref. [52]. Copyright American Chemical Society, 2023.

Using a similar approach, Zhang et al. successfully identified potential drugs for triple-negative breast cancer (TNBC) [53]. They focused on the derivatives of ZINCO3830212, which was selected through molecular docking studies involving pocket U of SIRT3, encompassing crucial amino acids, such as Phe157, Arg158, Ser159, Pro176, Glu177, and Glu323. This particular pocket is known for its significant role in modulating autophagy and its associations with various human cancers. To synthesize these derivatives, the researchers utilized alkanes as linkers and introduced halogen substituents to obtain compounds **117a–r** through an O-alkylation reaction. Subsequently, all synthesized compounds **117a–r** were evaluated against four human cancer cell lines: HL60, U937, MCF-7, and MDA-MB-231, using the MTT method and ZINCO3830212 as the reference compound (Table 26). Among these derivatives, compounds **117c–e** exhibited the most promising results in activating SIRT3. The researchers also investigated the impact of alkane length on SIRT3 activation by varying the linker length. Interestingly, they observed a decrease in activity for compound **117o** ($n = 0$, $R^1 = I$, $R^2 = 1$-pyrrolidinyl) with a shorter linker. Moreover, when substituents other than iodine (i.e., H, Cl, Br, and Me) were used, the activity was not favored in compounds **117n** ($n = 0$, $R^1 = H$, $R^2 = 1$-pyrrolidinyl) and **117p–r** ($n = 1$, $R^1 = $ Cl, Br, Me, $R^2 = 1$-pyrrolidinyl). However, in the case of compound **117c** ($n = 1$, $R^1 = I$, $R^2 = 1$-pyrrolidinyl) containing iodine, its activity surpassed that of compounds **117p–r**. Remarkably, compound **117c** showed the highest antiproliferative activity against the breast carcinoma line MDA-MB-231, with an IC_{50} value of 2.19 µM, compared to ZINCO3830212 ($IC_{50} = 33.43$ µM). This significant finding highlights its potential as a promising SIRT3 activator for TNBC treatment (Table 26).

Table 26. Antiproliferative activity of compounds 117a–r.

Compound	R¹	R²	n	10 μM of DMSO SIRT3	IC$_{50}$ (μM) [a]			
					HL60	U937	MCF-7	MDA-MB-231
117a	I	N(Me)Me	0	116.35 ± 1.34	8.22 ± 1.42	8.68 ± 0.48	39.37 ± 1.28	6.60 ± 0.27
117b	I	piperidine	0	121.12 ± 2.76	>50	>50	12.78 ± 0.52	>50
117c	I	pyrrolidine	1	144.69 ± 0.73	2.80 ± 0.64	2.52 ± 2.32	12.46 ± 0.85	2.19 ± 0.16
117d	I	piperidine	1	134.73 ± 0.93	11.46 ± 1.45	18.19 ± 1.42	37.94 ± 0.37	15.66 ± 0.95
117e	I	morpholine	1	136.73 ± 1.24	>50	>50	8.05 ± 0.61	6.81 ± 1.69
117f	I	N-Me piperazine	1	114.52 ± 0.71	>50	>50	5.37 ± 0.23	8.76 ± 1.25
117g	I	N-Et piperazine	1	126.52 ± 2.19	28.55 ± 1.56	13.65 ± 0.98	39.47 ± 1.26	17.96 ± 0.94
117h	I	N-Bn piperazine	1	93.63 ± 1.97	>50	41.36 ± 1.30	28.67 ± 1.03	15.68 ± 1.52
117i	I	N(Pr)	1	113.64 ± 0.17	16.02 ± 1.56	30.04 ± 2.45	21.54 ± 1.92	26.80 ± 0.87
117j	I	N(Bu)₂	1	123.40 ± 1.50	24.32 ± 1.49	7.44 ± 0.34	18.52 ± 0.79	11.27 ± 0.58
117k	I	NHBn	1	107.25 ± 1.53	4.40 ± 0.46	2.30 ± 1.57	8.05 ± 0.16	11.57 ± 0.83
117l	I	N(Me)Bn	1	92.43 ± 3.64	8.51 ± 1.30	6.86 ± 1.40	8.02 ± 0.29	53.97 ± 1.89
117m	H	N(Pr)	1	92.44 ± 1.53	2.77 ± 3.62	3.47 ± 0.56	5.26 ± 0.68	19.4 ± 0.78
117n	H	pyrrolidine	1	93.53 ± 3.14	6.38 ± 0.74	6.82 ± 0.16	3.73 ± 0.26	10.54 ± 0.88
117o	I	pyrrolidine	0	111.48 ± 1.83	23.15 ± 2.23	31.66 ± 1.86	>50	20.43 ± 2.06
117p	Cl	pyrrolidine	1	121.31 ± 2.92	11.34 ± 1.09	25.24 ± 0.95	27.66 ± 1.42	28.16 ± 1.53
117q	Br	pyrrolidine	1	119.09 ± 1.06	38.27 ± 1.36	>50	31.34 ± 1.22	21.64 ± 1.53
117r	Me	pyrrolidine	1	108.47 ± 1.95	41.33 ± 2.27	48.29 ± 1.98	15.58 ± 0.96	29.87 ± 0.91
ZINCO3830212 [b]	I	N(Et)	0	115.23 ± 2.20	5.27 ± 1.06	9.31 ± 1.45	>50	33.43 ± 2.84

[a] The data indicate the mean ± SD of at least three independent experiments. [b] The reference compound for the study.

This study compared the deacetylation activity of the compound **117c** with resveratrol and honokiol, known SIRT3 activators, using molecular coupling and evaluating their antiproliferative effects [53]. Molecular docking revealed that, unlike the compound **117c**, resveratrol and honokiol tended to bind to the acetylated substrate recognition site, indicating no allosteric effect (Figure 38a). Also, compound **117c** showed significantly higher potency than the reference activators, with an IC_{50} value of 2.19 µM against the MDA-MB-231 cell line. In contrast, resveratrol and honokiol showed IC_{50} values of 98.89 µM and 44.89 µM, respectively, with normalized E_{max} values of 0.25 for resveratrol, 0.91 for honokiol, and 1.00 for compound **117c** (Table 27). Additionally, compound **117c** induced the deacetylation of two tested SIRT3 substrates, MnSOD2 and p53 (Figure 38b), further confirming its potent deacetylation and antiproliferative effects.

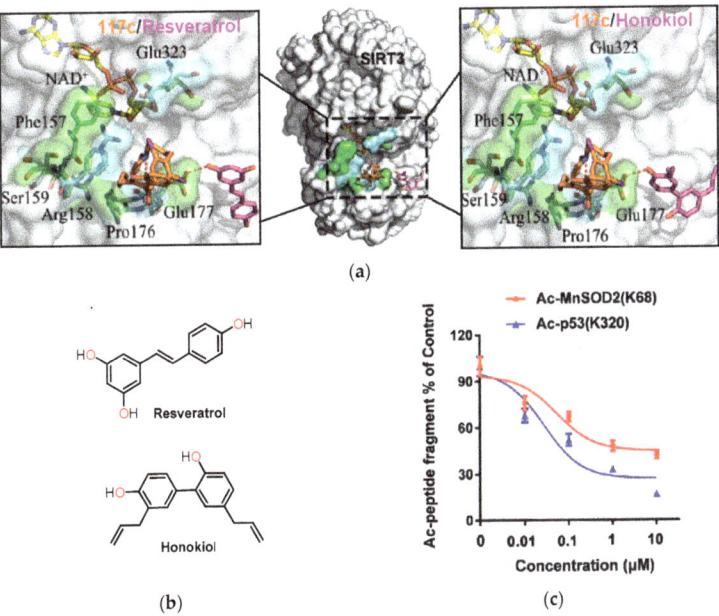

Figure 38. Effect of compound **117c** on SIRT3 activation. (**a**) The molecular coupling of compound **117c** with resveratrol and honokiol in SIRT3. (**b**) The chemical structures of resveratrol and honokiol. (**c**) SIRT3-dependent substrate deacetylation with compound **117c**. Reprinted (adapted) with permission from ref. [53]. Copyright American Chemical Society, 2023.

Table 27. E_{max}, EC_{50}, and IC_{50} values of compound **117c**.

Compound	E_{max} [a]	EC_{50} (µM)	IC_{50} (µM) [a] MDA-MB-231
117c	1.00 ± 0.07	0.21	2.19 ± 0.16
Resveratrol [b]	0.25 ± 0.03	55.19	98.89 ± 2.16
Honokiol [b]	0.91 ± 0.06	0.17	44.89 ± 1.86

[a] The data indicate the mean ± SD of at least three independent experiments. [b] The reference compounds for the study.

The conducted studies provided significant insights, demonstrating the activation capability of compound **117c** on SIRT3 [53]. To assess its selectivity toward SIRT3, a CETSA cellular thermal shift assay was performed on all SIRTs (SIRT1, SIRT2, SIRT3, and SIRT5). The collected data clearly revealed a direct interaction between compound **117c** and SIRT3, while SIRT1, SIRT2, and SIRT5 remained unaffected in their thermal stability. This distinct result confirmed the specific binding affinity of compound **117c** for SIRT3 in the MDA-MB-231 cell line (Figure 39). With its selectivity established, the researchers explored the influence of SIRT3 on both short- and long-term effects of compound **117c** in the MDA-MB-231 cell line, confirming a concentration- and time-dependent inhibition of tumor cell proliferation (Figure 40a,b). Next, the antiproliferative activity was examined in the presence of SIRT3, and the findings indicated a significant impairment in the inhibitory effect of compound **117c** when SIRT3 was absent. As a result, the antiproliferative effect displayed a marked reduction after SIRT3 knockdown, firmly establishing the dependence of compound **117c** on SIRT3 for its activity (Figure 40c).

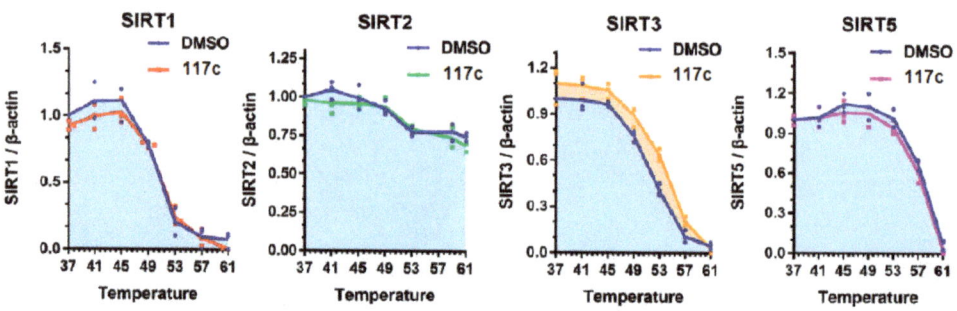

Figure 39. Selectivity of compound **117c** binding to SIRT3 in MDA-MB-231 cell line. Reprinted (adapted) with permission from ref. [53]. Copyright American Chemical Society, 2023.

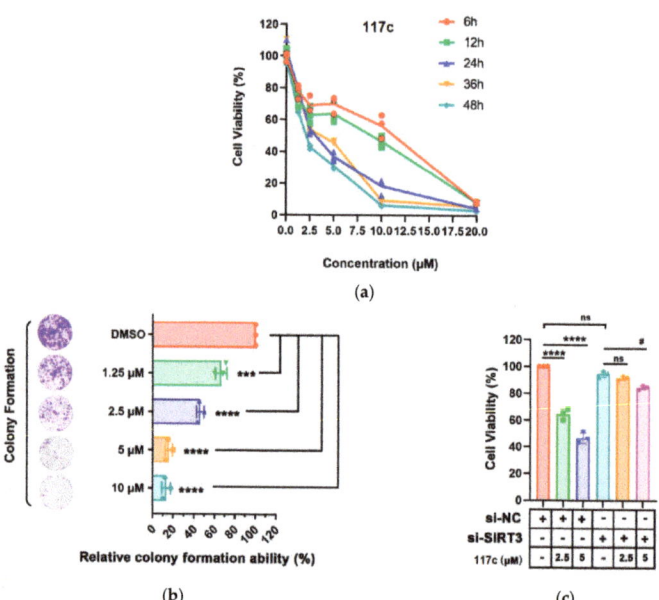

Figure 40. Inhibition of MDA-MB-231 tumor cell proliferation after treatment with compound **117c**. (a) MTT cell viability assay for 6–48 h with **117c**. (b) Colony formation assay with concentrations ranging from 1.25 to 10 μM. *** $p < 0.001$ and **** $p < 0.0001$ compared with the control group. (c) Transfection of MDA-MB-231 cells with siRNA SIRT3 and treated with **117c** at 2.5 and 5 μM. ns means no significance, **** $p < 0.0001$ compared with the si-NC group, and # $p < 0.05$ compared with the si-SIRT3 group. Reprinted (adapted) with permission from ref. [53]. Copyright American Chemical Society, 2023.

On the other hand, the regulation of autophagy by SIRT3 could have a suppressive effect on tumor growth and elimination [53]. As depicted in Figure 41a, compound **117c** inhibits autophagy, as evidenced by the increase in LC3-II expression and the decrease in p56 expression following treatment. Moreover, the inhibition of autophagy resulted in a significant increase in cell viability and a notable attenuation of tumor migration (Figure 41b,c). This indicates that compound **117c** exerts its antiproliferative effect on MDA-MB-231 cells by inducing autophagy. To delve into the role of SIRT3 in this process, the researchers employed SIRT3-specific siRNA to attenuate SIRT3 expression. Intriguingly, the attenuation of SIRT3 resulted in a significant decrease in the induction of autophagy, which is evident from the reduced levels of LC3-II and p62, along with the down-regulation of E-cadherin (Figure 41d). These findings suggest that compound **117c** regulates autophagy and tumor migration by activating SIRT3.

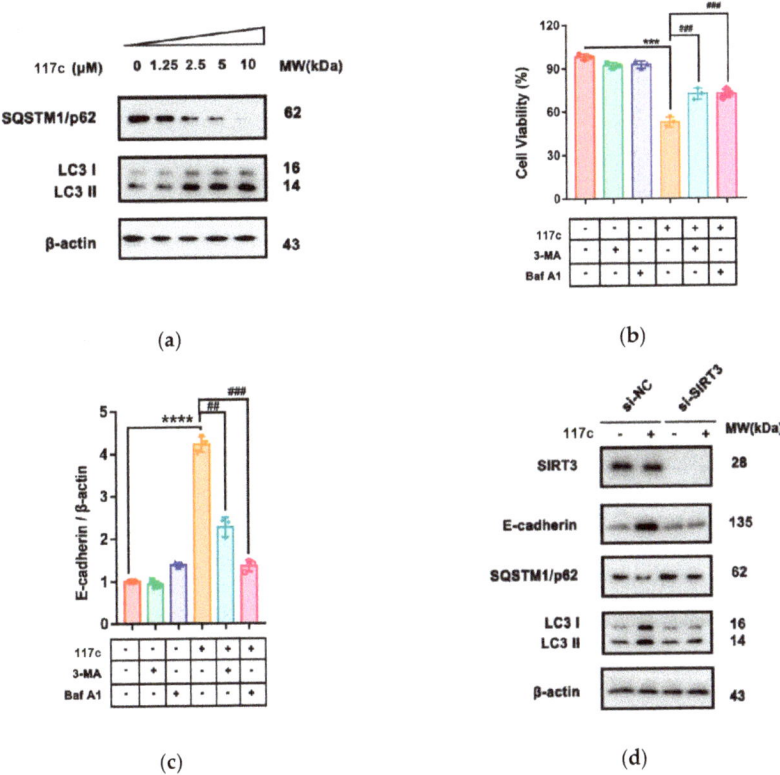

Figure 41. Effect of compound **117c** on autophagy-induced cell death in MDA-MB-231 cells by activation of SIRT3. (**a**) Western blot analysis of MDA-MB-231 cells treated with **117c**, showing expression levels of p62 and LC3. (**b**) MTT assay evaluating cellular capacity with **117c** in the presence or absence of autophagy inhibitors. (**c**) Cell migration study with **117c** in the presence or absence of autophagy inhibitors. (**d**) Western blot analysis of MDA-MB-231 cells treated with **117c**, measuring expression levels of SIRT3, E-cadherin, MMP-2, and LC3. NS means no significance, *** $p < 0.001$, **** $p < 0.0001$, ## $p < 0.01$, and ### $p < 0.001$. Reprinted (adapted) with permission from ref. [53]. Copyright American Chemical Society, 2023.

Finally, in vivo studies were carried out to evaluate the antitumor activity of compound **117c** in TNBC mouse xenograft models at three different concentrations (25, 50, and 100 mg/kg) [53]. After 16 days of treatment, compound **117c** exhibited significant antiproliferative activity in a dose-dependent manner. Additionally, the administration of **117c** resulted in a notable decrease in tumor growth, tumor volume, and tumor weight compared to the control group (Figure 42a,b). However, when evaluating the toxicity of **117c**, certain degrees of toxicity were observed in vivo at high concentrations, leading to pulmonary septum widening and other cellular abnormalities.

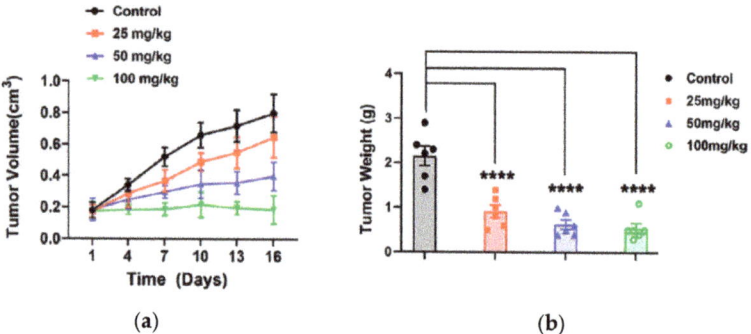

Figure 42. In vivo study conducted in a xenograft model with TNBC mice treated with compound **117c**. (**a**) Tumor volume and (**b**) tumor weight with compound **117c** treatment ($n = 10$ per group). Tumor volume data were collected every 3 days and presented as mean ± SD. Tumor weight data were collected on the final day and presented as mean ± SD. **** $p < 0.001$ compared with control group. Reprinted (adapted) with permission from ref. [53]. Copyright American Chemical Society, 2023.

In addition to the previous findings, complementary in vivo studies were conducted to assess the ability of compound **117c** to activate SIRT3 and induce autophagy in a mouse xenograft model using MDA-MB-231 TNBC cells [53]. The research team measured acetyl-lysine (Ac. K) acetylation levels, specifically AcK68-MnSOD2 and AcK122-MnSOD2, and examined the expression levels of SQSTM1/p62 and LC3 (Figure 43). The results provided compelling evidence that the acetylation levels of K68-MnSOD2 and K122-MnSOD2 precisely matched the deacetylation sites of SIRT3. Moreover, a noticeable decrease in the expression of p62 was observed, while LC3-II levels were significantly up-regulated compared to the positive control (β-actin). These findings convincingly demonstrate the activation of autophagy by compound **117c**. In conclusion, these promising in vivo results underscore the potential of compound **117c**, in combination with SIRT3, as a promising alternative for the treatment of TNBC-type cancer.

In their pursuit of employing eco-friendly synthetic strategies and achieving favorable yields, Irfan et al. conducted the synthesis of benzo[*b*]furan-based oxadiazole/triazole derivatives **120a–g** and **121a–h** using ultrasound and microwave irradiation, respectively (Table 28) [54]. In method A, benzo[*b*]furan–oxadiazole derivatives **120a–g** were obtained in 60–90% yields through *S*-nucleophilic substitution of 5-(benzofuran-2-yl)-1,3,4-oxadiazole-2-thiol **118a** with bromoacetanilides **119a–g** utilizing pyridine in acetonitrile under ultrasound irradiation at 40 °C for 30 min. Similarly, method B was employed to synthesize benzo[*b*]furan–triazole derivatives **121a–h** in 68–96% yields through a *S*-nucleophilic substitution of 5-(benzofuran-2-yl)-4-phenyl-4*H*-1,2,4-triazole-3-thiol **118b** with bromoacetanilides **119a–h**, utilizing pyridine in DMF under microwave irradiation for 60–70 s. In summary, the microwave-assisted method B led to better yields and shorter reaction times for benzo[*b*]furan derivatives than the ultrasound-assisted method A.

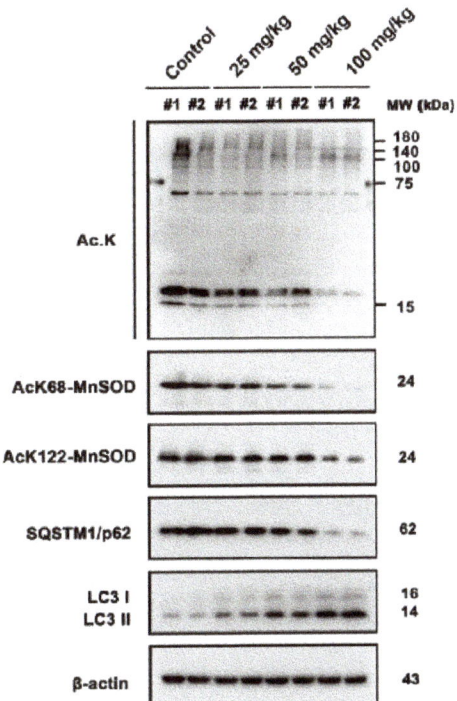

Figure 43. Western blot analysis of the expression levels of Ac. K, AcK68-MnSOD2, AcK122-MnSOD2, SQSTM1/p62, and LC3 in the MDA-MB-231 TNBC mouse xenograft model. β-Actin was used as the positive control. Reprinted (adapted) with permission from ref. [53]. Copyright American Chemical Society, 2023.

Table 28. Synthesis of benzo[*b*]furan-based oxadiazole/triazole derivatives **120a–g** and **121a–h** under ultrasound/microwave irradiation.

Compound	R¹	Yield (%)	
		Method A	Method B
120a	NH-Ph	70	75
120b	morpholine	66	70
120c	NH-(2-Cl-C₆H₄)	68	75
120d	NH-(2-OMe-C₆H₄)	86	94

Table 28. Cont.

Compound	R¹	Yield (%)	
		Method A	Method B
120e	4-EtO-phenyl-NH–	73	82
120f	2,6-dimethyl-phenyl-NH–	60	69
120g	diethylamino	66	74
121a	phenyl-NH–	80	90
121b	morpholinyl	64	73
121c	2-Cl-phenyl-NH–	60	68
121d	2-F-phenyl-NH–	63	74
121e	4-F-phenyl-NH–	69	77
121f	2,4-dimethyl-phenyl-NH–	61	70
121g	4-Cl-phenyl-NH–	77	89
121h	3,4-dichloro-phenyl-NH–	90	96

(a) Reagents and conditions: (i) method A: pyridine, CH3CN, 30 min, 40 °C, USI and (ii) method B: pyridine, DMF, 60–70 s, MWI.

The hemolytic, thrombolytic, and anticancer activities of previously synthesized benzo[*b*]furan-based oxadiazole/triazole derivatives were assessed (Table 29) [54]. Among these compounds, **121b** (X = NPh, R¹ = N-morpholinyl) exhibited the lowest cytotoxicity (0.1%), while **121g** (X = NPh, R¹ = 4-chloro-N-anilinyl) and **120b** (X = O, R¹ = N-morpholinyl) showed the highest toxicity (23.4% and 22.12%, respectively) compared to ABTS (95.9%). In the thrombolysis assay, the majority of compounds demonstrated moderate activity when compared to the positive control (ABTS). Notably, compound **121f** (X = NPh, R¹ = 2,4-dimethyl-N-anilinyl) exhibited the highest thrombolytic potential with a value of 61.4% compared to the positive control ABTS (86%).

Table 29. Hemolysis, thrombolysis, and antiproliferative studies of benzo[b]furan-based oxadiazole/triazole derivatives **120a–g** and **121a–h**.

Compound	% Hemolysis [a]	% Thrombolysis [a]	% Cell Viability [a] A549	IC$_{50}$ (µM) [a] A549
120a	3.7 ± 0.008	56.8 ± 0.081	64.1 ± 1.72	ND
120b	22.12 ± 0.008	50.7 ± 0.081	45.99 ± 4.22	ND
120c	1.3 ± 0.008	52.8 ± 0.081	43.7 ± 0.94	ND
120d	5.02 ± 0.008	53.5 ± 0.081	27.49 ± 1.90	6.3 ± 0.7
120e	0.5 ± 0.008	52.4 ± 0.081	34.47 ± 2.19	17.9 ± 0.46
120f	0.74 ± 0.008	56.5 ± 0.081	43.67 ± 4.43	ND
120g	4.86 ± 0.047	48.3 ± 0.081	41.45 ± 4.10	ND
121a	9.6 ± 0.081	52.2 ± 0.081	49.8 ± 1.06	ND
121b	0.1 ± 0.004	52.5 ± 0.081	57.62 ± 4.94	ND
121c	2.15 ± 0.008	54 ± 0.081	44.52 ± 5.01	ND
121d	6.13 ± 0.047	56.2 ± 0.081	39.12 ± 2.21	ND
121e	3.11 ± 0.008	59.1 ± 0.008	44.72 ± 0.84	ND
121f	15.7 ± 0.081	61.4 ± 0.081	99.1 ± 5.04	ND
121g	23.4 ± 0.081	49.06 ± 0.047	36.26 ± 0.41	19.8 ± 0.54
121h	14.8 ± 0.081	48.1 ± 0.081	29.29 ± 3.98	10.9 ± 0.94
ABTS	95.9	86	ND	ND
DMSO	ND	ND	100 ± 0	ND
Crizotinib [b]	ND	ND	28.22 ± 3.88	8.54 ± 0.84
Cisplatin [b]	ND	ND	15.34 ± 2.98	3.88 ± 0.76

[a] The data indicate the mean ± SD of at least three independent experiments. [b] The standard drugs for the study. ND means not determined.

The IC$_{50}$ value and cell viability percentage against the A549 lung cancer cell line were determined using an MTT assay, with crizotinib and cisplatin as the standard drugs (Table 29). In summary, compound **120d** (X = O, R^1 = 2-methoxy-N-anilinyl) exhibited the highest potency with a cell viability of 27.49% and an IC$_{50}$ value of 6.3 µM, demonstrating greater activity than crizotinib (28.22% and 8.54 µM, respectively) and lower cytotoxicity than cisplatin (15.34% and 3.88 µM, respectively). In addition, compound **121h** (X = NPh, R^1 = 2,4-dichloro-N-anilinyl) showed slightly lower activity with a cell viability of 29.29% and an IC$_{50}$ value of 10.9 µM. Other oxadiazole/triazole derivatives, such as **120b–c**, **120e–g**, **121a**, **121c–e**, and **121g**, exhibited moderate anticancer activity, with cell viability ranging from 34.47% to 49.8%. Although compound **121f** (X = NPh, R^1 = 2,4-dimethyl-N-anilinyl) had the highest cell viability (99.1%), it did not demonstrate any activity against the A549 cell line.

A molecular docking simulation was performed to investigate the interactions of compound **120d** with anaplastic lymphoma kinase (ALK) in conjunction with crizotinib (PDB ID code: 2XP2) [54]. In summary, crizotinib displayed direct contact with ALK residues in the active site, while compound **120d** exhibited even more effective binding to these ALK residues (Figure 44a,b). Notably, the phenyl and heterocyclic rings of **120d** engaged in π-sigma interactions with Leu1256 and Val1130, while the NH group formed hydrogen bonds with Gly1201 and π-anion bonds with Glu1210.

Isatin, also known as 1H-indole-2,3-dione, is an important N-heterocycle in medicinal chemistry and drug discovery [55]. In their study, Mohammed et al. synthesized an isatin–benzofuran hybrid **126** and investigated its antiproliferative activity against HT29 and SW620 cancer cell lines, along with its impact on tumor and metastatic development involved in primary cellular processes [56]. Scheme 20 shows the three-step synthesis of the isatin–benzofuran hybrid **126**. Firstly, 2-hydroxyacetophenone **122** underwent cyclization with ethyl bromoacetate utilizing K$_2$CO$_3$ as a base in refluxing acetonitrile for 8 h to afford ethyl 3-methylbenzofuran-2-carboxylate **123**, which then reacted with hydrazine hydrate in refluxing methanol for 4 h to give 3-methylbenzofuran-2-carbohydrazide **124**. Finally, a condensation reaction between **124** and isatin **125** catalyzed by acetic acid in refluxing ethanol for 5 h afforded isatin–benzofuran hybrid **126** with an 80% yield.

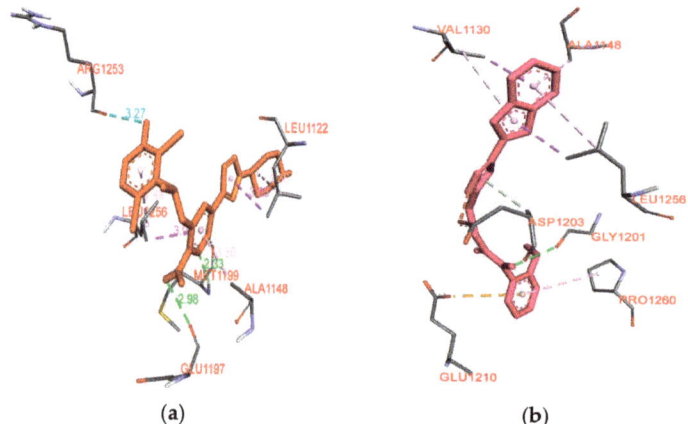

Figure 44. (**a**) Molecular docking of crizotinib in the ALK protein binding pocket (PDB ID code: 2XP2), and (**b**) 3D model of compound **120d** with its interacting amino acid residues. This is an open-access article distributed under the terms of the Creative Commons CC BY license [54].

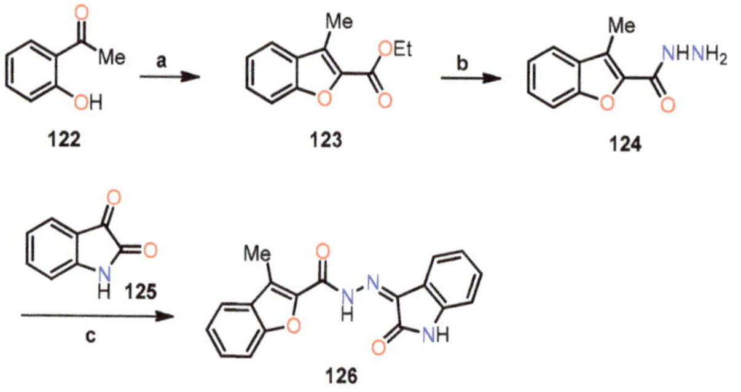

Scheme 20. Synthesis of benzofuran–isatin hybrid **126**, and reagents and conditions: (**a**) ethyl 2-bromoacetate, CH_3CN, K_2CO_3, reflux, 8 h; (**b**) hydrazine hydrate, MeOH, reflux, 4 h; (**c**) ethanol, AcOH, reflux, 5 h, 80%.

Next, comprehensive biological tests and analyses were carried out, specifically targeting cell viability, real-time migration, invasion studies, cell cycle assays related to apoptosis, and cytotoxicity evaluations [56]. In the initial phase, a cell viability, migration, and invasion assays were carried out using an xCELLigence Automated Dual Layer Real-time Cell Analyzer (RTCA-DP) with various concentrations of isatin–benzofuran hybrid **126**. The results revealed a noteworthy dose-dependent reduction in cell proliferation, migration, and invasion in both cancer cell lines compared to untreated cells (Figure 45a–c). Moreover, notable variations were observed in the inhibitory effects of hybrid **126** on the proliferation and migration of the SW620 line compared to the HT29 line, while in invasion, the HT29 line exhibited a more pronounced inhibitory effect. In addition, the impact of hybrid **126** on tumor suppression, based on the p53 protein, was evaluated. The results indicated a significant increase in p53 expression levels, 2.46-fold in HT29 cells, and 4.81-fold in SW260 cells at a concentration of 10 µM, demonstrating the potent inhibitory effect on tumor cell proliferation when utilizing hybrid **126** (Figure 45d).

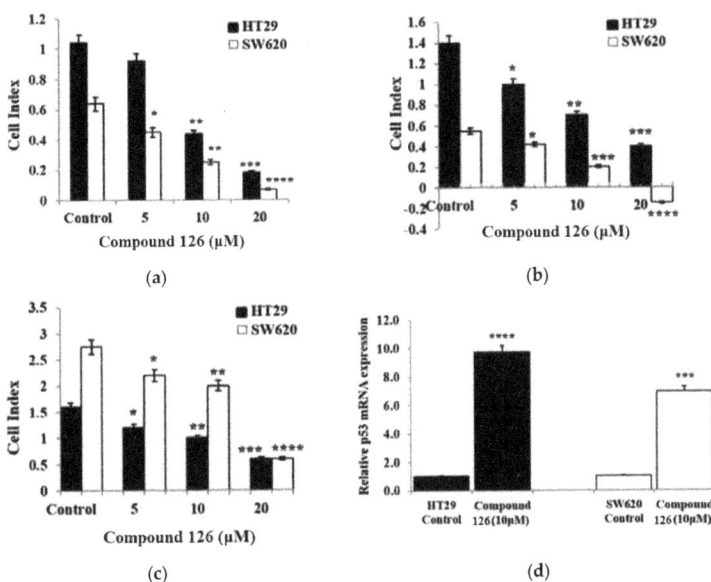

Figure 45. Inhibition of benzofuran–isatin hybrid **126** on (**a**) cell proliferation, (**b**) migration, (**c**) invasion in the HT29 and SW620 cancer cell lines, and (**d**) its influence on gene expression in the p53 protein. Data are shown as mean ± SD (n = 3). The data were considered significant when reporting * $p < 0.05$, ** $p < 0.01$, *** $p < 0.001$, and **** $p < 0.0001$ vs. control. This is an open-access article distributed under the terms of the Creative Commons CC BY license [56].

Apoptotic studies were performed to evaluate the pro-apoptotic effects of hybrid **126**, involving its role in suppressing the expression levels of mitochondrial proteins Bcl-x, Bax, and cytochrome C utilizing flow cytometry [56]. The results showed a significant suppression of the Bcl-x protein expression by 50% in the HT29 cells, and by 75% and 90% in the SW620 cells at concentrations of 5 and 10 μM, respectively, in comparison to untreated cells (Figure 46a,b). On the other hand, a significant increase in the expression of Bax and cytochrome C was observed in both HT29 and SW620 cell lines, showing an approximately two-fold increase compared to the basal expression of untreated cells (Figure 46a,b). Moreover, the down-regulation of Bcl-x and the up-regulation of Bax and cytochrome C in both cancer cell lines exposed to hybrid **126** were further confirmed in gene expression levels, when compared to untreated cells. These findings support the apoptotic effects of hybrid **126** on cell cycle disruption in both HT29 and SW620 cell lines (Figure 46c,d).

In this particular study, they assessed the cytotoxic effect of hybrid **126** both alone and in combination with three anticancer drugs: irinotecan (IRI), 5-fluorouracil (5-FU), and oxaliplatin (OXA) in the HT29 and SW620 cancer cell lines at different concentrations (Figure 47a,b) [56]. Notably, when combining hybrid **126** with IRI in the HT29 cell line (Figure 47a), there was a significant inhibition of cell proliferation at 10 μM (−75% vs. −50%) and 20 μM (−90% vs. −65%) compared to the single treatment with IRI alone. When combining 5-FU with hybrid **126**, cell proliferation was significantly inhibited at 5 μM (−50% vs. −20%) and 10 μM (−67% vs. −55%) compared to the single drug treatment. Similarly, OXA inhibited cell proliferation at 5 μM (−55% vs. −25%) and 10 μM (−75% vs. −45%) when used in conjunction with hybrid **126**. On the other hand, in the SW620 cell line (Figure 47b), the combined treatment of IRI with **126** resulted in a substantial inhibition of cell proliferation at 5 μM (−45% vs. −15%) and at 10 μM (−75% vs. −45%). Similarly, 5-FU combined with **126** showed comparable inhibition of cell proliferation to IRI at 5 μM (−55% vs. −15%) and at 10 μM (−70% vs. −50%). Moreover, when utilizing the treatment in conjunction with OXA, a more pronounced inhibition of cell proliferation was observed at 5 µM (−55% vs. −15%), 10 µM (−65% vs. −50%), and 20 µM (−90% vs. −75%) compared to the single drug treatment. These findings demonstrate a significant enhancement in the effectiveness of the anticancer drugs employed in this study when combined with hybrid **126**.

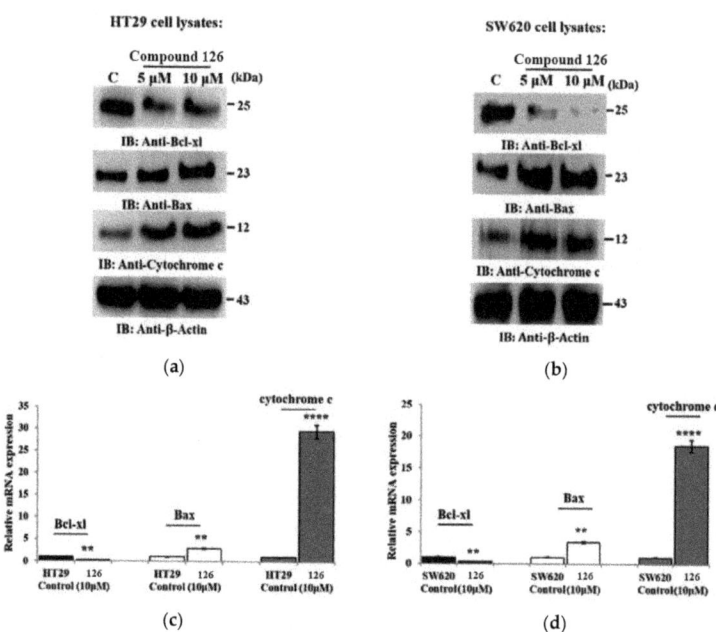

Figure 46. Flow cytometry analysis of hybrid **126** against cell lines (**a**) HT29 and (**b**) SW620 at Bcl-x, Bax, and cytochrome C expression levels with β-actin as the control. Additionally, bar graphs show the gene expression of these mitochondrial proteins for (**c**) HT29 and (**d**) SW620. Data are shown as mean ± SD (n = 3). ** $p < 0.01$ and **** $p < 0.0001$ vs. control. This is an open-access article distributed under the terms of the Creative Commons CC BY license [56].

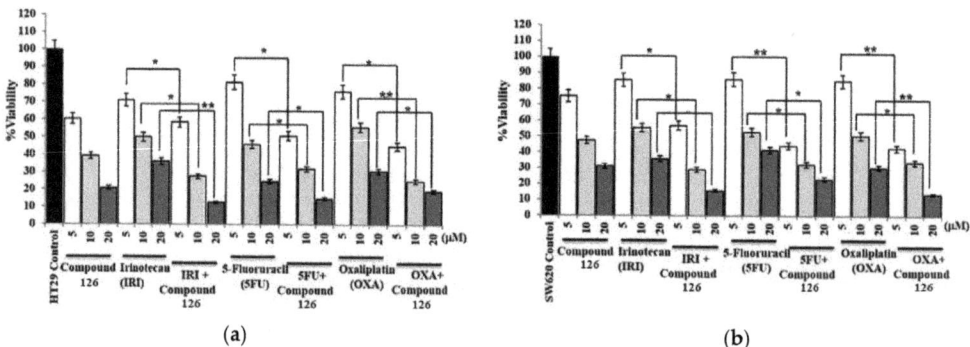

Figure 47. Effects of hybrid **126** on the cellular capacity of (**a**) HT29 and (**b**) SW620 cell lines in combination with three chemotherapeutic drugs: irinotecan (IRI), 5-fluorouracil (5-FU), and oxaliplatin (OXA) at concentrations of 5, 10, and 20 µM. Data are expressed as mean ± SD (n = 3). * $p < 0.001$ and ** $p < 0.0001$ vs. control. This is an open-access article distributed under the terms of the Creative Commons CC BY license [56].

2.2. Antibacterial Activity

Antibiotic resistance presents a worldwide concern, steadily escalating in severity, and posing substantial challenges to healthcare systems globally. Novel approaches are urgently needed to address this critical problem. Recently, benzo[b]furans and their derivatives have shown remarkable inhibitory potential against various Gram-positive bacteria, including *Staphylococcus aureus* (*S. aureus*), *Bacillus subtilis* (*B. subtilis*), and *Enterococcus* spp. (*E.* spp.), as well as Gram-negative bacteria, such as *Pseudomonas syringae* (*P. syringae*), *Klebsiella pneumoniae* (*K. pneumoniae*), *Salmonella typhi* (*S. typhi*), *Pseudomonas aeruginosa* (*P. aeruginosa*), and *Escherichia coli* (*E. coli*) [57]. In the province of Lampang, Thailand, a noteworthy medicinal discovery was made from the root extracts of *Stemona aphylla*, resulting in the isolation of alkaloids **127a–i** (Table 30) [58]. The extraction process involved drying 15.14 kg of ground root material, which was then extracted with 95% ethanol for 16 days at room temperature. After evaporating the extract, a portion of the residue was partitioned between 50% aqueous methanol and dichloromethane to give 8.86 g of the dichloromethane extract. The alkaloids were further isolated using column chromatography or preparative thin-layer chromatography through successive separations. The identified alkaloids and their corresponding masses are listed in Table 30.

Table 30. Isolation of alkaloids **127a–i** and their corresponding masses.

Compound	R^1	R^2	R^3	R^4	R^5	R^6	Mass (mg)
127a	H	H	Me	H	Me	H	13.2
127b	H	H	Me	H	Me	Me	1.8
127c	H	H	Me	Me	Me	H	33.9
127d	H	OMe	Me	H	H	H	6.4
127e	H	OMe	H	Me	H	H	0.9
127f	H	H	Me	Me	Me	Me	0.6
127g	H	OMe	Me	Me	H	H	5.8
127h	Me	OMe	Me	H	H	H	2.0
127i	H	OMe	Me	Me	Me	H	12.8

Antimicrobial studies were conducted on alkaloids **127a**, **127c–d**, **127g**, and **127i**, evaluating their MIC values against two Gram-negative bacteria (*Escherichia coli* and *Klebsiella pneumoniae*) and three Gram-positive bacteria (*Staphylococcus aureus*, methicillin-resistant *Staphylococcus aureus* (MRSA), and *Streptococcus pyogenes*), with gentamicin used as the standard drug (Table 31) [58]. The results indicated minimal activity against Gram-negative bacteria for all tested compounds. However, alkaloids **127a** (R^1 = H, R^2 = H, R^3 = Me, R^4 = H, R^5 = Me, R^6 = H), **127c** (R^1 = H, R^2 = H, R^3 = Me, R^4 = Me, R^5 = Me, R^6 = H), and **127i** (R^1 = H, R^2 = OMe, R^3 = Me, R^4 = Me, R^5 = Me, R^6 = H) showed significant activity against *MRSA*, exhibiting MIC values of 15.6 µg/mL, outperforming the control antibiotic (MIC = 45.0 µg/mL). Compounds **127d**, **127g**, and **127i** showed moderate activity against *S. aureus*, with MIC values of 31.3 µg/mL, in comparison to gentamicin (MIC = 22.5 µg/mL).

Table 31. MIC values (μg/mL) of some alkaloids of **127**.

Compound	Gram-Negative Bacteria		Gram-Positive Bacteria		
	E. coli	K. pneumoniae	S. aureus	MRSA	S. pyogenes
127a	62.5	62.5	62.5	15.6	ND
127c	62.5	62.5	62.5	15.6	62.5
127d	125.0	62.5	31.3	31.3	62.5
127g	62.5	125.0	31.3	62.5	ND
127i	62.5	62.5	31.3	15.6	ND
Gentamicin [a]	11.3	11.3	22.5	45.0	5.6

[a] The standard drug for the study. ND means not determined.

Next, Ashok et al. synthesized a series of *E*-(1)-(6-benzoyl-3,5-dimethylfuro [3′,2′:4,5]benzo[*b*]furan-2-yl)-3-(aryl)-2-propen-1-ones **130a–g** using both conventional and microwave heating protocols (Table 32) [59]. Firstly, they obtained 2-acetyl-3,5-dimethyl-6-benzoylbenzodifuran **129** by cyclizing 5-acetyl-2-benzoyl-6-hydroxy-3-methylbenzo[*b*]furan **128** with 2-chloroacetone employing K_2CO_3 as a base in refluxing acetone for 8 h (Method A), and microwave heating at 120 °C for 4 min under solvent-free conditions (Method B). Subsequently, *bis*-chalcones **130a–g** were synthesized through a Claisen–Schmidt condensation reaction of **129** with (hetero)aromatic aldehydes using NaOH as a base in refluxing ethanol for 6–8 h (Method A, 53–68%) and microwave-assisted aldol condensation at 90 °C for 4–5 min under solvent-free conditions (Method B, 87–94%). In summary, method B led to *bis*-chalcones in higher yields and shorter reaction times compared to method A, which utilized conventional heating.

Table 32. Synthesis of *bis*-chalcones **130a–g** via Claisen–Schmidt condensation using both conventional and microwave heating methods.

Compound	Ar	Method A		Method B	
		Time (h)	Yield (%)	Time (min)	Yield (%)
130a	Ph	6	68	5	92
130b	2-ClPh	7	64	5	90
130c	4-MeOPh	7	65	4	94
130d	2-Furyl	6	64	4	89
130e	α-Napthyl	8	66	5	92
130f	1,3-diPhenyl-1*H*-pyrazol-4-yl	8	56	5	87
130g	1-Phenyl-3-(4-bromophenyl)-1*H*-pyrazol-4-yl	8	53	5	93

(*i*) Method A: (**a**) K_2CO_3, $ClCH_2COCH_3$, CH_3COCH_3, reflux, 8 h; (**b**) NaOH, ArCHO, EtOH, reflux, 6–8 h, 53–68%, and (*ii*) Method B: (**a**) K_2CO_3, $ClCH_2COCH_3$, 120 °C, 4 min, MWI; (**b**) NaOH, ArCHO, 90 °C, 4–5 min, MWI, 87–94%.

The antimicrobial activities of *bis*-chalcones **130a–g** were evaluated using the plate count method with nutrient agar as the culture medium. The bacterial strains tested included two Gram-negative bacteria, *Escherichia coli* and *Pseudomonas aeruginosa*, and two Gram-positive bacteria, *Bacillus subtilis* and *Staphylococcus aureus*, using chloramphenicol, carbenicillin, streptomycin, and tetracycline as the standard drugs [59]. The inhibition zones (in mm) were measured after 24 h of incubation at 37 °C (Table 33). In summary, *bis*-chalcones **130a–g** showed inhibition zones ranging from 6 to 11 mm and 7 to 17 mm against Gram-negative and Gram-positive bacterial strains, respectively. In particular, compounds **130a** (Ar = Ph), **130b** (Ar = 2-ClPh), **130f** (Ar = 1,3-diphenyl-1*H*-pyrazol-4-yl), and **130g** (Ar = 1-phenyl-3-(4-bromophenyl)-1*H*-pyrazol-4-yl) showed good activity against all bacterial strains, with inhibition zones in a range of 9 to 17 mm. However, none of these *bis*-chalcones demonstrated higher activity than the control drugs for each bacterial strain (13–22 mm).

Table 33. Inhibition zones (in mm) of *bis*-chalcones 130a–g.

Compound	Gram-Negative Bacteria		Gram-Positive Bacteria	
	E. coli	P. aeruginosa	B. subtilis	S. aureus
130a	9	10	15	10
130b	10	9	16	9
130c	7	7	10	8
130d	8	6	10	8
130e	7	8	9	7
130f	10	9	16	10
130g	11	10	17	10
Chloramphenicol [a]	13	ND	ND	ND
Carbenicillin [a]	ND	13	ND	ND
Streptomycin [a]	ND	ND	22	ND
Tetracycline [a]	ND	ND	ND	15

[a] The standard drugs for the study. ND means not determined.

In a separate study, Ostrowska et al. utilized microwave irradiation to synthesize a collection of *O*-alkylamino benzo[*b*]furancarboxylates with good yields [60,61]. Firstly, the esterification reaction of 6-acetyl-5-hydroxy-2-methyl-3-benzo[*b*]furancarboxylic acid 131a with methanol catalyzed by sulfuric acid afforded benzo[*b*]furancarboxylate 132a (Scheme 21). Subsequently, microwave-assisted *O*-alkylation reaction of 132a with 2-chloroethyl-*N*,*N*-diethylamine in the presence of K_2CO_3 and Aliquat 336 in acetone gave *O*-alkylated benzofuran-3-carboxylate 133a. In an alternative approach, compounds 134a and 135a were synthesized under analogous conditions to the previous *O*-alkylation protocol, using carboxylic acid 131a instead of ester 132a. Notably, compounds 133b–g and 134b–g were obtained using the corresponding benzo[*b*]furancarboxylic acids 131b–g as the initial reactants (Table 34). Finally, the benzo[*b*]furancarboxylate derivatives 133a–d, 133f, 134a–d, 134f, and 135a were transformed into their respective hydrochloride salts.

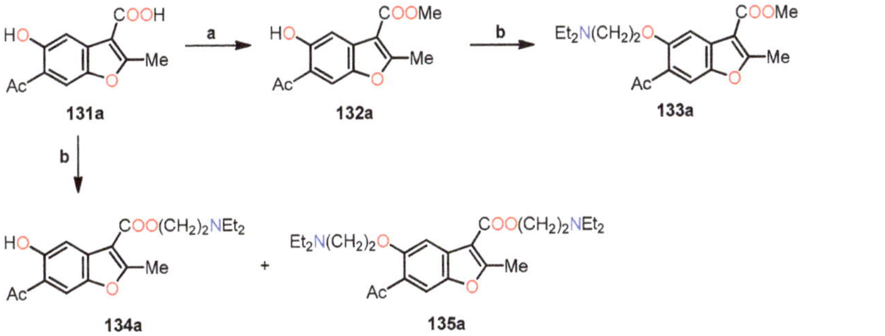

Scheme 21. Synthesis of *O*-alkylamino benzo[*b*]furancarboxylates 132–135, and reagents and conditions: (**a**) MeOH, H_2SO_4; (**b**) Cl(CH$_2$)$_2$NEt$_2$·HCl, K_2CO_3, Aliquat 336 (*N*-methyl-*N*,*N*-dioctyloctan-1-ammonium chloride), acetone, MWI (4–8 cycles: heating 6 min and cooling 2 min).

Table 34. Structures of benzo[b]furancarboxylic acids 131b–g, and benzo[b]furancarboxylates 133b–d, 133f, 134c–d, and 134f–g.

Compound	R^1	R^2	R^3	R^4	R^5
131b	COOH	Me	H	OH	Ac
131c	COOH	Me	OMe	OH	Ac
131d	COOH	Me	H	OH	4-Methoxycinnamaldehyde
131e	COOH	Me	H	OMe	Ac
131f	COOH	H	Br	OMe	OH
131g	COOH	H	H	H	OMe
133b	COOMe	Me	H	$O(CH_2)_2NEt_2$	Ac
133c	COOMe	Me	OMe	$O(CH_2)_2NEt_2$	Ac
133d	COOMe	Me	H	$O(CH_2)_2NEt_2$	4-Methoxycinnamaldehyde
133f	COOMe	H	Br	OMe	$O(CH_2)_2NEt_2$
134c	$COO(CH_2)_2NEt_2$	Me	OMe	OH	Ac
134d	$COO(CH_2)_2NEt_2$	Me	H	OH	4-Methoxycinnamaldehyde
134f	$COO(CH_2)_2NEt_2$	H	Br	OMe	OMe
134g	$COO(CH_2)_2NEt_2$	H	H	H	OMe

The antimicrobial activity of hydrochlorides of benzo[b]furancarboxylates 133a–d, 133f, 134a–d, 134f–g, and 135a was screened against six Gram-positive bacterial strains, including *Micrococcus luteus*, *Bacillus cereus*, *Bacillus subtilis*, *Staphylococcus epidermidis*, *Staphylococcus aureus*, and *Enterococcus hirae*, along with two Gram-negative bacterial strains, such as *Escherichia coli* and *Pseudomonas aeruginosa* (Table 35). The results unveiled that the O-alkyl-benzo[b]furancarboxylate 133f.HCl (R^1 = COOMe, R^2 = H, R^3 = Br, R^4 = OMe, R^5 = $O(CH_2)_2NEt_2$) showed the highest potency, with MIC values ranging from 0.003 to 0.012 µmol/cm^3 against all Gram-positive bacterial strains. Conversely, the hydroxy-benzo[b]furancarboxylate 134g.HCl (R^1 = COO(CH_2)$_2$NEt$_2$, R^2 = H, R^3 = H, R^4 = H, R^5 = OMe) showed the lowest potency, with an MIC value of 15.28 µmol/cm^3 in all Gram-positive bacterial strains. Additionally, it becomes evident that within the Gram-positive strains, compounds 135a.HCl and 133f.HCl exhibited the most remarkable activity against *E. coli* and *P. aeruginosa* with MIC values of 0.59 and 3.12 µmol/cm^3, respectively. Notably, the observed structure–activity relationships are as follows: (i) compound 133c.HCl exhibited higher activity than 133b.HCl due to the introduction of a methoxy group at the C–5 position, (ii) compound 133d.HCl displayed higher activity than 133b.HCl due to the presence of the 7-(4-methoxycinnamoyl) group, and (iii) 2-(N,N-diethylamino)ethyl esters 134a.HCl, 134c.HCl, and 134d.HCl exhibited superior activity against Gram-positive strains compared to compounds 133a.HCl, 133c.HCl, and 133d.HCl. These variations underscore the influential role of substituents in the benzo[b]furan moiety, as evidenced by their distinct activities against diverse bacterial strains.

Concurrently, Kenchappa et al. synthesized a series of (5-substituted-1-benzofuran-2-yl)(2,4-phenyl-substituted)methanones 139a–i by incorporating a pharmacophore group at the 2 position of the benzo[b]furan ring, a response to the consistent trend of enhanced antimicrobial activity observed across various studies [62]. Initially, the synthesis commenced with the cyclization reaction of salicylaldehyde derivatives 136a–c with α-bromoacetophenones 137a–c utilizing potassium carbonate as a base in refluxing acetonitrile to afford benzo[b]furan derivatives 138a–i (Scheme 22). Subsequently, the Knoevenage condensation of compounds 138a–i with Meldrum's acid catalyzed by acetic acid at temperatures of 110–115 °C for a duration of 8–10 h resulted in the formation of (5-substituted-1-benzofuran-2-yl)(2,4-phenyl)methanones 139a–i in 75–91% yields. It is important to emphasize that the presence of acetic acid facilitated the generation of a carbanion in Meldrum's acid, thereby enhancing the nucleophilic addition and subsequent dehydration processes.

Table 35. Antimicrobial activity of hydrochlorides of methyl benzo[b]furancarboxylates **133**, **134**, and **135**.

Compound	MIC (µmol/cm^3)							
	M. luteus	B. cereus	B. subtilis	S. epidermidis	S. aureus	E. hirae	E. coli	P. aeruginosa
133a.HCl	0.05	0.05	0.05	0.05	0.05	0.39	6.51	13.02
133b.HCl	0.75	1.49	1.49	1.49	3.11	3.11	12.44	ND
133c.HCl	0.05	0.36	0.18	0.18	0.18	0.36	6.04	ND
133d.HCl	0.04	0.30	0.04	0.04	0.15	0.60	ND	ND
133f.HCl	0.003	0.012	0.012	0.012	0.012	0.012	1.50	3.12
134a.HCl	0.01	0.10	0.01	ND	0.41	ND	ND	ND
134c.HCl	0.01	0.05	0.09	0.19	0.09	0.75	ND	ND
134d.HCl	0.01	0.04	0.04	0.01	0.04	0.08	ND	ND
134f.HCl	0.09	0.71	0.35	0.09	0.18	ND	ND	ND
134g.HCl	15.28	15.28	15.28	15.28	15.28	>30.56	>30.56	30.56
135a.HCl	0.04	0.04	0.04	0.04	0.07	0.30	0.59	ND

ND means not determined.

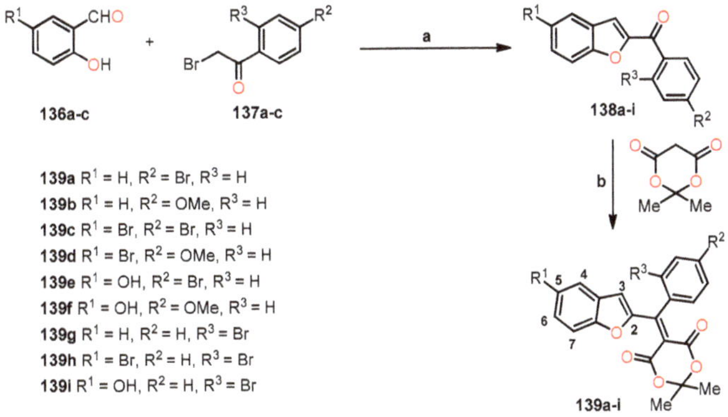

139a R^1 = H, R^2 = Br, R^3 = H
139b R^1 = H, R^2 = OMe, R^3 = H
139c R^1 = Br, R^2 = Br, R^3 = H
139d R^1 = Br, R^2 = OMe, R^3 = H
139e R^1 = OH, R^2 = Br, R^3 = H
139f R^1 = OH, R^2 = OMe, R^3 = H
139g R^1 = H, R^2 = H, R^3 = Br
139h R^1 = Br, R^2 = H, R^3 = Br
139i R^1 = OH, R^2 = H, R^3 = Br

Scheme 22. Synthesis of benzo[b]furan derivatives **139a–i**, and reagents and conditions: (**a**) K$_2$CO$_3$, CH$_3$CN, reflux; (**b**) AcOH, reflux, 8–10 h, 110–115 °C, 75–91%.

The antimicrobial activity of benzo[b]furan derivatives **139a–i** was screened against one Gram-positive bacterial strain, including *Bacillus subtilis*, as well as four Gram-negative bacterial strains, including *Pseudomonas syringae*, *Salmonella typhi*, *Klebsiella pneumoniae*, and *Escherichia coli*, using the agar well diffusion method [62]. Streptomycin was employed as the standard reference. The minimum inhibitory concentration (MIC) studies were performed using a serial broth-dilution method at different concentrations, including 1, 10, 25, 50, and 100 mol/L [62]. Based on the findings presented in Table 36, compound **139c** (R^1 = Br, R^2 = Br, R^3 = H) showed the most potent activity against the Gram-positive strain with an inhibition zone of 13 mm, closely approximating the effectiveness of streptomycin (16 mm) at a concentration of 0.5 mg/mL. Interestingly, compound **139c** also emerged as the most effective against all Gram-negative strains with inhibition zones ranging from 10 to 14 mm, akin to the performance of streptomycin (13–17 mm), at a concentration of 0.5 mg/mL. As shown in Table 37, compounds **139c** (R^1 = Br, R^2 = Br, R^3 = H) and **139a** (R^1 = H, R^2 = Br, R^3 = H) demonstrated remarkable activity across all bacterial strains, displaying MIC values in ranges of 14.80–16.00 µg/mL and 15.50–16.50 µg/mL, aligning closely with the efficacy of streptomycin (MIC = 14.8–16.0 µg/mL). In contrast, compounds **134b** (R^1 = H, R^2 = OMe, R^3 = H) and **139f** (R^1 = OH, R^2 = OMe, R^3 = H) showed reduced activity, possibly attributed to the presence of electron-donating groups at the C–5 position of the benzo[b]furan ring. Furthermore, the introduction of a bromine group at C–4 of the benzo[b]furan ring did not increase the activity of compound **139i** (R^1 = OH, R^2 = H, R^3 = Br) (Table 37).

Table 36. Zone of inhibition of benzo[b]furan derivatives **139a–i**.

Compound	Concentration (mg/mL)	Zone of Inhibition (mm) [a]				
		P. syringae	S. typhi	B. subtilis	K. pneumoniae	E. coli
139a	1.0	9 ± 0.3	10 ± 0.2	12 ± 0.1	13 ± 0.1	10 ± 0.3
	0.5	7 ± 0.1	9 ± 0.3	10 ± 0.3	11 ± 0.1	8 ± 0.1
139b	1.0	7 ± 0.2	6 ± 0.1	7 ± 0.2	7 ± 0.3	6 ± 0.2
	0.5	6 ± 0.3	5 ± 0.2	5 ± 0.2	5 ± 0.1	4 ± 0.2
139c	1.0	12 ± 0.2	12 ± 0.1	15 ± 0.2	16 ± 0.3	13 ± 0.2
	0.5	10 ± 0.3	10 ± 0.2	13 ± 0.1	14 ± 0.1	11 ± 0.3
139d	1.0	10 ± 0.2	10 ± 0.1	10 ± 0.2	11 ± 0.3	10 ± 0.2
	0.5	8 ± 0.3	7 ± 0.2	8 ± 0.1	10 ± 0.1	8 ± 0.3
139e	1.0	11 ± 0.2	11 ± 0.1	12 ± 0.2	13 ± 0.3	11 ± 0.2
	0.5	8 ± 0.3	8 ± 0.2	10 ± 0.1	12 ± 0.1	9 ± 0.3
139f	1.0	9 ± 0.2	8 ± 0.1	9 ± 0.2	8 ± 0.3	7 ± 0.2
	0.5	6 ± 0.3	6 ± 0.1	6 ± 0.2	ND	6 ± 0.3
139g	1.0	10 ± 0.2	9 ± 0.1	10 ± 0.2	9 ± 0.3	9 ± 0.2
	0.5	8 ± 0.3	8 ± 0.2	7 ± 0.1	9 ± 0.1	7 ± 0.3
139h	1.0	11 ± 0.2	12 ± 0.1	13 ± 0.2	14 ± 0.3	12 ± 0.2
	0.5	8 ± 0.3	9 ± 0.2	11 ± 0.1	13 ± 0.1	9 ± 0.3
139i	1.0	8 ± 0.2	7 ± 0.1	8 ± 0.2	8 ± 0.3	6 ± 0.2
	0.5	7 ± 0.3	6 ± 0.2	6 ± 0.2	5 ± 0.1	4 ± 0.3
Streptomycin [b]	1.0	15 ± 0.2	16 ± 0.1	18 ± 0.2	19 ± 0.3	16 ± 0.2
	0.5	13 ± 0.3	15 ± 0.2	16 ± 0.1	17 ± 0.1	14 ± 0.3

[a] The data indicate the mean ± SD of at least three independent experiments. [b] The standard drug for the study. ND means not determined.

Table 37. Minimum inhibitory concentration (MIC) of benzo[b]furan derivatives **139a–i**.

Compound	MIC (µg/mL)				
	P. syringae	S. typhi	B. subtilis	K. pneumoniae	E. coli
139a	15.50	15.80	16.50	16.50	16.00
139b	120.35	125.40	195.50	125.55	185.30
139c	14.80	15.25	15.75	16.00	15.75
139d	60.25	55.75	ND	32.85	38.65
139e	30.25	25.25	20.25	ND	40.25
139f	ND	70.35	85.25	75.35	85.35
139g	105.35	110.35	95.25	155.35	185.35
139h	16.25	16.75	17.50	17.25	17.50
139i	125.35	115.35	195.25	120.35	175.35
Streptomycin [a]	14.50	14.50	15.50	15.25	15.25

[a] The standard drug for the study. ND means not determined.

Naftifine, a topical allylamine, exhibits effectiveness across an extensive spectrum of dermatophytic fungi, including *Trichophyton* and *Microsporum* spp., and has also shown significant efficacy against *Candida* and *Aspergillus* spp. [63]. In 2016, Wang et al. undertook the synthesis of naphthalene hydrochloride (NFT) derivatives, previously recognized as potent inhibitors of the diapophytoene desaturase (CrtN) enzyme, which is a crucial molecular target against infections caused by pigmented *Staphylococcus aureus* [64]. The process of molecular design comprised several sequential stages. It commenced with an analysis of the naphthalene moiety of NFT, which served as a potential pharmacophore group. Subsequently, modifications were introduced in the N-methyl group, involving various steric groups (region A). Concurrently, the synthesis of specific analogs was undertaken to explore the impact of different linker types within the allyl portion on inhibitory activity (region B). Finally, a meticulous design approach led to the synthesis of 21 analogs, each featuring distinct substituents (region C), as illustrated in Scheme 23. In pursuit of this goal, several syntheses were undertaken to generate the varied analogs portrayed in Scheme 24.

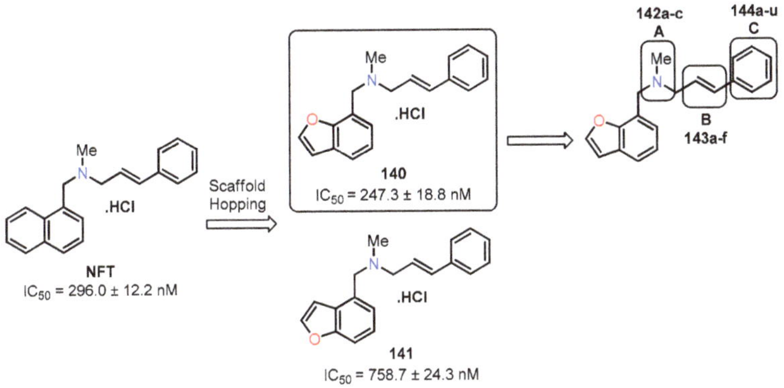

Scheme 23. Scaffold hopping from the active compound NTF, featuring a central nucleotide modification (**140–141**) along with three structural modifications (**142–144**).

Scheme 24 sowed synthetic procedures to synthesize a series of naftifine analogs—**140**, **142a–c**, **143a–b**, and **144a–t** [64]. The synthesis began with the nucleophilic substitution of 2-iodophenol **145** utilizing 1-bromo-2,2-diethoxyethane and NaH in DMF at 90 °C to give compound **146**, which was cyclized in the presence of polyphosphoric acids under refluxing toluene to yield 7-iodobenzofuran **147**. Further progression involved the substitution of the iodine atom in **147** with a cyano group in DMF at 130 °C for 4 h to furnish benzofuran-7-carbonitrile **148**. The subsequent reduction of the cyano group in **148** was performed with LiAlH$_4$ under mild reaction conditions to obtain benzofuran-7-ylmethanamine **149**. Later, allylamine **142a** was synthesized with an overall yield of 95% through a reductive amination reaction of compound **149** with *trans*-cinnamaldehyde utilizing NaBH$_4$ as the reducing agent. Subsequently, the N-alkylation of allylamine **142a** was conducted using iodoethane or 2-iodopropane in the presence of NaH as a base in DMF at ambient temperature to deliver compounds **142b–c** with a purity ≥ 95%. In another synthetic strategy, amine **149** was initially protected with di-*tert*-butyl dicarbonate, followed by reduction using LiAlH$_4$ to afford 1-(benzofuran-7-yl)-N-methylmethanamine **151** (Scheme 24). Simultaneously, the α,β-unsaturated aldehyde **152** were reduced using NaBH$_4$ to afford allylic alcohol **153**, which was subjected to an Appel reaction utilizing PBr$_3$ in Et$_2$O at ambient temperature to afford allylic bromide **154**. Finally, aliphatic nucleophilic substitution between compounds **151** and **154** gave a series of naftifine analogs, which were converted into their corresponding hydrochloride salts **140**, **143a–b**, and **144a–t**.

In a similar manner, the synthesis of naftifine analog **141** involved an O-alkylation reaction of 3-bromophenol **155** with 1-bromo-2,2-diethoxyethane to give compound **156**, which was then cyclized using polyphosphoric acids, leading to the formation of two isomeric products **157**, namely, 4-bromobenzo[*b*]furan and 6-bromobenzo[*b*]furan (Scheme 25). Subsequently, the bromine atom within isomers **157** was substituted with a cyano group to afford isomers **158**. Finally, through a series of consecutive reactions involving reduction and Boc$_2$O protection, reduction, and nucleophilic substitution, the naftifine analog **141** was successfully synthesized.

After obtaining the desired analogs, their inhibitory potential against *S. aureus* Newman was systematically evaluated using naftifine (NFT) as the reference drug [64]. The results indicated that compound **140** (R^3 = Me, R^5 = 7-benzofuranyl) showed the highest potency with an IC$_{50}$ value of 247.3 nM, in comparison to NFT as the standard drug (IC$_{50}$ = 296.0 nM). In contrast, the isomeric compound **141** demonstrated low activity with an IC$_{50}$ value of 758.7 nM. In the case of analogs featuring N-methyl substitutions **142a–c**, the incorporation of an ethyl or *iso*-propyl group markedly diminished activity (IC$_{50}$ > 1000 nm), as shown in Table 38. While the incorporation of cycloalkyl substituents **144a–b** and the 2-furanyl group **144c** did not improve activity (IC$_{50}$ > 1000 nM), the incorporation of 1- and 2-naphthalenyl groups **144d** and **144e** led to better activity (IC$_{50}$ = 887.7

and 17.1 nM, respectively). It is worth noting that the presence of electron-donating and electron-withdrawing groups attached to the aromatic ring significantly influenced the activity profile. Also, the position of the substituent on the aromatic ring affected the activity, such as compounds **144h** (R^1 = 4-FPh, IC_{50} = 31.2 nM) vs. **144o** (R^1 = 2-FPh, IC_{50} = 288.3 nM) vs. **144r** (R^1 = 3-FPh, IC_{50} = 513.0 nM). The same behavior was observed for compounds **144k** (R^1 = 4-NO$_2$Ph, IC_{50} = 71.1 nM) vs. **144p** (R^1 = 2-NO$_2$Ph, IC_{50} > 1000 nM) vs. **144s** (R^1 = 3-NO$_2$Ph, IC_{50} > 1000 nM). A key highlight is the exceptional activity displayed by compound **144l** (R^1 = 4-CF$_3$Ph), showcasing an impressive IC_{50} value of 4.0 nM, which is 74 times lower than NFT (IC_{50} = 296.0 nM), as shown in Table 39.

153a, 154a, 140 R^1 = Ph, R^2 = H
153b, 154b, 143a R^1 = Ph, R^2 = Me
153c, 154c, 143b R^1 = Styryl, R^2 = H
153d, 154d, 144a R^1 = Cyclopentyl, R^2 = H
153e, 154e, 144b R^1 = Cyclohexyl, R^2 = H
153f, 154f, 144c R^1 = 2-Furanyl, R^2 = H
153g, 154g, 144d R^1 = 1-Napthyl, R^2 = H
153h, 154h, 144e R^1 = 2-Napthyl, R^2 = H
153i, 154i, 144f R^1 = 4-MeOPh, R^2 = H
153j, 154j, 144g R^1 = 4-(COOMe)Ph, R^2 = H
153k, 154k, 144h R^1 = 4-FPh, R^2 = H
153l, 154l, 144i R^1 = 4-ClPh, R^2 = H
153m, 154m, 144j R^1 = 4-BrPh, R^2 = H
153n, 154n, 144k R^1 = 4-NO$_2$Ph, R^2 = H
153o, 154o, 144l R^1 = 4-CF$_3$Ph, R^2 = H
153p, 154p, 144m R^1 = 4-(t-Butyl)Ph, R^2 = H
153q, 154q, 144n R^1 = 4-MePh, R^2 = H
153r, 154r, 144o R^1 = 2-FPh, R^2 = H
153s, 154s, 144p R^1 = 2-NO$_2$Ph, R^2 = H
153t, 154t, 144q R^1 = 3-MePh, R^2 = H
153u, 154u, 144r R^1 = 3-FPh, R^2 = H
153v, 154v, 144s R^1 = 3-NO$_2$Ph, R^2 = H
153w, 154w, 144t R^1 = 2,4-diClPh, R^2 = H

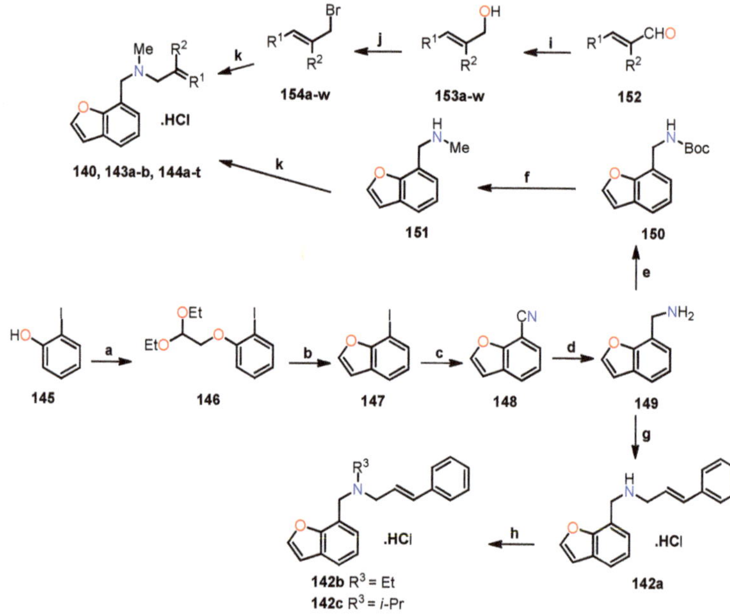

Scheme 24. Synthesis of benzo[b]furan derivatives **140**, **142a–c**, **143a–b**, and **144a–t**, and reagents and conditions: (**a**) 1-bromo-2,2-diethoxyethane, NaH, DMF, 90 °C, overnight, 93%; (**b**) polyphosphoric acids, toluene, reflux, overnight, 55%; (**c**) CuCN, DMF, 130 °C, 4 h, 90%; (**d**) LiAlH$_4$, THF, −78 °C to r.t, overnight, N$_2$, 95%; (**e**) di-*tert*-butyl dicarbonate, NaOH, THF, 0 °C to r.t, 1 h, 85%; (**f**) LiAlH$_4$, THF, 0 °C to r.t, overnight, N$_2$, 95%; (**g**) (*i*) *trans*-cinnamaldehyde, molecular sieves, CH$_2$Cl$_2$, reflux, 17 h; (*ii*) NaBH$_4$, methanol, 0 °C, 30 min, 95% (2 steps); (**h**) iodoethane or 2-iodopropane, NaH, DMF 0 °C to r.t, overnight, N$_2$, 50%; (**i**) NaBH$_4$, methanol, 0 °C, 30 min; (**j**) PBr$_3$, Et$_2$O, 0 °C to r.t, overnight, N$_2$, 57–84% (2 steps); (**k**) (*i*) K$_2$CO$_3$, DMF, r.t, overnight, 43–85%, (*ii*) bubbled into hydrogen chloride gas.

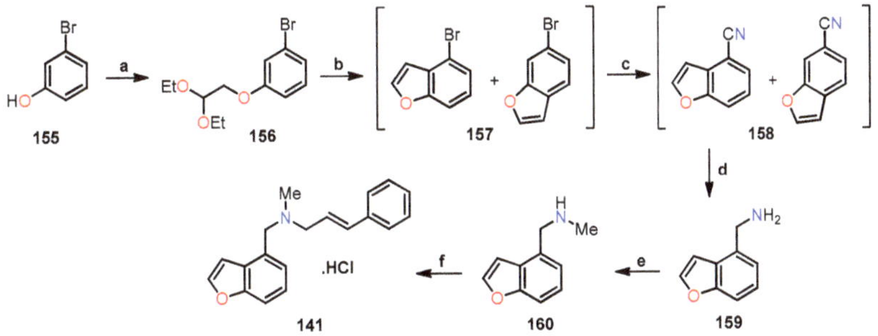

Scheme 25. Synthesis of benzo[b]furan derivative **141**, and reagents and reaction conditions: (**a**) 1-bromo-2,2-diethoxyethane, NaH, DMF, 90 °C, overnight, 95%; (**b**) polyphosphoric acids, toluene, reflux, overnight; (**c**) CuCN, DMF, 130 °C, 4 h, 30% (2 steps); (**d**) LiAlH$_4$, THF, −78 °C to r.t, overnight, under N$_2$, 94%; (**e**) (*i*) di-*tert*-butyl dicarbonate, NaOH, THF, 0 °C, r.t, 1 h; (*ii*) LiAlH$_4$, THF, 0 °C to r.t, overnight, N$_2$, 70% (2 steps); (**f**) (*i*) cinnamyl bromide, K$_2$CO$_3$, DMF, r.t, overnight, 44%, (*ii*) bubbled into hydrogen chloride gas.

Table 38. Antibacterial activity of naftifine analogs **140**, **141**, and **142a–c** against *S. aureus* Newman.

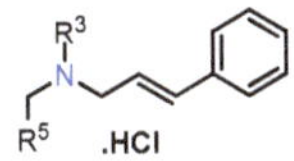

Compound	R^3	R^5	*S. aureus* Newman IC$_{50}$ (nM) [a]
140	Me	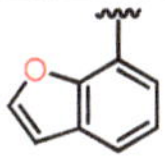 benzofuran-7-yl	247.3 ± 18.8
141	Me	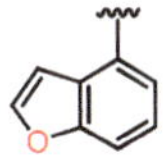 benzofuran-4-yl	758.7 ± 24.3
142a	H	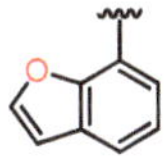 benzofuran-7-yl	>1000
142b	Et	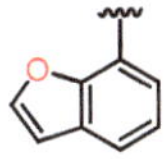 benzofuran-7-yl	>1000
142c	*i*-Propyl	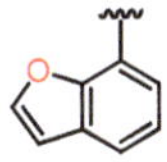 benzofuran-7-yl	>1000

[a] The data indicate the mean ± SD of at least three independent experiments.

Initial investigations unveiled the most prospective analogs with the potential to inhibit diapophytoene desaturase (CrtN) enzyme in the Staphyloxanthin (STX) biosynthesis pathway. STX is a notable golden carotenoid pigment synthesized by *S. aureus*, which opens up a novel avenue for treating *S. aureus* or *methicillin-resistant S. aureus* (MRSA) infections [64]. As shown in Table 40, a selection process identified five analogs (**144f**, **144i**, **144j**, **144l**, and **144t**) with the highest activity against *S. aureus* Newman, which were subjected to an evaluation of their capacity to inhibit the CrtN enzyme. The results revealed that five analogs displayed a remarkable 40-fold increase in inhibitory potency against CrtN compared to NFT. Interestingly, this potent inhibition of CrtN stands in contrast to their comparatively weaker impact on the enzymatic activity of pigmented *S. aureus* Newman. Furthermore, an assessment was conducted on the water solubility of the five analogs, revealing NFT to possess low solubility (6.2 mg/mL). This investigation facilitated the clarification of the interplay between solubility and chemical structure. Indeed, the replacement of the naphthalene ring with a benzo[b]furan ring generated an elevation in solubility, particularly evident in the cases of analogs **144f** and **144l**, showcasing solubilities 2–3 times greater than that of NFT, measuring 19.7 and 10.0 mg/mL, respectively. These findings served as the basis for advancing the assessment of the compound **144l**, both in vitro and in vivo.

Table 39. Antibacterial activity of naftifine analogs **144a–t** against *S. aureus* Newman.

Compound	R^1	S. aureus Newman IC$_{50}$ (nM) [a]	Compound	R^1	S. aureus Newman IC$_{50}$ (nM) [a]
144a	cyclopentyl	>1000	144k	4-NO$_2$-C$_6$H$_4$	71.1 ± 2.5
144b	cyclohexyl	>1000	144l	4-CF$_3$-C$_6$H$_4$	4.0 ± 0.2
144c	2-furyl	>1000	144m	4-tBu-C$_6$H$_4$	46.7 ± 3.2
144d	1-naphthyl	887.7 ± 60.0	144n	4-Me-C$_6$H$_4$	35.2 ± 3.8
144e	2-naphthyl	17.1 ± 1.5	144o	2-F-C$_6$H$_4$	288.3 ± 18.2
144f	4-OMe-C$_6$H$_4$	11.2 ± 1.1	144p	2-NO$_2$-C$_6$H$_4$	>1000
144g	4-CO$_2$Me-C$_6$H$_4$	16.9 ± 1.5	144q	3-Me-C$_6$H$_4$	>1000
144h	4-F-C$_6$H$_4$	31.2 ± 2.1	144r	3-F-C$_6$H$_4$	513.0 ± 11.1
144i	4-Cl-C$_6$H$_4$	6.3 ± 0.1	144s	3-NO$_2$-C$_6$H$_4$	>1000
144j	4-Br-C$_6$H$_4$	6.7 ± 0.1	144t	2,4-Cl$_2$-C$_6$H$_3$	8.3 ± 0.3

[a] The data indicate the mean ± SD of at least three independent experiments.

Table 40. CrtN enzymatic inhibition and water solubility of naftifine analogs 144.

Compound	CrtN IC$_{50}$ (nM) [a]	S. aureus Newman IC$_{50}$ (nM) [a]	Solubility (mg/mL)
144f	683.7 ± 68.1	11.08 ± 1.06	19.7
144i	219.0 ± 16.8	6.20 ± 1.02	7.4
144j	355.1 ± 26.1	6.44 ± 1.02	3.9
144l	338.8 ± 28.3	3.93 ± 1.02	10.0
144t	740.2 ± 55.6	8.21 ± 1.03	7.2
NFT [b]	8830.0 ± 109.1	296.0 ± 12.2	6.1

[a] The data indicate the mean ± SD of at least three independent experiments. [b] The standard drug for the study.

In their in vitro investigations, Wang et al. examined the impact of compound 144l on three MRSA strains: USA400 MW2, USA300 LAC, and Mu50 [64]. The results revealed a reduction in color due to the inhibitory effects of 144l, evidenced by IC$_{50}$ values of 5.45, 3.39, and 0.38 nM. These findings mirrored the observations made with the S. aureus Newman strain (Figure 48).

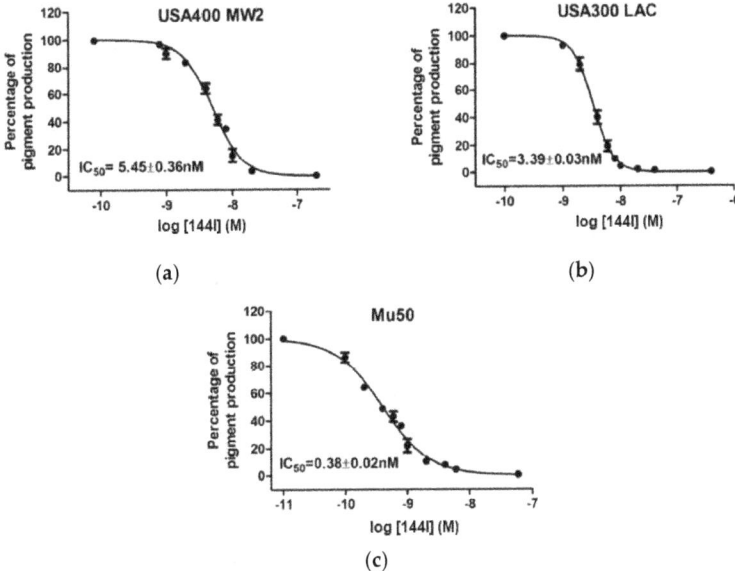

Figure 48. Dose–response curves in pigment formation across three MRSA strains: (a) USA400 MW2, (b) USA300 LAC, and (c) Mu50, with their corresponding IC$_{50}$ values using the compound 144l. This is an open-access article distributed under the terms of the Creative Commons CC BY license [64].

In vivo studies allowed the evaluation of virulence reduction in three of four colonies. Mice were infected with mock or treated with the compound 144l with S. aureus Newman, USA400 MW2, and Mu50 strains by retro-orbital injection [64]. Bacterial survival within host organs was then measured. Notably, in the S aureus Newman strain, the group treated with compound 144l displayed a reduction in bacterial survival. Kidneys and hearts showed decreases of 0.85 and 1.01 log$_{10}$CFU/organ, respectively (Figure 49).

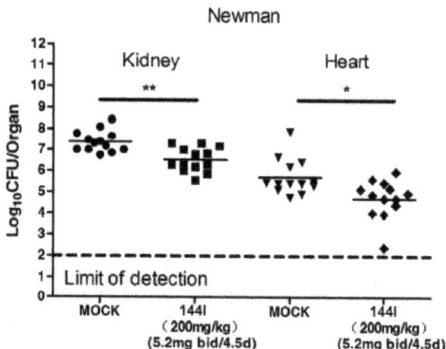

Figure 49. Effect of the compound **144l** on the virulence of *S. aureus* Newman in mouse kidneys and hearts ($n = 13$), quantified in terms of colony-forming units (Log_{10}CFUs/organ). Statistical significance determined by the Mann–Whitney test (two-tailed): * $p < 0.05$ and ** $p < 0.01$. This is an open-access article distributed under the terms of the Creative Commons CC BY license [64].

Regarding the MRSA strains, BPH-652 served as a reference CrtN inhibitor [64]. For the USA400 MW2 strain, the administration of a 200 mg/kg dose of compound **144l** to mice resulted in a remarkable 99.6% reduction in survival rates within hepatic organs (2.35 $\log_{10}$CFU). Impressively, this outcome surpassed that of the BPH-652-treated group (1.58 $\log_{10}$CFU). Furthermore, with the dosage scaled down to 50 mg/kg, the bacterial survival rate showed only a marginal increase in both scenarios while still maintaining superiority over BPH-652 by 0.71 $\log_{10}$CFU in the **144l**-treated group and by 0.25 $\log_{10}$CFU in the BPH-652-treated group (Figure 50a). Within the renal organs, the administration of the 200 mg/kg dose resulted in a notable 96.6% decrease in staphylococcal loads of the **144l**-treated group (1.47 $\log_{10}$CFU), surpassing the BPH-652-treated group outcome of 1.14 $\log_{10}$CFU. Upon reducing the dose to 50 mg/kg, slight increases in bacterial survival rates were noted compared to the high-dose treatment groups (Figure 50b).

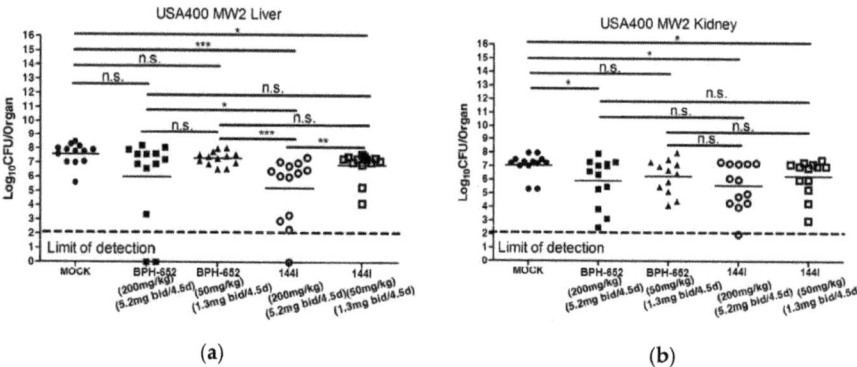

Figure 50. Effect of BPH-652 and compound **144l** on the virulence of USA400 MW2 in the livers (**a**) and kidneys (**b**) of mice ($n = 13$) as colony-forming units (Log_{10}CFUs/organ). Statistical significance determined by the Mann–Whitney test (two-tailed): * $p < 0.05$, ** $p < 0.01$, *** $p < 0.001$, and n.s. indicates no significant difference. This is an open-access article distributed under the terms of the Creative Commons CC BY license [64].

Employing the Mu50 strain, a parallel pattern emerged. Within liver organs, a dose of 200 mg/kg led to a reduction in survival rates by 3.58 $\log_{10}$CFU (**144l**-treated) and 2.84 $\log_{10}$CFU (BPH-652-treated). Similarly, at 50 mg/kg, the survival rate diminished by 2.94 $\log_{10}$CCFU (**144l**-treated) and 1.87 $\log_{10}$CFU (BPH-652-treated), as depicted in Figure 51a. However, outcomes in renal organs yielded inconclusive results. With a dosage of 200 mg/kg, the decrease amounted to 1.11 $\log_{10}$CFU (**144l**-treated) and merely 0.30 $\log_{10}$CFU (BPH-652-treated). At 50 mg/kg, survival was reduced to 0.68 $\log_{10}$CFU (**144l**-treated) and 0.25 $\log_{10}$CFU (BPH-652-treated), as shown in Figure 51b.

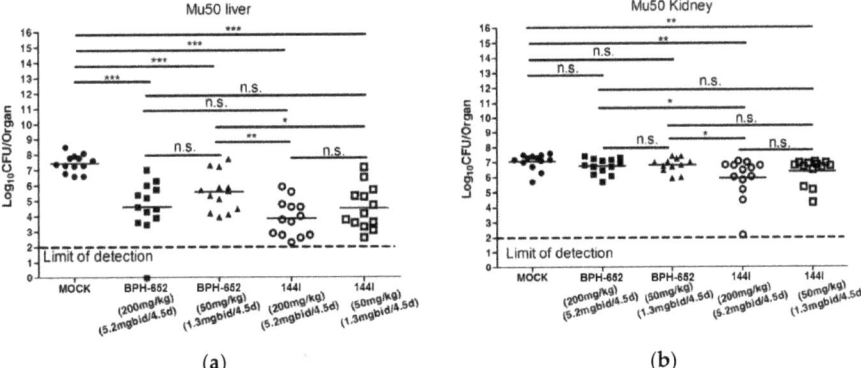

Figure 51. Effect of BPH-652 and compound **144l** on Mu50 virulence in the livers (**a**) and kidneys (**b**) of mice ($n = 13$) as colony-forming units ($\log_{10}$CFUs/organ). Statistical significance determined by the Mann–Whitney test (two-tailed): * $p < 0.05$, ** $p < 0.01$, *** $p < 0.001$, and n.s. indicates no significant difference. This is an open-access article distributed under the terms of the Creative Commons CC BY license [64].

On the other hand, pyrrolobenzodiazepines (PBDs) have garnered considerable attention as promising antibacterial agents derived from natural sources. In line with this, Andriollo et al. undertook the synthesis of a series of PBDs incorporating C–8 linkers (Scheme 26), with the primary objective of evaluating their bioactivity and elucidating the structure–activity relationship (SAR) [65]. The synthetic methodology entails the synthesis of benzo[*b*]furan-based pyrroles **163a–b** and **167**, achieved through an amidation reaction between *N*-methylpyrrole derivatives and benzo[*b*]furans in DMF, utilizing the EDCI/DMAP coupling system. Subsequently, nitrile **163b** underwent hydrolysis under reflux conditions using dioxane and H_2SO_4 to afford carboxylic acid **164** in a modest 9% yield (Scheme 26). After acquiring these intermediate components, the synthesis of PBD derivatives **169**, **171**, and **172** with C–8 linkers was undertaken (Scheme 27). In this stage, the deprotection of BOC-protected amines **163a** and **170** was achieved by treating them with an acidic solution (TFA in DCM). Additionally, derivatives containing methyl esters, identified as **167**, underwent hydrolysis using an aqueous NaOH solution. Subsequent to this, an amide coupling reaction was facilitated utilizing the EDCI/DMAP coupling system, effectively linking the PBD core and side chains. Lastly, employing pyrrolidine and Pd(PPh$_3$)$_4$ in DCM, the conjugates underwent deprotection, leading to the generation of PDB derivatives **169**, **171**, and **172**, each exhibiting standard and reverse orientations of the amide bond (Scheme 27).

The previously obtained PBD derivatives **169**, **171**, and **172** underwent an assessment to determine their capacity to bind to DNA and impart stability using a Förster resonance energy transfer (FRET)-based DNA fusion assay, utilizing netropsin as the positive control [65]. To achieve this objective, two oligonucleotide sequences labeled with distinct fluorophores were utilized: sequence F1'-FAM-TAT-ATA-TAG-ATA-TTT-TTT-TAT-CTA-TAT-ATA-3'-TAMRA and sequence F2'-FAM-TAT-AGA-TAT-AGA-TAT-TAT-TTT-ATA-TCT-ATA-TCT-ATA-TCT-ATA-3'-TAMRA. Here, FAM corresponds to 6-carboxyfluorescein, and TAMRA represents 5-carboxytetramethylrhodamine. The results revealed that compound **169**, featuring a conventional orientation, adeptly conferred substantial stability to both

DNA sequences, akin to the notable effect seen with netropsin. In contrast, compounds **171** and **172**, characterized by reversed orientations, displayed a clear inability to confer stability upon either DNA sequence, as evidenced by ΔT_m values below 1 °C. This compelling observation strongly suggests that the inversion of one or more amide bonds within these compounds markedly curtailed their DNA stabilizing efficacy, as firmly corroborated by the comprehensive data outlined in Table 41.

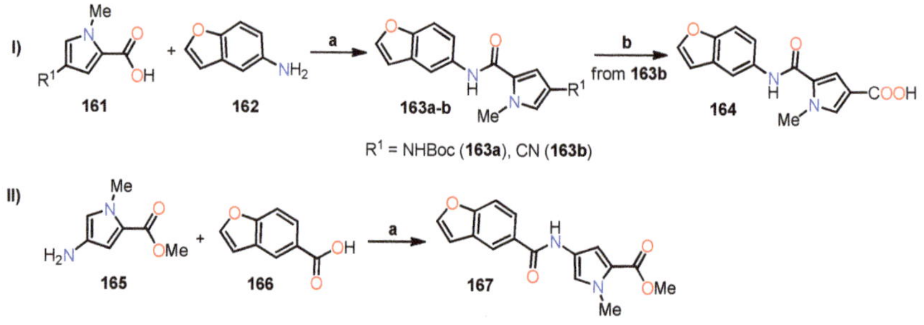

Scheme 26. Synthesis of benzo[b]furan-based pyrroles, and reagents and reaction conditions: (**I**) (**a**) EDCI, DMAP, DMF, 67–79%; (**b**) dioxane, H$_2$SO$_4$, reflux, 9%; (**II**) (**a**) EDCI, DMAP, DMF, 20%.

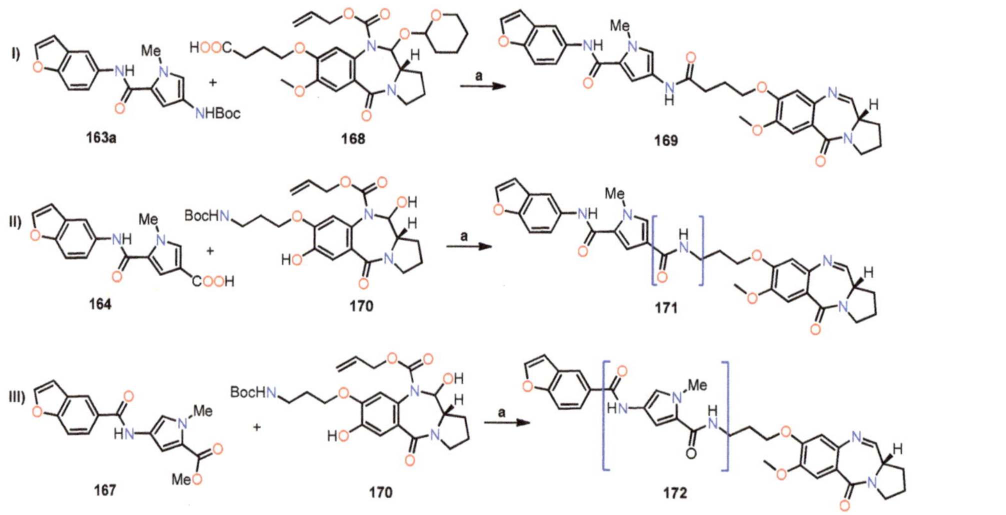

Scheme 27. Synthesis of benzo[b]furan-based pyrrolobenzodiazepines, and reagents and reaction conditions: (**I**) and (**II**) (**a**) (*i*) TFA, DCM, (*ii*) EDCI, DMAP, DMF, (*iii*) pyrrolidine, Pd(PPh$_3$)$_4$, PPh$_3$, DCM, 14–38%; (**III**) (**a**) (*i*) NaOH, H$_2$O, dioxane, (*ii*) TFA, DCM, (*iii*) EDCI, DMAP, DMF, (*iv*) pyrrolidine, Pd(PPh$_3$)$_4$, PPh$_3$, DCM, 22%. Bracketed structures denote compounds with an inverted orientation.

Table 41. Evaluation of DNA binding ability for PDB derivatives **169**, **171**, and **172**.

Compound	ΔT_m (°C) at 1 µM [a]	
	Sequence F1	Sequence F2
169	20.8 ± 0.2	9.5 ± 0.5
171	0.0 ± 0.2	0.4 ± 0.1
172	0.1 ± 0.1	0.5 ± 0.3
Netropsin [b]	13.8 ± 0.3	11.0 ± 0.2
Ciprofloxacin [b]	0.0 ± 0.4	0.2 ± 0.3

[a] The data indicate the mean ± SD of three independent experiments. [b] The standard drugs for the study.

Subsequently, PBD derivatives **169**, **171**, and **172** were subjected to antimicrobial testing against diverse Gram-positive bacterial strains, including methicillin-sensitive *S. aureus* (MSSA) strain ATCC 9144, as well as two methicillin-resistant *S. aureus* (MRSA) strains, namely, EMRSA-15 (strain HO 5096 0412) and EMRSA-16 (strain MRSA 252) [65]. Additionally, the study incorporated vancomycin-sensitive *Enterococcus faecalis* (VSE) strain NCTC 755, vancomycin-resistant *E. faecalis* (VRE) strain NCTC 12201, and vancomycin-resistant *Enterococcus faecium* (VRE) strain NCTC 12204. The results are presented in Table 42. Compound **169**, characterized by its standard orientation, exhibited remarkable antibacterial efficacy, with MIC values of 0.125 µg/mL against all assessed Gram-positive strains. In contrast, compound **171**, characterized by its inverted amide bond orientation, exhibited a significant reduction in activity across all bacterial strains, particularly evident in MRSA strains, where MIC values exceeded 32 µg/mL. Notably, compound **172**, characterized by its inversion of both amide bonds linked to the *N*-methylpyrrole ring, exhibited enhanced antibacterial activity compared to compound **171**. However, its activity was inferior to that of compound **169**. It is worth considering that the assessment of the orientation's impact on antibacterial activity is at the forefront of our analysis.

Table 42. Antibacterial activity of PBD derivatives **169**, **171**, and **172**.

Compound	Gram-Positive Strains MIC (µg/mL)					
	VRE		VSE	MRSA		MSSA
	NCTC 12201	NCTC 12204	NCTC 775	EMRSA 15	EMRSA 16	ATCC 9144
169	≤0.125	≤0.125	≤0.125	≤0.125	≤0.125	≤0.125
171	2	2	16	>32	>32	16
172	0.5	1	4	2	4	2
Ciprofloxacin [a]	0.5	2	1	32	32	≤0.125

[a] The standard drug for the study.

The mechanism of action of compound **172** was investigated via time–kill assays conducted on MRSA (EMRSA-15) and VRE (NCTC 12201) strains. These strains were exposed to compound **172** at a concentration of 4 × MIC for 24 h (Figure 52). Compound **172** demonstrated a rapid and robust bactericidal effect, leading to a reduction in cell counts below the detectable limit within a 2 h period. In contrast, ciprofloxacin exhibited bacteriostatic activity (Figure 52). Despite a modest cell population persisting in EMRSA-15 after 24 h of exposure to compound **172**, it exhibited no resistance to the compound, suggesting the presence of a potential persister population. This trend was more pronounced in EMRSA-15 than in NCTC 12201, underscoring the potential role of fluoroquinolone resistance.

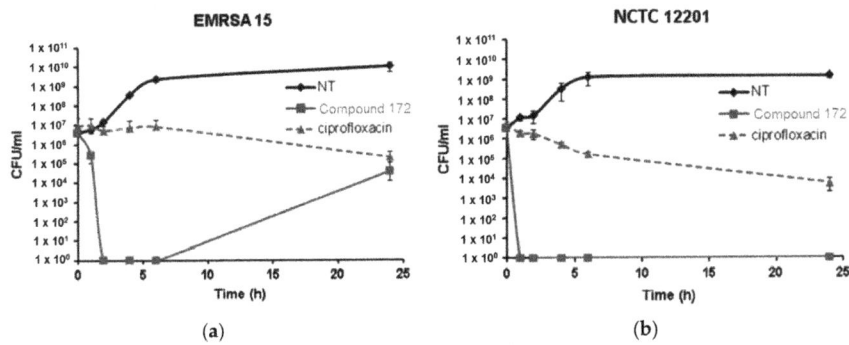

Figure 52. Mechanism of action of compound **172** against two bacterial strains: (**a**) EMRSA-15 and (**b**) NCTC 12201. This is an open-access article distributed under the terms of the Creative Commons CC BY license [65].

Moreover, an extensive in silico analysis was performed, employing ESI mutagenesis to unravel the mechanism of action inherent in the PBD derivatives. The gathered data unveiled that compound **172** exhibited a distinct interaction pattern with the ligand-binding domain of DNA gyrase, showcasing a notably robust binding affinity for both subunits of the bacterial DNA gyrase complex. The visualization of these interactions involving the bacterial gyrase from *Staphylococcus aureus* (PDB ID code: 2XCT) is visually presented in Figure 53A,B. As delineated in the 2D patterns depicted in Figure 53C, it becomes apparent that compound **172** forms three conventional hydrogen bonds, establishing connections with serine 98, arginine 92, and glutamine 95 within DNA gyrase A subunit 1. Analogously, the interaction extends to subunit 2 of DNA gyrase A, where it interacts with serine 85, arginine 92, and serine 98 (Figure 53D). The implications of the interaction underscore that the antibacterial activity attributed to compound **172** stems from its direct modulation of gyrase A via enzyme interaction, as opposed to its ability to stabilize DNA.

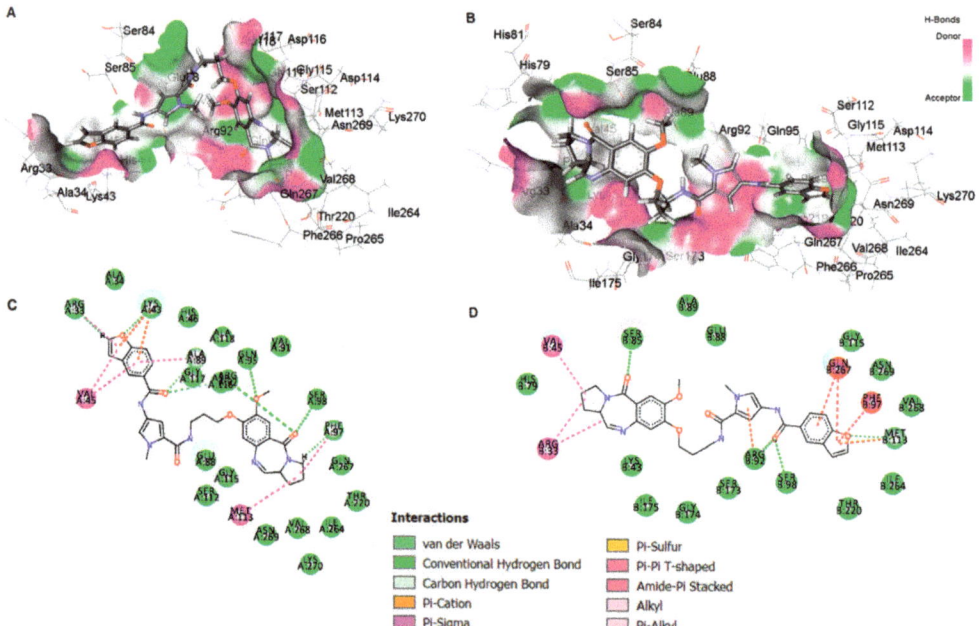

Figure 53. Molecular model depicting the interaction between compound **172** and (**A**) subunit 1 and (**B**) subunit 2 of the gyrase A in *Staphylococcus aureus* (PDB ID code: 2XCT). 2D model of the interaction of compound **172** with (**C**) subunit 1 and (**D**) subunit 2 of gyrase A in *Staphylococcus aureus*. This is an open-access article distributed under the terms of the Creative Commons CC BY license [65].

Considering the significance of Sortase A (SrtA) as a cysteine transpeptidase prevalent in most Gram-positive bacteria, its pivotal role in the infection process of these organisms is well-established. Inhibition of this enzyme has been shown to exert a discernible impact on the virulence of Gram-positive bacteria, thereby rendering them more resistant to antibiotics. Acknowledging this premise, Lei et al. embarked upon extending these insights to *Staphylococcus aureus* (*S. aureus*). Given its susceptibility to detection and elimination by the immune system due to its lower viscosity, inhibiting SrtA emerged as a pertinent strategy [66]. To this end, a series of derivatives of 2-(4-(1-cyano-2-phenylvinyl)phenyl)-*N*-isobutylbenzofuran-3-carboxamide **175a–z**, **175a$_2$–i$_2$**, and **176** were synthesized from intermediates **173** [67], through substitution reactions involving cyanide, the compounds **174** were obtained (Scheme 28). Ultimately, via condensation reactions with diverse aldehydes, the cyano derivatives of benzo[*b*]furan **175a–z**, **175a₂–i₂**, and **176** were successfully generated.

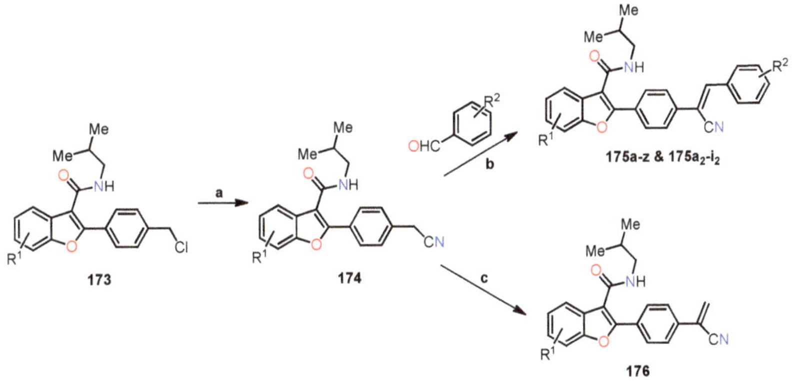

Scheme 28. Synthesis of benzo[*b*]furan cyanide derivatives **175a–z**, **175a₂-i₂**, and **176**. Reagents and conditions: (**a**) (Me)₃Si-CN, TBAF, MeCN, reflux, 2 h, 79–85%; (**b**) EtONa, EtOH, reflux, 1 h, 40–87%; (**c**) formaldehyde, EtONa, EtOH, reflux, 1 h.

Additionally, three benzofuran-3-carboxamide derivatives were synthesized to evaluate the effect of the olefin cyanide group in inhibiting SrtA activity (Scheme 29). For this, a double reduction was carried out using Pd/C under a hydrogen atmosphere to reduce the olefinic double bond adjacent to the cyanide, followed by the removal of the benzyl group, thus obtaining the compounds **177a–b** and **178a–b**, respectively.

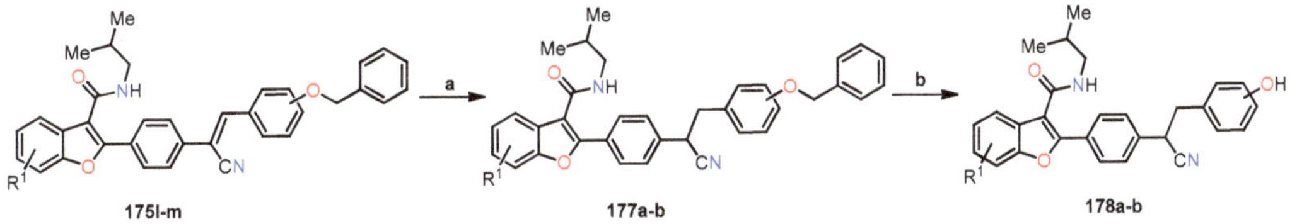

Scheme 29. Synthesis of benzo[*b*]furan cyanide derivatives **177a–b** and **178a–b**. Reagents and conditions: (**a**) Pd/C, H₂, MeOH, r.t, 1 h, 41–80%; (**b**) Pd/C, H₂, MeOH, r.t, 12 h, 37–40%.

A comprehensive synthesis of 39 benzofuran cyanide derivatives was conducted, followed by their rigorous evaluation for in vitro inhibitory potential against SrtA in *S. aureus*, with pHMB serving as a positive control. The findings, detailed in Table 43, underscored the noteworthy activity exhibited by most of the synthesized analogs, with IC_{50} values spanning the range of 3–100 μM. Among these, compounds **175a**, **175e**, **175g**, **175i**, **175m–o**, **175w**, and **175h₂** emerged as particularly significant performers. This performance differential might be attributed to the intricate interplay of structure–activity relationships, favoring the potency of analog **175a** (R^1 = H, R^2 = H, IC_{50} = 8.8 μM) over counterparts **175p** (R^1 = 7-OMe, R^2 = H, IC_{50} = 11.9 μM) and **175z** (R^1 = 5-Cl, R^2 = H, IC_{50} = 29.9 μM), primarily due to the strategic introduction of a substituent within the benzo[*b*]furan ring. The influence of R^2 substitution on inhibitory activity was distinctly pronounced, as evidenced by the superiority of electron-withdrawing groups over donor groups, **175b** (R^1 = H, R^2 = 4-Me) vs. **175i** (R^1 = H, R^2 = 4-Cl), **175q** (R^1 = 7-OMe, R^2 = 4-Me) vs. **175w** (R^1 = 7-OMe, R^2 = 4-Cl), and **175b₂** (R^1 = 5-Cl, R^2 = 3,4-Me) vs. **175h₂** (R^1 = 5-Cl, R^2 = 4-Cl). Furthermore, the size of substituents emerged as another pivotal determinant, favoring the chlorine group in compounds **175i**, **175w**, and **175h₂** (IC_{50} = 9.8, 5.9, and 6.8 μM, respectively), in contrast to the bromide group in compounds **175j** (R^1 = H, R^2 = 4-Br), **175x** (R^1 = 7-OMe, R^2 = 4-Br), and **175i$_2$** (R^1 = 5-Cl, R^2 = 4-Br) with IC$_{50}$ values of 19.1, 15.5, and 16.4 µM, respectively.

Table 43. IC$_{50}$ values of benzo[b]furan cyanide derivatives in Sortase A inhibition.

Compound	R^1	R^2	IC$_{50}$ (µM) [a]
175a	H	H	8.8 ± 1.4
175b	H	4-Me	14.1 ± 0.9
175c	H	3,4-diMe	17.5 ± 2.2
175d	H	2,4,6-triMe	32.2 ± 1.6
175e	H	4-Et	9.4 ± 0.9
175f	H	4-OMe	11.9 ± 0.7
175g	H	3,4-diOMe	9.6 ± 0.3
175h	H	2,5-diOMe	21.8 ± 1.4
175i	H	4-Cl	9.8 ± 1.5
175j	H	4-Br	19.1 ± 0.6
175k	H	4-F	19.7 ± 1.0
175l	H	4-CF$_3$	12.2 ± 1.8
175m	H	2-COOH	9.7 ± 0.3
175n	H	4-OBn	6.0 ± 0.8
175o	H	3,4-diOBn	3.3 ± 0.3
175p	7-OMe	H	11.9 ± 1.8
175q	7-OMe	4-Me	10.1 ± 0.7
175r	7-OMe	3,4-diMe	>100
175s	7-OMe	4-Et	48.6 ± 5.9
175t	7-OMe	4-OMe	12.1 ± 0.1
175u	7-OMe	3,4-diOMe	14.5 ± 1.1
175v	7-OMe	2,5-diOMe	52.8 ± 7.9
175w	7-OMe	4-Cl	5.9 ± 1.5
175x	7-OMe	4-Br	15.5 ± 1.3
175y	7-OMe	2-COOH	28.4 ± 0.4
175z	5-Cl	H	29.9 ± 1.2
175a$_2$	5-Cl	4-Me	47.1 ± 7.3
175b$_2$	5-Cl	3,4-diMe	>100
175c$_2$	5-Cl	2,4,6-triMe	37.9 ± 1.5
175d$_2$	5-Cl	4-Et	13.8 ± 0.6
175e$_2$	5-Cl	4-OMe	64.1 ± 4.4
175f$_2$	5-Cl	3,4-diOMe	49.1 ± 4.4
175g$_2$	5-Cl	2,5-diOMe	15.5 ± 0.9
175h$_2$	5-Cl	4-Cl	6.8 ± 0.1
175i$_2$	5-Cl	4-Br	16.4 ± 0.8
176	--	--	15.2 ± 1.7
177a	H	4-OBn	15.0 ± 0.9
178a	H	4-OH	7.9 ± 0.7
178b	H	3,4-diOH	13.7 ± 1.2
pHMB [b]	-	-	130.0 ± 4.3

[a] The data indicate the mean ± SD of at least three independent experiments. [b] The standard reference for the study.

Furthermore, the selection process extended to the identification of four compounds **175a, 175o, 175h$_2$, and 178a** boasting the most noteworthy SrtA inhibition activity. Subsequently, the impact of these compounds on impeding the formation of bacterial biofilms, a significant contributor to drug resistance, was diligently evaluated. To facilitate this assessment, SPSS Software was harnessed to compute the inhibitory potential of the chosen compounds, as outlined in Table 44. The findings underscored the inhibitory efficacy of all four compounds against the development of *S. aureus* biofilms, with IC$_{50}$ values spanning a range of 2.1–54.2 µM. Remarkably, compound **175o** emerged as the most potent inhibitor, displaying an impressive IC$_{50}$ of 2.1 µM. These results harmoniously align with the outcomes derived from the SrtA inhibition study. In addition, a further assay was performed to investigate the potential of the four compounds to disrupt the invasion of 293T cells (human embryonic kidney cells) by *S. aureus*, utilizing a group of drug-free FITC-labeled bacteria as the control reference. Intriguingly, all four compounds demonstrated a pronounced reduction in the invasiveness of *S. aureus* strains within 293T cells. Notably, compound **175o** exhibited the most substantial interference, diminishing bacterial invasion by a noteworthy 24.0% at 100 μM, as contrasted with the control blank (Table 44 and Figure 54).

Table 44. Inhibition of biofilm formation and bacterial invasion using four cyano derivatives of benzo[*b*]furan.

Compound	Biofilm Inhibition IC$_{50}$ (μM) [a]	% Reduction [a]
175a	19.7 ± 0.3	17.9 ± 3.2
175o	2.1 ± 0.1	24.0 ± 6.4
175h$_2$	54.2 ± 0.5	4.0 ± 1.9
178a	14.3 ± 0.4	2.8 ± 1.6

[a] The data indicate the mean ± SD of at least three independent experiments.

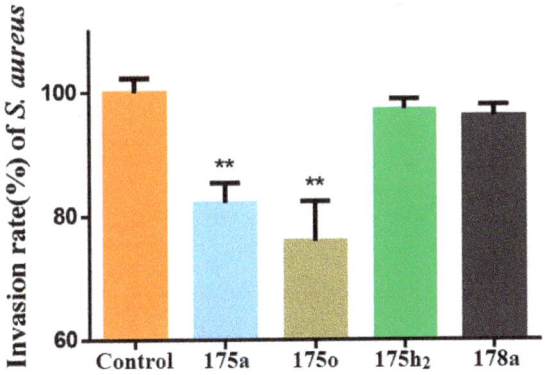

All the compounds with 100 μM

Figure 54. Quantitative analysis of *Staphylococcus aureus* invasion in 293T cells using flow cytometry. The evaluation encompasses the effect of compounds **175a, 175o, 175h$_2$**, and **178a** on the invasion process. Data are reported as mean ± SD of at least three independent experiments performed in duplicate. ** $p < 0.05$ vs. control. Reproduced with permission from ref. [66]. Copyright Elsevier Inc., 2023.

2.3. Antifungal Activity

Furthermore, benzo[*b*]furans have demonstrated pronounced antifungal efficacy against a diverse spectrum of fungal strains, encompassing *Candida albicans* (*C. albicans*), *Candida neoformans* (*C. neoformans*), *Candida parapsilosis* (*C. parapsilosis*), *Aspergillus flavus* (*A. flavus*), and *Microspora griseus* (*M. griseus*), among others. A seminal study by Sastraruji et al. has yielded a pivotal series of alkaloids **127a–i** extracted from *Stemona aphylla*, meticulously documented in Table 30 [58]. Within this set of nine isolated alkaloid derivatives, a judicious selection of five underwent rigorous examination to assess their antifungal prowess, with Amphotericin B employed as a comparative benchmark. Strikingly, the results unveiled that compounds **127g** (R^1 = H, R^2 = OMe, R^3 = Me, R^4 = Me, R^5 = H, R^6 = H) and **127i** (R^1 = H, R^2 = OMe, R^3 = Me, R^4 = Me, R^5 = Me, R^6 = H) exhibited performance tantamount to Amphotericin B against *C. albicans*, each recording an MIC value of 15.6 μg/mL. In contrast, with regard to *C. neoformans*, none of the evaluated alkaloids surpassed the efficacy of the control drug. Noteworthily, compounds **127a, 127c**, and **127g** emerged as the exclusive contributors to antifungal activity against the *C. neoformans* strain, with recorded MIC values of 7.8 μg/mL (Table 45).

Table 45. Minimum inhibitory concentration (MIC) values in μg/mL for alkaloids against two fungal strains.

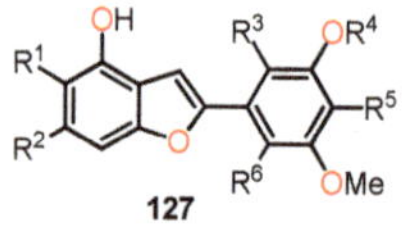

127

Compound	C. albicans	C. neoformans
127a	31.3	7.8
127c	31.3	7.8
127d	31.3	31.3
127g	15.6	7.8
127i	15.6	15.6
Amphotericin B [a]	15.6	3.9

[a] The standard drug for the study.

In a subsequent study conducted in 2013, Ostrowska et al. synthesized a series of chlorinated 2- and 3-benzo[b]furan carboxylates (Scheme 21 and Table 34). These derivatives underwent meticulous in vitro assessments to gauge their antimicrobial efficacy against both bacterial strains, as previously discussed, and fungal strains [60,61]. In terms of antifungal activity, they were scrutinized across strains encompassing *Aspergillus brasiliensis*, *Candida albicans*, *Candida parapsilosis*, *Saccharomyces cerevisiae*, and *Zygosaccharomyces rouxii*. The reference benchmark was fluconazole, spanning a range of MIC values from 3.9×10^{-4} to 8.4×10^{-1} μmol/cm^3. The outcomes, as presented in Table 46, align with the antibacterial findings (Table 35), underscoring that the most potent derivative across all tested fungal strains was the compound **133f.HCl** (R^1 = COOMe, R^2 = H, R^3 = Br, R^4 = OMe, R^5 = O(CH$_2$)$_2$NEt$_2$).

Table 46. Antifungal activity of hydrochlorides of methyl benzo[b]furan carboxylates against five fungal strains.

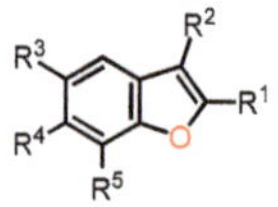

133, 134, and 135

Compound	MIC (μmol/cm^3)				
	C. albicans	C. parapsilosis	S. cerevisiae	Z. rouxii	A. brasiliensis
133a.HCl	0.39	0.39	ND	0.39	0.78
133b.HCl	1.49	1.49	ND	ND	1.49
133c.HCl	0.36	0.72	ND	0.36	0.72
133d.HCl	4.98	0.2987	ND	ND	1.19
133f.HCl	0.09	0.094	0.187	0.023	0.094
134a.HCl	0.41	ND	1.62	ND	ND
134c.HCl	ND	ND	3.13	ND	ND
134d.HCl	0.08	0.15	0.08	0.04	ND
134f.HCl	ND	ND	1.42	5.91	ND
134g.HCl	>30.56	>30.56	15.28	>30.56	15.28
135a.HCl	4.96	4.96	4.96	0.59	4.96

ND means not determined.

In a similar vein, Kenchappa et al. embarked on the synthesis of a series of (5-substituted-1-benzofuran-2-yl)(2,4-phenyl-substituted)methanone derivatives **139a–i** (Scheme 22), which were subsequently subjected to comprehensive evaluation for their antifungal potential [62]. The outcomes are meticulously presented in Table 47, revealing noteworthy observations. Specifically, compounds **139a** (R^1 = H, R^2 = Br, R^3 = H) and **139c** (R^1 = Br, R^2 = Br, R^3 = H) exhibited a comparably potent activity to fluconazole (control drug) across all tested fungal strains, displaying MIC values ranging from 6.30 to 12.75 µg/mL. Similarly, compounds **139e** (R^1 = OH, R^2 = Br, R^3 = H) and **139h** (R^1 = Br, R^2 = H, R^3 = Br) demonstrated robust antifungal efficacy, recording MIC values spanning from 6.40 to 13.70 µg/mL. In contrast, compounds **139d** and **139g** showcased minimal activity against the tested fungal strains.

Table 47. Antifungal activity of compounds **139a–i** against various fungal strains.

Compound	MIC (µg/mL)			
	C. albicans	A. flavus	M. griseus	A. terus
139a	12.75	12.70	6.35	6.50
139b	35.50	ND	22.50	25.55
139c	12.60	12.60	6.30	6.35
139d	15.80	16.85	9.75	7.55
139e	13.70	12.85	8.35	8.55
139f	23.50	21.65	18.50	21.55
139g	31.45	ND	24.70	24.75
139h	12.80	12.75	6.45	6.40
139i	32.80	35.75	20.50	28.50
Fluconazole [a]	12.50	12.50	6.25	6.25

[a] The standard drug for the study. ND means not determined.

In the same year, Wang et al. undertook the synthesis of a series of analogs derived from naftifine hydrochloride (NFT), specifically designed to target the CrtN enzyme for the treatment of MRSA infections [64]. Among this array, compound **144l** emerged as the most promising analog (Table 39). Recognizing NFT's established antifungal properties, a comparative analysis was conducted, gauging whether compound **144l** exhibited analogous effects. The outcomes, as presented in Table 48, unveiled that the compound **144l** analog displayed modest antifungal activity across all tested fungal strains, including *Trichophyton rubrum*, with MIC values reaching 16 µg/mL (128 times less potent than NFT). This variance in efficacy might be attributed to the structural difference between the benzofuran core of **144l** and the naphthalene moiety of NFT.

Table 48. Antifungal activity of the NFT analogue **144l**.

Compound	MIC (µg/mL)		
	T. rubrum	M. gypseum	T. barbae
144l	16	32	>64
NFT [a]	0.125	0.25	0.125
Fluconazole [a]	1	8	2
Voriconazole [a]	0.03125	0.25	0.03125
Ketoconazole [a]	0.5	2	0.0625

[a] The standard drugs for the study.

2.4. Analysis of the Structure–Activity Relationship in Anticancer Results

Interactions within the α- and β-tubulin structure have surfaced as a remarkable topic for exploring novel chemotherapeutic agents. Among the key domains, the colchicine binding site has emerged as a pivotal target for potential inducers of destabilization in tubulin polymerization. Colchicine binding site inhibitors (CBSIs) exert their biological impact by impeding tubulin assembly and limiting microtubule formation. In this context, benzo[b]furans have taken center stage, being designed and synthesized into a wide range of molecules interacting with the colchicine binding site, imparting marked structural diversity. Many compounds in this review demonstrated anticancer properties due to their inhibition of tubulin polymerization, achieved primarily through binding to the colchicine binding site on tubulin [32,34,37,39,40,43,46,49,51] (Figure 55). This is mainly achieved by the incorporation of the characteristic recognition fragment analogous to colchicine at positions C–2 or C–3 of benzo[b]furan, (3,4,5-trimethoxyphenyl)methanone [32,34,43,46,51], and their corresponding analogs [37,39,40], which interact through hydrogen bonds between methoxy groups and Cys241 as well as van der Waals interactions between phenyl ring and aromatic residues at the colchicine binding site on tubulin [40]. On the other hand, the methyl group located at positions C–2 or C–3 occupies hydrophobic pockets, while the necessity of hydrogen bonding interactions of methoxy or hydroxyl groups at positions C–6 and C–7 with the tubulin structure is evident. In particular, the hydrogen bonding interaction of the methoxy group at C–6 with Asnβ258 has remarkable significance [32,39]. This interaction not only enhances the antiproliferative activity in breast, renal, and melanoma cancer cell lines [36,46] but also contributes to the suppression of breast cancer metastasis [42]. Furthermore, additional hydrogen bonding interactions with donating groups like amino, hydroxyl, and unsaturations at C–5, along with alkyl ester derivatives [42] and alkoxy derivatives [45], play a role in stabilizing the active site of α- and β-tubulin [32], leading to the inhibition of different cancer cell lines [36]. Regarding substitution at position C–2, the diversity of functional groups allows the formation of hydrogen bonding interactions at active sites, and they also encompass aromatic groups that possess the ability to engage in π–π stacking with aromatic residues [34,37–39,42,43,46,48]. Notably, modifications such as the loss of aromaticity in the furan ring and the fusion of aliphatic chains on the d-side of benzo[b]furan enable the maintenance of activity against diverse cancer cell lines.

Some benzo[b]furan derivatives discussed in this review have undergone comprehensive studies analyzing their impact on the apoptotic pathway of cell death across various cell lines, including nuclear condensation [37], increased activation of caspase-3 [37,48,49,52], down-regulation of procaspase-9 [41,49], and negative regulation of the Bcl-2 protein family [37,40,41,48,49,52], as well as Bcl-x expression [54], PARP-1 cleavage [48,49], phosphorylation and stabilization of p53 [37,48,50], and a reduction in the expression of Bubr1 and Mad-2 proteins [51], among other effects. Furthermore, some benzo[b]furan derivatives demonstrated anti-vascular properties [40,51], a significant aspect in tumoral angiogenesis. Several researchers have confirmed the predominant cell cycle arrest in the G2/M phase [37,41,44,46,48,49,51]. Notably, effects on the PI3K/Akt/mTOR signaling pathway have been observed [41], along with the inhibition of Top 1 DNA relaxation activity [52], suppression of SIRT3 histone deacetylase [53], mitigation of tumor cell metastasis as demonstrated in in vivo mouse assays [42], facilitation of the tumor resection process via bioluminescence monitoring [42,51–53], and anti-metastatic effects underscored by interleukin-25 secretion from tumor-associated fibroblasts (TAF) [42]. In terms of innovation, this review highlights the advancement of novel compounds through the utilization of comparative molecular field analysis (CoMFA) and comparative molecular similarity indices analysis (CoMSIA) methodologies [43]. What is particularly remarkable is the versatility exhibited by benzo[b]furans, which extends to the exploration of triple-negative breast cancer [52,53], an exceptionally aggressive cancer type known for its rapid recurrence and challenging prognosis.

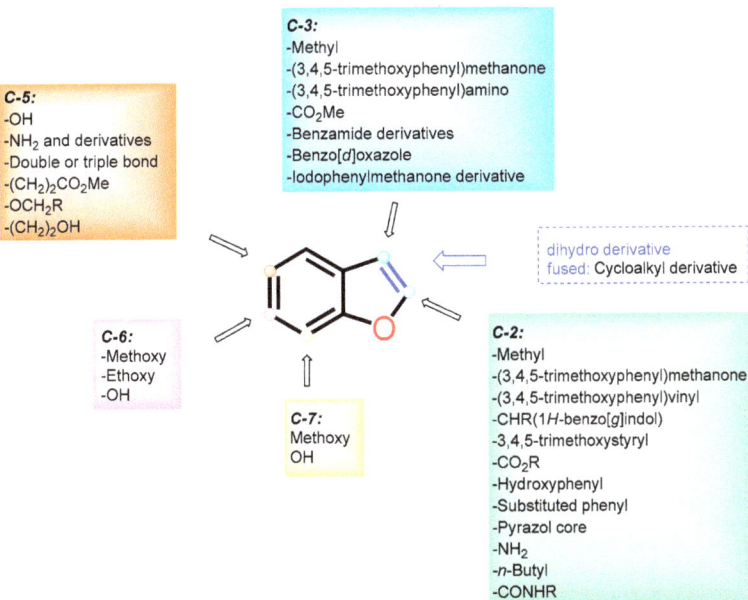

Figure 55. Analysis of the structure–activity relationship in anticancer results.

2.5. Analysis of the Structure–Activity Relationship in Antibacterial and Antifungal Results

To date the capacity to regulate the proliferation of bacteria and fungi holds considerable significance. The minimum inhibitory concentration (MIC) enables the identification of chemical structures capable of effectively managing visible growth inhibition in vitro against bacterial and fungal strains. Notably, the investigation into novel compounds possessing antibacterial activity is one of the most significant focuses within the microbial field, owing to the emergence of extensive resistance to conventionally employed antibacterial drugs.

In this review, the adaptability and versatility of benzo[b]furans toward both Gram-positive and Gram-negative bacteria are evident. This adaptability is apparent in the various structural modifications involving the incorporation of different functional groups at the six carbons of the structure (Figure 56). Beginning with the former, significant activity was observed against Gram-positive bacteria, including *methicillin-resistant Staphylococcus aureus* (MRSA) [58,64,65], *Staphylococcus aureus Newman* [64], *Staphylococcus aureus* [59,61], *Bacillus subtilis* [59,61,62], *Micrococcus luteus* [61], *Bacillus cereus* [61], *Staphylococcus epidermidis* [61], *Enterococcus hirae* [61], *vancomycin-sensitive Enterococcus faecalis* (VSE) [65], *vancomycin-resistant Enterococcus faecalis* (VRE) [65], and *vancomycin-resistant Enterococcus faecium* (VRE). Additionally, Gram-negative bacteria were targeted, including *Escherichia coli* [59,61,62], *Pseudomonas aeruginosa* [59,61], *Pseudomonas syringae* [62], *Salmonella typhi* [62], and *Klebsiella pneumoniae* [62]. Furthermore, enzymatic assays were performed against typical *Staphylococcus aureus* enzymes, specifically 4,4′-diapophytoene desaturase (CrtN) [64] and cysteine transpeptidase Sortase A (SrtA) [66]. Moreover, dose–response analyses of benzo[b]furans against pigment formation in various bacterial strains were performed [64]. In addition, in vivo assays were conducted to evaluate virulence in mice through retro-orbital injection [64], while incidence measurements were taken in organs such as the kidney and heart [64]. The potential of benzo[b]furans to bind to DNA and provide stability was investigated using the Förster resonance energy transfer (FRET) technique [65]. Additionally, in silico analyses were performed utilizing ESI mutagenesis in gyrase A in *Staphylococcus aureus* [65]. Among other tests, the disruption of invasion in 293T cells (human embryonic kidney cells) was also examined [66]. It is crucial to emphasize that, in some cases, the benzo[*b*]furans compiled in this review exhibit superior activity when compared to reference antibacterial drugs, such as Chloramphenicol [59], Carbenicillin [59], Streptomycin [59,62], Tetracycline [59], Netropsin [65], Ciprofloxacin [65], and polyhexamethylene biguanide (pHMB) [66].

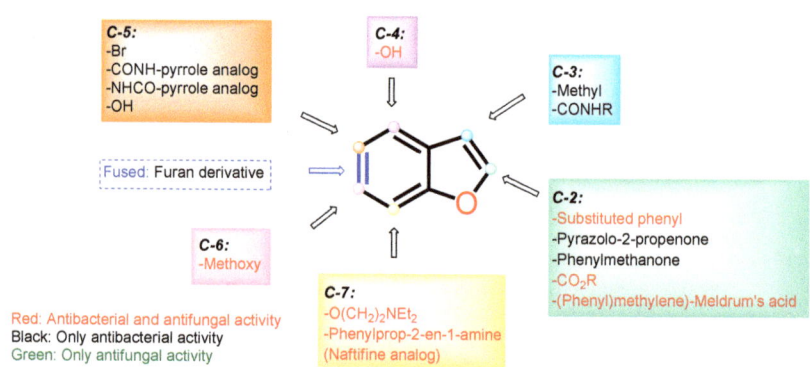

Figure 56. Analysis of the structure–activity relationship in antibacterial and antifungal results.

Conversely, a significant number of antibacterial compounds also demonstrate antifungal properties. Just like their antibacterial counterparts, the demand for novel compounds is pressing due to the emergence of multiple resistances. This review reveals that at a structural level, numerous benzo[*b*]furan derivatives exhibit both antibacterial and antifungal properties, as illustrated in Figure 56. The fungal strains examined within the studies discussed in this review displayed significant efficacy against *Candida albicans* (*C. albicans*) [58,61], *Candida neoformans* (*C. neoformans*) [58], *Candida parapsilosis* (*C. parapsilosis*) [61], *Saccharomyces cerevisiae* (*S. cerevisiae*) [61], *Zygosaccharomyces rouxii* (*Z. rouxii*) [61], *Aspergillus flavus* (*A. flavus*), *Microspora griseus* (*M. griseus*), and *Trichophyton rubrum* (*T. rubrum*) [64]. Notably, in some cases, their antifungal activity is comparable to reference drugs, such as Amphotericin B [58], Fluconazole [62,64], Voriconazole [64], and Ketoconazole [64].

3. Conclusions

Benzo[*b*]furans play a significant role in drug discovery and medicinal chemistry due to their unique structural features and wide range of biological and pharmaceutical properties. Most of the synthetic approaches for the synthesis and functionalization of these bioactive compounds involve multistep sequences. This comprehensive review provides a detailed overview of the synthesis and biological studies conducted on benzo[*b*]furans from 2011 to 2022, mainly focusing on their potential as anticancer, antibacterial, and antifungal agents. The selected articles provide invaluable insights related to chemical synthesis, discussions on molecular docking, and both in vitro and in vivo studies. By underscoring the importance of comprehending mechanisms of action, which encompass important processes such as inhibiting the cell cycle, inducing tumor regression, and demonstrating potent antibacterial and antifungal effects. Notably, the anticancer activity of benzo[*b*]furans correlates with their binding to tubulin's colchicine site, leading to polymerization inhibition. Colchicine-like fragments on specific positions of benzo[*b*]furan engage in interactions that impede cancer cell growth and metastasis. These compounds also influence apoptotic pathways, activate caspases, regulate Bcl-2, stabilize p53, affect angiogenesis, impact cell cycle and signaling pathways, modulate DNA activity, and influence metastasis in vivo. Conversely, benzo[*b*]furans exhibited a dual effectiveness against antibacterial and antifungal infections. In particular, benzo[*b*]furans exhibited significant efficacy against Gram-positive and Gram-negative bacterial strains, positioning them as potential superior antibacterial agents compared to reference drugs. These findings hold utmost importance in propelling scientific research and drug discovery, possibly paving the way for novel and potent therapeutic agents containing the benzo[*b*]furan ring.

Author Contributions: Conceptualization, writing—original draft preparation, and writing—review and editing, L.A.-R., J.-C.C. and D.B.; supervision, D.B. All authors have read and agreed to the published version of the manuscript.

Funding: This research received no external funding.

Institutional Review Board Statement: Not applicable.

Informed Consent Statement: Not applicable.

Data Availability Statement: Data sharing is not applicable.

Acknowledgments: The authors give thanks to the Universidad Pedagógica y Tecnológica de Colombia.

Conflicts of Interest: The authors declare no conflict of interest.

References

1. Taylor, A.P.; Robinson, R.P.; Fobian, Y.M.; Blakemore, D.C.; Jones, L.H.; Fadeyi, O. Modern advances in heterocyclic chemistry in drug discovery. *Org. Biomol. Chem.* **2016**, *14*, 6611–6637. [CrossRef] [PubMed]
2. Kerru, N.; Gummidi, L.; Maddila, S.; Gangu, K.K.; Jonnalagadda, S.B. A review on recent advances in nitrogen-containing molecules and their biological applications. *Molecules* **2020**, *25*, 1909. [CrossRef] [PubMed]
3. Kabir, E.; Uzzaman, M. A review on biological and medicinal impact of heterocyclic compounds. *Results Chem.* **2022**, *4*, 100606. [CrossRef]
4. Vitaku, E.; Smith, D.T.; Njardarson, J.T. Analysis of the structural diversity, substitution patterns, and frequency of nitrogen heterocycles among U.S. FDA-approved pharmaceuticals. *J. Med. Chem.* **2014**, *57*, 10257–10274. [CrossRef] [PubMed]
5. Heravi, M.M.; Zadsirjan, V. Prescribed drugs containing nitrogen heterocycles: An overview. *RSC Adv.* **2020**, *10*, 44247–44311. [CrossRef] [PubMed]
6. Nishanth Rao, R.; Jena, S.; Mukherjee, M.; Maiti, B.; Chanda, K. Green synthesis of biologically active heterocycles of medicinal importance: A review. *Environ. Chem. Lett.* **2021**, *19*, 3315–3358. [CrossRef]
7. Becerra, D.; Abonia, R.; Castillo, J.C. Recent applications of the multicomponent synthesis for bioactive pyrazole derivatives. *Molecules* **2022**, *27*, 4723. [CrossRef]
8. Martins, P.; Jesus, J.; Santos, S.; Raposo, L.R.; Roma-Rodrigues, C.; Baptista, P.V.; Fernandes, A.R. Heterocyclic anticancer compounds: Recent advances and the paradigm shift towards the use of nanomedicine's tool box. *Molecules* **2015**, *20*, 16852–16891. [CrossRef]
9. Kumar, D.; Jain, S.K. A comprehensive review of N-heterocycles as cytotoxic agents. *Curr. Med. Chem.* **2016**, *23*, 4338–4394. [CrossRef]
10. Serrano-Sterling, C.; Becerra, D.; Portilla, J.; Rojas, H.; Macías, M.; Castillo, J.-C. Synthesis, biological evaluation and X-ray crystallographic analysis of novel (*E*)-2-cyano-3-(het)arylacrylamides as potential anticancer agents. *J. Mol. Struct.* **2021**, *1244*, 130944. [CrossRef]
11. Hurtado-Rodríguez, D.; Salinas-Torres, A.; Rojas, H.; Becerra, D.; Castillo, J.-C. Bioactive 2-pyridone-containing heterocycle syntheses using multicomponent reactions. *RSC Adv.* **2022**, *12*, 35158–35176. [CrossRef] [PubMed]
12. Leeson, P.; Springthorpe, B. The influence of drug-like concepts on decision-making in medicinal chemistry. *Nat. Rev. Drug Discov.* **2007**, *6*, 881–890. [CrossRef] [PubMed]
13. Katritzky, A.R.; Ramsden, C.A.; Joule, J.A.; Zhdankin, V.V. *Handbook of Heterocyclic Chemistry*; Elsevier: Amsterdam, The Netherlands, 2010.
14. Cravotto, G.; Cintas, P. Power ultrasound in organic synthesis: Moving cavitational chemistry from academia to innovative and large-scale applications. *Chem. Soc. Rev.* **2006**, *35*, 180–196. [CrossRef] [PubMed]
15. Insuasty, D.; Castillo, J.; Becerra, D.; Rojas, H.; Abonia, R. Synthesis of biologically active molecules through multicomponent reactions. *Molecules* **2020**, *25*, 505. [CrossRef]
16. DeSimone, R.W.; Currie, K.S.; Mitchell, S.A.; Darrow, J.W.; Pippin, D.A. Privileged structures: Applications in drug discovery. *Comb. Chem. High Throughput Screen.* **2004**, *7*, 473–494. [CrossRef]
17. Karras, G.I.; Kustatscher, G.; Buhecha, H.R.; Allen, M.D.; Pugieux, C.; Sait, F.; Bycroft, M.; Ladurner, A.G. The macro domain is an ADP-ribose binding module. *EMBO J.* **2005**, *24*, 1911–1920. [CrossRef]
18. Zhao, K.; Harshaw, R.; Chai, X.; Marmorstein, R. Structural basis for nicotinamide cleavage and ADP-ribose transfer by NAD(+)-dependent Sir2 histone/protein deacetylases. *Proc. Natl. Acad. Sci. USA* **2004**, *101*, 8563–8568. [CrossRef]
19. Spaniol, M.; Bracher, R.; Ha, H.R.; Follath, F.; Kralhenbühl, S. Toxicity of amiodarone and amiodarone analogues on isolated rat liver mitochondria. *J. Hepatol.* **2001**, *35*, 628–636. [CrossRef]
20. Narimatsu, S.; Takemi, C.; Kuramoto, S.; Tsuzuki, D.; Hichiya, H.; Tamagake, K.; Yamamoto, S. Stereoselectivity in the oxidation of Bufuralol, a chiral substrate, by human cyctochrome P450s. *Chirality* **2003**, *15*, 333–339. [CrossRef]

21. Tseng, T.-H.; Lee, H.-J.; Lee, Y.-J.; Lee, K.-C.; Shen, C.-H.; Kuo, H.-C. Ailanthoidol, a neolignan, suppresses TGF-β1-induced HepG2 hepatoblastoma cell progression. *Biomedicines* **2021**, *9*, 1110. [CrossRef]
22. Proksch, P.; Rodríguez, E. Chromenes and benzofurans of the asteraceae, their chemistry and biological significance. *Phytochemistry* **1983**, *22*, 2335–2348. [CrossRef]
23. Miao, Y.-H.; Hu, Y.-H.; Yang, J.; Liu, T.; Sun, J.; Wang, X.-J. Natural source, bioactivity and synthesis of benzofuran derivatives. *RSC Adv.* **2019**, *9*, 27510–27540. [CrossRef] [PubMed]
24. Hwang, S.M.; Lee, H.-J.; Jung, J.-H.; Sim, D.Y.; Hwang, J.; Park, J.E.; Shim, B.S.; Kim, S.-H. Inhibition of Wnt3a/FOXM1/β-catenin axis and activation of GSK3β and caspases are critically involved in apoptotic effect of moracin D in breast cancers. *Int. J. Mol. Sci.* **2018**, *19*, 2681. [CrossRef] [PubMed]
25. Yoon, J.S.; Lee, H.-J.; Sim, D.Y.; Im, E.; Park, J.E.; Park, W.Y.; Koo, J.I.; Shim, B.S.; Kim, S.-H. Moracin D induces apoptosis in prostate cancer cells via activation of PPAR gamma/PKC delta and inhibition of PKC alpha. *Phytother. Res.* **2021**, *35*, 6944–6953. [CrossRef]
26. Aslam, S.N.; Stevenson, P.C.; Kokubun, T.; Hall, D.R. Antibacterial and antifungal activity of cicerfuran and related 2-arylbenzofurans and stilbenes. *Microbiol. Res.* **2009**, *164*, 191–195. [CrossRef]
27. Kadieva, M.G.; Oganesyan, É.T. Methods for the synthesis of benzofuran derivatives. *Chem. Heterocycl. Compd.* **1997**, *33*, 1245–1258. [CrossRef]
28. Shamsuzzaman, H.K. Bioactive benzofuran derivatives: A review. *Eur. J. Med. Chem.* **2015**, *97*, 483–504. [CrossRef]
29. Khodarahmi, G.; Asadi, P.; Hassanzadeh, F.; Khodarahmi, E. Benzofuran as a promising scaffold for the synthesis of antimicrobial and antibreast cancer agents: A review. *J. Res. Med. Sci.* **2015**, *20*, 1094–1104. [CrossRef]
30. Dawood, K.M. An update on benzofuran inhibitors: A patent review. *Expert Opin. Ther. Pat.* **2019**, *29*, 841–870. [CrossRef]
31. Dwarakanath, D.; Gaonkar, S.L. Advances in synthetic strategies and medicinal importance of benzofurans: A review. *Asian J. Org. Chem.* **2022**, *11*, e202200282. [CrossRef]
32. Flynn, B.L.; Gill, G.S.; Grobelny, D.W.; Chaplin, J.H.; Paul, D.; Leske, A.F.; Lavranos, T.C.; Chalmers, D.K.; Charman, S.A.; Kostewicz, E.; et al. Discovery of 7-hydroxy-6-methoxy-2-methyl-3-(3,4,5-trimethoxybenzoyl)benzo[b]furan (BNC105), a tubulin polymerization inhibitor with potent antiproliferative and tumor vascular disrupting properties. *J. Med. Chem.* **2011**, *54*, 6014–6027. [CrossRef] [PubMed]
33. Larock, R.C.; Yum, E.K.; Doty, M.J.; Sham, K.K.C. Synthesis of aromatic heterocycles via palladium-catalyzed annulation of internal alkynes. *J. Org. Chem.* **1995**, *60*, 3270–3271. [CrossRef]
34. Romagnoli, R.; Baraldi, P.G.; Lopez-Cara, C.; Cruz-Lopez, O.; Carrion, M.D.; Kimatrai Salvador, M.; Bermejo, J.; Estévez, S.; Estévez, F.; Balzarini, J.; et al. Synthesis and antitumor molecular mechanism of agents based on amino 2-(3′,4′,5′-trimethoxybenzoyl)benzo[b]furan: Inhibition of tubulin and induction of apoptosis. *ChemMedChem* **2011**, *6*, 1841–1853. [CrossRef] [PubMed]
35. Vincent, L.; Kermani, P.; Young, L.M.; Cheng, J.; Zhang, F.; Shido, K.; Lam, G.; Bompais-Vincent, H.; Zhu, Z.; Hicklin, D.J.; et al. Combretastatin A4 phosphate induces rapid regression of tumor neovessels and growth through interference with vascular endothelial-cadherin signaling. *J. Clin. Investig.* **2005**, *115*, 2992–3006. [CrossRef] [PubMed]
36. Wellington, K.W.; Qwebani-Ogunleye, T.; Kolesnikova, N.I.; Brady, D.; De Koning, C.B. One-pot laccase-catalysed synthesis of 5,6-dihydroxylated benzo[b]furans and catechol derivatives, and their anticancer activity. *Arch. Pharm.* **2013**, *346*, 266–277. [CrossRef]
37. Kamal, A.; Reddy, N.V.S.; Nayak, V.L.; Reddy, V.S.; Prasad, B.; Nimbarte, V.D.; Srinivasulu, V.; Vishnuvardhan, M.V.P.S.; Reddy, C.S. Synthesis and biological evaluation of benzo[b]furans as inhibitors of tubulin polymerization and inducers of apoptosis. *ChemMedChem* **2014**, *9*, 117–128. [CrossRef] [PubMed]
38. Frías, M.; Padrón, J.M.; Alemán, J. Dienamine and Friedel-Crafts one-pot synthesis, and antitumor evaluation of diheteroarylalkanals. *Chem. Eur. J.* **2015**, *21*, 8237–8241. [CrossRef]
39. Penthala, N.R.; Zong, H.; Ketkar, A.; Madadi, N.R.; Janganati, V.; Eoff, R.L.; Guzman, M.L.; Crooks, P.A. Synthesis, anticancer activity and molecular docking studies on a series of heterocyclic trans-cyanocombretastatin analogues as antitubulin agents. *Eur. J. Med. Chem.* **2015**, *92*, 212–220. [CrossRef]
40. Romagnoli, R.; Baraldi, P.G.; Salvador, M.K.; Prencipe, F.; Lopez-Cara, C.; Schiaffino Ortega, S.; Brancale, A.; Hamel, E.; Castagliuolo, I.; Mitola, S.; et al. Design, synthesis, in vitro, and in vivo anticancer and antiangiogenic activity of novel 3-arylaminobenzofuran derivatives targeting the colchicine site on tubulin. *J. Med. Chem.* **2015**, *58*, 3209–3222. [CrossRef]
41. Kamal, A.; Lakshma Nayak, V.; Nagesh, N.; Vishnuvardhan, M.V.P.S.; Subba Reddy, N.V. Benzo[b]furan derivatives induces apoptosis by targeting the PI3K/Akt/MTOR signaling pathway in human breast cancer cells. *Bioorg. Chem.* **2016**, *66*, 124–131. [CrossRef]
42. Yin, S.Y.; Jian, F.Y.; Chen, Y.H.; Chien, S.C.; Hsieh, M.C.; Hsiao, P.W.; Lee, W.H.; Kuo, Y.H.; Yang, N.S. Induction of IL-25 secretion from tumour-associated fibroblasts suppresses mammary tumour metastasis. *Nat. Commun.* **2016**, *7*, 11311. [CrossRef]
43. Quan, Y.P.; Cheng, L.P.; Wang, T.C.; Pang, W.; Wu, F.H.; Huang, J.W. Molecular modeling study, synthesis, and biological evaluation of combretastatin A-4 analogues as anticancer agents and tubulin inhibitors. *Medchemcomm* **2018**, *9*, 316–327. [CrossRef] [PubMed]

44. Lauria, A.; Gentile, C.; Mingoia, F.; Palumbo Piccionello, A.; Bartolotta, R.; Delisi, R.; Buscemi, S.; Martorana, A. Design, synthesis, and biological evaluation of a new class of benzo[b]furan derivatives as antiproliferative agents, with in silico predicted antitubulin activity. *Chem. Biol. Drug Des.* **2018**, *91*, 39–49. [CrossRef] [PubMed]
45. Pervaram, S.; Ashok, D.; Sarasija, M.; Reddy, C.V.R.; Sridhar, G. Synthesis and anticancer activity of 1,2,4-oxadiazole fused benzofuran derivatives. *Russ. J. Gen. Chem.* **2018**, *88*, 1219–1223. [CrossRef]
46. Kwiecień, H.; Perużyńska, M.; Stachowicz, K.; Piotrowska, K.; Bujak, J.; Kopytko, P.; Droździk, M. Synthesis and biological evaluation of 3-functionalized 2-phenyl- and 2-alkylbenzo[b]furans as antiproliferative agents against human melanoma cell line. *Bioorg. Chem.* **2019**, *88*, 102930. [CrossRef] [PubMed]
47. Abd El-Karim, S.S.; Anwar, M.M.; Mohamed, N.A.; Nasr, T.; Elseginy, S.A. Design, synthesis, biological evaluation and molecular docking studies of novel benzofuran–pyrazole derivatives as anticancer agents. *Bioorg. Chem.* **2015**, *63*, 1–12. [CrossRef] [PubMed]
48. Anwar, M.M.; Abd El-Karim, S.S.; Mahmoud, A.H.; Amr, A.E.-G.E.; Al-Omar, M.A. A comparative study of the anticancer activity and PARP-1 inhibiting effect of benzofuran–pyrazole scaffold and its nano-sized particles in human breast cancer cells. *Molecules* **2019**, *24*, 2413. [CrossRef]
49. Shao, Y.Y.; Yin, Y.; Lian, B.P.; Leng, J.F.; Xia, Y.Z.; Kong, L.Y. Synthesis and biological evaluation of novel shikonin-benzo[b]furan derivatives as tubulin polymerization inhibitors targeting the colchicine binding site. *Eur. J. Med. Chem.* **2020**, *190*, 112105. [CrossRef]
50. Sivaraman, A.; Kim, J.S.; Harmalkar, D.S.; Min, K.H.; Park, J.W.; Choi, Y.; Kim, K.; Lee, K. Synthesis and cytotoxicity studies of bioactive benzofurans from lavandula agustifolia and modified synthesis of ailanthoidol, homoegonol, and egonol. *J. Nat. Prod.* **2020**, *83*, 3354–3362. [CrossRef]
51. Oliva, P.; Romagnoli, R.; Manfredini, S.; Brancale, A.; Ferla, S.; Hamel, E.; Ronca, R.; Maccarinelli, F.; Giacomini, A.; Rruga, F.; et al. Design, synthesis, in vitro and in vivo biological evaluation of 2-amino-3-aroylbenzo[b]furan derivatives as highly potent tubulin polymerization inhibitors. *Eur. J. Med. Chem.* **2020**, *200*, 112448. [CrossRef]
52. Xu, S.; Yao, H.; Qiu, Y.; Zhou, M.; Li, D.; Wu, L.; Yang, D.H.; Chen, Z.S.; Xu, J. Discovery of novel polycyclic heterocyclic derivatives from evodiamine for the potential treatment of triple-negative breast cancer. *J. Med. Chem.* **2021**, *64*, 17346–17365. [CrossRef]
53. Zhang, J.; Zou, L.; Shi, D.; Liu, J.; Zhang, J.; Zhao, R.; Wang, G.; Zhang, L.; Ouyang, L.; Liu, B. Structure-guided design of a small-molecule activator of sirtuin-3 that modulates autophagy in triple negative breast cancer. *J. Med. Chem.* **2021**, *64*, 14192–14216. [CrossRef]
54. Irfan, A.; Faiz, S.; Rasul, A.; Zafar, R.; Zahoor, A.F.; Kotwica-Mojzych, K.; Mojzych, M. Exploring the synergistic anticancer potential of benzofuran–oxadiazoles and triazoles: Improved ultrasound- and microwave-assisted synthesis, molecular docking, hemolytic, thrombolytic and anticancer evaluation of furan-based molecules. *Molecules* **2022**, *27*, 1023. [CrossRef] [PubMed]
55. Becerra, D.; Castillo, J.; Insuasty, B.; Cobo, J.; Glidewell, C. Synthesis of N-substituted 3-(2-aryl-2-oxoethyl)-3-hydroxyindolin-2-ones and their conversion to N-substituted (E)-3-(2-aryl-2-oxo-ethylidene)indolin-2-ones: Synthetic sequence, spectroscopic characterization and structures of four 3-hydroxycompounds and five oxo-ethylidene products. *Acta Cryst.* **2020**, *C76*, 433–445. [CrossRef]
56. Vaali-Mohammed, M.-A.; Abdulla, M.-H.; Matou-Nasri, S.; Eldehna, W.M.; Meeramaideen, M.; Elkaeed, E.B.; El-Watidy, M.; Alhassan, N.S.; Alkhaya, K.; Al Obeed, O. The anticancer effects of the pro-apoptotic benzofuran-isatin conjugate (5a) are associated with P53 upregulation and enhancement of conventional chemotherapeutic drug efficiency in colorectal cancer cell lines. *Front. Pharmacol.* **2022**, *13*, 923398. [CrossRef]
57. Hiremathad, A.; Patil, M.R.; Chethana, K.R.; Chand, K.; Santos, M.A.; Keri, R.S. Benzofuran: An emerging scaffold for antimicrobial agents. *RSC Adv.* **2015**, *5*, 1–65. [CrossRef]
58. Sastraruji, T.; Chaiyong, S.; Jatisatienr, A.; Pyne, S.G.; Ung, A.T.; Lie, W. Phytochemical studies on *Stemona aphylla*: Isolation of a new stemofoline alkaloid and six new stemofurans. *J. Nat. Prod.* **2011**, *74*, 60–64. [CrossRef]
59. Ashok, D.; Sudershan, K.; Khalilullah, M. Solvent-free microwave-assisted synthesis of E-(1)-(6-benzoyl-3,5-dimethylfuro[3′,2′:4,5] benzo[b]furan-2-yl)-3-(aryl)-2-propen-1-ones and their antibacterial activity. *Green Chem. Lett. Rev.* **2012**, *5*, 121–125. [CrossRef]
60. Kossakowski, J.; Ostrowska, K.; Hejchman, E.; Wolska, I. Synthesis and structural characterization of derivatives of 2- and 3-benzo[b]furan carboxylic acids with potential cytotoxic activity. *Farmaco* **2005**, *60*, 519–527. [CrossRef]
61. Ostrowska, K.; Hejchman, E.; Wolska, I.; Kruszewska, H.; Maciejewska, D. Microwave-assisted preparation and antimicrobial activity of O-alkylamino benzofurancarboxylates. *Monatsh. Chem.* **2013**, *144*, 1679–1689. [CrossRef]
62. Kenchappa, R.; Bodke, Y.D.; Telkar, S.; Sindhe, M.A.; Giridhar, M. Synthesis, characterization, and antimicrobial activity of new benzofuran derivatives. *Russ. J. Gen. Chem.* **2016**, *86*, 2827–2836. [CrossRef]
63. Abonia, R.; Garay, A.; Castillo, J.-C.; Insuasty, B.; Quiroga, J.; Nogueras, M.; Cobo, J.; Butassi, E.; Zacchino, S. Design of two alternative routes for the synthesis of naftifine and analogues as potential antifungal agents. *Molecules* **2018**, *23*, 520. [CrossRef]
64. Wang, Y.; Chen, F.; Di, H.; Xu, Y.; Xiao, Q.; Wang, X.; Wei, H.; Lu, Y.; Zhang, L.; Zhu, J.; et al. Discovery of potent benzofuran-derived diapophytoene desaturase (CrtN) inhibitors with enhanced oral bioavailability for the treatment of methicillin-resistant *Staphylococcus aureus* (MRSA) infections. *J. Med. Chem.* **2016**, *59*, 3215–3230. [CrossRef] [PubMed]
65. Andriollo, P.; Hind, C.K.; Picconi, P.; Nahar, K.S.; Jamshidi, S.; Varsha, A.; Clifford, M.; Sutton, J.M.; Rahman, K.M. C8-linked pyrrolobenzodiazepine monomers with inverted building blocks show selective activity against multidrug resistant gram-positive bacteria. *ACS Infect. Dis.* **2018**, *4*, 158–174. [CrossRef] [PubMed]

66. Lei, S.; Hu, Y.; Yuan, C.; Sun, R.; Wang, J.; Zhang, Y.; Zhang, Y.; Lu, D.; Fu, L.; Jiang, F. Discovery of sortase A covalent inhibitors with benzofuranene cyanide structures as potential antibacterial agents against *Staphylococcus aureus*. *Eur. J. Med. Chem.* **2022**, *229*, 114032. [CrossRef] [PubMed]
67. He, W.; Zhang, Y.; Bao, J.; Deng, X.; Batara, J.; Casey, S.; Guo, Q.; Jiang, F.; Fu, L. Synthesis, biological evaluation and molecular docking analysis of 2-phenyl-benzofuran-3-carboxamide derivatives as potential inhibitors of *Staphylococcus aureus* sortase A. *Bioorg. Med. Chem.* **2017**, *25*, 1341–1351. [CrossRef]

Disclaimer/Publisher's Note: The statements, opinions and data contained in all publications are solely those of the individual author(s) and contributor(s) and not of MDPI and/or the editor(s). MDPI and/or the editor(s) disclaim responsibility for any injury to people or property resulting from any ideas, methods, instructions or products referred to in the content.

Article

Synthesis, DFT Study, and In Vitro Evaluation of Antioxidant Properties and Cytotoxic and Cytoprotective Effects of New Hydrazones on SH-SY5Y Neuroblastoma Cell Lines

Diana Tzankova [1], Hristina Kuteva [2], Emilio Mateev [1], Denitsa Stefanova [2], Alime Dzhemadan [2], Yordan Yordanov [2], Alexandrina Mateeva [1], Virginia Tzankova [2], Magdalena Kondeva-Burdina [2], Alexander Zlatkov [1] and Maya Georgieva [1,*]

[1] Department "Pharmaceutical Chemistry", Faculty of Pharmacy, Medical University-Sofia, 2 Dunav Str., 1000 Sofia, Bulgaria; d.tsankova@pharmfac.mu-sofia.bg (D.T.); e.mateev@pharmfac.mu-sofia.bg (E.M.); a.dineva@pharmfac.mu-sofia.bg (A.M.); azlatkov@pharmfac.mu-sofia.bg (A.Z.)

[2] Laboratory "Drug metabolism and Drug Toxicity", Department "Pharmacology, Pharmacotherapy and Toxicology", Faculty of Pharmacy, Medical University-Sofia, 2 Dunav Str., 1000 Sofia, Bulgaria; hristina.kuteva@abv.bg (H.K.); denitsa.stefanova@pharmfac.mu-sofia.bg (D.S.); yyordanov@pharmfac.mu-sofia.bg (Y.Y.); vtzankova@pharmfac.mu-sofia.bg (V.T.); mkondeva@pharmfac.mu-sofia.bg (M.K.-B.)

* Correspondence: mgeorgieva@pharmfac.mu-sofia.bg

Citation: Tzankova, D.; Kuteva, H.; Mateev, E.; Stefanova, D.; Dzhemadan, A.; Yordanov, Y.; Mateeva, A.; Tzankova, V.; Kondeva-Burdina, M.; Zlatkov, A.; et al. Synthesis, DFT Study, and In Vitro Evaluation of Antioxidant Properties and Cytotoxic and Cytoprotective Effects of New Hydrazones on SH-SY5Y Neuroblastoma Cell Lines. *Pharmaceuticals* 2023, 16, 1198. https://doi.org/10.3390/ph16091198

Academic Editor: Valentina Noemi Madia

Received: 29 July 2023
Revised: 17 August 2023
Accepted: 18 August 2023
Published: 23 August 2023

Copyright: © 2023 by the authors. Licensee MDPI, Basel, Switzerland. This article is an open access article distributed under the terms and conditions of the Creative Commons Attribution (CC BY) license (https://creativecommons.org/licenses/by/4.0/).

Abstract: A series of ten new hydrazide–hydrazone derivatives bearing a pyrrole ring were synthesized and structurally elucidated through appropriate spectral characteristics. The target hydrazones were assessed for radical scavenging activity through 1,1-diphenyl-2-picrylhydrazyl (DPPH) and 2,2′-azino-bis(3-ethylbenzothiazoline-6-sulphonic acid) (ABTS) tests, with ethyl 5-(4-bromophenyl)-1-(2-(2-(4-hydroxy-3,5-dimethoxybenzylidene)hydrazine-yl)-2-oxoethyl)-2-methyl-1H-pyrrole-3-carboxylate (**7d**) and ethyl 5-(4-bromophenyl)-1-(3-(2-(4-hydroxy-3,5-dimethoxybenzylidene) hydrazine-yl)-3-oxopropyl)-2-methyl-1H-pyrrole-3-carboxylate (**8d**) highlighted as the best radical scavengers from the series. Additional density functional theory (DFT) studies have indicated that the best radical scavenging ligands in the newly synthesized molecules are stable, do not decompose into elements, are less polarizable, and with a hard nature. The energy of the highest occupied molecular orbital (HOMO) revealed that both compounds possess good electron donation capacities. Overall, **7d** and **8d** can readily scavenge free radicals in biological systems via the donation of hydrogen atoms and single electron transfer. The performed in vitro assessment of the compound's protective activity on the H_2O_2-induced oxidative stress model on human neuroblastoma cell line SH-SY5Y determined **7d** as the most perspective representative with the lowest cellular toxicity and the highest protection.

Keywords: pyrrole; synthesis; DFT; antioxidant; SH-SY5Y; cellular toxicity; cell protection

1. Introduction

The generation of reactive oxygen species (ROS) leads to oxidative damage to biomolecules, such as cellular proteins, lipids, and DNA, and, as a result, to disruption to normal cellular functions [1]. A gradual decline in cellular antioxidant defense mechanisms due to the generation of ROS during aging leads to increases in oxidative stress. Therefore, aging, various genetic mutations, the influence of the environment, and the associated increase in oxidative stress are a prerequisite for the development of many neurodegenerative diseases, including Parkinson's disease (PD), Alzheimer's disease (AD), Huntington's disease (HD), amyotrophic lateral sclerosis (ALS), and many others [2]. Accordingly, the discovery of novel active substances with antioxidant potential is the basis of many programs for providing effective neuroprotection in the therapy of neurodegenerative diseases [3].

A lot of market drugs containing pyrrole ring show different pharmacological activities, such as antipsychotics [4], antidepressants [5], anticonvulsants [6], anti-inflammatories [7], and many others. There is increased interest in pyrroles and pyrrole derivatives exhibiting proven antioxidant effects [8,9]. Many heterocyclic hydrazones have shown antioxidant and neuroprotective activity, too [10].

For example, Boulebd et al. [11] successfully synthesized a series of new phenolic hydrazide-hydrazone derivatives and tested them for radical scavenging activity using DPPH and ABTS methods. The effects were compared with the action of ascorbic acid and Trolox, compounds with proven antioxidant activity. It was found that the synthesized new derivatives exhibited antioxidant activity comparable to those of ascorbic acid and Trolox [11]. In another study [12], twelve heterocyclic compounds bearing hydrazone functional groups were evaluated for antioxidant activity using DPPH, ABTS, and DMSO alkaline assays. The results showed that these heterocyclic compounds are potent antiradical agents [12]. In addition, a preliminary study performed by Tzankova et al. [13] identified the promising radical scavenging potential of pyrrole-based hydrazones, as determined through DPPH and ABTS assays, which pointed our attention toward the synthesis of analogous representatives in an attempt to enrich the variety of molecules.

In the current study, the initial hydrazides were synthesized in our laboratory, as previously explained in detail by Bijev et al. [14] for hydrazide **7** and Georgieva et al. [15] for hydrazide **8**, with the general structure presented in Figure 1.

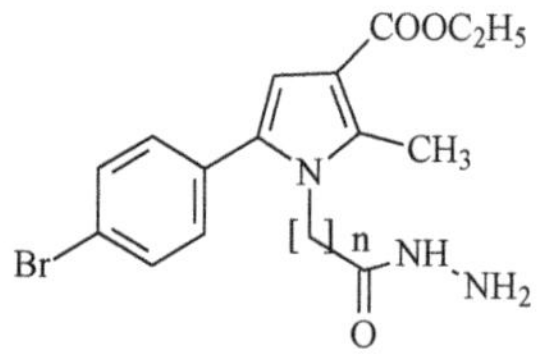

n = 1 (**7**) and n = 2 (**8**)

Figure 1. General structure of the initial pyrrole hydrazides.

The following study is focused on the synthesis of new hydrazones containing a pyrrole ring system and the evaluation of their in vitro safety profile on human neuroblastoma SH-SY5Y cells, radical scavenging activity through 1,1-diphenyl-2-picrylhydrazyl (DPPH) and 2,2′-azino-bis(3-ethylbenzothiazoline-6-sulphonic acid) (ABTS) assays and in silico assessment of possible antioxidant mechanisms through DFT calculations. The most promising structures are evaluated for antioxidative protective properties in a H_2O_2-induced oxidative stress model on SH-SY5Y cells.

2. Results

2.1. Chemistry

2.1.1. Synthesis of the N-pyrrolyl hydrazides 7 (ethyl 5-(4-bromophenyl)-1-(2-hydrazinyl-2-oxoethyl)-2-methyl-1H-pyrrole-3-carboxylate) and 8 (ethyl 5-(4-bromophenyl)-1-(3-hydrazinyl-3-oxopropyl)-2-methyl-1H-pyrrole-3-carboxylate)

The pyrrole ring in the target hydrazone molecules was formed through a Paal–Knorr cyclization reaction based on the C-alkylation of a 1,3-dicarbonyl compound to the corresponding 1,4-dicarbonyl derivative (**2**), cycled afterward with the corresponding amino acids L-glycine (n = 1) and L-β-alanine (n = 2) to give the following N-substituted pyrrole carboxylic acids (**3** and **4**, respectively). An esterification of the obtained acids was conducted, followed by a hydrazinolysis reaction with hydrazine hydrate, to give the target hydrazides (**7** and **8**, respectively) according to the procedure presented in Scheme 1 and described in detail in [14,15].

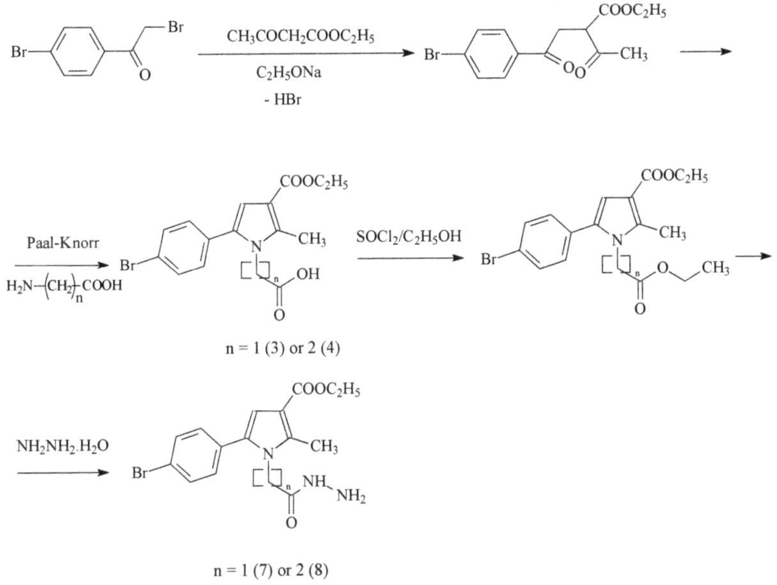

Scheme 1. Synthesis of the initial hydrazides **7** and **8** [14,15].

2.1.2. Synthesis of the New N–pyrrolylhydrazide–hydrazones 7a–e and 8a–e

The novel series of N–pyrrolylhydrazide–hydrazone derivatives were prepared under microsynthesis scale conditions through condensation reaction from the previously synthesized in our laboratory hydrazides **7** and **8** and the selected carbonyl partners (Figure 2), assuring about 64–86% yield of the purified product. The new compounds were synthesized according to the procedure presented in Scheme 2.

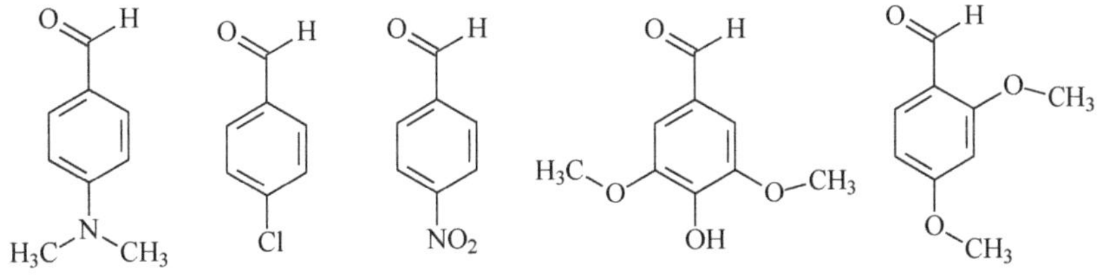

a **b** **c** **d** **e**

Figure 2. Used aldehydes (**a–e**).

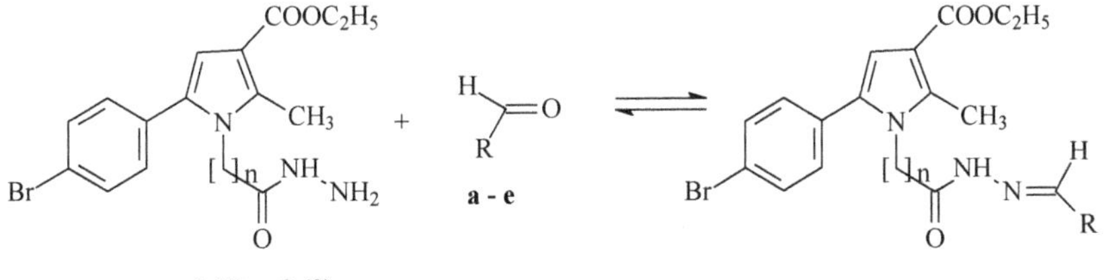

n = 1 (**7**) or 2 (**8**) **7a-e** or **8a-e**

Scheme 2. Synthesis of the new hydrazone compounds.

The reaction conditions were altered, as presented in Table 1, wherein the reflux in glacial acetic acid was selected as the most appropriate. All the new derivatives were obtained according to that presented in the Materials and Methods section procedures.

Table 1. Reaction conditions, reaction times, and yields.

Reaction Media	Reaction Temperature °C	Reaction Time (min)	Yields %
Ethanol	Heating	50–60	46–80
Ethanol + HCl	Room temperature	20–30	56–94
Methanol + conc. HCl	Room temperature	1440	15–56
Ethanol + glacial acetic acid	100 °C	30–50	26–64
Glacial acetic acid	**100 °C**	**20–30**	**68–84**

The synthesized pyrrole hydrazones are stable at room temperature for long periods of time. The structures of the new compounds were elucidated by melting points, TLC characteristics, IR, and ^{1}H-NMR spectral data, followed by MS data. The results from the spectral analysis confirmed that the new compounds are consistent with the expected structure. The corresponding experimental IR, ^{1}H-NMR, and LC-MS spectra of the target derivatives are supplied as Supplementary Materials. The purity of the obtained compounds was proven by corresponding elemental analyses.

The IR data determine the presence of new signals at 3245 cm^{-1} for the valence asymmetric (ν^{as}) vibrations for the amide NH group in the molecule of the new hydrazones and a band at 1666 cm^{-1} (Amide I) and 1533 cm^{-1} for deformational (δ) vibrations of the amide NH group (Amide II). In addition, the appearance of a band at around 810 cm^{-1} determines the presence of p-substituted phenyl residue. The ester group (COOC$_2$H$_5$) in the third position in the central pyrrole ring is pointed by the appearance of a band at around 1693–1698 cm^{-1}. A confirmation of the structural elucidation may be seen in the relevant ^{1}H NMR spectra, where the corresponding groups are available at 7.86 ppm for the CH=N group and at 11.38 ppm for the CONH group, respectively. The obtained values from the experimental spectral data are consistent with the theoretical ones.

The corresponding IDs, melting points, TLC characteristics, MS data, and yields are given in Table 2. The respective IR characteristics and ^{1}H-NMR spectral data are presented in the Materials and Methods section.

Table 2. IDs, melting points, TLC characteristics (Rf), MS data, and yields for the new N-pyrrolyl-hydrazones.

IDs	m.p. °C	Rf	MS Data [M+H]$^+$ (m/z)	Yields %
7a	211.4–213.6	0.38	511.13	84
7b	229.4–231.2	0.33	502.05	78
7c	245.9–247.2	0.33	512.07	76
7d	191.9–194.4	0.29	544.10	68
7e	206.0–207.6	0.40	528.11	72
8a	212.0–213.3	0.35	525.15	86
8b	181.4–184.6	0.31	517.06	74
8c	214.4–217.1	0.33	527.09	82
8d	196.6–197.6	0.28	558.12	64
8e	170.0–171.9	0.34	542.13	74

2.2. Antioxidant Assays

2.2.1. DPPH Radical Scavenging Assay

The 1,1-diphenyl-2-picrylhydrazyl (DPPH) radical scavenging activity of the newly synthesized derivatives was determined at one concentration—1 mg/mL. Trolox was used as a standard. The obtained results are presented in Figure 3.

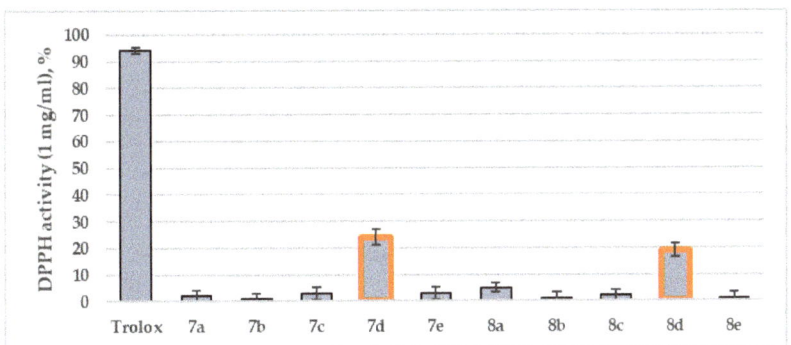

Figure 3. DPPH radical capacities of the newly synthesized compounds **7a–e** and **8a–e** at concentrations of 1 mg/mL. Trolox is used as an internal standard. Data are presented as means from three independent experiments. Standard deviation (SD) (n = 3).

The highest DPPH scavenging activity was achieved by compound **7d** (24%). The β-alanine hydrazide–hydrazone condensed with the same aldehyde (**8d**) demonstrated similar radical scavenging activity (19%). The standard, Trolox, showed 94% DPPH activity. The rest of the newly synthesized molecules demonstrated weak to no antioxidant effects.

2.2.2. ABTS Radical Scavenging Assay

During the 2,2′-azino-bis(3-ethylbenzothiazoline-6-sulphonic acid) (ABTS) assay, the discoloration of the initial color could be detected at 734 nm. The ABTS antioxidant assay of the title compounds is provided in Figure 4.

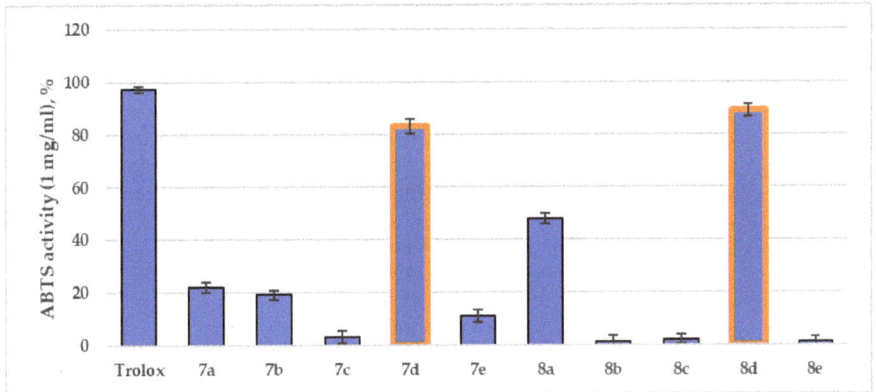

Figure 4. ABTS test of the newly synthesized compounds **7a–e** and **8a–e** at concentrations of 1 mg/mL. Trolox is used as an internal standard. Data are presented as means from three independent experiments. Standard deviation (SD) (n = 3).

2.3. DFT Calculations

To rationalize the antioxidant assays and to determine the favored mechanism involved in free radical scavenging, density functional theory (DFT) calculations were carried out at the B3LYP/6-311++(d,p) level of theory.

2.3.1. Optimized Geometries

The most prominent newly synthesized hydrazide–hydrazones (**7d** and **8d**) were selected for full optimization calculations. An initial conformational search has been performed by setting 2500 iterations with the OPLS4 force field. The best solutions were further optimized by full DFT geometry optimization at B3LYP/6-311++ (d,p) level of theory. The final geometries of **7d** and **8d** are visualized in Figure 5.

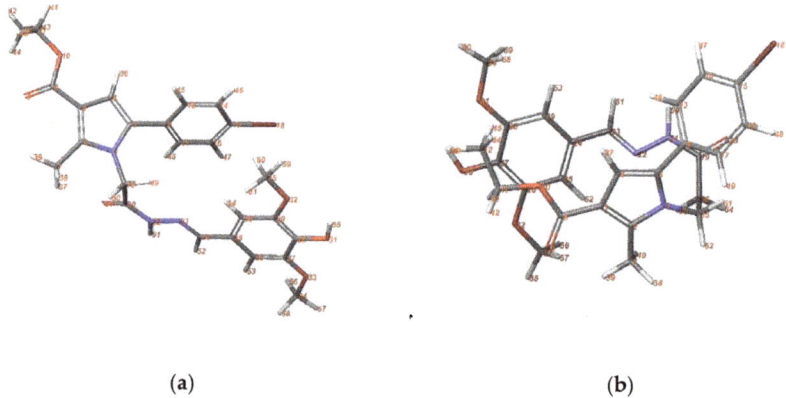

(a) (b)

Figure 5. Final optimized geometries of the structures with 2500 iterations with OPLS4 force field. The best solutions for (**a**) **7d**; (**b**) **8d** are further optimized at B3LYP/6-311++(d,p).

The overall minimized energies of **7d** and **8d** are −4197 and −4157 Hartree, respectively. It is important to note that *p*-bromophenyl moiety in structure **7d** was located in close proximity to the 3,5-dimethoxy-4-hydroxyphenyl fragment, while the same groups in **8d** were in different positions. In the latter case, the *p*-bromophenyl moiety faced away from the stable conformation. The most stable geometries have been selected as input geometries for further DFT studies.

2.3.2. Analysis of Frontier Molecular Orbitals FMOs and Global Reactivity Descriptors

A quantitative conceptual DFT analysis of the stability and reactivity of the investigated molecules (**7d** and **8d**) was carried out by calculations of the highest occupied molecular orbital (HOMO), the lowest unoccupied molecular orbital (LUMO), and subsequently, the global reactivity descriptors, such as ionization potential (IP), electron affinity (EA), molecular hardness and softness, electronegativity, and electrophilicity. The HOMO-LUMO electronic densities of **7d** and **8d** obtained from the optimized structures with DFT/B3LYP/6-311++(d,p) calculations are given in Figure 6. The blue color was used for a positive phase, and red was used for a negative phase.

It was found that the HOMO of both ligands is localized majorly on the pyrrole ring in **7d** and **8d**, while the LUMO is centered around the hydrazide–hydrazone moiety and the 3,5-dimethoxy-4-hydroxyphenyl fragment. The more energetically favorable conformation of **8d** (-4157 Hartree) led to a more compact structure and, therefore, the localization of HOMO is further spread to the hydrazide–hydrazone and the 3,5-dimethoxy-4-hydroxyphenyl moieties.

The energies (in atomic units) of the FMOs and the global reactivity descriptors (hardness (η), softness (S), electronegativity (χ), chemical potential (μ), and electrophilicity index (ω)) were calculated by applying the Koopman's theorem for closed-shell compounds [16]. The data is reported in Table 3.

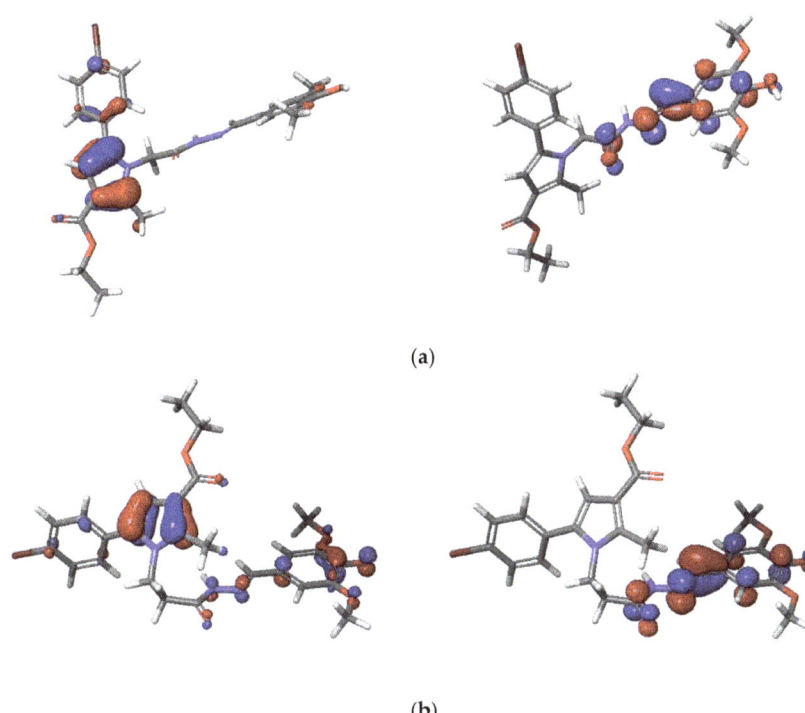

Figure 6. FMOs of **7d** and **8d**, calculated at DFT/B3LYP/6–311++(d,p): (**a**) HOMO and LUMO of **7d**; (**b**) HOMO and LUMO of **8d**. The HOMO-LUMO electronic densities are obtained from the optimized structures with DFT/B3LYP/6–311++(d,p) calculations. Blue color indicates a positive phase, and red indicates a negative phase.

Table 3. FMOs energies and global reactivity descriptors for **7d** and **8d** at B3LYP/6–311++(d,p) level of theory.

Electronic Parameter	7d	8d
E_{HOMO}	−0.2013	−0.1976
E_{LUMO}	−0.0564	−0.0470
$\Delta E_{HOMO-LUMO}$	0.1449	0.1506
Ionization Energy (IP)	0.2013	0.1976
Electron Affinity	0.0564	0.0470
Chemical Hardness	0.0724	0.0753
Softness	6.9060	6.6401
Electronegativity	0.1288	0.1223
Chemical Potential	−0.1288	−0.1223
Electrophilicity Index	0.1145	0.0992

2.3.3. Descriptors of the Antioxidant Properties

The calculated values for the dissociation of hydrogen-connected bonds and the IPs are provided in Table 4.

Table 4. Calculated bond dissociation energies (BDEs) and ionization potentials (IPs) of the corresponding compounds in gas phase.

Compound	Bond	BDE (Kcal/mol)	IP (Kcal/mol)
7d	O_{31}-H	83.55	126.31
	C_7-H	90.4	
	N_{22}-H	95.10	
8d	O_{30}-H	83.09	123.99
	C_7-H	88.9	
	N_{21}-H	96.2	

2.4. In Vitro Evaluations of the Cytotoxicity and Antioxidative Protective Activity on the SH-SY5Y Neuroblastoma Cell Line

2.4.1. Effects of the Newly Synthesized Derivatives **7a–e** and **8a–e** on the SH-SY5Y Cell Viability

The effects of the newly synthesized N-pyrrolyl-hydrazone derivatives on the cellular viability of neuronal SH-SY5Y cells were evaluated, and the corresponding IC50 values were calculated. In the current study, the cells were seeded at density 2×10^4 and were treated with the test compounds at concentrations 1–500 μM for 24 h. The calculated values of IC50 are presented in Table 5.

Table 5. In vitro cytotoxicity evaluation of the newly synthesized derivatives **7a–e** and **8a–e** on SH-SY5Y cells (IC50 values).

Compound IDs	IC50 [μM]	95% Confidence Intervals (CI)
7a	55.75	55.65–61.37
7b	67.60	54.31–79.64
7c	>500	NA
7d	99.56	88.87–103.23
7e	63.08	55.56–73.23
8a	56.33	45.25–67.63
8b	57.36	46.26–68.36
8c	58.23	47.36–69.23
8d	57.26	46.29–68.39
8e	91.07	83.65–105-36
Melatonin	>500	NA

The results demonstrated that compounds **7c**, **7d**, and **8e** possess lackluster or low toxicity on human neuronal SH-SY5Y cells, with IC50 of >500 μM, 99.56 μM, and 91.07 μM, respectively.

2.4.2. Effects of the Newly Synthesized Derivatives **7a–e** and **8a–e** in a Model of H_2O_2-Induced Oxidative Stress In Vitro

The potential cell protective effects of **7c**, **7d**, and **8e** (the less cytotoxic compounds from the series) were evaluated against H_2O_2-induced oxidative stress in SH-SY5Y cells (Figure 7).

The cells were pre-incubated with the tested compounds at concentrations 1, 10, and 20 μM for 90 min. Melatonin was used as a reference compound because of its well-established neuroprotective effects in various in vitro and in vivo models [17]. After 90 min pre-incubation with the test substances, SH-SY5Y cells were exposed to H_2O_2 (1 mM, 15 min), as described in the Materials and Methods section.

The most potent protection was observed with compound **7c**, followed by **7d** and **8e**. Interestingly, both **7c** and **7d** showed better protection in all tested concentrations than the reference compound melatonin. It should be noted that compounds **7e** and **7d** showed statistically significant protection even at the lowest tested concentration (1 μM). Compound **8e** showed lower antioxidant protection.

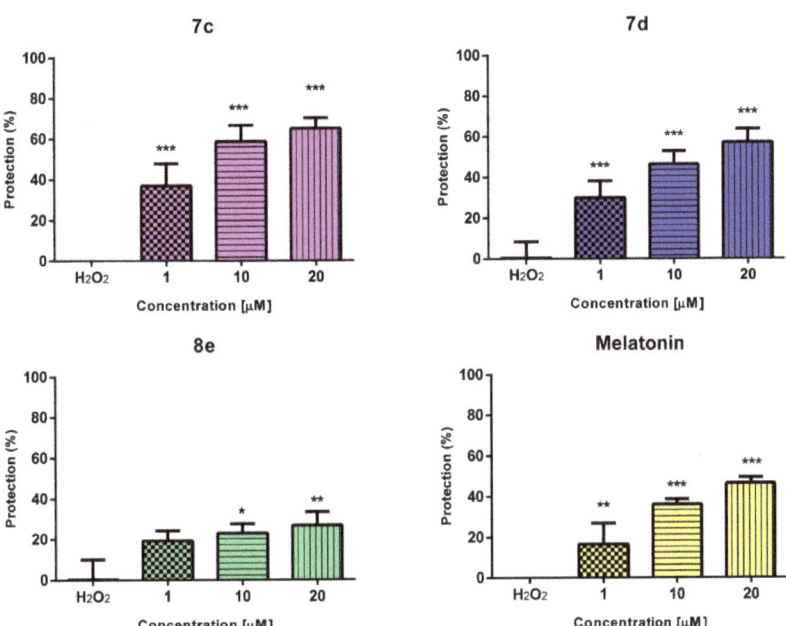

Figure 7. Protective effects of **7c**, **7d**, **8e**, and melatonin in a model of H_2O_2-induced oxidative damage in human neuroblastoma SH-SY5Y cells. Data are presented as means from three independent experiments ± SD (n = 8). * $p < 0.05$; ** $p < 0.01$; *** $p < 0.001$, vs. H_2O_2 group (one-way analysis of variance with Dunnet's post hoc test).

3. Discussion

3.1. Synthesis

The selection of the applied synthetic path was based on literary data showing the relation between the acidity of the media and the condensation time. Initially, the synthesis was carried out in an ethanol medium without a catalyst, with reaction times of 50–60 min [18]. In an attempt to speed up the process, the same reactions were carried out in an ethanol medium with a catalyst of glacial acetic acid. Under these conditions, the hydrazones were obtained in lower yields of 26–64%, with no significant alteration of the reaction time [19]. Thus, Koopaei presents some data on the interaction of carbohydrazides with carbonyl compounds in an ethanol medium with hydrochloric acid as the catalyst at 100 °C. Applying these conditions, it took 20–30 min to complete the reaction [20,21]. The obtained products were isolated and recrystallized. These conditions did not change either the yield or the time of the condensation.

Thus, in the current work, we conducted the synthesis entirely in glacial acetic acid media. This allowed good yields and optimal time for obtaining pure products (Table 1).

3.2. Antioxidant Assays

Two of the most common assays for the assessment of antioxidant capacities of natural and/or synthetic molecules are the well-applied 2,2′-azino-bis(3-ethylbenzothiazoline-6-sulphonic acid) (ABTS) and 1,1-diphenyl-2-picrylhydrazyl (DPPH) assays. Both techniques are based on spectrophotometric measurement of the quenching of stable colored radicals (ABTS$^{•+}$ or DPPH), which, in turn, allows evaluation of the radical scavenging ability of antioxidants even in complex mixtures [22].

3.2.1. DPPH Radical Scavenging Assay

One of the most common assays for the evaluation of antioxidant effects is the DPPH test. Its frequent applications are mainly due to its relative inexpensiveness [23].

The scavenging activity in the 1,1-diphenyl-2-picrylhydrazyl (DPPH) assay is attributed to the hydrogen-donating ability of antioxidants with a following decrease in the absorbance at 517 nm. The most probable mechanisms of the DPPH·assay are based on hydrogen atom abstraction (HAT) [24], single electron transfer (SET) [25], or mixed mode [26]. The following reaction schemes are related to these mechanisms:

$$\text{DPPH (violet at 515 nm)} + \text{ArOH} \rightarrow \text{DPPHH (colorless)} + \text{ArO· HAT} \quad (1)$$

$$\text{DPPH (violet at 515 nm)} + \text{ArOH} \rightarrow \text{DPPH}- \text{ (colorless)} + \text{ArO·} + \text{SET} \quad (2)$$

as suggested in [27].

This allowed us to apply this method in order to evaluate the possible radical scavenging effects of the newly synthesized pyrrole-based hydrazones. The obtained results for the target hydrazones are presented in Figure 3.

3.2.2. ABTS Radical Scavenging Assay

In the ABTS$^{•+}$ radical scavenging assay (an electron-transfer-based assay), the 2,2′-azino-bis(3-ethylbenzothiazoline-6-sulfonate) radical cation (ABTS$^{•+}$), which has a dark blue color, is reduced by an antioxidant into colorless ABTS, which can be measured spectrophotometrically at 734 nm [28].

A literary-based comparative analysis suggested two possible mechanisms of radical scavenging reactions related to the ABTS assay: formation of coupling adducts with ABTS$^{•+}$—reaction considered as more characteristic for antioxidants of phenolic nature and oxidation without coupling. In addition, the data suggested the possibility for further oxidative degradation of the obtained coupling adducts, leading to hydrazindyilidene-like and/or imine-like adducts with 3-ethyl-2-oxo-1,3-benzothiazoline-6-sulfonate and 3-ethyl-2-imino-1,3-benzothiazoline-6-sulfonate as marker compounds, respectively [29].

Considering the presence of a phenyl residue in the structure of the new pyrrole-based hydrazones, it was of interest to compare the possible radical scavenging activity of the target compounds when determined through both the most common in vitro antioxidant assays (Figures 3 and 4). The results from our evaluations demonstrated that when compared to the DPPH test, the ABTS assay revealed significantly better antioxidant capacities for the title compounds. The most prominent antioxidant was **8d**, with 89% scavenging activity, while **7d** showed 83%. Moreover, the hydrazide–hydrazones condensed with 4-dimethylamino benzaldehyde and **7a** and **8a** demonstrated moderate ABTS scavenging capacities of 22% and 48%, respectively. The drastic differences between the two assays could be related to the ability of the ABTS test to examine both hydrophobic and hydrophilic compounds, as well as to test bulky structures [30]. However, the application of the DPPH assay is still used by numerous research groups to produce significant and comparable antioxidant results [31].

In general, the best radical scavenging effect of compounds **7d** and **8d** is quite expected due to the presence of a free *p*-hydroxyl group in the phenyl part of the structure. The appearance of weak radical scavenging properties in the ABTS assay for compounds **7a** and **8a** is probably due to the free electron pair in the N-atom at the *p*-dimethylamino fragment of the phenyl residue. In addition, a negligible effect is observed for **7b**, containing an electron-withdrawing halogen Cl. The presence of *o*-methoxy groups in the phenyl residue is related to the lack of radical scavenging effects in **7e** and **8e**, probably due to the less likely possibility of these groups creating intermolecular hydrogen bonds.

It should also be mentioned that, in general, the shorter series **7a–e** is related to better radical scavenging activity in comparison to the one methylene longer series **8a–e**. This may be due to additional conformational changes leading to the hindrance of the possibility of hydrogen atom transfer and/or electron transfer ability of the active functional groups.

3.3. DFT Calculations

Several mechanisms for free-radical scavenging action of phenyl-containing molecules are proposed, where it is reported that phenolic O–H bond dissociation enthalpy (BDE), adiabatic ionization potential (IP), proton dissociation enthalpy (PDE), proton affinity (PA), and electron-transfer enthalpy (ETE) are important factors used to evaluate the preferred free-radical scavenging pathways of such structures thermodynamically [32]. In an attempt to clarify these mechanisms, some theoretical approaches are applied, with density functional theory (DFT) being most helpful in the calculation of these physicochemical descriptors [33].

The energies of the frontier molecular orbitals and the global reactivity descriptors were calculated with the B3LYP hybrid functional considering recent findings [32]. The corresponding HOMO and LUMO values provide information about the energy distribution and the energetic behavior of the **7d** and **8d**. The comparatively high negative value of HOMO emphasizes the high ability of both compounds to donate electrons to empty molecular orbital energy. Moreover, the negative magnitude of E_{HOMO} and E_{LUMO} establishes the stability of the compounds [34]. The energy gap of the FMOs ($E_{HOMO}-E_{LUMO}$) corresponds to the chemical reactivity and the kinetic stability. The calculated frontier molecular energy gaps are 0.1449 a.u and 0.1506 a.u for **7d** and **8d**, respectively. The large gap accounts for hard, unreactive, and less polarizable molecules.

The value of chemical softness (S) and hardness (η) identifies whether the compounds are reactive and can donate electrons. These descriptors are directly correlated with molecular stability and chemical reactivity. When compared, **7d** possesses slightly lower chemical hardness, which corresponds to lower stability. The calculated value of softness revealed that **8d** is more favorable in the charge–transfer mechanism. The negative value of the chemical potential of both ligands (−0.1288 and −0.1223) establishes good stabilities and resistance to sudden decomposition into their elements. The calculated electrophilicity index values (0.1145 and 0.0992 Ha), which are related to chemical potential and hardness, are indicative of the nucleophilicity power [35].

The DFT studies have indicated that the best radical scavenging ligands (**7d** and **8d**) in the newly synthesized molecules are stable, which do not decompose into elements, are less polarizable, and with hard nature. The energy of HOMO revealed that both compounds possess good electron donation capacities.

To examine the most probable mechanism of antioxidant activity of **7d** and **8d**, two processes were observed: hydrogen atom transfer (HAT) and single electron transfer (SET).

The bond dissociation energies (BDE) characterize the viability of hydrogen donation to the free radical. It is considered one of the most important descriptors in the process of determining the antioxidant effects [36]. Thus, the BDE was used as the best reliable thermodynamic parameter describing the hydrogen atom transfer (HAT) mechanism. Because this route involves the H atom transferring from a hydroxyl group of an antioxidant compound to the free radical, the weakest O–H bond with the lowest BDE is expected to be abstracted easier, revealing its higher antiradical (antioxidant) activity. Therefore, a computational calculation of the BDEs of available −OH groups in **7d** and **8d** will benefit in characterizing the theoretical antioxidant mechanism of the compounds. The activity of the −OH groups in both active ligands and the formation of stable radicals was conducted by calculating the dissociation energies.

In addition, apart from the HAT mechanism, another possible pathway for antioxidant molecules is single electron transfer followed by proton transfer (SET-PT). In this route, an electron is transferred from the antioxidant to the free radical, leading to radical cation formation, which subsequently deprotonates. Hence, adiabatic ionization proton (IP) and PDE are the most important parameters in describing the feasibility of the mechanism. In general, lower IPs are more subject to ionization and easier in electron-transfer rate between free radicals and antioxidants [33]. SET is related to the conceptual DFT parameters, mainly the IP energies [37]. This order reveals that compounds **7d** and **8d** are the most active antioxidants, with all the other calculated IPs being larger than the reference compound Trolox, suggesting their lower activity than Trolox.

The computationally calculated BDEs values of all possible dissociation sites in **7d** and **8d** revealed that radical formation occurs at O_{31}, followed by C_7 and N_{21}. A formation of possible intramolecular stabilization forces could conclude the result of strong N–H bond dissociation in the hydrazide moiety. The O–H bond is the weakest one in both **7d** and **8d**, which describes the potential HAT mechanism in the exerted antioxidant activity. The IPs energies are responsible for the electron donation abilities of the title ligands. The former values of both compounds were higher when compared to the BDEs. Moreover, the relatively low IP energies of the pyrrole-based ligands [38] imply the substantial contribution of the SET mechanism in the antioxidant capacities. Overall, the DFT calculations demonstrated that **7d** and **8d** possess similar values of BDEs and IPs, which provides a good theoretical explanation for the experimental, radical scavenging assays.

3.4. In Vitro Cytotoxicity and Antioxidative Protective Activity on SH-SY5Y Cells

3.4.1. Effects of the Newly Synthesized Derivatives **7a–e** and **8a–e** on SH-SY5Y Cell Viability

Toxicity evaluation of newly synthetized substances in different in vitro models is an important issue in the drug development process.

In the current study, we selected the human neuroblastoma cell line SH-SY5Y as an appropriate model for in vitro neurotoxicity evaluation due to its frequent use in experimental neuroscience [39]. This cell line is also an appropriate model for mimicking various neurodegenerative disorders since its cells are able to convert to numerous types of functional neurons by the introduction of specific compounds. In addition, the SH-SY5Y cells possess tyrosine hydroxylase and dopamine-β-hydroxylase activities because of their sympathetic adrenergic ganglia origin [39], thus making it appropriate for evaluation of dopamine and serotonin modulations in neurodegeneration.

We aimed to evaluate the effects of the newly synthesized N-pyrrolyl hydrazone derivatives **7a–e** and **8a–e** on the cellular viability of the neuronal SH-SY5Y cells and calculated the corresponding IC50 values (Table 5). The results indicated three compounds with low toxicity; among them, compound **7c** may be considered non-toxic with effects comparable to the ones of melatonin. Interestingly, the results showed that the prolongation of the methylene bridge between the central pyrrole ring and the azomethine fragment of the structure with one carbon atom is related to a slight increase in cellular toxicity. The obtained IC50 values showed that the overall cytotoxicity of compounds from series **8a–e** is slightly higher compared to series **7a–e**. Particularly, the toxicity of derivative **8d** is 1.7 times higher than the one of its monomethyl analogue **7d**. Based on their promising safety profile, the compounds **7c**, **7d**, and **8e** were chosen for the further antioxidative protection assay in vitro.

3.4.2. Effects of the Newly Synthesized Derivatives **7a–e** and **8a–e** in a Model of H_2O_2-Induced Oxidative Stress In Vitro

Oxidative stress plays a key role in the damage to the central nervous system and in the pathogenesis of neurodegenerative diseases [40]. When an imbalance between the production of free radicals and detoxification occurs, the high production of ROS can overcome the body's antioxidant defenses, leading to harmful conditions such as oxidative stress and damage to cellular structures and functions.

In addition, free radical overproduction is often considered to be involved in the acute damage of the central nervous system and is strictly related to the development of neurodegenerative diseases. There are a number of facts that determine the free radical overproduction as a direct cause of immature cultured cortical neurons apoptosis [41], induction of DNA damage [42], and necrotic cell death [43], among others [44].

H_2O_2 is one of the most important ROS generated through oxidative stress. It causes oxidative damage in different cell structures such as nucleic acids, proteins, and membrane lipids; therefore, it is commonly used as an in vitro model for inducing oxidative cell damage [45]. The main mechanism of induction of oxidative stress relates to the production of reactive hydroxyl radicals and the formation of Fenton's reaction, which interacts directly with proteins, lipids, and DNA.

The less cytotoxic compounds **7c**, **7d**, and **8e** were tested for antioxidant protection on the H_2O_2-induced oxidative stress model in neuroblastoma SH-SY5Y cells. As expected, H_2O_2-treatment (1 mM, 15 min) induced a significant decrease in SH-SY5Y cell viability. Our results demonstrated that the pre-treatment with compounds **7c**, **7d**, and **8e** significantly increased the SH-SY5Y cell survival compared to H_2O_2 treatment. Both melatonin and all three test compounds show dose-dependent protective effects. Compared to the reference compound melatonin, at all the applied concentrations between 1 and 20 µM, compounds **7c** and **7d** both exert stronger protective effects, and **8e**'s effect is lower than melatonin's. Noteworthy, despite the stronger radical scavenging activity of **7d** compared to **7c**, shown by both the DPPH and ABTS assay, both compounds cause similar in vitro protective effects on peroxide-damaged SH-SY5Y cells, which may indicate that in the cells, the compounds have pleiotropic modes of antioxidant action.

The overall results from the performed in vitro assessment of the radical scavenging activity and antioxidative protective effects on neuronal SH-SY5Y cells of the newly synthesized N-pyrrolyl hydrazones determined the ethyl 5-(4-bromophenyl)-1-(2-(2-(4-hydroxy-3,5-dimethoxybenzylidene)hydrazine-yl)-2-oxoethyl)-2-methyl-1H-pyrrole-3-carboxylate (**7d**) as the most perspective structure with lowest cytotoxicity, highest radical scavenging activity and best antioxidative protection on H_2O_2-induced oxidative stress model.

3.5. Limitations and Implications Section

Some limitations of the applied assays are related to the specificity of the radical scavenging evaluations used, such as the fact that DPPH and ABTS are not physiological radicals; thus, they are not similar to peroxyl-radicals or other oxygen-based radicals. This makes the test an indirect method based on the reduction of persistent radicals. Thus, some discrepancies may be observed since light, oxygen, and pH have an influence on the final absorbance. In this relation, we plan to add some additional cellular evaluations on the effect of the newly obtained substances on cell-defined oxidative mechanisms, which will add to the performed cell protective experiments on H_2O_2-induced oxidative stress applied here. These evaluations include 6-hydroxydopamine (6-OHDA)-, tert-butyl hydroperoxide (tBuOOH)-, and Fe^{2+} ascorbate-induced oxidative stress assays on cellular and sub-cellular levels, which will be published in a following study.

4. Materials and Methods

4.1. Materials

The purity of the obtained compounds and the progress of the reactions were controlled by thin layer chromatography (TLC) on TLC-CardsSilicagel 60 F254, 1.05554, Merck, Darmstadt, Germany, using $CHCl_3/CH_3CH_2OH$ as a mobile phase. Melting points were determined in open capillary tubes on an IA 9200 ELECTROTHERMAL apparatus, Southend-on-Sea, England. All chemical names are given according to IUPAC by the program ChemBioDraw Ultra software, Version 11.0, CambridgeSoft. The IR spectra were performed in the range 400–4000 cm^{-1} on a NicoletiS10 FT-IR spectrophotometer, using ATR technique with Smart iTR adapter. ^{1}H-NMR spectra were recorded on a Bruker-Spectrospin WM250 MHz, Faelanden, Switzerland, operating at 250 MHz, as δ (ppm) relative to TMS as internal standard. Mass spectra were registered on a 6410 Agilent LCMS triple quadrupole mass spectrometer (LCMS) with an electrospray ionization (ESI) interface. Elemental analyses were performed on a Euro EA 3000-Single, EUROVECTOR SpAanalyser. All chemicals and reagents used as starting materials were purchased from Merck (Darmstadt, Germany).

The necessary for the pharmacological evaluations RPMI cell culture medium, heat-inactivated fetal bovine serum (FBS), L-glutamine, (3-(4,5-dimethylthiazol-2-yl)-2,5- diphenyltetrazolium bromide (MTT), hydrogen peroxide (H_2O_2), and dimethyl sulfoxide (DMSO) were acquired from Sigma-Aldrich (Merck KGaA, Darmstadt, Germany).

4.2. General Synthesis of the New Compounds

The corresponding N-pyrrolyl-carbohydrazide **7** or **8** and any of the carbonyl compounds **a, b, c, d,** or **e** (Figure 2) were incubated in a glacial acetic acid in a round bottom flask of 50 mL and stirred at 100 °C to complete the reaction under TLC-control. The obtained products were isolated, washed with diethyl ether, and recrystallized, where necessary, by ethanol.

4.2.1. (E)-ethyl 5-(4-bromophenyl)-1-(2-(2-(4-(dimethylamino)benzylidene)hydrazinyl)-2-oxoethyl)-2-methyl-1H-pyrrole-3-carboxylate (**7a**)

Yield: 84% as an orange powder; m.p. 211.4–213.6; IR (cm^{-1}): 3245 (NH), 2976 (CH$_3$ and CH$_2$), 1701 (COOC$_2$H$_5$), 1666 (Amide I), 1533 (Amide II), 1232 (C-O), 1073 (C-N), 810 (p-substituted C$_6$H$_4$), 552 (C-Br); ^{1}HNMR (δH, 250 MHz, CDCl$_3$): 1.99 [s, 3H, CH$_2$CH$_3$], 2.54 [s, 3H, CH$_3$(2)], 3.12–3.15 [m, 6H, N(CH$_3$)2], 4.20–4.24 [m, 2H, CH$_2$CH$_3$], 4.82 [s, 2H, CH$_2$CO], 6.53 [s, 1H, H(4)], 7.19 [s, 2H, H(3″), H(5″)], 7.21 [s, 2H, H(2″), H(6″)], 7.68–7.69 [d, 2H, H(3′), H(5′)], 7.88–7.90 [d, 2H, H(2′), H(6′)], 9.99 [s, 2H, NH-N=CH]; LC-MS (ESI): Calc. for C$_{25}$H$_{28}$O$_3$N$_4$Br [M+H]$^+$: 511.1339; Found: 511.1339.; Anal. Calc. for C$_{25}$H$_{27}$BrN$_4$O$_3$: C, 58.71; H, 5.32. Found: C, 58.68; H, 5.34.

4.2.2. (E)-ethyl 5-(4-bromophenyl)-1-(2-(2-(4-chlorobenzylidene)hydrazinyl)-2-oxoethyl)-2-methyl-1h-pyrrole-3-carboxylate (**7b**)

Yield: 78% as a white powder; m.p. 229.4–231.2; IR (cm^{-1}): 3179 (NH), 2998 (CH$_3$ and CH$_2$), 1697 (COOC$_2$H$_5$), 1667 (Amide I), 1570 (Amide II), 1240 (C-O), 816 (p-substituted C$_6$H$_4$), 772 (C-Cl), 557 (C-Br); ^{1}HNMR (δH, 250 MHz, DMSO-d$_6$): 1.25–1.29 [m, 3H, CH$_2$CH$_3$], 2.44–2.48 [m, 3H, CH$_3$(2)], 4.14–4.23 [m, 2H, CH$_2$CH$_3$], 5.10 [s, 2H, CH$_2$CO], 6.49 [s, 1H, H(4)], 7.23–7.27 [m, 1H, H(3′)], 7.31–7.36 [m, 1H, H(5″)], 7.46–7.47 [d, 1H, H(6′)], 7.50–7.51 [d, 1H, H(2′)], 7.58 [s, 1H, H2″], 7.59 [s, 1H, H(6″)], 7.68 [s, 1H, H(3′)], 7.71 [s, 1H, H(5′)], 8.0 [s, 1H, CH=N], 11.81 [s, 1H, CONH]; LC-MS (ESI): Calc. for C$_{23}$H$_{22}$O$_3$N$_3$BrCl [M+H]$^+$: 502.0527; Found: 502.0528.; Anal. Calc. for C$_{23}$H$_{21}$BrClN$_3$O$_3$: C, 54.94; H, 4.21. Found: C, 54.84; H, 4.29.

4.2.3. (E)-ethyl 5-(4-bromophenyl)-2-methyl-1-(2-(2-(4-nitrobenzylidene)hydrazinyl)-2-oxo-ethyl)-1H-pyrrole-3-carboxylate (**7c**)

Yield: 76% as a white powder; m.p. 245.9–247.2; IR (cm^{-1}): 3201 (NH), 3064 (CH$_3$ and CH$_2$), 1678 (COOC$_2$H$_5$), 1592 (Amide I), 1558 (Amide II), 1348 (NO$_2$), 1240 (C-O), 1205 (C-N), 814 (p-substituted C$_6$H$_4$), 556 (C-Br); ^{1}HNMR (δH, 250 MHz, CDCl$_3$): 1.35–1.37 [m, 3H, CH$_2$CH$_3$], 2.58 [s, 3H, CH$_3$(2)], 4.92 [s, 2H, CH$_2$CH$_3$], 5.05 [s, 2H, CH$_2$CO], 6.65 [s, 1H, H(4)], 7.58 [s, 1H, H(5′)], 7.59 [s, 1H, H(3′)], 7.78–7.93 [m, 2H, H(2′), H(6′)], 8.28 [s, 1H, H(2″)], 8.30 [s, 1H, H(6″)], 8.40 [s, 1H, H(3″)], 8.41 [s, 1H, H(5″)], 10.18 [s, 2H, NH-N=CH]; LC-MS (ESI): Calc. for C$_{23}$H$_{22}$O$_5$N$_4$Br [M+H]$^+$: 513.0768; Found: 513.0766; Anal. Calc. for C$_{23}$H$_{21}$BrN$_4$O$_5$: C, 53.81; H, 4.12. Found: C, 53.78; H, 4.10.

4.2.4. (E)-ethyl 5-(4-bromophenyl)-1-(2-(2-(4-hydroxy-3,5-dimethoxybenzylidene)hydrazine-yl)-2-oxoethyl)-2-methyl-1H-pyrrole-3-carboxylate (**7d**)

Yield: 68% as a white powder; m.p. 191.9- 194.4; IR (cm^{-1}): 3424 (OH), 3208 (NH), 3067 (CH$_3$ and CH$_2$), 1694 (COOC$_2$H$_5$), 1673 (Amide I), 1570 (Amide II), 1236 (C-O), 814 (p-substituted C$_6$H$_4$), 554 (C-Br); ^{1}HNMR (δH, 250 MHz, DMSO-d$_6$): 1.29 [t, 3H, CH$_2$CH$_3$], 2.47 [s, 3H, CH$_3$(2)], 3.77 [s, 6H, OCH$_3$(3″), OCH$_3$(5″)], 4.15–4.21 [m, 2H, CH$_2$CH$_3$], 5.05 [s, 1H, OH], 5.09 [s, 2H, CH$_2$CO], 6.50 [s, 1H, H(4)], 6.93 [s, 1H, H(6″)], 6.98 [s, 1H, H(2″)], 7.28–7.38 [m, 2H, H(3′), H(5′)], 7.59–7.63 [m, 2H, H(2′), H(6′)], 7.89 [s, 1H, CH=N], 11.67 [s, 1H,CONH]; LC-MS (ESI): Calc. for C$_{25}$H$_{27}$O$_6$N$_3$Br [M+H]$^+$: 544.1078; Found: 544.1078; Anal. Calc. for C$_{25}$H$_{26}$BrN$_3$O$_6$: C, 55.16; H, 4.81. Found: C, 55.15; H, 4.85.

4.2.5. (E)-ethyl 5-(4-bromophenyl)-1-(2-(2-(2,4-dimethoxybenzylidene)hydrazinyl)-2-oxoethyl)-2-methyl-1H-pyrrole-3-carboxylate (**7e**)

Yield: 72% as a white powder; m.p. 206.0–207.6; IR (cm^{-1}): 3234 (NH), 2984 (CH$_3$ and CH$_2$), 1698 (COOC$_2$H$_5$), 1668 (Amide I), 1520 (Amide II), 1240 (C-O), 822 (*p*-substituted C$_6$H$_4$), 554 (C-Br); ^{1}HNMR (δH, 250 MHz, CDCl$_3$): 1.36 [t, 3H, CH$_2$CH$_3$], 2.54 [s, 3H, CH$_3$(2)], 3.93 [s, 3H, OCH$_3$(2″)], 3.94 [s, 3H, OCH$_3$(4″)], 4.29–4.31 [m, 2H, CH$_2$CH$_3$], 4.92 [s, 2H, CH$_2$CO], 6.64 [s, 1H, H(4)], 6.47 [s, 1H, H(3″)], 6.48 [s, 1H, H(5″)], 7.27 [s, 1H, H(6″)], 7.56 [s, 1H, H(2′)], 7.57 [s, 1H, H(6′)], 7.82 [s, 1H, H(3′)], 7.85 [s, 1H, H(5′)], 10.31 [s, 1H, CH=N], 10.32 [s, 1H, CONH]; LC-MS (ESI): Calc. for C$_{25}$H$_{27}$O$_5$N$_3$Br [M+H]$^+$: 528.1129; Found: 528.1132; Anal. Calc. for C$_{25}$H$_{26}$BrN$_3$O$_5$: C, 56.83; H, 4.96. Found: C, 56.84; H, 4.98.

4.2.6. (E)-ethyl 5-(4-bromophenyl)-1-(3-(2-(4-(dimethylamino)benzylidene)hydrazinyl)-3-oxopropyl)-2-methyl-1H-pyrrole-3-carboxylate (**8a**)

Yield: 86% as a white powder; m.p. 212.0–213.3; IR (cm^{-1}): 3266 (NH), 2907 (CH$_3$ and CH$_2$), 1693 (COOC$_2$H$_5$), 1668 (Amide I), 1570 (Amide II), 1265 (C-N), 1249 (C-O), 831 (*p*-substituted C$_6$H$_4$), 554 (C-Br); ^{1}HNMR (δH, 250 MHz, CDCl$_3$): 1.27 [s, 3H, CH$_2$CH$_3$], 2.54 [s, 3H, CH$_3$(2)], 2.59–2.62 [m, 2H, CH$_2$CH$_2$CO], 3.12 [s, 3H, N(CH$_3$)], 3.14 [s, 3H, N(CH$_3$)], 4.17–4.19 [m, 2H, CH$_2$CH$_2$CO], 4.27–4.28 [m, 2H, CH$_2$CH$_3$], 6.49 [s, 1H, H(4)], 7.18–7.19 [m, 2H, H(3″), H(5″)], 7.45–7.49 [m, 2H, H(2″), H(6″)], 7.67–7.69 [d, 2H, H(3′), H(5′)], 7.95–7.96 [d, 2H, H(2′), H(6′)], 8.60 [s, 1H, CH=N], 9.97 [s, 1H, CONH]; LC-MS (ESI): Calc. for C$_{26}$H$_{30}$O$_3$N$_4$Br [M+H]$^+$: 525.1496; Found: 525.1504; Anal. Calc. for C$_{26}$H$_{29}$BrN$_4$O$_3$: C, 59.43; H, 5.56. Found: C, 59.23; H, 5.66.

4.2.7. (E)-ethyl 5-(4-bromophenyl)-1-(3-(2-(4-chlorobenzylidene)hydrazinyl)-3-oxopropyl)-2-methyl-1H-pyrrole-3-carboxylate (**8b**)

Yield: 74% as a white powder; m.p. 181.4–184.6; IR (cm^{-1}): 3282 (NH), 2931 (CH$_3$ and CH$_2$), 1698 (COOC$_2$H$_5$), 1668 (Amide I), 1571 (Amide II), 1252 (C-O), 813 (*p*-substituted C$_6$H$_4$), 771 (C-Cl), 560 (C-Br); ^{1}HNMR (δH, 250 MHz, DMSO-d$_6$): 1.23–1.27 [m, 3H, CH$_2$CH$_3$], 2.61 [s, 3H, CH$_3$(2)], 2.83 [t, 2H, CH$_2$CH$_2$CO], 4.13–4.16 [q, 2H, CH$_2$CH$_2$CO], 4.24–4.33 [q, 2H, CH$_2$CH$_3$], 6.41 [s, 1H, H(4)], 7.35–7.37 [d, 2H, H(3″), H(5″)], 7.51–7.53 [d, 2H, H(2″), H(6″)], 7.56–7.60 [m, 2H, H(2′), H(6′)], 7.61–7.63 [d, 1H, H(3′)], 7.69–7.71 [d, 1H, H(5′)], 7.86 [s, 1H, CH=N], 11.38 [s, 1H, CONH]; LC-MS (ESI): Calc. for C$_{24}$H$_{24}$O$_3$N$_3$BrCl [M+H]$^+$: 516.0684; Found: 516.0684; Anal. Calc. for C$_{24}$H$_{23}$BrClN$_3$O$_3$: C, 55.78; H, 4.49. Found: C, 55.76; H, 4.48.

4.2.8. (E)-ethyl 5-(4-bromophenyl)-2-methyl-1-(3-(2-(4-nitrobenzylidene)hydrazinyl)-3-oxopropyl)-1H-pyrrole-3-carboxylate (**8c**)

Yield: 82% as a yellow powder; m.p. 214.4–217.1; IR (cm^{-1}): 3294 (NH), 2929 (CH$_3$ and CH$_2$), 1689 (COOC$_2$H$_5$), 1658 (Amide I), 1572 (Amide II), 1512 (NO$_2$), 1340 (C-N), 1249 (C-O), 814 (*p*-substituted C$_6$H$_4$), 558 (C-Br); ^{1}HNMR (δH, 250 MHz, CDCl$_3$): 1.26–1.29 [m, 3H, CH$_2$CH$_3$], 2.55 [s, 3H, CH$_3$(2)], 2.61 [t, 2H, CH$_2$CH$_2$CO], 4.18–4.20 [m, 2H, CH$_2$CH$_2$CO], 4.21–4.25 [m, 2H, CH$_2$CH$_3$], 6.49 [s, 1H, H(4)], 7.17 [s, 1H, H(3′)], 7.18 [s, 1H, H(5′)], 7.48 [s, 1H, H(6′)], 7.49 [s, 1H, H(2′)], 8.00 [s, 1H, H(2″)], 8.02 [s, 1H, H(6″)], 8.34 [s, 2H, H(3″), H(5″)], 10.09 [s, 2H, NH-N=CH]; LC-MS (ESI): Calc. for C$_{24}$H$_{24}$O$_5$N$_4$Br [M+H]$^+$: 527.0925; Found: 527.0927; Anal. Calc. for C$_{24}$H$_{23}$BrN$_4$O$_5$: C, 54.66; H, 4.40. Found: C, 54.64; H, 4.39.

4.2.9. (E)-ethyl 5-(4-bromophenyl)-1-(3-(2-(4-hydroxy-3,5-dimethoxybenzylidene) hydrazine-yl)-3-oxopropyl)-2-methyl-1H-pyrrole-3-carboxylate (**8d**)

Yield: 64% as a white powder; m.p. 196.6–197.6; IR (cm^{-1}): 3422 (OH), 3278 (NH), 2971 (CH$_3$ and CH$_2$), 1693 (COOC$_2$H$_5$), 1666 (Amide I), 1574 (Amide II), 1250 (C-O), 814 (*p*-substituted C$_6$H$_4$), 551 (C-Br); ^{1}HNMR (δH, 250 MHz, DMSO-d$_6$): 1.22–1.27 [m, 3H, CH$_2$CH$_3$], 2.61 [t, 2H, CH$_2$CH$_2$CO], 2.64 [s, 3H, CH$_3$(2)], 3.78 [s, 6H, OCH3(3″), OCH3(5″)], 4.12–4.16 [m, 2H, CH$_2$CH$_2$CO], 4.20–4.26 [m, 2H, CH$_2$CH$_3$], 6.41 [s, 1H, H(4)], 6.80 [s, 1H,

OH], 7.36–7.38 [m, 2H, H(2″), H(6″)], 7.54–7.56 [m, 2H, H(3′), H(5′)], 7.62–7.64 [m, 2H, H(2′), H(6′)], 7.78 [s, 1H, CH=N], 11.20 [s, 1H,CONH]; LC-MS (ESI): Calc. for $C_{26}H_{29}O_6N_3Br$ [M+H]$^+$: 558.1234; Found: 558.1243; Anal. Calc. for $C_{26}H_{28}BrN_3O_6$: C, 55.92; H, 5.05. Found: C, 55.98; H, 5.02.

4.2.10. (E)-ethyl 5-(4-bromophenyl)-1-(3-(2-(2,4-dimethoxybenzylidene)hydrazinyl)-3-oxopropyl)-2-methyl-1H-pyrrole-3-carboxylate (**8e**)

Yield: 74% as a white powder; m.p. 170.0–171.9; IR (cm^{-1}): 3298 (NH), 2968 (CH$_3$ and CH$_2$), 1692 (COOC$_2$H$_5$), 1673 (Amide I), 1571 (Amide II), 1253 (C-O), 813 (*p*-substituted C$_6$H$_4$), 549 (C-Br); ^{1}HNMR (δH, 250 MHz, CDCl$_3$): 1.24–1.28 [m, 3H, CH$_2$CH$_3$], 2.54 [s, 3H, CH$_3$(2)], 2.58–2.63 [m, 2H, CH$_2$CH$_2$CO], 3.18 [s, 3H, OCH$_3$(4″)], 3.83 [s, 3H, OCH$_3$(2″)], 4.18–4.20 [m, 2H, CH$_2$CH$_2$CO], 4.21–4.23 [m, 2H, CH$_2$CH$_3$], 6.37–6.68 [d, 1H, H(3″)], 6.47–6.48 [m, 1H, H(5″)], 6.49 [s, 1H, H(4)], 7.14 [s, 1H, H(3′)], 7.18 [s, 1H, H(5′)], 7.48 [s, 1H, H(6′)], 7.49 [s, 1H, H(2′)], 7.73–7.76 [d, 1H, H(6″)], 8.34 [s, 1H, CH=N], 10.22 [s, 1H, CONH]; LC-MS (ESI): Calc. for $C_{26}H_{29}O_5N_3Br$ [M+H]$^+$: 542.1285; Found: 542.1293; Anal. Calc. for $C_{26}H_{28}BrN_3O_5$: C, 57.57; H, 5.20. Found: C, 57.55; H, 5.22.

4.3. Antioxidant Activity Evaluation

4.3.1. DPPH Radical Scavenging Assay

The scavenging assay of the title compounds against DPPH radical was carried out by the widely employed protocol of Brand-Williams et al. [46]. Briefly, a single concentration of 1 mg/mL for each synthesized compound in methanol was obtained. Subsequent addition of 1 mL of the methanol solution of DPPH (1 mmol/L) was performed. The reaction mixtures were incubated in the dark for 30 min. The absorbance was measured at 517 nm. Three measurements were carried out for each sample; 6-Hydroxy-2,5,7,8-tetramethylchroman-2-carboxylic acid (Trolox) was applied as a standard. The percentage inhibition of the tested samples was calculated by the following Formula (3):

$$\text{DPPH}_{\text{scavenging activity}} = \text{Abs}_{\text{control}} - \text{Abs}_{\text{sample}} / \text{Abs}_{\text{control}} \times 100\% \quad (3)$$

where Abs$_{\text{control}}$ is the absorbance of the DPPH radical in methanol, and Abs$_{\text{sample}}$ is the absorbance of the DPPH radical solution mixed with the sample.

4.3.2. ABTS Radical Scavenging Assay

The ABTS radical tests were measured according to a modified method of Arnao et al. [47]. The test solutions were dissolved in methanol (1 mg/mL) at ambient temperature. The radical cation of ABTS (ABTS$^{+\bullet}$) was created by mixing 7 mmol/L solution of ABTS and 2.4 mmol/L solution of potassium persulphate, which were set to react for 14 h in the dark at room temperature. The working solutions consisted of 2 mL of the stock solution diluted in 50 mL of methanol with an absorbance of 0.294 ± 0.05 units at 734 nm. Then, 1 mL of the ABTS working solution was allowed to react with the title compounds for 10 min, with a subsequent absorbance determination. The inhibition percentages were evaluated by applying the same formula as the DPPH assay (4).

$$\text{ABTS}_{\text{scavenging activity}} = \text{Abs}_{\text{control}} - \text{Abs}_{\text{sample}} / \text{Abs}_{\text{control}} \times 100\% \quad (4)$$

where Abs$_{\text{control}}$ is the absorbance of the ABTS radical in methanol, and Abs$_{\text{sample}}$ is the absorbance of the ABTS radical solution mixed with the sample.

4.4. DFT Theoretical Calculations

All of the theoretical computations were carried out by applying Jaguar [48]. The initial geometries for the DFT calculations were achieved after a conformational search with 2500 iterations and OPLS4 force field. Subsequently, full geometry optimizations of the best conformations of compounds **7d** and **8d** with Becke's three-parameter hybrid exchange–correlation functional (B3LYP) and 6-311++G (d,p) basis set in the gas phase were performed. It was found that out of 11 functions and 14 basis sets, the most optimal combination in terms of both accuracy and resource usage for bond dissociation energies calculations is M06-2X/6-311G(d,p) [49]. The frontier molecular orbitals and the global reactivity descriptors were calculated at the same level of theory. The bond dissociation energies were obtained with M06–2X and 6–311G basis set considering a recent report [49].

4.5. In Vitro Pharmacological Evaluations

4.5.1. Cell Line

Human neuroblastoma cell line SH-SY5Y was purchased from European Collection of Cell Cultures (ECACC, Salisbury, UK). SH-SY5Y cells were cultivated in an RPMI 1640 medium supplemented with 10% heat-inactivated FBS, 2 mM L-glutamine, and 1% antibiotics (penicillin/streptomycin). The cell line was incubated at 37 °C with 5% of CO_2, and the culture's medium was replaced with a time interval of 2–3 days.

4.5.2. Cell Viability Assay

The SH-SY5Y cells were plated for 24 h at 37 °C on 96-well plates at a density of 2×10^4 cells per well to confluence. After 24 h, the cells were treated with the compounds (1–500 µM). MTT assay was performed to measure the metabolic activity in living cells. The mitochondria succinate dehydrogenase system of living cells metabolized yellow-colored, water-soluble, tetrazolium salt MTT (3-(4,5-dimethylthiazol2-yl)-2,5-diphenyltetrazolium bromide to a purple insoluble formazan crystal. The cells were incubated with the test solutions for 24 h, the MTT solution (10 mg/mL in PBS) was then added to each well, and the mixture was incubated at 37 °C for 3 h. Thereafter, the MTT solution was carefully aspirated, and the obtained formazan crystals were dissolved by the addition of 100 mL DMSO. The cell viability was determined by measuring the optical density −570–690 nm in a multiplate reader Synergy 2 (BioTek Instruments, Inc., Highland Park, Winooski, VT, USA) [50].

4.5.3. H_2O_2-Induced Oxidative Stress Model in SH-SY5Y Cells

For the H_2O_2-induced oxidative stress model, SH-SY5Y cells were seeded in 96-well plates at a density of 3.5×10^4/well in 100 µL for 24 h. Afterwards, the cell medium was aspirated, and the cells were treated with different concentrations of tested compounds (1, 10, 20 µM) for 90 min; then, the SH-SY5Y cells were washed with phosphate-buffered saline (PBS) and were exposed to the oxidative stress (H_2O_2 1 mM in PBS, 15 min). The contents of all wells were changed with culture medium. After 24 h, the amount of attached viable cells was evaluated by MTT assay. Negative controls (cells without hydrogen peroxide treatment) were considered as 100% protection, and hydrogen-peroxide-treated cells as 0% protection.

4.5.4. Statistical Analysis

Statistical analysis of data has been carried out on GraphPad Prism 6 Software. All experiments were carried out in triplicate, and results were presented as mean ± SD (n = 8). Comparisons between groups have been performed by applying one-way ANOVA with Dunnet's multiple comparisons post-test. We performed statistical analyses on the data with the aim of validating the significance of the observed differences. Differences between groups were considered significant for $p < 0.05$, $p < 0.01$ and $p < 0.001$.

5. Conclusions

Ten new N-pyrrolyl hydrazide–hydrazones were synthesized. The structures of the new compounds were elucidated through appropriate IR, ^{1}H-NMR, and MS spectral data. The purity of the compounds was proven by the corresponding melting points, TLC characteristics, and elemental analyses.

The DFT studies have indicated that the best radical scavenging ligands (**7d** and **8d**) in the newly synthesized compounds are stable molecules that do not decompose into elements, are less polarizable, and with hard nature. The energy of HOMO revealed that both compounds possess good electron donation capacities. Overall, **7d** and **8d** can readily scavenge free radicals in the biological system by donation of hydrogen atoms and single electron transfer.

The performed in vitro neurotoxicity and cellular toxicity and cell protection evaluations on human neuroblastoma cell line SH-SY5Y determined **7d** as the most prospective with the lowest toxicity and highest antioxidant protection.

Supplementary Materials: The following supporting information can be downloaded at https://www.mdpi.com/article/10.3390/ph16091198/s1.

Author Contributions: D.T. contributed to the synthesis of the new molecules; H.K. contributed to the chemical characterization of the new compounds; E.M. contributed to the DFT simulations and analysis; D.S. contributed to the in vitro neuroprotection evaluation; A.D. contributed to the in vitro neuroprotection evaluation; Y.Y. contributed to the in vitro neurotoxicity evaluation; A.M. contributed to in vitro evaluation of the antioxidant activity; V.T. contributed to in vitro neurotoxicity evaluation; M.K.-B. contributed to the in vitro neurotoxicity and neuroprotection evaluation and composing of the manuscript; A.Z. contributed to the composing of the manuscript and the overall data evaluation; M.G. contributed to the initiation of the scientific idea and composing of the manuscript. All authors have read and agreed to the published version of the manuscript.

Funding: This study was financed by the European Union NextGenerationEU through the National Recovery and Resilience Plan of the Republic of Bulgaria, project No. BG-RRP-2.004-0004-C01.

Institutional Review Board Statement: Not applicable.

Informed Consent Statement: Not applicable.

Data Availability Statement: Data is contained within the article and supplementary material.

Acknowledgments: The authors would like to thank Paraskev Nedjalkov for their support in conducting the LC-MS evaluations for structure elucidation.

Conflicts of Interest: The authors declare no conflict of interest.

References

1. Singh, S.; Singh, R.P. In vitro methods of assay of antioxidants: An overview. *Food Rev. Int.* **2008**, *24*, 392–415. [CrossRef]
2. Linseman, D.A. Targeting oxidative stress for neuroprotection. *Antioxid. Redox Signal.* **2009**, *11*, 421–424. [CrossRef] [PubMed]
3. Cho, K.S.; Shin, M.; Kim, S.; Lee, S.B. Recent advances in studies on the therapeutic potential of dietary carotenoids in neurodegenerative diseases. *Oxid. Med. Cell. Longev.* **2018**, *2018*, 4120458. [CrossRef] [PubMed]
4. Thurkauf, A.; Yuan, J.; Chen, N.; Wasley, J.W.; Meade, R.; Woodruff, K.H.; Ross, P.C. 1-Phenyl-3-(aminomethyl) pyrroles as potential antipsychotic agents. Synthesis and dopamine receptor binding. *J. Med. Chem.* **1995**, *38*, 4950–4952. [CrossRef]
5. de Oliveira, K.N.; Costa, P.; Santin, J.R.; Mazzambani, L.; Bürger, C.; Mora, C.; Nunes, R.J.; de Souza, M.M. Synthesis and antidepressant-like activity evaluation of sulphonamides and sulphonyl-hydrazones. *Bioorg. Med. Chem.* **2011**, *19*, 4295–4306. [CrossRef]
6. Ragavendran, J.V.; Sriram, D.; Patel, S.K.; Reddy, I.V.; Bharathwajan, N.; Stables, J.; Yogeeswari, P. Design and synthesis of anticonvulsants from a combined phthalimide–GABA–anilide and hydrazone pharmacophore. *Eur. J. Med. Chem.* **2007**, *42*, 146–151. [CrossRef]
7. Zlatanova, H.; Vladimirova, S.; Kandilarov, I.; Kostadinov, I.; Delev, D.; Kostadinova, I.; Bijev, A. Analgesic effect of a newly synthesized N-pyrrolylcarboxylic acid in experimental conditions. *Europ. Neuropsychopharmacol.* **2019**, *29*, S468–S469. [CrossRef]
8. MacLean, P.D.; Chapman, E.E.; Dobrowolski, S.L.; Thompson, A.; Barclay, L.R.C. Pyrroles as antioxidants: Solvent effects and the nature of the attacking radical on antioxidant activities and mechanisms of pyrroles, dipyrrinones, and bile pigments. *J. Org. Chem.* **2008**, *73*, 6623–6635. [CrossRef]
9. Bhosale, J.D.; Dabur, R.; Jadhav, G.P.; Bendre, R.S. Facile syntheses and molecular-docking of novel substituted 3,4-dimethyl-1H-pyrrole-2-carboxamide/carbohydrazide analogues with antimicrobial and antifungal properties. *Molecules* **2018**, *23*, 875. [CrossRef]
10. Peng, Z.; Wang, G.; Zeng, Q.H.; Li, Y.; Wu, Y.; Liu, H.; Wang, J.J.; Zhao, Y. Synthesis, antioxidant and anti-tyrosinase activity of 1,2,4-triazole hydrazones as antibrowning agents. *Food Chem.* **2021**, *341*, 128265. [CrossRef]
11. Boulebd, H.; Zine, Y.; Khodja, I.A.; Mermer, A.; Demir, A.; Debache, A. Synthesis and radical scavenging activity of new phenolic hydrazone/hydrazide derivatives: Experimental and theoretical studies. *J. Mol. Struct.* **2022**, *1249*, 131546. [CrossRef]

12. Khodja, A.; Bensouici, C.; Bouleb, H. Combined experimental and theoretical studies of the structure-antiradical activity relationship of heterocyclic hydrazone compounds. *J. Mol. Struct.* **2020**, *1221*, 128858. [CrossRef]
13. Tzankova, D.G.; Vladimirova, S.P.; Peikova, L.P.; Georgieva, M.B. Synthesis and preliminary antioxidant activity evaluation of new pyrrole based aryl hydrazones. *Bulg. Chem. Commun.* **2019**, *51*, 179–185.
14. Bijev, A.; Georgieva, M. Pyrrole-based hydrazones synthesized and evaluated in vitro as potential tuberculostatics. *Lett. Drug Des. Discov.* **2010**, *7*, 430–437. [CrossRef]
15. Georgieva, M.; Bijev, A.; Prodanova, P. Synthesis and comparative study of tuberculostatic activity of pyrrole-based hydrazones related to structural variations. *Pharmacia* **2010**, *52*, 3–14.
16. Tsuneda, T.; Song, J.W.; Suzuki, S.; Hirao, K. On Koopmans' theorem in density functional theory. *J. Chem. Phys.* **2010**, *133*, 174101. [CrossRef]
17. Zhang, Y.; Chen, D.; Wang, Y.; Wang, X.; Zhang, Z.; Xin, Y. Neuroprotective effects of melatonin-mediated mitophagy through nucleotide-binding oligomerization domain and leucine-rich repeat-containing protein X1 in neonatal hypoxic–ischemic brain damage. *FASEB J.* **2023**, *37*, e22784. [CrossRef]
18. Joshi, S.D.; Kumar, D.; Dixit, S.R.; Tigadi, N.; More, U.A.; Lherbet, C.; Aminabhavi, T.M.; Yang, K.S. Synthesis, characterization and antitubercular activities of novel pyrrolyl hydrazones and their Cu-complexes. *Eur. J. Med. Chem.* **2016**, *121*, 21–39. [CrossRef]
19. Manvar, A.; Bavishi, A.; Radadiya, A.; Patel, J.; Vora, V.; Dodia, N.; Rawal, K.; Shah, A. Diversity oriented design of various hydrazides and their in vitro evaluation against Mycobacterium tuberculosis H37Rv strains. *Bioorg. Med. Chem. Lett.* **2011**, *21*, 4728–4731. [CrossRef]
20. Nassiri, K.M.; Assarzadeh, M.J.; Almasirad, A.; Ghasemi-Niri, S.F.; Amini, M.; Kebriaeezadeh, A.; Nassiri, K.N.; Ghadimi, M.; Tabei, A. Synthesis and analgesic activity of novel hydrazide and hydrazine derivatives. *Iran. J. Pharm. Res.* **2013**, *12*, 721–727.
21. Rawat, P.; Singh, R.N. Synthesis, spectral and chemical reactivity analysis of 2,4-dinitrophenyl hydrazone having pyrrole moiety. *J. Mol. Struct.* **2015**, *1097*, 214–225. [CrossRef]
22. Sujarwo, W.; Keim, A.P. Spondias pinnata (L. f.) Kurz. (Anacardiaceae): Profiles and Applications to Diabetes. In *Bioactive Food as Dietary Interventions for Diabetes*, 2nd ed.; Elsevier Inc.: Amsterdam, The Netherlands, 2019; ISBN 978-0-12-813822-9. [CrossRef]
23. Kedare, S.B.; Singh, R.P. Genesis and development of DPPH method of antioxidant assay. *J. Food Sci. Technol.* **2011**, *48*, 412–422. [CrossRef] [PubMed]
24. Dawidowicz, A.L.; Olszowy, M.; Jóźwik-Dolęba, M. Importance of solvent association in the estimation of antioxidant properties of phenolic compounds by DPPH method. *J. Food Sci. Technol.* **2015**, *52*, 4523–4529. [CrossRef]
25. Brizzolari, A.; Foti, M.C.; Saso, L.; Ciuffreda, P.; Lazarević, J.; Santaniello, E. Evaluation of the radical scavenging activity of some representative isoprenoid and aromatic cytokinin ribosides (N6-substituted adenosines) by in vitro chemical assays. *Nat. Prod. Res.* **2022**, *36*, 6443–6447. [CrossRef]
26. Schaich, K.M. Lipid oxidation in specialty oils. In *Nutraceutical and Specialty Oils*; Shahidi., F., Ed.; CRC Press/Taylor & Francis: London, UK, 2006; pp. 401–448.
27. Amarowicz, R.; Pegg, R.B. Functional Food Ingredients from Plants. In *Advances in Food and Nutrition Research*; Elsevier Inc.: Amsterdam, The Netherlands, 2019.
28. Dasgupta, A.; Klein, K. Methods for Measuring Oxidative Stress in the Laboratory. In *Antioxidants in Food, Vitamins and Supplements*; Elsevier Inc.: Amsterdam, The Netherlands, 2014. [CrossRef]
29. Ilyasov, I.R.; Beloborodov, V.L.; Selivanova, I.A.; Terekhov, R.P. ABTS/PP Decolorization assay of antioxidant capacity reaction pathways. *Int. J. Mol. Sci.* **2020**, *21*, 1131. [CrossRef]
30. Mateev, E.; Georgieva, M.; Zlatkov, A. Design, microwave-assisted synthesis, biological evaluation, molecular docking, and adme studies of pyrrole-based hydrazide-hydrazones as potential antioxidant agents. *Maced. J. Chem. Chem. Eng.* **2022**, *41*, 175–186. [CrossRef]
31. Munteanu, I.G.; Apetrei, C. Analytical methods used in determining antioxidant activity: A Review. *Int. J. Mol. Sci.* **2021**, *22*, 3380. [CrossRef] [PubMed]
32. Leopoldini, M.; Russo, N.; Toscano, M. The molecular basis of working mechanism of natural polyphenolic antioxidants. *Food Chem.* **2011**, *125*, 288–306. [CrossRef]
33. Farrokhnia, M. Density Functional Theory Studies on the Antioxidant Mechanism and Electronic Properties of Some Bioactive Marine Meroterpenoids: Sargahydroquionic Acid and Sargachromanol. *ACS Omega* **2020**, *5*, 20382–20390. [CrossRef]
34. Safna Hussan, K.P.; Shahin Thayyil, M.; Rajan, V.K.; Muraleedharan, K. DFT studies on global parameters, antioxidant mechanism and molecular docking of amlodipine besylate. *Comput. Biol. Chem.* **2019**, *80*, 46–53. [CrossRef]
35. Choudhary, V.K.; Bhatt, A.K.; Dash, D.; Sharma, N. DFT calculations on molecular structures, HOMO–LUMO study, reactivity descriptors and spectral analyses of newly synthesized diorganotin (IV) 2-chloridophenylacetohydroxamate complexes. *J. Comput. Chem.* **2019**, *40*, 2354–2363. [CrossRef] [PubMed]
36. Dávalos, J.Z.; Valderrama-Negrón, A.C.; Barrios, J.R.; Freitas, V.L.S.; Ribeiro da Silva, M.D.M.C. Energetic and structural properties of two phenolic antioxidants: Tyrosol and hydroxytyrosol. *J. Phys. Chem. A* **2018**, *122*, 4130–4137. [CrossRef] [PubMed]
37. Liang, N.; Kitts, D.D. Antioxidant property of coffee components: Assessment of methods that define mechanisms of action. *Molecules* **2014**, *19*, 19180–19208. [CrossRef]
38. Lu, L.; Qiang, M.; Li, F.; Zhang, H.; Zhang, S. Theoretical investigation on the antioxidative activity of anthocyanidins: A DFT/B3LYP study. *Dyes Pigment.* **2014**, *103*, 175–182. [CrossRef]

39. Lopez-Suarez, L.; Al Awabdh, S.; Coumoul, X.; Chauvet, C. The SH-SY5Y human neuroblastoma cell line, a relevant in vitro cell model for investigating neurotoxicology in human: Focus on organic pollutants. *Neurotoxicology* **2022**, *92*, 131–155. [CrossRef]
40. Wang, X.; Zhou, Y.; Gao, Q.; Ping, D.; Wang, Y.; Wu, W.; Lin, X.; Fang, Y.; Zhang, J.; Shao, A. The Role of exosomal microRNAs and oxidative stress in neurodegenerative diseases. *Oxid. Med. Cell. Longev.* **2020**, *2020*, 3232869. [CrossRef]
41. Ratan, R.R.; Murphy, T.H.; Baraban, J.M. Oxidative stress induces apoptosis in embryonic cortical neurons. *J. Neurochem.* **1994**, *62*, 376–379. [CrossRef]
42. Floyd, R.A.; Carney, J.M. Free radical damage to protein and DNA: Mechanisms involved and relevant observations on brain undergoing oxidative stress. *Ann. Neurol.* **1992**, *32* (Suppl. S1), S22–S27. [CrossRef]
43. Kane, D.J.; Sarafian, T.A.; Anton, R.; Hahn, H.; Gralla, E.B.; Valentine, J.S.; Ord, T.; Bredesen, D.E. Bcl-2 inhibition of neural death: Decreased generation of reactive oxygen species. *Science* **1993**, *262*, 1274–1281. [CrossRef]
44. Amoroso, S.; Gioielli, A.; Cataldi, M.; Di Renzo, G.; Annunziato, L. In the neuronal cell line SH-SY5Y, oxidative stress-induced free radical overproduction causes cell death without any participation of intracellular Ca^{2+} increase. *Biochim. Biophys. Acta Mol. Cell Res.* **1999**, *1452*, 151–160. [CrossRef]
45. Salminen, A.; Kaarniranta, K.; Kauppinen, A. Crosstalk between oxidative stress and SIRT1: Impact on the aging process. *Int. J. Mol. Sci.* **2013**, *14*, 3834–3859. [CrossRef] [PubMed]
46. Brand-Williams, W.; Cuvelier, M.E.; Berset, C. Use of a free radical method to evaluate antioxidant activity. *LWT Food Sci. Technol.* **1995**, *28*, 25–30. [CrossRef]
47. Arnao, M.B.; Cano, A.; Hernández-Ruiz, J.; García-Cánovas, F.; Acosta, M. Inhibition by L-ascorbic acid and other antioxidants of the 2.2'-azino-bis(3-ethylbenzthiazoline-6-sulfonic acid) oxidation catalyzed by peroxidase: A new approach for determining total antioxidant status of foods. *Anal. Biochem.* **1996**, *236*, 255–261. [CrossRef] [PubMed]
48. Bochevarov, A.D.; Harder, E.; Hughes, T.F.; Greenwood, J.R.; Braden, D.A.; Philipp, D.M.; Rinaldo, D.; Halls, M.D.; Zhang, J.; Friesner, R.A. Jaguar: A high-performance quantum chemistry software program with strengths in life and materials sciences. *Int. J. Quantum Chem.* **2013**, *113*, 2110–2142. [CrossRef]
49. Spiegel, M.; Gamian, A.; Sroka, Z. A statistically supported antioxidant activity dft benchmark-the effects of hartree-fock exchange and basis set selection on accuracy and resources uptake. *Molecules* **2021**, *26*, 5058. [CrossRef]
50. Mosmann, T. Rapid colorimetric assay for cellular growth and survival: Application to proliferation and cytotoxicity assays. *J. Immunol. Methods* **1983**, *65*, 55–63. [CrossRef]

Disclaimer/Publisher's Note: The statements, opinions and data contained in all publications are solely those of the individual author(s) and contributor(s) and not of MDPI and/or the editor(s). MDPI and/or the editor(s) disclaim responsibility for any injury to people or property resulting from any ideas, methods, instructions or products referred to in the content.

Article

New Imadazopyrazines with CDK9 Inhibitory Activity as Anticancer and Antiviral: Synthesis, In Silico, and In Vitro Evaluation Approaches

Aisha A. Alsfouk [1], Hanan M. Alshibl [2], Najla A. Altwaijry [1], Ashwag Alanazi [1], Omkulthom AlKamaly [1], Ahlam Sultan [1] and Bshra A. Alsfouk [1,*]

1 Department of Pharmaceutical Sciences, College of Pharmacy, Princess Nourah bint Abdulrahman University, P.O. Box 84428, Riyadh 11671, Saudi Arabia; aaalsfouk@pnu.edu.sa (A.A.A.); naaltwaijry@pnu.edu.sa (N.A.A.); asalanzi@pnu.edu.sa (A.A.); omalkmali@pnu.edu.sa (O.A.); ahmsultan@pnu.edu.sa (A.S.)
2 Department of Pharmaceutical Chemistry, College of Pharmacy, King Saud University, P.O. Box 2457, Riyadh 11451, Saudi Arabia; halshibl@ksu.edu.sa
* Correspondence: baalsfouk@pnu.edu.sa

Abstract: This study describes the synthesis and biological activity of new imadazopyrazines as first-in-class CDK9 inhibitors. The inhibition of CDK9 is a well-established therapeutic target in cancer therapy. The new compounds were assessed using an in vitro kinase assay against CDK9. In this assay, compound **1d** exhibited the highest CDK9 inhibition with an IC$_{50}$ of 0.18 μM. The cytotoxicity effect of the novel compounds was evaluated in three cancer cell lines: HCT116, K652, and MCF7. The results of this assay showed a correlation between the antiproliferative effect of the inhibitors and their CDK9 inhibitory effect in the biochemical assay. This suggests CDK9 inhibition as a mechanistic pathway for their anticancer effect. Several compounds demonstrated potent cytotoxic effects with single-digit micromolar IC$_{50}$ values yielded through an MTT assay. The compounds with the most promising data were further assessed for their antiviral activity against human Coronavirus 229E. The results showed that compound **4a** showed the highest antiviral potency with an IC$_{50}$ of 63.28 μM and a selectivity index of 4.8. In silico target prediction data showed that **4a** displayed a good affinity to proteases. The result of the docking studies of **4a** with COVID-19 main protease revealed a high binding affinity, which confirmed the results obtained from in vitro study. The physiochemical and in silico pharmacokinetic parameters indicated reasonable drug-likeness properties of the new compounds, including solubility, lipophilicity, absorption, oral bioavailability, and metabolic stability. Further lead optimization of this novel scaffold could lead to a revolution of a new class of preclinical CDK9 agents.

Keywords: cyclin-dependent kinase; kinase inhibitor; COVID-19; 229E; coronavirus

Citation: Alsfouk, A.A.; Alshibl, H.M.; Altwaijry, N.A.; Alanazi, A.; AlKamaly, O.; Sultan, A.; Alsfouk, B.A. New Imadazopyrazines with CDK9 Inhibitory Activity as Anticancer and Antiviral: Synthesis, In Silico, and In Vitro Evaluation Approaches. Pharmaceuticals 2023, 16, 1018. https://doi.org/10.3390/ph16071018

Academic Editor: Valentina Noemi Madia

Received: 13 June 2023
Revised: 11 July 2023
Accepted: 17 July 2023
Published: 18 July 2023

Copyright: © 2023 by the authors. Licensee MDPI, Basel, Switzerland. This article is an open access article distributed under the terms and conditions of the Creative Commons Attribution (CC BY) license (https://creativecommons.org/licenses/by/4.0/).

1. Introduction

Cyclin-dependent kinases are an integral component of the cell cycle machinery. They are a family of serine/threonine kinases that are dimerized with cyclins to catalyze the phosphorylation of various endogenous substrates that result in cell-cycle progression at different phases [1–3]. CDK9 is a member of this family that, in collaboration with cyclin T, forms positive transaction elongation factor b (p-TEFB), which plays an important role in RNA transcription [4,5].

There is satisfactory evidence that supports the role of CDK9 in cancer via controlling proapoptotic proteins and the promotion of cell proliferation [6–8]. Its pathogenic function has been well-established in a number of malignancies such as breast, pancreatic, prostate, lymphomas, and others [6,9–14].

Since the recognition of CDK9 as a therapeutic possibility for treating cancer, large pharmaceutical companies have developed several scaffolds as CDK9 inhibitors. The majority of them were developed as ATP-competitive inhibitors with small molecular weights and good drug-likeness properties. Several of them were successfully progressed to clinical trials as antiproliferative agents targeting various types of cancers [2,9,15–23]. The initial clinical experience with the inhibition of CDK9 was with nonselective CDK9 inhibitors such as flavopiridol, dinaciclib, and SNS-032 (Table 1). The clinical results with these inhibitors showed good responses, but they also showed a high incidence of adverse effects [24–30]. Currently, there are three highly selective CDK9 inhibitors in clinical evaluation targeting hematological malignancies: atuveciclib, BAY-1251152, and AZD4573 (Table 2) [31–33]

Table 1. Old-generation CDK9 inhibitors were evaluated in clinical trials.

CDK9 Inhibitor		Flavopiridol	Dinaciclib	SNS-032
Structure				
IC_{50} (nM)	CDK9	11	4	4
	CDK2	282	1	48
	CDK4	132	-	925
	CDK5	110	1	340
	CDK7	514	-	62
References		[8,34,35]	[8,36]	[8,37]

Table 2. New-generation CDK9 inhibitors are currently in clinical trials.

CDK9 Inhibitor		Atuveciclib	BAY-1251152	AZD4573
Structure				
IC_{50} (nM)	CDK9	6	4	3
	CDK2	1000	2920	
	CDK4	-	50-fold greater than CDK9 IC50	10-fold greater than CDK9 IC50
	CDK5	1600		
	CDK7	>10,000		
References		[21]	[32]	[22]

In 2019, an imadazopyrazine **1** (Figure 1) was isolated from natural resources and identified by virtual screening as a first-in-class CDK9 inhibitor with an IC_{50} of 7.88 µM, and it exhibited an antiproliferative effect against a panel of human breast cancer cell lines [38,39]. Further optimization of this novel scaffold was conducted by our team with the identification of 2-phenylimidazopyrazin-3-amine **2** as a CDK9 inhibitor with submicromolar IC_{50} values and a potent antiproliferative effect against a panel of cancer cell lines with single-digit IC_{50} [40]. The current study aims to further develop the novel imidazopyrazine scaffold targeting CDK9 in cancer. The activity of the new compounds has been evaluated using an MTT assay against three cancer cell lines, the binding mode of the novel compounds has been studied in silico, and their pharmacokinetics and drug-likeness properties have been examined. In addition, based on the previous data, this scaffold possesses some antiviral activity against the human Coronavirus [40,41]. Therefore, the antiviral effect of the agents has been assessed in this study.

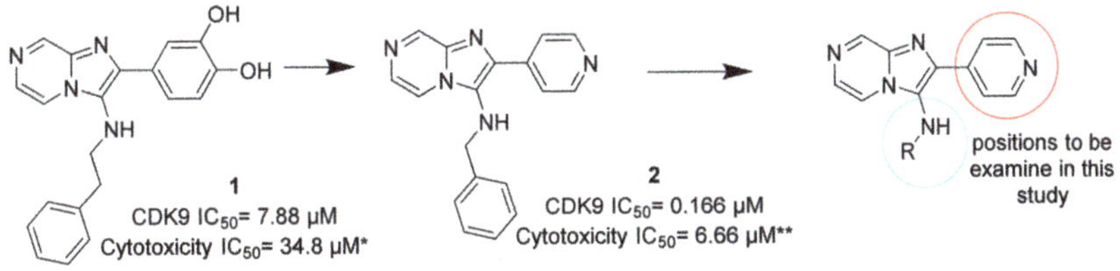

Figure 1. Development of imidazopyrazine scaffold used in this study (* average of six breast cancer cell lines, ** average of three cancer cell lines).

2. Results and Discussion

2.1. Chemistry

2.1.1. Synthetic Approach

The synthetic route used to synthesize the new compounds in this work, for **1a–4d**, is illustrated in Scheme 1. The novel compounds were synthesized using a Groebke–Blackburn–Bienaymé reaction; it is a one-pot multicomponent reaction where 2-aminopyrazine, aldehydes, and isocyanides were reacted in the presence of scandium (III)-triflate catalyst and a solvent mixture of dichloromethane and methanol (3:1). This reaction was conducted in a microwave using a temperature of 150° for 30 min. The purification of the final compounds was carried out using column chromatography that utilized the mobile phases of ethyl acetate and hexane.

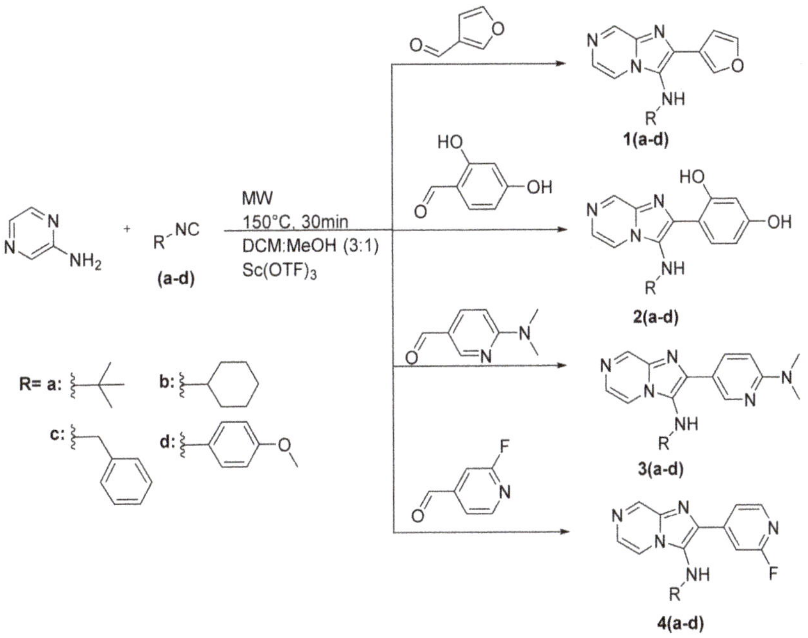

Scheme 1. The proposed synthetic approach used to synthesize compounds **1–4**.

Compounds **1a–1d** carry furan-3-yl in position 2 of imadazopyrazine and different amines in position 3, such as *t*-butyl, cyclohexyl, benzyl, and 4-methoxyphenyl. These compounds were obtained in a good yield of 86–91%. Compounds **2a–2d** carry phenyl-2,4 diol in position 2 of imadazopyrazine, and they were obtained in a good yield of 79–96%. Compounds **3a–3d** have 4-(dimethylamino)-pyridin-3-yl in position 2 of imadazopyrazine. The yield of this series ranged from 43 to 88%. Compounds **4a–4d** have 2-fluoro-pyridin-4-yl in position 2 of imadazopyrazine. The yield of these compounds was 79–97%.

The construction of the imadazopyrazine core was established using (Nuclear Magnetic Resonance (NMR). In the ^{1}H NMR spectrum of all compounds, three aromatic signals appeared that corresponded to the methylene protons of the pyrazine ring. For instance, the ^{1}H NMR spectrum of **3a** displayed the presence of two pair of doublets at 7.79 (d, J = 4.12 Hz, 1 H) and 8.34 (d, J = 4.12 Hz, 1 H) in addition to the presence of singlet signal at 8.84 (s, 1 H). These three protons represent imadazopyrazine's methylene protons (Figure 2).

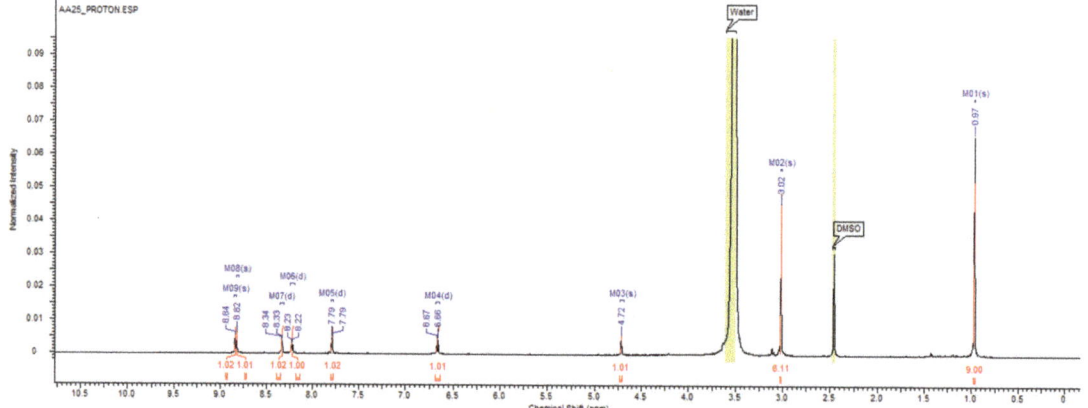

Figure 2. ^{1}H NMR spectrum of compound **3a**.

2.1.2. Rational of Molecule Design

A previous study by our team demonstrated that compounds with lipophilic substituents at position 3 of imadazopyrazine have optimum activity (compounds **1** and **2** in Figure 1) [40]. Moreover, docking studies showed that two to three carbons is the optimal distance between the lipophilic substituent and the imadazopyrazine core to occupy the G-rich pocket and form several hydrophobic interactions with Ile25, Gly26, and Val33 (Figure 3). Therefore, phenyl, benzyl, cyclohexyl, and t-butyl groups were selected to be tested at this position (R in Figure 1). The data from a previous study showed that compounds with 4-pyridinyl and 4-hydroxyphenyl in position 2 of imadazopyrazine have the optimum activity (compounds **1** and **2** in Figure 1). Also, the docking studies showed that the hydrogen bond donor/acceptor group of 4-pyridinyl and 4-hydroxyphenyl is pointed toward the solvent-exposed area and forms a hydrogen bond with Asp109. Therefore, several heterocycles with hydrogen bond donor/acceptor moieties were incorporated in 2-imadazopyrazine.

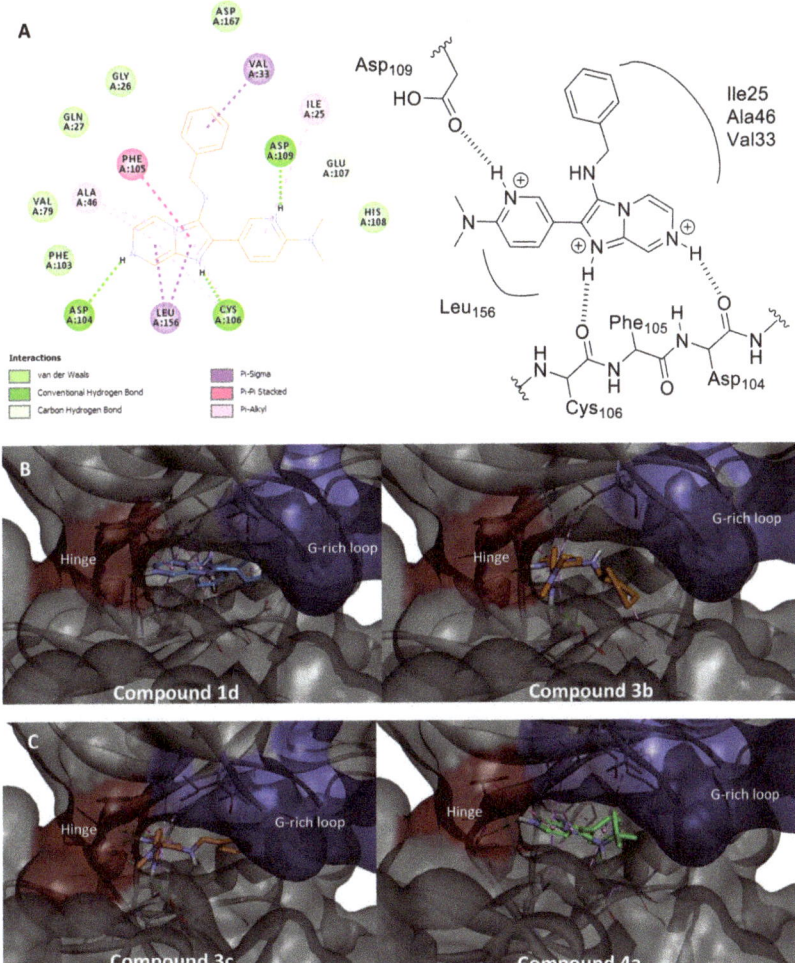

Figure 3. (**A**) A 2D illustration of **3c** displaying the main binding interactions with ATP binding site of CDK9 showing that the imadazopyrazine interacts with the hinge. (**B**) The proposed binding mode of **1d** showing that methoxyphenyl group occupies hydrophobic pocket (**left**) and **3b** showing the pyridyl pointed toward the solvent expose (**right**). (**C**) The proposed binding mode of **3c** showing the pyridyl pointed toward the solvent expose (**left**) and the proposed binding mode of **4a** where the t-butyl side chain occupied the area below the G-loop (**right**).

2.2. Anticancer Activity

2.2.1. CDK9 Activity

CDK9 is a member of the kinase family of enzymes that has a role in cell growth through the activation of RNA polymerase II via phosphorylation [4,5]. Its pathogenic function has been well-established in a number of malignancies such as acute myeloid leukemia, as well as pancreatic and prostate cancers. Its pathogenetic effects are mediated through the regulation of antiapoptotic proteins that are essential for tumor initiation and progression [6,9–13].

In this work, the synthesized compounds were assessed for their inhibition against an isolated CDK9 enzyme using the biochemical kinase assay. Standard dinaciclib (a well-established potent CDKs inhibitor) was used as a control in this experiment.

Table 3 displays the outcomes of the kinase assay. The data showed that these compounds revealed good CDK9 inhibition with an IC_{50} value of 0.18–1.78 µM. Compounds with furan-3-yl in position 2 of imadazopyrazine exhibited the highest CDK9 inhibitory activity with sub-µM IC_{50} values. In particular, **1d**, a compound with a 4-methoxyphenyl amine in position 3 of imadazopyrazine, exhibited the most potent CDK9 inhibitory activity among all the tested compounds with an IC_{50} of 0.18 µM.

Table 3. CDK9 inhibitory results of compounds **1a–4d**. Data are means (SD) of three individual trials.

Compound	CDK9 Inhibitory Activity IC_{50} (µM)
1a	0.19 ± 0.003
1b	1.78 ± 0.029
1c	0.46 ± 0.008
1d	0.18 ± 0.003
2a	0.45 ± 0.008
2b	0.30 ± 0.006
2c	0.65 ± 0.011
2d	0.89 ± 0.018
3a	0.30 ± 0.005
3b	0.23 ± 0.005
3c	0.33 ± 0.007
3d	1.66 ± 0.034
4a	0.24 ± 0.004
4b	1.22 ± 0.011
4c	1.11 ± 0.02
4d	0.57 ± 0.011
Dinaciclib	0.08 ± 0.002

In addition, the data indicated that derivatives with *t*-butylamine in position 3 of imadazopyrazine (**1a, 2a, 3a,** and **4a**) showed good CDK9 inhibition with a low IC_{50} range of 0.19–0.45 µM.

2.2.2. Cytotoxicity Assay

The cytotoxicity effect of the new compounds was assessed in an MTT assay against chronic myelogenous leukemia (K652), colorectal (HCT116), and breast (MCF7) cancer cell lines. Staurosporine standard was used as a control in this experiment. The data is presented in Table 4 as IC_{50} values (in µM).

The results of the MTT assay indicated that the new compounds described in this work have good cytotoxicity effects, with IC_{50} values in the three cell lines ranging from 10.65 to 143.79 µM. Furthermore, the cytotoxic effects of the inhibitors in this assay against several cell lines were allied to their CDK9 inhibitory effects in the biochemical assay. The derivatives with the highest potency in the CDK9 primary assay showed the most potent cytotoxic effects in the MTT assay (for example **1d, 1a, 3b,** and **4a** exhibited IC_{50} values of 0.18, 0.19, 0.23, and 0.24 µM in the biochemical assay, respectively, as well as average IC_{50} values of 11.62, 12.61, 10.65, and 20.73 µM in the MTT assay, respectively). In addition, the agents that exhibited the weakest CDK9 inhibition in the primary assay such as **1b, 3d,** and **4c** (with IC_{50} values of 1.78, 1.66, and 1.11 µM, respectively) showed the weakest cytotoxic effects in the antiproliferative assay (average IC_{50} values of 143.79, 140.04, and 92.79 µM, respectively), thus suggesting the CDK9 inhibition as a mechanistic pathway of their anticancer effects.

Table 4. Cytotoxicity results of compounds 1a–4d in the MTT assay against MCF7, HCT116, and K652 cancer cells. Data are means (SD) of three individual trials.

Compound	IC$_{50}$ (µM)			Average of the Three Cell Lines
	MCF7	HCT116	K652	
1a	2.85 ± 0.03	21.55 ± 0.29	13.43 ± 0.19	12.61
1b	99.16 ± 3.08	163.98 ± 2.1	168.23 ± 5.99	143.79
1c	17.39 ± 0.34	16.39 ± 0.69	24.83 ± 0.61	19.54
1d	12.35 ± 0.17	10.67 ± 0.16	11.83 ± 0.19	11.62
2a	48.26 ± 0.24	136.42 ± 0.25	71.05 ± 0.39	85.24
2b	45.93 ± 0.67	7.61 ± 2.12	38.84 ± 1.16	30.79
2c	63.48 ± 0.7	56.56 ± 0.13	50.24 ± 0.69	56.76
2d	71.24 ± 0.99	103.99 ± 0.98	25.30 ± 0.91	66.84
3a	22.81 ± 0.33	4.13 ± 0.07	78.21 ± 1.32	35.05
3b	16.42 ± 0.26	11 ± 0.19	4.5 ± 0.08	10.65
3c	9.52 ± 0.15	42.22 ± 0.68	26.38 ± 0.43	26.04
3d	119.91 ± 2.02	147.46 ± 2.75	152.74 ± 2.99	140.04
4a	4.38 ± 0.06	38.2 ± 0.51	19.62 ± 0.26	20.73
4b	16.21 ± 0.41	15.28 ± 0.13	23.15 ± 0.81	18.22
4c	67.32 ± 1.01	97.388 ± 1.46	113.67 ± 1.7	92.79
4d	25.07 ± 0.39	10.22 ± 0.16	47.11 ± 0.26	27.47
Staurosporine	18.41 ± 0.4	10.86 ± 0.26	22.08 ± 0.56	17.12

To ensure that the cytotoxicity effect of our compounds observed on the previous assay was selective to cancer cells, the most cytotoxic agents to cancer cells (**1a**, **1c**, **1d**, **3b**, and **4b**) were evaluated on normal cells. The data is presented in Table 5; our compounds exhibited weak cytotoxicity against normal noncancerous FHC cells, with IC$_{50}$ values ranging from 58.64–187.31 µM, as well as a selectivity index ranging from 2.5–20.5. This data indicated that the most promising agents showed acceptable safety for the normal cells.

Table 5. Cytotoxicity results of compounds **1a**, **1c**, **1d**, **3b**, and **4b** in MTT assay against FHC normal cells. Data are means (SD) of three individual trials(*SI = Selectivity Index).

Compound	FHC IC$_{50}$ (µM)	SI*		
		MCF7	HCT116	K652
1a	58.64 ± 0.8	20.5	2.7	4.3
1c	61.68 ± 0.96	15.16	17.55	15.8
1d	187.31 ± 2.92	3.5	3.7	2.5
3b	95.80 ± 1.72	5.8	8.7	21.3
4b	74.31 ± 1.26	4.5	4.8	3.2
Staurosporine	38.19 ± 0.97	2.07	3.5	1.7

2.2.3. Molecular Docking Study into CDK9 Active Site

Simulation docking studies into the ATP binding site of the CDK9 enzyme of compounds **1d**, **3b**, **3c**, and **4a** were conducted using AutoDock Vina embedded in PyRx 0.8 software. In these studies, the CDK9 complex with a UT5 ligand was used as a macromolecule (PDB: 7NWK). The validity of the docking studies was carried out initially by the redocking of a cocrystalized inhibitor (UT5). It showed a docking score of −8.5 kcal/mol with an RMSD value of 1.6 Å. The main hydrogen bond interactions between the imdazopyridine ring of UT5 and the Cys106 residue of the protein were obtained.

In general, the tested compounds fitted well into the ATP binding site and showed several interactions with key amino acid residues in the catalytic site of the CDK9 enzyme (Figure 3, Table 6). All tested compounds showed interactions between the hinge region of the CDK9 enzyme and imadazopyrazine by forming hydrogen bonds with Cys106 and Asp104.

Table 6. Data of docking studies of **1d**, **3b**, **3c**, and **4a** with CDK9 catalytic site.

Compound	Docking Score (Kcal/mol)	Interaction Residue (Type of Interaction)	Bond Length (Å)
1d	−8.3	Cys106 (HB)	1.83
		Val33 (pi–sigma)	3.63
		Leu156 (pi–sigma)	3.94
		Ile25 (pi–sigma)	3.63
		Phe103 (pi–pi stacked)	3.90
		Phe105 (pi–pi stacked)	5.25
		Al166 (pi–alkyl)	4.57
		Val79 (pi–alkyl)	3.93
3b	−8.4	Cys106 (HB)	2.00
		Asp109 (HB)	2.53
		Phe105 (pi–pi stacked)	5.26
		Leu156 (pi–sigma)	3.99
		Ile25 (pi–sigma)	3.65
		Ala46 (pi–alkyl)	3.80
3c	−8	Cys106 (HB)	1.93
		Asp104 (HB)	2.79
		Asp109 (HB)	2.29
		Phe105 (pi–pi stacked)	5.17
		Leu156 (pi–sigma)	3.75 and 3.48
		Val33 (pi–sigma)	3.78
		Ala46 (pi–alkyl)	4.63 and 3.05
4a	−8	Cys106 (HB)	2.25
		Asp167 (HB)	3.14
		Phe103 (pi–pi stacked)	4.30
		Phe105 (pi–pi stacked)	5.48
		Leu156 (pi–sigma)	4.67
		Ile25 (pi–sigma)	3.76
		Val33 (pi–alkyl)	5.46
		Al46 (pi–alkyl)	5.43 and 4.69

In compounds **3b**, **3c**, and **4a**, the lipophilic side chain of the alkylamine or benzylamine in position 3 of imadazopyrazine formed several hydrophobic interactions with Ile25, Gly26, and Val33 in the G-rich loop, while the pyridine ring in position 2 of imadazopyrazine pointed toward the solvent-exposed areas and formed a hydrogen bond with Asp109 (Figure 3A,C).

Compound **1d** adapted a flipped orientation in comparison to the poses of compounds **3b**, **3c**, and **4a**, where the furan ring of **1d** in position 2 of the imadazopyrazine occupied a region down the G-loop and interacted with Gly28, Gly26, and Ile25, while the methoxyphenyl ring in position 3 of the imadazopyrazine engaged the hydrophobic pocket and interacted with Phe103, Val166, and Val79 (Figure 3B).

In terms of biological activity, **1d** demonstrated the most potent activity among all the tested compounds in both the CDK9 and cytotoxicity assays (Sections 2.2.1 and 2.2.2). This may indicate that the lipophilic small group in position 2 of the imadazopyrazine is more favorable than the ionizable groups (such as pyridine and phenol). The docking studies showed that **1d** has a flipped orientation in comparison to the poses of compounds **3b**, **3c**, and **4a** (Figure 3). This flipped orientation, where the 2-imadazopyrazine substituent occupies a region down the G-loop and 3-imadazopyrazine moieties engaged in the hydrophobic pocket, seems to be more favorable for biological activity.

2.3. Antiviral Activity

2.3.1. 229E Inhibitory Assay

The antiviral effect of the compounds was examined in a human Coronavirus (HCoV-229E) inhibitory assay. The standard ribavirin (a well-established antiviral agent) was used as a positive control in this experiment.

The outcomes of this assay are displayed in Table 7. The data showed that compound **4a** showed good antiviral activity. In comparison to the standard antiviral agent ribavirin, **4a** demonstrated better antiviral activity against HCoV-229E with an IC_{50} of 63.28 µM. It also showed a weak cytotoxicity effect on the target cells at the concentrations that achieved its anticoronaviral effect (0.1–1000 µg/mL), with a 50% cellular cytotoxicity concentration (CC_{50}) of 303.15 µM and a selectivity index (SI) of 4.8 [42,43].

Table 7. 229E inhibitory effect of selected compounds in the primary assay. Data are means (SD) of three individual trials.

Compound	CC50 (µM)	IC50 (µM)	SI
1a	1319.90	1057.79	1.25
1c	652.80	516.45	2.1
1d	845.57	404.72	1.27
3a	186.14	392.83	0.48
3b	212.98	590.07	0.37
4a	303.15	63.28	4.8
4b	174.36	617.75	0.29
4c	270.64	330.56	0.9
Ribavirin	160.47	113.81	1.4

SI = selectivity index (CC_{50}/IC_{50}).

2.3.2. Antiviral Target Prediction and Molecular Docking Studies

A target prediction study was performed in silico using SwissTarget [44]. Figure 4 shows the result of the target prediction of compound **4a**, which revealed a good affinity to protease enzymes. This suggests the inhibition of protases enzyme as a mechanistic pathway of the observed anticoronaviral activity of **4a** in an in vitro assay.

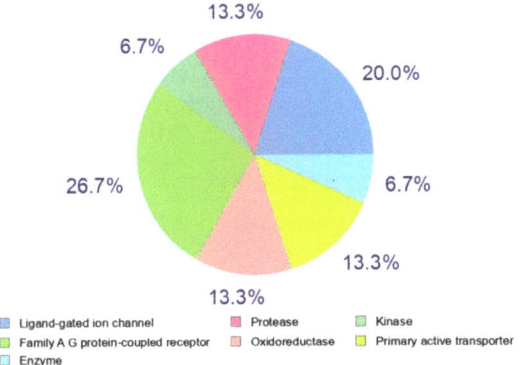

Figure 4. Prediction of the molecular target of compound **4a**.

Docking studies of the compound **4a** with a COVID-19 main protease were conducted using AutoDock Vina embedded in PyRx 0.8 software. HCoV-229E is an isoform of the Coronavirus that shows high homologous sequence similarity to the SARS-CoV-2 isoform; the virus caused the pandemic respiratory disease in 2019 (COVID-19) [45,46]. Therefore, the main protease was selected as a potential target for studying the binding mode of the novel antiviral agents in this work. In these studies, the main protease of COVID-19 in a complex with X77 was used as a macromolecule (PDB: 6W63). The validity test of our docking analysis was conducted initially through redocking of the X77 inhibitor, which was the cocrystalized ligand. It showed a docking score of −9.7 kcal/mol with an RMSD value of 1.7 Å. The main H-bonding between the Gly143 and Glu166 amino acids of the receptor and the cocrystalized X77 were gained.

The docking outcomes showed that the H-bonding between the pyridine ring of compound **4a** formed with the Glu166 amino acid residue. It also revealed a hydrogen bond interaction between the amino group of the imadazopyrazine at position 3 of **4a** and the Phe140 residue. The imadazopyrazine formed several hydrophobic interactions with Cy145, His41, and Met165. The fluorine atom at position 2 of the pyridine ring interacted with Asn142 (Figure 5, Table 8).

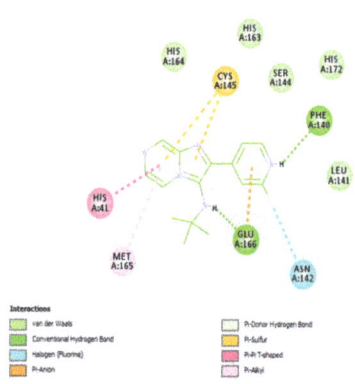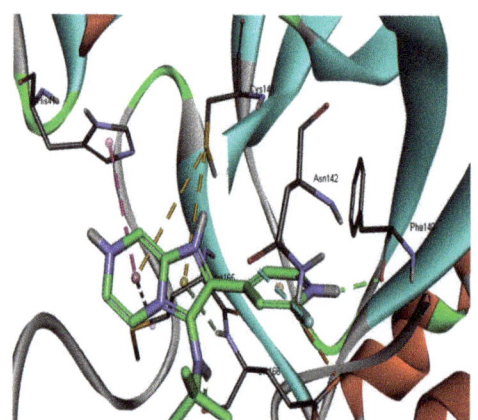

Figure 5. 2D and 3D diagrams showing the key binding residues of **4a** with COVID-19 main protease.

Table 8. Data of docking studies of **4a** with COVID-19 main protease.

Docking Score (Kcal/mol)	Interaction Residue (Type of Interaction)	Bond Length (Å)
−8.7	Glu166 (HB)	2.15
	Phe140 (HB)	2.16
	Cys145 (pi–sulfur, pi–anion)	4.57, 5.21
	His41 (pi–pi stacked)	5.02
	Met165 (pi–Alkyl)	5.12
	Asn142 (fluorine)	3.51

2.4. In Silico Prediction of Drug-Likeness Properties

2.4.1. Molecular Structured and Physicochemical Properties

The physiochemical properties of the selected inhibitors (**1a**, **1d**, **2c**, **3b**, **4a**, and **4b**) were estimated using SwissADME [47]. All the studied compounds displayed good physiochemical properties, as shown in Table 9. All the compounds were shown to be in line with Lipinski's rule of five, with zero violations, and are therefore expected to be orally bioavailable. Their molecular weights were shown to be <500, and they demonstrated optimal lipophilicity (Log P values ranged from 1–5), optimal polarity (tPSA values < 90 Å^2), and reasonable aqueous solubility (Log S > −4 mol/L).

The compounds with furan-3-yl in position 2 of the imadazopyrazine displayed the highest lipophilicity. In the biochemical and cell assays, they also exhibited the most potent activity among all the tested compounds. This could indicate that the lipophilic small group is more favorable than the ionizable groups (such as pyridine and phenol) in position 2 of the imadazopyrazine.

Table 9. Predicated molecular and physiochemical properties of selected compounds (calculated using SwissADME, tPSA = total polar surface area).

Compound	M. wt	Clog P	tPSA (Å^2)	Log S	HBA	HBD	Lipinski
1a	286.37	4.02	55.36	−4.44	3	1	Yes; with 0 violations
1d	306.32	2.79	64.59	−4.18	4	1	Yes; with 0 violations
2c	332.36	2.42	82.68	−4.44	4	3	Yes; with 0 violations
3b	336.43	3.15	58.35	−4.54	3	1	Yes; with 0 violations
4a	285.32	2.55	55.11	−3.84	4	1	Yes; with 0 violations
4b	313.37	2.90	55.10	−3.08	5	1	Yes; with 0 violations

2.4.2. ADMET Studies

In silico pharmacokinetic studies of the inhibitors **1d**, **3b**, and **4a** were conducted through the ADMETLab platform [48]. Table 10 shows the data of these studies. The three studied compounds displayed high oral absorption by demonstrating good intestinal permeability and low efflux liability (with a Papp index > −5.15, which is a positive test result for human intestinal absorption). None of these compounds are predicted to be a P-glycoprotein substrate or inhibitor (except for **1d**, which is expected to be an inhibitor of P-glycoprotein). All three compounds showed good bioavailability scores. With respect to distribution, the volume of distribution of the studied compounds was within the optimal range (0.04–20 L/kg), and plasma protein binding (PPB) was within the acceptable level (<90%). The three tested derivatives are expected to cross the blood–brain barrier (BBB). Regarding the metabolism, all three compounds are estimated to be an inhibitor of CYP3A4 (the main metabolizing CYP450 isoform), but none of them are expected to be metabolized by the same isoform. With respect to excretion, the compounds displayed half-lives > 0.5 h and a clearance that was <15 mL/min/kg, which are considered acceptable values in the drug development process. With respect to the toxicity profile, all three of the studied derivatives were expected to block the hERG channel and produce mutagenicity, as well as liver injury, but none of them were shown to be a skin sensitizer.

Table 10. Predicated ADMET properties of compounds **1d**, **3b**, and **4a** (estimated using ADMETLab; red color indicates unfavorable parameter).

Compound				1d	3b	4a
Property	Test		Recommended			
Absorption	Papp (Caco-2 permeability) (cm/s)		>−5.15	−4.871	−4.677	−4.525
	Pg protein inhibitor			Inhibitor	Noninhibitor	Noninhibitor
	Pg protein substrate			Nonsubstrate	Nonsubstrate	Nonsubstrate
	HIA (human intestinal absorption)			HIA+	HIA+	HIA+
	Bioavailability score		0.55	0.55	0.55	0.55
Distribution	PPB (Plasma protein binding) %		<90	81.421	82.13	78.804
	BBB (Blood–brain barrier)			BBB+	BBB+	BBB+
	VD (volume of distribution) L/kg		0.04–20	0.125	0.725	0.121
Metabolism	CYP3A4 inhibitor			Inhibitor	Inhibitor	Inhibitor
	CYP3A4 substrate			Nonsubstrate	Substrate	Substrate
Excretion	$T_{1/2}$ (Half live) h		>0.5	1.812	1.827	1.807
	Clearance mL/min/kg		<15	1.979	2.028	1.95
Toxicity	hERG blocker			Blocker	Blocker	Blocker
	Ames mutagenicity			Ames+	Ames+	Ames−
	Skin sensitivity			Nonsensitizer	Nonsensitizer	Nonsensitizer
	LD$_{50}$ of acute toxicity		>500 mg/kg	2.527	2.598	2.701
	DILI (drug-induced liver injury)			DILI+	DILI+	DILI+
	FDAMDD (maximum recommended daily dose)			FDAMDD+	FDAMDD−	FDAMDD+

3. Materials and Methods

3.1. Chemistry

All starting materials, building blocks, catalysts, solvents, and reagents, including 2-aminopyrazine, isocyanides, aldehydes, scandium (III)-trifluoromethanesulfonate, methanol, and dichloromethane, were purchased from Sigma-Aldrich and TCI chemical companies and used directly in the reactions without prior purification. Electrothermal melting point apparatus was used to measure the melting points of the compounds without correction. The ^{1}H NMR spectra at 700.17 MHz and ^{13}C spectra at 176.08 MHz were obtained using Bruker Ascend 700 NMR spectrometer (Fällanden, Switzerland) and DMSO-d_6 as solvents in all samples. TLC was used to monitor the progression of the reactions using aluminum sheets of precoated silica gel (60 F254, Merck) and visualized under UV light at 365 and 254.

General procedure:

In a microwave vial, a mixture of 2-aminopyrazine (0.52 mmol), required isocyanide (0.52 mmol), required aldehyde (0.4 mmol), scandium (III)-trifluoromethanesulfonate catalyst (0.015 mmol), and 2 mL of a solvent system of 3-to-1 dichloromethane-to-methanol were added. The vial was then sealed and heated to 150 °C for 30 min in the microwave. The reaction mixture was then allowed to cool down to room temperature and concentrated. The reaction residue was then extracted with 5 mL dichloromethane three times. The organic extract was then collected, dried over magnesium sulfate, and evaporated. The residue was then purified using column chromatography using hexane and ethyl acetate.

The N-(Tert-butyl)-2-(furan-3-yl) imidazo[1,2-a] pyrazin-3-amine (**1a**) *Yield*: 89.3%; yellow oil; ^{1}H NMR (700 MHz, DMSO-d_6) δ ppm 1.09 (br. s., 9 H), 4.76 (s, 1 H), 7.13 (s, 1 H), 7.76 (br. s., 1 H), 7.84 (br. s., 1 H), 8.32 (s, 1 H), 8.38 (br. s., 1 H), 8.89 (s, 1 H); ^{13}C NMR (176 MHz, DMSO-d_6) δ ppm 30.53, 56.85, 110.46, 117.60, 120.40, 125.57, 128.81, 135.55, 137.31, 141.81, 142.52, and 143.76. *m/z* (ESI-MS) [M]$^+$ 257.13.

The N-Cyclohexyl-2-(furan-3-yl) imidazo[1,2-a] pyrazin-3-amine (**1b**) *yield*: 86.4%; yellow oil; ^{1}H NMR (700 MHz, DMSO-d_6) δ ppm 1.09 (br. s., 9 H), 1.10–1.15 (m, 3 H), 1.28–1.33 (m, 2 H), 1.53 (br. s., 1 H), 1.66 (d, *J* = 11.83 Hz, 2 H), 1.75 (d, *J* = 12.69 Hz, 2 H), 2.86–2.90 (m, 1 H), 4.92 (d, *J* = 7.10 Hz, 1 H), 7.79 (s, 1 H), 7.85 (d, *J* = 4.52 Hz, 1 H), 7.87 (s, 1 H) 8.25 (s, 1 H), 8.30 (dd, *J* = 4.52, 1.08 Hz, 1 H), 8.88 (s, 1 H); ^{13}C NMR (176 MHz, DMSO-d_6) δ ppm 25.07, 25.81, 34.19, 56.82, 109.78, 116.72, 120.13, 127.17, 120.95, 132.89, 136.65, 141.06, 142.46, and 144.16. *m/z* (ESI-MS) [M]$^+$ 283.07.

The N-Benzyl-2-(furan-3-yl) imidazo[1,2-a]pyrazin-3-amine (**1c**) *yield*: 91.0%; yellow oil; ^{1}H NMR (700 MHz, DMSO-d_6) δ ppm 4.17 (d, *J* = 6.24 Hz, 2 H), 5.57 (t, *J* = 6.35 Hz, 1 H), 7.04 (s, 1 H), 7.23–7.31 (m, 5 H), 7.75 (d, *J* = 4.52 Hz, 1 H), 7.79 (s, 1 H), 8.12 (d, *J* = 4.30 Hz, 1 H), 8.20 (s, 1 H), 8.86 (s, 1 H); ^{13}C NMR (176 MHz, DMSO-d_6) δ ppm 51.09, 109.79, 116.46, 119.94, 127.63, 126.63, 126.75, 131.98, 132.05, 132.89, 136.47, 140.13, 141.06, 142.45, and 144.20. *m/z* (ESI-MS) [M]$^+$ 291.14.

The 2-(Furan-3-yl)-N-(4-methoxyphenyl) imidazo[1,2-a] pyrazin-3-amine (**1d**) *yield*: 88.4%; yellow oil; ^{1}H NMR (700 MHz, DMSO-d_6) δ ppm 3.64 (s, 3 H), 6.47 (d, *J* = 7.31 Hz, 2 H), 6.77 (d, *J* = 7.74 Hz, 2 H), 6.93 (s, 1 H), 7.75 (br. s., 1 H), 7.89 (br. s., 1 H), 7.99 (br. s., 1 H), 8.03 (br. s., 1 H), 8.04 (br. s., 1 H), 9.04 (s, 1 H); ^{13}C NMR (176 MHz, DMSO-d_6) δ ppm 55.69, 109.55, 114.76, 115.47, 132.04, 132.92, 134.91, 137.66, 138.87, 141.47, 142.31, 142.85, 144.41, 153.12, and 156.53. *m/z* (ESI-MS) [M]$^+$ 291.14.

The 4-(3-(Tert-butylamino) imidazo[1,2-a]pyrazin-2-yl)benzene-1,3-diol (**2a**) *yield*: 96.6%; white solid (MP: 174–176 °C); ^{1}H NMR (700 MHz, DMSO-d_6) δ ppm 1.00 (s, 9 H), 5.35 (s, 1 H), 6.28 (d, *J* = 2.75 Hz, 1 H), 6.34 (d, *J* = 2.06 Hz, 1 H), 7.86 (d, *J* = 4.81 Hz, 1 H), 7.89 (d, *J* = 8.94 Hz, 1 H), 8.39 (dd, *J* = 4.81, 1.37 Hz, 1 H), 8.91 (d, *J* = 1.37 Hz, 1 H), 9.51 (br. s., 1 H), 11.53 (br. s., 1 H); ^{13}C NMR (176 MHz, DMSO-d_6) δ ppm 29.77, 109.8, 116.7, 119.80, 127.61, 126.67, 126.79, 132.01, 132.70, 132.89, 136.54, 140.15, 141.67, 142.45, and 144.22. *m/z* (ESI-MS) [M]$^-$ 297.04.

The 4-(3-(Cyclohexylamino) imidazo[1,2-a] pyrazin-2-yl)benzene-1,3-diol (**2b**) yield: 79.5%; white solid (MP: 245–247 °C); ^{1}H NMR (700 MHz, DMSO-d_6) δ ppm 1.02–1.11 (m, 4 H), 1.20 (d, *J* = 11.00 Hz, 2 H), 1.58–1.61 (m, 2 H), 1.66 (d, *J* = 12.37 Hz, 2 H), 2.81 (tdt, *J* = 10.35, 10.35, 6.96, 3.69, 3.69 Hz, 1 H), 5.21 (d, *J* = 6.87 Hz, 1 H), 6.28 (d, *J* = 2.75 Hz, 1 H), 6.34 (dd, *J* = 8.59, 2.41 Hz, 1 H), 7.89 (d, *J* = 4.81 Hz, 1 H), 7.99 (d, *J* = 8.25 Hz, 1 H), 8.34 (dd, *J* = 4.12, 1.37 Hz, 1 H), 8.92 (d, *J* = 1.38 Hz, 1 H), 9.57 (s, 1 H), 12.15 (s, 1 H); ^{13}C NMR (176 MHz, DMSO-d_6) δ ppm 51.09, 109.79, 116.46, 119.94, 127.63, 126.63, 126.75, 131.98, 132.05, 132.89, 136.47, 140.13, 141.06, 142.45, and 144.20. *m/z* (ESI-MS) [M]$^+$ 325.17.

The 4-(3-(Benzylamino)imidazo[1,2-a]pyrazin-2-yl)benzene-1,3-diol (**2c**) yield: 91.8%; white solid (MP: 167–169 °C); ^{1}H NMR (700 MHz, DMSO-d_6) δ ppm 4.06 (d, *J* = 6.87 Hz, 2 H), 5.65 (t, *J* = 6.19 Hz, 1 H), 6.31 (d, *J* = 2.06 Hz, 1 H), 6.33–6.36 (m, 1 H), 7.21–7.23 (m, 4 H), 7.29–7.32 (m, 1 H), 7.79 (d, *J* = 4.12 Hz, 1 H), 7.98 (d, *J* = 8.94 Hz, 1 H), 8.14 (dd, *J* = 4.12, 1.37 Hz, 1 H), 8.89 (d, *J* = 1.37 Hz, 1 H), 9.60 (s, 1 H), 12.10 (s, 1 H)); ^{13}C NMR (176 MHz, DMSO-d_6) δ ppm 51.29, 103.64, 107.74, 109.38, 116.25, 126.32, 127.78, 128.78, 128.83, 129.32, 129.51, 134.65, 136.85, 139.77, 141.63, 158.34, and 159.44. *m/z* (ESI-MS) [M]$^+$ 331.06.

The 4-(3-((4-Methoxyphenyl) amino)imidazo[1,2-a]pyrazin-2-yl)benzene-1,3-diol (**2d**) yield: 79.9%; white solid (MP: 186–188 °C); ^{1}H NMR (700 MHz, DMSO-d_6) δ ppm 3.60 (s, 3 H), 6.20–6.22 (m, 1 H), 6.29 (d, *J* = 2.75 Hz, 1 H), 6.44 (m, *J* = 8.94 Hz, 2 H), 6.72 (m, *J* = 8.94 Hz, 2 H), 7.75 (d, *J* = 8.93 Hz, 1 H), 7.93 (d, *J* = 4.81 Hz, 1 H), 8.00 (s, 1 H) 8.02 (dd, *J* = 4.12, 1.37 Hz, 1 H), 9.09 (s, 1 H), 9.64 (s, 1 H), 12.49 (s, 1 H); ^{13}C NMR (176 MHz, DMSO-d_6) δ ppm 55.74, 103.63, 107.75, 107.96, 114.81, 115.58, 116.44, 119.63, 128.70, 130.62, 135.32, 138.60, 139.75, 141.89, 153.26, 159.15, and 159.81. *m/z* (ESI-MS) [M]$^+$ 349.12.

The N-(Tert-butyl)-2-(6-(dimethylamino) pyridin-3-yl)imidazo[1,2-a]pyrazin-3-amine (**3a**) yield: 98%; white solid (MP: 126–128 °C); ^{1}H NMR (700 MHz, DMSO-d_6) δ ppm 0.97 (s, 9 H), 3.02 (s, 6 H), 4.72 (s, 1 H), 6.67 (d, *J* = 8.94 Hz, 1 H), 7.79 (d, *J* = 4.12 Hz, 1 H), 8.22 (d, *J* = 8.93 Hz, 1 H), 8.34 (d, *J* = 4.12 Hz, 1 H), 8.84 (s, 1 H), 8.84 (s, 1H); ^{13}C NMR (176 MHz, DMSO-d_6) δ ppm 30.52, 38.16, 56.66, 105.57, 117.54, 118.36, 124.96, 128.85, 136.98, 137.20, 139.86, 142.22, 147.54, and 158.65. *m/z* (ESI-MS) [M]$^+$ 311.20.

The N-Cyclohexyl-2-(6-(dimethylamino) pyridin-3-yl)imidazo[1,2-a]pyrazin-3-amine (**3b**) yield: 85.1%; white solid (MP: 191–193 °C); ^{1}H NMR (700 MHz, DMSO-d_6) δ ppm 1.04 (br. s., 3 H), 1.19–1.24 (m, 2 H), 1.45 (br. s., 1 H), 1.58 (br. s., 2 H), 1.65 (d, *J* = 12.37 Hz, 2 H), 2.79 (dd, *J* = 10.31, 3.44 Hz, 1 H), 3.03 (s, 6 H), 4.91 (d, *J* = 6.87 Hz, 1 H), 6.70 (d, *J* = 8.94 Hz, 1 H), 7.78 (d, *J* = 4.81 Hz, 1 H), 8.23 (dd, *J* = 8.94, 2.75 Hz, 1 H), 8.26 (d, *J* = 4.12 Hz, 1 H), 8.81 (s, 1 H), 8.86 (s, 1 H); ^{13}C NMR (176 MHz, DMSO-d_6) δ ppm 25.02, 25.84, 34.07, 38.19, 56.84, 106.01, 116.66, 117.90, 118.35, 126.36, 128.99, 136.00, 136.59, 142.14, 146.69, and 158.64. *m/z* (ESI-MS) [M]$^+$ 336.96.

The N-Benzyl-2-(6-(dimethylamino) pyridin-3-yl)imidazo[1,2-a]pyrazin-3-amine (**3c**) yield: 43.3%; yellow oil; ^{1}H NMR (700 MHz, DMSO-d_6) δ ppm 3.05 (s, 6 H), 4.08 (d, *J* = 6.87 Hz, 2 H), 5.58 (t, *J* = 6.53 Hz, 1 H), 6.71 (d, *J* = 8.94 Hz, 1 H), 7.20–7.22 (m, 4 H), 7.68 (d, *J* = 4.81 Hz, 1 H), 7.76–7.88 (m, 1 H), 8.09 (dd, *J* = 4.81, 1.37 Hz, 1 H), 8.17 (dd, *J* = 8.94, 2.75 Hz, 1 H), 8.79 (d, *J* = 1.37 Hz, 1 H), 8.82 (d, *J* = 2.06 Hz, 1 H); ^{13}C NMR (176 MHz, DMSO-d_6) δ ppm 38.22, 51.22, 106.07, 106.44, 116.36, 117.94, 121.71, 127.67, 128.61, 128.78, 136.12, 136.42, 140.10, 142.22, 146.68, 158.64, and 189.82. *m/z* (ESI-MS) [M]$^+$ 345.22.

The 2-(6-(Dimethylamino) pyridin-3-yl)-N-(4-methoxyphenyl) imidazo[1,2-a]pyrazin-3-amine (**3d**) yield: 47.7%; white solid (MP: 180–182 °C); ^{1}H NMR (700 MHz, DMSO-d_6) δ ppm 3.00 (s, 6 H), 3.59 (s, 3 H), 6.40 (m, *J* = 8.94 Hz, 2 H), 6.67 (d, *J* = 8.94 Hz, 1 H), 6.72 (m, *J* = 8.94 Hz, 2 H), 7.82 (d, *J* = 4.12 Hz, 1 H), 7.96 (dd, *J* = 4.81, 1.37 Hz, 1 H), 7.97 (s, 1 H), 8.09 (dd, *J* = 8.94, 2.06 Hz, 1 H), 8.69 (d, *J* = 2.06 Hz, 1 H), 8.98 (d, *J* = 1.37 Hz, 1 H); ^{13}C NMR (176 MHz, DMSO-d_6) δ ppm 38.14, 55.75, 106.11, 114.59, 115.61, 116.60, 117.11, 120.28, 122.57, 129.61, 135.88, 137.70, 139.10, 142.66, 146.80, 135.11, and 158.87. *m/z* (ESI-MS) [M]$^+$ 361.31.

The N-(Tert-butyl)-2-(2-fluoropyridin-4-yl) imidazo[1,2-a] pyrazin-3-amine (**4a**) yield: 87.7%; yellow oil; ^{1}H NMR (700 MHz, DMSO-d_6) δ ppm 1.00 (s, 9 H), 5.04 (s, 1 H), 7.85 (s, 1 H), 7.87 (d, *J* = 4.12 Hz, 1 H), 8.11 (d, *J* = 4.81 Hz, 1 H), 8.27 (d, *J* = 4.81 Hz, 1 H), 8.43 (d, *J* = 4.81 Hz, 1 H), 8.98 (s, 1 H); ^{13}C NMR (176 MHz, DMSO-d_6) δ ppm 30.40, 57.26, 107.40, 118.13, 120.70, 128.34, 129.30, 136.70, 148.22, 163.34, and 164.89. *m/z* (ESI-MS) [M]$^+$ 286.04.

The N-Cyclohexyl-2-(2-fluoropyridin-4-yl) imidazo[1,2-a] pyrazin-3-amine (**4b**) yield: 81.2%; yellow oil; ^{1}H NMR (700 MHz, DMSO-d_6) δ ppm 1.04–1.10 (m, 3 H), 1.25–1.31 (m, 2 H), 1.47 (br. s., 1 H), 1.59–1.63 (m, 2 H), 1.70 (d, *J* = 12.37 Hz, 2 H), 2.79–2.86 (m, 1 H), 5.30 (d, *J* = 7.56 Hz, 1 H), 7.77 (s, 1 H), 7.86 (d, *J* = 4.81 Hz, 1 H), 8.06 (d, *J* = 5.50 Hz, 1 H), 8.29 (d, *J* = 5.50 Hz, 1 H), 8.37 (dd, *J* = 4.12, 1.37 Hz, 1 H), 8.97 (s, 1 H); ^{13}C NMR (176 MHz, DMSO-d_6) δ ppm 25.11, 25.76, 34.17, 57.59, 105.92, 117.29, 119.44, 129.34, 130.71, 134.14, 136.67, 144.22, 148.51, 163.64, and 165.18. *m/z* (ESI-MS) [M]$^+$ 312.11.

The N-Benzyl-2-(2-fluoropyridin-4-yl) imidazo[1,2-a] pyrazin-3-amine (**4c**) yield: 79.2%; yellow oil; ^{1}H NMR (700 MHz, DMSO-d_6) δ ppm 4.14 (d, *J* = 6.87 Hz, 2 H), 5.94 (t, *J* = 6.53 Hz, 1 H), 7.16–7.19 (m, 5 H), 7.65 (s, 1 H), 7.77 (d, *J* = 4.81 Hz, 1 H), 7.96 (d, *J* = 4.81 Hz, 1 H), 8.20 (dd, *J* = 4.47, 1.72 Hz, 1 H), 8.25 (d, *J* = 5.50 Hz, 1 H), 8.94 (d, *J* = 1.37 Hz, 1 H); ^{13}C NMR (176 MHz, DMSO-d_6) δ ppm 51.50, 106.11, 117.07, 119.59, 127.81, 128.70, 128.82, 129.11, 131.01, 134.15, 136.54, 139.65, 144.13, 148.38, 163.57, and 165.12. *m/z* (ESI-MS) [M]$^+$ 320.11.

The 2-(2-Fluoropyridin-4-yl)-N-(4-methoxyphenyl) imidazo[1,2-a]pyrazin-3-amine (**4d**) yield: 97.0%; yellow oil; ^{1}H NMR (700 MHz, DMSO-d_6) δ ppm 3.60 (s, 3 H), 6.48 (d, *J* = 8.94 Hz, 2 H), 6.74 (d, *J* = 8.94 Hz, 2 H), 6.99 (m, *J* = 8.94 Hz, 2 H), 7.37 (m, *J* = 8.94 Hz, 2 H), 7.82 (s, 1 H), 7.89 (d, *J* = 4.81 Hz, 1 H), 8.02–8.03 (m, 1 H), 8.27 (d, *J* = 1.37 Hz, 1 H); ^{13}C NMR (176 MHz, DMSO-d_6) δ ppm 55.91, 106.22, 115.09, 115.61, 117.30, 119.56, 120.76, 123.68, 130.03, 132.08, 132.94, 137.74, 142.37, 144.52, 149.04, and 155.55. *m/z* (ESI-MS) [M]$^+$ 336.18.

3.2. In Vitro CDK9 Kinase Assay

The in vitro kinase activity was measured using a CDK9 assay kit obtained from BPS Biosciences (San Diego, CA). The inhibitory effect of the tested compounds was assessed following the manufacturer's instructions as indicated in the kit. GraphPad Prism 5.0 software was used to analyze the results. DMSO was used as a negative standard in this assay, and dinaciclib was used as a positive standard.

3.3. MTT Cytotoxicity Assay

Cells were obtained from American Type Culture Collection. The cell culture DMEM was obtained from Life Technologies and Invitrogen and supplied with 10% FBS from Hyclone, 1% penicillin-streptomycin, and 10 µg/mL insulin from Sigma-Aldrich.

The MTT assay was used to monitor the in vitro cytotoxicity of our compounds. The cells were treated with serial concentrations of the compounds to be tested, which ranged from 0.1–10 µM at 37 °C for 48 h. Then, the cells were incubated with 10% *v/v* reconstituted MTT at 37 °C for 3 h. The multiwell plates were then read using Wallac Victor2 1420 multilabel counter, and the absorbance was measured at a wavelength of 450 (ex) and 590 nm (em).

3.4. Antiviral Assay

The cytopathic inhibition effect was used to assess the antiviral and cytotoxicity of the compounds using a method described by Choi et al. (2009) [49]. Coronavirus 229E cells were used in this assay. The antiviral activity of the tested compounds was measured as a percentage using a method described by Pauwels et al.'s (1988) research [50]. GraphPad Prism 5.0 software was used to analyze the results. Ribavirin was used as a positive standard in this assay, and DMSO was used as a negative standard.

3.5. Docking Studies

In these studies, the required proteins' crystal structures were gained from PDB (3LQ5: CDK9 costructure with CR8 ligand; 6W63: COVID-19 costructure with X77 ligand). The proteins were downloaded as PDB files and prepared using Discovery Studio by keeping one subunit and removing the water and other ligands. The prepared protein was then saved in PDB format and converted to PDBQT format using AutoDock. The required inhibitors (compounds **4a**, **3a**, **2c**, and **3c**) were prepared by ChemDraw Ultra 14.0 and then saved as a PDB file and used as ligands in the docking studies. The docking studies were performed using AutoDock Vina implemented in PyRx. The analysis of the obtained docking results was conducted using Discovery Studio.

4. Conclusions

In conclusion, new imadazopyrazines as first-in-class CDK9 inhibitors were synthesized and biologically evaluated for their anticancer and antiviral activity. The new derivatives were assessed in vitro against isolated CDK9, and the data of this assay showed that our compounds demonstrated a good CDK9 inhibition effect with an IC_{50} of 0.18–1.78 µM. In the MTT cytotoxicity assay against the MCF7, HCT116, and K652 cancer cell lines, our compounds demonstrated good antiproliferative effects, with an average IC_{50} in the three cell lines ranging from 10.65 to 143.79 µM. In addition, the results of this assay showed a correlation between the antiproliferative effects of the inhibitors and their inhibition of CDK9, which suggests the CDK9 inhibition as a mechanistic pathway for their anticancer effects. The physiochemical and pharmacokinetic parameters of the new agents were predicated in silico, and they exhibited reasonable drug-likeness properties. The compounds with the most promising data were further assessed for their antiviral activity against human Coronavirus 229E, and the most potent agent showed a good inhibitory effect with an IC_{50} of 63.28 µM and a selectivity index of 4.8. This data was supported by docking studies with a COVID-19 main protease, which showed a high binding affinity.

Author Contributions: Conceptualization, A.A.A. and B.A.A.; methodology, H.M.A., A.S. and O.A.; software, N.A.A.; validation, A.A.A. and B.A.A.; investigation, O.A., H.M.A. and A.S.; resources, A.A.A.; data curation, H.M.A.; writing—original draft preparation, A.A.A.; writing—review and editing, A.A.A.; visualization, A.A.; supervision, A.A.A.; project administration, A.A.A.; funding acquisition, A.A.A. All authors have read and agreed to the published version of the manuscript.

Funding: This research was funded by Princess Nourah bint Abdulrahman University Researchers Supporting Project, number (PNURSP2023R116), Princess Nourah bint Abdulrahman University, Riyadh, Saudi Arabia.

Institutional Review Board Statement: Not applicable.

Informed Consent Statement: Not applicable.

Data Availability Statement: Data is contained within the article.

Conflicts of Interest: The authors declare no conflict of interest.

References

1. Bruyère, C.; Meijer, L. Targeting Cyclin-Dependent Kinases in Anti-Neoplastic Therapy. *Curr. Opin. Cell Biol.* **2013**, *25*, 772–779. [CrossRef] [PubMed]
2. McInnes, C. Progress in the Evaluation of CDK Inhibitors as Anti-Tumor Agents. *Drug Discov. Today* **2008**, *13*, 875–881. [CrossRef] [PubMed]
3. Malumbres, M.; Barbacid, M. Cell Cycle, CDKs and Cancer: A Changing Paradigm. *Nat. Rev. Cancer* **2009**, *9*, 153–166. [CrossRef] [PubMed]
4. Morales, F.; Giordano, A. Overview of CDK9 as a Target in Cancer Research. *Cell Cycle* **2016**, *15*, 519–527. [CrossRef] [PubMed]
5. Alsfouk, A. Small Molecule Inhibitors of Cyclin-Dependent Kinase 9 for Cancer Therapy. *J. Enzym. Inhib. Med. Chem.* **2021**, *36*, 693–706. [CrossRef]
6. Ma, H.; Seebacher, N.A.; Hornicek, F.J.; Duan, Z. Cyclin-Dependent Kinase 9 (CDK9) Is a Novel Prognostic Marker and Therapeutic Target in Osteosarcoma. *EBioMedicine* **2019**, *39*, 182–193. [CrossRef]

7. Narita, T.; Ishida, T.; Ito, A.; Masaki, A.; Kinoshita, S.; Suzuki, S.; Takino, H.; Yoshida, T.; Ri, M.; Kusumoto, S.; et al. Cyclin-Dependent Kinase 9 Is a Novel Specific Molecular Target in Adult T-Cell Leukemia/Lymphoma. *Blood* **2017**, *130*, 1114–1124. [CrossRef]
8. Sonawane, Y.A.; Taylor, M.A.; Napoleon, J.V.; Rana, S.; Contreras, J.I.; Natarajan, A. Cyclin Dependent Kinase 9 Inhibitors for Cancer Therapy. *J. Med. Chem.* **2016**, *59*, 8667–8684. [CrossRef]
9. Krystof, V.; Baumli, S.; Furst, R. Perspective of Cyclin-Dependent Kinase 9 (CDK9) as a Drug Target. *Curr. Pharm. Des.* **2012**, *18*, 2883–2890. [CrossRef]
10. Wang, J.; Dean, D.C.; Hornicek, F.J.; Shi, H.; Duan, Z. Cyclin-Dependent Kinase 9 (CDK9) Is a Novel Prognostic Marker and Therapeutic Target in Ovarian Cancer. *FASEB J.* **2019**, *33*, 5990–6000. [CrossRef]
11. Kretz, A.L.; Schaum, M.; Richter, J.; Kitzig, E.F.; Engler, C.C.; Leithäuser, F.; Henne-Bruns, D.; Knippschild, U.; Lemke, J. CDK9 Is a Prognostic Marker and Therapeutic Target in Pancreatic Cancer. *Tumor Biol.* **2017**, *39*, 1010428317694304. [CrossRef] [PubMed]
12. Franco, L.C.; Morales, F.; Boffo, S.; Giordano, A. CDK9: A Key Player in Cancer and Other Diseases. *J. Cell Biochem.* **2018**, *119*, 1273–1284. [CrossRef] [PubMed]
13. Boffo, S.; Damato, A.; Alfano, L.; Giordano, A. CDK9 Inhibitors in Acute Myeloid Leukemia. *J. Exp. Clin. Cancer Res.* **2018**, *37*, 36. [CrossRef]
14. Alsfouk, A.A.; Alshibl, H.M.; Altwaijry, N.A.; Alsfouk, B.A.; Al-Abdullah, E.S. Synthesis and Biological Evaluation of Seliciclib Derivatives as Potent and Selective CDK9 Inhibitors for Prostate Cancer Therapy. *Monatsh. Chem.* **2021**, *152*, 109–120. [CrossRef]
15. Senderowicz, A.M. Flavopiridol: The First Cyclin-Dependent Kinase Inhibitor in Human Clinical Trials. *Investig. New Drugs* **1999**, *3*, 313–320. [CrossRef] [PubMed]
16. Kumar, S.K.; Fruth, B.; Roy, V.; Erlichman, C.; Stewart, A.K. Dinaciclib, a Novel CDK Inhibitor, Demonstrates Encouraging Single-Agent Activity in Patients with Relapsed Multiple Myeloma. *Blood* **2015**, *125*, 443–448. [CrossRef]
17. Tong, W.; Chen, R.; Plunkett, W.; Siegel, D.; Sinha, R.; Harvey, R.D.; Badros, A.Z.; Popplewell, L.; Coutre, S.; Fox, J.A.; et al. Phase I and Pharmacologic Study of SNS-032, a Potent and Selective Cdk2,7, and 9 Inhibitor, in Patients with Advanced Chronic Lymphocytic Leukemia and Multiple Myeloma. *J. Clin. Oncol.* **2010**, *28*, 3015–3022. [CrossRef]
18. Van der Biessen, D.A.J.; Burger, H.; Bruijn, P.D.; Lamers, C.H.J.; Naus, N.; Loferer, H.; Wiemer, E.A.C.; Mathijssen, R.H.J.; de Jonge, M.J.A. Phase I Study of RGB-286638, a Novel, Multitargeted Cyclin- Dependent Kinase Inhibitor in Patients with Solid Tumors. *Clin. Cancer Res.* **2014**, *20*, 4776–4784. [CrossRef]
19. Walsby, E.; Pratt, G.; Shao, H.; Abbas, A.Y.; Fischer, P.M.; Bradshaw, T.D.; Brennan, P.; Fegan, C.; Wang, S.; Pepper, C. A Novel Cdk9 Inhibitor Preferentially Targets Tumor Cells and Synergizes with Fludarabine. *Oncotarget* **2014**, *5*, 375–385. [CrossRef]
20. Zhai, S.; Senderowicz, A.M.; Sausville, E.A.; Figg, W.D. Flavopiridol, a Novel Cyclin-Dependent Kinase Inhibitor, in Clinical Development. *Ann. Pharmacother.* **2002**, *36*, 905–911. [CrossRef]
21. Lücking, U.; Scholz, A.; Lienau, P.; Siemeister, G.; Kosemund, D.; Bohlmann, R.; Briem, H.; Terebesi, I.; Meyer, K.; Prelle, K.; et al. Identification of Atuveciclib (BAY 1143572), the First Highly Selective, Clinical PTEFb/CDK9 Inhibitor for the Treatment of Cancer. *ChemMedChem* **2017**, *12*, 1776–1793. [CrossRef]
22. Cidado, J.; Boiko, S.; Proia, T.; Ferguson, D.; Criscione, S.W.; Martin, M.S.; Pop-Damkov, P.; Su, N.; Franklin, V.N.R.; Chilamakuri, C.S.R.; et al. AZD4573 Is a Highly Selective CDK9 Inhibitor That Suppresses Mcl-1 and Induces Apoptosis in Hematologic Cancer Cells. *Clin. Cancer Res.* **2020**, *26*, 922–934. [CrossRef] [PubMed]
23. Zhang, M.; Zhang, L.; Hei, R.; Li, X.; Cai, H.; Wu, X.; Zheng, Q.; Cai, C. CDK Inhibitors in Cancer Therapy, an Overview of Recent Development. *Am. J. Cancer Res.* **2021**, *11*, 1913–1935.
24. Gojo, I.; Sadowska, M.; Walker, A.; Feldman, E.J.; Iyer, S.P.; Baer, M.R.; Sausville, E.A.; Lapidus, R.G.; Zhang, D.; Zhu, Y.; et al. Clinical and Laboratory Studies of the Novel Cyclin-Dependent Kinase Inhibitor Dinaciclib (SCH 727965) in Acute Leukemias. *Cancer Chemother. Pharmacol.* **2013**, *72*, 897–908. [CrossRef]
25. Stephenson, J.J.; Nemunaitis, J.; Joy, A.A.; Martin, J.C.; Jou, Y.M.; Zhang, D.; Statkevich, P.; Yao, S.L.; Zhu, Y.; Zhou, H.; et al. Randomized Phase 2 Study of the Cyclin-Dependent Kinase Inhibitor Dinaciclib (MK-7965) versus Erlotinib in Patients with Non-Small Cell Lung Cancer. *Lung Cancer* **2014**, *83*, 219–223. [CrossRef] [PubMed]
26. Mita, M.M.; Joy, A.A.; Mita, A.; Sankhala, K.; Jou, Y.M.; Zhang, D.; Statkevich, P.; Zhu, Y.; Yao, S.L.; Small, K.; et al. Randomized Phase II Trial of the Cyclin-Dependent Kinase Inhibitor Dinaciclib (MK-7965) versus Capecitabine in Patients with Advanced Breast Cancer. *Clin. Breast Cancer* **2014**, *14*, 169–176. [CrossRef]
27. Conroy, A.; Stockett, D.E.; Walker, D.; Arkin, M.R.; Hoch, U.; Fox, J.A.; Hawtin, R.E. SNS-032 Is a Potent and Selective CDK 2, 7 and 9 Inhibitor That Drives Target Modulation in Patient Samples. *Cancer Chemother. Pharmacol.* **2009**, *64*, 723–732. [CrossRef]
28. Chen, R.; Wierda, W.G.; Chubb, S.; Hawtin, R.E.; Fox, J.A.; Keating, M.J.; Gandhi, V.; Plunkett, W. Mechanism of Action of SNS-032, a Novel Cyclin-Dependent Kinase Inhibitor, in Chronic Lymphocytic Leukemia. *Blood* **2009**, *113*, 4637–4645. [CrossRef]
29. Lin, T.S.; Ruppert, A.S.; Johnson, A.J.; Fischer, B.; Heerema, N.A.; Andritsos, L.A.; Blum, K.A.; Flynn, J.M.; Jones, J.A.; Hu, W.; et al. Phase II Study of Flavopiridol in Relapsed Chronic Lymphocytic Leukemia Demonstrating High Response Rates in Genetically High-Risk Disease. *J. Clin. Oncol.* **2009**, *27*, 6012–6018. [CrossRef] [PubMed]
30. Karp, J.E.; Garrett-Mayer, E.; Estey, E.H.; Rudek, M.A.; Douglas Smith, B.; Greer, J.M.; Michelle Drye, D.; Mackey, K.; Dorcy, K.S.; Gore, S.D.; et al. Randomized Phase II Study of Two Schedules of Flavopiridol given as Timed Sequential Therapy with Cytosine Arabinoside and Mitoxantrone for Adults with Newly Diagnosed, Poor-Risk Acute Myelogenous Leukemia. *Haematologica* **2012**, *97*, 1736–1742. [CrossRef]

31. Luecking, U.T.; Scholz, A.; Kosemund, D.; Bohlmann, R.; Briem, H.; Lienau, P.; Siemeister, G.; Terebesi, I.; Meyer, K.; Prelle, K.; et al. Abstract 984: Identification of Potent and Highly Selective PTEFb Inhibitor BAY 1251152 for the Treatment of Cancer: From p.o. to i.v. Application via Scaffold Hops. *Cancer Res.* **2017**, *77*, 984. [CrossRef]
32. Byrne, M.; Frattini, M.G.; Ottmann, O.G.; Mantzaris, I.; Wermke, M.; Lee, D.J.; Morillo, D.; Scholz, A.; Ince, S.; Valencia, R.; et al. Phase I Study of the PTEFb Inhibitor BAY 1251152 in Patients with Acute Myelogenous Leukemia. *Blood* **2018**, *132*, 4055. [CrossRef]
33. Cidado, J.; Proia, T.; Boiko, S.; Martin, M.S.; Criscione, S.; Ferguson, D.; Shao, W.; Drew, L. Abstract 310: AZD4573, a Novel CDK9 Inhibitor, Rapidly Induces Cell Death in Hematological Tumor Models through Depletion of Mcl1. *Cancer Res.* **2018**, *78*, 310. [CrossRef]
34. Byth, K.F.; Thomas, A.; Hughes, G.; Forder, C.; McGregor, A.; Geh, C.; Oakes, S.; Green, C.; Walker, M.; Newcombe, N.; et al. AZD5438, a Potent Oral Inhibitor of Cyclin-Dependent Kinases 1, 2, and 9, Leads to Pharmacodynamic Changes and Potent Antitumor Effects in Human Tumor Xenografts. *Mol. Cancer Ther.* **2009**, *8*, 1856–1866. [CrossRef] [PubMed]
35. Mariaule, G.; Belmont, P. Cyclin-Dependent Kinase Inhibitors as Marketed Anticancer Drugs: Where Are We Now? A Short Survey. *Molecules* **2014**, *19*, 14366–14382. [CrossRef]
36. Parry, D.; Guzi, T.; Shanahan, F.; Davis, N.; Prabhavalkar, D.; Wiswell, D.; Seghezzi, W.; Paruch, K.; Dwyer, M.P.; Doll, R.; et al. Dinaciclib (SCH 727965), a Novel and Potent Cyclin-Dependent Kinase Inhibitor. *Mol. Cancer Ther.* **2010**, *9*, 2344–2353. [CrossRef] [PubMed]
37. Heath, E.I.; Bible, K.; Martell, R.E.; Adelman, D.C.; LoRusso, P.M. A Phase 1 Study Of SNS-032 (Formerly BMS-387032), A Potent Inhibitor Of Cyclin-Dependent Kinases 2, 7 And 9 Administered As A Single Oral Dose And Weekly Infusion In Patients With Metastatic Refractory Solid Tumors. *Investig. New Drugs* **2008**, *26*, 59–65. [CrossRef]
38. Kim, I.K.; Nam, K.Y.; Kim, S.Y.; Park, S.J. Composition for Prevention and Treatment of Cancer Including CDK9 Inhibitor as Active Ingredient, University of Ulsan Foundation for Industry Cooperation. Patent KR1020180106188, 22 February 2019.
39. Nandi, S.; Dey, R.; Dey, S.; Samadder, A.; Saxena, A.K. Naturally Sourced CDK Inhibitors and Current Trends in Structure-Based Synthetic Anticancer Drug Design by Crystallography. *Anticancer Agents Med. Chem.* **2022**, *22*, 485–498. [CrossRef] [PubMed]
40. Alsfouk, A.A.; Alshibl, H.M.; Alsfouk, B.A.; Altwaijry, N.A.; Al-Abdullah, E.S. Synthesis and Biological Evaluation of Imadazo[1,2-a]Pyrazines as Anticancer and Antiviral Agents through Inhibition of CDK9 and Human Coronavirus. *Pharmaceuticals* **2022**, *15*, 859. [CrossRef] [PubMed]
41. Cheol-Gyu, H.; Jeong-Hyeok, Y. Novel Imidazole Pyrazine Derivative Compound, a Method for Preparing the Same, and a Pharmaceutical Composition for Antiviral Treatment Containing the Same as an Active Ingredient, Sihwa Industriy. Patent 1020110097448, 31 August 2011.
42. Nandi, S.; Kumar, M.; Saxena, M.; Saxena, A.K. The Antiviral and Antimalarial Drug Repurposing in Quest of Chemotherapeutics to Combat COVID-19 Utilizing Structure-Based Molecular Docking. *Comb. Chem. High Throughput Screen.* **2021**, *24*, 1055–1068. [CrossRef]
43. Nandi, S.; Roy, H.; Gummadi, A.; Saxena, A.K. Exploring Spike Protein as Potential Target of Novel Coronavirus and to Inhibit the Viability Utilizing Natural Agents. *Curr. Drug Targets* **2021**, *22*, 2006–2020. [CrossRef] [PubMed]
44. Daina, A.; Michielin, O.; Zoete, V. Swiss Target Prediction: Updated Data and New Features for Efficient Prediction of Protein Targets of Small Molecules. *Nucleic Acids Res.* **2019**, *47*, 357–364. [CrossRef] [PubMed]
45. Li, Y.C.; Bai, W.Z.; Hashikawa, T. The Neuroinvasive Potential of SARS-CoV2 May Play a Role in the Respiratory Failure of COVID-19 Patients. *J. Med. Virol.* **2020**, *92*, 552–555. [CrossRef]
46. V'kovski, P.; Kratzel, A.; Steiner, S.; Stalder, H.; Thiel, V. Coronavirus Biology and Replication: Implications for SARS-CoV-2. *Nat. Rev. Microbiol.* **2021**, *19*, 155–170. [CrossRef] [PubMed]
47. Daina, A.; Michielin, O.; Zoete, V. SwissADME: A Free Web Tool to Evaluate Pharmacokinetics, Drug-Likeness and Medicinal Chemistry Friendliness of Small Molecules. *Sci. Rep.* **2017**, *7*, 42717. [CrossRef]
48. Dong, J.; Wang, N.N.; Yao, Z.J.; Zhang, L.; Cheng, Y.; Ouyang, D.; Lu, A.P.; Cao, D.S. ADMETlab: A Platform for Systematic ADMET Evaluation Based on a Comprehensively Collected ADMET Database. *J. Cheminform.* **2018**, *10*, 29. [CrossRef]
49. Choi, H.J.; Song, J.H.; Park, K.S.; Kwon, D.H. Inhibitory Effects of Quercetin 3-Rhamnoside on Influenza A Virus Replication. *Eur. J. Pharm. Sci.* **2009**, *37*, 329–333. [CrossRef]
50. Pauwels, R.; Balzarini, J.; Baba, M.; Snoeck, R.; Schols, D.; Herdewijn, P.; Desmyter, J.; De Clercq, E. Rapid and Automated Tetrazolium-Based Colorimetric Assay for the Detection of Anti-HIV Compounds. *J. Virol. Methods* **1988**, *20*, 309–321. [CrossRef]

Disclaimer/Publisher's Note: The statements, opinions and data contained in all publications are solely those of the individual author(s) and contributor(s) and not of MDPI and/or the editor(s). MDPI and/or the editor(s) disclaim responsibility for any injury to people or property resulting from any ideas, methods, instructions or products referred to in the content.

MDPI AG
Grosspeteranlage 5
4052 Basel
Switzerland
Tel.: +41 61 683 77 34

Pharmaceuticals Editorial Office
E-mail: pharmaceuticals@mdpi.com
www.mdpi.com/journal/pharmaceuticals

Disclaimer/Publisher's Note: The statements, opinions and data contained in all publications are solely those of the individual author(s) and contributor(s) and not of MDPI and/or the editor(s). MDPI and/or the editor(s) disclaim responsibility for any injury to people or property resulting from any ideas, methods, instructions or products referred to in the content.

www.ingramcontent.com/pod-product-compliance
Lightning Source LLC
LaVergne TN
LVHW070244100526
838202LV00015B/2174